Formulas for
Stress and Strain

Formulas for Stress and Strain

FIFTH EDITION

RAYMOND J. ROARK

WARREN C. YOUNG

McGraw-Hill Book Company

New York St. Louis San Francisco Auckland Bogotá Düsseldorf Johannesburg London Madrid Mexico Montreal New Delhi Panama Paris São Paulo Singapore Sydney Tokyo Toronto

Library of Congress Cataloging in Publication Data

Roark, Raymond Jefferson, 1890–1966.
Formulas for stress and strain.

Includes bibliographical references and indexes.
1. Strength of materials—Tables, calculations, etc. 2. Strains and stresses—Tables, calculations, etc. I. Young, Warren Clarence, date, joint author. II. Title.
TA407.2.R6 1975 624'.176'0212 75-26612
ISBN 0-07-053031-9

67890 KPKP 784321098

The editors for this book were Tyler G. Hicks and Ross Kepler, the designer was Naomi Auerbach, and the production supervisor was George Oechsner. It was set in Baskerville by York Graphic Services. It was printed and bound by The Kingsport Press.

Contents

List of Tables

Preface to the Fifth Edition

A major change in format in the many expanded tables of formulas is the most obvious modification introduced in this edition. The choice of format was greatly influenced by the general availability to design engineers of high-speed digital computers and by the recent development of both hand- and desk-model scientific digital calculators. Where space permits, some equations are presented in reduced form for use with a slide rule; but for many of the more complex loadings, the evaluation of the necessary constants and variables requires calculator accuracy. Since trigonometric, exponential, and hyperbolic functions are now available with the necessary accuracy, the evaluation of these functions presents little difficulty. The general use of the singularity functions defined in Chapter 1 has measurably reduced the space required to present the formulas in this edition.

All the new formulas tabulated have been programmed for a digital computer, and numerical values of forces, moments, and deformations have been determined and checked against the assumed boundary conditions. Where space permits, representative values of coefficients are tabulated to enable the reader to verify his own programs or, by interpolation, use the tabulated coefficients directly.

The addition of applied deformation as a form of loading has greatly expanded the versatility of many of the tables. The applied deformations consist of externally applied concentrated angular rotations and lateral displacements to beams, plates, and shells as well as linear temperature variations through the thickness of these structural members.

The major additions in this edition are as follows: In Chapter 7 (Beams; Flexure of Straight Bars), in addition to a substantial increase in the loadings considered, all stable combinations of support conditions, including free, simply supported, guided, and fixed, are considered. This has been done in Table 3 for common straight beams, in Table 4 for rigid frames, in Table 7 for short beams on elastic foundations, in Table 8 for long beams on elastic foundations, and in Table 10 for beams under axial compression and transverse loading. Table 13 has been added to give correction coefficients to be applied to reactions and deformations of uniform beams to account for many types and degrees of taper in the beam dimensions. In Article 7-2 on composite beams and bimetallic strips, examples illustrate the use of formulas for equivalent material properties and stresses.

A separate chapter on curved beams (Chapter 8) has been included owing to the growth in information on this subject. Table 17 on circular rings now includes corrections for deflections due to hoop stress and transverse shear, a more detailed set of formulas for deformations, and tabulated force, moment, and deformation coefficients for several specific load locations. For many practical problems the formulas do not need to be solved. Table 18 for circular arches includes most of the loadings and all of the combinations of support conditions mentioned previously for straight beams and rigid frames. Table 19 is an extensive tabulation of formulas for curved beams loaded normal to the plane of curvature, including the majority of the stable combinations of support conditions. For beams with round cross sections or having any shape where the polar moment of inertia is twice the bending moment of inertia, numerical coefficients for forces, moments, and deformations are tabulated for several combinations of angular spans and load locations.

In Chapter 9 (Torsion) Table 21 has been added to give torsional stiffness and warping constants for thin-walled open cross sections; Table 22 uses these constants in formulas to determine angles of twist and stresses in thin-walled beams under many torsional loading conditions and most stable combinations of support conditions. In Chapter 10 (Flat Plates) Table 24 presents extensive tables of numerical coefficients as well as formulas for both solid circular and annular plates. Table 25 gives shear deflections for circular plates; Table 26 for plates with straight boundaries has been greatly expanded. Formulas for stresses and deformations in bimetallic circular plates have been added in Article 10-4.

In Chapter 12 (Shells of Revolution; Pressure Vessels; Pipes) the tables have been greatly expanded. Specifically, Table 30 has been added to cover both long and short thin-walled cylindrical shells under axisymmetric loading within the span rather than just at the ends. Table 31 covers bending stresses in thin-walled vessels, and the formulas for conical shells have been completely revised to give more accurate results for a much wider range of cone parameters; numerical data for short conical shells are also included. Body-force loadings for thick-walled vessels have been included in Table 32, in addition to the usual pressure loadings.

In Chapter 14 (Elastic Stability) there has been a substantial increase in the numerical data for buckling with loads applied at more than one position on a column simultaneously. In Chapter 15 (Dynamic and Temperature Stresses), Table 36 gives natural frequencies and nodal positions for many continuous members, including beams and flat plates.

It is the sincere hope of the author that the modifications made in this edition are in keeping with the philosophy practiced by Raymond J. Roark in previous editions. He assumed that good engineering judgment, coupled with sufficient information about the behavior of a similarly shaped structural member, should enable the reader to find an adequate solution to a given problem. This book was his continuing contribution to making such information readily available.

It should be understood that although many new references are included herein, there is no way to ensure that much excellent work has not been overlooked and omitted.

The author wishes to acknowledge the assistance of those persons who have carefully checked formulas in previous editions and called attention to errors and omissions, as well as those individuals, publishers, institutions, and corporations who have generously given permission to use material in this edition. Special thanks are due those who have volunteered unpublished material, which has added very measurably to the usefulness of this book.

Warren C. Young

Preface to the First Edition

This book was written for the purpose of making available a compact, adequate summary of the formulas, facts, and principles pertaining to strength of materials. It is intended primarily as a reference book and represents an attempt to meet what is believed to be a present need of the designing engineer.

This need results from the necessity for more accurate methods of stress analysis imposed by the trend of engineering practice. That trend is toward greater speed and complexity of machinery, greater size and diversity of structures, and greater economy and refinement of design. In consequence of such developments, familiar problems, for which approximate solutions were formerly considered adequate, are now frequently found to require more precise treatment, and many less familiar problems, once of academic interest only, have become of great practical importance. The solutions and data desired are often to be found only in advanced treatises or scattered through an extensive literature, and the results are not always presented in such form as to be suited to the requirements of the engineer. To bring together as much of this material as is likely to prove generally useful and to present it in convenient form have been the author's aim.

The scope and arrangement of the book are indicated by the Contents. In Part 1 are defined all terms whose exact meaning might otherwise not be clear. In Part 2 certain useful general principles are stated; analytical and experimental methods of stress analysis are briefly described, and information concerning the behavior of material under stress is given. In Part 3 the behavior of structural elements under various conditions of loading is discussed, and extensive tables of formulas for the calculation of stress, strain, and strength are given.

Because they are not believed to serve the purpose of this book, derivations of formulas and detailed explanations, such as are appropriate in a textbook, are omitted, but a sufficient number of examples are included to illustrate the application of the various formulas and methods. Numerous references to more detailed discussions are given, but for the most part these are limited to sources that are generally available and no attempt has been made to compile an exhaustive bibliography.

That such a book as this derives almost wholly from the work of others is self-evident, and it is the author's hope that due acknowledgment has been made of the immediate sources of all material here presented. To the publishers and others who have generously permitted the use of material, he wishes to express his thanks. The helpful criticisms and suggestions of his colleagues, Professors E. R. Maurer, M. O. Withey, J. B. Kommers, and K. F. Wendt, are gratefully acknowledged. A considerable number of the tables of formulas have been published from time to time in *Product Engineering,* and the opportunity thus afforded for criticism and study of arrangement has been of great advantage.

Finally, it should be said that, although every care has been taken to avoid errors, it would be oversanguine to hope that none had escaped detection; for any suggestions that readers may make concerning needed corrections the author will be grateful.

Raymond J. Roark

PART ONE

Definitions

ONE

Definitions

The definitions given here apply to the terms in question as they are used in this book. Some of these terms are defined differently by other authors; when this is the case, the fact is noted. When two or more terms with identical meaning are in general acceptance, they are given in the order of the present writers' preference. The references referred to by number are listed at the end of this section.

Allowable stress (working stress): If a member is so designed that the maximum stress as calculated for the expected conditions of service is less than some certain value, the member will have a proper margin of security against damage or failure. This certain value is the *allowable stress* of the kind and for the material and condition of service in question. The allowable stress is less than the damaging stress because of uncertainty as to the conditions of service, nonuniformity of material, and inaccuracy of stress analysis. The margin between the allowable stress and the damaging stress may be reduced in proportion to the certainty with which the conditions of service are known, the intrinsic reliability of the material, the accuracy with which the stress produced by the loading can be calculated, and the degree to which failure is unattended by danger or loss. (Compare *Damaging stress; Factor of safety; Factor of utilization; Margin of safety*. See Refs. 1, 2, and Table 40.)

Apparent elastic limit (useful limit point): The stress at which the rate of change of strain with respect to stress is 50 percent greater than at zero stress. It is more definitely determinable from the stress-strain diagram than is the proportional limit, and is useful for comparing materials of the same general class. (Compare *Elastic limit; Proportional limit; Yield point; Yield strength.*)

Apparent stress: The stress corresponding to a given unit strain on the assumption of uniaxial elastic stress. It is calculated by multiplying the unit strain by the modulus of elasticity and may differ from the true stress because the effect of transverse stresses is not taken into account.

Bending moment: Reference is to a beam, assumed for convenience to be horizontal and loaded and supported by forces, all of which lie in a vertical plane. The *bending moment* at any section of the beam is the moment of all forces that act on the beam to the left of that section, taken about the horizontal axis of the section. The bending moment is positive when clockwise and negative when counterclockwise; a positive bending moment therefore bends the beam so that it is concave upward, and a negative bending moment bends it so that it is concave downward. The *moment equation* is an expression for the bending moment at any section in terms of x, the distance to that section measured from a chosen origin, usually taken at the left end of the beam.

Boundary conditions: As used in strength of materials, the term usually refers to the condition of stress, displacement, or slope at the ends or edges of a member, where these conditions are apparent from the circumstances of the problem. Thus for a beam with fixed ends, the zero slope at each end is a boundary condition; for a pierced circular plate with freely supported edges, the zero radial stress at each edge is a boundary condition.

Brittle fracture: The tensile failure with negligible plastic deformation of an ordinarily ductile metal. (See Art. 2.7.)

Bulk modulus of elasticity: The ratio of a tensile or compressive stress, triaxial and equal in all directions (e.g., hydrostatic pressure), to the relative change it produces in volume.

Central axis (centroidal axis): A central axis of an area is one that passes through the centroid; it is understood to lie in the plane of the area unless the contrary is stated. When taken normal to the plane of the area, it is called the *central polar axis.*

Centroid of an area (center of gravity of an area): That point in the plane of the area about any axis through which the moment of the area is zero; it coincides with the center of gravity of the area materialized as an infinitely thin homogeneous and uniform plate.

Corrosion fatigue: Fatigue aggravated by corrosion, as in parts repeatedly stressed while exposed to salt water.

Creep: Continuous increase in deformation under constant or decreasing stress. The term is ordinarily used with reference to the behavior of metals under tension at elevated temperatures. The similar yielding of a material under compressive stress is usually called *plastic flow,* or *flow.* Creep at atmospheric temperature due to sustained elastic stress is sometimes called *drift,* or *elastic drift.* Another manifestation of creep, the diminution in *stress* when the deformation is maintained constant, is called *relaxation.*

Damaging stress: The least unit stress of a given kind and for a given material and condition of service that will render a member unfit for service before the end of its normal life. It may do this by producing excessive set, by causing creep to occur at an excessive rate, or by causing fatigue cracking, excessive strain hardening, or rupture.

Damping capacity: The amount of work dissipated into heat per unit of total strain energy present at maximum strain for a complete cycle.

Deformation (strain): Change in the form or dimensions of a body produced by stress. *Elongation* is often used for tensile strain, *compression* or *shortening* for compressive strain, and *detrusion* for shear strain. *Elastic deformation* is such deformation as disappears on removal of stress; *permanent deformation* is such deformation as remains on removal of stress. (Compare *Set*.)

Eccentricity: A load or component of a load normal to a given cross section of a member is eccentric with respect to that section if it does not act through the centroid. The perpendicular distance from the line of action of the load to either principal central axis is the *eccentricity* with respect to that axis.

Elastic: Capable of sustaining stress without permanent deformation; the term is also used to denote conformity to the law of stress-strain proportionality. An elastic stress or strain is a stress or strain within the elastic limit.

Elastic axis: The elastic axis of a beam is the line, lengthwise of the beam, along which transverse loads must be applied in order to produce bending only, with no torsion of the beam at any section. Strictly speaking, no such line exists except for a few conditions of loading. Usually the elastic axis is assumed to be the line that passes through the elastic center of every section. The term is most often used with reference to an airplane wing of either the shell or multiple-spar type. (Compare *Torsional center; Flexural center; Elastic center*. See Ref. 4.)

Elastic center: The elastic center of a given section of a beam is that point in the plane of the section lying midway between the flexural center and center of twist of that section. The three points may be identical and are usually assumed to be so. (Compare *Flexural center; Torsional center; Elastic axis*. See Refs. 4 and 5.)

Elastic curve: The curve assumed by the axis of a normally straight beam or column when bent by loads that do not stress it beyond the proportional limit.

Elastic limit: The least stress that will cause permanent set. (Compare *Proportional limit; Apparent elastic limit; Yield point; Yield strength*. See Art. 2.2 and Ref. 6.)

Elastic ratio: The ratio of the elastic limit to the ultimate strength.

Ellipsoid of strain: An ellipsoid that represents the state of strain at any given point in a body; it has the form assumed under stress by a sphere centered at the point in question (Ref. 7).

Ellipsoid of stress: An ellipsoid that represents the state of stress at a given point in a body; its semiaxes are vectors representing the principal stresses at the point, and any radius vector represents the resultant stress on a particular plane through the point. For a condition of plane stress (one principal stress zero) the ellipsoid becomes the *ellipse of stress* (see Ref. 8).

Endurance limit (fatigue strength): The maximum stress that can be reversed an indefinitely large number of times without producing fracture of a material (see Art. 2.8).

Endurance ratio: Ratio of the endurance limit to the ultimate static tensile strength.

Endurance strength: The highest stress that a material can withstand with repeated application or reversal without rupture for a given number of cycles is the endurance strength of that material for that number of cycles. Unless otherwise specified, reversed stressing is usually implied. (Compare *Endurance limit.*)

Energy of rupture: The work done per unit volume in producing fracture. It is not practicable to establish a definite energy of rupture value for a given material because the result obtained depends upon the form and proportions of the test specimen and the manner of loading. As determined by similar tests on similar specimens, the energy of rupture affords a criterion for comparing the toughness of different materials.

Equivalent bending moment: A bending moment that, acting alone, would produce in a circular shaft a normal (tensile or compressive) stress of the same magnitude as the maximum normal stress produced by a given bending moment and a given twisting moment acting simultaneously.

Equivalent twisting moment: A twisting moment that, acting alone, would produce in a circular shaft a shear stress of the same magnitude as the shear stress produced by a given twisting moment and a given bending moment acting simultaneously.

Factor of safety: The ratio of the load that would cause failure of a member or structure to the load that is imposed upon it in service. The term usually has this meaning; it may also be used to represent the ratio of breaking to service value of speed, deflection, temperature variation, or other stress-producing factor against possible increase in which the factor of safety is provided as a safeguard. (Compare *Allowable stress; Margin of safety.*)

Factor of strain concentration: In the presence of stress raisers, localized peak strains are developed. The factor of strain concentration is the ratio of the *localized maximum strain* at a given cross section to the *nominal average strain* on that cross section. The nominal average strain is computed from the average stress and a knowledge of the stress strain behavior of the material. In a situation where all stresses and strains are elastic, the factors of stress concentration and strain concentration are equal. (Compare *Factor of stress concentration.*)

Factor of stress concentration: Irregularities of form such as holes, screw threads, notches, and sharp shoulders, when present in a beam, shaft, or other member subject to loading, may produce high localized stresses. This phenomenon is called *stress concentration,* and the form irregularities that cause it are called *stress raisers.* The ratio of the true maximum stress to the stress calculated by the ordinary formulas of mechanics (flexure formula, torsion formula, etc.), using the net section but ignoring the changed distribution of stress, is the factor of stress concentration for the particular type of stress raiser in question. (See Art. 2.10.)

Factor of stress concentration in fatigue: At a specified number of loading cycles, the fatigue strength of a given geometry depends upon the stress concentration factor and upon material properties. The *factor of stress concentration in fatigue* is the ratio of the fatigue strength without a stress concentration to the fatigue strength with the given stress concentration. It may vary with the specified number of cycles as well as with the material. (See *Notch-sensitivity ratio.*)

Factor of utilization: The ratio of the allowable stress to the ultimate strength. For cases in which stress is proportional to load, the factor of utilization is the reciprocal of the factor of safety (see Ref. 1).

Fatigue: Tendency of materials to fracture under many repetitions of a stress considerably less than the ultimate static strength.

Fatigue-strength reduction factor: Alternative term for *factor of stress concentration in fatigue.*

Fiber stress: A term used for convenience to denote the longitudinal tensile or compressive stress in a beam or other member subject to bending. It is sometimes used to denote this stress at the point or points most remote from the neutral axis, but the term *stress in extreme fiber* is preferable for this purpose. Also, for convenience, the longitudinal elements or filaments of which a beam may be imagined as composed are called *fibers.*

Fixed (clamped, built-in, encastré): A condition of support at the ends of a beam or column or at the edges of a plate or shell that prevents *rotation and transverse displacement* of the edge of the neutral surface but permits *longitudinal displacement.* (Compare *Guided; Held; Supported.*)

Flexural center (shear center): With reference to a beam, the flexural center of any section is that point in the plane of the section through which a transverse load, applied at that section, must act if bending deflection only is to be produced, with no twist of the section. (Compare *Torsional center; Elastic center; Elastic axis.* See Refs. 4 and 9.)

Form factor: The term pertains to a beam section of a given shape and means the ratio of the modulus of rupture of a beam having that particular section to the modulus of rupture of a beam otherwise similar but having a section adopted as standard. This standard section is usually taken as rectangular or square; for wood it is a 2 by 2 in square, with edges horizontal and vertical (see Arts. 2.11 and 7.3). The term is also used to mean, for a given maximum fiber stress within the elastic limit, the ratio of the actual resisting moment of a wide-flanged beam to the resisting moment the beam would develop if the fiber stress were uniformly distributed across the entire width of the flanges. So used, the term expresses the strength-reducing effect of shear lag.

Fretting fatigue (chafing fatigue): Fatigue aggravated by surface rubbing, as in shafts with press-fitted collars.

Guided: A condition of support at the ends of a beam or column or at the edge of a plate or shell that prevents rotation of the edge of the neutral surface but permits longitudinal and transverse displacement. (Compare *Fixed; Held; Supported.*)

Held: A condition of support at the ends of a beam or column or at the edge of a plate or shell that prevents longitudinal and transverse displacement of the edge of the neutral surface but permits rotation in the plane of bending. (Compare *Fixed; Guided; Supported.*)

Influence line: Usually pertaining to a particular section of a beam, an influence line is a curve drawn so that its ordinate at any point represents the value of the reaction, vertical shear, bending moment, or deflection produced at the particular section by a unit load applied at the point where the ordinate is measured. An influence line may be used to show the effect of load position on any quantity dependent thereon, such as the stress in a given truss member, the deflection of a truss, or the twisting moment in a shaft.

Isoclinic: A line (in a stressed body) at all points on which the corresponding principal stresses have the same direction.

Isotropic: Having the same properties in all directions. In discussions pertaining to strengths of materials, isotropic usually means having the same strength and elastic properties (modulus of elasticity, modulus of rigidity, and Poisson's ratio) in all directions.

Kern (kernel): Reference is to some particular section of a member. The kern is that area in the plane of the section through which the line of action of a force must pass if that force is to produce, at all points in the given section, the same kind of normal stress, i.e., tension throughout or compression throughout.

Lüders lines: See *Slip lines.*

Margin of safety: As used in aeronautical design, margin of safety is the percentage by which the ultimate strength of a member exceeds the *design load.* The *design load* is the applied load, or maximum probable load, multiplied by a specified factor of safety. [The use of the terms margin of safety and design load in this sense is practically restricted to aeronautical engineering (see Ref. 10).]

Mechanical hysteresis: The dissipation of energy as heat during a stress cycle, which is revealed graphically by failure of the descending and ascending branches of the stress-strain diagram to coincide.

Member: Any part or element of a machine or structure such as a beam, column, or shaft.

Modulus of elasticity (Young's modulus): The rate of change of unit tensile or compressive stress with respect to unit tensile or compressive strain for the condition of uniaxial stress within the proportional limit. For most, but not all, materials, the modulus of elasticity is the same for tension and compression. For nonisotropic materials such as wood, it is necessary to distinguish between the moduli of elasticity in different directions.

Modulus of resilience: The strain energy per unit volume absorbed up to the elastic limit under the condition of uniform uniaxial stress.

Modulus of rigidity (modulus of elasticity in shear): The rate of change of unit shear stress with respect to unit shear strain for the condition of pure shear within the proportional limit. For nonisotropic materials such as wood, it is necessary to distinguish between the moduli of rigidity in different directions.

Modulus of rupture in bending (computed ultimate bending strength): The fictitious tensile or compressive stress in the extreme fiber of a beam computed by the flexure equation $\sigma = Mc/I$, where M is the bending moment that causes rupture.

Modulus of rupture in torsion (computed ultimate twisting strength): The fictitious shear stress at the surface of a circular shaft computed by the torsion formula $\tau = Tr/J$, where T is the twisting moment that causes rupture.

Moment of an area (first moment of an area, statical moment of an area): With respect to an axis, the sum of the products obtained by multiplying each element of the area dA by its distance from the axis y; it is therefore the quantity $\int dA\, y$. An axis in the plane of the area is implied.

Moment of inertia of an area (second moment of an area): The moment of inertia of an area with respect to an axis is the sum of the products obtained by multiplying each element of the area dA by the square of its distance from the axis y; it is therefore the quantity $\int dA\, y^2$. An axis in the plane of the area is implied; if the axis is normal to that plane, the term *polar moment of inertia* is used (see Chap. 5).

Neutral axis: The line of zero fiber stress in any given section of a member subject to bending; it is the line formed by the intersection of the neutral surface and the section.

Neutral surface: The longitudinal surface of zero fiber stress in a member subject to bending; it contains the neutral axis of every section.

Notch-sensitivity ratio: Used to compare *stress concentration factor* k_t and *fatigue-strength reduction factor* k_f, the notch-sensitivity ratio is commonly defined as the ratio $(k_f - 1)/(k_t - 1)$. It varies from 0, for some soft ductile materials, to 1, for some hard brittle materials.

Plastic moment; plastic hinge; plastic section modulus: These terms are explained in Art. 7.16.

Plasticity: The property of sustaining appreciable (visible to the eye) permanent deformation without rupture. The term is also used to denote the property of yielding or flowing under steady load (Ref. 11).

Poisson's ratio: The ratio of lateral unit strain to longitudinal unit strain under the condition of uniform and uniaxial longitudinal stress within the proportional limit.

Polar moment of inertia: See *Moment of inertia of an area.*

Principal axes: The principal axes of an area for a given point in its plane are the two mutually perpendicular axes, passing through the point and lying in the plane of the area, for one of which the moment of inertia is greater and for the other less than for any other coplanar axis passing through that point. If the point in question is the centroid of the area, these axes are called *principal central axes* (see Chap. 5).

Principal moment of inertia: The moment of inertia of an area about either principal axis (see Chap. 5).

Principal planes; principal stresses: Through any point in a stressed body there pass three mutually perpendicular planes, the stress on each of which is purely normal, tension, or compression; these are the *principal planes* for that point. The stresses on these planes are the *principal stresses;* one of them is the maximum stress at the point, and one of them is the minimum stress at the point. When one of the principal stresses is zero, the condition is one of *plane stress;* when two of them are zero, the condition is one of *uniaxial stress.*

Product of inertia of an area: With respect to a pair of rectangular axes in its plane, the sum of the products obtained by multiplying each element of the area dA by its coordinates with respect to those axes x and y; it is therefore the quantity $\int dA\, xy$ (see Chap. 5).

Proof stress: Pertaining to acceptance tests of metals, a specified tensile stress that must be sustained without deformation in excess of a specified amount.

Proportional limit: The greatest stress that a material can sustain without deviating from the law of stress-strain proportionality. (Compare *Elastic limit; Apparent elastic limit; Yield point; Yield strength.* See Art. 2.2 and Ref. 6.)

Radius of gyration: The radius of gyration of an area with respect to a given axis is the square root of the quantity obtained by dividing the moment of inertia of the area with respect to that axis by the area (see Chap. 5).

Reduction of area: The difference between the cross-sectional area of a tension specimen at the section of rupture before loading and after rupture, usually expressed as a percentage of the original area.

Rupture factor: Used in reference to brittle materials, i.e., materials in which failure occurs through tensile rupture rather than through excessive deformation. For a member of given form, size, and material, loaded and supported in a given manner, the *rupture factor* is the ratio of the fictitious maximum tensile stress at failure, as calculated by the appropriate formula for elastic stress, to the ultimate tensile strength of the material, as determined by a conventional tension test (Art. 2.11).

Section modulus (section factor): Pertaining to the cross section of a beam, the *section modulus* with respect to either principal central axis is the moment of inertia with respect to that axis divided by the distance from that axis to the most remote point of the section. The section modulus largely determines the flexural strength of a beam of given material.

Set (permanent set, permanent deformation, plastic strain, plastic deformation): Strain remaining after removal of stress.

Shakedown load (stabilizing load): The maximum load that can be applied to a beam or rigid frame and on removal leave such residual moments that subsequent applications of the same or a smaller load will cause only elastic stresses.

Shape factor: The ratio of the plastic section modulus to the elastic section modulus.

Shear center: See *Flexural center.*

Shear lag: On account of shear strain, the longitudinal tensile or compressive bending stress in wide beam flanges diminishes with the distance from the web or webs, and this stress diminution is called *shear lag.*

Singularity functions: A class of functions that, when used with some caution, permit expressing in one equation what would normally be expressed in several separate equations, with boundary conditions being matched at the ends of the intervals over which the several separate expressions are valid. *Singularity functions* are commonly employed in heat transfer and electromagnetic theory as well as in the evaluation of shears, bending moments, and deformations in beams, plates, and shells. The singularity function most often expressed in this reference is the *step function,* written with the bracket notation $\langle x - a \rangle^0$, which is defined as having a value of *zero* if $x < a$ and a value of *unity* if $x > a$. The indeterminate value when $x = a$ is of no consequence since herein it is always multiplied by another function that will have a value of *zero* when $x = a$. Integrals of the step function, such as the *ramp function* $\langle x - a \rangle^1$ and all others expressed as $\langle x - a \rangle^n$, are defined as having a value of zero if $x < a$ and the normal functional values $(x - a)^n$ if $x > a$.

Slenderness ratio: The ratio of the length of a uniform column to the least radius of gyration of the cross section.

Slip lines (Lüders lines): Lines that appear on the polished surface of a crystal or crystalline body that has been stressed beyond the elastic limit. They represent the intersection of the surface by planes on which shear stress has produced plastic slip or gliding (see Art. 2.5 and Ref. 11).

Step function: See *Singularity function.*

Strain: Any forced change in the dimensions of a body. A stretch is a *tensile strain;* a shortening is a *compressive strain;* and an angular distortion is a *shear strain.* The word *strain* is commonly used to connote *unit strain* (which see); this usage is followed in this book.

Strain concentration factor: See *Factor of strain concentration.*

Strain energy (elastic energy, potential energy of deformation): Mechanical energy stored up in stressed material. Stress within the elastic limit is implied; therefore, the strain energy is equal to the work done by the external forces in producing the stress and is recoverable.

Strain rosette: At any point on the surface of a stressed body, strains measured on each of three intersecting gage lines make possible the calculation of the principal stresses. Such gage lines and the corresponding strains are called a *strain rosette.*

Stress: Internal force exerted by either of two adjacent parts of a body upon the other across an imagined plane of separation. When the forces are parallel to the plane, the stress is called *shear stress;* when the forces are normal to the plane, the stress is called *normal stress;* when the normal stress is directed toward the part on which it acts, it is called *compressive stress;* and when it is directed away from the part on which it acts, it is called *tensile stress.* Shear, compressive, and tensile stresses, respectively, resist the tendency of the parts to mutually slide, approach, or separate under the action of applied forces. For brevity, the word *stress* is often used to connote *unit stress* (which see); this usage is followed in this book.

Stress concentration factor: See *Factor of stress concentration.*

Stress solid: The solid figure formed by surfaces bounding vectors drawn at all points of the cross section of a member and representing the unit normal stress at each such point. The stress solid gives a picture of the stress distribution on a section.

Stress-strain diagram (stress diagram): The curve obtained by plotting unit stresses as ordinates against corresponding unit strains as abscissas.

Stress trajectory (isostatic): A line (in a stressed body) tangent to the direction of one of the principal stresses at every point through which it passes.

Supported (simply supported): A condition of support at the ends of a beam or column or at the edge of a plate or shell, that prevents transverse displacement of the edge of the neutral surface but permits rotation and longitudinal displacement. (Compare *Fixed; Guided; Held.*)

System: Denotes any member or assemblage of members such as a composite column, coupling, truss, or other structure.

Torsional center (center of twist, center of torsion, center of shear): If a twisting couple is applied at a given section of a straight member, that section rotates about some point in its plane. This point, which does not move when the member twists, is the *torsional center* of that section. It is sometimes defined as though identical with the flexural center, but the two points do not always coincide. (Compare *Flexural center; Elastic Center; Elastic Axis.* See Refs. 4 and 5.)

True strain (natural strain, logarithmic strain): The integral, over the whole of a finite extension, of each infinitesimal elongation divided by the corresponding momentary length. It is equal to $\log_e (1 + \epsilon)$, where ϵ is the unit normal strain as ordinarily defined (Ref. 12).

True stress: For an axially loaded bar, the load divided by the corresponding actual cross-sectional area. It differs from the stress as ordinarily defined because of the change in area due to loading.

Twisting moment (torque): At any section of a member, the moment of all forces that act on the member to the left (or right) of that section, taken about a polar axis through the flexural center of that section. For sections that are symmetrical about each principal central axis, the flexural center coincides with the centroid (see Refs. 5 and 9).

Ultimate elongation: The percentage of permanent deformation remaining after tensile rupture measured over an arbitrary length including the section of rupture.

Ultimate strength: The *ultimate strength* of a material in tension, compression, or shear, respectively, is the maximum tensile, compressive, or shear stress that the material can sustain calculated on the basis of the ultimate load and the original or unstrained dimensions. It is implied that the condition of stress represents uniaxial tension, uniaxial compression, or pure shear, as the case may be.

Unit strain: *Unit tensile strain* is the elongation per unit length; *unit compressive strain* is the shortening per unit length; and *unit shear strain* is the change in angle (radians) between two lines originally at right angles to each other.

Unit stress: The amount of stress per unit of area. The unit stress (tensile, compressive, or shear) at any point on a plane is the limit, as ΔA approaches 0, of $\Delta P/\Delta A$, where ΔP is the total tension, compression, or shear on an area ΔA that lies in the plane and includes the point. In this book, *stress* connotes *unit stress,* in general.

Vertical shear: Refers to a beam, assumed for convenience to be horizontal and loaded and supported by forces that all lie in a vertical plane. The *vertical shear* at any section of the beam is the vertical component of all forces that act on the beam to the left of the section. The *shear equation* is an expression for the vertical shear at any section in terms of x, the distance to that section measured from a chosen origin, usually taken at the left end of the beam.

Yield point: The lowest stress at which strain increases without increase in stress. For some purposes it is important to distinguish between the *upper* yield point, which is the stress at which the stress-strain diagram first becomes horizontal, and the *lower* yield point, which is the somewhat lower and almost constant stress under which the metal continues to deform. Only a few materials exhibit a true yield point; for other materials the term is sometimes used synonymously with yield strength. (Compare *Yield strength; Elastic limit; Apparent elastic limit; Proportional limit.* See Ref. 6.)

Yield strength: The stress at which a material exhibits a specified permanent deformation or set. The set is usually determined by measuring the departure of the actual stress-strain diagram from an extension of the initial straight portion. The specified value is often taken as a unit strain of 0.002. (See Ref. 6.)

REFERENCES

1. Soderberg, C. R.: Working Stresses, *ASME Paper* A-106, *J. Appl. Mech.*, vol. 2, no. 3, 1935.
2. Unit Stress in Structural Materials (Symposium), *Trans. Am. Soc. Civil Eng.*, vol. 91, p. 388, 1927.
3. von Heydenkampf, G. S.: Damping Capacity of Materials, *Proc. ASTM,* vol. 21, part II, p. 157, 1931.
4. Kuhn, P.: Remarks on the Elastic Axis of Shell Wings, *Nat. Adv. Comm. Aeron., Tech. Note* 562, 1936.

5. Schwalbe, W. L.: The Center of Torsion for Angle and Channel Sections, *Trans. ASME, Paper* APM-54-11, vol. 54, no. 1, 1932.
6. Tentative Definitions of Terms Relating to Methods of Testing, *Proc. ASTM*, vol. 35, part I, p. 1315, 1935.
7. Morley, A.: "Strength of Materials," 5th ed., Longmans, Green & Co., Ltd., 1919.
8. Timoshenko, S.: "Theory of Elasticity," Engineering Societies Monograph, McGraw-Hill Book Company, 1934.
9. Griffith, A. A., and G. I. Taylor: The Problem of Flexure and Its Solution by the Soap Film Method, *Tech. Rep. Adv. Comm. Aeron.* (British), *Reports and Memoranda* no. 399, p. 950, 1917.
10. Airworthiness Requirements for Aircraft, *Aeron. Bull.* 7-A, U.S. Dept. of Commerce, 1934.
11. Nadai, A.: "Plasticity," Engineering Societies Monograph, McGraw-Hill Book Company, 1931.
12. Freudenthal, A. M.: "The Inelastic Behavior of Engineering Materials and Structures," John Wiley & Sons, Inc., 1950.

PART TWO

Facts; Principles; Methods

TWO

The Behavior of Bodies Under Stress

This discussion pertains to what are commonly designated as *structural materials,* that is, materials suitable for structures and members that must sustain loads without suffering damage. In this category are included most of the metals, concrete, wood, brick and tile, stone, glass, some plastics, etc. It is beyond the scope of this book to give more than a mere statement of a few important facts concerning the behavior of a material under stress; extensive literature is available on every phase of the subject, and the articles referred to will serve as an introduction. (*For numerical values of quantities discussed in this section, the reader is referred to Table 38.*)

2.1 *Methods of loading*

The mechanical properties of a material are usually determined by laboratory tests, and the commonly accepted values of ultimate strength, elastic limit, etc., are those found by testing a specimen of a certain form in a certain manner. To apply results so obtained in engineering design requires an understanding of the effects of many different variables, such as form and scale, temperature and other conditions of service, and method of loading.

The method of loading, in particular, affects the behavior of bodies under stress. There are an infinite number of ways in which stress may be applied to a body, but for most purposes it is sufficient to distinguish the types of loading now to be defined.

1. ***Short-time static loading.*** The load is applied so gradually that at any

instant all parts are essentially in equilibrium. In testing, the load is increased progressively until failure occurs, and the total time required to produce failure is not more than a few minutes. In service, the load is increased progressively up to its maximum value, is maintained at that maximum value for only a limited time, and is not reapplied often enough to make fatigue a consideration. The ultimate strength, elastic limit, yield point, yield strength, and modulus of elasticity of a material are usually determined by short-time static testing at room temperature.

2. ***Long-time static loading.*** The maximum load is applied gradually and maintained. In testing, it is maintained for a sufficient time to enable its probable final effect to be predicted; in service, it is maintained continuously or intermittently during the life of the structure. The creep, or flow characteristics, of a material and its probable permanent strength are determined by long-time static testing at the temperatures prevailing under service conditions. (See Art. 2.6.)

3. ***Repeated loading.*** Typically, a load or stress is applied and wholly or partially removed or reversed repeatedly. This type of loading is important if high stresses are repeated for a few cycles or if relatively lower stresses are repeated many times; it is discussed under *Fatigue*. (See Art. 2.8.)

4. ***Dynamic loading.*** The circumstances are such that the rate of change of momentum of the parts must be taken into account. One such condition may be that the parts are given definite *accelerations* corresponding to a controlled motion, such as the constant acceleration of a part of a rotating member or the repeated accelerations suffered by a portion of a connecting rod. As far as stress effects are concerned, these loadings are treated as virtually static and the *inertia forces* (Art. 15.2) are treated exactly as though they were ordinary static loads.

A second type of *quasi-static* loading, *quick static loading,* can be typified by the rapid burning of a powder charge in a gun barrel. Neither the powder, gas, nor any part of the barrel acquires appreciable radial momentum; therefore equilibrium may be considered to exist at any instant and the maximum stress produced in the gun barrel is the same as though the powder pressure had developed gradually.

In static loading and the two types of dynamic loading just described, the loaded member is required to resist a definite *force*. It is important to distinguish this from *impact loading,* where the loaded member is usually required to absorb a definite amount of *energy*.

Impact loading can be divided into two general categories. In the first case a relatively large slow-moving mass strikes a less massive beam or bar and the *kinetic energy* of the moving mass is assumed to be converted into *strain energy* in the beam. All portions of the beam and the moving mass are assumed to stop moving simultaneously. The shape of the elastic axis of the deflected beam or bar is thus the same as in static loading. A special case of this loading, generally called *sudden loading,* occurs when a mass that is not moving is released when in contact with a beam and falls through the distance the beam deflects. This produces approximately twice the stress and deflection that would have been produced had the mass been "eased" onto the beam (see Art. 15.4). The second case of impact loading involves the mass of the member being struck. *Stress waves* travel through the member during the impact and continue even after the impacting mass has rebounded (see Art. 15.3).

On consideration, it is obvious that methods of loading really differ only in degree. As the time required for the load to be applied increases, short-time static loading changes imperceptibly into long-time static loading; impact may be produced by a body moving so slowly that the resulting stress conditions are practically the same as though equal deflection had been produced by static loading; the number of stress repetitions at which fatigue becomes involved is not altogether definite. Furthermore, all these methods of loading may be combined or superimposed in various ways. Nonetheless, the classification presented is convenient because most structural and machine parts function under loading that may be classified definitely as one of the types described.

2.2 *Elasticity; proportionality of stress and strain*

In determining stress by mathematical analysis, it is customary to assume that material is elastic, isotropic, homogeneous, and infinitely divisible without change in properties and that it conforms to Hooke's law, which states that strain is proportional to stress. Actually, none of these assumptions is strictly true. A structural material is usually an aggregate of crystals, fibers, or cemented particles, the arrangement of which may be either random or systematic. When the arrangement is random, the material is essentially isotropic if the part considered is large in comparison with the constituent units; when the arrangement is systematic, the elastic properties and strength are usually different in different directions and the material is anisotropic. Again, when subdivision is carried to the point where the part under consideration comprises only a portion of a single crystal, fiber, or other unit, in all probability its properties will differ from those of a larger part that is an aggregate of such units. Finally, very careful experiments show that for all materials there is probably some set and some deviation from Hooke's law for any stress, however small.

These facts impose certain limitations upon the conventional methods of stress analysis and must often be taken into account, but formulas for stress and strain, mathematically derived and based on the assumptions stated, give satisfactory results for nearly all problems of engineering design. In particular, Hooke's law may be regarded as practically true up to a proportional limit, which, though often not sharply defined, can be established for most materials with sufficient definiteness. So, too, a fairly definite elastic limit is determinable; in most cases it is so nearly equal to the proportional limit that no distinction need be made between the two.

2.3 *Factors affecting elastic properties*

For ordinary purposes it may be assumed that the elastic properties of most metals, when stressed below a nominal proportional limit, are constant with respect to stress, unaffected by ordinary atmospheric variations of temperature, unaffected by prior applications of moderate stress, and independent of the rate of loading. When precise relations between stress and strain are important, as in the design or calibration of instruments, these assumptions cannot always be made. The fourth edition of this book (Ref. 1) discussed in detail the effects of strain rate, temperature, etc., on the elastic properties of many metals and gave references for the experiments performed. The relationships between atomic and molecular structure and the elastic properties are discussed in Ref. 2.

Wood exhibits a higher modulus of elasticity and much higher proportional limit when tested rapidly than when tested slowly. The standard impact test on a beam indicates a fiber stress at the proportional limit approximately twice as great as that found by the standard static bending test. Absorption of moisture up to the fiber saturation point greatly lowers both the modulus of elasticity and the proportional limit (Ref. 4).

Both concrete and cast iron have stress-strain curves more or less curved throughout, and neither has a definite proportional limit. For these materials it is customary to define E as the ratio of some definite stress (for example, the allowable stress or one-fourth the ultimate strength) to the corresponding unit strain; the quantity so determined is called the *secant* modulus since it represents the slope of the secant of the stress-strain diagram drawn from the origin to the point representing the stress chosen. The moduli of elasticity of cast iron are much more variable than those of steel, and the stronger grades are stiffer than the weaker ones. Cast iron suffers a distinct set from the first application of even a moderate stress; but after several repetitions of that stress, the material exhibits perfect elasticity up to, but not beyond, that stress. The modulus of elasticity is slightly less in tension than in compression (Ref. 5).

Concrete also shows considerable variation in modulus of elasticity, and in general its stiffness increases with its strength. Like cast iron, concrete can be made to exhibit perfect elasticity up to a moderate stress by repeated loading up to that stress. Because of its tendency to yield under continuous loading, the modulus of elasticity indicated by long-time loading is much less than that obtained by progressive loading at ordinary speeds (Refs. 3 and 18).

2.4 *Load-deformation relation for a body*

If Hooke's law holds for the material of which a member or structure is composed, the member or structure will usually conform to a similar law of load-deformation proportionality and the deflection of a beam or truss, the twisting of a shaft, the dilation of a pressure container, etc., may in most instances be assumed proportional to the magnitude of the applied load or loads.

There are two important exceptions to this rule. One is to be found in any case where the stresses due to the loading are appreciably affected by the deformation. Examples of this are: a beam subjected to axial and transverse loads; a flexible wire or cable held at the ends and loaded transversely; a thin diaphragm held at the edges and loaded normal to its plane; a ball pressed against a plate or against another ball; and a helical spring under severe extension.

The second exception is represented by any case in which failure occurs through elastic instability, as in a slender (Euler) column. Here, for loads less than the critical, elastic instability plays no part and the load deformation is linear. At the critical load the type of deformation changes, the column bending instead of merely shortening axially, and the deformation becomes indeterminate. For any load beyond the critical load, failure occurs through excessive deflection (see Art. 2.13).

2.5 *Plasticity*

Elastic deformation represents an actual change in the distance between atoms or molecules; plastic deformation represents a permanent change in their relative positions. In crystalline material this permanent rearrangement consists largely of group displacements of the atoms in the crystal lattice brought about by slip on planes of least resistance, parts of a crystal sliding past one another and in some instances suffering angular displacement. In amorphous material the rearrangement appears to take place through the individual shifting from positions of equilibrium of many atoms or molecules, the cause being thermal agitation due to external work and the result appearing as a more or less uniform flow like that of a viscous liquid. It should be noted that plastic deformation before rupture is much less for biaxial or triaxial tension than for one-way stress; for this reason metals that are ordinarily ductile may prove brittle when thus stressed.

The laws governing plastic deformation are less amenable to mathematical statement than those assumed to govern elastic behavior, but a mathematical theory of plastic action is being developed. Important applications are to the prediction of the ultimate strength and postbuckling behavior of structures and in the study of metal working, creep, and flow (see Refs. 6 to 8).

2.6 *Creep and rupture under long-time loading*

Most materials will creep or flow to some extent and eventually fail under a sustained stress less than the short-time ultimate strength. After a short time at load, the initial creep related to stress redistribution in the structure and strain hardening ccases and the *steady state,* or *viscous* creep, predominates. The viscous creep will continue until fracture unless the load is reduced sufficiently, but it is seldom important in materials at temperatures less than 40 to 50 percent of their absolute melting temperatures. Thus, creep and long-time strength at atmospheric temperatures must sometimes be taken into account in designing members of nonferrous metals and in selecting allowable stresses for wood, plastics, and concrete.

Metals. Creep is an important consideration in high-pressure steam and distillation equipment, gas turbines, nuclear reactors, supersonic vehicles, etc. Lubahn and Felgar (Ref. 8) relate creep to plasticity. Marin, Odqvist, and Finnie, in Ref. 9, give excellent surveys and list references on creep in metals and structures. Conway (Refs. 10 and 11) discusses the effectiveness of various parametric equations, and Conway and Flagella (Ref. 12) present extensive creep-rupture data for the refractory metals. Odqvist (Ref. 13) discusses the theory of creep and its application to large deformation and stability problems in plates, shells, membranes, and beams and tabulates creep constants for 15 common metals and alloys. Hult (Ref. 14) also discusses creep theory and its application to many structural problems. Penny and Marriott (Ref. 15) discuss creep theories and the design of experiments to verify them. They also discuss the development of several metals for increased resistance to creep at high temperatures as well as polymeric and composite materials at lower temperatures. Reference 16 is a series of papers with extensive references covering creep theory, material properties, and structural problems.

Plastics. The literature on the behavior of the many plastics being used for structural or machine applications is too extensive to list here. Reference 17 summarizes much of the data available and gives the necessary references.

Concrete. Under sustained compressive stress concrete suffers considerable plastic deformation and may flow for a very long time at stresses less than the ordinary working stress. Continuous flow has been observed over a period of 10 years, though ordinarily it ceases or becomes imperceptible within 1 or 2 years. The rate of flow is greater for air than for water storage, greater for small than for large specimens, and for moderate stresses increases approximately as the applied stress. On removal of stress, some elastic recovery occurs. Concrete also shows creep under tensile stress, the early creep rate being greater than the flow rate under compression (Refs. 18 and 23).

Under very gradually applied loading concrete exhibits an ultimate strength considerably less than that found under short-time loading; in certain compression tests it was found that increasing the time of testing from 1 s to 4 h decreased the unit stress at failure about 30 percent, most of this decrease occurring between the extremely quick (1 or 2 s) and the conventional (several minutes) testing. This indicates that the compressive stress that concrete can sustain indefinitely may be considerably less than the ultimate strength as determined by a conventional test. On the other hand, the long-time imposition of a moderate loading appears to have no harmful effect; certain tests show that after 10 years of constant loading equal to one-fourth the ultimate strength, the compressive strength of concrete cylinders is practically the same and the modulus of elasticity is considerably greater than for similar cylinders that were not kept under load (Ref. 22).

The modulus of rupture of plain concrete also decreases with the time of loading, and some tests indicate that the long-time strength in cross-breaking may be only 55 to 75 percent of the short-time strength (Ref. 19).

Ref. 24 is a compilation of 12 papers, each with extensive references, dealing with the effect of volumetric changes on concrete structures. Design modifications to accommodate these volumetric changes is the main thrust of the papers.

Wood. Wood also yields under sustained stress; the long-time (several years) strength is about 55 percent of the short-time (several minutes) strength in bending; for direct compression parallel to the grain the corresponding ratio is about 75 percent (Ref. 4).

2.7 *Criteria of elastic failure and of rupture*

For the purpose of this discussion it is convenient to divide metals into two classes: (1) *ductile* metals, in which marked plastic deformation commences at a fairly definite stress (yield point, yield strength, or possibly elastic limit) and which exhibit considerable ultimate elongation; and (2) *brittle* metals, for which the beginning of plastic deformation is not clearly defined and which exhibit little ultimate elongation. Mild steel is typical of the first class, and cast iron is typical of the second; an ultimate elongation of 5 percent has been suggested as the arbitrary dividing line between the two classes of metals (Ref. 20).

A ductile metal is usually considered to have failed when it has suffered *elastic failure,* i.e., when marked plastic deformation has begun. Under simple uniaxial tension this occurs when the stress reaches a value we will denote by σ_y, which represents the yield strength, yield point, or elastic limit, according to which one of these is the most satisfactory indication of elastic failure for the material in question. The question arises, when does elastic failure occur under other conditions of stress, such as compression, shear, or a combination of tension, compression, and shear?

The four theories of elastic failure that have received the widest acceptance at various times are: (1) the *maximum stress theory,* which states that elastic failure occurs when the maximum tensile stress becomes equal to σ_y; (2) the *maximum strain theory,* which states that elastic failure occurs when the maximum tensile strain becomes equal to σ_y/E; (3) the *maximum shear stress theory,* which states that elastic failure occurs when the maximum shear stress becomes equal to $\frac{1}{2}\sigma_y$; and (4) the *theory of constant energy of distortion,* which states that elastic failure occurs when the principal stresses σ_1, σ_2, and σ_3 satisfy the equation

$$(\sigma_1 - \sigma_2)^2 + (\sigma_2 - \sigma_3)^2 + (\sigma_3 - \sigma_1)^2 = 2\,\sigma_y^2$$

Of these four theories, the fourth is the one that agrees best with experimental evidence, but the third leads to results so nearly the same and is so much simpler in application that it is much more widely used as a basis for design (Refs. 20 and 21).

The criteria just discussed concern the elastic failure of *material.* Such failure may occur locally in a *member* and may do no real damage if the volume of material affected is so small or so located as to have only negligible influence on the form and strength of the member as a whole. Whether or not such local overstressing is significant depends upon the properties of the material and the conditions of service. Fatigue properties, resistance to impact, and mechanical functioning are much more likely to be affected than static strength, and a degree of local overstressing that would constitute failure in a high-speed machine part might be of no consequence whatever in a bridge member.

A brittle material cannot be considered to have definitely failed until it has broken, which can occur either through a *tensile fracture,* when the maximum tensile stress reaches the ultimate strength, or through what appears to be a *shear fracture,* when the maximum compressive stress reaches a certain value. The fracture occurs on a plane oblique to the maximum compressive stress but not, as a rule, on the plane of maximum shear stress, and so it cannot be considered to be purely a shear failure (see Ref. 21). The results of some tests on glass and Bakelite (Ref. 45) indicate that for these brittle materials either the maximum stress or the maximum strain theory affords a satisfactory criterion of rupture while neither the maximum shear stress nor the constant energy of distortion theory does. These tests also indicated that strength increases with rate of stress application and that the increase is more marked when the location of the most stressed zone changes during loading (pressure of a sphere on a flat surface) than when this zone is fixed (axial tension).

Another failure theory that is applicable to brittle materials is the *Mohr theory of failure.* Brittle materials have compressive strengths greater than their tensile strengths, and therefore both a uniaxial tensile test and a uniaxial compressive test must be run to use the Mohr theory. First we draw on a single plot both Mohr's stress circle for the tensile test at the instant of failure and Mohr's stress circle for the compressive test at the instant of failure; then

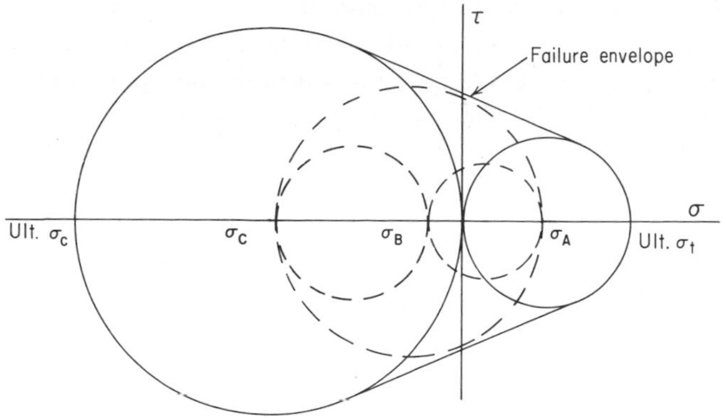

Fig. 2.1

we complete a failure envelope simply by drawing a pair of tangent lines to the two circles, as shown in Fig. 2.1.

Failure under a complex stress situation is expected if the largest of the three Mohr circles for the given situation touches or extends outside the envelope just described. If all normal stresses are tensile, the results coincide with the maximum stress theory. For a condition where the three principal stresses are σ_A, σ_B, and σ_C, as shown in Fig. 2.1, failure is being approached but will not take place unless the dotted circle passing through σ_A and σ_C reaches the failure envelope.

The accurate prediction of the breaking strength of a member composed of brittle metal requires a knowledge of the effect of *form and scale,* and these effects are expressed by the *rupture factor* (see Art. 2.11). In addition, what has been said here concerning brittle metals applies also to any essentially isotropic brittle material.

Thus far, our discussion of failure has been limited to *isotropic* materials. For wood, which is distinctly *anisotropic,* the possibility of failure in each of several ways and directions must be taken into account, viz.: (1) by tension parallel to the grain, which causes fracture; (2) by tension transverse to the grain, which causes fracture; (3) by shear parallel to the grain, which causes fracture; (4) by compression parallel to the grain, which causes gradual buckling of the fibers usually accompanied by a shear displacement on an oblique plane; (5) by compression transverse to the grain, which causes sufficient deformation to make the part unfit for service. The unit stress producing each of these types of failure must be ascertained by suitable tests (Ref. 4).

2.8 *Fatigue*

Practically all materials will break under numerous repetitions of a stress that is not as great as the stress required to produce immediate rupture. This phenomenon is known as *fatigue* and was formerly attributed to a supposed alteration in structure brought about by recurrent stressing. Metals, for instance, were said to crystallize, a belief probably arising from the facts that fatigue fractures often occur at points of weakness where the grain structure is abnormally coarse and, even in ductile material, are square and crystalline in appearance because of the absence of gross plastic flow.

Over the past 100 years the effects of surface condition, corrosion, temperature, etc., on fatigue properties have been well documented, but only in recent years has the microscopic cause of fatigue damage been attributed to *cyclic plastic flow* in the material at the source of a fatigue crack (*crack initiation*) or at the tip of an existing fatigue crack (*crack propagation;* Ref. 33). The development of extremely sensitive extensometers has permitted the separation of elastic and plastic strains when testing axially loaded specimens over short gage lengths. With this instrumentation it is possible to determine whether cyclic loading is accompanied by significant cyclic plastic strain and, if it is, whether the cyclic plastic strain continues at the same level, increases, or decreases. Sandor (Ref. 35) discusses this instrumentation and its use in detail.

It is not feasible to reproduce here even a small portion of the fatigue data available for various engineering materials. The reader should consult materials handbooks, manufacturers' literature, and design manuals such as Refs. 25 to 29, 47, 49, and 53. Some of the more important factors governing fatigue behavior in general will be outlined in the following material.

Number of cycles to failure. Most data concerning the number of cycles to failure are presented in the form of an S/N curve where the cyclic stress amplitude is plotted versus the number of cycles to failure. This generally leads to a straight line plot on log paper if we account for the scatter in the data. For the ferrous metals a lower limit exists on stress amplitude called the *fatigue limit,* or *endurance limit.* This generally occurs at a life of from 10^5 to 10^7 cycles of reversed stress and we assume that stresses below this limit will not cause failure regardless of the number of repetitions. With the ability to separate elastic and plastic strains accurately, there are instances when a plot of plastic-strain amplitudes versus N and elastic-strain amplitudes versus N will reveal more useful information (Refs. 30, 34, 35, 71, and 72; and Morrow in Ref. 31, p. 45).

Rate of loading. Up to a frequency of 10,000 c/min, the rate of stress repetition does not appear to affect the endurance limit of steel; for higher frequencies, the endurance limit increases somewhat, reaching a maximum value at from 1200 to 1800 Hz, beyond which point there is a decrease (Ref. 37).

Method of loading and size of specimen. Uniaxial stress can be produced by axial load, bending, or a combination of both. In flat-plate bending only the upper and lower surfaces are subjected to the full range of cyclic stress. In rotating bending all surface layers are similarly stressed, but in axial loading the entire cross section is subjected to the same average stress. Since fatigue properties of a material depend upon the statistical distribution of defects throughout the specimen, it is apparent that the three methods of loading will produce different results.

In a similar way the size of a bending specimen will affect the fatigue behavior while it will have little effect on an axially loaded specimen. Several empirical formulas have been proposed to represent the influence of size on a machine part or test specimen in bending. For steel, Moore (Ref. 69) suggests the equation

$$\sigma_e'\left(1 - \frac{0.016}{d'}\right) = \sigma_e''\left(1 - \frac{0.016}{d''}\right)$$

where σ_e' is the endurance limit for a specimen of diameter d' and σ_e'' is the endurance limit for a specimen of diameter d''. This formula was based on test results obtained with specimens from 0.125 to 1.875 in in diameter and shows good agreement within that size range. Obviously it cannot be used for predicting the endurance limit of very small specimens.

Heywood (Ref. 49) suggests an equation equivalent to

$$\sigma_d - \left(1 + \frac{0.014}{0.1 + d^2}\right)\sigma_D$$

where σ_d is the predicted endurance limit for a specimen of diameter d and σ_D is the experimentally determined endurance limit for a specimen of diameter $D > d$; in this reference $D = 2$ in. This formula also shows reasonably good agreement with test results up to a diameter of 2 in. The few relevant test results available indicate a considerable decrease in endurance limit for very large diameters (Refs. 39 to 41).

Stress concentrations. Fatigue failures occur at stress levels less than those necessary to produce the gross yielding which would blunt the sharp rise in stress at a stress concentration. It is necessary, therefore, to apply the fatigue strengths of a smooth specimen to the peak stresses expected at the stress concentrations unless the size of the stress-concentrating notch or fillet approaches the grain size or the size of an anticipated defect in the material itself (see *Factor of stress concentration in fatigue* in Chap. 1 and in Art. 2.10). References 74 and 75 discuss the effect of notches on low-cycle fatigue.

Surface conditions. Surface roughness constitutes a kind of stress raiser, and its effect is indicated in Table 37. Discussion of the effect of surface coatings and platings is beyond the scope of this book (see Refs. 49 and 67).

Corrosion fatigue. Under the simultaneous action of *corrosion* and repeated stress, the fatigue strength of most metals is drastically reduced, sometimes to a small fraction of the strength in air, and a true endurance limit can no longer be said to exist. Liquids and gases not ordinarily thought of as especially conducive to corrosion will often have a very deleterious effect on fatigue properties, and resistance to corrosion is more important than normal fatigue strength in determining the relative rating of different metals (Refs. 41, 42, 48, and 53).

Range of stress. Stressing a ductile material beyond the elastic limit or yield point in tension will raise the elastic limit for subsequent cycles but lower the elastic limit for compression. For a more detailed description of the Bauschinger effect, see Ref. 2. The consequence of this effect on fatigue is apparent if one accepts the statement that fatigue damage is a result of cyclic plastic flow; i.e., if the range of cyclic stress is reduced sufficiently, higher peak stresses can be accepted without suffering continuing damage.

Various empirical formulas for the endurance limit corresponding to any given range of stress variation have been suggested, the most generally accepted of which is expressed by the *Goodman diagram* or some modification thereof. Figure 2.2 shows one method of constructing this diagram. In each cycle the stress varies from a maximum value σ_{max} to a minimum value σ_{min},

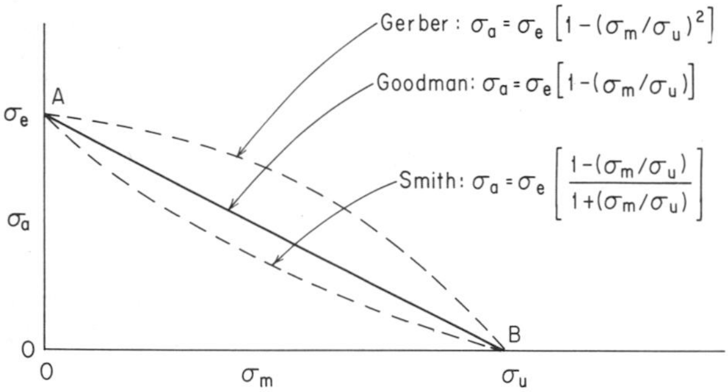

Fig. 2.2

either of which is plus or minus according to whether it is tensile or compressive. The *mean stress* is

$$\sigma_m = \tfrac{1}{2}(\sigma_{max} + \sigma_{min})$$

and the *alternating stress* is

$$\sigma_a = \tfrac{1}{2}(\sigma_{max} - \sigma_{min})$$

the addition and subtraction being algebraic. With reference to rectangular axes, σ_m is measured horizontally and σ_a vertically. Obviously when $\sigma_m = 0$, the limiting value of σ_a is the endurance limit for fully reversed stress, denoted here by σ_e. When $\sigma_a = 0$, the limiting value of σ_m is the ultimate tensile strength, denoted here by σ_u. Points *A* and *B* on the axes are thus located.

According to the Goodman theory, the ordinate to a point on the straight line *AB* represents the maximum alternating stress σ_a that can be imposed in conjunction with the corresponding mean stress σ_m. Any point above *AB* represents a stress condition that would eventually cause failure; any point below *AB* represents a stress condition with more or less margin of safety.

A more conservative construction, suggested by Soderberg (Ref. 20), is to move B back to σ_y, the yield strength. A less conservative but sometimes preferred construction, proposed by Gerber, is to replace the straight line by the parabola.

The Goodman diagrams described can be used for steel and for aluminum and titanium alloys, but for cast iron many test results fall below the straight line AB and the lower curved line, suggested by Smith (Ref. 36), is preferred. Test results for magnesium alloys also sometimes fall below the straight line.

Figure 2.2 represents conditions where σ_m is tensile. If σ_m is compressive, σ_a is increased; and for values of σ_m less than the compression yield strength, the relationship is represented approximately by the straight line AB extended to the left with the same slope. When the mean stress and alternating stress are both torsional, σ_a is practically constant until σ_m exceeds the yield strength in shear; and for alternating bending combined with mean torsion, the same thing is true. But when σ_m is tensile and σ_a is torsional, σ_a diminishes as σ_m increases in almost the manner represented by the Goodman line. When stress concentration is to be taken into account, the accepted practice is to apply k_f (or k, if k_f is not known) to σ_a only, not to σ_m.

Residual stress. Since residual stresses, whether deliberately introduced or merely left over from manufacturing processes, will influence the *mean* stress, their effects can be accounted for. One should be careful, however, not to expect the beneficial effects of a residual stress if during the expected life of a structure it will encounter overloads sufficient to change the residual-stress distribution. Sandor (Ref. 35) discusses this in detail and points out that an occasional overload might be beneficial in some cases.

Combined stress. No one of the theories of failure discussed in Art. 2.7 can be applied to all fatigue loading conditions. The *constant energy of distortion theory* seems to be conservative in most cases, however. Reference 25 gives a detailed description of an acceptable procedure for designing for fatigue under conditions of combined stress. The procedure described also considers the effect of mean stress on the cyclic stress range. Three criteria for failure are discussed: *gross yielding*, *crack initiation*, and *crack propagation*. An extensive discussion of fatigue under combined stress is found in Refs. 38, 47, and 53.

Stress history. A very important question and one that has been given much attention is the influence of previous stressing on fatigue strength. One theory that has had considerable acceptance is the *linear damage law* (Miner in Ref. 47); here the assumption is made that the damage produced by repeated stressing at any level is directly proportional to the number of cycles. Thus, if the number of cycles producing failure (100 percent damage) at a stress range σ_1 is N_1, then the proportional damage produced by N cycles of the stress is N/N_1 and stressing at various stress levels for various numbers of cycles causes cumulative damage equal to the summation of such fractional values. Failure occurs, therefore, when $\Sigma N/N_1 = 1$. The formula implies that the effect of a given number of cycles is the same, whether they are applied continuously or intermittently, and does not take into account the fact that for some metals understressing (stressing below the endurance limit) raises the endurance limit. The linear damage law is not reliable for all stress conditions, and various modifications have been proposed, such as replacing 1 in the formula by a quantity x whose numerical value, either more or less than unity, must be determined experimentally. Attempts have been made to develop a better theory (e.g., Corten and Dolan, Freudenthal and Gumbel, in Ref. 54). Though all the several theories are of value when used knowledgeably, it does not appear that as yet any generally reliable method is available for predicting the life of a stressed part under variable or random loading. (See Refs. 32 and 73.)

A modification of the foil strain gage called an *S/N fatigue life gage* (Refs. 55 and 56) measures accumulated plastic deformation in the form of a permanent change in resistance. A given total change in resistance can be correlated with the damage necessary to cause a fatigue failure in a given material.

Structural joints. The fatigue properties of riveted, bolted, and welded structural joints have been studied extensively. Reference 26 discusses extensively the effects of various welding techniques on steel structures and gives references to the work done. Reference 34 gives information on riveted, bolted, and welded joints in materials used in aircraft (see also Refs. 43, 44, and 49).

Wood and concrete. The fatigue properties of wood and concrete have been less thoroughly investigated than those of metals. Neither material can be said to have a true endurance limit. Tests in reversed bending on both solid wood and plywood have indicated an endurance strength at 30 million cycles of about 30 percent of the static modulus of rupture. The weakening influence of sloping grain, knots, and drilled holes is greater for repeated than for static loading (Ref. 4).

For concrete, the endurance strength for 1 to 2 million cycles of compression from zero to a maximum is 50 to 55 percent of the ultimate compressive strength, and the endurance strength determined by either one-way or reversed bending of plain beams is about 50 percent of the static modulus of rupture. The endurance strength of concrete, like that of steel, can be slightly raised by *understressing* (Ref. 3).

2.9 *Brittle fracture*

Brittle fracture is a term applied to an unexpected brittle failure of a material such as low-carbon steel where large plastic strains are usually noted before actual separation of the part. For a brittle fracture to take place the material must be subjected to a tensile stress at a location where a crack or other very sharp notch or defect is present and the temperature must be lower than the so-called *transition temperature*. To determine a transition temperature for a given material, a series of notched specimens is tested under impact loading, each at a different temperature, and the ductility or the energy required to cause fracture is noted. There will be a limited range of temperatures over which the ductility or fracture energy will drop significantly. Careful examination of the fractured specimens will show that the material at the root of the notch has tried to contract laterally. Where the fracture energy is large, there is evidence of a large lateral contraction; and where the fracture energy is small, the lateral contraction is essentially zero. In all cases the lateral contraction is resisted by the adjacent less stressed material. The deeper and sharper cracks have relatively more material to resist lateral contraction. Thicker specimens have a greater distance over which to build up the necessary *triaxial* tensile stresses that lead to a tensile failure without producing enough shear stress to cause yielding. Thus, the term *transition temperature* is somewhat relative since it depends upon notch geometry as well as specimen size and shape. Since yielding is a flow phenomenon, it is apparent that rate of loading is also important. Static loading of sufficient intensity may start a brittle fracture, but it can continue under much lower stress levels owing to the higher rate of loading.

A safe design where brittle fracture is a possibility requires a consideration of defect size and shape such as voids and inclusions in castings, forgings, and weldments, residual stresses due to processing and assembly, rate of loading, material thickness in the structure, and ambient temperature when under load. A wealth of good practical information can be found in Refs. 57 to 64, 70, and 76. Pellini (Refs. 62 and 63) discusses the theory of brittle fracture extensively and gives many more references to specific problems and applications.

2.10 *Stress concentration*

The distribution of elastic stress across the section of a member may be nominally uniform or may vary in some regular manner, as illustrated by the linear distribution of stress in flexure. When the variation is abrupt so that within a very short distance the intensity of stress increases greatly, the condition is described as *stress concentration.* It is usually due to local irregularities of form such as small holes, screw threads, scratches, and similar *stress raisers.* There is obviously no hard and fast line of demarcation between the rapid variation of stress brought about by a stress raiser and the variation that occurs in such members as sharply curved beams, but in general the term *stress concentration* implies some form irregularity not inherent in the member as such but accidental (tool marks) or introduced for some special purpose (screw thread).

The maximum intensity of elastic stress produced by many of the common kinds of stress raisers can be ascertained by mathematical analysis, photoelastic analysis, or direct strain measurement and is usually expressed by the *factor of stress concentration.* This term has been defined (Chap. 1), but its meaning may be made clearer by an example. Consider a straight rectangular beam, originally of uniform breadth b and depth D, which has had cut across the lower face a fairly sharp transverse V notch of uniform depth h, making the net depth of the beam section at that point $D - h$. If now the beam is subjected to a uniform bending moment M, the *nominal* fiber stress at the root of the notch may be calculated by the ordinary flexure formula $\sigma = Mc/I$, which here reduces to $\sigma = M/\frac{1}{6}b(\text{D} - h)^2$. But the actual stress σ' is very much greater than this because of the stress concentration that occurs at the point of the notch. The ratio σ'/σ, actual stress divided by nominal stress, is the factor of stress concentration k for this particular case. Values of k (or k_t, to use the more common current notation) for a number of common stress raisers are given in Table 37. The most complete single source for numerical values of stress concentration factors is Peterson (Ref. 77). It also contains an extensive bibliography.

The abrupt variation and high local intensity of stress produced by stress raisers are characteristics of *elastic behavior.* The plastic yielding that occurs on overstressing greatly mitigates stress concentration even in relatively brittle materials and causes it to have much less influence on breaking strength than might be expected from a consideration of the elastic stresses only. The practical significance of stress concentration therefore depends on circumstances. For ductile metal under static loading it is usually (though not always) of little or no importance; for example, the high stresses that occur at the edges of rivet holes in structural steel members are safely ignored, the stress due to a tensile load being assumed uniform on the net section. (In the case of eyebars and other pin-connected members, however, a reduction of 25 percent in allowable stress on the net section is recommended.) For brittle material under static loading, stress concentration is often a serious consideration, but its effect varies widely and cannot be predicted either from k or from the brittleness of the material (see Ref. 65).

What may be termed the *factor of stress concentration at rupture,* or the *factor of strength reduction,* represents the significance of stress concentration for static loading. This factor, which will be denoted by k_r, is the ratio of the computed stress at rupture for a plain specimen to the computed stress at rupture for the specimen containing the stress raiser. For the case just described, it would be the ratio of the modulus of rupture of the plain beam to that of the notched beam, the latter being calculated for the net section. k_r is therefore a ratio of stresses, one or both of which may be fictitious, but is nonetheless a measure of the strength-reducing effect of stress concentration. Some values of k_r are given in Table 17 of Ref. 1.

It is for conditions involving fatigue that stress concentration is most important. Even the most highly localized stresses, such as those produced by small surface scratches, may greatly lower the apparent endurance limit, but materials vary greatly in *notch sensitivity,* as susceptibility to this effect is sometimes called. Contrary to what might be expected, ductility (as ordinarily determined by axial testing) is not a measure of immunity to stress concentration in fatigue; for example, steel is much more susceptible than cast iron. What may be termed the *factor of stress concentration for fatigue* k_f is the practical measure of notch sensitivity. It is the ratio of the endurance limit of a plain specimen to the nominal stress at the endurance limit of a specimen containing the stress raiser.

A study of available experimental data shows that k_f is almost always less, and often significantly less, than k, and various methods for estimating k_f from k have been proposed. Neuber (Ref. 68) proposes the formula

$$k_f = 1 + \frac{k - 1}{1 + \pi \sqrt{\rho'/\rho}/(\pi - w)}$$

where ω is the flank angle of the notch (called θ in Table 37), ρ is the radius of curvature (in inches) at the root of the notch (called r in Table 37), and ρ' is a dimension related to the grain size, or size of some type of basic building block, of the material and may be taken as 0.0189 in for steel. Moore (Ref. 69) states that for steel both the size effect and ductility can be taken into account by taking

$$\rho' = 0.2\left(1 - \frac{\sigma_y}{\sigma_u}\right)^3\left(1 - \frac{0.05}{d}\right)$$

where σ_y = yield strength, σ_u = ultimate tensile strength, and d = minimum diameter or depth of specimen (in inches). This formula yields results in good agreement with tests over a limited range of sizes. Heywood (Ref. 49) suggests a relationship that can be expressed by the formula

$$k_f = \frac{k}{1 + 2\,[(k - 1)/k]\,\sqrt{a/R}}$$

where R is the root radius of the notch and $\sqrt{a} = 5000/\sigma_u$, σ_u again is the ultimate tensile strength. This formula also shows good agreement with a large number of test results.

It has been suggested (Ref. 46) that the relationship between k and k_f can be expressed by the equation $(k_f - 1)/(k - 1) = q$, a constant called the *notch-sensitivity index*. For a given *material* q does seem to be approximately constant for all ordinary types of stress raisers except those that are very sharp (for example, V notch with root radius nominally zero), and so by ascertaining the value of q for a case where both k and k_f are known, the value of k_f can be estimated for a different case where only k is known. But a common value of q cannot be used for different materials.

All the methods described are valuable and applicable within certain limitations, but none can be applied with confidence to all situations (Ref. 50). Probably none of them gives sufficient weight to the effect of scale in the larger size range. There is abundant evidence to show that the significance of stress concentration increases with size for both static and repeated loading, especially the latter. Of geometrically similar specimens, the larger indicate higher values for both k_r and k_f (Refs. 51, 65, 66, and 69). For this reason, when giving values of these factors that have been experimentally determined, it is important to state the dimensions of the specimens tested. On the whole, it seems doubtful that any completely reliable criterion of notch sensitivity is available except tests on specimens of the material, shape, and approximate size of the part under consideration, and for this reason it is a growing practice to make fatigue tests on the actual part or assembly of parts with which the design problem is concerned.

An important fact concerning stress concentration is that a single isolated notch or hole has a worse effect than have a number of similar stress raisers placed close together; thus, a single V groove reduces the strength of a part more than does a continuous screw thread of almost identical form. The deleterious effect of an unavoidable stress raiser can, therefore, be mitigated sometimes by juxtaposing additional form irregularities of like nature, but the actual superposition of stress raisers, such as the introduction of a small notch in a fillet, may result in a stress concentration factor equal to or even exceeding the product of the factors for the individual stress raisers (Refs. 52 and 78).

2.11 *Effect of form and scale on strength; rupture factor*

It has been pointed out (Art. 2.7) that a member composed of brittle material breaks in tension when the maximum tensile stress reaches the ultimate strength or in shear when the maximum compressive stress reaches a certain value. In calculating the stress at rupture in such a member it is customary to employ an *elastic-stress formula;* thus the ultimate fiber stress in a beam is usually calculated by the ordinary flexure formula. It is known that the result (modulus of rupture) is not a true stress, but it can be used to predict the strength of a similar beam of the same material. However, if another beam of the same material but of different cross section, span/depth ratio, size, or manner of loading and support is tested, the modulus of rupture will be found to be different. (The effect of the shape of the section is often taken into account by the *form factor,* and the effects of the span/depth ratio and manner of loading are recognized in the testing procedure.) Similarly, the *calculated* maximum stress at rupture in a curved beam, flat plate, or torsion member is not equal to the ultimate strength of the material, and the magnitude of the disparity will vary greatly with the material, form of the member, manner of loading, and absolute scale. In order to predict accurately the breaking load for such a member, it is necessary to take this variation into account, and the *rupture factor* (defined in Chap. 1) provides a convenient means of doing so. Values of the rupture factor for a number of materials and types of member are given in Table 18 of Ref. 1.

On the basis of many experimental determinations of the rupture factor (Ref. 65) the following generalizations may be made:

1. The smaller the proportional part of the member subjected to high stress, the larger the rupture factor. This is exemplified by the facts that a beam of circular section exhibits a higher modulus of rupture than a rectangular beam and that a flat plate under a concentrated center load fails at a higher computed stress than one uniformly loaded. The extremes in this respect are, on the one hand, a uniform bar under axial tension for which the rupture factor is unity and, on the other hand, a case of severe stress concentration such as a sharply notched bar for which the rupture factor may be indefinitely large.

2. In the flexure of statically indeterminate members, the redistribution of bending moments that occurs when plastic yielding starts at the most highly stressed section increases the rupture factor. For this reason a flat plate gives a higher value than a simple beam, and a circular ring gives a higher value than a portion of it tested as a statically determinate curved beam.

3. The rupture factor seems to vary inversely with the absolute scale for conditions involving abrupt stress variation, which is consistent with the fact (already noted) that for cases of stress concentration both k_r and k_f diminish with the absolute scale.

4. As a rule, the more brittle the material, the more nearly all rupture factors approach unity. There are, however, many exceptions to this rule. It has been pointed out (Art. 2.10) that immunity to notch effect even under static loading is not always proportional to ductility.

The practical significance of these facts is that for a given material and given factor of safety, some members may be designed with a much higher allowable stress than others. This fact is often recognized in design; for example, the allowable stress for wooden airplane spars varies according to the form factor and the proportion of the stress that is flexural.

What has been said here pertains especially to comparatively brittle materials, i.e., materials for which failure consists in fracture rather than in the beginning of plastic deformation. The effect of form on the ultimate strength of ductile members is less important, although even for steel the allowable unit stress is often chosen with regard to circumstances such as those discussed previously. For instance, in gun design the maximum stress is allowed to approach and even exceed the nominal elastic limit, the volume of material affected being very small, and in structural design extreme fiber stresses in bending are permitted to exceed the value allowed for axial loading. In testing, account must be taken of the fact that some ductile metals exhibit a higher ultimate strength when fracture occurs at a reduced section such as would be formed in a tensile specimen by a concentric groove or notch. Whatever effect of stress concentration may remain during plastic deformation is more than offset by the supporting action of the shoulders, which tends to prevent the normal "necking down."

2.12 *Prestressing*

Parts of an elastic system, by accident or design, may have introduced into them stresses that cause and are balanced by opposing stresses in other parts, so that the system reaches a state of stress without the imposition of any external load. Examples of such initial, or locked-up, stresses are the temperature stresses in welded members, stresses in a statically indeterminate truss due to tightening or "rigging" some of the members by turnbuckles, and stresses in the flange couplings of a pipeline caused by screwing down the nuts. The effects of such prestressing upon the rigidity and strength of a system will now be considered, the assumption being made that prestressing is not so severe as to affect the properties of the *material.*

In discussing this question it is necessary to distinguish two types of systems, viz., one in which the component parts can sustain reversal of stress and one in which at least some of the component parts cannot sustain reversal of stress. Examples of the first type are furnished by a solid bar and by a truss, all members of which can sustain either tension or compression. Examples of the second type are furnished by the bolt-flange combination mentioned and by a truss with wire diagonals that can take tension only.

For the first type of system, prestressing has no effect on initial rigidity. Thus a plain bar with locked-up temperature stresses will exhibit the same modulus of elasticity as a similar bar from which these stresses have been removed by annealing; two prestressed helical springs arranged in parallel, the tension in one balancing the compression in the other, will deflect neither more nor less under load than the same two springs similarly placed without prestressing.

Prestressing will lower the elastic limit (or allowable load, or ultimate strength) provided that in the absence of prestressing all parts of the system reach their respective elastic limits (or allowable loads, or ultimate strengths) simultaneously. But if this relation between the parts does not exist, then prestressing may raise any or all of these quantities. One or two examples illustrating each condition may make this clear.

Consider first a plain bar that is to be loaded in axial tension. If there are no locked-up stresses, then (practically speaking) all parts of the bar reach their allowable stress, elastic limit, and ultimate strength simultaneously. But if there are locked-up stresses present, then the parts in which the initial tension is highest reach their elastic limit before other parts and the elastic limit of the bar as a whole is thus lowered. The load at which the allowable unit stress is first reached is similarly lowered, and the ultimate strength may also be reduced; although if the material is ductile, the equalization of stress that occurs during elongation will largely prevent this.

As an example of the second condition (all parts do not simultaneously reach the elastic limit or allowable stress) consider a thick cylinder under internal pressure. If the cylinder is not prestressed, the stress at the interior surface reaches the elastic limit first and so governs the pressure that may be applied. But if the cylinder is prestressed by shrinking on a jacket or wrapping with wire under tension, as is done in gun construction, then the walls are put into an initial state of compression. This compressive stress also is greatest at the inner surface, and the pressure required to reverse it and produce a tensile stress equal to the elastic limit is much greater than before. As another example, consider a composite member comprising two rods of equal length, one aluminum and the other steel, that are placed side by side to jointly carry a tensile load. For simplicity, it will be assumed that the allowable unit stresses for the materials are the same. Because the modulus of elasticity of the steel is about three times that of the aluminum, it will reach the allowable stress first and at a total load less than the sum of the allowable loads for the bars acting separately. But if the composite bar is properly prestressed, the steel being put into initial compression and the aluminum into initial tension (the ends being in some way rigidly connected to permit this), then on the application of a tensile load the two bars will reach the allowable stress simultaneously and the load-carrying capacity of the combination is thus greater than before. Similarly the elastic limit and sometimes the ultimate strength of a composite member may be raised by prestressing.

In a system of the second type (in which all parts *cannot* sustain stress reversal) prestressing increases the rigidity for any load less than that required to produce stress reversal. The effect of prestressing up to that point is to make the rigidity of the system the same as though all parts were effective. Thus in the case of the truss with wire diagonals it is as though the counterwires were taking compression; in the case of the flange-bolt combination it is as though the flanges were taking tension. (If the flanges are practically rigid in comparison with the bolts, there is no deformation until the applied load exceeds the bolt tension and so the system is rigid.) When the applied load becomes large enough to cause stress reversal (to make the counterwires go slack or to separate the flanges), the effect of prestressing disappears and the system is neither more nor less rigid than a similar one not prestressed provided, of course, none of the parts has been overstressed.

The elastic limit (or allowable load, or ultimate strength) of a system of this type is not affected by prestressing unless the elastic limit (or allowable load, or ultimate strength) of one or more of the parts is reached before stress reversal occurs. In effect, a system of this type is exactly like a system of the first type until stress reversal occurs, after which all effects of prestressing vanish.

The effects of prestressing are often taken advantage of, notably in bolted joints (flanges, cylinder heads, etc.), where high initial tension in the bolts prevents stress fluctuation and consequent fatigue, and in prestressed reinforced-concrete members, where the initially compressed concrete is enabled, in effect, to act in tension without cracking up to the point of stress reversal. The example of the prestressed thick cylinder has already been mentioned.

2.13 *Elastic stability*

Under certain circumstances the maximum load a member will sustain is determined, not by the strength of the material, but by the stiffness of the member. This condition arises when the load produces a bending or a twisting moment that is proportional to the corresponding deformation. The most familiar example is the *Euler column.* When a straight slender column is loaded axially, it remains straight and suffers only axial compressive deformation under small loads. If while thus loaded it is slightly deflected by a transverse force, it will straighten after removal of that force. But there is obviously some axial load that will just hold the column in the deflected position, and since both the bending moment due to the load and the resisting moment due to the stresses are directly proportional to the deflection, the load required thus to hold the column is independent of the amount of the deflection. If this condition of balance obtains at stresses less than the elastic limit, the condition is called *elastic stability* and the load that produces this condition is called the *critical* load. Any increase of the load beyond this critical value is usually attended by immediate collapse of the member.

Other examples of elastic stability are afforded by a thin cylinder under external pressure, a thin plate under edge compression or edge shear, and a deep thin cantilever beam under a transverse end load applied at the top surface. Some such elements, unlike the simple column described previously, do not fail under the load that initiates elastic buckling but demonstrate increasing resistance as the buckling progresses. Such *postbuckling* behavior is important in many problems of shell design. Elastic stability is discussed further in Chap. 14, and formulas for the critical loads for various members and types of loading are given in Tables 34 and 35.

REFERENCES

1. Roark, R. J.: "Formulas for Stress and Strain," 4th ed., McGraw-Hill Book Company, 1965.
2. Richards, C. W.: "Engineering Materials Science," Wadsworth Publishing Company, Inc., 1961.
3. Withey, M. O., and G. W. Washa: "Materials of Construction," John Wiley & Sons, Inc., 1954.
4. "Wood Handbook," Forest Products Laboratory, U.S. Dept. of Agriculture, 1974.
5. Symposium on Cast Iron, *Proc. ASTM,* vol. 33, part II, p. 115, 1933.
6. Nádai, A.: "Theory of Flow and Fracture of Solids," 2d ed., vol. I, Engineering Societies Monograph, McGraw-Hill Book Company, 1950.
7. Hencky, H.: The New Theory of Plasticity, Strain Hardening and Creep and the Testing

of the Inelastic Behavior of Metals, *ASME Paper* APM-55-18; *J. Appl. Mech.*, vol. 1, no. 4, 1933.
8. Lubahn, J. D., and R. P. Felgar: "Plasticity and Creep of Metals," John Wiley & Sons, Inc., 1961.
9. Abramson, H. N., H. Leibowitz, J. M. Crowley, and S. Juhasz (eds.): "Applied Mechanics Surveys," Spartan Books, 1966.
10. Conway, J. B.: "Stress-rupture Parameters: Origin, Calculation, and Use," Gordon and Breach, Science Publishers, Inc., 1969.
11. Conway, J. B.: "Numerical Methods for Creep and Rupture Analyses," Gordon and Breach, Science Publishers, Inc., 1967.
12. Conway, J. B., and P. N. Flagella: "Creep-rupture Data for the Refractory Metals to High Temperatures," Gordon and Breach, Science Publishers, Inc., 1971.
13. Odqvist, F. K. G.: "Mathematical Theory of Creep and Creep Rupture," Oxford University Press, 1966.
14. Hult, J. A. H.: "Creep in Engineering Structures," Blaisdell Publishing Company, 1966.
15. Penny, R. K., and D. L. Marriott: "Design for Creep," McGraw-Hill Book Company, 1971.
16. Smith, A. I., and A. M. Nicolson (eds.): "Advances in Creep Design, The A. E. Johnson Memorial Volume," Applied Science Publishers, Ltd., 1971.
17. "Encyclopedia of Polymer Science and Technology," Interscience Publishers, a division of John Wiley & Sons, Inc., 1964.
18. Davis, R. E., H. E. Davis, and J. S. Hamilton: Plastic Flow of Concrete under Sustained Stress, *Proc. ASTM,* vol. 34, part II, p. 354, 1934.
19. Report of Committee on Materials of Construction, *Bull. Assoc. State Eng. Soc.*, July 1934.
20. Soderberg, R.: Working Stresses, *ASME Paper* A-106, *J. Appl. Mech.*, vol. 2, no. 3, 1935.
21. Nádai, A.: Theories of Strength, *ASME Paper* APM 55-15, *J. Appl. Mech.*, vol. 1, no. 3, 1933.
22. Washa, G. W., and P. G. Fluck: Effect of Sustained Loading on Compressive Strength and Modulus of Elasticity of Concrete, *J. Am. Concr. Inst.,* vol. 46, May 1950.
23. Neville, A. M.: "Creep of Concrete: Plain, Reinforced, and Prestressed," North-Holland Publishing Company, 1970.
24. Designing for Effects of Creep, Shrinkage, Temperature in Concrete Structures, *Am. Concr. Inst. Publ. SP*-27, 1971.
25. "Fatigue Design Handbook," Society of Automotive Engineers, Inc., 1968.
26. Munse, W. H. (L. M. Grover, ed.): "Fatigue of Welded Steel Structures," Welding Research Council, 1964.
27. Sors, L.: "Fatigue Design of Machine Components," Pergamon Press, 1971 (English transl., S. E. Mitchell).
28. ASME Boiler and Pressure Vessel Code, Section VIII, Rules for Construction of Pressure Vessels, Division 2—Alternative Rules, July 1971.
29. ASME Boiler and Pressure Vessel Code, Section III, Rules for Construction of Nuclear Power Plant Components, July 1971.
30. Tavernelli, J. F., and L. F. Coffin, Jr.: A Compilation and Interpretation of Cyclic Strain Fatigue Tests on Metals, *Trans. Am. Soc. Met.*, vol. 51, 1959.
31. Internal Friction, Damping, and Cyclic Plasticity, *ASTM Spec. Tech. Publ.* 378, 1964.
32. Structural Fatigue in Aircraft, *ASTM Spec. Tech. Publ.* 404, 1965.
33. Fatigue Crack Propagation, *ASTM Spec. Tech. Publ.* 415, 1966.
34. Grover, H. J.: "Fatigue of Aircraft Structures," NAVAIR 01-1A-13, U.S. Government Printing Office, 1966.
35. Sandor, B. I.: "Fundamentals of Cyclic Stress and Strain," The University of Wisconsin Press, 1972.
36. Smith, J. O.: The Effect of Range of Stress on Fatigue Strength, *Univ. Ill., Eng. Exp. Sta. Bull.* 334, 1942.

37. Lomas, T. W., J. O. Wardt, J. R. Rait, and E. W. Colbeck: The Influence of Frequency of Vibration on the Endurance Limit of Ferrous Alloys at Speeds up to 150,000 Cycles per Minute (in Ref. 54).
38. Bowman, C. E., and T. J. Dolan: Biaxial Fatigue Properties of Pressure Vessel Steels, *Welding J. Res. Suppl.*, November 1953 and January 1955.
39. Horger, O. J., and H. R. Neifert: Fatigue Strength of Machined Forgings 6 to 7 Inches in Diameter, *Proc. ASTM,* vol. 39, 1939.
40. Eaton, F. C.: Fatigue Tests of Large Alloy Steel Shafts; Symposium on Large Fatigue Testing Machines and their Results, *ASTM Spec. Tech. Publ.* 216, 1957.
41. Jiro, H., and A. Junich: Studies on Rotating Beam Fatigue of Large Mild Steel Specimens, *Proc. 9th Jap. Natl. Congr. Appl. Mech.*, 1959.
42. Gould, A. J.: Corrosion Fatigue (in Ref. 54).
43. Nordfin, L.: Some Problems of Fatigue of Bolts and Bolted Joints in Aircraft Applications, *NBS Tech. Note* 136, 1962.
44. Munse, W. H., D. T. Wright, and N. M. Newmark: *Am. Soc. Civil Eng. Proc. Separate* 441, 1954.
45. Weibull, W.: Investigations into Strength Properties of Brittle Materials, *Proc. R. Swed. Inst. Eng. Res.,* no. 149, 1938.
46. Wilson, W. K.: "Practical Solution of Torsional Vibration Problems," 2d ed., vol. II, John Wiley & Sons, Inc., 1941.
47. Sines, George, and J. L. Waisman (eds.): "Metal Fatigue," McGraw-Hill Book Company, 1959.
48. Gough, H. J.: Corrosion Fatigue of Metals, *J. Inst. Met.*, vol. 49, no. 2, 1932.
49. Heywood, R. B.: "Designing against Fatigue of Metals," Reinhold Publishing Corporation, 1962.
50. Yen, C. S., and T. J. Dolan: A Critical Review of the Criteria for Notch Sensitivity in Fatigue of Metals, *Univ. Ill., Exp. Sta. Bull.* 398, 1952.
51. Phillips, C. E., and R. B. Heywood: Size Effect in Fatigue of Steel Specimens under Reversed Direct Stress, *Proc. Inst. Mech. Eng., Lond.*, vol. 165, 1951.
52. Mowbray, A. Q., Jr.: The Effect of Superposition of Stress Raisers on Members Subjected to Static or Repeated Loads, *Proc. Soc. Exp. Stress Anal.*, vol. 10, no. 2, 1953.
53. Forrest, P. G.: "Fatigue of Metals," Pergamon Press, Addison-Wesley Series in Metallurgy and Materials, 1962.
54. International Conference on Fatigue of Metals. Institution of Mechanical Engineers, London, and American Society of Mechanical Engineers, New York, 1956.
55. Harting, D. R.: The -S/N- Fatigue Life Gage: A Direct Means of Measuring Cumulative Fatigue Damage, *Exp. Mech.*, vol. 6, no. 2, February 1966.
56. Descriptive literature, Micro-Measurements, Inc., Romulus, Mich.
57. Tipper, C. F.: "The Brittle Fracture Story," Cambridge University Press, 1962.
58. Kobayashi, A. S. (ed.): "Experimental Techniques in Fracture Mechanics," Society for Experimental Stress Analysis, The Department of Publications, State University of Iowa, 1974.
59. Hall, W. J., H. Kihara, W. Soete, and A. A. Wells: "Brittle Fracture of Welded Plate," Prentice-Hall, Inc., 1967.
60. Boyd, G. M. (ed.): "Brittle Fracture in Steel Structures," Butterworth & Co., Ltd., 1970.
61. Richards, K. G.: "Brittle Fracture of Welded Structures," The Welding Institute, 1971.
62. Pellini, W. S.: Evolution of Engineering Principles for Fracture-Safe Design of Steel Structures, *Nav. Res. Lab. NRL Rept.* 6957, September 23, 1969.
63. Pellini, W. S.: Integration of Analytical Procedures for Fracture-Safe Design of Metal Structures, *Nav. Res. Lab. NRL Rept.* 7251, March 26, 1971.
64. Pellini, W. S.: Criteria for Fracture Control Plans, *Nav. Res. Lab. NRL Rept.* 7406, May 11, 1972.

65. Roark, R. J., R. S. Hartenberg, and F. Z. Williams: The Influence of Form and Scale on Strength, *Univ. Wis. Exp. Sta. Bull.* 84, 1938.
66. Peterson, R. E.: Model Testing as Applied to Strength of Materials, *Am. Soc. Mech. Eng., Paper* APM-55-11; *J. Appl. Mech.*, vol. 1, no. 2, 1933.
67. Battelle Memorial Institute: "Prevention of Fatigue of Metals," John Wiley & Sons, Inc., 1941.
68. Neuber, H.: "Theory of Notch Stresses," J. W. Edwards, Publisher, Incorporated, 1946.
69. Moore, H. F.: A Study of Size Effect and Notch Sensitivity in Fatigue Tests of Steel, *Proc. Am. Soc. Test. Mater.*, vol. 45, 1945.
70. Shannon, J. L., Jr., and W. F. Brown, Jr.: Progress in Fracture Mechanics, *Mach. Des.*, March 5, 1970.
71. Manson, S. S.: "Thermal Stress and Low-Cycle Fatigue," McGraw-Hill Book Company, 1966.
72. Manual on Low Cycle Fatigue Testing, *ASTM Spec. Tech. Publ.* 465, 1969.
73. Metal Fatigue Damage: Mechanism, Detection, Avoidance, and Repair, *ASTM Spec. Tech. Publ.* 495, 1971.
74. Cyclic Stress-Strain Behavior: Analysis, Experimentation, and Failure Prediction, *ASTM Spec. Tech. Publ.* 519, 1973.
75. Effect of Notches on Low-Cycle Fatigue: A Literature Survey, *ASTM Spec. Tech. Publ.* 490, 1972.
76. Wells, A. A.: Fracture Control: Past, Present and Future, *Exp. Mech.*, vol. 13, no. 10, October 1973.
77. Peterson, R. E.: "Stress Concentration Factors," John Wiley & Sons, Inc., 1974.
78. Vicentini, V.: Stress Concentration Factors for Superposed Notches, *Exp. Mech.*, vol. 7, no. 3, March 1967.

THREE

Principles and Analytical Methods

Most of the formulas of strength of materials express the relations among the form and dimensions of a member, the loads applied thereto, and the resulting stress or deformation. Any such formula is valid only within certain limitations and is applicable only to certain problems. An understanding of these limitations and of the way in which formulas may be combined and extended for the solution of problems to which they do not immediately apply requires a knowledge of certain principles and methods that are stated briefly in the following articles. The significance and use of these principles and methods are illustrated in Part 3 by examples that accompany the discussion of specific problems.

3.1 *Equations of motion and of equilibrium*

The relations that exist at any instant between the motion of a body and the forces acting on it may be expressed by these two equations: (1) F_x (the component along any line x of all forces acting on a body) $= m\bar{a}_x$ (the product of the mass of the body and the x component of the acceleration of its mass center); (2) T_x (the torque about any line x of all forces acting on the body) $= dh_x/dt$ (the time rate at which its angular momentum about that line is changing). If the body in question is in equilibrium, these equations reduce to (1) $F_x = 0$ and (2) $T_x = 0$.

These equations, Hooke's law, and experimentally determined values of the elastic constants E, G, and ν constitute the basis for the mathematical analysis of most problems of strength of materials. The majority of the common formulas for stress are derived by considering a portion of the loaded member as a body in equilibrium under the action of forces that include the stresses sought and then solving for these stresses by applying the equations of equilibrium.

3.2 *Principle of superposition*

With certain exceptions, the effect (stress, strain, or deflection) produced on an elastic system by any final state of loading is the same whether the forces that constitute that loading are applied simultaneously or in any given sequence and is the result of the effects that the several forces would produce if each acted singly.

An exception to this principle is afforded by any case in which some of the forces cause a deformation that enables other forces to produce an effect they would not have otherwise. A beam subjected to transverse and axial loading is an example; the transverse loads cause a deflection that enables the longitudinal load to produce a bending effect it would not produce if acting alone. In no case does the principle apply if the deformations are so large as to alter appreciably the geometrical relations of the parts of the system.

The principle of superposition is important and has many applications. It often makes it possible to resolve or break down a complex problem into a number of simple ones, each of which can be solved separately for stresses, deformations, etc., that are then algebraically added to yield the solution of the original problem.

3.3 *Principle of reciprocal deflections*

Let A and B be any two points of an elastic system. Let the displacement of B in any direction U due to a force P acting in any direction V at A be u; and let the displacement of A in the direction V due to a force Q acting in the direction U at B be v. Then $Pv = Qu$.

This is the general statement of the *principle of reciprocal deflections*. If P and Q are equal and parallel and u and v are parallel, the statement can be simplified greatly. Thus, for a horizontal beam with vertical loading and deflection understood, the principle expresses the following relation: A load applied at any point A produces the same deflection at any other point B as it would produce at A if applied at B.

The principle of reciprocal deflections is a corollary of the principle of superposition and so can be applied only to cases for which that principle is valid. It can be used to advantage in many problems involving deformation and is the basis of certain mechanical methods of structural analysis (see Refs. 1 and 2). Examples of the application of the principle are given in Chaps. 7 and 10.

3.4 *Method of consistent deformations (strain compatibility)*

Many statically indeterminate problems are easily solved by utilizing the obvious relations among the deformations of the several parts or among the deformations produced by the several loads. Thus the division of load between the parts of a composite member is readily ascertained by expressing the deformation or deflection of each part in terms of the load it carries and then equating these deformations or deflections. Or the reaction at the supported end of a beam with one end fixed and the other supported can be found by regarding the beam as a cantilever, acted on by the downward loads and an upward end load (the reaction), and setting the resultant deflection at the end equal to zero.

The method of consistent deformations is based on the principle of superposition; it can be applied only to cases for which that principle is valid.

3.5 *Principles and methods involving strain energy*

Strain energy has been defined (Chap. 1) as the mechanical energy stored up in an elastically stressed system; formulas for the amount of strain energy developed in members under various conditions of loading are given in Part 3. It is the purpose of this article to state certain relations between strain energy and external forces that are useful in the analysis of stress and deformation. For convenience, external forces with points of application that do not move will here be called *reactions,* and external forces with points of application that move will be called *loads.*

External work equal to strain energy. When an elastic system is subjected to static loading, the external work done by the loads as they increase from zero to their maximum value is equal to the strain energy acquired by the system.

This relation may be used directly to determine the deflection of a system under a single load; for such a case it shows that the deflection at the point of loading in the direction of the load is equal to twice the strain energy divided by the load. The relationship also furnishes a means of determining the critical load that produces elastic instability in a member. A reasonable form of curvature, compatible with the boundary conditions, is assumed, and the corresponding critical load found by equating the work of the load to the strain energy developed, both quantities being calculated for the curvature assumed. For each such assumed curvature, a corresponding approximate critical load will be found and the least load so found represents the closest approximation to the true critical load (see Ref. 3).

Method of unit loads. During the static loading of an elastic system the external work done by a *constant* force acting thereon is equal to the internal work done by the stresses caused by that constant force. This relationship is the basis of the following method for finding the deflection of any given point of an elastic system: A unit force is imagined to act at the point in question and in the direction of the deflection that is to be found. The stresses produced by such a unit force will do a certain amount of internal work during the application of the actual loads. This work, which can be readily found, is equal to the work done by the unit force; but since the unit force is constant, this work is equal to the deflection sought.

If the direction of the deflection cannot be ascertained in advance, its horizontal and vertical components can be determined separately in the way described and the resultant deflection found therefrom (see Refs. 4 and 5). Examples of application of the method are given in Art. 6.4.

Deflection, the partial derivative of strain energy. When an elastic system is statically loaded, the partial derivative of the strain energy with respect to any one of the applied forces is equal to the movement of the point of application of that force in the direction of that force. This relationship, known as *Castigliano's first theorem*, provides a means of finding the deflection of a beam or truss under several loads (see Ref. 4).

Theorem of least work.[1] When an elastic system is statically loaded, the distribution of stress is such as to make the strain energy a minimum consistent with equilibrium and the imposed boundary conditions. This principle is used extensively in the solution of statically indeterminate problems. In the simpler type of problem (beams with redundant supports or trusses with redundant members) the first step in the solution consists in arbitrarily selecting certain reactions or members to be considered redundant, the number and identity of these being such that the remaining system is just determinate. The strain energy of the entire system is then expressed in terms of the unknown redundant reactions or stresses. The partial derivative of the strain energy with respect to each of the redundant reactions or stresses is then set equal to zero and the resulting equations solved for the redundant

[1] By *theorem of least work* is usually meant only so much of the theorem as is embodied in the first application here described, and so understood it is often referred to as *Castigliano's second theorem.* But, as originally stated by Castigliano, it had a somewhat different significance. (See his "Théorème de l'équilibre des systèmes élastiques et ses applications," Paris, 1879, or the English translation "Elastic Stresses in Structures," by E. S. Andrews, Scott, Greenwood, London. See also R. V. Southwell, Castigliano's Principle of Minimum Strain-energy, *Proc. Roy. Soc. Lond., Ser. A,* vol. 154, 1936.) The more general theory stated is called *theorem of minimum energy* by Love (Chap. 15, Ref. 6) and *theorem of minimum resilience* by Morley (Ref. 59).

reactions or stresses. The remaining reactions or stresses are then found by the equations of equilibrium. An example of the application of this method is given in Art. 6.4.

As defined by this procedure, the *theorem of least work* is implicit in Castigliano's first theorem: It furnishes a method of solution identical with the method of consistent deflections, the deflection used being zero and expressed as a partial derivative of the strain energy. In a more general type of problem, it is necessary to determine which of an infinite number of possible stress distributions or configurations satisfies the condition of minimum strain energy. The electronic computer has made practicable the solution of many problems of this kind—shell analysis, elastic and plastic buckling, etc.—that formerly were relatively intractable (Refs. 6 and 9 to 11).

3.6 *Dimensional analysis*

Most physical quantities can be expressed in terms of mass, length, and time conveniently represented by the symbols M, L, and T, respectively. Thus velocity is LT^{-1}; acceleration is LT^{-2}; force is MLT^{-2}; unit stress is $ML^{-1}T^{-2}$; etc. A formula in which the several quantities are thus expressed is a dimensional formula, and the various applications of this system of representation constitute *dimensional analysis.*

Dimensional analysis may be used to check formulas for homogeneity, check or change units, derive formulas, and establish the relationships between similar physical systems that differ in scale (e.g., a model and its prototype). In strength of materials, dimensional analysis is especially useful in checking formulas for homogeneity. To do this, it is not always necessary to express *all* quantities dimensionally since it may be possible to cancel some terms. Thus it is often convenient to express force by some symbol, as F, until it is ascertained whether or not all terms representing force can be canceled.

For example, consider the formula for the deflection y at the free end of a cantilever beam of length l carrying a uniform load W. This formula (Table 3) is

$$y = \frac{1}{8}\frac{Wl^3}{EI}$$

To test for homogeneity, omit the coefficient $\frac{1}{8}$ (which is dimensionless) and write the formula

$$L = \frac{FL^3}{(F/L^2)\,L^4}$$

It is seen that F cancels and the equation reduces at once to $L = L$, showing that the original equation was homogeneous.

Instead of the symbols M, L, T, and F, we can use the names of the *units* in which the quantities are to be expressed. Thus the above equation may be written

$$\text{Inches} = \frac{(\text{pounds})\,(\text{inches}^3)}{(\text{pounds/inches}^2)\,(\text{inches}^4)} = \text{inches}$$

This practice is especially convenient if it is desired to change units. Thus it might be desired to write the above formula so that y is given in inches when l is expressed in feet. It is only necessary to write

$$\text{Inches} = \frac{1}{8}\frac{\text{pounds}\,(\text{feet} \times 12)^3}{(\text{pounds/inches}^2)\ \text{inches}^4}$$

and the coefficient is thus found to be 216 instead of $\frac{1}{8}$.

By what amounts to a reversal of the checking process described, it is often possible to determine the way in which a certain term or terms should appear in a formula provided the other terms involved are known. For example, consider the formula for the critical load on the Euler column. Familiarity with the theory of flexure suggests that this load will be directly proportional to E and I. It is evident that the length l will be involved in some way as yet unknown. It is also reasonable to assume that the load is independent of the deflection since both the bending moment and the resisting moment would be expected to vary in direct proportion to the deflection. We can then write: $P = kEIl^a$, where k is a dimensionless constant that must be found in some other way, and the exponent a shows how l enters the expression. Writing the equation dimensionally and omitting k, we have

$$F = \frac{F}{L^2} L^4 L^a \qquad \text{or} \qquad L^2 = L^{4+a}$$

Equating the exponents of L (as required for homogeneity) we find $a = -2$, showing that the original formula should be $P = kEI/l^2$. Note that the derivation of a formula in this way requires at least a partial knowledge of the relationship that is to be expressed.

The use of dimensional analysis in planning and interpreting model tests is of greater importance in aerodynamics and hydrodynamics than in strength of materials, although it has important applications in problems involving dynamic loading and vibrations (Refs. 7 and 8).

3.7 *Remarks on the use of formulas*

No calculated value of stress, strength, or deformation can be regarded as exact. The formulas used are based on certain assumptions as to properties of materials, regularity of form, and boundary conditions that are only approximately true, and they are derived by mathematical procedures that often involve further approximations. In general, therefore, great precision in numerical work is not justified. Each individual problem requires the exercise of judgment, and it is impossible to lay down rigid rules of procedure; but the following suggestions concerning the use of formulas may be of value.

1. For most cases, slide-rule calculations giving results to three significant figures are sufficiently precise. An exception is afforded by any calculation that involves the algebraic addition of quantities that are large in comparison with the final result (e.g., some of the formulas for beams under axial and transverse loading, some of the formulas for circular rings, and any case of superposition in which the effects of several loads tend to counteract each other). For such cases a calculator should be used.

2. In view of uncertainties as to actual conditions, many of the formulas may appear to be unnecessarily elaborate and include constants given to more significant figures than is warranted. For this reason, we may often be inclined to simplify a formula by dropping unimportant terms, "rounding off" constants, etc. It is sometimes advantageous to do this, but it is usually better to use the formula as it stands, bearing in mind that the result is at best only a close approximation. The only disadvantage of using an allegedly "precise" formula is the possibility of being misled into thinking that the result it yields corresponds exactly to a real condition. So far as the time required for calculation is concerned, little is saved by simplification and it is as easy to operate with large numbers as with small ones when a slide rule or calculator is used. (If a complicated formula is to be used frequently, it is of course advantageous to represent it by charts or graphs.)

3. When using an unfamiliar formula, we may be uncertain as to the correctness of the numerical substitutions made and mistrustful of the result. It is nearly always possible to effect some sort of check by analogy, superposition, reciprocal deflections, comparison, or merely by judgment and common sense. Thus the membrane analogy (Art. 4.4) shows that the torsional stiffness of any irregular section is greater than that of the largest inscribed circular section and less than that of the smallest circumscribed section. Superposition shows that the deflection and bending moment at the center of a beam under triangular loading (Table 3, case 2e) is the same as under an equal load uniformly distributed. The principle of reciprocal deflections shows that the stress and deflection at the center of a circular flat plate under eccentric concentrated load (Table 24, case 18) are the same as for an equal load uniformly distributed along a concentric circle with radius equal to the eccentricity (case 9a). Comparison shows that the critical unit compressive stress is greater for a thin plate under edge loading than for a strip of that plate regarded as an Euler column. Common sense and judgment should generally serve to prevent the acceptance of grossly erroneous calculations.

4. A difficulty frequently encountered is uncertainty as to boundary conditions—whether a beam or flat plate should be calculated as freely supported or fixed, whether a load should be assumed uniformly or otherwise distrib-

uted, etc. In any such case it is a good plan to make *bracketing assumptions,* i.e., to calculate the desired quantity on the basis of each of two assumptions representing limits between which the actual conditions must lie. Thus for a beam with ends having an unknown degree of fixity, the bending moment at the center cannot be more than if the ends were freely supported and the bending moments at the ends cannot be more than if the ends were truly fixed. If so designed as to be safe for either extreme condition, the beam will be safe for any intermediate degree of fixity.

5. Formulas concerning the validity of which there is reason for doubt, especially empirical formulas, should be checked dimensionally. If such a formula expresses the results of some intermediate condition, it should be checked for extreme or terminal conditions; thus an expression for the deflection of a beam carrying a uniform load over a portion of its length should agree with the corresponding expression for a fully loaded beam when the loaded portion becomes equal to the full length and should vanish when the loaded portion becomes zero.

REFERENCES

1. Timoshenko, S., and J. M. Lessells: "Applied Elasticity," Westinghouse Technical Night School Press, 1925.
2. Beggs, G. E.: The Use of Models in the Solution of Indeterminate Structures, *J. Franklin Inst.*, March 1927.
3. Timoshenko, S.: "Theory of Elastic Stability," Engineering Societies Monograph, McGraw-Hill Book Company, 1936.
4. Spofford, C. M.: "Theory of Structures," McGraw-Hill Book Company, 1928.
5. Niles, A. S., and J. S. Newell: "Airplane Structures," 3d ed., John Wiley & Sons, Inc., 1943.
6. Jakobsen, B. F.: Stresses in Gravity Dams by Principle of Least Work, *Trans. Am. Soc. Civil Eng.*, vol. 96, p. 489, 1932.
7. Robertson, B. L.: Dimensional Analysis, *Gen. Elec. Rev.*, April, 1930.
8. Bridgman, P. W.: "Dimensional Analysis," Yale University Press, 1922.
9. Langhaar, H. L.: "Energy Methods in Applied Mechanics," John Wiley & Sons, Inc., 1962.
10. Southwell, R. V.: "Relaxation Methods in Engineering Science," Oxford University Press, 1940.
11. Berg, G. V.: "Computer Analysis of Structures," College of Engineering, University of Michigan, 1963.

FOUR

Experimental Methods

A structural member or part may be of such a form or may be loaded in such a way that calculation of the stresses and strains produced in it is impracticable. When this is the case, we may resort to experimental methods, which can be applied to the actual member, or to a model thereof, or to a conventionalized specimen. Some of the more important methods used for this purpose are described briefly in this chapter. The interested reader is referred to extensive literature on experimental stress determination (see Refs. 1 to 11).

4.1 *Measurement of strain*

The most direct way of determining the stress produced under given circumstances is to measure the accompanying strain. Measurement is comparatively easy when the stress is fairly uniform over a considerable length of the part in question, but becomes difficult when the stress is localized or varies abruptly since measurement must then be made over a very short gage length and with great precision. When localized stress occurs at the edge of a plate or similar member, its value can sometimes best be determined by measuring the *lateral strain* across the thickness of the plate and computing the corresponding *longitudinal strain* from Poisson's ratio. Under conditions of uniaxial tension or compression, it is sufficient to measure strains in one direction; under conditions of combined stress, it is necessary to measure strains in each of two or three directions, preferably in the directions of the principal stresses. When a strain rosette (see Chap. 1, Definitions) is to be analyzed, either graphical or mathematical methods may be used (see Refs. 2 to 5).

Extensometers and strain gages commonly used in the laboratory testing of materials, where the form of specimen facilitates attachment and permits the use of almost any desired gage length, are often not suitable for field use or even for laboratory tests on parts of irregular form. Of the many instruments and techniques that have been successfully used for the measurement of strain, the following are among the most important. They are listed in general order of mechanical, optical, and electrical and in order of sophistication; many of them are commercially available.

1. ***Mechanical direct measurement.*** A direct measurement of strain is simple if the gage length is great enough, and indeed steel tapes have been used to measure strains over several feet of gage length on such as truss members or columns. Optical magnification of the motion of the fiduciary mark makes it possible to use short gage lengths also, and the Prewitt scratch strain gage (Ref. 12) is a commercially available gage using this principle. Reasonably large strains are utilized to move the target by a ratchet and therefore provide a permanent record of the strain in the form of scratches on the polished target over many hundreds of strain cycles. This gage is especially valuable for impact and vibration studies.

2. ***Brittle coatings.*** Easily applied surface coatings formulated to crack at strain levels well within the elastic limit of most structural materials provide a means of locating points of maximum strain and directions of principal strains. When suitably calibrated by means of calibration bars and fixtures, quantitative results can be obtained. Resin-based coatings are commonly used (Refs. 21 and 25), but *ceramic* coatings (Ref. 21), useful under higher temperature and other adverse ambient conditions, can be employed if the structure can be baked at 1000°F to cure the coating.

3. ***Mechanical lever.*** The addition of mechanical levers to increase the motion of the fiduciary mark is the principle used in several commercially available gages, but the added mass, the friction of the contacts, and the added contact pressure on the specimen needed to move the levers make them generally useful only for *static* or slowly varying *monotonic* strains (Refs. 16 and 17).

4. ***Optical direct measurement.*** Most of the optical measurement techniques are based on the principles of optical interference.

Moire techniques. Moire techniques use grids of alternate equally wide bands of relatively transparent or light-colored material and opaque or dark-colored material in order to observe the relative motion of two such grids. The most common technique (Refs. 2, 5, 8, and 11) uses an alternate transparent and opaque grid to produce photographically a matching grid on the flat surface of the specimen. Then the full-field relative motion is observed between the reproduction and the original when the specimen is loaded. Similarly, the original may be used with a projector to produce the photographic image on the specimen and then produce interference with the projected image after loading (Ref. 25). These methods can use ordinary white light, and the interference is due merely to *geometric* blocking of the light as it passes through or is reflected from the grids.

Another similar technique, *shadow Moire,* produces interference patterns due to motion of the specimen at right angles to its surface between an alternately transparent and opaque grid and the shadow of the grid on the specimen (Refs. 14 and 15).

Interferometry. The availability of the laser as an intense source of monochromatic, coherent light has made it possible to use the development and motion of interference fringes as a means of strain measurement (Refs. 29 and 57).

Holographic interferometry permits full-field measurement of changes in surface contour of any specimen (Refs. 27 and 28) or changes in thickness of plane transparent specimens (Refs. 30 to 33). The thickness change of a plane specimen can be related to the sum of the principal stresses $\Delta t = (-\nu/E)(\sigma_1 + \sigma_2)$. With a knowledge of the difference in the principal stresses and the direction of the principal stresses from a photoelastic analysis (see Art. 4.2), the individual principal stresses can then be determined.

A *diffraction grating* consisting of equally spaced parallel lines ruled on the surface of a specimen or on a piece of material subsequently bonded to the surface of a specimen can be used as a strain gage. The sensitivity is controlled by the line spacing, the wavelength of light, and the angle of incidence of the light beam to the grating. Thus, gages can be designed to measure small elastic strains or large plastic strains. See Refs. 34 to 36, where gage lengths from 0.001 to 0.25 in are discussed with static and dynamic strains measured up to 10 percent. In Ref. 34 it is stated that gratings remained usable at elevated temperatures to within 100°F of the melting point of the aluminum specimen; in Ref. 36 full-field contour lines are produced for constant strain in a specific direction.

X-rays. X-ray diffraction makes possible the determination of changes in interatomic distance and thus the measurement of elastic strain. The method has the particular advantages that it can be used at points of high stress concentration and to determine residual stresses without cutting the object of investigation (Ref. 39).

5. ***Optical-lever extensometer.*** The use of an optical lever in place of a mechanical lever increases the sensitivity and reduces the mass and friction but still requires a moving mirror or prism coupled to the specimen. The Tuckermann optical strain gage (Ref. 18) uses an autocollimated light source and image so that many strain gages can be read with one instrument.

6. ***Optical remote sensing.*** Electro-optical measurement of displacement or strain can be accomplished by rapid scanning of the optical image of a fiduciary mark on a stationary or moving specimen (Ref. 20). The change in light intensity as the scan passes the fiduciary mark is electronically converted to a displacement measurement or, by using two sensors, to a strain measurement. The distance from the sensors to the specimen can be as little

as 0.1 in or as far as the quality of the optical system will permit. Some instruments will measure displacement along two axes simultaneously.

7. *Electric strain gages.* Strain gages have been developed using the changes in resistance, capacitance, and inductance as a measure of strain or displacement.

Capacitance strain gage. Capacitance strain gages measure the change in spacing between two surfaces by the change in capacitance. They are generally large but can be made quite sensitive with proper instrumentation (Refs. 2 and 3).

Inductance strain gage. The change in air gap in a magnetic circuit can create a large change in *inductance* depending upon the design of the rest of the circuit. The large change in inductance is accompanied by a large change in *force* across the gap, and so the very sensitive inductance strain gages can be used only on more massive structures. They have been used as overload indicators on presses with no electronic amplification necessary. The linear relationship between core motion and output voltage of a *linear differential transformer* makes possible accurate measurement of displacements under a wide variety of conditions (Refs. 3, 4, and 19).

Electric resistance strain gage. The change in resistance of an electric conductor due to strain-induced changes in the size and specific sensitivity of the conductor has led to a wide variety of strain gages. Very light and suitable for both static and dynamic loading, the filament may be in the form of a wire, a wire grid, a foil grid, or, for maximum sensitivity, a filament of silicon. These gages are available in single gages and in many forms of rosettes (Refs. 21 to 24).

4.2 *Photoelastic analysis*

When a beam of polarized light passes through an elastically stressed transparent isotropic material, the beam may be treated as having been decomposed into two rays polarized in the planes of the principal stresses in the material. In birefringent materials the indexes of refraction of the material encountered by these two rays will depend upon the principal stresses. Therefore, interference patterns will develop which are proportional to the differences in the principal stresses.

Two-dimensional analysis. With suitable optical elements—polarizers and wave plates of specific relative retardation—both the principal stress differences and the directions of principal stresses may be determined at every point in a two-dimensional specimen (Refs. 2 to 6, 10 and 38). Many suitable photoelastic plastics are available, but the most commonly used are the epoxies and polyester resins. The material properties that must be considered are transparency, sensitivity (relative index of refraction change with stress), optical and mechanical creep, modulus of elasticity, ease of machining, cost, and stability (freedom from stresses developing with time).

Three-dimensional analysis. Several photoelastic techniques are used to determine stresses in three-dimensional specimens. If information is desired at a single point only, the optical polarizers, wave plates, and photoelastically sensitive material can be *embedded* in a transparent model (Ref. 2) and two-dimensional techniques used. A modification of this technique, *stress freezing,* is possible in some biphase materials. By heating, loading, cooling, and unloading, it is possible to lock permanently into the specimen, on a *molecular* level, stresses proportional to those present under load (Refs. 2 to 4). Since equilibrium exists at a molecular level, the specimen can be cut into two-dimensional slices and all *secondary principal stress differences* determined. The secondary principal stresses at a point are defined as the largest and smallest normal stresses in the plane of the slice; these in general will not correspond with the principal stresses at that same point in the three-dimensional structure. If desired, the specimen can be cut into cubes and the three principal stress differences determined. The individual principal stresses at a given point cannot be determined from photoelastic data taken at that point alone since the addition of a hydrostatic stress to any cube of material would not be revealed by differences in the indexes of refraction. Mathematical integration techniques, which start at a point where the hydrostatic stress component is known, can be used with photoelastic data to determine all individual principal stresses (Refs. 3 and 6).

A third method, *scattered light photoelasticity,* uses a thin pencil beam of intense monochromatic polarized light or a similar thin sheet of light (Refs. 3, 13, and 25) passing through photoelastically sensitive transparent models that have the additional property of being able to scatter uniformly a small portion of the light from any point on the beam or sheet. The same general restrictions apply to this analysis as applied to the stress-frozen three-dimensional analysis except that the specimen does not have to be cut. However, the amount of light available for analysis is much less, the specimen must be immersed in a fluid with an index of refraction that very closely matches that of the specimen, and in general the data are much more difficult to analyze.

Photoelastic coating. Photoelastic coatings have been sprayed, bonded in the form of thin sheets, or cast directly in place on the surface of models or structures to determine the two-dimensional surface strains. The surface is made reflective before bonding the plastics in place so the effective thickness of the photoelastic plastic is doubled and all two-dimensional techniques can be applied with suitable instrumentation (Refs. 3 to 5, 25, and 26).

4.3 *Detection of plastic yielding*

In parts made of ductile metal, sometimes a great deal can be learned concerning the location of the most highly stressed region and the load that produces elastic failure by noting the first signs of plastic yielding. Such yielding may be detected in the following ways.

Observation of slip lines. If yielding occurs first at some point on the surface, it can be detected by the appearance of slip lines if the surface is suitably polished. If yielding occurs first at some interior point, it can sometimes be discovered by cutting a section that includes the zone of high stress and then etching the cut surfaces with a solution of cupric chloride in hydrochloric acid and water to make the slip lines visible. However, this method is applicable only to certain low-carbon steels, and there appears to be no way to ascertain (except by trial) whether or not a given steel can thus be made to reveal overstrain (Ref. 40).

Brittle coating. If a member is coated with some material that will flake off easily, this flaking will indicate local yielding of the member. A coating of rosin or a wash of lime or white portland cement, applied and allowed to dry, is best for this purpose, but chalk or mill scale will often suffice. By this method zones of high stress such as those that occur in pressure vessels around openings and projections can be located and the load required to produce local yielding can be determined approximately (Ref. 37).

Photoelastic coatings. Thin photoelastic coatings show very characteristic patterns analogous to slip lines when the material beneath the coating yields (Refs. 5, 25, and 26).

4.4 *Analogies*

Certain problems in elasticity involve equations that cannot be solved but that happen to be mathematically identical with the equations that describe some other physical phenomenon which can be investigated experimentally. Among the more useful of such analogies are the following.

Membrane analogy This is especially useful in determining the torsion properties of bars having noncircular sections. If in a thin flat plate holes are cut having the outlines of various sections and over each of these holes a soap film (or other membrane) is stretched and slightly distended by pressure from one side, the volumes of the bubbles thus formed are proportional to the torsional rigidities of the corresponding sections and the slope of a bubble surface at any point is proportional to the stress caused at that point of the corresponding section by a given twist per unit length of bar. By cutting in the plate one hole the shape of the section to be studied and another hole that is circular, the torsional properties of the irregular section can be determined by comparing the bubble formed on the hole of that shape with the bubble formed on the circular hole since the torsional properties of the circular section are known. This method has been used successfully to determine the torsional stiffness of various members employed in airplane construction (Refs. 41 to 43).

A similar membrane analogy may be used to determine the distribution of shear stress and the position of the center of flexure in a beam having any given section. Here, however, the procedure is much less simple than the one described above. The soap film, instead of being inflated, is stretched over a hole cut in a curved sheet of metal. The shape of the hole and the curvature of the sheet must be such that a projection of the hole on a fixed reference plane has the form of the section being studied while distances of points along the edge of the hole from that reference plane represent a given function of the shape of the section (see Ref. 44).

Hydrodynamic analogy for torsion. If a prismatic vessel or tube having a cross section like that of a twisted member is filled with fluid, set with its axis vertical and rotated by a horizontal couple the velocity of the fluid at any point relative to the vessel will be proportional to and have the same direction as the shear stress at a corresponding point in the twisted member.

This hydrodynamic analogy and others are not well adapted to quantitative investigation of torsion problems, but they may be used to give a good representation of the shear distribution and warping of sections (see Refs. 45 and 46).

Electrical analogy for torsion. This analogy is especially applicable to the study of stresses in a circular shaft having circumferential grooves or other somewhat abrupt changes in section. Consider a plate having the outline of a longitudinal half-section of the shaft and a thickness at any point proportional to the cube of the distance from that edge which represents the axis of the shaft. If the ends of this plate, which correspond to the ends of the shaft, are maintained at a constant difference of electric potential, the rate of drop of potential at any point along the thick edge of the plate is proportional to the shear stress at the corresponding point on the surface of the shaft.

This analogy has been used to determine the factor of torsional-stress concentration at fillets, etc. (Refs. 43, 46, and 47).

Electrical analogy for isopachic lines. *Isopachic lines* are lines along which the sums of the principal stresses are equal in a two-dimensional plane stress problem. The voltage at any point on a uniform two-dimensional conducting surface is governed by the same form of equation as is the principal stress sum. Teledeltos paper is a uniform layer of graphite particles on a paper backing and makes an excellent material from which to construct the electrical analog. The paper is cut to a geometric outline corresponding to the shape of the two-dimensional structure or part, and boundary potentials are applied by an adjustable power supply. The required boundary potentials are obtained from a photoelastic study of the part where the principal stress sums can be found from the principal stress differences on the boundaries (Refs. 2 and 3). A similar membrane analogy has the height of a nonpressurized membrane proportional to the principal stress sum (Refs. 2 and 3).

4.5 *Models*

In addition to being used in photoelastic analysis, models may be employed in other ways to investigate stress in members of peculiar form and to predict strength.

Rubber models. By ruling a grid or lattice of fine lines on the surface of a model and studying the distortion of this grid under load, it is possible to ascertain something concerning the distribution and magnitude of the stresses in a member or structure. The material used must be one that can sustain very large deformations without ceasing to be approximately elastic. Models of rubber and of gelatin have been used to study the stresses in dams (Ref. 48), flat slabs (Ref. 49), and members subject to stress concentration (Ref. 50).

Reinforced plastic models. An aluminum-filled plastic model material is available either in liquid form to be cast or in block form for machining (Refs. 25 and 58). The material is linearly elastic and available in two different compositions, the stiffer one having twice the modulus of elasticity of the other. It can easily be cast in inexpensive molds and shaped by sawing, filing, and bonding on extra materials. Much lower loads are needed to create measurable strains that can be measured by all the conventional techniques, including strain gages, brittle coatings, and photoelastic coatings.

Brittle models. If a brittle material were available for which the proportional limit and ultimate strength are identical, then for members made thereof all stresses would be proportional to the applied load and formulas for elastic stress would be valid clear up to rupture. By testing to failure models made of such material, it would therefore be possible to determine the relation between load and maximum stress in a member of any form simply by determining the relation between the breaking load and the ultimate tensile strength.

There is probably no material available for which Hooke's law applies clear up to fracture, but there are materials which almost conform to this requirement. Models made of pottery plaster, plaster of paris, and Bakelite have been used to check formulas for the maximum stress in torsion members, curved beams, flat plates, etc., and to determine factors of stress concentration (Refs. 51 to 54).

This method has the advantages of being cheap, quick, and applicable to any form of member and any condition of stress. It has the disadvantages of yielding information concerning the maximum stress only and being only fairly dependable because of the fact that there is some plastic action even in the most brittle of available materials.

Proportionate-strength models. The breaking strength of members and even of structures of some complexity can be predicted with fair accuracy from tests of plaster models provided allowance is made for all factors involved. In principle, this method comprises the following steps: (1) determination by test of the load P_m carried by a scale model $1/n$ times as large as the structure; (2) determination, by tests of suitable coupons, of R, the ratio of the ultimate strength of the structural material to the ultimate strength of the model material; (3) calculation of the predicted load P that the structure will carry by the relation $P = P_m \times n^2 \times R$.

To secure satisfactory results from so simple a procedure would require certain relationships between the structural and model materials; in particular, the strength ratio R should be the same for all kinds of stress likely to prove critical, and the materials should respond similarly to the influence of form on strength. Differences in this latter respect can be taken into account if values of k_r, rupture factors, form factors, etc., are known for each material. This method has been used to predict the strength of castings of irregular form and of reinforced concrete structures with reasonable accuracy (Ref. 53).

Models for investigation of elastic stability. The critical loading for thin-walled structures—submarine hulls, pipes, monocoque fuselages, etc.—is sometimes determined by tests on small models. Since elastic stability is influenced greatly by small departures from geometrical regularity of form, the models so used must be constructed with great care and of material similar to that used to make the prototype. Paper models have been used to check basic theories and secure information as to the general behavior of thin-walled cylinders, but it is generally agreed that for reliable quantitative results the model material should be identical with that used in the actual structure (Refs. 55 and 56).

REFERENCES

1. *Proc. Soc. Exp. Stress Anal.,* 1943–1960; and *Exp. Mech., J. Soc. Exp. Stress Anal.,* after Jan. 1, 1961.
2. Hetenyi, M.: "Handbook of Experimental Stress Analysis," John Wiley & Sons, Inc., 1950.
3. Dally, J. W., and W. F. Riley: "Experimental Stress Analysis," McGraw-Hill Book Company, 1965.
4. Dove, R. C., and P. H. Adams: "Experimental Stress Analysis and Motion Measurement," Charles E. Merrill Books, Inc., 1955.
5. Holister, G. S.: "Experimental Stress Analysis: Principles and Methods," Cambridge University Press, 1967.
6. Frocht, M. M.: "Photoelasticity," vols. 1 and 2, John Wiley & Sons, Inc., 1941, 1948.
7. Durelli, A. J.: "Applied Stress Analysis," Prentice-Hall, Inc., 1967.
8. Durelli, A. J., and V. J. Parks,: "Moire Analysis of Strain," Prentice-Hall, Inc., 1970.
9. Durelli, A. J., E. A. Phillips, and C. H. Tsao: "Introduction to the Theoretical and Experimental Analysis of Stress and Strain," Prentice-Hall, Inc., 1958.
10. Durelli, A. J., and W. F. Riley,: "Introduction to Photomechanics," Prentice-Hall, Inc., 1965.
11. Theocaris, P. S.: "Moire Fringes in Strain Analysis," Pergamon Press, 1969.
12. Tatnall, F. G.: Development of the Scratch Gage, *Exp. Mech., J. Soc. Exp. Stress Anal.*, vol. 9, no. 6, June 1969.
13. Tuppeny, W. H., Jr., and A. S. Kobayashi (eds.): "Manual on Experimental Stress Analysis," 2d. ed., Society for Experimental Stress Analysis, 1965.
14. Theocaris, P. S.: Moire Topography of Curved Surfaces. *Exp. Mech., J. Soc. Exp. Stress Anal.,* vol. 7, no. 7, July 1967.
15. Chiang, F. P., and G. Jaisingh: Dynamic Moire Methods for the Bending of Plates, *Exp. Mech., J. Soc. Exp. Stress Anal.,* vol. 13, no. 4, April 1973.
16. Descriptive literature: Tinius Olsen Testing Machine Co., Willow Grove, Pa.; Wiedemann

Machine Co., King of Prussia, Pa.; P. L. Porter, Los Angeles, Calif.; Baldwin-Lima-Hamilton Corp., Philadelphia, Pa.
17. Descriptive literature: Federal Products Corp., Providence, R.I.; Standard Gage Co., Poughkeepsie, N.Y.; B. C. Ames Co., Waltham, Mass.; The L. S. Starrett Co., Athol, Mass.
18. Descriptive literature: American Instrument Co., Inc., Silver Springs, Md.
19. Descriptive literature: Bourns, Inc., Riverside, Calif.; Columbia Research Labs., Inc., Woodlyn, Pa.; Daytronic Corp., Dayton, Ohio; Honeywell, Minneapolis, Minn.; Pickering & Co., Inc., Plainview, N.Y.; Sanborn Co., Waltham, Mass.; Schaevitz Engineering, Pennsauken, N.J.; Tresco, Inc., Philadelphia, Pa.
20. Descriptive literature: Optron Corp., New Haven, Conn.; PhysiTech, Inc., Willow Grove, Pa.
21. Descriptive literature, Magnaflux Corp., Chicago, Ill.
22. Descriptive literature: Micro-Measurements, Inc., Romulus, Mich.; BLH Electronics, Inc., Waltham, Mass.
23. Descriptive literature, Ailtech, A Cutler-Hammer Co., City of Industry, Ca.
24. Descriptive literature, William T. Bean, Inc., Lampasas, Texas.
25. Descriptive literature, Photoelastic Inc., Malvern, Pa.
26. Descriptive literature, Photostress Corporation, Philadelphia, Pa.
27. Fagan, W. F., P. Waddell, and W. McCracken: The Study of Vibration Patterns Using Real-Time Hologram Interferometry, *Opt. & Laser Tech.*, vol. 4, no. 3, August 1972.
28. Hazell, C. R., and S. D. Liem: Vibration Analysis of Plates by Real-time Stroboscopic Holography, *Exp. Mech., J. Soc. Exp. Stress Anal.,* vol. 13, no. 8, August 1973.
29. Post, D.: Photoelastic Evaluation of Individual Principal Stresses by Large Field Absolute Retardation Measurements, *Proc. Soc. Exp. Stress Anal.*, vol. 13, no. 2, 1956.
30. Fourney, M. E.: Application of Holography to Photoelasticity, *Exp. Mech., J. Soc. Exp. Stress Anal.*, vol. 8, no. 1, January 1968.
31. Hovanesian, J. D., V. Brcic, and R. L. Powell: A New Experimental Stress-Optic Method: Stress-Holo-Interferometry, *Exp. Mech., J. Soc. Exp. Stress Anal.,* vol. 8, no. 8, August 1968.
32. Sanford, R. J., and A. J. Durelli: Interpretation of Fringes in Stress-Holo-Interferometry, *Exp. Mech., J. Soc. Exp. Stress Anal.*, vol. 11, no. 4, April 1971.
33. O'Regan, R., and T. D. Dudderar: A New Holographic Interferometer for Stress Analysis, *Exp. Mech., J. Soc. Exp. Stress Anal.,* vol. 11, no. 6, June 1971.
34. Bell, J. F.: Diffraction Grating Strain Gauge, *Proc. Soc. Exp. Stress Anal.*, vol. 17, no. 2, December 1959.
35. Sharpe, W. N., Jr.: The Interferometric Strain Gage, *Exp. Mech., J. Soc. Exp. Stress Anal.*, vol. 8, no. 4, April 1968.
36. Boone, P. M.: A Method for Directly Determining Surface Strain Fields Using Diffraction Gratings, *Exp. Mech., J. Soc. Exp. Stress Anal.,* vol. 11, no. 11, November 1971.
37. Radaj, D.: Detection of Lueders' Lines by Means of Brittle Coating, *Exp. Mech., J. Soc. Exp. Stress Anal.*, vol. 9, no. 3, March 1969.
38. Heywood, R. B.: "Photoelasticity for Designers," Pergamon Press, 1969.
39. Norton, J. T., and D. Rosenthal: Stress Measurement by X-ray Diffraction, and Applications of the X-ray Diffraction Method of Stress Measurement to Problems Involving Residual Stresses in Metals, *Proc. Soc. Exp. Stress Anal.*, vol. 1, no. 2, 1944.
40. Fry, A.: *Stahl und Eisen,* Aug. 11, 1921; also *Iron Age,* Dec. 1, 1921.
41. Trayer, G. W., and H. W. March: The Torsion of Members Having Sections Common in Aircraft Construction, *Adv. Comm. Aeron. Rept.* 334, 1930.
42. Taylor, G. I., and A. A. Griffith: The Use of Soap Films in Solving Torsion Problems, *Tech. Rept. Adv. Comm. Aeron.* (British), *Reports and Memoranda,* no. 333, p. 920, 1917.
43. Higgins, T. J.: Analogic Experimental Methods in Stress Analysis as Examplified by Saint-Venant's Torsion Problem, *Proc. Soc. Exp. Stress Anal.*, vol. 2, no. 2, p. 17, 1945.
44. Griffith, A. A., and G. I. Taylor: The Problem of Flexure and Its Solution by the Soap

Film Method, *Tech. Report Adv. Comm. Aeron.* (British), *Reports and Memoranda,* no. 399, p. 950, 1917.

45. Den Hartog, J. P.: On the Hydrodynamic Analogy of Torsion, *ASME Paper* A-46; *J. Appl. Mech.,* vol. 2, no. 2, 1935.
46. Timoshenko, S.: "Theory of Elasticity," Engineering Societies Monograph, McGraw-Hill Book Company, 1934.
47. Jacobsen, L. S.: Torsional-Stress Concentrations in Shafts of Circular Cross-section and Variable Diameter, *Trans. ASME,* vol. 47, p. 619, 1925.
48. "An Experimental Study of the Stresses in Masonry Dams," Research Memoirs, Technical Series V, Cambridge University Press. (See also *Trans. Am. Soc. Civil Eng.*, vol. 98, p. 1022, 1933.)
49. Trelease, F. J.: "The Design of Concrete Flat Slabs," *Proc. Natl. Assoc. Cement Users,* vol. 8, p. 218, 1912.
50. Chiles, G. S., and R. G. Kelley: The Resistance of Materials; the Effect of Sudden or Abrupt Changes in the Section on the Distribution of Unit-stress, *Railway Mech. Eng.*, March, April, and May 1919.
51. Peterson, R. E.: An Investigation of Stress Concentration by Means of Plaster of Paris Specimens, *Mech. Eng.*, p. 1449, December 1926.
52. Seely, F. B., and R. V. James: The Plaster-model Method of Determining Stresses Applied to Curved Beams, *Univ. Ill. Eng. Exp. Sta., Bull.* 195, 1928.
53. Roark, R. J., and R. S. Hartenberg: Predicting the Strength of Structures from Tests of Plaster Models, *Univ. Wis. Eng. Exp. Sta., Bull.* 81, 1935.
54. Frocht, M. M.: The Behavior of Brittle Materials at Failure, *ASME Paper* A-99; *J. Appl. Mech.*, vol. 3, no. 3, 1931.
55. Rhode, R. V., and E. E. Lundquist: Strength Tests of Paper Cylinders in Compression, Bending and Shear, *Nat. Adv. Comm. Aeron., Tech. Note* 370, 1931.
56. Saunders, H. E., and D. F. Windenburg: The Use of Models in Determining the Strength of Thin-walled Structures, *Trans. ASME,* vol. 54, no. 23, p. 263, 1932.
57. Descriptive literature, Diffracto Ltd., Troy, Michigan.
58. Riegner, E. I., and A. E. Scotese: Use of Reinforced Epoxy Models to Design and Analyze Aircraft Structures, *AIAA, J. Aircraft,* vol. 8, no. 10, October 1971.
59. Morley, A.: "Theory of Structures," 5th ed., Longmans, Green and Co., 1948.

FIVE

Properties of a Plane Area

Because of their importance in connection with the analysis of bending and torsion, certain relations between the moments of inertia and product of inertia of a plane area (here called *section*) are indicated in the following paragraphs. The equations are given with reference to Fig. 5.1, and the notation is as follows: A is the area of the section; X and Y are rectangular

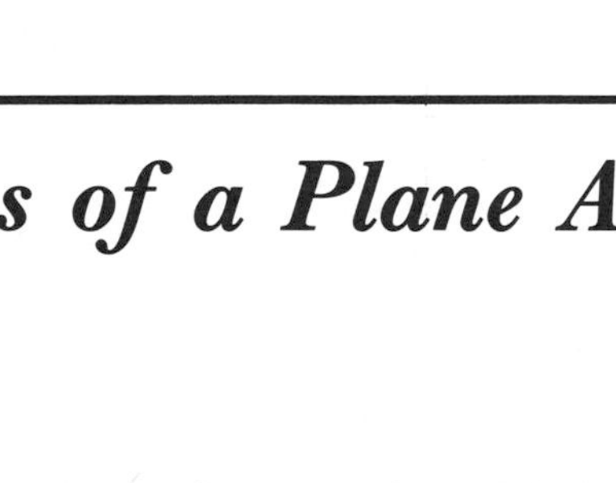

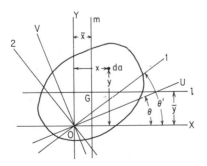

Fig. 5.1

axes in the plane of the section intersecting at any point 0; Z is a polar axis through 0; U and V are rectangular axes through 0 inclined at an angle θ to X and Y; 1 and 2 are the *principal axes* at the point 0; l and m are rectangular axes parallel to X and Y, respectively, and intersecting at G, the centroid of the area; $\bar{x}$ and $\bar{y}$ are the distances of X and Y from l and m, respectively; r is the distance from 0 to dA.

By definition (see Chap. 1)

$$I_x = \int dA \cdot y^2$$
$$I_y = \int dA \cdot x^2$$
$$J_z = \int dA \cdot r^2$$
$$H_{xy} = \int dA \cdot xy$$
$$k_x = \sqrt{\frac{I_x}{A}}$$
$$k_y = \sqrt{\frac{I_y}{A}}$$

The following equations and statements hold true:

$$J_z = I_x + I_y = I_u + I_v = I_1 + I_2$$
$$I_x = I_l + A\bar{y}^2$$
$$I_u = I_x \cos^2 \theta + I_y \sin^2 \theta - H_{xy} \sin 2\theta$$
$$\theta' = \frac{1}{2} \text{arc tan} \frac{2H_{xy}}{I_y - I_x}$$
$$I_{1,2} = \tfrac{1}{2}(I_x + I_y) \pm \sqrt{\tfrac{1}{4}(I_y - I_x)^2 + H_{xy}{}^2} \qquad \begin{cases} (+ \text{ for } I_1 \text{ the max}) \\ (- \text{ for } I_2 \text{ the min}) \end{cases}$$
$$H_{xy} = H_{lm} + A\overline{xy}$$
$$H_{uv} = H_{xy} \cos 2\theta - \tfrac{1}{2}(I_y - I_x) \sin 2\theta$$
$$H_{12} = 0$$

Which of the axes, 1 or 2, is the axis of maximum moment of inertia and which is the axis of minimum moment of inertia must be ascertained by calculation, unless the shape of the section is such as to make this obvious.

Any axis of symmetry is one of the principal axes for every point thereon.

The product of inertia is zero for any pair of axes, one of which is an axis of symmetry.

If the moment of inertia for one of the principal axes through a point is equal to that for any other axis through that point, it follows that the moments of inertia for all axes through that point are equal. (This refers to axes in the plane of the section.) Thus the moment of inertia of a square, an equilateral triangle, or any section having two or more axes of identical symmetry is the same for any central axis.

The moment of inertia and radius of gyration of a section with respect to a central axis are less than for any other axis parallel thereto.

The moment of inertia of a composite section (one regarded as made up of rectangles, triangles, etc.) is equal to the sum of the moments of inertia of its components parts. Voids are taken into account by subtracting the moment of inertia of the corresponding area.

Expressions for the area, distance of centroid from edges, moment of inertia, and radius of gyration are given in Table 1 for each of a number of representative sections. I and H for composite areas can be found by addition; the centroid of composite areas can be found by using the relation that the statical moment about any line of the entire area is equal to the sum of the statical moments of its component parts. Properties of structural sections—I-beams, channels, angles, etc.—are given in the structural-steel handbooks.

A closely approximate formula (due to Steinitz, *Aero. Digest,* November 1939) for the section modulus I/c of any solid section of compact form (e.g., approximately square, circular, triangular, or trapezoidal) is $I/c = A^2/6.15b$, where A is the area and b is the maximum width of the section. The formula gives the least I/c for nonsymmetrical sections and makes it unnecessary to locate the position of the neutral axis. The formula applies (though with less accuracy) to closed hollow sections of uniform wall thickness for which b should be taken as the net width equal to twice the wall thickness. It cannot be used for flanged or otherwise "spread" sections.

TABLE 1 ***Properties of sections***

Form of section	Area and distances from centroid to extremities	Moments and products of inertia about central axes	Radii of gyration about central axes
1. Square 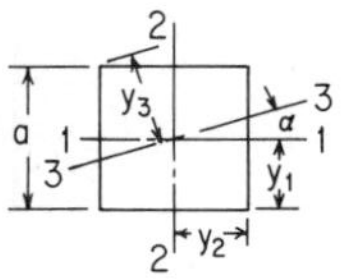	$A = a^2$ $y_1 = y_2 = \frac{a}{2}$ $y_3 = 0.707a \cos\left(\frac{\pi}{4} - \alpha\right)$	$I_1 = I_2 = I_3 = \frac{1}{12}a^4$	$r_1 = r_2 = r_3 = 0.2887a$
2. Rectangle 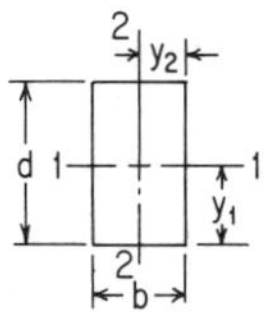	$A = bd$ $y_1 = \frac{d}{2}$ $y_2 = \frac{b}{2}$	$I_1 = \frac{1}{12}bd^3$ $I_2 = \frac{1}{12}db^3$ $I_1 > I_2$ if $d > b$	$r_1 = 0.2887d$ $r_2 = 0.2887b$
3. Equilateral triangle 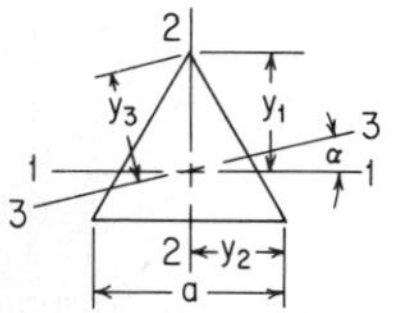	$A = 0.4330a^2$ $y_1 = 0.5774a$ $y_2 = 0.5000a$ $y_3 = 0.5774a \cos \alpha$	$I_1 = I_2 = I_3 = 0.01804a^4$	$r_1 = r_2 = r_3 = 0.2041a$
4. Isosceles triangle 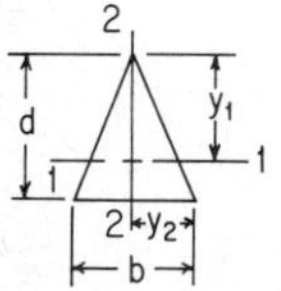	$A = \frac{bd}{2}$ $y_1 = \frac{2}{3}d$ $y_2 = \frac{b}{2}$	$I_1 = \frac{1}{36}bd^3$ $I_2 = \frac{1}{48}db^3$ $I_1 > I_2$ if $d > 0.866b$	$r_1 = 0.2357d$ $r_2 = 0.2041b$

5. Triangle 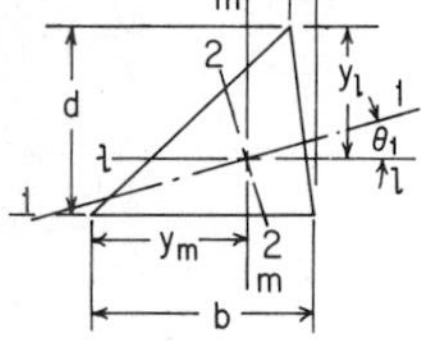	$A = \frac{bd}{2}$ $y_l = \frac{2}{3}d$ $y_m = \frac{2}{3}b - \frac{1}{3}a$	$I_l = \frac{1}{36}bd^3$ $I_m = \frac{1}{36}bd(b^2 - ab + a^2)$ $H_{lm} = \frac{1}{72}bd^2(b - 2a)$ $\theta_{1,2} = \frac{1}{2}\tan^{-1}\frac{d(b-2a)}{b^2 - ab + a^2 - d^2}$	$r_l = 0.2357d$ $r_m = 0.2357\sqrt{b^2 - ab + a^2}$
6. Parallelogram 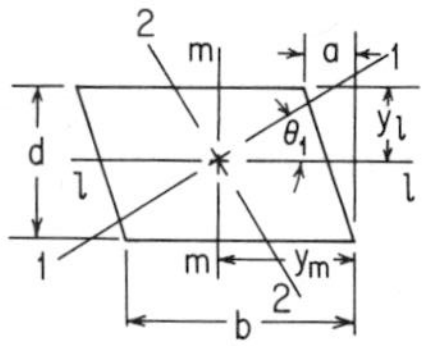	$A = bd$ $y_l = \frac{d}{2}$ $y_m = \frac{1}{2}(b + a)$	$I_l = \frac{1}{12}bd^3$ $I_m = \frac{1}{12}bd(b^2 + a^2)$ $H_{lm} = \frac{-1}{12}abd^2$ $\theta_{1,2} = \frac{1}{2}\tan^{-1}\frac{-2ad}{b^2 + a^2 - d^2}$	$r_l = 0.2887d$ $r_m = 0.2887\sqrt{b^2 + a^2}$
7. Diamond 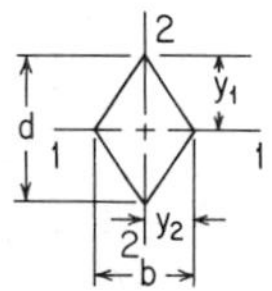	$A = \frac{bd}{2}$ $y_1 = \frac{d}{2}$ $y_2 = \frac{b}{2}$	$I_1 = \frac{1}{48}bd^3$ $I_2 = \frac{1}{48}db^3$	$r_1 = 0.2041d$ $r_2 = 0.2041b$
8. Trapezoid 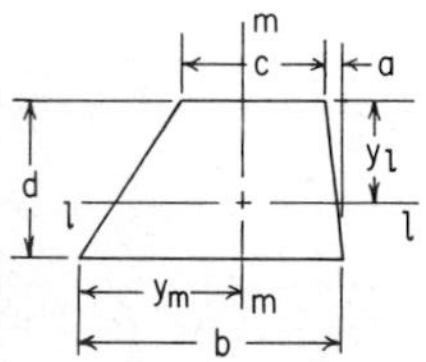	$A = \frac{d}{2}(b + c)$ $y_l = \frac{d}{3}\frac{2b + c}{b + c}$ $y_m = \frac{2b^2 + 2bc - ab - 2ac - c^2}{3(b + c)}$	$I_l = \frac{d^3}{36}\frac{b^2 + 4bc + c^2}{b + c}$ $I_m = \frac{d}{36(b+c)}[b^4 + c^4 + 2bc(b^2 + c^2) - a(b^3 + 3b^2c - 3bc^2 - c^3) + a^2(b^2 + 4bc + c^2)]$ $H_{lm} = \frac{d^2}{72(b+c)}[c(3b^2 - 3bc - c^2) + b^3 - a(2b^2 + 8bc + 2c^2)]$	$r_l = \sqrt{\frac{I_l}{A}}$ $r_m = \sqrt{\frac{I_m}{A}}$

TABLE 1 ***Properties of sections (Cont.)***

Form of section	Area and distances from centroid to extremities	Moments and products of inertia about central axes	Radii of gyration about central axes
9. Regular polygon with n sides 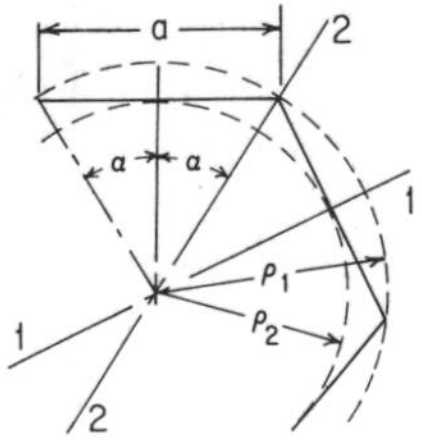	$A = \dfrac{a^2 n}{4 \tan \alpha}$ $\rho_1 = \dfrac{a}{2 \sin \alpha}$ $\rho_2 = \dfrac{a}{2 \tan \alpha}$ If n is odd: $y_1 = y_2 = \rho_1 \cos\left[\alpha\left(\dfrac{n+1}{2}\right) - \dfrac{\pi}{2}\right]$ If $n/2$ is odd: $y_1 = \rho_1 \qquad y_2 = \rho_2$ If $n/2$ is even: $y_1 = \rho_2 \qquad y_2 = \rho_1$	$I_1 = I_2 = \frac{1}{24}A(6\rho_1^2 - a^2)$	$r_1 = r_2 = \sqrt{\frac{1}{24}(6\rho_1^2 - a^2)}$
10. Solid circle 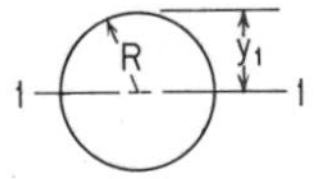	$A = \pi R^2$ $y_1 = R$	$I_1 = \dfrac{\pi}{4} R^4$	$r_1 = \dfrac{R}{2}$
11. Hollow circle 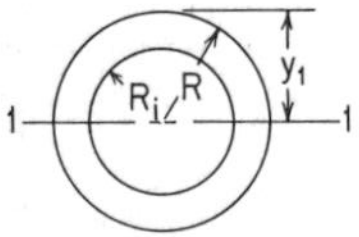	$A = \pi(R^2 - R_i^2)$ $y_1 = R$	$I_1 = \dfrac{\pi}{4}(R^4 - R_i^4)$	$r_1 = \frac{1}{2}\sqrt{R^2 + R_i^2}$
12. Very thin annulus 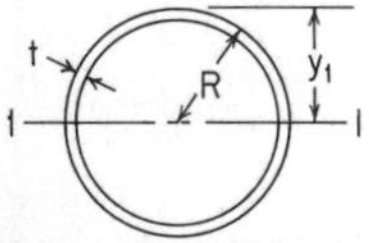	$A = 2\pi R t$ $y_1 = R$	$I_1 = \pi R^3 t$	$r_1 = 0.707R$

13. Solid ellipse 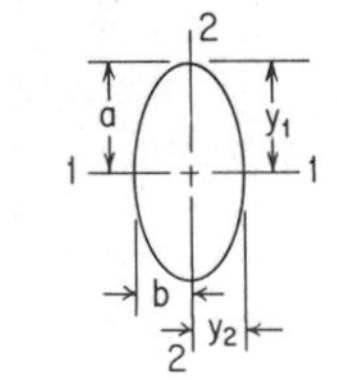	$A = \pi ab$ $y_1 = a$ $y_2 = b$	$I_1 = \frac{\pi}{4} ba^3$ $I_2 = \frac{\pi}{4} ab^3$	$r_1 = \frac{a}{2}$ $r_2 = \frac{b}{2}$
14. Hollow ellipse 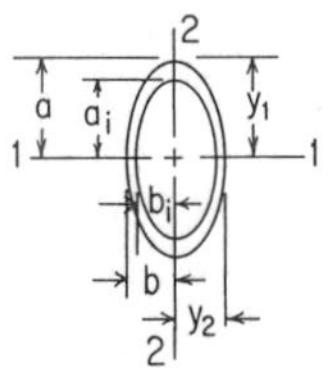	$A = \pi(ab - a_i b_i)$ $y_1 = a$ $y_2 = b$	$I_1 = \frac{\pi}{4}(ba^3 - b_i a_i^3)$ $I_2 = \frac{\pi}{4}(ab^3 - a_i b_i^3)$	$r_1 = \frac{1}{2}\sqrt{\frac{ba^3 - b_i a_i^3}{ab - a_i b_i}}$ $r_2 = \frac{1}{2}\sqrt{\frac{ab^3 - a_i b_i^3}{ab - a_i b_i}}$
15. Solid semicircle 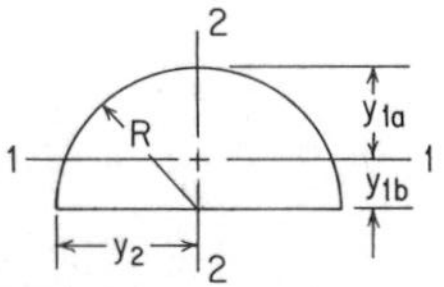	$A = \frac{\pi}{2} R^2$ $y_{1a} = 0.5756R$ $y_{1b} = 0.4244R$ $y_2 = R$	$I_1 = 0.1098R^4$ $I_2 = \frac{\pi}{8} R^4$	$r_1 = 0.2643R$ $r_2 = \frac{R}{2}$
16. Sector of solid circle 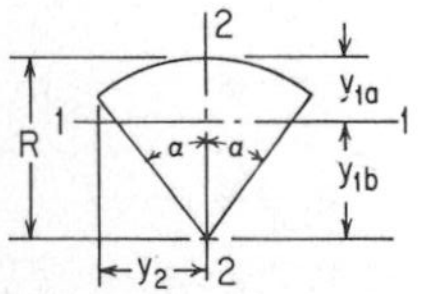	$A = \alpha R^2$ $y_{1a} = R\left(1 - \frac{2 \sin \alpha}{3\alpha}\right)$ $y_{1b} = \frac{2R \sin \alpha}{3\alpha}$ $y_2 = R \sin \alpha$	$I_1 = \frac{R^4}{4}\left(\alpha + \sin \alpha \cos \alpha - \frac{16 \sin^2 \alpha}{9\alpha}\right)$ $I_2 = \frac{R^4}{4}(\alpha - \sin \alpha \cos \alpha)$ (*Note:* If α is small, $\alpha - \sin \alpha \cos \alpha = \frac{2}{3}\alpha^3 - \frac{2}{15}\alpha^5$)	$r_1 = \frac{R}{2}\sqrt{1 + \frac{\sin \alpha \cos \alpha}{\alpha} - \frac{16 \sin^2 \alpha}{9\alpha^2}}$ $r_2 = \frac{R}{2}\sqrt{1 - \frac{\sin \alpha \cos \alpha}{\alpha}}$

TABLE 1 *Properties of sections (Cont.)*

Form of section	Area and distances from centroid to extremities	Moments and products of inertia about central axes	Radii of gyration about central axes
17. Segment of solid circle (*Note:* If $\alpha < \frac{\pi}{4}$, use expressions from case 18) 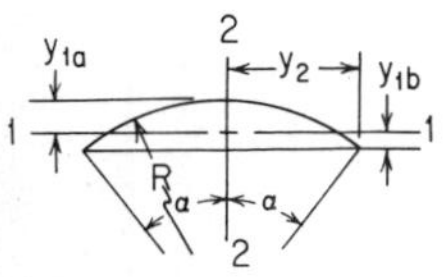	$A = R^2(\alpha - \sin\alpha\cos\alpha)$ $y_{1a} = R\left[1 - \frac{2\sin^3\alpha}{3(\alpha - \sin\alpha\cos\alpha)}\right]$ $y_{1b} = R\left[\frac{2\sin^3\alpha}{3(\alpha - \sin\alpha\cos\alpha)} - \cos\alpha\right]$ $y_2 = R\sin\alpha$	$I_1 = \frac{R^4}{4}\left[\alpha - \sin\alpha\cos\alpha + 2\sin^3\alpha\cos\alpha - \frac{16\sin^6\alpha}{9(\alpha - \sin\alpha\cos\alpha)}\right]$ $I_2 = \frac{R^4}{12}(3\alpha - 3\sin\alpha\cos\alpha - 2\sin^3\alpha\cos\alpha)$	$r_1 = \frac{R}{2}\sqrt{1 + \frac{2\sin^3\alpha\cos\alpha}{\alpha - \sin\alpha\cos\alpha} - \frac{16\sin^6\alpha}{9(\alpha - \sin\alpha\cos\alpha)^2}}$ $r_2 = \frac{R}{2}\sqrt{1 - \frac{2\sin^3\alpha\cos\alpha}{3(\alpha - \sin\alpha\cos\alpha)}}$
18. Segment of solid circle (*Note:* Do not use if $\alpha > \frac{\pi}{4}$) 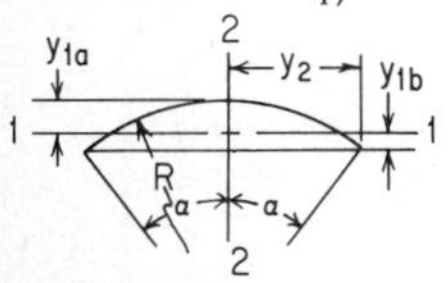	$A = \frac{2}{3}R^2\alpha^3(1 - 0.2\alpha^2 + 0.019\alpha^4)$ $y_{1a} = 0.3R\alpha^2(1 - 0.0976\alpha^2 + 0.0028\alpha^4)$ $y_{1b} = 0.2R\alpha^2(1 - 0.0619\alpha^2 + 0.0027\alpha^4)$ $y_2 = R\alpha(1 - 0.1667\alpha^2 + 0.0083\alpha^4)$	$I_1 = 0.01143R^4\alpha^7(1 - 0.3491\alpha^2 + 0.0450\alpha^4)$ $I_2 = 0.1333R^4\alpha^5(1 - 0.4762\alpha^2 + 0.1111\alpha^4)$	$r_1 = 0.1309R\alpha^2(1 - 0.0745\alpha^2)$ $r_2 = 0.4472R\alpha(1 - 0.1381\alpha^2 + 0.0184\alpha^4)$

19. Sector of hollow circle (*Note:* If t/R is small, α can exceed π to form an overlapped annulus) 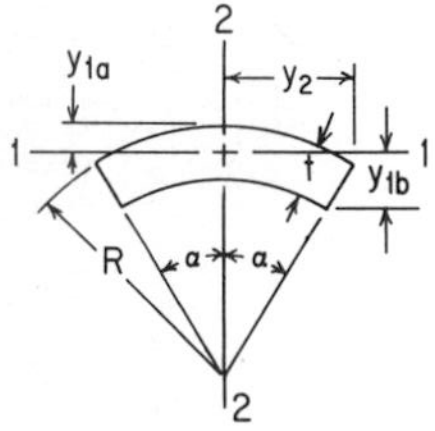	$A = \alpha t(2R - t)$ $y_{1a} = R\left[1 - \frac{2\sin\alpha}{3\alpha}\left(1 - \frac{t}{R} + \frac{1}{2 - t/R}\right)\right]$ $y_{1b} = R\left[\frac{2\sin\alpha}{3\alpha(2 - t/R)} + \left(1 - \frac{t}{R}\right)\frac{2\sin\alpha - 3\alpha\cos\alpha}{3\alpha}\right]$ $y_2 = R\sin\alpha$	$I_1 = R^3 t\left[\left(1 - \frac{3t}{2R} + \frac{t^2}{R^2} - \frac{t^3}{4R^3}\right) \times \left(\alpha + \sin\alpha\cos\alpha - \frac{2\sin^2\alpha}{\alpha}\right) + \frac{t^2\sin^2\alpha}{3R^2\alpha(2 - t/R)}\left(1 - \frac{t}{R} + \frac{t^2}{6R^2}\right)\right]$ $I_2 = R^3 t\left(1 - \frac{3t}{2R} + \frac{t^2}{R^2} - \frac{t^3}{4R^3}\right) \times (\alpha - \sin\alpha\cos\alpha)$	$r_1 = \sqrt{\frac{I_1}{A}}$ $r_2 = \sqrt{\frac{I_2}{A}}$
		Note: If α is small: $\frac{\sin\alpha}{\alpha} = 1 - \frac{\alpha^2}{6} + \frac{\alpha^4}{120}$ $\cos\alpha = 1 - \frac{\alpha^2}{2} + \frac{\alpha^4}{24}$ $\frac{\sin^2\alpha}{\alpha} = \alpha\left(1 - \frac{\alpha^2}{3} + \frac{2\alpha^4}{45}\right)$	$\alpha - \sin\alpha\cos\alpha = \frac{2}{3}\alpha^3\left(1 - \frac{\alpha^2}{5} + \frac{2\alpha^4}{105}\right)$ $\alpha + \sin\alpha\cos\alpha - \frac{2\sin^2\alpha}{\alpha} = \frac{2\alpha^5}{45}\left(1 - \frac{\alpha^2}{7} + \frac{\alpha^4}{105}\right)$
20. Hollow regular polygon with n sides 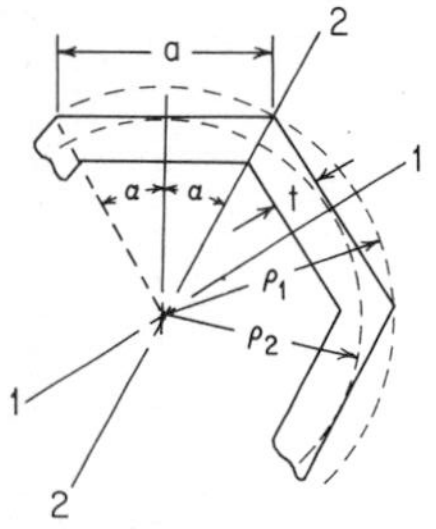	$A = nat\left(1 - \frac{t\tan\alpha}{a}\right)$ $\rho_1 = \frac{a}{2\sin\alpha}$ $\rho_2 = \frac{a}{2\tan\alpha}$ If n is odd: $y_1 = y_2 = \rho_1\cos\left(\alpha\frac{n+1}{2} - \frac{\pi}{2}\right)$ If $n/2$ is odd: $y_1 = \rho_1 \quad y_2 = \rho_2$ If $n/2$ is even: $y_1 = \rho_2 \quad y_2 = \rho_1$	$I_1 = I_2 = \frac{na^3 t}{8}\left(\frac{1}{3} + \frac{1}{\tan^2\alpha}\right) \times \left[1 - 3\frac{t\tan\alpha}{a} + 4\left(\frac{t\tan\alpha}{a}\right)^2 - 2\left(\frac{t\tan\alpha}{a}\right)^3\right]$	$r_1 = r_2 = \frac{a}{\sqrt{8}} \times \sqrt{\left(\frac{1}{3}\right) + \frac{1}{\tan^2\alpha}\left[1 - 2\frac{t\tan\alpha}{a} + 2\left(\frac{t\tan\alpha}{a}\right)^2\right]}$

PART THREE

Formulas and Examples

Each of the following chapters deals with a certain type of structural member or a certain condition of stress. What may be called the common, or typical, case is usually discussed first; special cases, representing peculiarities of form, proportions, or circumstances of loading, are considered subsequently. In the discussion of each case the underlying assumptions are stated, the general behavior of the loaded member is described, and formulas for the stress and deformation are given. The more important of the general equations are numbered consecutively throughout each section to facilitate reference, but, wherever possible, formulas applying to specific cases are tabulated for convenience and economy of space.

In all formulas which contain numerical constants having dimensions, unless other units are specified, the unit of distance is the inch and the unit of force is the pound. Therefore all areas are in square inches, all moments of inertia are in inches to the fourth, all distributed loads are in pounds per linear inch or pounds per square inch, all moments are in inch-pounds, and all stresses are in pounds per square inch.

Most formulas contain only dimensionless constants and can be evaluated in any consistent system of units.

SIX

Tension, Compression, Shear, and Combined Stress

6.1 *Bar under axial tension (or compression); common case*

The bar is straight, of any uniform cross section, of homogeneous material, and (if under compression) short or constrained against lateral buckling. The loads are applied at the ends, centrally, and in such a manner as to avoid nonuniform stress distribution at any section of the part under consideration. The stress does not exceed the proportional limit.

Behavior. Parallel to the load the bar elongates (under tension) or shortens (under compression), the unit longitudinal strain being ϵ and the total longitudinal strain in the length l being e. At right angles to the load the bar contracts (under tension) or expands (under compression); the unit lateral strain ϵ' is the same in all transverse directions, and the total lateral strain e' in any direction is proportional to the lateral dimension d measured in that direction. Both longitudinal and lateral strains are proportional to the

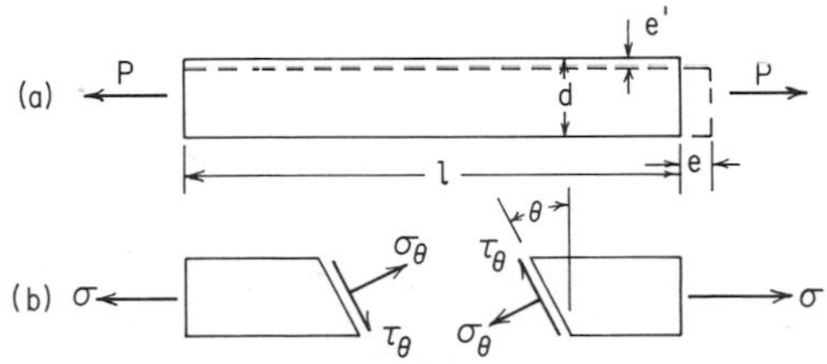

Fig. 6.1

applied load. On any right section there is a uniform tensile (or compressive) stress σ; on any oblique section there is a uniform tensile (or compressive) normal stress σ_θ and a uniform shear stress τ_θ. The deformed bar under tension is represented in Fig. 6.1*a*, and the stresses in Fig. 6.1*b*.

Formulas Let

P = applied load
A = cross-sectional area (before loading)
l = length (before loading)
E = modulus of elasticity
ν = Poisson's ratio

Then

$$\sigma = \frac{P}{A} \qquad (1)$$

$$\sigma_\theta = \frac{P}{A}\cos^2\theta \qquad \max \sigma_\theta = \sigma \text{ (when } \theta = 0°)$$

$$\tau_\theta = \frac{P}{2A}\sin 2\theta \qquad \max \tau_\theta = \tfrac{1}{2}\sigma \text{ (when } \theta = 45 \text{ or } 135°)$$

$$\epsilon = \frac{\sigma}{E} \qquad (2)$$

$$e = l\epsilon = \frac{Pl}{AE} \qquad (3)$$

$$\epsilon' = -\nu\epsilon \qquad (4)$$

$$e' = \epsilon' d \qquad (5)$$

$$\text{Strain energy per unit volume} = \frac{1}{2}\frac{\sigma^2}{E} \text{ (in-lb in}^3) \qquad (6)$$

$$\text{Total strain energy} = \frac{1}{2}\frac{\sigma^2}{E} lA = \frac{1}{2} Pe \text{ (in-lb)} \qquad (7)$$

Each square inch of cross section changes by $(-2\nu\epsilon)$ in^2 under load, and each cubic inch of volume changes by $(1 - 2\nu)\,\varepsilon$ in^3 under load.

In some discussions it is convenient to refer to the *stiffness* of a member, which is a measure of the resistance it offers to being deformed. The stiffness of a uniform bar under axial load is shown by Eq. 3 to be proportional to A and E directly and to l inversely, i.e., proportional AE/l.

EXAMPLE

A cylindrical specimen of steel 4 in long and $1\frac{1}{2}$ in in diameter has applied to it a compressive load of 20,000 lb. For this steel $\nu = 0.285$ and $E = 30{,}000{,}000$. It is required to find (*a*) the unit compressive stress σ; (*b*) the total longitudinal deformation e; (*c*) the total transverse deformation e'; (*d*) the change in volume ΔV; and (*e*) the total energy, or work done in applying the load.

(*a*) $\sigma = \dfrac{P}{A} = \dfrac{-20{,}000}{1.77} = -11{,}300 \text{ lb/in}^2$

(*b*) $\epsilon = \dfrac{\sigma}{E} = \dfrac{-11{,}300}{30{,}000{,}000} = -0.000377$

$e = l\epsilon = 4\,(-0.000377) = -0.00151$ in (shortening)

(*c*) $\epsilon' = -\nu\epsilon = 0.285\,(0.000377) = 0.0001075$

$e' = \epsilon' d = (0.0001075)\,1.5 = 0.000161$ in (expansion)

(*d*) Change in volume/in.$^3 = (1 - 2\nu)\,\varepsilon = -0.000162$ in^3

Total change in volume $\Delta V = 4\,(1.77)\,(-0.000162) = -0.00115$ in^3 (decrease)

(*e*) Strain energy $= \frac{1}{2} Pe = \frac{1}{2}\,(20{,}000)\,(0.00151) = 15.1$ in-lb

6.2 *Bar under tension (or compression); special cases*

If the bar is not straight, it is subject to bending; formulas for this case are given in Art. 11.4.

If the load is applied eccentrically, the bar is subject to bending; formulas for this case are given in Arts. 7.7 and 11.4. If the load is compressive and the bar is long and not laterally constrained, it must be analyzed as a column by the methods of Chap. 11.

If the stress exceeds the proportional limit, the formulas for stress given in Art. 6.1 still hold but the deformation and work done in producing it can be determined only from experimental data relating unit strain to unit stress.

If the section is not uniform but changes *gradually,* the stress at any section can be found by dividing the load by the area of that section; the total longitudinal deformation over a length l is given by $\int_0^l \frac{P}{AE}\,dl$, and the strain energy is given by $\int_0^l \frac{1}{2}\frac{P^2}{AE}\,dl$. If the change in section is *abrupt,* stress concentration may have to be taken into account, values of k being used to find elastic stresses and values of k_r being used to predict the breaking load. Stress concentration may also have to be considered if the end attachments for loading involve pinholes, screw threads, or other stress raisers (see Art. 2.10 and Table 37).

If instead of being applied at the ends of a uniform bar the load is applied at an intermediate point, both ends being held, the *method of consistent deformations* shows that the load is apportioned to the two parts of the bar in inverse proportion to their respective lengths.

If a uniform bar is supported at one end in a vertical position and loaded only by its own weight, the maximum stress occurs at the supported end and is equal to the weight divided by the cross-sectional area. The total elongation is *half* as great and the total strain energy *one-third* as great as if a load equal to the weight were applied at the unsupported end. A bar supported at one end and loaded by its own weight and an axial downward load P (in pounds) applied at the unsupported end will have the same unit stress σ (in pounds per square inch) at all sections if it is tapered so that all sections are similar in form but vary in scale according to the formula

$$y = \frac{\sigma}{w} \log_e \frac{A\sigma}{P}$$

where y is the distance (in inches) from the free end of the bar to any section, A is the area (in square inches) of that section, and w is the density of the material (in pounds per cubic inch).

If a bar is stressed by having both ends rigidly held while a change in temperature is imposed, the resulting stress is found by calculating the longitudinal expansion (or contraction) that the change in temperature would produce if the bar were not held and then calculating the load necessary to shorten (or lengthen) it by that amount (principle of superposition). If the bar is uniform, the unit stress produced is independent of the length of the bar if restraint against buckling is provided. If a bar is stressed by being struck an axial blow at one end, the case is one of *impact* loading, discussed in Art. 15.3.

EXAMPLES

1. Figure 6.2 represents a uniform bar rigidly held at the ends A and D and axially loaded at the intermediate points B and C. It is required to determine the total force in each portion of the bar AB, BC, CD.

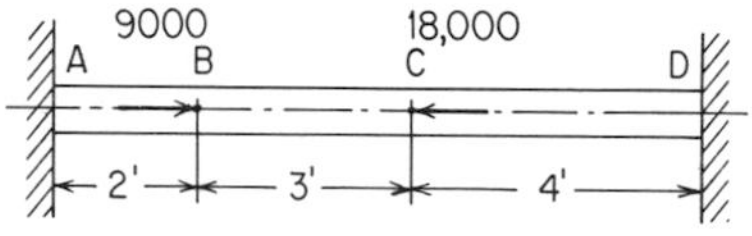

Fig. 6.2

Solution. Each load is divided between the portions of the bar to right and left in inverse proportion to the lengths of these parts (consistent deformations), and the total force sustained by each part is the algebraic sum of the forces imposed by the individual loads (superposition). Of the 9000-lb load, therefore, $\frac{7}{9}$, or 7000 lb, is carried in tension by the part AB, and $\frac{2}{9}$, or 2000 lb, is carried in compression by the part BD. Of the 18,000-lb load, $\frac{4}{9}$, or 8000 lb, is carried in compression by the part AC, and $\frac{5}{9}$, or 10,000 lb, is carried in tension by the part CD. Denoting tension by the plus sign and compression by the minus sign, and adding algebraically, the actual stresses in the parts are found to be

$$\begin{aligned}
&(\text{In } AB)\ +7000 - 8000 &&= -1000 \text{ lb}\\
&(\text{In } BC)\ -2000 - 8000 &&= -10{,}000 \text{ lb}\\
&(\text{In } CD)\ -2000 + 10{,}000 &&= +8000 \text{ lb}
\end{aligned}$$

The results are quite independent of the diameter of the bar and of E.

If instead of being *held* at the ends, the bar is prestressed by wedging it between rigid walls under an initial compression of, say, 10,000 lb and the loads at B and C are then applied, the results secured above would represent the *changes* in force the several parts would undergo. The final forces in the bar would therefore be 11,000 lb compression in AB, 20,000 lb compression in BC, and 2000 lb compression in CD. But if the initial compression were less than 8000 lb, the bar would break contact with the wall at D (no tension possible); there would be no force at all in CD, and the forces in AB and BC, now statically determinate, would be 9000 and 18,000 lb compression, respectively.

2. A steel bar 24 in long has the form of a truncated cone, being circular in section with a diameter at one end of 1 in and at the other of 3 in. For this steel, $E = 30{,}000{,}000$ and the coefficient of thermal expansion is 0.0000065/°F. This bar is rigidly held at both ends and subjected to a drop in temperature of 50°F. It is required to determine the maximum tensile stress thus caused.

Solution. Using the principle of superposition, the solution is effected in three steps: (*a*) the shortening e due to the drop in temperature is found, assuming the bar free to contract; (*b*) the force P required to produce an elongation equal to e, that is, to stretch the bar back to its original length, is calculated; (*c*) the maximum tensile stress produced by this force P is calculated.

(*a*) $e = 50(0.0000065)(24) = 0.00780$ in.

(*b*) Let d denote the diameter and A the area of any section a distance x in from the small end of the bar. Then

$$d = 1 + \frac{1}{12}x \qquad A = \frac{\pi}{4}\left(1 + \frac{1}{12}x\right)^2$$

and

$$e = \int_0^{24} \frac{P}{(E\,\pi/4)\,(1 + \frac{1}{12}x)^2}\,dx = 0.00000034P = 0.00780$$

or

$$P = 22{,}900 \text{ lb}$$

(*c*) The maximum stress occurs at the smallest section and is

$$\sigma = \frac{22{,}900}{0.785} = 29{,}200 \text{ lb/in}^2$$

The result can be accepted as correct only if the proportional limit of the steel is known to be as great as or greater than the maximum stress and if the concept of a rigid support can be accepted. (See page 146 for deformations of elastic bodies.)

6.3 *Composite members*

A tension or compression member may be made up of parallel elements or parts which jointly carry the applied load. The essential problem is to determine how the load is apportioned among the several parts, and this is easily done by the method of consistent deformations. If the parts are so arranged that all undergo the same total elongation or shortening, then each will carry a portion of the load proportional to its stiffness, i.e., proportional to AE/l if each is a uniform bar and proportional to AE if all these uniform bars are of equal length. It follows that if there are n bars, with section areas $A_1, A_2, \ldots, A_n$, lengths $l_1, l_2, \ldots, l_n$, and moduli $E_1, E_2, \ldots, E_n$, then the loads on the several bars $P_1, P_2, \ldots, P_n$ are given by

$$P_1 = P \frac{\dfrac{A_1E_1}{l_1}}{\dfrac{A_1E_1}{l_1} + \dfrac{A_2E_2}{l_2} + \cdots + \dfrac{A_nE_n}{l_n}} \tag{8}$$

$$P_2 = P \frac{\dfrac{A_2E_2}{l_2}}{\dfrac{A_1E_1}{l_1} + \dfrac{A_2E_2}{l_2} + \cdots + \dfrac{A_nE_n}{l_n}} \tag{9}$$

....

A composite member of this kind can be *prestressed.* P_1, P_2, etc., then represent the *increments* of force in each member due to the applied load, and can be found by Eqs. 8 and 9, provided all bars can sustain reversal of stress, or provided the applied load is not great enough to cause such reversal in any bar which cannot sustain it. As explained in Art. 2.12, by proper prestressing, all parts of a composite member can be made to reach their allowable loads, elastic limits, or ultimate strengths simultaneously (Example 2).

EXAMPLES

1. A ring is suspended by three vertical bars *A, B,* and *C* of unequal lengths. The upper ends of the bars are held at different levels, so that as assembled none of the bars is stressed. *A* is 4 ft long, has a section area of 0.3 in^2, and is of steel for which $E = 30{,}000{,}000$; *B* is 3 ft long, has a section area of 0.2 in^2, and is of copper for which $E = 17{,}000{,}000$; *C* is 2 ft long, has a section area of 0.4 in^2, and is of aluminum for which $E = 10{,}000{,}000$. A load of 10,000 lb is hung on the ring. It is required to determine how much of this load is carried by each bar.

Solution. Denoting by P_A, P_B, and P_C the loads carried by *A, B,* and *C,* respectively, and expressing the moduli of elasticity in millions of pounds per square inch and the lengths in feet, we substitute in Eq. 8 and find

$$P_A = 10{,}000 \left[\frac{\dfrac{(0.3)\,(30)}{4}}{\dfrac{(0.3)\,(30)}{4} + \dfrac{(0.2)\,(17)}{3} + \dfrac{(0.4)\,(10)}{2}} \right] = 4180 \text{ lb}$$

Similarly

$$P_B = 2100 \text{ lb} \qquad \text{and} \qquad P_C = 3720 \text{ lb}$$

2. A composite member is formed by passing a steel rod through an aluminum tube of the same length and fastening the two parts together at both ends. The fastening is accomplished by adjustable nuts, which make it possible to assemble the rod and tube so that one is under initial tension and the other is under an equal initial compression. For the steel rod the section area is 1.5 in^2, the modulus of elasticity 30,000,000, and the allowable stress 15,000 lb/in^2. For the aluminum tube the section area is 2 in^2, the modulus of elasticity 10,000,000, and the allowable stress 10,000 lb/in^2. It is desired to prestress the composite member so that under a tensile load both parts will reach their allowable stresses simultaneously.

Solution. When the allowable stresses are reached, the force in the steel rod will be 1.5(15,000) = 22,500 lb; the force in the aluminum tube will be 2(10,000) = 20,000 lb; and the total load on the member will be 22,500 + 20,000 = 42,500 lb. Let P_i denote the initial tension or compression in the members, and, as before, let tension be considered positive and compression negative. Then, since Eq. 9 gives the *increment* in force, we have for the aluminum tube

$$P_i + 42{,}500\frac{(2)(10)}{(2)(10) + (1.5)(30)} = 20{,}000$$

or

$$P_i = +6920 \text{ lb} \qquad \text{(initial tension)}$$

For the steel rod, we have

$$P_i + 42{,}500\frac{(1.5)(30)}{(2)(10) + (1.5)(30)} = 22{,}500$$

or

$$P_i = -6920 \text{ lb} \qquad \text{(initial compression)}$$

If the member were not prestressed, the unit stress in the steel would always be just three times as great as that in the aluminum because it would sustain the same unit deformation and its modulus of elasticity is three times as great. Therefore, when the steel reached its allowable stress of 15,000 lb/in^2, the aluminum would be stressed to only 5000 lb/in^2 and the allowable load on the composite member would be only 32,500 lb instead of 42,500 lb.

6.4 *Trusses*

A conventional truss is essentially an assemblage of straight uniform bars that are subjected to axial tension or compression when the truss is loaded at the joints. The deflection of any joint of a truss is easily found by the *method of unit loads* (Art. 3.5). Let p_1, p_2, p_3, etc., denote the forces produced in the several members by an *assumed unit load* acting in the direction x at the joint whose deflection is to be found, and let e_1, e_2, e_3, etc., denote the longitudinal deformations produced in the several members by the *actual applied loads.* The deflection D_x in the direction x of the joint in question is given by

$$D_x = p_1e_1 + p_2e_2 + p_3e_3 + \cdots = \Sigma pe \tag{10}$$

The deflection in the direction y, at right angles to x, can be found similarly by assuming the unit load to act in the y direction; the resultant deflection is then determined by combining the x and y deflections. Attention must be given to the *signs* of p and e; p is positive if a member is subjected to tension and negative if under compression, and e is positive if it represents an elongation and negative if it represents a shortening. A positive value for Σpe means that the deflection is in the direction of the assumed unit load, and a negative value means that it is in the opposite direction. (This procedure is illustrated in Example 1 at the end of this article.)

A statically indeterminate truss can be solved by the *method of least work* (Art. 3.5). To do this, it is necessary to write down the expression for the total strain energy in the structure, which, being simply the sum of the strain energies of the constituent bars, is given by

$$\frac{1}{2}P_1e_1 + \frac{1}{2}P_2e_2 + \frac{1}{2}P_3e_3 + \cdots = \sum \frac{1}{2}Pe = \sum \frac{1}{2}\frac{P^2l}{AE} \tag{11}$$

Here P_1, P_2, etc., denote the forces in the individual members due to the applied loads and e has the same meaning as above. It is necessary to express each force P as the sum of the two forces; one of these is the force the applied loads would produce with the redundant member removed, and the other is the force due to the unknown force (say, F) exerted by this redundant member on the rest of the structure. The total strain energy is thus expressed as a function of F, the force in the redundant member. The partial derivative with respect to F of this expression for strain energy is then set equal to zero and solved for F. If there are two or more redundant members, the expression for strain energy with all the redundant forces, F_1, F_2, etc., represented is differentiated once with respect to each. The equations thus obtained are then solved simultaneously for the unknown forces. (The procedure is illustrated in Example 2.)

EXAMPLES

1. The truss shown in Fig. 6.3 is composed of tubular steel members, for which $E = 30{,}000{,}000$. The section areas of the members are given in the table below. It is required to determine D_x and D_y, the horizontal and vertical components of the displacement of joint A produced by the indicated loading.

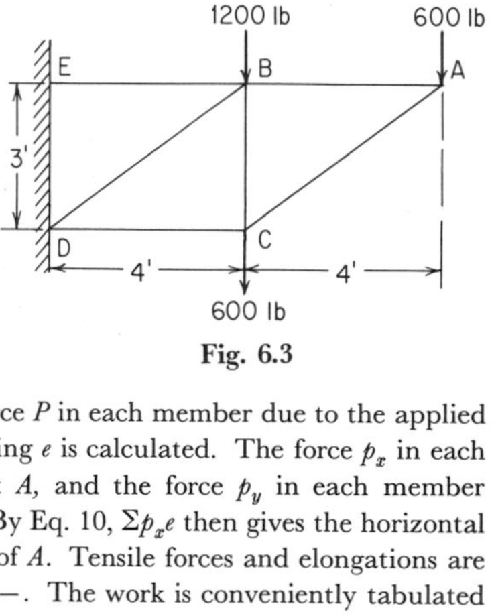

Fig. 6.3

Solution. The method of unit loads is used. The force P in each member due to the applied loads is found, and the resulting elongation or shortening e is calculated. The force p_x in each member due to a load of 1 lb acting to the right at A, and the force p_y in each member due to a load of 1 lb acting down at A are calculated. By Eq. 10, $\Sigma p_x e$ then gives the horizontal and $\Sigma p_y e$ gives the vertical displacement or deflection of A. Tensile forces and elongations are denoted by +, compressive forces and shortenings by −. The work is conveniently tabulated as follows:

Member	Area A, in^2	Length l, in	P, lb	$e = \frac{Pl}{AE}$, in	p_x, lb/lb	$p_x e$, in	p_y, lb/lb	$p_y e$, in
AB	0.07862	48	+ 800	+0.0163	+1.00	+0.0163	+1.33	+0.0217
AC	0.07862	60	−1000	−0.0254	0	0	−1.67	+0.0424
BC	0.1464	36	+1200	+0.0098	0	0	+1.00	+0.0098
BE	0.4142	48	+4000	+0.0154	+1.00	+0.0154	+2.67	+0.0411
BD	0.3318	60	−4000	−0.0241	0	0	−1.67	+0.0402
CD	0.07862	48	− 800	−0.0163	0	0	−1.33	+0.0217
							D_x =	+0.0317
							D_y =	+0.1769

D_x and D_y are both found to be positive, which means that the displacements are in the directions of the assumed unit loads—to the right and down. Had either been found to be negative, it would have meant that the displacement was in a direction opposite to that of the corresponding unit load.

2. Assume a diagonal member, running from A to D and having a section area 0.3318 in^2, is to be added to the truss of Example 1; the structure is now statically indeterminate. It is required to determine the force in each member of the altered truss due to the loads shown.

Solution. We use the method of least work. The truss has one redundant member; any member except BE may be regarded as redundant, since if any one were removed, the remaining structure would be stable and statically determinate. We select AD to be regarded as redundant, denote the unknown force in AD by F, and assume F to be compression. We find the force in each member assuming AD to be removed, then find the force in each member due to a push F exerted at A by AD, and then add these forces, thus getting an expression for the force in each member of the actual truss in terms of F. The expression for the strain energy can then be written out, differentiated with respect to F, equated to zero, and solved for F. F being known, the force in each member of the truss is easily found. The computations are conveniently tabulated as follows:

Member	Stress due to				
	Applied loads, AD out	Push F exerted by AD		Applied loads, AD in place	
		$F = F$	$F =$ value found	In terms of F	Actual value
	(1)	(2)	(3)	(4) = (1) + (2) = P	(5) = (1) + (3)
AB	+ 800	+0.470F	+494	+ 800 + 0.470F	+1290(T)
AC	−1000	+0.584F	+612	−1000 + 0.584F	− 390(C)
BC	+1200	−0.351F	−369	+1200 − 0.351F	+ 830(T)
BE	+4000	0	0	+4000	+4000(T)
BD	−4000	+0.584F	+612	−4000 + 0.584F	−3390(C)
CD	− 800	+0.470F	+494	− 800 + 0.470F	− 306(C)
AD	0	F	1050	F	1050(C)

$$U = \sum \frac{1}{2}\frac{P^2 l}{AE} = \frac{1}{2E}\left[\frac{(800 + 0.470F)^2(48)}{0.07862} + \frac{(-1000 + 0.584F)^2(60)}{0.07862}\right.$$

$$+ \frac{(1200 - 0.351F)^2(36)}{0.1464} + \frac{(4000^2)\,(48)}{0.4142} + \frac{(-4000 + 0.584F)^2(60)}{0.3318}$$

$$\left. + \frac{(-800 + 0.470F)^2(48)}{0.07862} + \frac{F^2(102.6)}{0.3318}\right]$$

$$\frac{\partial U}{\partial F} = \frac{1}{2E}\left[\frac{2(800 + 0.470F)\,(48)\,(0.470)}{0.07862} + \cdots\right]$$

$$= 0$$

which gives $\quad F = +1050$

The plus sign here means simply that F is *compression, as assumed.* If F had mistakenly been assumed to be tension, a negative result would have been obtained, showing that the assumption was incorrect and that F was really compression.

6.5 *Body under pure shear stress*

A condition of pure shear may be produced by any one of the methods of loading shown in Fig. 6.4. In Fig. 6.4*a* a rectangular block of length a, height b, and uniform thickness t is shown loaded by forces P_1 and P_2, uniformly distributed over the surfaces to which they are applied and satisfying the equilibrium equation $P_1 b = P_2 a$. There are equal shear stresses on all vertical and horizontal planes, so that any contained cube oriented like $ABCD$ has on each of four faces the shear stress $\tau = P_1/at = P_2/bt$ and no other stress.

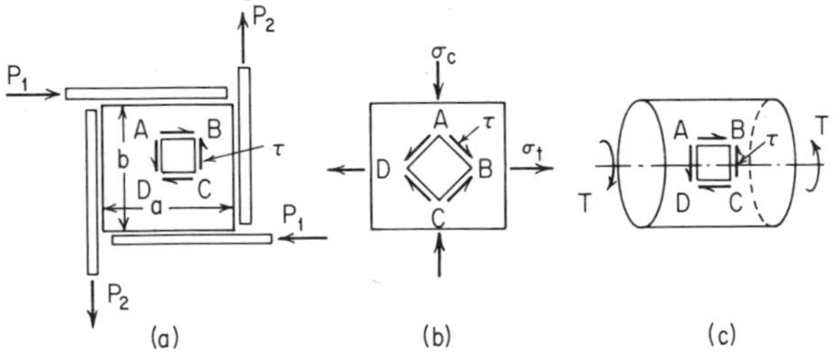

Fig. 6.4

In Fig. 6.4*b* a rectangular block is shown under equal and opposite biaxial stresses σ_t and σ_c. There are equal shear stresses on all planes inclined at 45° to the top and bottom faces, so that a contained cube oriented like $ABCD$ has on each of four faces the shear stress $\tau = \sigma_t = \sigma_c$ and no other stress.

In Fig. 6.4*c* a circular shaft is shown under a twisting moment T; a cube of infinitesimal dimensions, a distance z from the axis and oriented like $ABCD$ has on each of four faces an essentially uniform shear stress $\tau = Tz/J$ (Art. 9.1) and no other stress.

In whatever way the loading is accomplished, the result is to impose on an elementary cube of the loaded body the condition of stress represented in Fig. 6.5, that is, shearing stress alone on each of four faces, these stresses being equal and so directed as to satisfy the equilibrium condition $T_x = 0$ (Art. 3.1).

Behavior. The strains produced by pure shear are as shown in Fig. 6.5; the cube $ABCD$ is deformed into the rhombohedron $A'B'C'D'$. On any vertical plane and on any horizontal plane there is a shear stress τ; on any oblique plane there is a normal stress σ_θ and a shear stress τ_θ.

Formulas. Let all stresses be expressed in pounds per square inch, let γ denote unit shear strain, and let G denote the modulus of rigidity in pounds

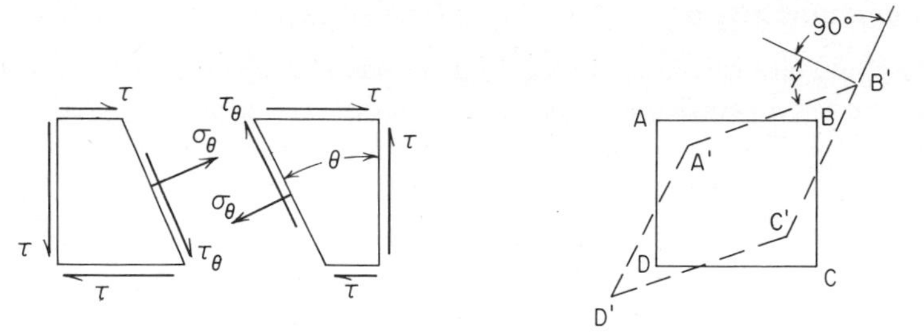

Fig. 6.5

per square inch. Then

$$\sigma_\theta = \tau \sin 2\theta; \ \max \sigma_\theta = \tau \qquad (\text{when } \theta = 45°)$$

$$\tau_\theta = -\tau \cos 2\theta$$

$$\gamma = \frac{\tau}{G} \tag{12}$$

$$\text{Strain energy per unit volume} = \frac{1}{2}\frac{\tau^2}{G} \qquad (\text{in-lb/in}^3) \tag{13}$$

The relations between τ, σ, and the strains represented in Fig. 6.5 make it possible to express G in terms of E and Poisson's ratio ν. The formula is

$$G = \frac{E}{2(1 + \nu)} \tag{14}$$

From known values of E (determined by a tensile test) and G (determined by a torsion test) it is thus possible to calculate ν.

6.6 *Cases of direct shear loading*

By *direct shear loading* is meant any case in which a member is acted on by equal, parallel, and opposite forces so nearly colinear that the material between them is subjected primarily to shear stress, with negligible bending. Examples of this are provided by rivets, bolts, and pins, shaft splines and keys, screw threads, short lugs, etc. These are not really cases of pure shear; the actual stress distribution is complex and usually indeterminate because of the influence of fit and other factors. In designing such parts, however, it is usually assumed that the shear is uniformly distributed on the critical section, and since working stresses are selected with due allowance for the approximate nature of this assumption, the practice is usually permissible. In *beams* subject to transverse shear, this assumption cannot be made as a rule.

Shear and other stresses in rivets, pins, keys, etc., are discussed more fully in Chap. 13, shear stresses in beams in Chap. 7, and shear stresses in torsion members in Chap. 9.

6.7 *Combined stress*

Under certain circumstances of loading, a body is subjected to a combination of tensile and compressive stresses (usually designated as *biaxial* or *triaxial stress*) or to a combination of tensile, compressive, and shear stresses (usually designated as *combined stress*). For example, the material at the inner surface of a thick cylindrical pressure vessel is subjected to triaxial stress (radial compression, longitudinal tension, and circumferential tension), and a shaft simultaneously bent and twisted is subjected to combined stress (longitudinal tension or compression, and torsional shear).

In most instances the normal and shear stresses on each of three mutually perpendicular planes are due to torsion, beam shear, flexure, axial loading, or some combination of these and can be readily calculated by the appropriate formulas. The principal stresses, the maximum shear stress, and the normal and shear stresses on any given plane can then be found by the formulas given in Table 2.

The *strains* produced by any combination of stresses not exceeding the proportional limit can be found by superposition. Consideration of the strains produced by equal triaxial stress leads to an expression for the bulk modulus of elasticity:

$$K = \frac{E}{3(1 - 2\nu)} \tag{15}$$

EXAMPLES

1. A rectangular block 12 in long, 4 in high, and 2 in thick is subjected to a longitudinal tensile stress $\sigma_x = 12{,}000$ lb/in^2, a vertical compressive stress $\sigma_y = 15{,}000$ lb/in^2, and a lateral compressive stress $\sigma_z = 9000$ lb/in^2. The material is steel, for which $E = 30{,}000{,}000$ and $\nu = 0.30$. It is required to find the total change in length.

Solution. The longitudinal deformation is found by superposition: The unit strain due to each stress is computed separately by Eq. 2; these results are added to give the resultant longitudinal unit strain, which is multiplied by the length to give the total elongation. Denoting unit longitudinal strain by ϵ_x and total longitudinal strain by e_x, we have

$$\epsilon_x = +\frac{12{,}000}{E} + \nu\frac{15{,}000}{E} + \nu\frac{9000}{E}$$

$$= +0.000400 + 0.000150 + 0.000090 = +0.00064$$

$$e_x = 12(0.00064) = 0.00768 \text{ in}$$

The lateral dimensions have nothing to do with the result since the lateral stresses, not the lateral loads, are given.

2. A piece of "standard extra-strong" pipe, 2 in nominal diameter, is simultaneously subjected to an internal pressure of 2000 lb/in^2 and to a twisting moment of 5000 in-lb caused by tightening a cap screwed on at one end. It is required to determine the maximum tensile stress and the maximum shear stress thus produced in the pipe.

Solution. The calculations will be made, first, for a point at the outer surface and, second, for a point at the inner surface. The dimensions of the pipe and properties of the cross section are as follows: Inner radius $R_0 = 0.9695$; outer radius $R_1 = 1.1875$; cross-sectional area of bore

$A_b = 2.955$; cross-sectional area of pipe wall $A_w = 1.475$; and polar moment of inertia $J = 1.735$.

We take axis x parallel to the axis of the pipe, axis y tangent to the cross section, and axis z radial. For a point at the outer surface the stress conditions are those of Table 2, case 4, where σ_x is the longitudinal tensile stress due to pressure, σ_y is the circumferential stress due to pressure, and τ_{xy} is the shear stress due to torsion. Using the formula for stress in thick cylinders (Table 13, case 33) to calculate σ_y, the formula for axial stress (Eq. 1) to calculate σ_x; and the formula for torsional stress (Chap. 9, Eq. 2) to calculate τ_{xy}, we have

$$\sigma_x = \frac{pA_b}{A_w} = \frac{(2000)(2.955)}{1.475} = 4000$$

$$\sigma_y = p\frac{R_0^2(R_1^2 + R_1^2)}{R_1^2(R_1^2 - R_0^2)} = 2000\frac{(0.9695^2)(1.1875^2 + 1.1875^2)}{(1.1875^2)(1.1875^2 - 0.9695^2)} = 8000$$

$$\tau_{xy} = \frac{TR_1}{J} = \frac{(5000)(1.1875)}{1.735} = 3420$$

$$\text{Max } \sigma = \frac{1}{2}(\sigma_x + \sigma_y) + \sqrt{\left(\frac{\sigma_x - \sigma_y}{2}\right)^2 + \tau_{xy}^2}$$

$$= \frac{1}{2}(4000 + 8000) + \sqrt{\left(\frac{4000 - 8000}{2}\right)^2 + 3420^2} = 9970$$

$$\text{Max } \tau = \sqrt{\left(\frac{4000 - 8000}{2}\right)^2 + 3420^2} = 3970$$

For a point at the inner surface the stress conditions are those of Table 2, case 6, with σ_x the longitudinal tension due to pressure, σ_y the circumferential tension due to pressure, σ_z the radial compression due to the direct pressure of the contained liquid, τ_{xy} the shear stress due to torsion, and τ_{xz} and τ_{yz} equal to zero. We have

$$\sigma_x = 4000$$

$$\sigma_y = p\frac{R_1^2 + R_0^2}{R_1^2 - R_0^2} = 2000\frac{1.1875^2 + 0.9695^2}{1.1875^2 - 0.9695^2} = 10{,}000$$

$$\sigma_z = -p = -2000$$

$$\tau_{xy} = \frac{TR_0}{J} = \frac{(5000)(0.9695)}{1.735} = 2790$$

$$\tau_{xz} = \tau_{yz} = 0$$

$$\sigma^3 - (4000 + 10{,}000 - 2000)\sigma^2 + [(4000)(10{,}000) + (10{,}000)(-2000) + (4000)(-2000) - 0 - 0 - 2790^2]\sigma - [(4000)(10{,}000)(-2000) + 0 - 0 - 0 - (-2000)(2790^2)] = 0$$

Solving, $\sigma = +11{,}100, +2900, -2000$. These are the three principal stresses. Obviously the maximum stress is 11,100 tension. The maximum shear stress is $\frac{1}{2}[11{,}100 - (-2000)] = 6550$. For this particular case, with τ_{xz} and $\tau_{yz} = 0$, the principal stresses $+11{,}100$ and $+2900$ are in the plane of axes x and y and the third principal stress -2000 is σ_z.

3. Stress calculations have revealed that an element of material is subjected to the following six stress components: $\sigma_x = 8000$ lb/in^2; $\sigma_y = -4000$ lb/in^2; $\sigma_z = 9000$ lb/in^2; $\tau_{xy} = -2000$ lb/in^2; $\tau_{xz} = 5000$ lb/in^2; and $\tau_{yz} = -3000$ lb/in^2. The negative signs indicate stresses in directions opposed to those shown in Table 2, case 6. Find the magnitude and directions of the maximum tensile stress and the magnitude of the maximum shear stress.

Solution. The three principal stresses σ_1, σ_2, and σ_3 are found from the roots of the following

TABLE 2 ***Formulas for combined stress***

All stresses are unit stresses and are positive when acting as shown. θ is positive when measured counterclockwise from the X face

Condition of applied stress	Formulas for σ_θ, τ_θ, the principal stresses, and the maximum shear stress
1. Axial stress 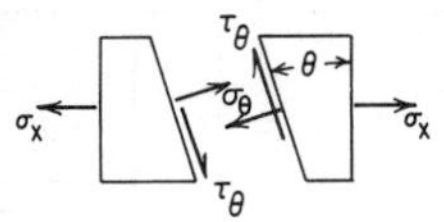	$\sigma_\theta = \sigma_x \cos^2 \theta$ $\tau_\theta = \frac{1}{2}\sigma_x \sin 2\theta$ Principal stresses $= \sigma_x$ and 0 (when $\theta = 0°$ and $90°$) Max $\tau = \frac{1}{2}\sigma_x$ (when $\theta = 45°$ and $135°$)
2. Biaxial stress 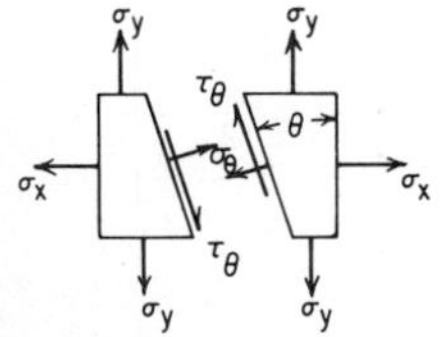	$\sigma_\theta = \frac{1}{2}(\sigma_x + \sigma_y) + \frac{1}{2}(\sigma_x - \sigma_y)\cos 2\theta$ $\tau_\theta = \frac{1}{2}(\sigma_x - \sigma_y)\sin 2\theta$ Principal stresses $= \sigma_x$ and σ_y (when $\theta = 0°$ and $90°$) Max $\tau =$ the numerically largest of the three values $\frac{1}{2}(\sigma_x - \sigma_y)$, $\frac{1}{2}\sigma_x$, or $\frac{1}{2}\sigma_y$ (on planes at 45° to X and Y faces, 45° to X face and 90° to Y face, or 90° to X face and 45° to Y face, respectively)
3. Pure shear 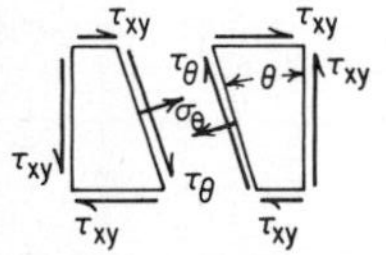	$\sigma_\theta = \tau_{xy} \sin 2\theta$ $\tau_\theta = -\tau_{xy} \cos 2\theta$ Principal stresses $= +\tau_{xy}$ and $-\tau_{xy}$ (when $\theta = 45°$ and $135°$) Max $\tau = \tau_{xy}$ (when $\theta = 0°$ and $90°$)
4. Biaxial stress combined with shear stress 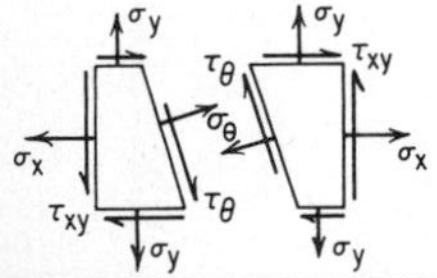	$\sigma_\theta = \frac{1}{2}(\sigma_x + \sigma_y) + \frac{1}{2}(\sigma_x - \sigma_y)\cos 2\theta + \tau_{xy}\sin 2\theta$ $\tau_\theta = \frac{1}{2}(\sigma_x - \sigma_y)\sin 2\theta - \tau_{xy}\cos 2\theta$ Principal stresses $= \dfrac{1}{2}(\sigma_x + \sigma_y) \pm \sqrt{\dfrac{1}{4}(\sigma_x - \sigma_y)^2 + \tau_{xy}^2}$ $\left(\text{when } \theta = \dfrac{1}{2}\tan^{-1}\dfrac{2\tau_{xy}}{\sigma_x - \sigma_y}\right)$ Max $\tau = \sqrt{\dfrac{1}{4}(\sigma_x - \sigma_y)^2 + \tau_{xy}^2}$ $\left(\text{when } \theta = \dfrac{1}{2}\tan^{-1}\dfrac{\sigma_x - \sigma_y}{-2\tau_{xy}}\right)$

5. Triaxial stress

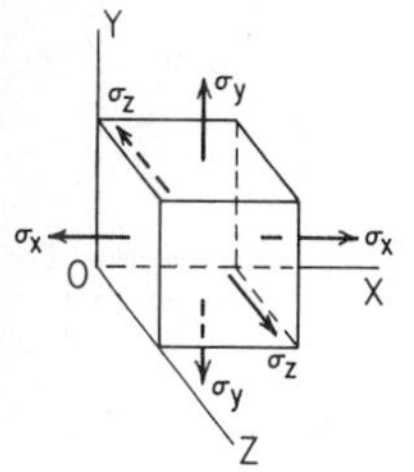

Let *OX'* be an axis making with *OX*, *OY*, and *OZ* angles whose cosines are $l_{x'x}$, $l_{x'y}$, and $l_{x'z}$, respectively. Then on a plane normal to *OX'*

$$\sigma_{x'} = \sigma_x l_{x'x}^2 + \sigma_y l_{x'y}^2 + \sigma_z l_{x'z}^2$$

Resultant shear stress $= \sqrt{\sigma_x^2 l_{x'x}^2 + \sigma_y^2 l_{x'y}^2 + \sigma_z^2 l_{x'z}^2 - \sigma_{x'}^2}$

Principal stresses $= \sigma_x, \sigma_y, \sigma_z$

The maximum shear stress occurs on each of two planes inclined at 45° to the two principal stresses whose algebraic difference is greatest and is equal to one-half that algebraic difference

6. Triaxial stress combined with shear stress (general case of stress)

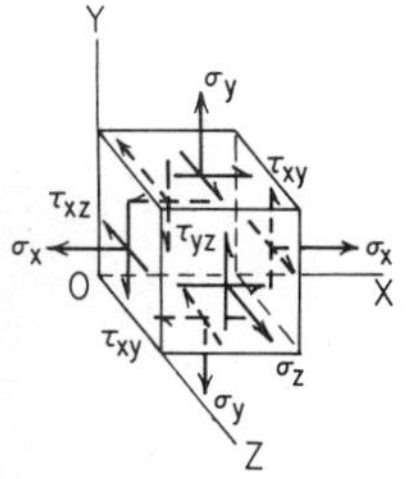

Let *OX'*, *OY'*, and *OZ'* be rectangular axes with the origin at *O*. *OX'* makes with *OX*, *OY*, *OZ* angles whose cosines are $l_{x'x}$, $l_{x'y}$, and $l_{x'z}$; *OY'* makes with *OX*, *OY*, and *OZ* angles whose cosines are $l_{y'x}$, $l_{y'y}$, and $l_{y'z}$; and *OZ'* makes with *OX*, *OY*, and *OZ* angles whose cosines are $l_{z'x}$, $l_{z'y}$, and $l_{z'z}$. Then on a plane normal to *OX*[1]

$$\sigma_{x'} = \sigma_x l_{x'x}^2 + \sigma_y l_{x'y}^2 + \sigma_z l_{x'z}^2 + 2\tau_{xy} l_{x'x} l_{x'y} + 2\tau_{xz} l_{x'x} l_{x'z} + 2\tau_{yz} l_{x'y} l_{x'z}$$
$$\tau_{x'y'} = \sigma_x l_{x'x} l_{y'x} + \sigma_y l_{x'y} l_{y'y} + \sigma_z l_{x'z} l_{y'z} + \tau_{xy}(l_{x'x} l_{y'y} + l_{x'y} l_{y'x}) + \tau_{xz}(l_{x'x} l_{y'z} + l_{x'z} l_{y'x}) + \tau_{yz}(l_{x'y} l_{y'z} + l_{x'z} l_{y'y})$$
$$\tau_{x'z'} = \sigma_x l_{x'x} l_{z'x} + \sigma_y l_{x'y} l_{z'y} + \sigma_z l_{x'z} l_{z'z} + \tau_{xy}(l_{x'x} l_{z'y} + l_{x'y} l_{z'x}) + \tau_{xz}(l_{x'x} l_{z'z} + l_{x'z} l_{z'x}) + \tau_{yz}(l_{x'y} l_{z'z} + l_{x'z} l_{z'y})$$

Resultant shear stress $= \sqrt{\tau_{x'y'}^2 + \tau_{x'z'}^2}$

The three principal stresses σ are given by the three roots of the equation

$$\sigma^3 - (\sigma_x + \sigma_y + \sigma_z)\sigma^2 + (\sigma_x\sigma_y + \sigma_y\sigma_z + \sigma_x\sigma_z - \tau_{xy}^2 - \tau_{xz}^2 - \tau_{yz}^2)\sigma - (\sigma_x\sigma_y\sigma_z + 2\tau_{xy}\tau_{xz}\tau_{yz} - \sigma_x\tau_{yz}^2 - \sigma_y\tau_{xz}^2 - \sigma_z\tau_{xy}^2) = 0$$

The direction of each principal stress is defined by the cosines of the angles it makes with *OX*, *OY*, and *OZ*. These direction cosines are found by substituting the value of σ in question into the following three equations and solving for the direction cosines. For the principal stress σ_1:

$$(\sigma_x - \sigma_1)l_{1x} + \tau_{xy} l_{1y} + \tau_{xz} l_{1z} = 0$$
$$\tau_{xy} l_{1x} + (\sigma_y - \sigma_1)l_{1y} + \tau_{yz} l_{1z} = 0$$
$$l_{1x}^2 + l_{1y}^2 + l_{1z}^2 = 1$$

Similar calculations are made for the directions of σ_2 and σ_3.

The maximum shear stress occurs on each of two planes inclined at 45° to the two principal stresses whose algebraic difference is greatest and is equal to one-half that algebraic difference.

equation from case 6:

$$\sigma^3 - [8000 - 4000 + 9000]\sigma^2 + [(8000)(-4000) + (-4000)(9000) + (8000)(9000) - (-2000^2) - (5000^2) - (-3000^2)]\sigma - [(8000)(-4000)(9000) + 2(-2000)(5000)(-3000) - (8000)(-3000^2) - (-4000)(5000^2) - (9000)(-2000^2)] = 0$$

$$\sigma^3 - (13)(10^3)\sigma^2 - (34)(10^6)\sigma + (236)(10^9) = 0$$

$$\sigma_1 = 14{,}220 \text{ lb/in}^2 \qquad \sigma_2 = 3{,}510 \text{ lb/in}^2 \qquad \sigma_3 = -4{,}730 \text{ lb/in}^2$$

Substituting σ_1 into the three equations for evaluating the direction cosines gives

$$(8000 - 14{,}220)\, l_{1x} - 2000\, l_{1y} + 5000\, l_{1z} = 0$$

$$-2000\, l_{1x} + (-4000 - 14{,}220)\, l_{1y} - 3000\, l_{1z} = 0$$

$$l_{1x}^2 + l_{1y}^2 + l_{1z}^2 = 1$$

$$l_{1x} = 0.652 \qquad l_{1y} = -0.189 \qquad l_{1z} = 0.735$$

The maximum shear stress is the largest value of $\frac{1}{2}(14{,}220 - 3510)$, $\frac{1}{2}(14{,}220 + 4730)$, or $\frac{1}{2}(3510 + 4730)$. Max $\tau = 9{,}475$ lb/in^2.

SEVEN

Beams; Flexure of Straight Bars

7.1 *Straight beams (common case) elastically stressed.*

The formulas of this article are based on the following assumptions: (1) The beam is of homogeneous material that has the same modulus of elasticity in tension and compression. (2) The beam is straight or nearly so; if it is slightly curved, the curvature is in the plane of bending and the radius of curvature is at least 10 times the depth. (3) The cross section is uniform. (4) The beam has at least one longitudinal plane of symmetry. (5) All loads and reactions are perpendicular to the axis of the beam and lie in the same plane, which is a longitudinal plane of symmetry. (6) The beam is long in proportion to its depth, the span/depth ratio being 8 or more for metal beams of compact section, 15 or more for beams with relatively thin webs, and 24 or more for rectangular timber beams. (7) The beam is not disproportionately wide. (8) The maximum stress does not exceed the proportional limit.

Applied to any case for which these assumptions are not valid, the formulas given yield results that at best are approximate and that may be grossly in error; such cases are discussed in subsequent articles. The limitations stated here with respect to straightness and proportions of the beam correspond to a maximum error in calculated results of about 5 percent.

In the following discussion, it is assumed for convenience that the beam is horizontal and the loads and reactions vertical.

Behavior. The beam bends; fibers on the convex side lengthen, and fibers on the concave side shorten. The neutral surface is normal to the plane of the loads and contains the centroids of all sections, hence the neutral axis of any section is the horizontal central axis. Plane sections remain plane, and hence unit fiber strains and stresses are proportional to distance from the neutral surface. Longitudinal displacements of points on the neutral surface are negligible. Vertical deflection is largely due to bending, that due to shear being usually negligible under the conditions stated.

There is at any point a longitudinal fiber stress σ, which is tensile if the point lies between the neutral and convex surfaces of the beam and compressive if the point lies between the neutral and concave surfaces of the beam. This fiber stress σ usually may be assumed uniform across the width of the beam (see Arts. 7.11 and 7.12).

There is at any point a longitudinal shear stress τ on the horizontal plane and an equal vertical shear stress on the transverse plane. These shear stresses may be assumed uniform across the width of the beam (see page 93).

Figures 7.1*a* and *b* represent a beam under load and show the various dimensions that appear in the formulas; Fig. 7.1*c* shows a small prism at a point *q* acted on by the stresses σ and τ.

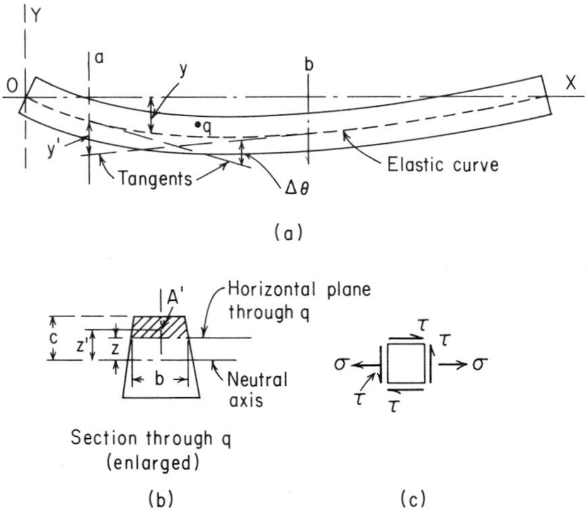

Fig. 7.1

Formulas. Let I = the moment of inertia of the section of the beam with respect to the neutral axis and E = modulus of elasticity of the material.

The fiber stress σ at any point q is

$$\sigma = \frac{Mz}{I} \tag{1}$$

where M is the bending moment at the section containing q, and z is the distance from the neutral axis to q.

The shear stress τ at any point q is

$$\tau = \frac{VA'z'}{Ib} \tag{2}$$

where V is the vertical shear at the section containing q, A' is the area of that part of the section above (or below) q, z' is the distance from the neutral axis to the centroid of A', and b is the net breadth of the section measured through q.

The strain energy of flexure U_f (inch-pounds) is

$$U_f = \int \frac{M^2}{2EI}\,dx \tag{3}$$

where M represents the bending moment equation in terms of x, the distance (inches) from the left end of the beam to any section.

The radius of curvature R of the elastic curve at any section is

$$R = \frac{EI}{M} \tag{4}$$

where M is the bending moment at the section in question.

The general differential equation of the elastic curve is

$$EI\frac{d^2y}{dx^2} = M \tag{5}$$

where M has the same meaning as in Eq. 3. Solution of this equation for the vertical deflection y (inches) is effected by writing out the expression for M, integrating twice, and determining the constants of integration by the boundary conditions.

By the method of unit loads the vertical deflection y at any point is found to be

$$y = \int \frac{Mm}{EI}\,dx \tag{6}$$

or by Castigliano's first theorem it is found to be

$$y = \frac{\partial U}{\partial P} \tag{7}$$

where M has the same meaning as in Eq. 3 and m is the equation of the bending moment due to a load of 1 lb acting vertically at the section where y is to be found. The integration indicated must be performed over each portion of the beam for which either M or m is expressed by a different equation. A positive result for y means that the deflection is in the direction of the assumed unit load; a negative result means it is in the opposite direction (see Example 2 at the end of this article).

In Eq. 7, U is given by Eq. 3 and P is a vertical load, real or imaginary, applied at the section where y is to be found. It is most convenient to perform the differentiation within the integral sign; as with Eq. 6, the integration must extend over the entire length of the beam, and the sign of the result is interpreted as before.

The change in slope of the elastic curve $\Delta\theta$ (radians) between any two sections a and b is

$$\Delta\theta = \int_a^b \frac{M}{EI}\,dx \tag{8}$$

where M has the same meaning as in Eq. 3.

The deflection y' at any section a, *measured vertically from a tangent drawn to the elastic curve at any section* b, is

$$y' = \int_a^b \frac{M}{EI}\,x\,dx \tag{9}$$

where x is the distance from a to any section dx between a and b.

Important relations between the bending moment and shear equations are

$$V = \frac{dM}{dx} \tag{10}$$

$$M = \int V\,dx \tag{11}$$

These relations are useful in constructing shear and moment diagrams and locating the section or sections of maximum bending moment since Eq. 10 shows that the maximum moment occurs when V, its first derivative, passes through zero and Eq. 11 shows that the increment in bending moment that occurs between any two sections is equal to the area under the shear diagram between those sections.

Maximum fiber stress. The maximum fiber stress at any section occurs at the point or points most remote from the neutral axis and is given by Eq. 1 when $z = c$; hence

$$\text{Max}\ \sigma = \frac{Mc}{I} = \frac{M}{I/c} \tag{12}$$

The maximum fiber stress in the beam occurs at the section of greatest bending moment; if the section is not symmetrical about the neutral axis, the stresses should be investigated at both the section of greatest positive moment and the section of greatest negative moment.

Maximum shear stress. The maximum shear stress in the beam occurs at the section of greatest vertical shear. The maximum shear stress at any section occurs at the neutral axis, provided the net width b is as small there as anywhere else; if the section is narrower elsewhere, the maximum shear stress may not occur at the neutral axis. This maximum shear stress can be expressed conveniently by the formula

$$\text{Max } \tau = \alpha \frac{V}{A} \tag{13}$$

where V/A is the *average* shear stress on the section and α is a factor that depends on the form of the section. For a rectangular section, $\alpha = \frac{3}{2}$ and the maximum stress is at the neutral axis; for a solid circular section, $\alpha = \frac{4}{3}$ and the maximum stress is at the neutral axis; for a triangular section, $\alpha = \frac{3}{2}$ and the maximum stress is halfway between the top and bottom of the section; for a diamond-shaped section of depth d, $\alpha = \frac{9}{8}$ and the maximum stress is at points that are a distance of $\frac{1}{8}d$ above and below the neutral axis.

In the derivation of Eq. 2 and in the preceding discussion, it is assumed that the shear stress is uniform across the width of the beam; i.e., it is the same at all points on any transverse line parallel to the neutral axis. Actually this is not the case; exact analysis (Ref. 1) shows that the shear stress varies across the width and that for a rectangle the maximum intensity occurs at the ends of the neutral axis where, for a wide beam, it is twice the average intensity. Photoelastic investigation of beams under concentrated loading shows that localized shearing stresses about four times as great as the maximum stress given by Eq. 2 occur near the points of loading and support (Ref. 2), but experience shows that this variation may be ignored and design based on the average value as determined by Eq. 2.

For some sections the greatest horizontal shear stress at a given point occurs, not on a horizontal plane, but on an inclined longitudinal plane which cuts the section so as to make b a minimum. Thus, for a circular tube or pipe the greatest horizontal shear stress at any point occurs on a *radial* plane; the corresponding shear stress in the plane of the section is not vertical but tangential, and in computing τ by Eq. 2, b should be taken as twice the thickness of the tube instead of the net horizontal breadth of the member. (See Table 17, cases 18 and 19, for examples of where this shear stress in a tube is of importance.)

In an I, T, or box section there is a horizontal shear stress on any vertical longitudinal plane through the flange, and this stress is usually a maximum at the juncture of flange and web. It may be calculated by Eq. 2, taking A' as the area outside of the vertical plane (for outstanding flanges) or between the vertical plane and the center of the beam section (for box girders), and b as the flange thickness (see the solution to Example 1*b*). The other terms have the same meanings explained previously.

Shear stresses are not often of controlling importance except in wood or metal beams that have thin webs or a small span/depth ratio. For beams that conform to the assumptions stated previously, strength will practically always be governed by fiber stress.

Change in projected length due to bending. The apparent shortening of a beam due to bending, i.e., the difference between its original length and the horizontal projection of the elastic curve, is given by

$$\Delta l = \frac{1}{2}\int_0^l \left(\frac{dy}{dx}\right)^2 dx \tag{14}$$

To evaluate Δl, dy/dx is expressed in terms of x (Eq. 5) and the square of this is integrated as indicated.

The extreme fibers of the beam undergo a change in actual length due to stress given by

$$e = \int_0^l \frac{Mc}{EI}\, dx \tag{15}$$

By means of these equations the actual relative horizontal displacement of points on the upper or lower surface of the beam can be predicted and the necessary allowances made in the design of rocker bearings, clearance for the corners, etc.

Tabulated formulas. Table 3 gives formulas for the reactions, moments, slopes, and deflections at each end of single-span beams supported and loaded in various ways. The table also gives formulas for the vertical shears, bending moments, slopes, and deflections at any distance x from the left end of the span.

In these formulas, the unit step function is used by itself and in combination with ordinary functions.

The unit step function is denoted by $\langle x - a\rangle^0$ where the use of the angle brackets $\langle\ \rangle$ is defined as follows: If $x < a$, $\langle x - a\rangle^0 = 0$; if $x > a$, $\langle x - a\rangle^0 = 1$. At $x = a$ the unit step function is undefined just as vertical shear is undefined directly beneath a concentrated load. The use of the angle brackets $\langle\ \rangle$ is extended to other cases involving powers of the unit step function and the ordinary function $(x - a)^n$. Thus the quantity $(x - a)^n \langle x - a\rangle^0$ is shortened to $\langle x - a\rangle^n$ and again is given a value of zero if $x < a$ and is treated in the normal fashion if $x > a$.

In addition to the usual concentrated vertical loads, concentrated couples, and distributed loads, Table 3 also presents cases where the loading is essentially a development of reactions due to deformations created within the span. These cases include the concentrated angular displacement, concentrated transverse deflection, and linear temperature differential across the beam from top to bottom. In all three cases it can be assumed that initially the beam was deformed but free of stress and then is forced to conform to the end conditions. (In many cases no forces are created, and the formulas give only the deformed shape.)

Hetényi (Ref. 29) discusses in detail the Maclaurin series form of the equations used in this article and suggests (Ref. 53) that the deformation type of loading might be useful in solving beam problems. Thomson (Ref. 65) describes the use of the unit step function in the determination of beam deflections. By superposition, the formulas can be made to apply to almost any type of loading and support. The use of the tabulated and fundamental formulas given in this article is illustrated in the following examples.

EXAMPLES

1. For a beam supported and loaded as shown in Fig. 7.2, it is required to determine the maximum tensile stress, maximum shear stress, and maximum compressive stress, assuming, first, that the beam is made of wood with section as shown in Fig. 7.2*a*; second, that the beam is made of wood with section as shown in Fig. 7.2*b*; and third, that the beam is a 4-in, 7.7-lb steel I-beam.

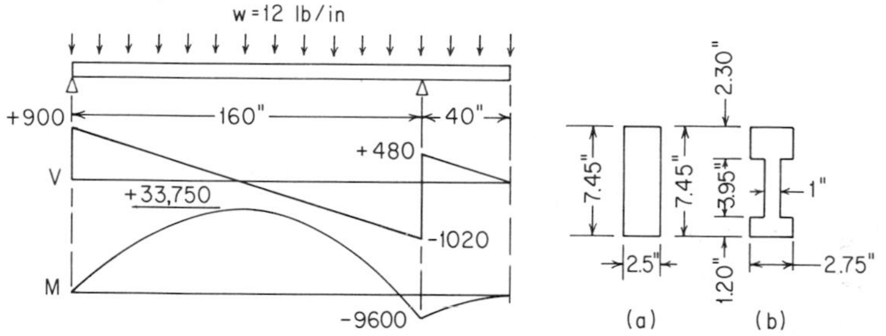

Fig. 7.2

Solution. By using the equations of equilibrium (Art. 3.1), the left and right reactions are found to be 900 and 1500 lb, respectively. The shear and moment equations are therefore

$$(x = 0 \text{ to } x = 160): \quad V = 900 - 12x$$
$$M = 900x - 12x(\tfrac{1}{2}x)$$
$$(x = 160 \text{ to } x = 200): \quad V = 900 - 12x + 1500$$
$$M = 900x - 12x(\tfrac{1}{2}x) + 1500(x - 160)$$

Using the step function described previously, these equations can be reduced to

$$V = 900 - 12x + 1500\, \langle x - 160 \rangle^0$$
$$M = 900x - 6x^2 + 1500\, \langle x - 160 \rangle$$

These equations are plotted, giving the shear and moment diagrams shown in Fig. 7.2. The maximum positive moment evidently occurs between the supports; the exact location is found by setting the first shear equation equal to zero and solving for x, which gives $x = 75$. Substitution of this value of x in the first moment equation gives $M = 33{,}750$ in-lb. The maximum negative moment occurs at the right support where the shear diagram again passes through zero and is 9600 in-lb.

The results obtained so far are independent of the cross section of the beam. The stresses will now be calculated for each of the sections (*a*), (*b*), and (*c*).

(*a*) For the rectangular section: $I = \frac{1}{12}bd^3 = 86.2$; $I/c = \frac{1}{6}bd^2 = 23.1$; and $A = bd = 18.60$. Therefore

$$\text{Max } \sigma = \frac{\max M}{I/c} = \frac{33{,}750}{23.1} = 1460 \text{ lb/in}^2 \qquad \text{(by Eq. 12)}$$

This stress occurs at $x = 75$ and is tension in the bottom fibers of the beam and compression in the top.

$$\text{Max } \tau = \frac{3}{2}\frac{\max V}{A} = \frac{3}{2}\frac{1020}{18.60} = 82 \text{ lb/in}^2 \qquad \text{(by Eq. 13)}$$

TABLE 3 ***Shear, moment, slope, and deflection formulas for beams***

NOTATION: W = load (pounds); w = unit load (pounds per linear inch); M_o = applied couple (inch-pounds); θ_o = externally created concentrated angular displacement (radians); Δ_o = externally created concentrated lateral displacement (inches); T_1 and T_2 = temperatures on the top and bottom surfaces, respectively (degrees). R_A and R_B are the vertical end reactions at the left and right, respectively, and are positive upward. M_A and M_B are the reaction end moments at the left and right, respectively. All moments are positive when producing compression on the upper portion of the beam cross section. The transverse shear force V is positive when acting upward on the left end of a portion of the beam. All applied loads, couples, and displacements are positive as shown. All forces are in pounds, all moments in inch-pounds, all deflections and beam dimensions in inches, all slopes in radians, and all temperatures in degrees. All deflections are positive upward, and all slopes are positive when up and to the right. E is the modulus of elasticity of the beam material, and I is the area moment of inertia about the centroidal axis of the beam cross section. γ is the temperature coefficient of expansion (inches per inch per degree)

1. Concentrated intermediate load

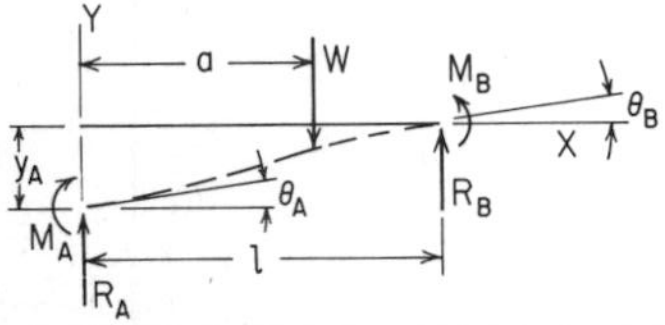

Transverse shear $= V = R_A - W\langle x - a\rangle^0$
Bending moment $= M = M_A + R_A x - W\langle x - a\rangle$

$$\text{Slope} = \theta = \theta_A + \frac{M_A x}{EI} + \frac{R_A x^2}{2EI} - \frac{W}{2EI}\langle x - a\rangle^2$$

$$\text{Deflection} = y = y_A + \theta_A x + \frac{M_A x^2}{2EI} + \frac{R_A x^3}{6EI} - \frac{W}{6EI}\langle x - a\rangle^3$$

(*Note:* see page 94 for a definition of the term $\langle x - a\rangle^n$.)

End restraints, reference no.	Boundary values	Selected maximum values of moments and deformations
1a. Left end free, right end fixed (cantilever) 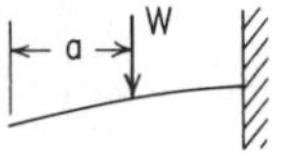	$R_A = 0 \quad M_A = 0 \quad \theta_A = \frac{W(l-a)^2}{2EI}$ $y_A = \frac{-W}{6EI}(2l^3 - 3l^2a + a^3)$ $R_B = W \quad M_B = -W(l-a)$ $\theta_B = 0 \quad y_B = 0$	Max $M = M_B$; max possible value $= -Wl$ when $a = 0$ Max $\theta = \theta_A$; max possible value $= \frac{Wl^2}{2EI}$ when $a = 0$ Max $y = y_A$; max possible value $= \frac{-Wl^3}{3EI}$ when $a = 0$
1b. Left end guided, right end fixed 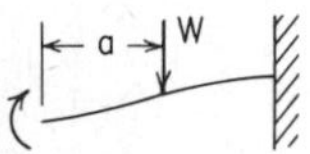	$R_A = 0 \quad M_A = \frac{W(l-a)^2}{2l} \quad \theta_A = 0$ $y_A = \frac{-W}{12EI}(l-a)^2(l+2a)$ $R_B = W \quad M_B = \frac{-W(l^2 - a^2)}{2l}$ $\theta_B = 0 \quad y_B = 0$	Max $+M = M_A$; max possible value $= \frac{Wl}{2}$ when $a = 0$ Max $-M = M_B$; max possible value $= \frac{-Wl}{2}$ when $a = 0$ Max $y = y_A$; max possible value $= \frac{-Wl^3}{12EI}$ when $a = 0$

1c. Left end simply supported, right end fixed 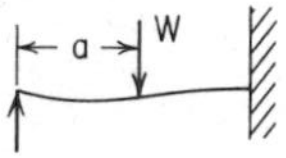	$R_A = \frac{W}{2l^3}(l-a)^2(2l+a) \qquad M_A = 0$ $\theta_A = \frac{-Wa}{4EIl}(l-a)^2 \qquad y_A = 0$ $R_B = \frac{Wa}{2l^3}(3l^2-a^2) \qquad \theta_B = 0$ $M_B = \frac{-Wa}{2l^2}(l^2-a^2) \qquad y_B = 0$	Max $+M = \frac{Wa}{2l^3}(l-a)^2(2l+a)$ at $x = a$; max possible value $= 0.174Wl$ when $a = 0.366l$ Max $-M = M_B$; max possible value $= -0.1924Wl$ when $a = 0.5773l$ Max $y = \frac{-Wa}{6EI}(l-a)^2\left(\frac{a}{2l+a}\right)^{1/2}$ at $x = l\left(\frac{a}{2l+a}\right)^{1/2}$ when $a > 0.414l$ Max $y = \frac{-Wa(l^2-a^2)^3}{3EI(3l^2-a^2)^2}$ at $x = \frac{l(l^2+a^2)}{3l^2-a^2}$ when $a < 0.414l$; max possible $y = -0.0098\frac{Wl^3}{EI}$ when $x = a = 0.414l$
1d. Left end fixed, right end fixed 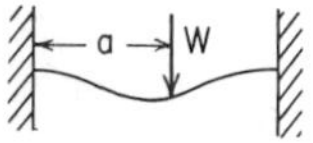	$R_A = \frac{W}{l^3}(l-a)^2(l+2a)$ $M_A = \frac{-Wa}{l^2}(l-a)^2$ $\theta_A = 0 \qquad y_A = 0$ $R_B = \frac{Wa^2}{l^3}(3l-2a)$ $M_B = \frac{-Wa^2}{l^2}(l-a)$ $\theta_B = 0 \qquad y_B = 0$	Max $+M = \frac{2Wa^2}{l^3}(l-a)^2$ at $x = a$; max possible value $= \frac{Wl}{8}$ when $a = \frac{l}{2}$ Max $-M = M_A$ if $a < \frac{l}{2}$; max possible value $= -0.1481Wl$ when $a = \frac{l}{3}$ Max $y = \frac{-2W(l-a)^2a^3}{3EI(l+2a)^2}$ at $x = \frac{2al}{l+2a}$ if $a > \frac{l}{2}$; max possible value $= \frac{-Wl^3}{192EI}$ when $x = a = \frac{l}{2}$
1e. Left end simply supported, right end simply supported 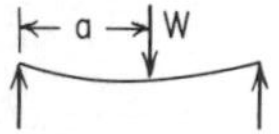	$R_A = \frac{W}{l}(l-a) \qquad M_A = 0$ $\theta_A = \frac{-Wa}{6EIl}(2l-a)(l-a) \qquad y_A = 0$ $R_B = \frac{Wa}{l} \qquad M_B = 0$ $\theta_B = \frac{Wa}{6EIl}(l^2-a^2) \qquad y_B = 0$	Max $M = R_A a$ at $x = a$; max possible value $= \frac{Wl}{4}$ when $a = \frac{l}{2}$ Max $y = \frac{-Wa}{3EIl}\left(\frac{l^2-a^2}{3}\right)^{3/2}$ at $x = l - \left(\frac{l^2-a^2}{3}\right)^{1/2}$ when $a < \frac{l}{2}$; max possible value $= \frac{-Wl^3}{48EI}$ at $x = \frac{l}{2}$ when $a = \frac{l}{2}$ Max $\theta = \theta_A$ when $a < \frac{l}{2}$; max possible value $= -0.0642\frac{Wl^2}{EI}$ when $a = 0.423l$
1f. Left end guided, right end simply supported 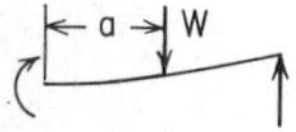	$R_A = 0 \qquad M_A = W(l-a) \qquad \theta_A = 0$ $y_A = \frac{-W(l-a)}{6EI}(2l^2+2al-a^2)$ $R_B = W \qquad M_B = 0$ $\theta_B = \frac{W}{2EI}(l^2-a^2) \qquad y_B = 0$	Max $M = M_A$ for $0 < x < a$; max possible value $= Wl$ when $a = 0$ Max $\theta = \theta_B$; max possible value $= \frac{Wl^2}{2EI}$ when $a = 0$ Max $y = y_A$; max possible value $= \frac{-Wl^3}{3EI}$ when $a = 0$

TABLE 3 *Shear, moment, slope, and deflection formulas for beams (Cont.)*

2. Partial distributed load

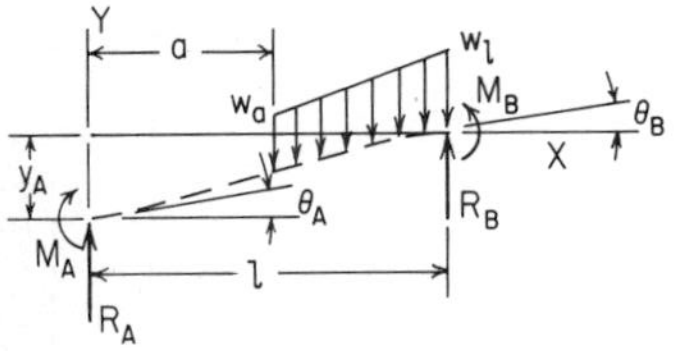

Transverse shear $= V = R_A - w_a\langle x-a\rangle - \dfrac{w_l - w_a}{2(l-a)}\langle x-a\rangle^2$

Bending moment $= M = M_A + R_A x - \dfrac{w_a}{2}\langle x-a\rangle^2 - \dfrac{w_l - w_a}{6(l-a)}\langle x-a\rangle^3$

Slope $= \theta = \theta_A + \dfrac{M_A x}{EI} + \dfrac{R_A x^2}{2EI} - \dfrac{w_a}{6EI}\langle x-a\rangle^3 - \dfrac{w_l - w_a}{24EI(l-a)}\langle x-a\rangle^4$

Deflection $= y = y_A + \theta_A x + \dfrac{M_A x^2}{2EI} + \dfrac{R_A x^3}{6EI} - \dfrac{w_a}{24EI}\langle x-a\rangle^4 - \dfrac{(w_l - w_a)}{120EI(l-a)}\langle x-a\rangle^5$

End restraints, reference no.	Boundary values	Selected maximum values of moments and deformations
2a. Left end free, right end fixed (cantilever) 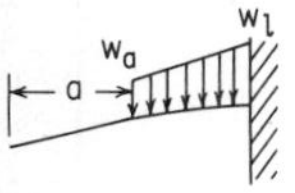	$R_A = 0 \qquad M_A = 0$ $\theta_A = \dfrac{w_a}{6EI}(l-a)^3 + \dfrac{w_l - w_a}{24EI}(l-a)^3$ $y_A = \dfrac{-w_a}{24EI}(l-a)^3(3l+a) - \dfrac{w_l - w_a}{120EI}(l-a)^3(4l+a)$ $R_B = \dfrac{w_a + w_l}{2}(l-a)$ $M_B = \dfrac{-w_a}{2}(l-a)^2 - \dfrac{w_l - w_a}{6}(l-a)^2$ $\theta_B = 0 \qquad y_B = 0$	If $a = 0$ and $w_l = w_a$ (uniform load on entire span), then: Max $M = M_B = \dfrac{-w_a l^2}{2}$ Max $\theta = \theta_A = \dfrac{w_a l^3}{6EI}$ Max $y = y_A = \dfrac{-w_a l^4}{8EI}$ If $a = 0$ and $w_a = 0$ (uniformly increasing load), then: Max $M = M_B = \dfrac{-w_l l^2}{6}$ Max $\theta = \theta_A = \dfrac{w_l l^3}{24EI}$ Max $y = y_A = \dfrac{-w_l l^4}{30EI}$ If $a = 0$ and $w_l = 0$ (uniformly decreasing load), then: Max $M = M_B = \dfrac{-w_a l^2}{3}$ Max $\theta = \theta_A = \dfrac{w_a l^3}{8EI}$ Max $y = y_A = \dfrac{-11 w_a l^4}{120EI}$

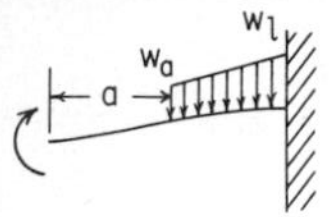 	$M_A = \frac{w_a}{6l}(l-a)^3 + \frac{w_l - w_a}{24l}(l-a)^3$ $y_A = \frac{-w_a}{24EI}(l-a)^3(l+a) - \frac{w_l - w_a}{240EI}(l-a)^3(3l+2a)$ $R_B = \frac{w_a + w_l}{2}(l-a)$ $M_B = \frac{-w_a}{6l}(l-a)^2(2l+a) - \frac{w_l - w_a}{24l}(l-a)^2(3l+a)$ $\theta_B = 0 \qquad y_B = 0$	$\text{Max} - M = M_B = \frac{-w_a l^2}{3} \qquad \text{Max} + M = M_A = \frac{w_a l^2}{6}$ $\text{Max } y = y_A = \frac{-w_a l^4}{24EI}$ If $a = 0$ and $w_a = 0$ (uniformly increasing load), then: $\text{Max} - M = M_B = \frac{-w_l l^2}{8} \qquad \text{Max} + M = M_A = \frac{w_l l^2}{24}$ $\text{Max } y = y_A = \frac{-w_l l^4}{80EI}$ If $a = 0$ and $w_l = 0$ (uniformly decreasing load), then: $\text{Max} - M = M_B = \frac{-5w_a l^2}{24} \qquad \text{Max} + M = M_A = \frac{w_a l^2}{8}$ $\text{Max } y = y_A = \frac{-7w_a l^4}{240EI}$
2c. Left end simply supported, right end fixed 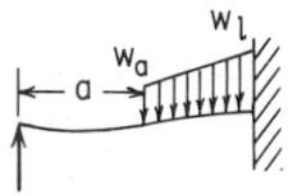 	$R_A = \frac{w_a}{8l^3}(l-a)^3(3l+a) + \frac{w_l - w_a}{40l^3}(l-a)^3(4l+a)$ $\theta_A = \frac{-w_a}{48EIl}(l-a)^3(l+3a) - \frac{w_l - w_a}{240EIl}(l-a)^3(2l+3a)$ $M_A = 0 \qquad y_A = 0$ $R_B = \frac{w_a + w_l}{2}(l-a) - R_A$ $M_B = R_A l - \frac{w_a}{2}(l-a)^2 - \frac{w_l - w_a}{6}(l-a)^2$ $\theta_B = 0 \qquad y_B = 0$	If $a = 0$ and $w_l = w_a$ (uniform load on entire span), then: $R_A = \frac{3}{8}w_a l \qquad R_B = \frac{5}{8}w_a l \qquad \text{Max} - M = M_B = \frac{-w_a l^2}{8}$ $\text{Max} + M = \frac{9w_a l^2}{128} \text{ at } x = \frac{3}{8}l \qquad \text{Max } \theta = \theta_A = \frac{-w_a l^3}{48EI}$ $\text{Max } y = -0.0054\frac{w_a l^4}{EI} \text{ at } x = 0.4215l$ If $a = 0$ and $w_a = 0$ (uniformly increasing load), then: $R_A = \frac{w_l l}{10} \qquad R_B = \frac{2w_l l}{5} \qquad \text{Max} - M = M_B = \frac{-w_l l^2}{15}$ $\text{Max} + M = 0.0298 w_l l^2 \text{ at } x = 0.4472l \qquad \text{Max } \theta = \theta_A = \frac{-w_l l^3}{120EI}$ $\text{Max } y = -0.00239\frac{w_l l^4}{EI} \text{ at } x = 0.4472l$ If $a = 0$ and $w_l = 0$ (uniformly decreasing load), then: $R_A = \frac{11}{40}w_a l \qquad R_B = \frac{9}{40}w_a l \qquad \text{Max} - M = M_B = \frac{-7}{120}w_a l^2$ $\text{Max} + M = 0.0422 w_a l^2 \text{ at } x = 0.329l$ $\text{Max } \theta = \theta_A = \frac{-w_a l^3}{80EI} \qquad \text{Max } y = -0.00304\frac{w_a l^4}{EI} \text{ at } x = 0.4025l$

TABLE 3 *Shear, moment, slope, and deflection formulas for beams* (*Cont.*)

End restraints, reference no.	Boundary values	Selected maximum values of moments and deformations
2d. Left end fixed, right end fixed 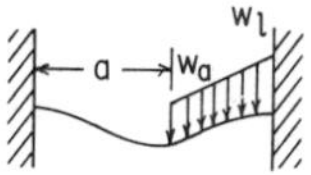	$R_A = \frac{w_a}{2l^3}(l-a)^3(l+a) + \frac{w_l - w_a}{20l^3}(l-a)^3(3l+2a)$ $M_A = \frac{-w_a}{12l^2}(l-a)^3(l+3a) - \frac{w_l - w_a}{60l^2}(l-a)^3(2l+3a)$ $\theta_A = 0 \qquad y_A = 0$ $R_B = \frac{w_a + w_l}{2}(l-a) - R_A$ $M_B = R_A l + M_A - \frac{w_a}{2}(l-a)^2 - \frac{w_l - w_a}{6}(l-a)^2$ $\theta_B = 0 \qquad y_B = 0$	If $a = 0$ and $w_l = w_a$ (uniform load on entire span), then: $\text{Max} - M = M_A = M_B = \frac{-w_a l^2}{12} \qquad \text{Max} + M = \frac{w_a l^2}{24}$ at $x = \frac{l}{2}$ $\text{Max } y = \frac{-w_a l^4}{384EI}$ at $x = \frac{l}{2}$ If $a = 0$ and $w_a = 0$ (uniformly increasing load), then: $R_A = \frac{3w_l l}{20} \qquad M_A = \frac{-w_l l^2}{30} \qquad R_B = \frac{7w_l l}{20} \qquad \text{Max} - M = M_B = \frac{-w_l l^2}{20}$ $\text{Max} + M = 0.0215 w_l l^2$ at $x = 0.548l$ $\text{Max } y = -0.001309 \frac{w_l l^4}{EI}$ at $x = 0.525l$
2e. Left end simply supported, right end simply supported 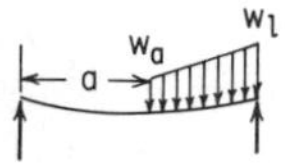	$R_A = \frac{w_a}{2l}(l-a)^2 + \frac{w_l - w_a}{6l}(l-a)^2$ $M_A = 0 \qquad y_A = 0$ $\theta_A = \frac{-w_a}{24EIl}(l-a)^2(l^2 + 2al - a^2)$ $- \frac{w_l - w_a}{360EIl}(l-a)^2(7l^2 + 6al - 3a^2)$ $R_B = \frac{w_a + w_l}{2}(l-a) - R_A$ $\theta_B = \frac{w_a}{24EIl}(l^2 - a^2)^2$ $+ \frac{w_l - w_a}{360EIl}(l-a)^2(8l^2 + 9al + 3a^2)$	If $a = 0$ and $w_l = w_a$ (uniform load on entire span), then: $R_A = R_B = \frac{w_a l}{2} \qquad \text{Max } M = \frac{w_a l^2}{8}$ at $x = \frac{l}{2}$ $\text{Max } \theta = \theta_B = \frac{w_a l^3}{24EI} \qquad \text{Max } y = \frac{-5w_a l^4}{384EI}$ at $x = \frac{l}{2}$ If $a = 0$ and $w_a = 0$ (uniformly increasing load), then: $R_A = \frac{w_l l}{6} \qquad R_B = \frac{w_l l}{3} \qquad \text{Max } M = 0.0641 w_l l^2$ at $x = 0.5773l$ $\theta_A = \frac{-7w_l l^3}{360EI} \qquad \theta_B = \frac{w_l l^3}{45EI}$ $\text{Max } y = -0.00653 \frac{w_l l^4}{EI}$ at $x = 0.5195l$

end simply supported 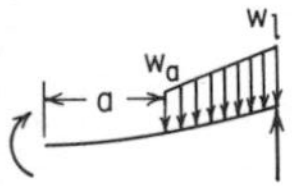	$M_A = \frac{w_a}{2}(l-a)^2 + \frac{w_l - w_a}{6}(l-a)^2$ $y_A = \frac{-w_a}{24EI}(l-a)^2(5l^2+2al-a^2) - \frac{w_l - w_a}{120EI}(l-a)^2(9l^2+2al-a^2)$ $R_B = \frac{w_a + w_l}{2}(l-a)$ $\theta_B = \frac{w_a}{6EI}(l-a)^2(2l+a) + \frac{w_l - w_a}{24EI}(l-a)^2(3l+a)$ $M_B = 0 \qquad y_B = 0$	If $a = 0$ and $w_l = w_a$ (uniform load on entire span), then: Max $M = M_A = \frac{w_a l^2}{2}$ Max $\theta = \theta_B = \frac{w_a l^3}{3EI}$ Max $y = y_A = \frac{-5w_a l^4}{24EI}$ If $a = 0$ and $w_a = 0$ (uniformly increasing load), then: Max $M = M_A = \frac{w_l l^2}{6}$ Max $\theta = \theta_B = \frac{w_l l^3}{8EI}$ Max $y = y_A = \frac{-3w_l l^4}{40EI}$ If $a = 0$ and $w_l = 0$ (uniformly decreasing load), then: Max $M = M_A = \frac{w_a l^2}{3}$ Max $\theta = \theta_B = \frac{5w_a l^3}{24EI}$ Max $y = y_A = \frac{-2w_a l^4}{15EI}$

3. Concentrated intermediate moment

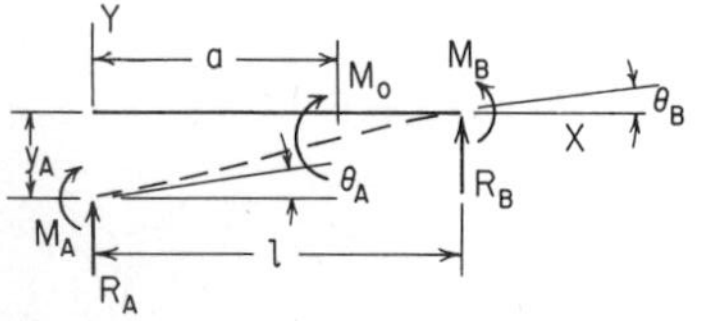

Transverse shear $= V = R_A$

Bending moment $= M = M_A + R_A x + M_o\langle x-a\rangle^0$

Slope $= \theta = \theta_A + \frac{M_A x}{EI} + \frac{R_A x^2}{2EI} + \frac{M_o}{EI}\langle x-a\rangle$

Deflection $= y = y_A + \theta_A x + \frac{M_A x^2}{2EI} + \frac{R_A x^3}{6EI} + \frac{M_o}{2EI}\langle x-a\rangle^2$

End restraints, reference no.	Boundary values	Selected maximum values of moments and deformations
3a. Left end free, right end fixed (cantilever) 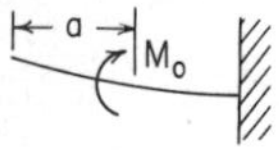	$R_A = 0 \qquad M_A = 0$ $\theta_A = \frac{-M_o(l-a)}{EI}$ $y_A = \frac{M_o(l^2-a^2)}{2EI}$ $R_B = 0 \qquad M_B = M_o$ $\theta_B = 0 \qquad y_B = 0$	Max $M = M_o$ Max $\theta = \theta_A$; max possible value $= \frac{-M_o l}{EI}$ when $a = 0$ Max $y = y_A$; max possible value $= \frac{M_o l^2}{2EI}$ when $a = 0$

TABLE 3 ***Shear, moment, slope, and deflection formulas for beams*** **(Cont.)**

End restraints, reference no.	Boundary values	Selected maximum values of moments and deformations
3b. Left end guided, right end fixed 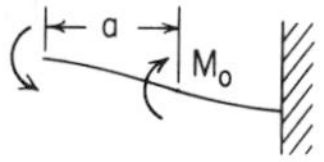	$R_A = 0 \quad \theta_A = 0$ $M_A = \frac{-M_o(l-a)}{l}$ $y_A = \frac{M_o a(l-a)}{2EI}$ $R_B = 0 \quad \theta_B = 0$ $M_B = \frac{M_o a}{l} \quad y_B = 0$	Max $+M = M_B$; max possible value $= M_o$ when $a = l$ Max $-M = M_A$; max possible value $= -M_o$ when $a = 0$ Max $y = y_A$; max possible value $= \frac{M_o l^2}{8EI}$ when $a = \frac{l}{2}$
3c. Left end simply supported, right end fixed 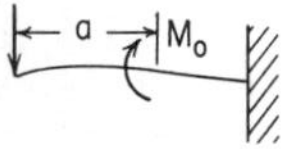	$R_A = \frac{-3M_o}{2l^3}(l^2 - a^2)$ $\theta_A = \frac{M_o}{4EIl}(l-a)(3a-l)$ $M_A = 0 \quad y_A = 0$ $R_B = \frac{3M_o}{2l^3}(l^2 - a^2)$ $M_B = \frac{M_o}{2l^2}(3a^2 - l^2)$ $\theta_B = 0 \quad y_B = 0$	Max $+M = M_o + R_A a$ just right of $x = a$; max possible value $= M_o$ when $a = 0$ or $a = l$; min possible value $= 0.423M_o$ when $a = 0.577l$ Max $-M = \frac{-3M_o a}{2l^3}(l^2 - a^2)$ just left of $x = a$ if $a > 0.282l$; max possible value $= -0.577M_o$ when $a = 0.577l$ Max $-M = \frac{-M_o}{2l^2}(l^2 - 3a^2)$ at B if $a < 0.282l$; max possible value $= -0.5M_o$ when $a = 0$ Max $+y = \frac{M_o(l-a)}{6\sqrt{3(l+a)}\,EI}(3a-l)^{3/2}$ at $x = l\sqrt{\frac{3a-l}{3l+3a}}$; max possible value $= 0.0257\frac{M_o l^2}{EI}$ at $x = 0.474l$ when $a = 0.721l$ $\left(\textit{Note:}\text{ There is no positive deflection if } a < \frac{l}{3}\right)$ Max $-y$ occurs at $x = \frac{2l^3}{3(l^2-a^2)}\left[1 - \frac{1}{2}\sqrt{1 - 6\left(\frac{a}{l}\right)^2 + 9\left(\frac{a}{l}\right)^4}\right]$; max possible value $= \frac{-M_o l^2}{27EI}$ at $x = \frac{l}{3}$ when $a = 0$

3d. Left end fixed, right end fixed 	$R_A = \frac{-6M_o a}{l^3}(l - a)$ $M_A = \frac{-M_o}{l^2}(l^2 - 4al + 3a^2)$ $\theta_A = 0 \quad y_A = 0$ $R_B = -R_A$ $M_B = \frac{M_o}{l^2}(3a^2 - 2al)$ $\theta_B = 0 \quad y_B = 0$	Max $+M = \frac{M_o}{l^3}(4al^2 - 9a^2l + 6a^3)$ just right of $x = a$; max possible value $= M_o$ when $a = l$ Max $-M = \frac{M_o}{l^3}(4al^2 - 9a^2l + 6a^3 - l^3)$ just left of $x = a$; max possible value $= -M_o$ when $a = 0$ Max $+y = \frac{2M_A^3}{3R_A^2 EI}$ at $x = \frac{l}{3a}(3a - l)$; max possible value $= 0.01617\frac{M_o l^2}{EI}$ at $x = 0.565l$ when $a = 0.767l$ $\left(\textit{Note:} \text{ There is no positive deflection if } a < \frac{l}{3}\right)$
3e. Left end simply supported, right end simply supported 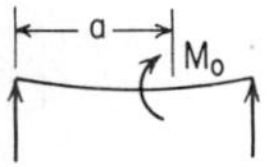	$R_A = \frac{-M_o}{l}$ $\theta_A = \frac{-M_o}{6EIl}(2l^2 - 6al + 3a^2)$ $M_A = 0 \quad y_A = 0$ $R_B = \frac{-M_o}{l}$ $\theta_B = \frac{M_o}{6EIl}(l^2 - 3a^2)$ $M_B = 0 \quad y_B = 0$	Max $+M = \frac{M_o}{l}(l - a)$ just right of $x = a$; max possible value $= M_o$ when $a = 0$ Max $-M = \frac{-M_o a}{l}$ just left of $x = a$; max possible value $= -M_o$ when $a = l$ Max $+y = \frac{M_o(6al - 3a^2 - 2l^2)^{3/2}}{9\sqrt{3}EIl}$ at $x = (2al - a^2 - \frac{2}{3}l^2)^{1/2}$ when $a > 0.423$; max possible value $= 0.0642\frac{M_o l^2}{EI}$ at $x = 0.577l$ when $a = l$ (*Note:* There is no positive deflection if $a < 0.423l$)
3f. Left end guided, right end simply supported 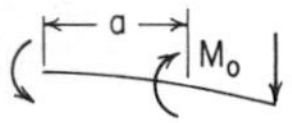	$R_A = 0 \quad \theta_A = 0$ $M_A = -M_o$ $y_A = \frac{M_o a}{2EI}(2l - a)$ $R_B = 0 \quad M_B = 0 \quad y_B = 0$ $\theta_B = \frac{-M_o a}{EI}$	Max $M = -M_o$ for $0 < x < a$ Max $\theta = \theta_B$; max possible value $= \frac{-M_o l}{EI}$ when $a = l$ Max $y = y_A$; max possible value $= \frac{M_o l^2}{2EI}$ when $a = l$

TABLE 3 *Shear, moment, slope, and deflection formulas for beams* (*Cont.*)

4. Intermediate externally created angular deformation

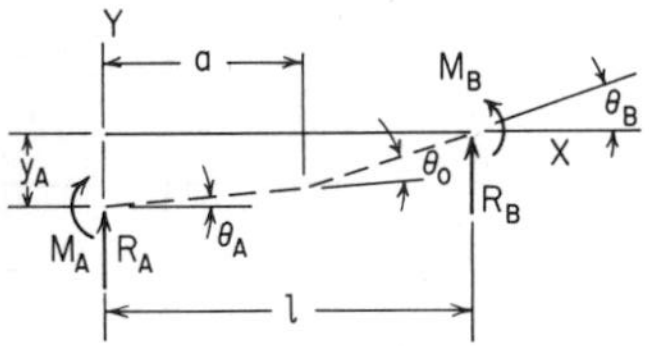

Transverse shear $= V = R_A$

Bending moment $= M = M_A + R_A x$

Slope $= \theta = \theta_A + \dfrac{M_A x}{EI} + \dfrac{R_A x^2}{2EI} + \theta_o \langle x - a \rangle^0$

Deflection $= y = y_A + \theta_A x + \dfrac{M_A x^2}{2EI} + \dfrac{R_A x^2}{6EI} + \theta_o \langle x - a \rangle$

End restraints, reference no.	Boundary values	Selected maximum values of moments and deformations
4a. Left end free, right end fixed 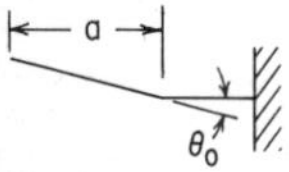	$R_A = 0$ $M_A = 0$ $\theta_A = -\theta_o$ $y_A = \theta_o a$ $R_B = 0$ $M_B = 0$ $\theta_B = 0$ $y_B = 0$	Max $y = y_A$; max possible value $= \theta_o l$ when $a = l$
4b. Left end guided, right end fixed 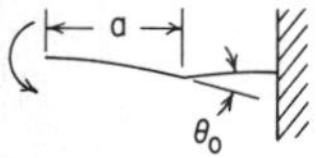	$R_A = 0$ $M_A = \dfrac{-EI\theta_o}{l}$ $\theta_A = 0$ $y_A = \theta_o\left(a - \dfrac{l}{2}\right)$ $R_B = 0$ $M_B = \dfrac{EI\theta_o}{l}$ $\theta_B = 0$ $y_B = 0$	Max $M = M_A$ Max $+ y = y_A$ when $a > \dfrac{l}{2}$; max possible value $= \dfrac{\theta_o l}{2}$ when $a = l$ Max $- y = \dfrac{-\theta_o}{2l}(l - a)^2$ at $x = a$; max possible value $= \dfrac{-\theta_o l}{2}$ when $a = 0$
4c. Left end simply supported, right end fixed 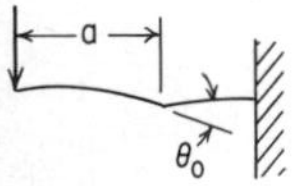	$M_A = 0$ $y_A = 0$ $R_A = \dfrac{-3EIa\theta_o}{l^3}$ $\theta_A = -\theta_o\left(1 - \dfrac{3a}{2l}\right)$ $R_B = -R_A$ $M_B = \dfrac{-3EIa\theta_o}{l^2}$	Max $M = M_B$; max possible value $= \dfrac{-3EI\theta_o}{l}$ when $a = l$ Max $+ y = \theta_o a\left(1 - \dfrac{2l}{3a}\right)^{3/2}$ at $x = l\left(1 - \dfrac{2l}{3a}\right)^{1/2}$ when $a \geq \frac{2}{3}l$; max possible value $= 0.1926\theta_o l$ at $x = 0.577l$ when $a = l$ (*Note:* There is no positive deflection if $a < \frac{2}{3}l$) Max $- y = -\theta_o a\left(1 - \dfrac{3a}{2l} + \dfrac{a^3}{2l^3}\right)$ at $x = a$; max possible value $= -0.232\theta_o l$ at $x = 0.366l$ when $a = 0.366l$

4d. Left end fixed, right end fixed

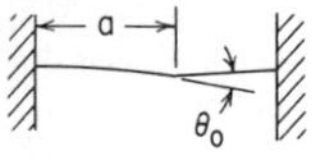

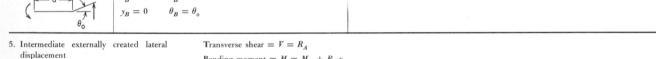

$R_A = \frac{6EI\theta_o}{l^3}(l - 2a)$

$M_A = \frac{2EI\theta_o}{l^2}(3a - 2l)$

$\theta_A = 0 \qquad y_A = 0$

$R_B = -R_A$

$M_B = \frac{2EI\theta_o}{l^2}(l - 3a)$

$\theta_B = 0 \qquad y_B = 0$

Max $+M = M_B$ when $a < \frac{l}{2}$; max possible value $= \frac{2EI\theta_o}{l}$ when $a = 0$

Max $-M = M_A$ when $a < \frac{l}{2}$; max possible value $= \frac{-4EI\theta_o}{l}$ when $a = 0$

Max $+y$ occurs at $x = \frac{l^2}{3(l - 2a)}$ if $a < \frac{l}{3}$; max possible value $= \frac{4}{27}l\theta_o$ when $a = 0$

$\left(\textit{Note:}\ \text{There is no positive deflection if } \frac{l}{3} < a < \frac{2}{3}l\right)$

Max $-y = \frac{-2\theta_o a^2}{l^3}(l - a)^2$ at $x = a$; max possible value $= \frac{-\theta_o l}{8}$ when $a = \frac{l}{2}$

4e. Left end simply supported, right end simply supported

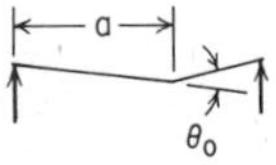

$R_A = 0 \qquad M_A = 0$

$\theta_A = \frac{-\theta_o}{l}(l - a) \qquad y_A = 0$

$R_B = 0 \qquad M_B = 0$

$y_B = 0 \qquad \theta_B = \frac{\theta_o a}{l}$

Max $y = \frac{-\theta_o a}{l}(l - a)$ at $x = a$; max possible value $= \frac{-\theta_o l}{4}$ when $a = \frac{l}{2}$

4f. Left end guided, right end simply supported

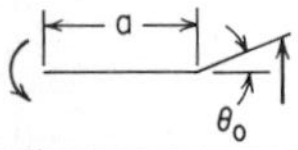

$R_A = 0 \qquad M_A = 0$

$\theta_A = 0 \qquad y_A = -\theta_o(l - a)$

$R_B = 0 \qquad M_B = 0$

$y_B = 0 \qquad \theta_B = \theta_o$

Max $y = y_A$; max possible value $= -\theta_o l$ when $a = 0$

---

5. Intermediate externally created lateral displacement

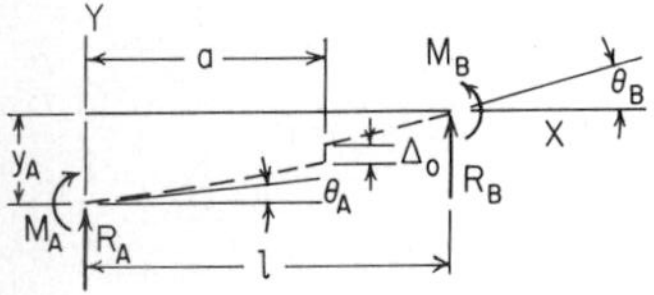

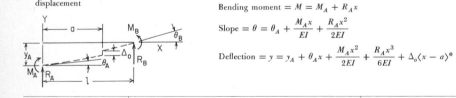

Transverse shear $= V = R_A$

Bending moment $= M = M_A + R_A x$

Slope $= \theta = \theta_A + \frac{M_A x}{EI} + \frac{R_A x^2}{2EI}$

Deflection $= y = y_A + \theta_A x + \frac{M_A x^2}{2EI} + \frac{R_A x^3}{6EI} + \Delta_o\langle x - a\rangle^0$

5a. Left end free, right end fixed

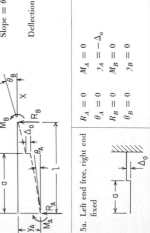

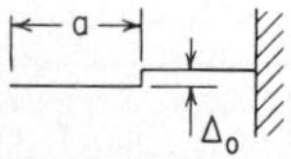

$R_A = 0 \qquad M_A = 0$

$\theta_A = 0 \qquad y_A = -\Delta_o$

$R_B = 0 \qquad M_B = 0$

$\theta_B = 0 \qquad y_B = 0$

Max $y = y_A$ when $x < a$

TABLE 3 *Shear, moment, slope, and deflection formulas for beams* (*Cont.*)

End restraints, reference no.	Boundary values	Selected maximum values of moments and deformations
5b. Left end guided, right end fixed 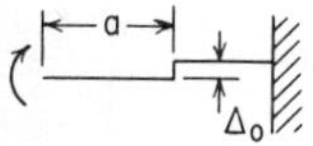	$R_A = 0$ $M_A = 0$ $\theta_A = 0$ $y_A = -\Delta_o$ $R_B = 0$ $M_B = 0$ $\theta_B = 0$ $y_B = 0$	Max $y = y_A$ when $x < a$
5c. Left end simply supported, right end fixed 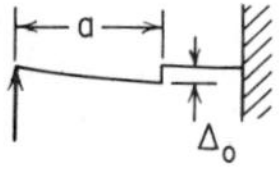	$R_A = \frac{3EI\Delta_o}{l^3}$ $M_A = 0$ $\theta_A = \frac{-3\Delta_o}{2l}$ $y_A = 0$ $R_B = -R_A$ $M_B = \frac{3EI\Delta_o}{l^2}$ $\theta_B = 0$ $y_B = 0$	Max $M = M_B$ Max $\theta = \theta_A$ Max $+ y = \frac{\Delta_o}{2l^3}(2l^3 + a^3 - 3l^2a)$ just right of $x = a$; max possible value $= \Delta_o$ when $a = 0$ Max $- y = \frac{-\Delta_o a}{2l^3}(3l^2 - a^2)$ just left of $x = a$; max possible value $= -\Delta_o$ when $a = l$
5d. Left end fixed, right end fixed 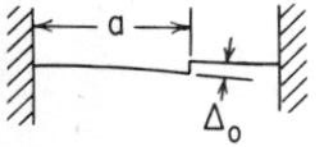	$R_A = \frac{12EI\Delta}{l^3}$ $\theta_A = 0$ $M_A = \frac{-6EI\Delta_o}{l^2}$ $y_A = 0$ $R_B = -R_A$ $M_B = -M_A$ $\theta_B = 0$ $y_B = 0$	Max $+ M = M_B$ Max $- M = M_A$ Max $\theta = \frac{-3\Delta_o}{2l}$ at $x = \frac{l}{2}$ Max $+ y = \frac{\Delta_o}{l^3}(l^3 + 2a^3 - 3a^2l)$ just right of $x = a$; max possible value $= \Delta_o$ when $a = 0$ Max $- y = \frac{-\Delta_o a^2}{l^3}(3l - 2a)$ just left of $x = a$; max possible value $= -\Delta_o$ when $a = l$
5e. Left end simply supported, right end simply supported 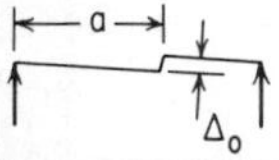	$R_A = 0$ $M_A = 0$ $y_A = 0$ $\theta_A = \frac{-\Delta_o}{l}$ $R_B = 0$ $M_B = 0$ $y_B = 0$ $\theta_B = \frac{-\Delta_o}{l}$	Max $+ y = \frac{\Delta_o}{l}(l - a)$ just right of $x = a$; max possible value $= \Delta_o$ when $a = 0$ Max $- y = \frac{-\Delta_o a}{l}$ just left of $x = a$; max possible value $= -\Delta_o$ when $a = l$
5f. Left end guided, right end simply supported 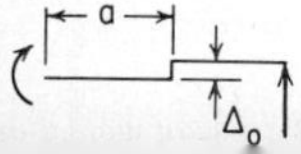	$R_A = 0$ $M_A = 0$ $\theta_A = 0$ $y_A = -\Delta_o$ $R_B = 0$ $M_B = 0$ $\theta_B = 0$ $y_B = 0$	Max $y = y_A$ when $x < a$

to bottom from a to l

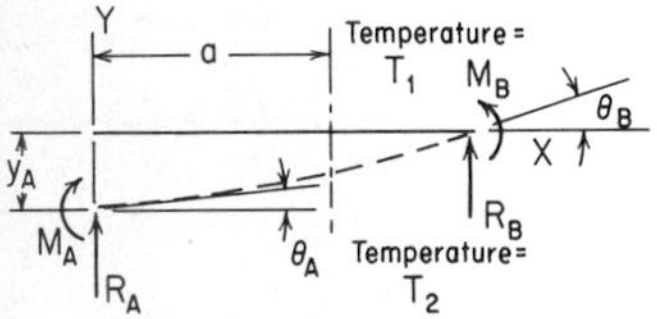

Bending moment $= M = M_A + R_A x$

Slope $= \theta = \theta_A + \dfrac{M_A x}{EI} + \dfrac{R_A x^2}{2EI} + \dfrac{\gamma}{t}(T_2 - T_1)\langle x - a\rangle$

Deflection $= y = y_A + \theta_A x + \dfrac{M_A x^2}{2EI} + \dfrac{R_A x^3}{6EI} + \dfrac{\gamma}{2t}(T_2 - T_1)\langle x - a\rangle^2$

γ = temperature coefficient of expansion (in/in/°) t = depth of beam

End restraints, reference no.	Boundary values	Selected maximum values of moments and deformations
6a. Left end free, right end fixed 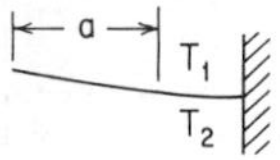	$R_A = 0 \quad M_A = 0$ $\theta_A = \dfrac{-\gamma}{t}(T_2 - T_1)(l - a)$ $y_A = \dfrac{\gamma}{2t}(T_2 - T_1)(l^2 - a^2)$ $R_B = 0 \quad M_B = 0$ $\theta_B = 0 \quad y_B = 0$	$M = 0$ everywhere Max $\theta = \theta_A$; max possible value $= \dfrac{-\gamma l}{t}(T_2 - T_1)$ when $a = 0$ Max $y = y_A$; max possible value $= \dfrac{\gamma l^2}{2t}(T_2 - T_1)$ when $a = 0$
6b. Left end guided, right end fixed 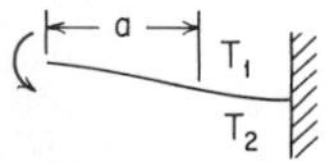	$R_A = 0 \quad \theta_A = 0$ $M_A = \dfrac{-EI\gamma}{t}(T_2 - T_1)(l - a)$ $y_A = \dfrac{a\gamma}{2t}(T_2 - T_1)(l - a)$ $R_B = 0 \quad M_B = M_A$ $\theta_B = 0 \quad y_B = 0$	$M = M_A$ everywhere; max possible value $= \dfrac{-EI\gamma}{t}(T_2 - T_1)$ when $a = 0$ Max $\theta = \dfrac{-a\gamma}{lt}(T_2 - T_1)(l - a)$ at $x = a$; max possible value $= \dfrac{-\gamma l}{4t}(T_2 - T_1)$ when $a = \dfrac{l}{2}$ Max $y = y_A$; max possible value $= \dfrac{\gamma l^2}{8t}(T_2 - T_1)$ when $a = \dfrac{l}{2}$
6c. Left end simply supported, right end fixed 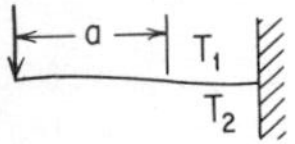	$M_A = 0 \quad y_A = 0$ $R_A = \dfrac{-3EI\gamma}{2tl^3}(T_2 - T_1)(l^2 - a^2)$ $\theta_A = \dfrac{\gamma}{4tl}(T_2 - T_1)(l - a)(3a - l)$ $R_B = -R_A \quad M_B = R_A l$ $\theta_B = 0 \quad y_B = 0$	Max $M = M_B$; max possible value $= \dfrac{-3EI\gamma}{2t}(T_2 - T_1)$ when $a = 0$ Max $+y = \dfrac{\gamma(T_2 - T_1)(l - a)}{6t\sqrt{3(l + a)}}(3a - l)^{3/2}$ at $x = l\left(\dfrac{3a - l}{3l + 3a}\right)^{1/2}$; max possible value $= 0.0257\dfrac{\gamma l^2}{t}(T_2 - T_1)$ at $x = 0.474l$ when $a = 0.721l$ (*Note:* There is no positive deflection if $a < l/3$) Max $-y$ occurs at $x = \dfrac{2l^3}{3(l^2 - a^2)}\left[1 - \dfrac{1}{2}\sqrt{1 - 6\left(\dfrac{a}{l}\right)^2 + 9\left(\dfrac{a}{l}\right)^4}\right]$; max possible value $= \dfrac{-\gamma l^2}{27t}(T_2 - T_1)$ at $x = \dfrac{l}{3}$ when $a = 0$

TABLE 3 ***Shear, moment, slope, and deflection formulas for beams*** **(*Cont.*)**

End restraints, reference no.	Boundary values	Selected maximum values of moments and deformations
6d. Left end fixed, right end fixed 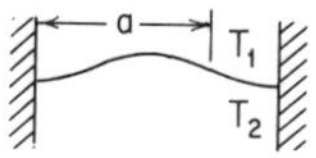	$R_A = \frac{-6EIa\gamma}{tl^3}(T_2 - T_1)(l - a)$ $M_A = \frac{EI\gamma}{tl^2}(T_2 - T_1)(l - a)(3a - l)$ $\theta_A = 0 \quad y_A = 0$ $R_B = -R_A$ $M_B = \frac{-EI\gamma}{tl^2}(T_2 - T_1)(l - a)(3a + l)$ $\theta_B = 0 \quad y_B = 0$	Max $+ M = M_A$; max possible value $= \frac{EI\gamma}{3t}(T_2 - T_1)$ when $a = \frac{2}{3}l$ $\left(\textit{Note: } \text{There is no positive moment if } a < \frac{l}{3}\right)$ Max $- M = M_B$; max possible value $= \frac{-4EI\gamma}{3t}(T_2 - T_1)$ when $a = \frac{l}{3}$ Max $+ y = \frac{2M_A^3}{3R_A^2EI}$ at $x = \frac{l}{3a}(3l - a)$; max possible value $= 0.01617\frac{\gamma l^2}{t}(T_2 - T_1)$ at $x = 0.565l$ when $a = 0.767l$ $\left(\textit{Note: } \text{There is no positive deflection if } a < \frac{l}{3}\right)$
6e. Left end simply supported, right end simply supported 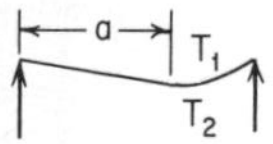	$R_A = 0 \quad M_A = 0 \quad y_A = 0$ $\theta_A = \frac{-\gamma}{2tl}(T_2 - T_1)(l - a)^2$ $R_B = 0 \quad M_B = 0 \quad y_B = 0$ $\theta_B = \frac{\gamma}{2tl}(T_2 - T_1)(l^2 - a^2)$	$M = 0$ everywhere Max $+ \theta = \theta_B$; max possible value $= \frac{\gamma l}{2t}(T_2 - T_1)$ when $a = 0$ Max $- \theta = \theta_A$; max possible value $= \frac{-\gamma l}{2t}(T_2 - T_1)$ when $a = 0$ Max $y = \frac{-\gamma}{8tl^2}(T_2 - T_1)(l^2 - a^2)^2$; max possible value $= \frac{-\gamma l^2}{8t}(T_2 - T_1)$ at $x = \frac{l}{2}$ when $a = 0$
6f. Left end guided, right end simply supported 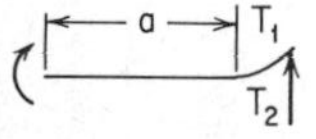	$R_A = 0 \quad M_A = 0 \quad \theta_A = 0$ $y_A = \frac{-\gamma}{2t}(T_2 - T_1)(l - a)^2$ $R_B = 0 \quad M_B = 0 \quad y_B = 0$ $\theta_B = \frac{\gamma}{t}(T_2 - T_1)(l - a)$	$M = 0$ everywhere Max $\theta = \theta_B$; max possible value $= \frac{\gamma l}{t}(T_2 - T_1)$ when $a = 0$ Max $y = y_A$; max possible value $= \frac{-\gamma l^2}{2t}(T_2 - T_1)$ when $a = 0$

which is the horizontal and vertical shear stress at the neutral axis of the section just to the left of the right support.

(*b*) For the routed section it is found (Chap. 5) that the neutral axis is 4 in from the base of the section and $I = 82.6$. The maximum shear stress on a horizontal plane occurs at the neutral axis since b is as small there as anywhere, and so in Eq. 2 the product $A'z'$ represents the statical moment about the neutral axis of all that part of the section above the neutral axis. Taking the moment of the flange and web portions separately, we find $A'z' = (2.75)(2.3)(2.30) + (1)(1.15)(0.575) = 15.2$. Also, $b = 1.00$.

Since the section is not symmetrical about the neutral axis, the fiber stresses will be calculated at the section of maximum positive moment and at the section of maximum negative moment. We have

At $x = 75$:

$$\sigma = \frac{(33750)(4)}{82.6} = 1630 \text{ lb/in}^2 \qquad \text{(tension in bottom fiber)}$$

$$\sigma = \frac{(33750)(3.45)}{82.6} = 1410 \text{ lb/in}^2 \qquad \text{(compression in top fiber)}$$

At $x = 160$:

$$\sigma = \frac{(9600)(4)}{82.6} = 465 \text{ lb/in}^2 \qquad \text{(compression in bottom fiber)}$$

$$\sigma = \frac{(9600)(3.45)}{82.6} = 400 \text{ lb/in}^2 \qquad \text{(tension in top fibers)}$$

It is seen that for this beam the maximum fiber stresses in both tension and compression occur at the section of maximum positive bending moment:

$$\text{Max } \tau = \frac{(1020)(15.2)}{(82.6)(1)} = 188 \text{ lb/in}^2 \qquad \text{(by Eq. 2)}$$

This is the maximum shear stress on a horizontal plane and occurs at the neutral axis of the section just to the left of the right support.

(*c*) For the steel I-beam, the structural steel handbook gives $I/c = 3.00$ and $t = 0.190$. Therefore

$$\text{Max } \sigma = \frac{33{,}750}{3} = 11{,}250 \text{ lb/in}^2$$

This stress occurs at $x = 75$, and is subject to tension in the bottom fibers and compression in the top.

$$\text{Approx max } \tau = \frac{1020}{(4)(0.19)} = 1340 \text{ lb/in}^2$$

Although this method of calculating τ (where the shear force is assumed to be carried entirely by the web) is only approximate, it is usually sufficiently accurate to show whether or not the shear stress is important. If it indicates that shear stress may govern, then the stress at the neutral axis may be calculated by Eq. 2. For standard I-beams, the allowable vertical shear is given by the structural-steel handbooks, making computation unnecessary.

2. The beam shown in Fig. 7.3 has a rectangular section 2 in wide and 4 in deep and is made of spruce, where $E = 1{,}300{,}000$. It is required to determine the deflection of the left end.

Solution. The solution will be effected first by superposition, using the formulas of Table 3. The deflection y of the left end is the sum of the deflection y_1 produced by the distributed

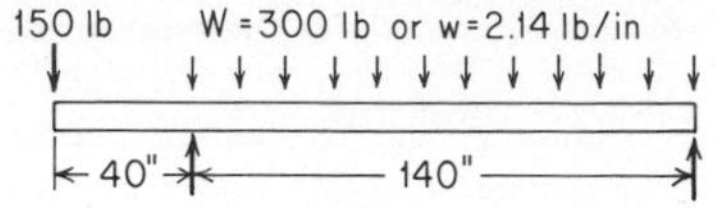

Fig. 7.3

load and the deflection y_2 produced by the concentrated load. Each of these is computed independently of the other. Thus

$$y_1 = -40\,\theta = (-40)\left(-\frac{1}{24}\frac{(2.14)(140^3)}{EI}\right) = +\frac{9{,}800{,}000}{EI}$$

by formula for θ at A, case 2e, where $a = 0$, $l = 140$, and $w_a = w_l = 2.14$. y_2 is calculated as the sum of the deflection the 150-lb load would produce if the beam were *fixed* at the left support and the deflection produced by the fact that it actually assumes a slope there. The first part of the deflection is given by the formula for max y (case 1a), and the second part is found by multiplying the overhang (40 in) by the slope produced at the left end of the 140-in span by a couple equal to $150(40) = 6000$ applied at that point (formula for θ at A, case 3e, where $a = 0$).

$$y_2 = -\frac{1}{3}\frac{(150)(40^3)}{EI} + (-40)\left[-\frac{1}{3}\frac{(-6000)(140)}{EI}\right] = -\frac{14{,}400{,}000}{EI}$$

Adding algebraically,

$$y = y_1 + y_2 = -\frac{4{,}600{,}000}{EI} = -0.33 \text{ in} \qquad \text{(deflection is downward)}$$

The solution of this problem can also be effected readily by using Eq. 6. The reaction at the left support due to the actual loads is 343, and the reaction due to a unit load acting down at the left end is 1.286. If x is measured from the extreme left end of the beam,

$$M = -150x + 343\,\langle x - 40\rangle - 2.14\frac{\langle x - 40\rangle^2}{2} \qquad \text{and} \qquad m = -x + 1.286\,\langle x - 40\rangle$$

Simplifying the equations, we have

$$y = \int \frac{Mm}{EI}\,dx = \frac{1}{EI}\left[\int_0^{40} (-150x)(-x)\,dx + \int_{40}^{180} (-1.071x^2 + 278.8x - 15{,}430)(0.286x - 51.6)\,dx\right] = +0.33 \text{ in}$$

(Here the plus sign means that y is in the direction of the assumed unit load, i.e., downward.)

This second solution involves much more labor than the first, and the calculations must be carried out with great accuracy to avoid the possibility of making a large error in the final result.

3. An aluminum beam is simply supported at the left end and fixed at the right end on a span of 100 in. The cross section is 4 in wide and 6 in deep ($I = 72 \text{ in}^4$). The modulus of elasticity of aluminum is 10^7 lb/in^2, and the coefficient of expansion is 0.000012 in/in/°F. It is desired to determine the locations and magnitudes of the maximum vertical deflection and the maximum bending stress in the beam. The loading consists of a uniformly increasing distributed load starting at 0 at 40 in from the left end and increasing to 200 lb/in at the right end. In addition, the beam, which was originally 70°F, is heated to 100°F on the top and 150°F on the bottom with the temperature assumed to vary linearly from top to bottom.

Solution. Superimposing cases 2c and 6c, the following reactions are determined. (*Note:* For case 2c, $w_a = 0$, $a = 40$, $w_l = 200$, and $l = 100$; for case 6*b*, $T_1 = 100$, $T_2 = 150$, $\gamma = 0.000012$, $t = 6$, and $a = 0$.)

$$R_A = \frac{200(100-40)^3(4\cdot 100+40)}{40(100^3)} - \frac{3(10^7)(72)(0.000012)(150-100)}{2(6)(100)} = -604.8 \text{ lb}$$

$$M_A = 0, \qquad y_A = 0$$

$$\theta_A = \frac{-200(100-40)^3(2\cdot 100+3\cdot 40)}{240(10^7)(72)(100)} + \frac{0.000012(150-100)(-100)}{4(6)} = -0.0033$$

Therefore

$$y = -0.0033x - \frac{604.8x^3}{6EI} - \frac{200\,\langle x-40\rangle^5}{(100-40)(120)\,EI} + \frac{0.000012(150-100)x^2}{2(6)}$$

or

$$y = -0.0033x - 1.4(10^{-7})x^3 - 3.86(10^{-11})\,\langle x-40\rangle^5 + 5.0(10^{-5})x^2$$

and

$$\frac{dy}{dx} = -0.0033 - 4.2(10^{-7})x^2 - 19.3(10^{-11})\,\langle x-40\rangle^4 + 10(10^{-5})x$$

The maximum deflection will occur at a position x_1 where the slope dy/dx is zero. At this time an assumption must be made as to which span, $x_1<40$ or $x_1>40$, will contain the maximum deflection. Assuming that x_1 is less than 40 and setting the slope equal to zero,

$$0 = -0.0033 - 4.2(10^{-7})x_1^2 + 10(10^{-5})x_1$$

Of the two solutions for x_1, 39.7 and 198, only the value of 39.7 is valid since it is less than 40. Substituting $x = 39.7$ into the deflection equation gives the maximum deflection of -0.061 in. Similarly,

$$M = -604.8x - \frac{200}{6(100-40)}\langle x-40\rangle^3$$

which has a maximum value where x is a maximum, i.e., at the right end:

$$\text{Max } M = -604.8(100) - \frac{200}{6(100-40)}(100-40)^3 = -180{,}480 \text{ in-lb}$$

$$\text{Max } \sigma = \frac{Mc}{I} = \frac{180{,}480(3)}{72} = 7520 \text{ lb/in}^2$$

4. The aluminum beam in Example 3 is to be simply supported at both ends and carry a concentrated load of 10,000 lb at a point 30 in from the right end. It is desired to determine the relative displacement of the lower edges of the end sections. For this example, case 1e can be used with $a = 70$ in:

$$R_A = \frac{10{,}000(100-70)}{100} = 3000 \text{ lb} \qquad M_A = 0$$

$$\theta_A = \frac{-10{,}000(70)}{6(10^7)(72)(100)}(200-70)(100-70) = -0.00632 \text{ rad} \qquad y_A = 0$$

$$\theta_B = \frac{10{,}000(70)}{6(10^7)(72)(100)}(100^2-70^2) = 0.00828 \text{ rad}$$

Then

$$\frac{dy}{dx} = -0.00632 + 2.083(10^{-6})x^2 - \frac{10{,}000}{2(10^7)(72)}\langle x-70\rangle^2$$

or

$$\frac{dy}{dx} = -0.00632 + 2.083(10^{-6})x^2 - 6.94(10^{-6})\,\langle x-70\rangle^2$$

The shortening of the neutral surface of the beam is given by Eq. 14 to be

$$\Delta l = \frac{1}{2}\int_0^l \left(\frac{dy}{dx}\right)^2 dx$$

$$= \frac{1}{2}\int_0^{70} \left[-0.00632 + 2.083(10^{-6})x^2\right]^2 dx$$

$$+ \frac{1}{2}\int_{70}^{100} [-0.04033 + 9.716(10^{-4})x - 4.857(10^{-6})x^2]^2 dx$$

or $$\Delta l = 0.00244 \text{ in} \qquad \text{(a shortening)}$$

In addition to the shortening of the neutral surface, the lower edge of the left end moves to the left by an amount $\theta_A c$ or $0.00632(3) = 0.01896$ in. Similarly, the lower edge of the right end moves to the right by an amount $\theta_B c$ or $0.00828(3) = 0.02484$ in. Evaluating the motion of the lower edges in this manner is equivalent to solving Eq. 15 for the total strain in the lower fibers of the beam.

The total relative motion of the lower edges of the end sections is therefore a moving apart by an amount $0.01896 + 0.02484 - 0.00244 = 0.04136$ in.

7.2 *Composite beams and bimetallic strips*

Beams that are constructed of more than one material can be treated by using an equivalent width technique if the maximum stresses in each of the several materials remain within the proportional limit. An equivalent cross section is developed in which the width of each component parallel to the principal axis of bending is increased in the same proportion that the modulus of elasticity of that component makes with the modulus of the assumed material of the equivalent beam.

EXAMPLE

The beam cross section shown in Fig. 7.4*a* is composed of three portions of equal width and depth. The top portion is made of aluminum for which $E_A = 10 \cdot 10^6$ lb/in^2; the center is made of brass for which $E_B = 15 \cdot 10^6$ lb/in^2; and the bottom is made of steel for which $E_S = 30 \cdot 10^6$ lb/in^2. Figure 7.4*b* shows the equivalent cross section, which is assumed to be made of aluminum. For this equivalent cross section the centroid must be located and the moment of inertia determined for the centroidal axis.

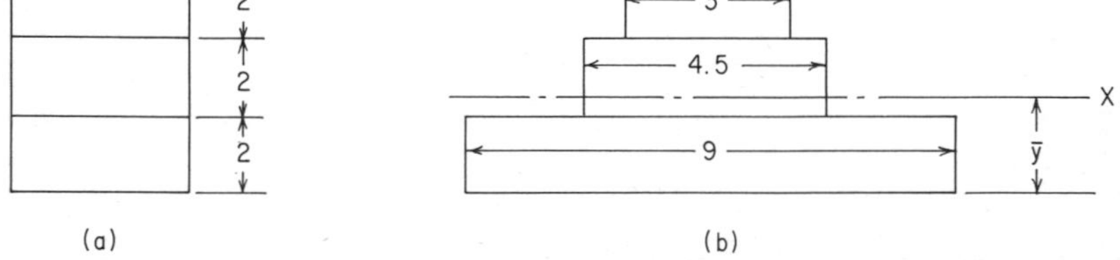

Fig. 7.4

Solution

$$\bar{y} = \frac{3(2)5 + 4.5(2)(3) + 9(2)(1)}{6 + 9 + 18} = 2.27 \text{ in}$$

$$I_x = \frac{3(2^3)}{12} + 6(5 - 2.27)^2 + \frac{4.5(2^3)}{12} + 9(3 - 2.27)^2 + \frac{9(2^3)}{12} + 18(2.27 - 1)^2$$

$$= 89.5 \text{ in}^4$$

The equivalent stiffness EI of this beam is therefore $10 \cdot 10^6(89.5)$, or $895 \cdot 10^6$ lb in^2.

A flexure stress computed by $\sigma = Mc/I_x$ will give a stress in the equivalent beam which can thereby be converted into the stress in the actual composite beam by multiplying by the modulus ratio. If a bending moment of 300,000 in-lb were applied to a beam with the cross section shown, the stress at the top surface on the equivalent beam would be $\sigma = 300{,}000(6 - 2.27)/89.5$, or 12,500 lb/in^2. Since the material at the top is the same in both the actual and equivalent beams, this is also the maximum stress in the aluminum portion of the actual beam. The stress at the bottom of the equivalent beam would be $\sigma = 300{,}000(2.27)/89.5 = 7{,}620$ lb/in^2. The actual stress at the bottom of the steel portion of the beam would be $\sigma = 7{,}620(30)/10 = 22{,}900$ lb/in^2.

Bimetallic strips are widely used in instruments to sense or control temperatures. The following formula gives the equivalent properties of the strip for which the cross section is shown in Fig. 7.5.

$$\text{Equivalent } EI = \frac{w t_b^3 t_a E_b E_a}{12(t_a E_a + t_b E_b)} K_1$$

$$\text{where } K_1 = 4 + 6\frac{t_a}{t_b} + 4\left(\frac{t_a}{t_b}\right)^2 + \frac{E_a}{E_b}\left(\frac{t_a}{t_b}\right)^3 + \frac{E_b}{E_a}\frac{t_b}{t_a}$$

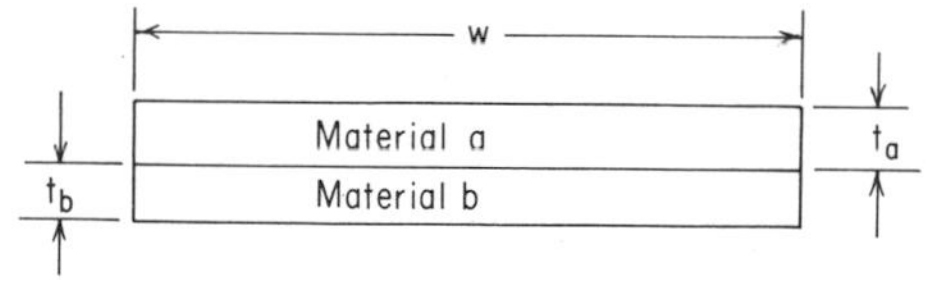

Fig. 7.5

All of the formulas in Table 3, cases 1 to 5, can be applied to the bimetallic beam by using this equivalent value of EI. Since a bimetallic strip is designed to deform when its temperature differs from T_o, the temperature at which the strip is straight, Table 3, case 6, can be used to solve for reaction forces and moments as well as deformations of the bimetallic strip under a uniform temperature T. To do this, the term $\gamma(T_2 - T_1)/t$ is replaced by the term $6(\gamma_b - \gamma_a)(T - T_o)(t_a + t_b)/(t_b^2 K_1)$ and EI is replaced by the equivalent EI given above.

After the moments and deformations have been determined, the flexure stresses can be computed. The stresses due to the bending moments caused by the restraints and any applied loads are given by the following expressions:

In the top surface of material a:

$$\sigma = \frac{-6M}{wt_b^2K_1}\left(2 + \frac{t_b}{t_a} + \frac{E_at_a}{E_bt_b}\right)$$

In the bottom surface of material b:

$$\sigma = \frac{6M}{wt_b^2K_1}\left(2 + \frac{t_a}{t_b} + \frac{E_bt_b}{E_at_a}\right)$$

If there are no restraints imposed, the distortion of a bimetallic strip due to a temperature change is accompanied by flexure stresses in the two materials. This differs from a beam made of a single material which deforms free of stress when subjected to a linear temperature variation through the thickness if there are no restraints. Therefore the following stresses must be added algebraically to the stresses caused by the bending moments, if any:
In the top surface of material a:

$$\sigma = \frac{-(\gamma_b - \gamma_a)(T - T_o)E_a}{K_1}\left[3\frac{t_a}{t_b} + 2\left(\frac{t_a}{t_b}\right)^2 - \frac{E_bt_b}{E_at_a}\right]$$

In the bottom surface of material b:

$$\sigma = \frac{(\gamma_b - \gamma_a)(T - T_o)E_b}{K_1}\left[3\frac{t_a}{t_b} + 2 - \frac{E_a}{E_b}\left(\frac{t_a}{t_b}\right)^3\right]$$

EXAMPLES

1. A bimetallic strip is made by bonding a piece of titanium alloy $\frac{1}{4}$ in wide by 0.030 in thick to a piece of stainless steel $\frac{1}{4}$ in wide by 0.060 in thick. For the titanium, $E = 17 \cdot 10^6$ lb/in², and $\gamma = 5.7 \cdot 10^{-6}$ in/in/°F; for the stainless steel, $E = 28 \cdot 10^6$ lb/in², and $\gamma = 9.6 \cdot 10^{-6}$ in/in/°F. It is desired to find the length of bimetal required to develop a reaction force of 5 oz at a simply supported left end when the right end is fixed and the temperature is raised 50°F; also the maximum stresses must be determined.

Solution. First find the value of K_1 and then evaluate the equivalent stiffness:

$$K_1 = 4 + 6\frac{0.03}{0.06} + 4\left(\frac{0.03}{0.06}\right)^2 + \frac{17}{28}\left(\frac{0.03}{0.06}\right)^3 + \frac{28}{17}\frac{0.06}{0.03} = 11.37$$

$$\text{Equivalent } EI = \frac{0.25(0.06^3)(0.03)(28 \cdot 10^6)(17 \cdot 10^6)}{12[0.03(17 \cdot 10^6) + 0.06(28 \cdot 10^6)]}11.37 = 333 \text{ lb/in}^2$$

The bimetallic strip curves under a temperature rise over the entire length just as a single material strip would curve under a temperature differential. In order to use case 6c in Table 3, the equivalent to $\gamma(T_2 - T_1)/t$ must be found. This equivalent value is given by

$$\frac{6(9.6 \cdot 10^{-6} - 5.7 \cdot 10^{-6})(50)(0.03 + 0.06)}{(0.06^2)(11.37)} = 0.00257$$

The expression for R_A can now be obtained from case 6c in Table 3, and, noting that $a = 0$, the value of the length l can be determined:

$$R_A = \frac{-3(l^2 - a^2)}{2l^3}EI\frac{\gamma}{t}(T_2 - T_1) = \frac{-3}{2l}(333)(0.00257) = \frac{-5}{16}\text{ lb}$$

Therefore $l = 4.11$ in.

The maximum bending moment is found at the fixed end and is equal to $R_A l$:

$$\text{Max } M = -\tfrac{5}{16}(4.11) = -1.285 \text{ in-lb}$$

The flexure stress on the top of the titanium is

$$\sigma = \frac{-6(-1.285)}{0.25(0.06^2)(11.37)}\left(2 + \frac{0.06}{0.03} + \frac{17}{28}\frac{0.03}{0.06}\right)$$
$$- \frac{(9.6 \cdot 10^{-6} - 5.7 \cdot 10^{-6})(50)(17 \cdot 10^6)}{11.37}\left[3\frac{0.03}{0.06} + 2\left(\frac{0.03}{0.06}\right)^2 - \frac{28}{17}\frac{0.06}{0.03}\right]$$
$$= 3242 + 378 = 3620 \text{ lb/in}^2$$

The flexure stress on the bottom of the stainless steel is

$$\sigma = \frac{6(-1.285)}{0.25(0.06^2)(11.37)}\left[2 + \frac{0.03}{0.06} + \frac{28}{17}\frac{0.06}{0.03}\right]$$
$$+ \frac{(9.6 \cdot 10^{-6} - 5.7 \cdot 10^{-6})(50)(28 \cdot 10^6)}{11.37}\left[3\frac{0.03}{0.06} + 2 - \frac{17}{28}\left(\frac{0.03}{0.06}\right)^3\right]$$
$$= -4365 + 1644 = -2720 \text{ lb/in}^2$$

7.3 *Three-moment equation*

The *three-moment equation*, which expresses the relationship between the bending moments found at three *consecutive* supports in a *continuous* beam, can be readily derived for any loading shown in Table 3. This is accomplished by taking any two consecutive spans and evaluating the slope for each span at the end where the two spans join. These slopes, which are expressed in terms of the three moments and the loads on the spans, are then equated and the equation reduced to its usual form.

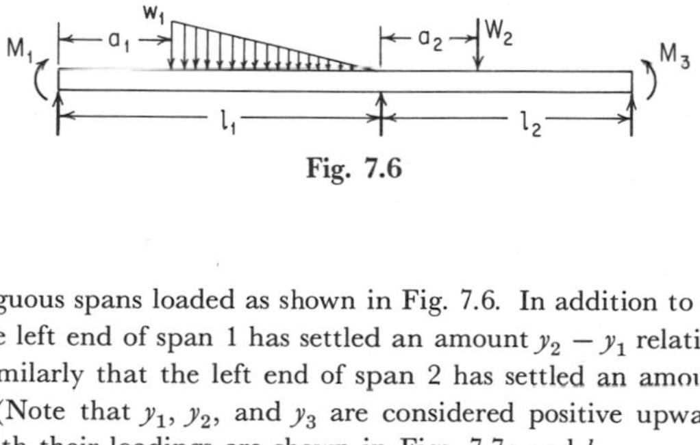

Fig. 7.6

EXAMPLE

Consider two contiguous spans loaded as shown in Fig. 7.6. In addition to the loading shown, it is known that the left end of span 1 has settled an amount $y_2 - y_1$ relative to the right end of the span, and similarly that the left end of span 2 has settled an amount $y_3 - y_2$ relative to the right end. (Note that y_1, y_2, and y_3 are considered positive upward as usual.) The individual spans with their loadings are shown in Figs. 7.7*a* and *b*.

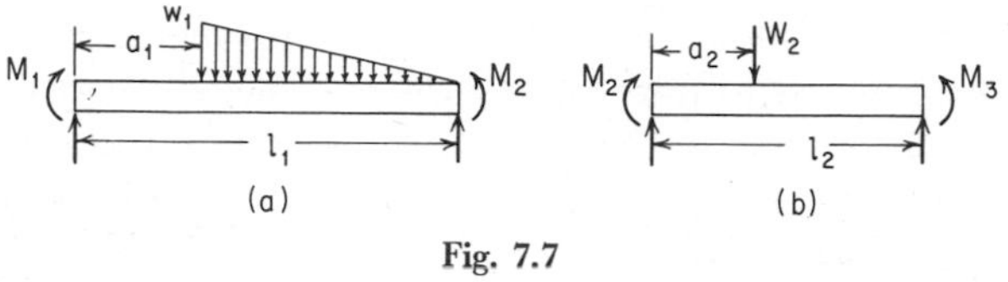

Fig. 7.7

Solution. Using cases 2e and 3e from Table 3 and noting the relative deflections mentioned above, the expression for the slope at the right end of span 1 is

$$\theta_2 = \frac{w_1(l_1^2 - a_1^2)^2}{24E_1I_1l_1} - \frac{w_1(l_1 - a_1)^2}{360E_1I_1l_1}(8l_1^2 + 9a_1l_1 + 3a_1^2)$$

$$+ \frac{M_1l_1^2}{6E_1I_1l_1} + \frac{-M_2(l_1^2 - 3l_1^2)}{6E_1I_1l_1} + \frac{y_2 - y_1}{l_1}$$

Similarly, using cases 1e and 3e from Table 3, the expression for the slope at the left end of span 2 is

$$\theta_2 = \frac{-W_2a_2}{6E_2I_2l_2}(2l_2 - a_2)(l_2 - a_2) - \frac{M_2}{6E_2I_2l_2}(2l_2^2) - \frac{-M_3}{6E_2I_2l_2}(2l_2^2 - 6l_2^2 + 3l_2^2) + \frac{y_3 - y_2}{l_2}$$

Equating these slopes gives

$$\frac{M_1l_1}{6E_1I_1} + \frac{M_2l_1}{3E_1I_1} + \frac{M_2l_2}{3E_2I_2} + \frac{M_3l_2}{6E_2I_2} = \frac{-w_1(l_1 - a_1)^2}{360E_1I_1l_1}(7l_1^2 + 21a_1l_1 + 12a_1^2)$$

$$- \frac{y_2 - y_1}{l_1} - \frac{W_2a_2}{6E_2I_2l_2}(2l_2 - a_2)(l_2 - a_2) + \frac{y_3 - y_2}{l_2}$$

If M_1 and M_3 are known, this expression can be solved for M_2; if not, similar expressions for the adjacent spans must be written and the set of equations solved for the moments.

The three-moment equation can also be applied to beams carrying axial tension or compression in addition to transverse loading. The procedure is exactly the same as that described above except the slope formulas to be used are those given in Tables 10 and 11.

7.4 *Rigid frames*

By superposition and the matching of slopes and deflections, the formulas in Table 3 can be used to solve for the indeterminate reactions in rigid frames or to determine the deformations where the support conditions permit such deformations to take place.

In Table 4 formulas are given for the indeterminate reactions and end deformations for rigid frames consisting of three members. Very extensive compilations of formulas for rigid frames are available, notably those of Kleinlogel (Ref. 56) and Leontovich (Ref. 57).

When the number of members is large, as in a framed building, a relaxation method such as moment distribution might be used or a digital computer could be programmed to solve the large number of equations. In all rigid frames, corner or knee design is important; much information and experimental data relating to this problem are to be found in the reports published by the Fritz Engineering Laboratories of Lehigh University. The frames given in Table 4 are assumed to have rigid corners; however, corrections can be made easily once the rigidity of a given corner design is known by making use of the concentrated angular displacement loading with the displacement positioned at the corner.

EXAMPLES

1. The frame shown in Fig. 7.8*a* is fixed at the lower end of the right-hand member and guided at the lower end of the left-hand member in such a way as to prevent any rotation of this end but permitting horizontal motion if any is produced by the loading. The configuration could represent the upper half of the frame shown in Fig. 7.8*b*; for this frame the material properties and physical dimensions are given as $l_1 = 40$ in, $l_2 = 20$ in, $l_3 = 15$ in, $E_1 = E_2 = E_3 = 30 \cdot 10^6$ lb/in², $I_1 = 8$ in⁴, $I_2 = 10$ in⁴, and $I_3 = 4$ in⁴. In addition to the load P of 1000 lb, the frame has been subjected to a temperature rise of 50°F since it was assembled in a stress-free condition. The coefficient of expansion of the material used in all three portions is 0.0000065 in/in/°F.

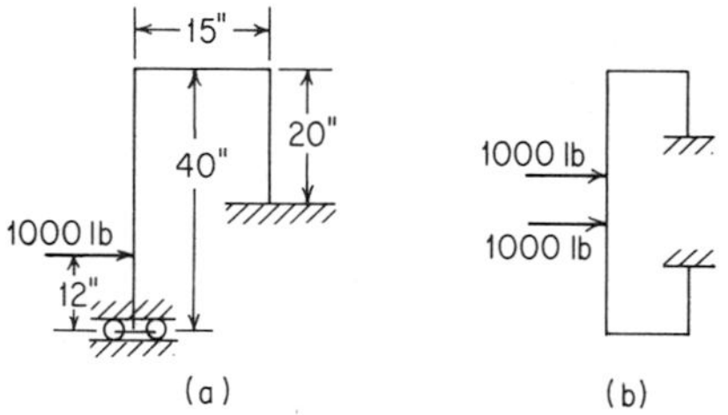

Fig. 7.8

Solution. An examination of Table 4 shows the required end or support conditions in case 7 with the loading cases f and q listed under case 5. For cases 5 to 12 the frame constants are evaluated as follows:

$$C_{HH} = \frac{l_1^3}{3E_1I_1} + \frac{l_1^3 - (l_1 - l_2)^3}{3E_2I_2} + \frac{l_1^2 l_3}{E_3 I_3} = \frac{\dfrac{40^3}{3(8)} + \dfrac{[40^3 - (40 - 20)^3]}{3(10)} + \dfrac{(40^2)(15)}{4}}{30(10^6)}$$

$$= \frac{2666.7 + 1866.7 + 6000}{30(10^6)} = 0.0003511$$

Similarly

$$C_{HV} = C_{VH} = 0.0000675$$
$$C_{HM} = C_{MH} = 0.00001033$$
$$C_{VV} = 0.0000244$$
$$C_{VM} = C_{MV} = 0.00000194$$
$$C_{MM} = 0.000000359$$

For case 7f

$$LF_H = W\left(C_{HH} - aC_{HM} + \frac{a^3}{6E_1I_1}\right) = 1000\left[0.0003511 - 12(0.00001033) + \frac{12^3}{6(30 \cdot 10^6)(8)}\right]$$

$$= 1000(0.0003511 - 0.000124 + 0.0000012) = 0.2282$$

Similarly

$$LF_V = 0.0442 \qquad \text{and} \qquad LF_M = 0.00632$$

For case 7q

$$LF_H = -(T - T_o)\gamma_3 l_3 = -50(0.0000065)(15) = -0.004875$$

$$LF_V = 0.0065 \qquad \text{and} \qquad LF_M = 0$$

TABLE 4 ***Reaction and deflection formulas for rigid frames***

NOTATION: W = load (pounds); w = unit load (pounds per linear inch); M_o = applied couple (inch-pounds); θ_o = externally created concentrated angular displacement (radians); Δ_o = externally created concentrated lateral displacement (inches); $T - T_o$ = uniform temperature rise (degrees); T_1 and T_2 = temperatures on outside and inside, respectively (degrees). H_A and H_B are the horizontal end reactions at the left and right, respectively, and are positive to the left (pounds); V_A and V_B are the vertical end reactions at the left and right, respectively, and are positive upward (pounds); M_A and M_B are the reaction moments at the left and right, respectively, and are positive clockwise (inch-pounds). I_1, I_2, and I_3 are the respective area moments of inertia for bending in the plane of the frame for the three members (inches to the fourth); E_1, E_2, and E_3 are the respective moduli of elasticity (pounds per square inch); γ_1, γ_2, and γ_3 are the respective temperature coefficients of expansion (inches per inch per degree)

General reaction and deformation expressions for cases 1 to 4, right end pinned in all four cases

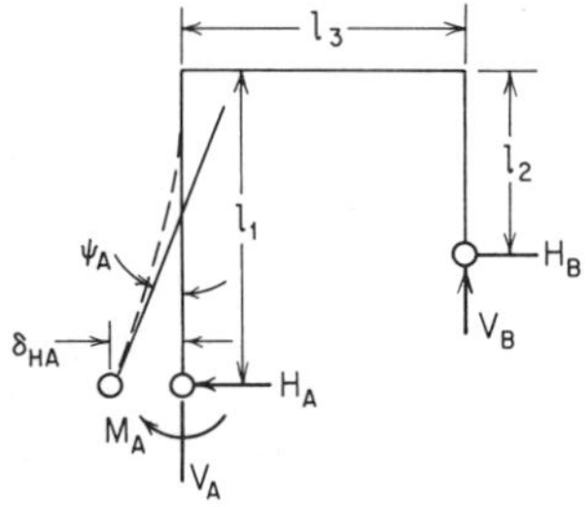

Deformation equations:

Horizontal deflection at $A = \delta_{HA} = A_{HH}H_A + A_{HM}M_A - LP_H$ (1)

Angular rotation at $A = \psi_A = A_{MH}H_A + A_{MM}M_A - LP_M$ (2)

where $$A_{HH} = \frac{l_1^3}{3E_1I_1} + \frac{l_2^3}{3E_2I_2} + \frac{l_3}{3E_3I_3}(l_1^2 + l_1l_2 + l_2^2)$$

$$A_{HM} = A_{MH} = \frac{l_1^2}{2E_1I_1} + \frac{l_3}{6E_3I_3}(2l_1 + l_2)$$

$$A_{MM} = \frac{l_1}{E_1I_1} + \frac{l_3}{3E_3I_3}$$

and where LP_H and LP_M are loading terms given below for several types of load

(*Note:* V_A, V_B, and H_B are to be evaluated from equilibrium equations after calculating H_A and M_A)

1. Left end pinned, right end pinned

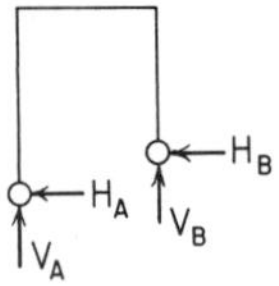

Since $\delta_{HA} = 0$ and $M_A = 0$,

$$H_A = \frac{LP_H}{A_{HH}} \quad \text{and} \quad \psi_A = A_{MH}H_A - LP_M$$

The loading terms are as follows.

Reference no., loading	Loading terms
1a. Concentrated load on the horizontal member 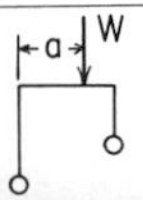	$LP_H = \frac{Wa}{6E_3I_3}\left[3l_1a - 2l_1l_3 - l_2l_3 - \frac{a^2}{l_3}(l_1 - l_2)\right]$ $LP_M = \frac{Wa}{6E_3I_3}\left(3a - 2l_3 - \frac{a^2}{l_3}\right)$

1b. Distributed load on the horizontal member	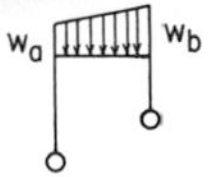	$LP_H = \dfrac{-w_a l_3^3}{24E_3I_3}(l_1 + l_2) - \dfrac{(w_b - w_a)l_3^3}{360E_3I_3}(7l_1 + 8l_2)$ $LP_M = \dfrac{-7w_a l_3^3}{72E_3I_3} - \dfrac{11(w_b - w_a)l_3^3}{180E_3I_3}$
1c. Concentrated moment on the horizontal member	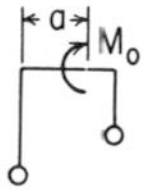	$LP_H = \dfrac{M_o}{6E_3I_3}\left[6l_1a - 2l_1l_3 - l_2l_3 - \dfrac{3a^2}{l_3}(l_1 - l_2)\right]$ $LP_M = \dfrac{M_o}{6E_3I_3}\left(4a - 2l_3 - \dfrac{3a^2}{l_3}\right)$
1d. Concentrated angular displacement on the horizontal member	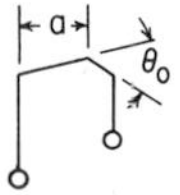	$LP_H = \theta_o\left[l_1 - \dfrac{a}{l_3}(l_1 - l_2)\right]$ $LP_M = \theta_o\left(\dfrac{2}{3} - \dfrac{a}{l_3}\right)$
1e. Concentrated lateral displacement on the horizontal member	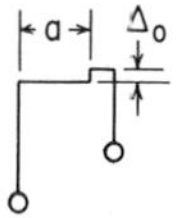	$LP_H = \dfrac{-\Delta_o(l_1 - l_2)}{l_3}$ (*Note:* Δ_o could also be an increase in the length l_1 or a decrease in the length l_2) $LP_M = \dfrac{-\Delta_o}{l_3}$
1f. Concentrated load on the left vertical member	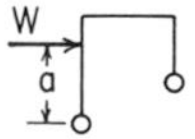	$LP_H = W\left(A_{HH} - aA_{HM} + \dfrac{a^3}{6E_1I_1}\right)$ $LP_M = W\left(A_{MH} - aA_{MM} + \dfrac{a^2}{2E_1I_1}\right)$
1g. Distributed load on the left vertical member	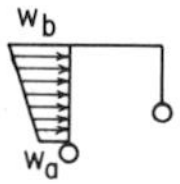	$LP_H = w_a\left(A_{HH}l_1 - A_{HM}\dfrac{l_1^2}{2} + \dfrac{l_1^4}{24E_1I_1}\right) + (w_b - w_a)\left(A_{HH}\dfrac{l_1}{2} - A_{HM}\dfrac{l_1^2}{3} + \dfrac{l_1^4}{30E_1i_1}\right)$ $LP_M = w_a\left[\dfrac{l_1^3}{6E_1I_1} + \dfrac{l_1l_3}{6E_3I_3}(l_1 + l_2)\right] + (w_b - w_a)\left[\dfrac{l_1^3}{24E_1I_1} + \dfrac{l_1l_3}{36E_3I_3}(2l_1 + 3l_2)\right]$
1h. Concentrated moment on the left vertical member	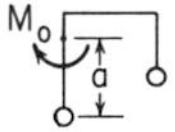	$LP_H = M_o\left(\dfrac{a^2}{2E_1I_1} - A_{HM}\right)$ $LP_M = M_o\left(\dfrac{a}{E_1I_1} - A_{MM}\right)$

TABLE 4 ***Reaction and deflection formulas for rigid frames (Cont.)***

Reference no., loading		Loading terms
1i. Concentrated angular displacement on the left vertical member	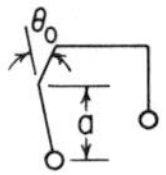	$LP_H = \theta_o(a)$ $LP_M = \theta_o(1)$
1j. Concentrated lateral displacement on the left vertical member	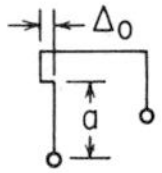	$LP_H = \Delta_o(1)$ (*Note:* Δ_o could also be a decrease in the length l_3) $LP_M = 0$
1k. Concentrated load on the right vertical member	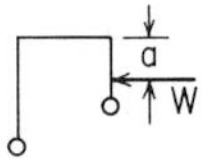	$LP_H = W\left[\frac{1}{6E_2I_2}(3l_2a^2 - 2l_2^3 - a^3) - \frac{l_3}{6E_3I_3}(l_2 - a)(l_1 + 2l_2)\right]$ $LP_M = W\left[\frac{-l_3}{6E_3I_3}(l_2 - a)\right]$
1l. Distributed load on the right vertical member	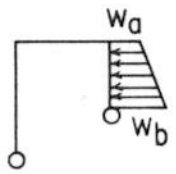	$LP_H = w_a\left[\frac{-5l_2^4}{24E_2I_2} - \frac{l_2^2l_3}{12E_3I_3}(l_1 + 2l_2)\right] + (w_b - w_a)\left[\frac{-3l_2^4}{40E_2I_2} - \frac{l_2^2l_3}{36E_3I_3}(l_1 + 2l_2)\right]$ $LP_M = w_a\left(\frac{-l_2^2l_3}{12E_3I_3}\right) + (w_b - w_a)\left(\frac{-l_2^2l_3}{36E_3I_3}\right)$
1m. Concentrated moment on the right vertical member	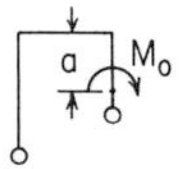	$LP_H = M_o\left[\frac{a}{2E_2I_2}(2l_2 - a) + \frac{l_3}{6E_3I_3}(l_1 + 2l_2)\right]$ $LP_M = M_o\frac{l_3}{6E_3I_3}$
1n. Concentrated angular displacement on the right vertical member	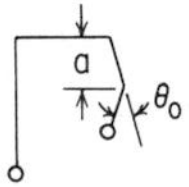	$LP_H = \theta_o(l_2 - a)$ $LP_M = 0$
1p. Concentrated lateral displacement on the right vertical member	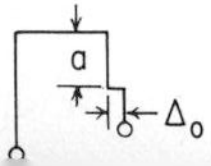	$LP_H = \Delta_o(-1)$ (*Note:* Δ_o could also be an increase in the length l_3) $LP_M = 0$

1q. Uniform temperature rise: T = uniform temperature; T_o = unloaded temperature	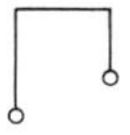	$LP_H = (T - T_o)\left[\gamma_3 l_3 - \frac{(l_1 - l_2)}{l_3}(\gamma_1 l_1 - \gamma_2 l_2)\right]$ γ = temperature coefficient of expansion (inches/inch/degree) $LP_M = (T - T_o)\left[\frac{-1}{l_3}(\gamma_1 l_1 - \gamma_2 l_2)\right]$
1r. Uniform temperature differential from outside to inside. Average temperature is T_o	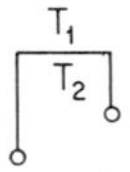	$LP_H = \frac{(T_1 - T_2)}{2}\left[\frac{\gamma_1 l_1^2}{t_1} + \frac{\gamma_2 l_2^2}{t_2} + \frac{\gamma_3 l_3(l_1 + l_2)}{t_3}\right]$ t_1, t_2, and t_3 are beam thicknesses from inside to outside $LP_M = \frac{(T_1 - T_2)}{2}\left(\frac{4\gamma_1 l_1}{t_1} + \frac{\gamma_3 l_3}{3t_3}\right)$
2. Left end guided horizontally, right end pinned	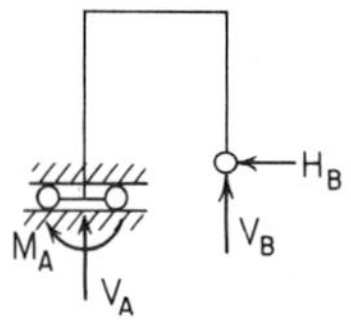	Since $\psi_A = 0$ and $H_A = 0$, $M_A = \frac{LP_M}{A_{MM}}$ and $\delta_{HA} = A_{HM}M_A - LP_H$ Use the loading terms for cases 1a to 1r
3. Left end roller supported along the horizontal, right end pinned	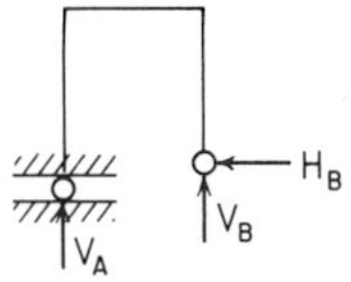	Since H_A and M_A are both zero, this is a statically determinate case. Equations 1 and 2 can be used to evaluate the deflections: $\delta_{HA} = -LP_H$ and $\psi_A = -LP_M$ Use the loading terms for cases 1a to 1r
4. Left end fixed, right end pinned	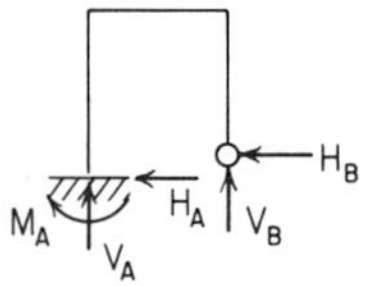	Since $\delta_{HA} = 0$ and $\psi_A = 0$, $H_A = \frac{A_{MM}LP_H - A_{HM}LP_M}{A_{HH}A_{MM} - (A_{HM})^2}$ and $M_A = \frac{A_{HH}LP_M - A_{HM}LP_H}{A_{HH}A_{MM} - (A_{HM})^2}$ Use the loading terms for cases 1a to 1r

TABLE 4 ***Reaction and deflection formulas for rigid frames*** **(Cont.)**

General reaction and deformation expressions for cases 5 to 12, right end fixed in all eight cases:

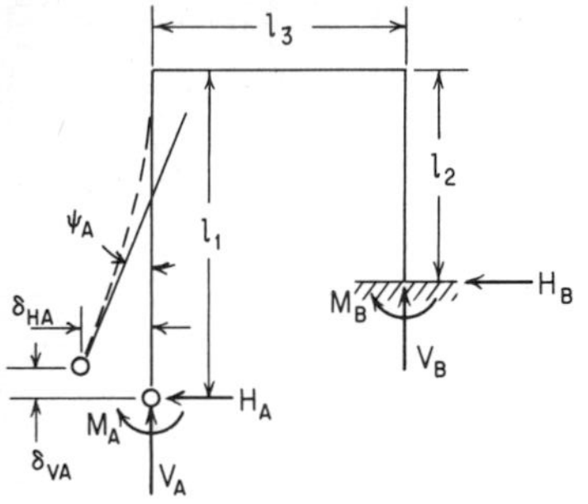

Deformation equations:

Horizontal deflection at $A = \delta_{HA} = C_{HH}H_A + C_{HV}V_A + C_{HM}M_A - LF_H$ (3)

Vertical deflection at $A = \delta_{VA} = C_{VH}H_A + C_{VV}V_A + C_{VM}M_A - LF_V$ (4)

Angular rotation at $A = \psi_A = C_{MH}H_A + C_{MV}V_A + C_{MM}M_A - LF_M$ (5)

where $C_{HH} = \dfrac{l_1^3}{3E_1I_1} + \dfrac{l_1^3 - (l_1 - l_2)^3}{3E_2I_2} + \dfrac{l_1^2 l_3}{E_3I_3}$

$$C_{HV} = C_{VH} = \frac{l_2 l_3}{2E_2I_2}(2l_1 - l_2) + \frac{l_1 l_3^2}{2E_3I_3}$$

$$C_{HM} = C_{MH} = \frac{l_1^2}{2E_1I_1} + \frac{l_2}{2E_2I_2}(2l_1 - l_2) + \frac{l_1 l_3}{E_3I_3}$$

$$C_{VV} = \frac{l_2 l_3^2}{E_2I_2} + \frac{l_3^3}{3E_3I_3}$$

$$C_{VM} = C_{MV} = \frac{l_2 l_3}{E_2I_2} + \frac{l_3^2}{2E_3I_3}$$

$$C_{MM} = \frac{l_1}{E_1I_1} + \frac{l_2}{E_2I_2} + \frac{l_3}{E_3I_3}$$

and where LF_H, LF_V, and LF_M are loading terms given below for several types of load

(*Note:* If desired, H_B, V_B, and M_B are to be evaluated from equilibrium equations after calculating H_A, V_A, and M_A)

5. Left end fixed, right end fixed

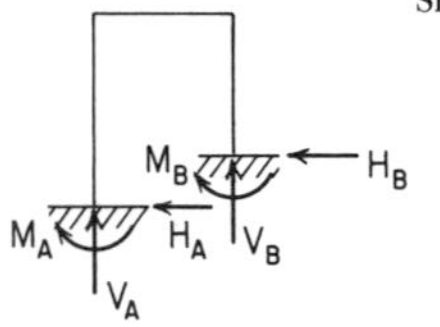

Since $\delta_{HA} = 0$, $\delta_{VA} = 0$, and $\psi_A = 0$, Eqs. 3 to 5 reduce to a set of three equations to be solved simultaneously for H_A, V_A, and M_A:

$$C_{HH}H_A + C_{HV}V_A + C_{HM}M_A = LF_H$$

$$C_{VH}H_A + C_{VV}V_A + C_{VM}M_A = LF_V$$

$$C_{MH}H_A + C_{MV}V_A + C_{MM}M_A = LF_M$$

The loading terms are given below.

Reference no., loading	Loading terms
5a. Concentrated load on the horizontal member 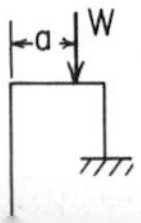	$LF_H = W\left[\dfrac{l_2}{2E_2I_2}(2l_1 - l_2)(l_3 - a) + \dfrac{l_1}{2E_3I_3}(l_3 - a)^2\right]$ $LF_V = W\left(C_{VV} - aC_{VM} + \dfrac{a^3}{6E_3I_3}\right)$

5b. Distributed load on the horizontal member	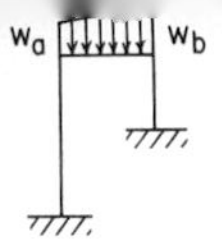	$LF_H = w_a\left[\frac{l_2 l_3^2}{4E_2I_2}(2l_1 - l_2) + \frac{l_1 l_3^3}{6E_3I_3}\right] + (w_b - w_a)\left[\frac{l_2 l_3^2}{12E_2I_2}(2l_1 - l_2) + \frac{l_1 l_3^3}{24E_3I_3}\right]$ $LF_V = w_a\left(\frac{l_2 l_3^3}{2E_2I_2} + \frac{l_3^4}{8E_3I_3}\right) + (w_b - w_a)\left(\frac{l_2 l_3^3}{6E_2I_2} + \frac{l_3^4}{30E_3I_3}\right)$ $LF_M = w_a\left(\frac{l_2 l_3^2}{2E_2I_2} + \frac{l_3^3}{6E_3I_3}\right) + (w_b - w_a)\left(\frac{l_2 l_3^2}{6E_2I_2} + \frac{l_3^3}{24E_3I_3}\right)$
5c. Concentrated moment on the horizontal member	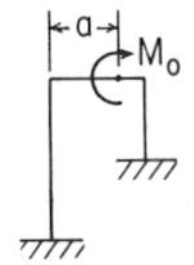	$LF_H = M_o\left[\frac{-l_2}{2E_2I_2}(2l_1 - l_2) - \frac{l_1}{E_3I_3}(l_3 - a)\right]$ $LF_V = M_o\left(-C_{VM} + \frac{a^2}{2E_3I_3}\right)$ $LF_M = M_o\left[\frac{-l_2}{E_2I_2} - \frac{1}{E_3I_3}(l_3 - a)\right]$
5d. Concentrated angular displacement on the horizontal member	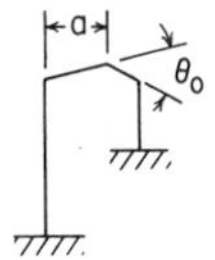	$LF_H = \theta_o(l_1)$ $LF_V = \theta_o(a)$ $LF_M = \theta_o(1)$
5e. Concentrated lateral displacement on the horizontal member	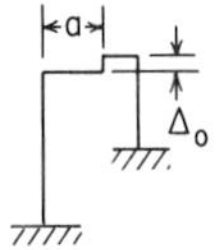	$LF_H = 0$ $LF_V = \Delta_o(1)$ $LF_M = 0$
5f. Concentrated load on the left vertical member	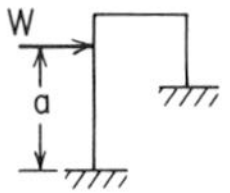	$LF_H = W\left(C_{HH} - aC_{HM} + \frac{a^3}{6E_1I_1}\right)$ $LF_V = W(C_{VH} - aC_{VM})$ $LF_M = W\left(C_{MH} - aC_{MM} + \frac{a^2}{2E_1I_1}\right)$
5g. Distributed load on the left vertical member	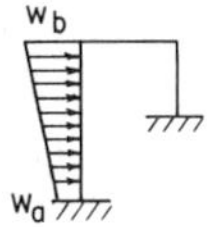	$LF_H = w_a\left(C_{HH}l_1 - C_{HM}\frac{l_1^2}{2} + \frac{l_1^4}{24E_1I_1}\right) + (w_b - w_a)\left(C_{HH}\frac{l_1}{2} - C_{HM}\frac{l_1^2}{3} + \frac{l_1^4}{30E_1I_1}\right)$ $LF_V = w_a\left(C_{VH}l_1 - C_{VM}\frac{l_1^2}{2}\right) + (w_b - w_a)\left(C_{VH}\frac{l_1}{2} - C_{VM}\frac{l_1^2}{3}\right)$ $LF_M = w_a\left(C_{MH}l_1 - C_{MM}\frac{l_1^2}{2} - \frac{l_1^3}{6E_1I_1}\right) + (w_b - w_a)\left(C_{MH}\frac{l_1}{2} - C_{MM}\frac{l_1^2}{3} + \frac{l_1^3}{8E_1I_1}\right)$

TABLE 4 *Reaction and deflection formulas for rigid frames* (*Cont.*)

Reference no., loading		Loading terms
5h. Concentrated moment on the left vertical member	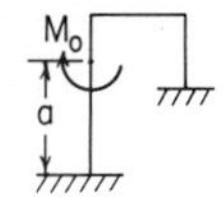	$LF_H = M_o\left(-C_{HM} + \frac{a^2}{2E_1I_1}\right)$ $LF_V = M_o(-C_{VM})$ $LF_M = M_o\left(-C_{MM} + \frac{a}{E_1I_1}\right)$
5i. Concentrated angular displacement on the left vertical member	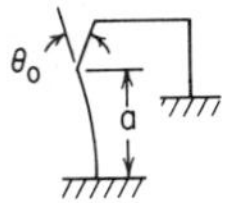	$LF_H = \theta_o(a)$ $LF_V = 0$ $LF_M = \theta_o(1)$
5j. Concentrated lateral displacement on the left vertical member	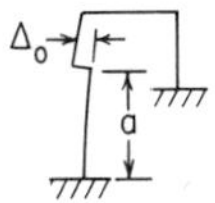	$LF_H = \Delta_o(1)$ $LF_V = 0$ $LF_M = 0$
5k. Concentrated load on the right vertical member	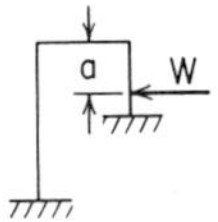	$LF_H = \frac{W}{6E_2I_2}[3l_1(l_2 - a)^2 - 2l_2^3 - a^3 + 3al_2^2]$ $LF_V = \frac{W}{2E_2I_2}[l_3(l_2 - a)^2]$ $LF_M = \frac{W}{2E_2I_2}(l_2 - a)^2$
5l. Distributed load on the right vertical member	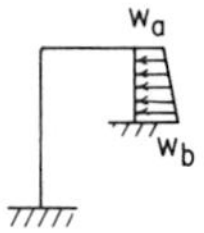	$LF_H = w_a\left[\frac{l_2^3}{24E_2I_2}(4l_1 - 3l_2)\right] + (w_b - w_a)\left[\frac{l_2^3}{120E_2I_2}(5l_1 - 4l_2)\right]$ $LF_V = w_a\frac{l_2^3l_3}{6E_2I_2} + (w_b - w_a)\frac{l_2^3l_3}{24E_2I_2}$ $LF_M = w_a\frac{l_2^3}{6E_2I_2} + (w_b - w_a)\frac{l_2^3}{24E_2I_2}$
5m. Concentrated moment on the right vertical member	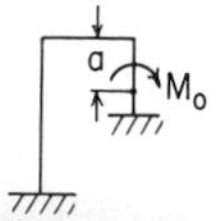	$LF_H = \frac{M_o}{2E_2I_2}[-2l_1(l_2 - a) - a^2 + l_2^2]$ $LF_V = \frac{M_o}{E_2I_2}[-l_3(l_2 - a)]$ $LF_M = M_o$ [illegible]

... displacement on the right vertical member	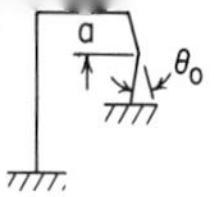	... $LF_V = \theta_o(l_3)$ $LF_M = \theta_o(1)$
5p. Concentrated lateral displacement on the right vertical member	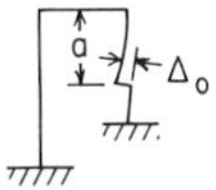	$LF_H = \Delta_o(-1)$ (*Note:* Δ_o could also be an increase in the length l_3) $LF_V = 0$ $LF_M = 0$
5q. Uniform temperature rise: T = uniform temperature T_o = unloaded temperature	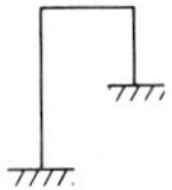	$LF_H = (T - T_o)(-\gamma_3 l_3)$ γ = temperature coefficient of expansion (inches/inch/degree) $LF_V = (T - T_o)(\gamma_1 l_1 - \gamma_2 l_2)$ $LF_M = 0$
5r. Uniform temperature differential from outside to inside; average temperature is T_o	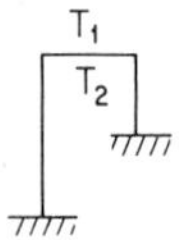	$LF_H = (T_1 - T_2)\left[\frac{l_1^2\gamma_1}{2t_1} + \frac{l_2\gamma_2}{2t_2}(2l_1 - l_2) - \frac{l_1 l_3 \gamma_3}{t_3}\right]$ $LF_V = (T_1 - T_2)\left(\frac{l_2 l_3 \gamma_2}{t_2} + \frac{l_3^2\gamma_3}{2t_3}\right)$ t_1, t_2, and t_3 are beam thicknesses from inside to outside $LF_M = (T_1 - T_2)\left(\frac{l_1\gamma_1}{t_1} + \frac{l_2\gamma_2}{t_2} + \frac{l_3\gamma_3}{t_3}\right)$
6. Left end pinned, right end fixed	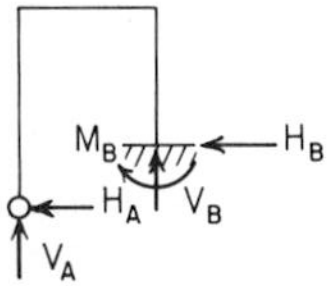	Since $\delta_{HA} = 0$, $\delta_{VA} = 0$ and $M_A = 0$, $H_A = \frac{LF_H C_{VV} - LF_V C_{HV}}{C_{HH}C_{VV} - (C_{HV})^2}$ $V_A = \frac{LF_V C_{HH} - LF_H C_{HV}}{C_{HH}C_{VV} - (C_{HV})^2}$ $\psi_A = C_{MH}H_A + C_{HV}V_A - LF_M$ Use the loading terms for cases 5a to 5r
7. Left end guided horizontally, right end fixed	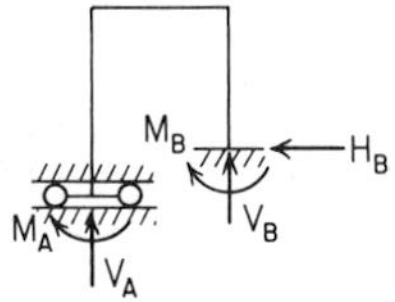	Since $\delta_{VA} = 0$, $\psi_A = 0$, and $H_A = 0$, $V_A = \frac{LF_V C_{MM} - LF_M C_{VM}}{C_{VV}C_{MM} - (C_{VM})^2}$ $M_A = \frac{LF_M C_{VV} - LF_V C_{VM}}{C_{VV}C_{MM} - (C_{VM})^2}$ $\delta_{HA} = C_{HV}V_A + C_{HM}M_A - LF_H$ Use the loading terms for cases 5a to 5r

TABLE 4 *Reaction and deflection formulas for rigid frames* (Cont.)

8.	Left end guided vertically, right end fixed	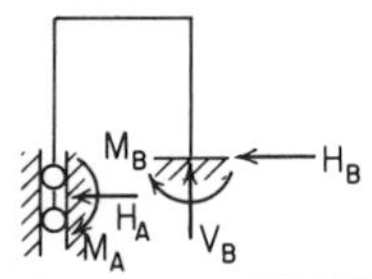	Since $\delta_{HA} = 0$, $\psi_A = 0$, and $V_A = 0$, $H_A = \dfrac{LF_H C_{MM} - LF_M C_{HM}}{C_{HH} C_{MM} - (C_{HM})^2}$ $M_A = \dfrac{LF_M C_{HH} - LF_H C_{HM}}{C_{HH} C_{MM} - (C_{HM})^2}$ $\delta_{VA} = C_{VH} H_A + C_{VM} M_A - LF_V$ Use the loading terms for cases 5a to 5r
9.	Left end roller supported along the horizontal, right end fixed	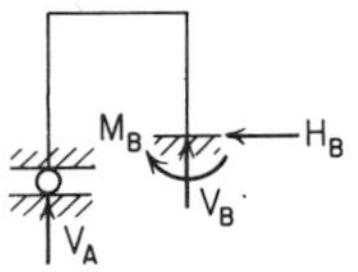	Since $\delta_{VA} = 0$, $H_A = 0$, and $M_A = 0$, $V_A = \dfrac{LF_V}{C_{VV}}$ $\delta_{HA} = C_{HV} V_A - LF_H$ and $\psi_A = C_{MV} V_A - LF_M$ Use the loading terms for cases 5a to 5r
10.	Left end roller supported along the vertical, right end fixed	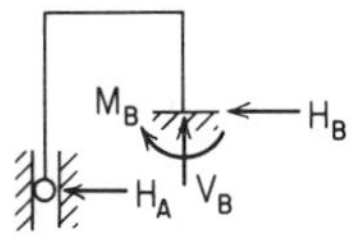	Since $\delta_{HA} = 0$, $V_A = 0$, and $M_A = 0$, $H_A = \dfrac{LF_H}{C_{HH}}$ $\delta_{VA} = C_{VH} H_A - LF_V$ and $\psi_A = C_{MH} H_A - LF_M$ Use the loading terms for cases 5a to 5r
11.	Left end guided by moment only (zero slope at the left end), right end fixed	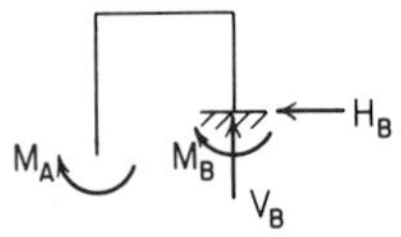	Since $\psi_A = 0$, $H_A = 0$, and $V_A = 0$, $M_A = \dfrac{LF_M}{C_{MM}}$ $\delta_{HA} = C_{HM} M_A - LF_H$ and $\delta_{VA} = C_{VM} M_A - LF_V$ Use the loading terms for cases 5a to 5r
12.	Left end free, right end fixed	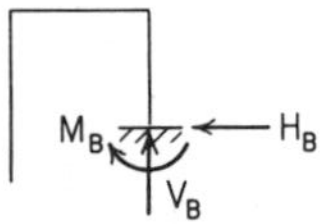	Since $H_A = 0$, $V_A = 0$, and $M_A = 0$, this is a statically determinate case. The deflections are given by $\delta_{HA} = -LF_H$ $\delta_{VA} = -LF_V$ and $\psi_A = -LF_M$ Use the loading terms for cases 5a to 5r

For the combined loading

$$LF_H = 0.2282 - 0.004875 = 0.2233$$
$$LF_V = 0.0507$$
$$LF_M = 0.00632$$

Now the left end force, moment, and displacement can be evaluated:

$$V_A = \frac{LF_V C_{MM} - LF_M C_{VM}}{C_{VV}C_{MM} - C_{VM}^2} = \frac{0.0507(0.359 \cdot 10^{-6}) - 0.00632(1.94 \cdot 10^{-6})}{(24.4 \cdot 10^{-6})(0.359 \cdot 10^{-6}) - (1.94 \cdot 10^{-6})^2} = 1{,}189 \text{ lb}$$

$$M_A = 11{,}179 \text{ in-lb}$$
$$\delta_{HA} = -0.0274 \text{ in}$$

Figure 7.9 shows the moment diagram for the entire frame.

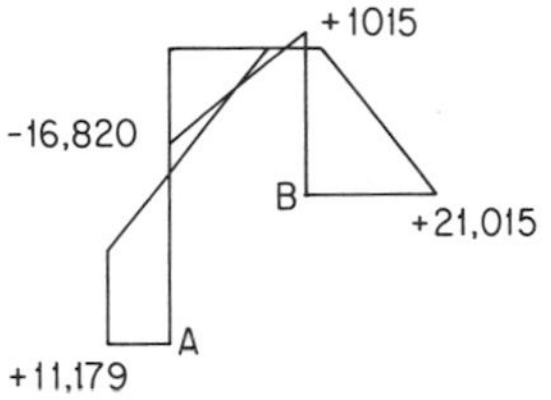

Fig. 7.9

2. If the joint at the top of the left vertical member in Example 1 had not been rigid but instead had been found to be deformable by 10^{-7} rad for every inch-pound of bending moment applied to it, the solution can be modified as follows.

Solution. The bending moment at the corner in question would be given by $M_A - 28(1000)$, and so the corner rotation would be $10^{-7}(M_A - 28{,}000)$ rad in a direction opposite to that shown by θ_o in case 5*a*. Note that the position of θ_o is at the corner, and so *a* would be 40 in. Therefore, the following load terms due to the corner deformation can be added to the previously determined load terms:

$$LF_H = -10^{-7}(M_A - 28{,}000)(40)$$
$$LF_V = 0$$
$$LF_M = -10^{-7}(M_A - 28{,}000)$$

Thus the resultant load terms become

$$LF_H = 0.2233 - 4 \cdot 10^{-6} M_A + 0.112 = 0.3353 - 4 \cdot 10^{-6} M_A$$
$$LF_V = 0.0507$$
$$LF_M = 0.00632 - 10^{-7} M_A + 0.0028 = 0.00912 - 10^{-7} M_A$$

Again, the left end force, moment, and displacement are evaluated:

$$V_A = \frac{0.0507(0.359 \cdot 10^{-6}) - (0.00912 - 10^{-7}M_A)(1.94 \cdot 10^{-6})}{4.996 \cdot 10^{-12}} = 100 + 0.0388 M_A$$

$$M_A = \frac{(0.00912 - 10^{-7}M_A)(24.4 \cdot 10^{-6}) - 0.0507(1.94 \cdot 10^{-6})}{4.996 \cdot 10^{-12}} = 24{,}800 - 0.488 M_A$$

or $\quad M_A = 16{,}670$ in-lb

$$\delta_{HA} = -0.0460 \text{ in}$$
$$V_A = 747 \text{ lb}$$

7.5 *Beams on elastic foundations*

There are cases in which beams are supported on foundations which develop essentially continuous reactions that are proportional at each position along the beam to the deflection of the beam at that position. This is the reason for the name *elastic foundation.* Solutions are available (Refs. 41 and 42) which consider that the foundation transmits shear forces within the foundation such that the reaction force is not directly proportional to the deflection at a given location but instead is proportional to a function of the deflections near the given location; these solutions are much more difficult to use and are not justified in many cases since the linearity of most foundations is open to question anyway.

It is not necessary, in fact, that a foundation be continuous. If a discontinuous foundation, such as is encountered in the support provided a rail by the cross ties, is found to have at least three concentrated reaction forces in every half wavelength of the deflected beam, then the solutions provided in this section are adequate.

Table 7 provides formulas for the reactions and deflections at the left end of a finite-length beam on an elastic foundation as well as formulas for the shear, moment, slope, and deflection at any point x along the length. The format used in presenting the formulas is designed to facilitate programming for use on a digital computer; but regardless of the form of the equations, a great deal of effort is needed to make the calculations with a slide rule. In some cases a slide rule will not provide sufficient accuracy to obtain a satisfactory answer unless the tabulated values of functions in Tables 5 and 6 are used. With a pocket or desk calculator having trigonometric and exponential functions, the calculations become routine, however. If only simple loads are encountered at the ends or centers of beams, tables and graphs of the necessary coefficients are given in Refs. 36 and 53.

In theory the equations in Table 7 are correct for any finite-length beam or for any finite foundation modulus, but for practical purposes they should not be used when βl exceeds a value of 6 because the roundoff errors that are created where two very nearly equal large numbers are subtracted will make the accuracy of the answer questionable. For this reason, Table 8 has been provided also. Table 8 contains formulas for semi-infinite- and infinite-length beams on elastic foundations. These formulas are of a much simpler form since the far end of the beam is assumed to be far enough away so as to have no effect on the response of the left end to the loading. If $\beta l > 6$ and the load is nearer the left end, this is the case.

Hetényi (Ref. 53) discusses this problem of a beam supported on an elastic foundation extensively and shows how the solutions can be adapted to other elements such as hollow cylinders. Hetényi (Ref. 51) has also developed a series solution for beams supported on elastic foundations in which the stiffness parameters of the beam and foundation are not incorporated in the arguments of trigonometric or hyperbolic functions. He gives tables of coefficients derived for specific boundary conditions from which deformation, moments, or shears can be found at any specific point along the beam. Any degree of accuracy can be obtained by using enough terms in the series.

Tables of numerical values, Tables 5 and 6, are provided to assist in the solution of the formulas in Table 7. Interpolation is possible for values that are not included but should be used with caution if it is noted that differences of large and nearly equal numbers are being encountered. A far better method of interpolation for a beam with a single load is to solve the problem twice. For the first solution move the load to the left until $\beta(l - a)$ is a value found in Table 5, and for the second solution move the load similarly to the right. A linear interpolation from these solutions should be very accurate.

Presenting the formulas for end reactions and displacements in Table 7 in terms of the constants C_i and C_{ai} is advantageous since it permits one to solve directly for loads anywhere on the span. If the loads are at the left end such that $C_i = C_{ai}$, then the formulas can be presented in a simpler form as is done in Ref. 6 of Chap. 12 for cylindrical shells. In order to facilitate the use of Table 7 when a concentrated load, moment, angular rotation, or lateral displacement is at the left end (that is, $a = 0$), the following equations are presented to simplify the numerators:

$$C_1C_2 + C_3C_4 = C_{12} \qquad 2C_1^2 + C_2C_4 = 2 + C_{11}$$
$$C_2C_3 - C_1C_4 = C_{13} \qquad C_2^2 - 2C_1C_3 = C_{14}$$
$$C_1^2 + C_3^2 = 1 + C_{11} \qquad 2C_3^2 - C_2C_4 = C_{11}$$
$$C_2^2 + C_4^2 = 2C_{14} \qquad 2C_1C_3 + C_4^2 = C_{14}$$

EXAMPLES

1. A 6-in, 12.5-lb I-beam 20 ft long is used as a rail for an overhead crane and is in turn being supported every 2 ft of its length by being bolted to the bottom of a 5-in, 10-lb I-beam at midlength. The supporting beams are each 21.5 ft long and are considered to be simply supported at the ends. This is a case of a discontinuous foundation being analyzed as a continuous foundation. It is desired to determine the maximum bending stresses in the 6-in beam as well as in the supporting beams when a load of 1 ton is supported at one end of the crane.

Solution. The spring constant for each supporting beam is $48EI/l^3$, or $(48)(30 \cdot 10^6)(12.1)/(21.5 \cdot 12)^3 = 1{,}013$ lb/in. If this is assumed to be distributed over a 2-ft length of the rail, the equivalent value of $b_o k_o$ is $1.013/24 = 42.2$ lb/in per inch of deflection. Therefore

$$\beta = \left(\frac{b_o k_o}{4EI}\right)^{\frac{1}{4}} = \left[\frac{42.2}{4(30 \cdot 10^6)(21.8)}\right]^{\frac{1}{4}} = 0.01127 \text{ in}^{-1}$$

and $$\beta l = (0.01127)(240) = 2.70$$

An examination of the deflection of a beam on an elastic foundation shows that it varies cyclicly in amplitude with the sine and cosine of βx. A half-wavelength of this cyclic variation would occur over a span l_1, where $\beta l_1 = \pi$, or $l_1 = \pi/0.01127 = 279$ in. There is no question about there being at least three supporting forces over this length, and so the use of the solution for a continuous foundation is entirely adequate.

Since βl is less than 6, Table 7 will be used. Refer to case 1 where both ends are free. It must be pointed out that a simple support refers to a reaction force, developed by a support other than the foundation, which is large enough to prevent any vertical deflection of the end of the beam. From the table we find that $R_A = 0$ and $M_A = 0$; and since the load is at the left end, $a = 0$. When $a = 0$, the C_a terms are equal to the C terms, and so the four terms C_1, C_2, C_3, and C_4 are calculated:

$$C_1 = \text{Cosh}\,\beta l \cos\beta l = 7.47(-0.904) = -6.76$$

$$C_2 = \text{Cosh}\,\beta l \sin\beta l + \text{Sinh}\,\beta l \cos\beta l = 7.47(0.427) + 7.41(-0.904) = -3.50$$

Similarly $C_3 = 3.17$, $C_4 = 9.89$, and $C_{11} = 54.7$. (See Tables 5 and 6.) Therefore,

$$\theta_A = \frac{2000}{2(30\cdot10^6)(21.8)(0.01127^2)}\,\frac{(-3.50^2) - (2)(3.17)(-6.76)}{54.7} = 0.01216 \text{ rad}$$

$$y_A = \frac{2000}{2(30\cdot10^6)(21.8)(0.01127^3)}\,\frac{(9.89)(-6.76) - (3.17)(-3.50)}{54.7} = -1.092 \text{ in}$$

With the deformations at the left end known, the expression for the bending moment can be written:

$$M = -y_A 2EI\beta^2 F_3 - \theta_A EI\beta F_4 - \frac{W}{2\beta}F_{a2}$$

$$= 1.092(2)(30\cdot10^6)(21.8)(0.01127^2)F_3 - 0.01216(30\cdot10^6)(21.8)(0.01127)F_4 - \frac{2000}{2(0.01127)}F_{a2}$$

$$= 181{,}400F_3 - 89{,}600F_4 - 88{,}700F_{a2}$$

Now substituting the expressions for F_{a2}, F_3, and F_4 gives

$$M = 181{,}400\,\text{Sinh}\,\beta x \sin\beta x - 89{,}600(\text{Cosh}\,\beta x \sin\beta x - \text{Sinh}\,\beta x \cos\beta x)$$
$$- 88{,}700(\text{Cosh}\,\beta x \sin\beta x + \text{Sinh}\,\beta x \cos\beta x)$$

or

$$M = 181{,}400\,\text{Sinh}\,\beta x \sin\beta x - 178{,}300\,\text{Cosh}\,\beta x \sin\beta x + 900\,\text{Sinh}\,\beta x \cos\beta x$$

The maximum value of M can be found by trying values of x in the neighborhood of $x = \pi/4\beta = \pi/4(0.01127) = 69.7$ in, which would be the location of the maximum moment if the beam were infinitely long (see Table 8). This procedure reveals that the maximum moment occurs at $x = 66.5$ in and has a value of $-55{,}400$ in-lb.

The maximum stress in the 6-in I-beam is therefore $55{,}400(3)/21.8 = 7{,}620$ lb/in^2. The maximum stress in the supporting 5-in I-beams is found at the midspan of the beam directly above the load. The deflection of this beam is known to be 1.092 in, and the spring constant is 1013 lb/in, so that the center load on the beam is $1.092(1013) = 1107$ lb. Therefore the maximum bending moment is $Pl/4 = 1107(21.5)(12)/4 = 71{,}400$ in-lb, and the maximum stress is $71{,}400(2.5)/12.1 = 14{,}780$ lb/in^2.

2. If the 6-in I-beam in Example 1 had been much longer but supported in the same manner, Table 8 could have been used. Case 8 reveals that for an end load the end deflection is $-W/2EI\beta^3 = -2000/2(30\cdot10^6)(21.8)(0.01127^3) = -1.070$ in and the maximum moment would have equaled $-0.3225\,W/\beta = -0.3225(2000)/0.01127 = -57{,}200$ in-lb at 69.7 in from the left end. We should not construe from this example that increasing the length will always increase the stresses; if the load had been placed elsewhere on the span, the longer beam could have had the lower maximum stress.

3. An aluminum alloy beam 3 in wide, 2 in deep, and 60 in long is manufactured with an initial concentrated angular deformation of 0.02 rad at midlength; this initial shape is shown in Fig. 7.10*a*. In use, the beam is placed on an elastic foundation which develops 500 lb/in^2 vertical upward pressure for every 1 in it is depressed. The beam is loaded by two concentrated loads of 4000 lb each and a uniformly distributed load of 80 lb/in over the portion between the concentrated loads. The loading is shown in Fig. 7.10*b*. It is desired to determine the maximum bending stress in the aluminum beam.

Solution. First determine the beam and foundation parameters:

$$E = 9.5\cdot10^6 \text{ lb/in}^2 \qquad I = \tfrac{1}{12}(3)(2^3) = 2 \text{ in}^4 \qquad k_o = 500 \text{ lb/in}^2\text{/in} \qquad b_o = 3 \text{ in}$$

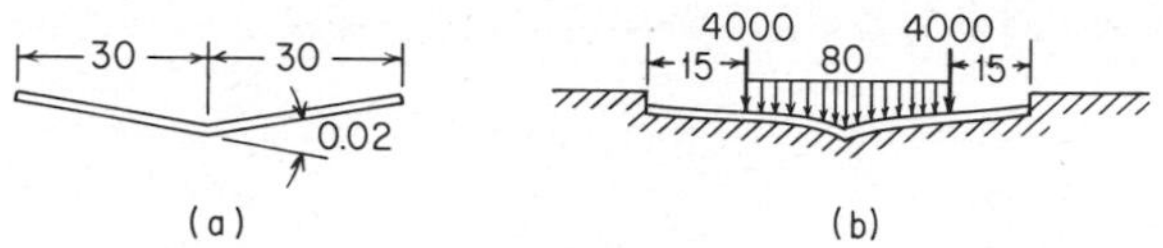

Fig. 7.10

$$\beta = \left[\frac{3(500)}{4(9.5 \cdot 10^6)(2)}\right]^{\frac{1}{4}} = 0.0666 \qquad l = 60 \text{ in} \qquad \beta l = 4.0$$

$$C_1 = -17.85 \qquad C_2 = -38.50 \qquad C_3 = -20.65 \qquad C_4 = -2.83 \qquad C_{11} = 744$$

An examination of Table 7 shows the loading conditions covered by the superposition of three cases in which both ends are free: case 1 used twice with $W_1 = 4000$ lb and $a_1 = 15$ in, and $W_2 = 4000$ lb and $a_2 = 45$ in; case 2 used twice with $w_3 = 80$ lb/in and $a_3 = 15$ in, and $w_4 = -80$ lb/in and $a_4 = 45$ in; case 5 used once with $\theta_o = 0.02$ and $a = 30$ in.

The loads and deformations at the left end are now evaluated by summing the values for the five different loads, which is done in the order in which the loads are mentioned. But before actually summing the end values, a set of constants involving the load positions must be determined for each case. For case 1, load 1:

$$C_{a1} = \text{Cosh } \beta(60 - 15) \cos \beta(60 - 15) = 10.068(-0.99) = -9.967$$

$$C_{a2} = -8.497$$

For case 1, load 2:

$$C_{a1} = 0.834 \qquad C_{a2} = 1.933$$

For case 2, load 3:

$$C_{a2} = -8.497 \qquad C_{a3} = 1.414$$

For case 2, load 4:

$$C_{a2} = 1.933 \qquad C_{a3} = 0.989$$

For case 5:

$$C_{a3} = 3.298 \qquad C_{a4} = 4.930$$

Therefore $R_A = 0$ and $M_A = 0$.

$$\theta_A = \frac{4000}{2EI\beta^2} \frac{(-38.50)(-8.497) - (2)(-20.65)(-9.967)}{744}$$

$$+ \frac{4000}{2EI\beta^2} \frac{(-38.50)(1.933) - (2)(-20.65)(0.834)}{744}$$

$$+ \frac{80}{2EI\beta^3} \frac{(-38.50)(1.414) - (-20.65)(-8.497)}{744}$$

$$+ \frac{-80}{2EI\beta^3} \frac{(-38.50)(0.989) - (-20.65)(1.933)}{744}$$

$$+ 0.02 \frac{(-38.50)(4.93) - (2)(-20.65)(3.298)}{744}$$

$$= \frac{4000}{EI\beta^2}(-0.0568) + \frac{4000}{EI\beta^2}(-0.02688) + \frac{80}{EI\beta^3}(-0.1545)$$

$$- \frac{80}{EI\beta^3}(0.00125) + 0.02(-0.0721)$$

$$= -0.007582 \text{ rad}$$

TABLE 5 ***Numerical values for functions used in Table 7***

βx	F_1	F_2	F_3	F_4
0.00	1.00000	0.00000	0.00000	0.00000
0.10	0.99998	0.20000	0.01000	0.00067
0.20	0.99973	0.39998	0.04000	0.00533
0.30	0.99865	0.59984	0.08999	0.01800
0.40	0.99573	0.79932	0.15995	0.04266
0.50	0.98958	0.99792	0.24983	0.08331
0.60	0.97841	1.19482	0.35948	0.14391
0.70	0.96001	1.38880	0.48869	0.22841
0.80	0.93180	1.57817	0.63709	0.34067
0.90	0.89082	1.76067	0.80410	0.48448
1.00	0.83373	1.93342	0.98890	0.66349
1.10	0.75683	2.09284	1.19034	0.88115
1.20	0.65611	2.23457	1.40688	1.14064
1.30	0.52722	2.35341	1.63649	1.44478
1.40	0.36558	2.44327	1.87659	1.79593
1.50	0.16640	2.49714	2.12395	2.19590
1.60	−0.07526	2.50700	2.37456	2.64573
1.70	−0.36441	2.46387	2.62358	3.14562
1.80	−0.70602	2.35774	2.86523	3.69467
1.90	−1.10492	2.17764	3.09266	4.29076
2.00	−1.56563	1.91165	3.29789	4.93026
2.10	−2.09224	1.54699	3.47170	5.60783
2.20	−2.68822	1.07013	3.60355	6.31615
2.30	−3.35618	0.46690	3.68152	7.04566
2.40	−4.09766	−0.27725	3.69224	7.78428
2.50	−4.91284	−1.17708	3.62088	8.51709
2.60	−5.80028	−2.24721	3.45114	9.22607
2.70	−6.75655	−3.50179	3.16529	9.88981
2.80	−7.77591	−4.95404	2.74420	10.48317
2.90	−8.84988	−6.61580	2.16749	10.97711
3.00	−9.96691	−8.49687	1.41372	11.33837
3.20	−12.26569	−12.94222	−0.71484	11.50778
3.40	−14.50075	−18.30128	−3.82427	10.63569
3.60	−16.42214	−24.50142	−8.09169	8.29386
3.80	−17.68744	−31.35198	−13.66854	3.98752
4.00	−17.84985	−38.50482	−20.65308	−2.82906
4.20	−16.35052	−45.41080	−29.05456	−12.72446
4.40	−12.51815	−51.27463	−38.74857	−26.24587
4.60	−5.57927	−55.01147	−49.42334	−43.85518
4.80	5.31638	−55.21063	−60.51809	−65.84195
5.00	21.05056	−50.11308	−71.15526	−92.21037
5.20	42.46583	−37.61210	−80.07047	−122.53858
5.40	70.26397	−15.28815	−85.54576	−155.81036
5.60	104.86818	19.50856	−85.35442	−190.22206
5.80	146.24469	69.51236	−76.72824	−222.97166
6.00	193.68136	137.31651	−56.36178	−250.04146

TABLE 6 *Numerical values for denominators used in Table 7*

βl	C_{11}	C_{12}	C_{13}	C_{14}
0.00	0.00000	0.00000	0.00000	0.00000
0.10	0.00007	0.20000	0.00133	0.02000
0.20	0.00107	0.40009	0.01067	0.08001
0.30	0.00540	0.60065	0.03601	0.18006
0.40	0.01707	0.80273	0.08538	0.32036
0.50	0.04169	1.00834	0.16687	0.50139
0.60	0.08651	1.22075	0.28871	0.72415
0.70	0.16043	1.44488	0.45943	0.99047
0.80	0.27413	1.68757	0.68800	1.30333
0.90	0.44014	1.95801	0.98416	1.66734
1.00	0.67302	2.26808	1.35878	2.08917
1.10	0.98970	2.63280	1.82430	2.57820
1.20	1.40978	3.07085	2.39538	3.14717
1.30	1.95606	3.60512	3.08962	3.81295
1.40	2.65525	4.26345	3.92847	4.59748
1.50	3.53884	5.07950	4.93838	5.52883
1.60	4.64418	6.09376	6.15213	6.64247
1.70	6.01597	7.35491	7.61045	7.98277
1.80	7.70801	8.92147	9.36399	9.60477
1.90	9.78541	10.86378	11.47563	11.57637
2.00	12.32730	13.26656	14.02336	13.98094
2.10	15.43020	16.23205	17.10362	16.92046
2.20	19.21212	19.88385	20.83545	20.51946
2.30	23.81752	24.37172	25.36541	24.92967
2.40	29.42341	29.87747	30.87363	30.33592
2.50	36.24681	36.62215	37.58107	36.96315
2.60	44.55370	44.87496	45.75841	45.08519
2.70	54.67008	54.96410	55.73686	55.03539
2.80	66.99532	67.29005	67.92132	67.21975
2.90	82.01842	82.34184	82.80645	82.13290
3.00	100.33792	100.71688	100.99630	100.37775
3.20	149.95828	150.51913	150.40258	149.96510
3.40	223.89682	224.70862	224.21451	224.02742
3.60	334.16210	335.25438	334.46072	334.55375
3.80	498.67478	500.03286	499.06494	499.42352
4.00	744.16690	745.73416	744.74480	745.31240
4.20	1110.50726	1112.19410	1111.33950	1112.02655
4.40	1657.15569	1658.85362	1658.26871	1658.96679
4.60	2472.79511	2474.39393	2474.17104	2474.76996
4.80	3689.70336	3691.10851	3691.28284	3691.68805
5.00	5505.19766	5506.34516	5506.88918	5507.03673
5.20	8213.62683	8214.49339	8215.32122	8215.18781
5.40	12254.10422	12254.71090	12255.69184	12255.29854
5.60	18281.71463	18282.12354	18283.10271	18282.51163
5.80	27273.73722	27274.04166	27274.86449	27274.16893
6.00	40688.12376	40688.43354	40688.97011	40688.27990

TABLE 7 ***Shear, moment, slope, and deflection formulas for finite-length beams on elastic foundations***

NOTATION: W = load (pounds); w = unit load (pounds per linear inch); M_o = applied couple (inch-pounds); θ_o = externally created concentrated angular displacement (radians); Δ_o = externally created concentrated lateral displacement (inches); γ = temperature coefficient of expansion (inches per inch per degree); T_1 and T_2 = temperatures on top and bottom surfaces, respectively (degrees). R_A and R_B are the vertical end reactions at the left and right, respectively, and are positive upward. M_A and M_B are the reaction end moments at the left and right, respectively, and all moments are positive when producing compression on the upper portion of the beam cross section. The transverse shear force V is positive when acting upward on the left end of a portion of the beam. All applied loads, couples, and displacements are positive as shown. All forces are in pounds; all moments are in inch-pounds; all deflections and beam dimensions are in inches; all slopes are in radians; and all temperatures are in degrees. All deflections are positive upward and slopes positive when up and to the right. Note that M_A and R_A are reactions, not applied loads. They exist only when necessary end restraints are provided.

The following constants and functions, involving both beam constants and foundation constants, are hereby defined in order to permit condensing the tabulated formulas which follow

k_o = foundation modulus (pounds per square inch per inch); b_o = beam width (inches); and $\beta = (b_o k_o / 4EI)^{\frac{1}{4}}$. (*Note:* See page 94 for a definition of $\langle x - a \rangle^n$.) The functions Cosh $\beta\langle x - a\rangle$, Sinh $\beta\langle x - a\rangle$, cos $\beta\langle x - a\rangle$, and sin $\beta\langle x - a\rangle$ are also defined as having a value of zero if $x < a$.

$F_1 = \text{Cosh}\,\beta x \cos \beta x$	$C_1 = \text{Cosh}\,\beta l \cos \beta l$	$C_{11} = \text{Sinh}^2\,\beta l - \sin^2 \beta l$
$F_2 = \text{Cosh}\,\beta x \sin \beta x + \text{Sinh}\,\beta x \cos \beta x$	$C_2 = \text{Cosh}\,\beta l \sin \beta l + \text{Sinh}\,\beta l \cos \beta l$	$C_{12} = \text{Cosh}\,\beta l\, \text{Sinh}\,\beta l + \cos \beta l \sin \beta l$
$F_3 = \text{Sinh}\,\beta x \sin \beta x$	$C_3 = \text{Sinh}\,\beta l \sin \beta l$	$C_{13} = \text{Cosh}\,\beta l\, \text{Sinh}\,\beta l - \cos \beta l \sin \beta l$
$F_4 = \text{Cosh}\,\beta x \sin \beta x - \text{Sinh}\,\beta x \cos \beta x$	$C_4 = \text{Cosh}\,\beta l \sin \beta l - \text{Sinh}\,\beta l \cos \beta l$	$C_{14} = \text{Sinh}^2\,\beta l + \sin^2 \beta l$
$F_{a1} = \langle x - a\rangle^0\, \text{Cosh}\,\beta\langle x - a\rangle \cos \beta\langle x - a\rangle$	$C_{a1} = \text{Cosh}\,\beta(l - a) \cos \beta(l - a)$	
$F_{a2} = \text{Cosh}\,\beta\langle x - a\rangle \sin \beta\langle x - a\rangle + \text{Sinh}\,\beta\langle x - a\rangle \cos \beta\langle x - a\rangle$	$C_{a2} = \text{Cosh}\,\beta(l - a) \sin \beta(l - a) + \text{Sinh}\,\beta(l - a) \cos \beta(l - a)$	
$F_{a3} = \text{Sinh}\,\beta\langle x - a\rangle \sin \beta\langle x - a\rangle$	$C_{a3} = \text{Sinh}\,\beta(l - a) \sin \beta(l - a)$	
$F_{a4} = \text{Cosh}\,\beta\langle x - a\rangle \sin \beta\langle x - a\rangle - \text{Sinh}\,\beta\langle x - a\rangle \cos \beta\langle x - a\rangle$	$C_{a4} = \text{Cosh}\,\beta(l - a) \sin \beta(l - a) - \text{Sinh}\,\beta(l - a) \cos \beta(l - a)$	
$F_{a5} = \langle x - a\rangle^0 - F_{a1}$	$C_{a5} = 1 - C_{a1}$	
$F_{a6} = 2\beta(x - a)\langle x - a\rangle^0 - F_{a2}$	$C_{a6} = 2\beta(l - a) - C_{a2}$	

1. Concentrated intermediate load

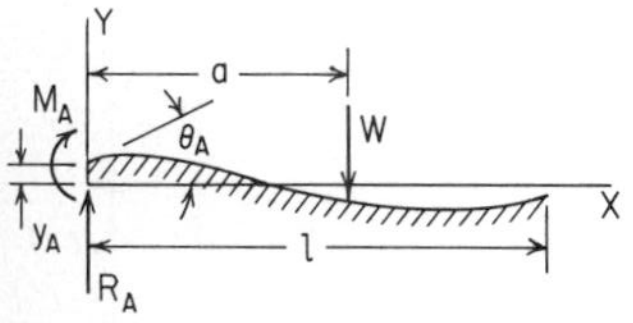

Transverse shear $= V = R_A F_1 - y_A 2EI\beta^3 F_2 - \theta_A 2EI\beta^2 F_3 - M_A \beta F_4 - W F_{a1}$

Bending moment $= M = M_A F_1 + \dfrac{R_A}{2\beta} F_2 - y_A 2EI\beta^2 F_3 - \theta_A EI\beta F_4 - \dfrac{W}{2\beta} F_{a2}$

Slope $= \theta = \theta_A F_1 + \dfrac{M_A}{2EI\beta} F_2 + \dfrac{R_A}{2EI\beta^2} F_3 - y_A \beta F_4 - \dfrac{W}{2EI\beta^2} F_{a3}$

Deflection $= y = y_A F_1 + \dfrac{\theta_A}{2\beta} F_2 + \dfrac{M_A}{2EI\beta^2} F_3 + \dfrac{R_A}{4EI\beta^3} F_4 - \dfrac{W}{4EI\beta^3} F_{a4}$

If $\beta l > 6$, see Table 8

Expressions for R_A, M_A, θ_A, and y_A are found below for several combinations of end restraints

Left end \ Right end	Free	Guided	Simply supported	Fixed
Free	$R_A = 0 \quad M_A = 0$ $\theta_A = \frac{W}{2EI\beta^2}\frac{C_2C_{a2} - 2C_3C_{a1}}{C_{11}}$ $y_A = \frac{W}{2EI\beta^3}\frac{C_4C_{a1} - C_3C_{a2}}{C_{11}}$	$R_A = 0 \quad M_A = 0$ $\theta_A = \frac{W}{2EI\beta^2}\frac{C_2C_{a3} - C_4C_{a1}}{C_{12}}$ $y_A = \frac{-W}{2EI\beta^3}\frac{C_1C_{a1} + C_3C_{a3}}{C_{12}}$	$R_A = 0 \quad M_A = 0$ $\theta_A = \frac{W}{2EI\beta^2}\frac{C_1C_{a2} + C_3C_{a4}}{C_{13}}$ $y_A = \frac{-W}{4EI\beta^3}\frac{C_4C_{a4} + C_2C_{a2}}{C_{13}}$	$R_A = 0 \quad M_A = 0$ $\theta_A = \frac{W}{2EI\beta^2}\frac{2C_1C_{a3} + C_4C_{a4}}{2 + C_{11}}$ $y_A = \frac{W}{2EI\beta^3}\frac{C_1C_{a4} - C_2C_{a3}}{2 + C_{11}}$
Guided	$R_A = 0 \quad \theta_A = 0$ $M_A = \frac{W}{2\beta}\frac{C_2C_{a2} - 2C_3C_{a1}}{C_{12}}$ $y_A = \frac{-W}{4EI\beta^3}\frac{2C_1C_{a1} + C_4C_{a2}}{C_{12}}$	$R_A = 0 \quad \theta_A = 0$ $M_A = \frac{W}{2\beta}\frac{C_2C_{a3} - C_4C_{a1}}{C_{14}}$ $y_A = \frac{-W}{4EI\beta^3}\frac{C_2C_{a1} + C_4C_{a3}}{C_{14}}$	$R_A = 0 \quad \theta_A = 0$ $M_A = \frac{W}{2\beta}\frac{C_1C_{a2} + C_3C_{a4}}{1 + C_{11}}$ $y_A = \frac{W}{4EI\beta^3}\frac{C_1C_{a4} - C_3C_{a2}}{1 + C_{11}}$	$R_A = 0 \quad \theta_A = 0$ $M_A = \frac{W}{2\beta}\frac{2C_1C_{a3} + C_4C_{a4}}{C_{12}}$ $y_A = \frac{W}{4EI\beta^3}\frac{C_2C_{a4} - 2C_3C_{a3}}{C_{12}}$
Simply supported	$M_A = 0 \quad y_A = 0$ $R_A = W\frac{C_3C_{a2} - C_4C_{a1}}{C_{13}}$ $\theta_A = \frac{W}{2EI\beta^2}\frac{C_1C_{a2} - C_2C_{a1}}{C_{13}}$	$M_A = 0 \quad y_A = 0$ $R_A = W\frac{C_1C_{a1} + C_3C_{a3}}{1 + C_{11}}$ $\theta_A = \frac{W}{2EI\beta^2}\frac{C_1C_{a3} - C_3C_{a1}}{1 + C_{11}}$	$M_A = 0 \quad y_A = 0$ $R_A = \frac{W}{2}\frac{C_2C_{a2} + C_4C_{a4}}{C_{14}}$ $\theta_A = \frac{W}{4EI\beta^2}\frac{C_2C_{a4} - C_4C_{a2}}{C_{14}}$	$M_A = 0 \quad y_A = 0$ $R_A = W\frac{C_2C_{a3} - C_1C_{a4}}{C_{13}}$ $\theta_A = \frac{W}{2EI\beta^2}\frac{C_3C_{a4} - C_4C_{a3}}{C_{13}}$
Fixed	$\theta_A = 0 \quad y_A = 0$ $R_A = W\frac{2C_1C_{a1} + C_4C_{a2}}{2 + C_{11}}$ $M_A = \frac{W}{\beta}\frac{C_1C_{a2} - C_2C_{a1}}{2 + C_{11}}$	$\theta_A = 0 \quad y_A = 0$ $R_A = W\frac{C_4C_{a3} + C_2C_{a1}}{C_{12}}$ $M_A = \frac{W}{\beta}\frac{C_1C_{a3} - C_3C_{a1}}{C_{12}}$	$\theta_A = 0 \quad y_A = 0$ $R_A = W\frac{C_3C_{a2} - C_1C_{a4}}{C_{13}}$ $M_A = \frac{W}{2\beta}\frac{C_2C_{a4} - C_4C_{a2}}{C_{13}}$	$\theta_A = 0 \quad y_A = 0$ $R_A = W\frac{2C_3C_{a3} - C_2C_{a4}}{C_{11}}$ $M_A = \frac{W}{\beta}\frac{C_3C_{a4} - C_4C_{a3}}{C_{11}}$

TABLE 7 ***Shear, moment, slope, and deflection formulas for finite-length beams on elastic foundations*** ***(Cont.)***

2. Partial uniformly distributed load

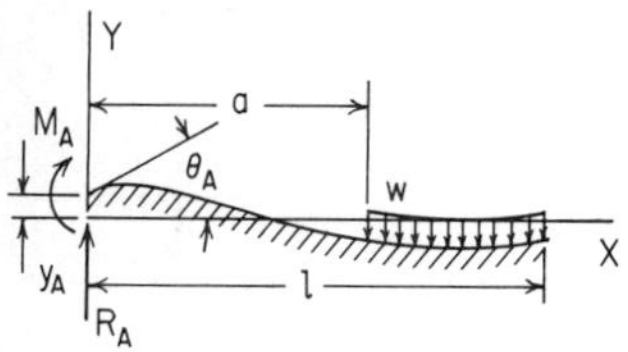

Transverse shear $= V = R_A F_1 - y_A 2EI\beta^3 F_2 - \theta_A 2EI\beta^2 F_3 - M_A \beta F_4 - \dfrac{w}{2\beta} F_{a2}$

Bending moment $= M = M_A F_1 + \dfrac{R_A}{2\beta} F_2 - y_A 2EI\beta^2 F_3 - \theta_A EI\beta F_4 - \dfrac{w}{2\beta^2} F_{a3}$

Slope $= \theta = \theta_A F_1 + \dfrac{M_A}{2EI\beta} F_2 + \dfrac{R_A}{2EI\beta^2} F_3 - y_A \beta F_4 - \dfrac{w}{4EI\beta^3} F_{a4}$

Deflection $= y = y_A F_1 + \dfrac{\theta_A}{2\beta} F_2 + \dfrac{M_A}{2EI\beta^2} F_3 + \dfrac{R_A}{4EI\beta^3} F_4 - \dfrac{w}{4EI\beta^4} F_{a5}$

If $\beta l > 6$, see Table 8

Expressions for R_A, M_A, θ_A, and y_A are found below for several combinations of end restraints

Left end \ Right end	Free	Guided	Simply supported	Fixed
Free	$R_A = 0 \quad M_A = 0$ $\theta_A = \dfrac{w}{2EI\beta^3} \dfrac{C_2 C_{a3} - C_3 C_{a2}}{C_{11}}$ $y_A = \dfrac{w}{4EI\beta^4} \dfrac{C_4 C_{a2} - 2C_3 C_{a3}}{C_{11}}$	$R_A = 0 \quad M_A = 0$ $\theta_A = \dfrac{w}{4EI\beta^3} \dfrac{C_2 C_{a4} - C_4 C_{a2}}{C_{12}}$ $y_A = \dfrac{-w}{4EI\beta^4} \dfrac{C_1 C_{a2} + C_3 C_{a4}}{C_{12}}$	$R_A = 0 \quad M_A = 0$ $\theta_A = \dfrac{w}{2EI\beta^3} \dfrac{C_1 C_{a3} + C_3 C_{a5}}{C_{13}}$ $y_A = \dfrac{-w}{4EI\beta^4} \dfrac{C_4 C_{a5} + C_2 C_{a3}}{C_{13}}$	$R_A = 0 \quad M_A = 0$ $\theta_A = \dfrac{w}{2EI\beta^3} \dfrac{C_1 C_{a4} + C_4 C_{a5}}{2 + C_{11}}$ $y_A = \dfrac{w}{4EI\beta^4} \dfrac{2C_1 C_{a5} - C_2 C_{a4}}{2 + C_{11}}$
Guided	$R_A = 0 \quad \theta_A = 0$ $M_A = \dfrac{w}{2\beta^2} \dfrac{C_2 C_{a3} - C_3 C_{a2}}{C_{12}}$ $y_A = \dfrac{-w}{4EI\beta^4} \dfrac{C_1 C_{a2} + C_4 C_{a3}}{C_{12}}$	$R_A = 0 \quad \theta_A = 0$ $M_A = \dfrac{w}{4\beta^2} \dfrac{C_2 C_{a4} - C_4 C_{a2}}{C_{14}}$ $y_A = \dfrac{-w}{8EI\beta^4} \dfrac{C_2 C_{a2} + C_4 C_{a4}}{C_{14}}$	$R_A = 0 \quad \theta_A = 0$ $M_A = \dfrac{w}{2\beta^2} \dfrac{C_1 C_{a3} + C_3 C_{a5}}{1 + C_{11}}$ $y_A = \dfrac{w}{4EI\beta^4} \dfrac{C_1 C_{a5} - C_3 C_{a3}}{1 + C_{11}}$	$R_A = 0 \quad \theta_A = 0$ $M_A = \dfrac{w}{2\beta^2} \dfrac{C_1 C_{a4} + C_4 C_{a5}}{C_{12}}$ $y_A = \dfrac{w}{4EI\beta^4} \dfrac{C_2 C_{a5} - C_3 C_{a4}}{C_{12}}$
Simply supported	$M_A = 0 \quad y_A = 0$ $R_A = \dfrac{w}{2\beta} \dfrac{2C_3 C_{a3} - C_4 C_{a2}}{C_{13}}$ $\theta_A = \dfrac{w}{4EI\beta^3} \dfrac{2C_1 C_{a3} - C_2 C_{a2}}{C_{13}}$	$M_A = 0 \quad y_A = 0$ $R_A = \dfrac{w}{2\beta} \dfrac{C_1 C_{a2} + C_3 C_{a4}}{1 + C_{11}}$ $\theta_A = \dfrac{w}{4EI\beta^3} \dfrac{C_1 C_{a4} - C_3 C_{a2}}{1 + C_{11}}$	$M_A = 0 \quad y_A = 0$ $R_A = \dfrac{w}{2\beta} \dfrac{C_2 C_{a3} + C_4 C_{a5}}{C_{14}}$ $\theta_A = \dfrac{w}{4EI\beta^3} \dfrac{C_2 C_{a5} - C_4 C_{a3}}{C_{14}}$	$M_A = 0 \quad y_A = 0$ $R_A = \dfrac{w}{2\beta} \dfrac{C_2 C_{a4} - 2C_1 C_{a5}}{C_{13}}$ $\theta_A = \dfrac{w}{4EI\beta^3} \dfrac{2C_3 C_{a5} - C_4 C_{a4}}{C_{13}}$
Fixed	$\theta_A = 0 \quad y_A = 0$ $R_A = \dfrac{w}{\beta} \dfrac{C_1 C_{a2} + C_4 C_{a3}}{2 + C_{11}}$ $M_A = \dfrac{w}{[illegible]} \dfrac{2C_1 C_{a3} - C_2 C_{a2}}{[illegible]}$	$\theta_A = 0 \quad y_A = 0$ $R_A = \dfrac{w}{2\beta} \dfrac{C_4 C_{a4} + C_2 C_{a2}}{C_{12}}$ $M_A = \dfrac{w}{[illegible]} \dfrac{C_1 C_{a4} - C_3 C_{a2}}{[illegible]}$	$\theta_A = 0 \quad y_A = 0$ $R_A = \dfrac{w}{\beta} \dfrac{C_3 C_{a3} - C_1 C_{a5}}{C_{13}}$ $M_A = \dfrac{w}{2\beta^2} \dfrac{C_2 C_{a5} - C_4 C_{a3}}{C_{13}}$	$\theta_A = 0 \quad y_A = 0$ $R_A = \dfrac{w}{\beta} \dfrac{C_3 C_{a4} - C_2 C_{a5}}{C_{11}}$ $M_A = \dfrac{w}{2\beta^2} \dfrac{2C_3 C_{a5} - C_4 C_{a4}}{C_{11}}$

3. Partial uniformly increasing load

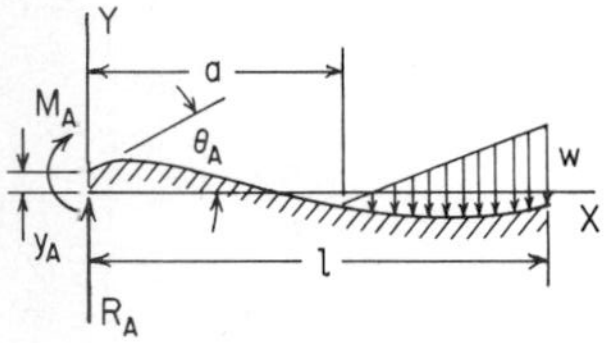

Transverse shear $= V = R_A F_1 - y_A 2EI\beta^3 F_2 - \theta_A 2EI\beta^2 F_3 - M_A \beta F_4 - \dfrac{wF_{a3}}{2\beta^2(l-a)}$

Bending moment $= M = M_A F_1 + \dfrac{R_A}{2\beta} F_2 - y_A 2EI\beta^2 F_3 - \theta_A EI\beta F_4 - \dfrac{wF_{a4}}{4\beta^3(l-a)}$

Slope $= \theta = \theta_A F_1 + \dfrac{M_A}{2EI\beta} F_2 + \dfrac{R_A}{2EI\beta^2} F_3 - y_A \beta F_4 - \dfrac{wF_{a5}}{4EI\beta^4(l-a)}$

Deflection $= y = y_A F_1 + \dfrac{\theta_A}{2\beta} F_2 + \dfrac{M_A}{2EI\beta^2} F_3 + \dfrac{R_A}{4EI\beta^3} F_4 - \dfrac{wF_{a6}}{8EI\beta^5(l-a)}$

If $\beta l > 6$, see Table 8

Expressions for R_A, M_A, θ_A, and y_A are found below for several combinations of end restraints

Left end \ Right end	Free	Guided	Simply supported	Fixed
Free	$R_A = 0 \quad M_A = 0$ $\theta_A = \dfrac{w(C_2C_{a4} - 2C_3C_{a3})}{4EI\beta^4(l-a)C_{11}}$ $y_A = \dfrac{w(C_4C_{a3} - C_3C_{a4})}{4EI\beta^5(l-a)C_{11}}$	$R_A = 0 \quad M_A = 0$ $\theta_A = \dfrac{w(C_2C_{a5} - C_4C_{a3})}{4EI\beta^4(l-a)C_{12}}$ $y_A = \dfrac{-w(C_1C_{a3} + C_3C_{a5})}{4EI\beta^5(l-a)C_{12}}$	$R_A = 0 \quad M_A = 0$ $\theta_A = \dfrac{w(C_1C_{a4} + C_3C_{a6})}{4EI\beta^4(l-a)C_{13}}$ $y_A = \dfrac{-w(C_2C_{a4} + C_4C_{a6})}{8EI\beta^5(l-a)C_{13}}$	$R_A = 0 \quad M_A = 0$ $\theta_A = \dfrac{w(2C_1C_{a5} + C_4C_{a6})}{4EI\beta^4(l-a)(2 + C_{11})}$ $y_A = \dfrac{w(C_1C_{a6} - C_2C_{a5})}{4EI\beta^5(l-a)(2 + C_{11})}$
Guided	$R_A = 0 \quad \theta_A = 0$ $M_A = \dfrac{w(C_2C_{a4} - 2C_3C_{a3})}{4\beta^3(l-a)C_{12}}$ $y_A = \dfrac{-w(2C_1C_{a3} + C_4C_{a4})}{8EI\beta^5(l-a)C_{12}}$	$R_A = 0 \quad \theta_A = 0$ $M_A = \dfrac{w(C_2C_{a5} - C_4C_{a3})}{4\beta^3(l-a)C_{14}}$ $y_A = \dfrac{-w(C_2C_{a3} + C_4C_{c5})}{8EI\beta^5(l-a)C_{14}}$	$R_A = 0 \quad \theta_A = 0$ $M_A = \dfrac{w(C_1C_{a4} + C_3C_{a6})}{4\beta^3(l-a)(1 + C_{11})}$ $y_A = \dfrac{w(C_1C_{a6} - C_3C_{a4})}{8EI\beta^5(l-a)(1 + C_{11})}$	$R_A = 0 \quad \theta_A = 0$ $M_A = \dfrac{w(2C_1C_{a5} + C_4C_{a6})}{4\beta^3(l-a)C_{12}}$ $y_A = \dfrac{w(C_2C_{a6} - 2C_3C_{a5})}{8EI\beta^5(l-a)C_{12}}$
Simply supported	$M_A = 0 \quad y_A = 0$ $R_A = \dfrac{w(C_3C_{a4} - C_4C_{a3})}{2\beta^2(l-a)C_{13}}$ $\theta_A = \dfrac{w(C_1C_{a4} - C_2C_{a3})}{4EI\beta^4(l-a)C_{13}}$	$M_A = 0 \quad y_A = 0$ $R_A = \dfrac{w(C_1C_{a3} + C_3C_{a5})}{2\beta^2(l-a)(1 + C_{11})}$ $\theta_A = \dfrac{w(C_1C_{a5} - C_3C_{a3})}{4EI\beta^4(l-a)(1 + C_{11})}$	$M_A = 0 \quad y_A = 0$ $R_A = \dfrac{w(C_2C_{a4} + C_4C_{a6})}{4\beta^2(l-a)C_{14}}$ $\theta_A = \dfrac{w(C_2C_{a6} - C_4C_{a4})}{8EI\beta^4(l-a)C_{14}}$	$M_A = 0 \quad y_A = 0$ $R_A = \dfrac{w(C_2C_{a5} - C_1C_{a6})}{2\beta^2(l-a)C_{13}}$ $\theta_A = \dfrac{w(C_3C_{a6} - C_4C_{a5})}{4EI\beta^4(l-a)C_{13}}$
Fixed	$\theta_A = 0 \quad y_A = 0$ $R_A = \dfrac{w(2C_1C_{a3} + C_4C_{a4})}{2\beta^2(l-a)(2 + C_{11})}$ $M_A = \dfrac{w(C_1C_{a4} - C_2C_{a3})}{2\beta^3(l-a)(2 + C_{11})}$	$\theta_A = 0 \quad y_A = 0$ $R_A = \dfrac{w(C_4C_{a5} + C_2C_{a3})}{2\beta^2(l-a)C_{12}}$ $M_A = \dfrac{w(C_1C_{a5} - C_3C_{a3})}{2\beta^3(l-a)C_{12}}$	$\theta_A = 0 \quad y_A = 0$ $R_A = \dfrac{w(C_3C_{a4} - C_1C_{a6})}{2\beta^2(l-a)C_{13}}$ $M_A = \dfrac{w(C_2C_{a6} - C_4C_{a4})}{4\beta^3(l-a)C_{13}}$	$\theta_A = 0 \quad y_A = 0$ $R_A = \dfrac{w(2C_3C_{A5} - C_2C_{a6})}{2\beta^2(l-a)C_{11}}$ $M_A = \dfrac{w(C_3C_{a6} - C_4C_{a5})}{2\beta^3(l-a)C_{11}}$

TABLE 7 ***Shear, moment, slope, and deflection formulas for finite-length beams on elastic foundations*** ***(Cont.)***

4. Concentrated intermediate moment

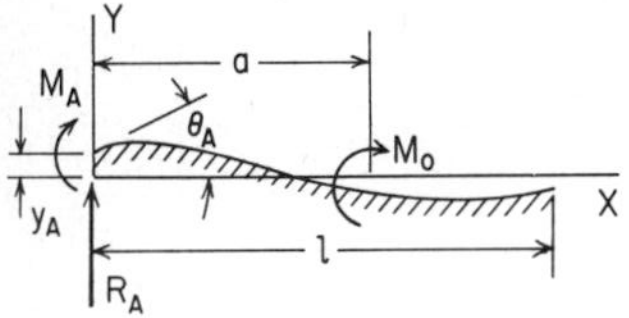

Transverse shear $= V = R_A F_1 - y_A 2EI\beta^3 F_2 - \theta_A 2EI\beta^2 F_3 - M_A \beta F_4 - M_o \beta F_{a4}$

Bending moment $= M = M_A F_1 + \frac{R_A}{2\beta} F_2 - y_A 2EI\beta^2 F_3 - \theta_A EI\beta F_4 + M_o F_{a1}$

Slope $= \theta = \theta_A F_1 + \frac{M_A}{2EI\beta} F_2 + \frac{R_A}{2EI\beta^2} F_3 - y_A \beta F_4 + \frac{M_o}{2EI\beta} F_{a2}$

Deflection $= y = y_A F_1 + \frac{\theta_A}{2\beta} F_2 + \frac{M_A}{2EI\beta^2} F_3 + \frac{R_A}{4EI\beta^3} F_4 + \frac{M_o}{2EI\beta^2} F_{a3}$

If $\beta l > 6$, see Table 8

Expressions for R_A, M_A, θ_A, and y_A are found below for several combinations of end restraints

Left end \ Right end	Free	Guided	Simply supported	Fixed
Free	$R_A = 0 \quad M_A = 0$ $\theta_A = \frac{-M_o}{EI\beta}\frac{C_3C_{a4} + C_2C_{a1}}{C_{11}}$ $y_A = \frac{M_o}{2EI\beta^2}\frac{2C_3C_{a1} + C_4C_{a4}}{C_{11}}$	$R_A = 0 \quad M_A = 0$ $\theta_A = \frac{-M_o}{2EI\beta}\frac{C_2C_{a2} + C_4C_{a4}}{C_{12}}$ $y_A = \frac{M_o}{2EI\beta^2}\frac{C_3C_{a2} - C_1C_{a4}}{C_{12}}$	$R_A = 0 \quad M_A = 0$ $\theta_A = \frac{-M_o}{EI\beta}\frac{C_1C_{a1} + C_3C_{a3}}{C_{13}}$ $y_A = \frac{M_o}{2EI\beta^2}\frac{C_4C_{a3} + C_2C_{a1}}{C_{13}}$	$R_A = 0 \quad M_A = 0$ $\theta_A = \frac{-M_o}{EI\beta}\frac{C_1C_{a2} + C_4C_{a3}}{2 + C_{11}}$ $y_A = \frac{-M_o}{2EI\beta^2}\frac{2C_1C_{a3} - C_2C_{a2}}{2 + C_{11}}$
Guided	$R_A = 0 \quad \theta_A = 0$ $M_A = -M_o\frac{C_2C_{a1} + C_3C_{a4}}{C_{12}}$ $y_A = \frac{-M_o}{2EI\beta^2}\frac{C_1C_{a4} - C_4C_{a1}}{C_{12}}$	$R_A = 0 \quad \theta_A = 0$ $M_A = \frac{-M_o}{2}\frac{C_2C_{a2} + C_4C_{a4}}{C_{14}}$ $y_A = \frac{M_o}{4EI\beta^2}\frac{C_4C_{a2} - C_2C_{a4}}{C_{14}}$	$R_A = 0 \quad \theta_A = 0$ $M_A = -M_o\frac{C_1C_{a1} + C_3C_{a3}}{1 + C_{11}}$ $y_A = \frac{M_o}{2EI\beta^2}\frac{C_3C_{a1} - C_1C_{a3}}{1 + C_{11}}$	$R_A = 0 \quad \theta_A = 0$ $M_A = -M_o\frac{C_1C_{a2} + C_4C_{a3}}{C_{12}}$ $y_A = \frac{M_o}{2EI\beta^2}\frac{C_3C_{a2} - C_2C_{a3}}{C_{12}}$
Simply supported	$M_A = 0 \quad y_A = 0$ $R_A = -M_o\beta\frac{2C_3C_{a1} + C_4C_{a4}}{C_{13}}$ $\theta_A = \frac{-M_o}{2EI\beta}\frac{2C_1C_{a1} + C_2C_{a4}}{C_{13}}$	$M_A = 0 \quad y_A = 0$ $R_A = -M_o\beta\frac{C_3C_{a2} - C_1C_{a4}}{1 + C_{11}}$ $\theta_A = \frac{-M_o}{2EI\beta}\frac{C_1C_{a2} + C_3C_{a4}}{1 + C_{11}}$	$M_A = 0 \quad y_A = 0$ $R_A = -M_o\beta\frac{C_2C_{a1} + C_4C_{a3}}{C_{14}}$ $\theta_A = \frac{-M_o}{2EI\beta}\frac{C_2C_{a3} - C_4C_{a1}}{C_{14}}$	$M_A = 0 \quad y_A = 0$ $R_A = -M_o\beta\frac{C_2C_{a2} - 2C_1C_{a3}}{C_{13}}$ $\theta_A = \frac{-M_o}{2EI\beta}\frac{2C_3C_{a3} - C_4C_{a2}}{C_{13}}$
Fixed	$\theta_A = 0 \quad y_A = 0$ $R_A = -M_o 2\beta\frac{C_4C_{a1} - C_1C_{a4}}{2 + C_{11}}$ $M_A = -M_o\frac{2C_1C_{a1} + C_2C_{a4}}{[illegible]}$	$\theta_A = 0 \quad y_A = 0$ $R_A = -M_o\beta\frac{C_4C_{a2} - C_2C_{a4}}{C_{12}}$ $M_A = -M_o\frac{C_1C_{a2} + C_3C_{a4}}{[illegible]}$	$\theta_A = 0 \quad y_A = 0$ $R_A = -M_o 2\beta\frac{C_3C_{a1} - C_1C_{a3}}{C_{13}}$ $M_A = -M_o\frac{C_2C_{a3} - C_4C_{a1}}{C_{13}}$	$\theta_A = 0 \quad y_A = 0$ $R_A = -M_o 2\beta\frac{C_3C_{a2} - C_2C_{a3}}{C_{11}}$ $M_A = -M_o\frac{2C_3C_{a3} - C_4C_{a2}}{C_{11}}$

5. Externally created concentrated angular displacement

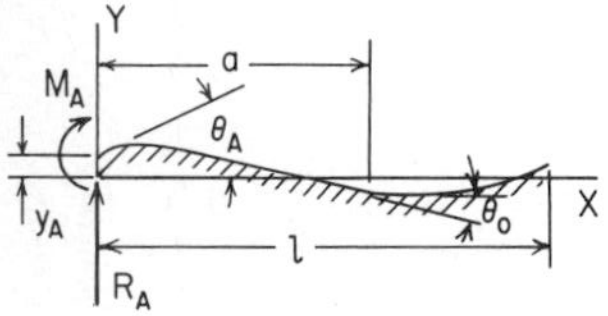

Transverse shear $= V = R_A F_1 - y_A 2EI\beta^3 F_2 - \theta_A 2EI\beta^2 F_3 - M_A \beta F_4 - \theta_o 2EI\beta^2 F_{a3}$

Bending moment $= M = M_A F_1 + \frac{R_A}{2\beta} F_2 - y_A 2EI\beta^2 F_3 - \theta_A EI\beta F_4 - \theta_o EI\beta F_{a4}$

Slope $= \theta = \theta_A F_1 + \frac{M_A}{2EI\beta} F_2 + \frac{R_A}{2EI\beta^2} F_3 - y_A \beta F_4 + \theta_o F_{a1}$

Deflection $= y = y_A F_1 + \frac{\theta_A}{2\beta} F_2 + \frac{M_A}{2EI\beta^2} F_3 + \frac{R_A}{4EI\beta^3} F_4 + \frac{\theta_o}{2\beta} F_{a2}$

If $\beta l > 6$, see Table 8

Expressions for R_A, M_A, θ_A, and y_A are found below for several combinations of end restraints

Left end \ Right end	Free	Guided	Simply supported	Fixed
Free	$R_A = 0 \quad M_A = 0$ $\theta_A = \theta_o \frac{C_2 C_{a4} - 2C_3 C_{a3}}{C_{11}}$ $y_A = \frac{\theta_o}{\beta} \frac{C_4 C_{a3} - C_3 C_{a4}}{C_{11}}$	$R_A = 0 \quad M_A = 0$ $\theta_A = -\theta_o \frac{C_2 C_{a1} + C_4 C_{a3}}{C_{12}}$ $y_A = \frac{\theta_o}{\beta} \frac{C_3 C_{a1} - C_1 C_{a3}}{C_{12}}$	$R_A = 0 \quad M_A = 0$ $\theta_A = \theta_o \frac{C_1 C_{a4} - C_3 C_{a2}}{C_{13}}$ $y_A = \frac{\theta_o}{2\beta} \frac{C_4 C_{a2} - C_2 C_{a4}}{C_{13}}$	$R_A = 0 \quad M_A = 0$ $\theta_A = -\theta_o \frac{2C_1 C_{a1} + C_4 C_{a2}}{2 + C_{11}}$ $y_A = \frac{-\theta_o}{\beta} \frac{C_1 C_{a2} - C_2 C_{a1}}{2 + C_{11}}$
Guided	$R_A = 0 \quad \theta_A = 0$ $M_A = \theta_o EI\beta \frac{C_2 C_{a4} - 2C_3 C_{a3}}{C_{12}}$ $y_A = \frac{-\theta_o}{2\beta} \frac{2C_1 C_{a3} + C_4 C_{a4}}{C_{12}}$	$R_A = 0 \quad \theta_A = 0$ $M_A = -\theta_o EI\beta \frac{C_2 C_{a1} + C_4 C_{a3}}{C_{14}}$ $y_A = \frac{\theta_o}{2\beta} \frac{C_4 C_{a1} - C_2 C_{a3}}{C_{14}}$	$R_A = 0 \quad \theta_A = 0$ $M_A = \theta_o EI\beta \frac{C_1 C_{a4} - C_3 C_{a2}}{1 + C_{11}}$ $y_A = \frac{-\theta_o}{2\beta} \frac{C_1 C_{a2} + C_3 C_{a4}}{1 + C_{11}}$	$R_A = 0 \quad \theta_A = 0$ $M_A = -\theta_o EI\beta \frac{2C_1 C_{a1} + C_4 C_{a2}}{C_{12}}$ $y_A = \frac{\theta_o}{2\beta} \frac{2C_3 C_{a1} - C_2 C_{a2}}{C_{12}}$
Simply supported	$M_A = 0 \quad y_A = 0$ $R_A = \theta_o 2EI\beta^2 \frac{C_3 C_{a4} - C_4 C_{a3}}{C_{13}}$ $\theta_A = \theta_o \frac{C_1 C_{a4} - C_2 C_{a3}}{C_{13}}$	$M_A = 0 \quad y_A = 0$ $R_A = \theta_o 2EI\beta^2 \frac{C_1 C_{a3} - C_3 C_{a1}}{1 + C_{11}}$ $\theta_A = -\theta_o \frac{C_1 C_{a1} + C_3 C_{a3}}{1 + C_{11}}$	$M_A = 0 \quad y_A = 0$ $R_A = \theta_o EI\beta^2 \frac{C_2 C_{a4} - C_4 C_{a2}}{C_{14}}$ $\theta_A = \frac{-\theta_o}{2} \frac{C_2 C_{a2} + C_4 C_{a4}}{C_{14}}$	$M_A = 0 \quad y_A = 0$ $R_A = \theta_o 2EI\beta^2 \frac{C_1 C_{a2} - C_2 C_{a1}}{C_{13}}$ $\theta_A = \theta_o \frac{C_4 C_{a1} - C_3 C_{a2}}{C_{13}}$
Fixed	$\theta_A = 0 \quad y_A = 0$ $R_A = \theta_o 2EI\beta^2 \frac{2C_1 C_{a3} + C_4 C_{a4}}{2 + C_{11}}$ $M_A = \theta_o 2EI\beta \frac{C_1 C_{a4} - C_2 C_{a3}}{2 + C_{11}}$	$\theta_A = 0 \quad y_A = 0$ $R_A = \theta_o 2EI\beta^2 \frac{C_2 C_{a3} - C_4 C_{a1}}{C_{12}}$ $M_A = -\theta_o 2EI\beta \frac{C_1 C_{a1} + C_3 C_{a3}}{C_{12}}$	$\theta_A = 0 \quad y_A = 0$ $R_A = \theta_o 2EI\beta^2 \frac{C_1 C_{a2} + C_3 C_{a4}}{C_{13}}$ $M_A = -\theta_o EI\beta \frac{C_2 C_{a2} + C_4 C_{a4}}{C_{13}}$	$\theta_A = 0 \quad y_A = 0$ $R_A = \theta_o 2EI\beta^2 \frac{C_2 C_{a2} - 2C_3 C_{a1}}{C_{11}}$ $M_A = \theta_o 2EI\beta \frac{C_4 C_{a1} - C_3 C_{a2}}{C_{11}}$

TABLE 7 ***Shear, moment, slope, and deflection formulas for finite-length beams on elastic foundations*** ***(Cont.)***

6. Externally created concentrated lateral displacement

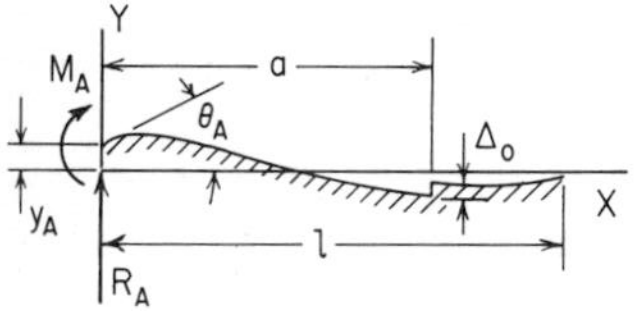

Transverse shear $= V = R_A F_1 - y_A 2EI\beta^3 F_2 - \theta_A 2EI\beta^2 F_3 - M_A \beta F_4 - \Delta_o 2EI\beta^3 F_{a2}$

Bending moment $= M = M_A F_1 + \frac{R_A}{2\beta} F_2 - y_A 2EI\beta^2 F_3 - \theta_A EI\beta F_4 - \Delta_o 2EI\beta^2 F_{a3}$

Slope $= \theta = \theta_A F_1 + \frac{M_A}{2EI\beta} F_2 + \frac{R_A}{2EI\beta^2} F_3 - y_A \beta F_4 - \Delta_o \beta F_{a4}$

Deflection $= y = y_A F_1 + \frac{\theta_A}{2\beta} F_2 + \frac{M_A}{2EI\beta^2} F_3 + \frac{R_A}{4EI\beta^3} F_4 + \Delta_o F_{a1}$

If $\beta l > 6$, see Table 8

Expressions for R_A, M_A, θ_A, and y_A are found below for several combinations of end restraints

Left end \ Right end	Free	Guided	Simply supported	Fixed
Free	$R_A = 0 \quad M_A = 0$ $\theta_A = \Delta_o 2\beta \frac{C_2C_{a3} - C_3C_{a2}}{C_{11}}$ $y_A = \Delta_o \frac{C_4C_{a2} - 2C_3C_{a3}}{C_{11}}$	$R_A = 0 \quad M_A = 0$ $\theta_A = \Delta_o \beta \frac{C_2C_{a4} - C_4C_{a2}}{C_{12}}$ $y_A = -\Delta_o \frac{C_1C_{a2} + C_3C_{A4}}{C_{12}}$	$R_A = 0 \quad M_A = 0$ $\theta_A = \Delta_o 2\beta \frac{C_1C_{a3} - C_3C_{a1}}{C_{13}}$ $y_A = \Delta_o \frac{C_4C_{a1} - C_2C_{a3}}{C_{13}}$	$R_A = 0 \quad M_A = 0$ $\theta_A = \Delta_o 2\beta \frac{C_1C_{a4} - C_4C_{a1}}{2 + C_{11}}$ $y_A = -\Delta_o \frac{2C_1C_{a1} + C_2C_{a4}}{2 + C_{11}}$
Guided	$R_A = 0 \quad \theta_A = 0$ $M_A = \Delta_o 2EI\beta^2 \frac{C_2C_{a3} - C_3C_{a2}}{C_{12}}$ $y_A = -\Delta_o \frac{C_1C_{a2} + C_4C_{a3}}{C_{12}}$	$R_A = 0 \quad \theta_A = 0$ $M_A = \Delta_o EI\beta^2 \frac{C_2C_{a4} - C_4C_{a2}}{C_{14}}$ $y_A = \frac{-\Delta_o}{2} \frac{C_2C_{a2} + C_4C_{a4}}{C_{14}}$	$R_A = 0 \quad \theta_A = 0$ $M_A = \Delta_o 2EI\beta^2 \frac{C_1C_{a3} - C_3C_{a1}}{1 + C_{11}}$ $y_A = -\Delta_o \frac{C_1C_{a1} + C_3C_{a3}}{1 + C_{11}}$	$R_A = 0 \quad \theta_A = 0$ $M_A = \Delta_o 2EI\beta^2 \frac{C_1C_{a4} - C_4C_{a1}}{C_{12}}$ $y_A = -\Delta_o \frac{C_2C_{a1} + C_3C_{a4}}{C_{12}}$
Simply supported	$M_A = 0 \quad y_A = 0$ $R_A = \Delta_o 2EI\beta^3 \frac{2C_3C_{a3} - C_4C_{a2}}{C_{13}}$ $\theta_A = \Delta_o \beta \frac{2C_1C_{a3} - C_2C_{a2}}{C_{13}}$	$M_A = 0 \quad y_A = 0$ $R_A = \Delta_o 2EI\beta^3 \frac{C_1C_{a2} + C_3C_{a4}}{1 + C_{11}}$ $\theta_A = \Delta_o \beta \frac{C_1C_{a4} - C_3C_{a2}}{1 + C_{11}}$	$M_A = 0 \quad y_A = 0$ $R_A = \Delta_o 2EI\beta^3 \frac{C_2C_{a3} - C_4C_{a1}}{C_{14}}$ $\theta_A = -\Delta_o \beta \frac{C_2C_{a1} + C_4C_{a3}}{C_{14}}$	$M_A = 0 \quad y_A = 0$ $R_A = \Delta_o 2EI\beta^3 \frac{C_2C_{a4} + 2C_1C_{a1}}{C_{13}}$ $\theta_A = -\Delta_o \beta \frac{2C_3C_{a1} + C_4C_{a4}}{C_{13}}$
Fixed	$\theta_A = 0 \quad y_A = 0$ $R_A = \Delta_o 4EI\beta^3 \frac{C_1C_{a2} + C_4C_{a3}}{2 + C_{11}}$	$\theta_A = 0 \quad y_A = 0$ $R_A = \Delta_o 2EI\beta^3 \frac{C_4C_{a4} + C_2C_{a2}}{C_{12}}$	$\theta_A = 0 \quad y_A = 0$ $R_A = \Delta_o 4EI\beta^3 \frac{C_3C_{a3} + C_1C_{a1}}{C_{13}}$	$\theta_A = 0 \quad y_A = 0$ $R_A = \Delta_o 4EI\beta^3 \frac{C_3C_{a4} + C_2C_{a1}}{C_{11}}$

7. Uniform temperature differential from top to bottom

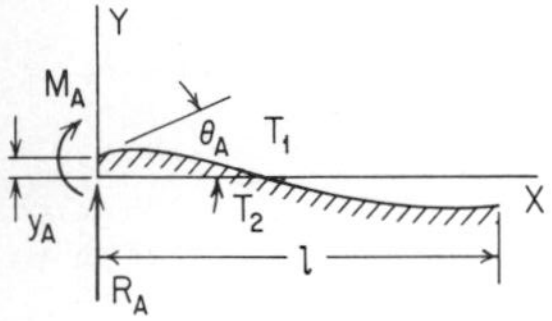

Transverse shear $= V = R_A F_1 - y_A 2EI\beta^3 F_2 - \theta_A 2EI\beta^2 F_3 - M_A \beta F_4 + \frac{T_1 - T_2}{t}\gamma EI\beta F_4$

Bending moment $= M = M_A F_1 + \frac{R_A}{2\beta}F_2 - y_A 2EI\beta^2 F_3 - \theta_A EI\beta F_4 - \frac{T_1 - T_2}{t}\gamma EI(F_1 - 1)$

Slope $= \theta = \theta_A F_1 + \frac{M_A}{2EI\beta}F_2 + \frac{R_A}{2EI\beta^2}F_3 - y_A \beta F_4 - \frac{T_1 - T_2}{2t\beta}\gamma F_2$

Deflection $= y = y_A F_1 + \frac{\theta_A}{2\beta}F_2 + \frac{M_A}{2EI\beta^2}F_3 + \frac{R_A}{4EI\beta^3}F_4 - \frac{T_1 - T_2}{2t\beta^2}\gamma F_3$

If $\beta l > 6$, see Table 8

Expressions for R_A, M_A, θ_A, and y_A are found below for several combinations of end restraints

Left end \ Right end	Free	Guided	Simply supported	Fixed
Free	$R_A = 0$ $M_A = 0$ $\theta_A = \frac{(T_1 - T_2)\gamma}{\beta t}\frac{C_1C_2 + C_3C_4 - C_2}{C_{11}}$ $y_A = \frac{-(T_1 - T_2)\gamma}{2\beta^2 t}\frac{C_4^2 + 2C_1C_3 - 2C_3}{C_{11}}$	$R_A = 0$ $M_A = 0$ $\theta_A = \frac{(T_1 - T_2)\gamma}{2\beta t}\frac{C_2^2 + C_4^2}{C_{12}}$ $y_A = \frac{-(T_1 - T_2)\gamma}{2\beta^2 t}\frac{C_2C_3 - C_1C_4}{C_{12}}$	$R_A = 0$ $M_A = 0$ $\theta_A = \frac{(T_1 - T_2)\gamma}{\beta t}\frac{C_1^2 + C_3^2 - C_1}{C_{13}}$ $y_A = \frac{-(T_1 - T_2)\gamma}{2\beta^2 t}\frac{C_1C_2 + C_3C_4 - C_2}{C_{13}}$	$R_A = 0$ $M_A = 0$ $\theta_A = \frac{(T_1 - T_2)\gamma}{\beta t}\frac{C_1C_2 + C_3C_4}{2 + C_{11}}$ $y_A = \frac{(T_1 - T_2)\gamma}{2\beta^2 t}\frac{2C_1C_3 - C_2^2}{2 + C_{11}}$
Guided	$R_A = 0$ $\theta_A = 0$ $M_A = \frac{(T_1 - T_2)\gamma EI}{t}\frac{C_1C_2 + C_3C_4 - C_2}{C_{12}}$ $y_A = \frac{-(T_1 - T_2)\gamma}{2\beta^2 t}\frac{C_4}{C_{12}}$	$R_A = 0$ $\theta_A = 0$ $M_A = \frac{(T_1 - T_2)\gamma EI}{t}$ $y_A = 0$	$R_A = 0$ $\theta_A = 0$ $M_A = \frac{(T_1 - T_2)\gamma EI}{t}\frac{C_1^2 + C_3^2 - C_1}{1 + C_{11}}$ $y_A = \frac{(T_1 - T_2)\gamma}{2\beta^2 t}\frac{C_3}{1 + C_{11}}$	$R_A = 0$ $\theta_A = 0$ $M_A = \frac{(T_1 - T_2)\gamma EI}{t}$ $y_A = 0$
Simply supported	$M_A = 0$ $y_A = 0$ $R_A = \frac{(T_1 - T_2)\gamma\beta EI}{t}\frac{2C_1C_3 + C_4^2}{C_{13}}$ $\theta_A = \frac{(T_1 - T_2)\gamma}{2\beta t}\frac{2C_1^2 + C_2C_4 - 2C_1}{C_{13}}$	$M_A = 0$ $y_A = 0$ $R_A = \frac{(T_1 - T_2)\gamma\beta EI}{t}\frac{C_2C_3 - C_1C_4}{1 + C_{11}}$ $\theta_A = \frac{(T_1 - T_2)\gamma}{2\beta t}\frac{C_1C_2 + C_3C_4}{1 + C_{11}}$	$M_A = 0$ $y_A = 0$ $R_A = \frac{(T_1 - T_2)\gamma\beta EI}{t}\frac{C_1C_2 + C_3C_4 - C_2}{C_{14}}$ $\theta_A = \frac{(T_1 - T_2)\gamma}{2\beta t}\frac{C_2C_3 - C_1C_4 + C_4}{C_{14}}$	$M_A = 0$ $y_A = 0$ $R_A = \frac{(T_1 - T_2)\gamma\beta EI}{t}\frac{C_2^2 - 2C_1C_3}{C_{13}}$ $\theta_A = \frac{(T_1 - T_2)\gamma}{2\beta t}\frac{2C_3^2 - C_2C_4}{C_{13}}$
Fixed	$\theta_A = 0$ $y_A = 0$ $R_A = \frac{(T_1 - T_2)\gamma 2\beta EI}{t}\frac{-C_4}{2 + C_{11}}$ $M_A = \frac{(T_1 - T_2)\gamma EI}{t}\frac{2C_1^2 + C_2C_4 - 2C_1}{2 + C_{11}}$	$\theta_A = 0$ $y_A = 0$ $R_A = 0$ $M_A = \frac{(T_1 - T_2)\gamma EI}{t}$	$\theta_A = 0$ $y_A = 0$ $R_A = \frac{(T_1 - T_2)\gamma\beta EI}{t}\frac{-2C_3}{C_{13}}$ $M_A = \frac{(T_1 - T_2)\gamma EI}{t}\frac{C_2C_3 - C_1C_4 + C_4}{C_{13}}$	$\theta_A = 0$ $y_A = 0$ $R_A = 0$ $M_A = \frac{(T_1 - T_2)\gamma EI}{t}$

TABLE 8 *Shear, moment, slope, and deflection formulas for semi-infinite beams on elastic foundations*

NOTATION: All notation is the same as that for Table 7. No length is defined since these beams are assumed to extend from the left end, for which restraints are defined, to a length beyond that portion affected by the loading. Note that M_A and R_A are reactions, not applied loads.

The following constants and functions, involving both beam constants and foundation constants, are hereby defined in order to permit condensing the tabulated formulas which follow

k_o = foundation modulus (pounds per square inch per inch); b_o = beam width (inches); and $\beta = \left(\frac{b_o k_o}{4EI}\right)^{\frac{1}{4}}$. (*Note:* See page 94 for a definition of $\langle x-a\rangle^n$)

$F_1 = \text{Cosh}\,\beta x \cos \beta x$
$F_2 = \text{Cosh}\,\beta x \sin \beta x + \text{Sinh}\,\beta x \cos \beta x$
$F_3 = \text{Sinh}\,\beta x \sin \beta x$
$F_4 = \text{Cosh}\,\beta x \sin \beta x - \text{Sinh}\,\beta x \cos \beta x$

$A_1 = 0.5e^{-\beta a} \cos \beta a$
$A_2 = 0.5e^{-\beta a} (\sin \beta a - \cos \beta a)$
$A_3 = -0.5e^{-\beta a} \sin \beta a$
$A_4 = 0.5e^{-\beta a} (\sin \beta a + \cos \beta a)$

$B_1 = 0.5e^{-\beta b} \cos \beta b$
$B_2 = 0.5e^{-\beta b} (\sin \beta b - \cos \beta b)$
$B_3 = -0.5e^{-\beta b} \sin \beta b$
$B_4 = 0.5e^{-\beta b} (\sin \beta b + \cos \beta b)$

$F_{a1} = \langle x-a\rangle^0 \,\text{Cosh}\,\beta\langle x-a\rangle \cos \beta\langle x-a\rangle$
$F_{a2} = \text{Cosh}\,\beta\langle x-a\rangle \sin \beta\langle x-a\rangle + \text{Sinh}\,\beta\langle x-a\rangle \cos \beta\langle x-a\rangle$
$F_{a3} = \text{Sinh}\,\beta\langle x-a\rangle \sin \beta\langle x-a\rangle$
$F_{a4} = \text{Cosh}\,\beta\langle x-a\rangle \sin \beta\langle x-a\rangle - \text{Sinh}\,\beta\langle x-a\rangle \cos \beta\langle x-a\rangle$
$F_{a5} = \langle x-a\rangle^0 - F_{a1}$
$F_{a6} = 2\beta(x-a)\langle x-a\rangle^0 - F_{a2}$

$F_{b1} = \langle x-b\rangle^0 \,\text{Cosh}\,\beta\langle x-b\rangle \cos \beta\langle x-b\rangle$
$F_{b2} = \text{Cosh}\,\beta\langle x-b\rangle \sin \beta\langle x-b\rangle + \text{Sinh}\,\beta\langle x-b\rangle \cos \beta\langle x-b\rangle$
$F_{b3} = \text{Sinh}\,\beta\langle x-b\rangle \sin \beta\langle x-b\rangle$
$F_{b4} = \text{Cosh}\,\beta\langle x-b\rangle \sin \beta\langle x-b\rangle - \text{Sinh}\,\beta\langle x-b\rangle \cos \beta\langle x-b\rangle$
$F_{b5} = \langle x-b\rangle^0 - F_{b1}$
$F_{b6} = 2\beta(x-a)\langle x-b\rangle^0 - F_{b2}$

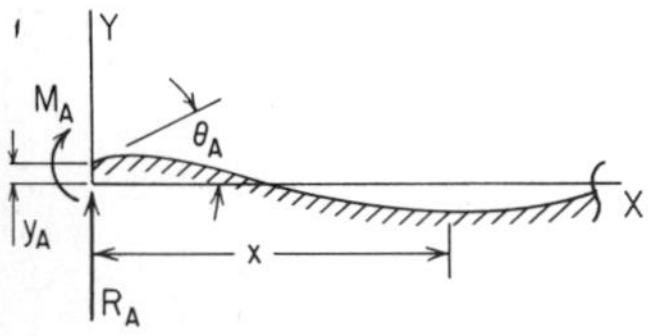

Transverse shear $= V = R_A F_1 - y_A 2EI\beta^3 F_2 - \theta_A 2EI\beta^2 F_3 - M_A \beta F_4 + LT_V$

Bending moment $= M = M_A F_1 + \frac{R_A}{2\beta} F_2 - y_A 2EI\beta^2 F_3 - \theta_A EI\beta F_4 + LT_M$

Slope $= \theta = \theta_A F_1 + \frac{M_A}{2EI\beta} F_2 + \frac{R_A}{2EI\beta^2} F_3 - y_A \beta F_4 + LT_\theta$

Deflection $= y = y_A F_1 + \frac{\theta_A}{2\beta} F_2 + \frac{M_A}{2EI\beta^2} F_3 + \frac{R_A}{4EI\beta^3} F_4 + LT_y$

Expressions for R_A, M_A, θ_A, and y_A are found below for several combinations of loading and left end restraints. The loading terms LT_V, LT_M, LT_θ, and LT_y are given for each loading condition.

Loading, reference no. \ Left end restraint	Free	Guided	Simply supported	Fixed	Loading terms
1. Concentrated intermediate load (if $a = 0$, see case 8; if $\beta a > 3$, see case 10) 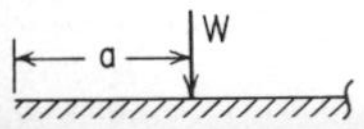	$R_A = 0 \quad M_A = 0$ $\theta_A = \frac{-W}{EI\beta^2} A_2$ $y_A = \frac{-W}{EI\beta^3} A_1$	$R_A = 0 \quad \theta_A = 0$ $M_A = \frac{-W}{\beta} A_2$ $y_A = \frac{-W}{2EI\beta^3} A_4$	$M_A = 0 \quad y_A = 0$ $R_A = 2WA_1$ $\theta_A = \frac{W}{EI\beta^2} A_3$	$\theta_A = 0 \quad y_A = 0$ $R_A = 2WA_4$ $M_A = \frac{2W}{\beta} A_3$	$LT_V = -WF_{a1}$ $LT_M = \frac{-W}{2\beta} F_{a2}$ $LT_\theta = \frac{-W}{2EI\beta^2} F_{a3}$ $LT_y = \frac{-W}{4EI\beta^3} F_{a4}$

from a to b 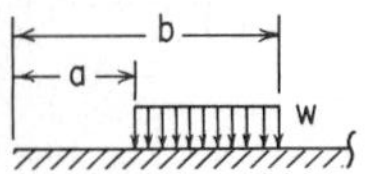	[illegible] $\theta_A = \frac{-w}{EI\beta^3}(B_3 - A_3)$ $y_A = \frac{-w}{2EI\beta^4}(B_2 - A_2)$	[illegible] $M_A = \frac{-w}{\beta^2}(B_3 - A_3)$ $y_A = \frac{w}{2EI\beta^4}(B_1 - A_1)$	$M_A = 0 \quad y_A = 0$ $R_A = \frac{w}{\beta}(B_2 - A_2$ $\theta_A = \frac{w}{2EI\beta^3}(B_4 - A_4)$	$\theta_A = 0 \quad y_A = 0$ $R_A = \frac{-2w}{\beta}(B_1 - A_1)$ $M_A = \frac{w}{\beta^2}(B_4 - A_4)$	$LT_V = \frac{w}{2\beta}(F_{a2} - F_{b2})$ $LT_M = \frac{-w}{2\beta^2}(F_{a3} - F_{b3})$ $LT_\theta = \frac{-w}{4EI\beta^3}(F_{a4} - F_{b4})$ $LT_y = \frac{-w}{4EI\beta^4}(F_{a5} - F_{b5})$
3. Uniformly increasing load from a to b 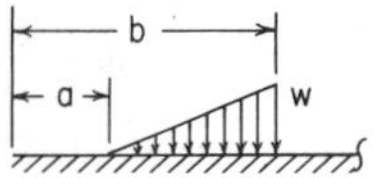	$R_A = 0 \quad M_A = 0$ $\theta_A = \frac{w}{2EI\beta^4}\left(\frac{B_4 - A_4}{b - a} - 2\beta B_3\right)$ $y_A = \frac{w}{2EI\beta^5}\left(\frac{B_3 - A_3}{b - a} - \beta B_2\right)$	$R_A = 0 \quad \theta_A = 0$ $M_A = \frac{w}{2\beta^3}\left(\frac{B_4 - A_4}{b - a} - 2\beta b_3\right)$ $y_A = \frac{-w}{4EI\beta^5}\left(\frac{B_2 - A_2}{b - a} - 2\beta B_1\right)$	$M_A = 0 \quad y_A = 0$ $R_A = \frac{-w}{\beta^2}\left(\frac{B_3 - A_3}{b - a} - \beta B_2\right)$ $\theta_A = \frac{w}{2EI\beta^4}\left(\frac{B_1 - A_1}{b - a} + \beta B_4\right)$	$\theta_A = 0 \quad y_A = 0$ $R_A = \frac{w}{\beta^2}\left(\frac{B_2 - A_2}{b - a} - 2\beta B_1\right)$ $M_A = \frac{w}{\beta^3}\left(\frac{B_1 - A_1}{b - a} + \beta B_4\right)$	$LT_V = \frac{-w}{2\beta^2}\left(\frac{F_{a3} - F_{b3}}{b - a} - \beta F_{b2}\right)$ $LT_M = \frac{-w}{4\beta^3}\left(\frac{F_{a4} - F_{b4}}{b - a} - 2\beta F_{b3}\right)$ $LT_\theta = \frac{-w}{4EI\beta^4}\left(\frac{F_{a5} - F_{b5}}{b - a} - \beta F_{b4}\right)$ $LT_y = \frac{-w}{8EI\beta^5}\left(\frac{F_{a6} - F_{b6}}{b - a} + 2\beta F_{b1}\right)$
4. Concentrated intermediate moment (if $a = 0$, see case 9; if $\beta a > 3$, see case 11) 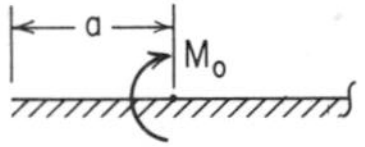	$R_A = 0 \quad M_A = 0$ $\theta_A = \frac{-2M_o}{EI\beta}A_1$ $y_A = \frac{M_o}{EI\beta^2}A_4$	$R_A = 0 \quad \theta_A = 0$ $M_A = -2M_oA_1$ $y_A = \frac{-M_o}{EI\beta^2}A_3$	$M_A = 0 \quad y_A = 0$ $R_A = -2M_o\beta A_4$ $\theta_A = \frac{M_o}{EI\beta}A_2$	$\theta_A = 0 \quad y_A = 0$ $R_A = 4M_o\beta A_3$ $M_A = 2M_oA_2$	$LT_V = -M_o\beta F_{a4}$ $LT_M = M_oF_{a1}$ $LT_\theta = \frac{M_o}{2EI\beta}F_{a2}$ $LT_y = \frac{M_o}{2EI\beta^2}F_{a3}$
5. Externally created concentrated angular displacement 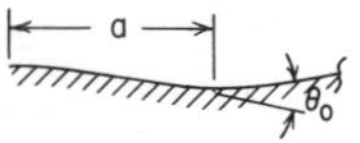	$R_A = 0 \quad M_A = 0$ $\theta_A = -2\theta_oA_4$ $y_A = \frac{-2\theta_o}{\beta}A_3$	$R_A = 0 \quad \theta_A = 0$ $M_A = -2\theta_oEI\beta A_4$ $y_A = \frac{\theta_o}{\beta}A_2$	$M_A = 0 \quad y_A = 0$ $R_A = 4\theta_oEI\beta^2A_3$ $\theta_A = -2\theta_oA_1$	$\theta_A = 0 \quad y_A = 0$ $R_A = -4\theta_oEI\beta^2A_2$ $M_A = -4\theta_oEI\beta A_1$	$LT_V = -2\theta_oEI\beta^2F_{a3}$ $LT_M = -\theta_oEI\beta F_{a4}$ $LT_\theta = \theta_oF_{a1}$ $LT_y = \frac{\theta_o}{2\beta}F_{a2}$
6. Externally created concentrated lateral displacement 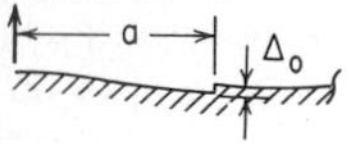	$R_A = 0 \quad M_A = 0$ $\theta_A = 4\Delta_o\beta A_3$ $y_A = 2\Delta_oA_2$	$R_A = 0 \quad \theta_A = 0$ $M_A = 4\Delta_oEI\beta^2A_3$ $y_A = -2\Delta_oA_1$	$M_A = 0 \quad y_A = 0$ $R_A = -4\Delta_oEI\beta^3A_2$ $\theta_A = -2\Delta_o\beta A_4$	$\theta_A = 0 \quad y_A = 0$ $R_A = 8\Delta_oEI\beta^3A_1$ $M_A = -4\Delta_oEI\beta^2A_4$	$LT_V = -2\Delta_oEI\beta^3F_{a2}$ $LT_M = -2\Delta_oEI\beta^2F_{a3}$ $LT_\theta = -\Delta_o\beta F_{a4}$ $LT_y = \Delta_oF_{a1}$

TABLE 8 ***Shear, moment, slope, and deflection formulas for semi-infinite beams on elastic foundations (Cont.)***

Loading, reference no. \ Left end restraint	Free	Guided	Simply supported	Fixed	Loading terms
7. Uniform temperature differential from top to bottom 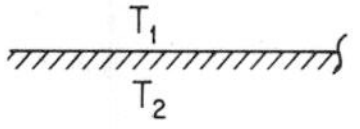	$R_A = 0 \quad M_A = 0$ $\theta_A = \frac{T_1 - T_2}{t\beta}\gamma$ $y_A = -\frac{T_1 - T_2}{2t\beta^2}\gamma$	$R_A = 0 \quad \theta_A = 0$ $M_A = \frac{T_1 - T_2}{t}\gamma EI$ $y_A = 0$	$M_A = 0 \quad y_A = 0$ $R_A = \frac{T_1 - T_2}{t}\gamma EI\beta$ $\theta_A = \frac{T_1 - T_2}{2t\beta}\gamma$	$\theta_A = 0 \quad y_A = 0$ $R_A = 0$ $M_A = \frac{T_1 - T_2}{t}\gamma EI$	$LT_V = \frac{T_1 - T_2}{t}\gamma EI\beta F_4$ $LT_M = \frac{T_1 - T_2}{t}\gamma EI(1 - F_1)$ $LT_\theta = -\frac{T_1 - T_2}{2t\beta}\gamma F_2$ $LT_y = -\frac{T_1 - T_2}{2t\beta^2}\gamma F_3$

Simple loads on semi-infinite and on infinite beams on elastic foundations

Loading, reference no.	Shear, moment, and deformation equations	Selected maximum values
8. Concentrated end load on a semi-infinite beam, left end free 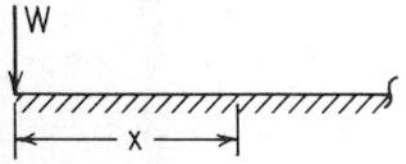	$V = -We^{-\beta x}(\cos\beta x - \sin\beta x)$ $M = -\frac{W}{\beta}e^{-\beta x}\sin\beta x$ $\theta = \frac{W}{2EI\beta^2}e^{-\beta x}(\cos\beta x + \sin\beta x)$ $y = -\frac{W}{2EI\beta^3}e^{-\beta x}\cos\beta x$	Max $V = -W$ at $x = 0$ Max $M = -0.3225\frac{W}{\beta}$ at $x = \frac{\pi}{4\beta}$ Max $\theta = \frac{W}{2EI\beta^2}$ at $x = 0$ Max $y = \frac{-W}{2EI\beta^3}$ at $x = 0$
9. Concentrated end moment on a semi-infinite beam, left end free 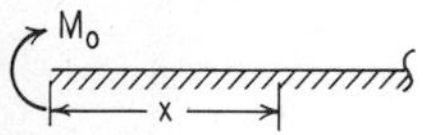	$V = -M_o\beta e^{-\beta x}\sin\beta x$ $M = M_o e^{-\beta x}(\cos\beta x + \sin\beta x)$ $\theta = -\frac{M_o}{EI\beta}e^{-\beta x}\cos\beta x$ $y = -\frac{M_o}{2EI\beta^2}e^{-\beta x}(\sin\beta x - \cos\beta x)$	Max $V = -0.3225M_o\beta$ at $x = \frac{\pi}{4\beta}$ Max $M = M_o$ at $x = 0$ Max $\theta = -\frac{M_o}{EI\beta}$ at $x = 0$ Max $y = \frac{M_o}{2EI\beta^2}$ at $x = 0$

10. Concentrated load on an infinite beam 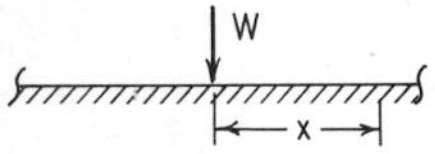	$V = -\frac{W}{2}e^{-\beta x}\cos\beta x$ $M = \frac{W}{4\beta}e^{-\beta x}(\cos\beta x - \sin\beta x)$ $\theta = \frac{W}{8EI\beta^2}e^{-\beta x}\sin\beta x$ $y = -\frac{W}{8EI\beta^3}e^{-\beta x}(\cos\beta x + \sin\beta x)$	Max $V = -\frac{W}{2}$ at $x = 0$ Max $M = \frac{W}{4\beta}$ at $x = 0$ Max $\theta = 0.0403\frac{W}{EI\beta^2}$ at $x = \frac{\pi}{4\beta}$ Max $y = -\frac{W}{8EI\beta^3}$ at $x = 0$
11. Concentrated moment on an infinite beam 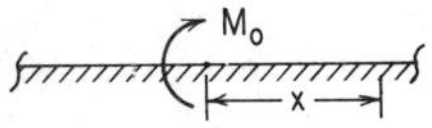	$V = -\frac{M_o\beta}{2}e^{-\beta x}(\cos\beta x + \sin\beta x)$ $M = \frac{M_o}{2}e^{-\beta x}\cos\beta x$ $\theta = -\frac{M_o}{4EI\beta}e^{-\beta x}(\cos\beta x - \sin\beta x)$ $y = -\frac{M_o}{8EI\beta^2}e^{-\beta x}\sin\beta x$	Max $V = -\frac{M_o\beta}{2}$ at $x = 0$ Max $M = \frac{M_o}{2}$ at $x = 0$ Max $\theta = -\frac{M_o}{4EI\beta}$ at $x = 0$ Max $y = -0.0403\frac{M_o}{EI\beta^2}$ at $x = \frac{\pi}{4\beta}$

Similarly,

$$y_A = -0.01172 \text{ in}$$

An examination of the equation for the transverse shear V shows that the value of the shear passes through zero at $x = 15$, 30, and 45 in. The maximum positive bending moment occurs at $x = 15$ in and is evaluated as follows, noting again that R_A and M_A are zero:

$$\begin{aligned} M_{15} = &-(-0.01172)(2)(9.5 \cdot 10^6)(2)(0.06666^2)[\text{Sinh } (0.06666)(15) \text{ sin } 1] \\ &-(-0.007582)(9.5 \cdot 10^6)(2)(0.06666)(\text{Cosh } 1 \text{ sin } 1 - \text{Sinh } 1 \text{ cos } 1) \\ = &\ 8{,}330 \text{ in-lb} \end{aligned}$$

Similarly, the maximum negative moment at $x = 30$ in is evaluated, making sure that the terms for the concentrated load at 15 in and the uniformly distributed load from 15 to 30 in are included:

$$M_{30} = -13{,}000 \text{ in-lb}$$

The maximum bending stress is given by $\sigma = Mc/I$ and is found to be 6500 lb/in^2

7.6 *Deformation due to the elasticity of fixed supports*

The formulas in Tables 3, 4, 7, 8, and 10 to 12 which apply to those cases where fixed or guided end supports are specified are based on the assumption that the support is rigid and holds the fixed or guided end truly horizontal or vertical. The slight deformation that actually occurs at the support permits the beam to assume there a slope $\Delta\theta$, which for the conditions represented in Fig. 7.11, that is, a beam integral with a semi-infinite supporting foundation, is given by

$$\Delta\theta = \frac{16.67M}{Eh_1^2} + \frac{(1-\nu)V}{Eh_1}$$

Here M is the bending moment and V is the shear per unit width of the beam at the support in inch-pounds per inch and pounds per inch, respectively; E is the modulus of elasticity, and ν is Poisson's ratio for the foundation material; and $h_1 = h + 1.5r$ (Ref. 64). The effect of this deformation is to increase the deflections of the beam. For a cantilever, this increase is simply $x\Delta\theta$, but for other support conditions the concept of the externally created angular deformation may be utilized (see Example 2 on page 127).

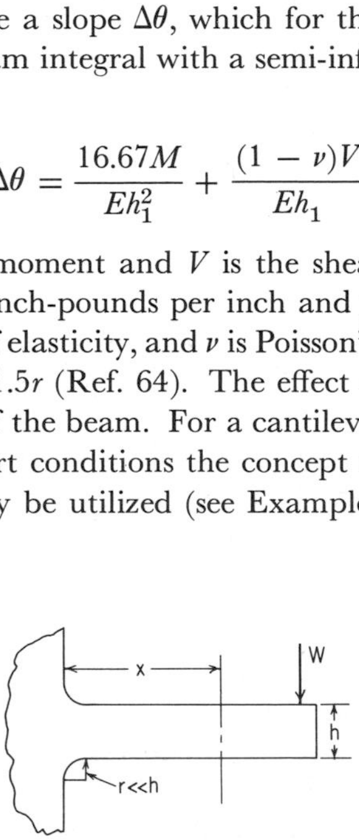

Fig. 7.11

7.7 *Beams under simultaneous axial and transverse loading*

Under certain conditions a beam may be subjected to axial tension or compression in addition to the transverse loads; examples are afforded by airplane-wing spars and the chord members of a bridge truss. Axial tension tends to straighten the beam and thus reduce the bending moments produced by the transverse loads, but axial compression has the opposite effect and may greatly increase the maximum bending moment and deflection. In either case solution cannot be effected by simple superposition but must be arrived at by methods that take into account the change in deflection produced by the axial load.

For any condition of loading, the maximum normal stress in an extreme fiber is given by

$$\text{Max } \sigma = \frac{P}{A} \pm \frac{Mc}{I} \tag{16}$$

where P is the axial load (positive if tensile and negative if compressive), A is the cross-sectional area of the beam, I/c is the section modulus, and M is the maximum bending moment due to the combined effect of axial and transverse loads. (Use the plus sign if M causes tension at the point in question and the minus sign if M causes compression.)

It is the determination of M that offers difficulty. For some cases, especially if P is small or tensile, it is permissible to ignore the small additional moment caused by P and to take M equal to M', the bending moment due to transverse loads only. Approximate formulas of the type (Ref. 33)

$$y_{\max} = \frac{y'_{\max}}{1 \pm \alpha_y P^2/EI} \qquad \theta_{\max} = \frac{\theta'_{\max}}{1 \pm \alpha_\theta P^2/EI} \qquad M_{\max} = \frac{M'_{\max}}{1 \pm \alpha_M P^2/EI} \tag{17}$$

have been used, but the values of α_y, α_θ, and α_M are different for each loading and each manner of supporting the beam.

Instead of tabulating the values of α, which give answers with increasing error as P increases, Table 9*a* to 9*d* gives values of the coefficient C_P which can be used in the expressions

$$y_A = C_P y'_A \qquad \theta_A = C_P \theta'_A \qquad M_A = C_p M'_A \qquad \text{etc.} \tag{18}$$

where the primed values refer to the laterally loaded beam without the axial load and can be evaluated from expressions found in Table 3. For those cases listed where the reactions are statically indeterminate, the reaction coefficients given will enable the remaining reactions to be evaluated by applying the principles of static equilibrium. The given values of C_P are exact, based on the assumption that deflections due to transverse shear are negligible. This same assumption was used in developing the equations for trans-

TABLE 9a ***Reaction and deflection coefficients for beams under simultaneous axial and transverse loading: cantilever end support***

Case no. in Table 3	Load location a/l	Coefficient listed for	Axial compressive load, $kl = \sqrt{Pl^2/EI}$					Axial tensile load, $kl = \sqrt{Pl^2/EI}$				
			0.2	0.4	0.6	0.8	1.0	0.5	1.0	2.0	4.0	8.0
1a. Conc. load	0	y_A	1.0163	1.0684	1.1686	1.3455	1.6722	0.9092	0.7152	0.3885	0.1407	0.0410
		θ_A	1.0169	1.0713	1.1757	1.3604	1.7016	0.9054	0.7039	0.3671	0.1204	0.0312
		M_B	1.0136	1.0570	1.1402	1.2870	1.5574	0.9242	0.7616	0.4820	0.2498	0.1250
	0.5	y_A	1.0153	1.0646	1.1589	1.3256	1.6328	0.9142	0.7306	0.4180	0.1700	0.0566
		θ_A	1.0195	1.0821	1.2026	1.4163	1.8126	0.8914	0.6617	0.2887	0.0506	0.0022
		M_B	1.0085	1.0355	1.0869	1.1767	1.3402	0.9524	0.8478	0.6517	0.4333	0.2454
2a. Uniform load	0	y_A	1.0158	1.0665	1.1638	1.3357	1.6527	0.9117	0.7228	0.4031	0.1552	0.0488
		θ_A	1.0183	1.0771	1.1900	1.3901	1.7604	0.8980	0.6812	0.3243	0.0800	0.0117
		M_B	1.0102	1.0427	1.1047	1.2137	1.4132	0.9430	0.8193	0.5969	0.3792	0.2188
	0.5	y_A	1.0150	1.0629	1.1548	1.3171	1.6161	0.9164	0.7373	0.4314	0.1851	0.0667
		θ_A	1.0198	1.0835	1.2062	1.4239	1.8278	0.8896	0.6562	0.2794	0.0447	0.0015
		M_B	1.0059	1.0248	1.0606	1.1229	1.2357	0.9666	0.8925	0.7484	0.5682	0.3773
2a. Uniformly increasing	0	y_A	1.0155	1.0652	1.1604	1.3287	1.6389	0.9135	0.7283	0.4137	0.1662	0.0552
		θ_A	1.0190	1.0799	1.1972	1.4051	1.7902	0.8942	0.6700	0.3039	0.0629	0.0057
		M_B	1.0081	1.0341	1.0836	1.1701	1.3278	0.9543	0.8543	0.6691	0.4682	0.2930
	0.5	y_A	1.0147	1.0619	1.1523	1.3118	1.6056	0.9178	0.7415	0.4400	0.1951	0.0740
		θ_A	1.0200	1.0843	1.2080	1.4277	1.8355	0.8887	0.6535	0.2748	0.0419	0.0012
		M_B	1.0046	1.0191	1.0467	1.0944	1.1806	0.9742	0.9166	0.8020	0.6489	0.4670
2a. Uniformly decreasing	0	y_A	1.0159	1.0670	1.1650	1.3382	1.6578	0.9110	0.7208	0.3992	0.1512	0.0465
		θ_A	1.0181	1.0761	1.1876	1.3851	1.7505	0.8992	0.6850	0.3311	0.0857	0.0136
		M_B	1.0112	1.0469	1.1153	1.2355	1.4559	0.9374	0.8018	0.5609	0.3348	0.1817
	0.5	y_A	1.0150	1.0633	1.1557	1.3189	1.6197	0.9159	0.7358	0.4284	0.1816	0.0642
		θ_A	1.0198	1.0833	1.2056	1.4226	1.8253	0.8899	0.6571	0.2809	0.0456	0.0016
		M_B	1.0066	1.0276	1.0676	1.1372	1.2632	0.9628	0.8804	0.7215	0.5279	0.3324
3a. Conc. moment	0	y_A	1.0169	1.0713	1.1757	1.3604	1.7016	0.9054	0.7039	0.3671	0.1204	0.0312
		θ_A	1.0136	1.0570	1.1402	1.2870	1.5574	0.9242	0.7616	0.4820	0.2498	0.1250
		M_B	1.0203	1.0857	1.2116	1.4353	1.8508	0.8868	0.6481	0.2658	0.0366	0.0007
	0.5	y_A	1.0161	1.0677	1.1668	1.3418	1.6646	0.9101	0.7180	0.3932	0.1437	0.0409
		θ_A	1.0186	1.0785	1.1935	1.3974	1.7747	0.8961	0.6754	0.3124	0.0664	0.0046
		M_B	1.0152	1.0641	1.1575	1.3220	1.6242	0.9147	0.7308	0.4102	0.1378	0.0183

TABLE 9b *Reaction and deflection coefficients for beams under simultaneous axial and transverse loading: simply supported ends*

Case no. in Table 3	Load location a/l	Coefficient listed for	Axial compressive load, $kl = \sqrt{Pl^2/EI}$					Axial tensile load, $kl = \sqrt{Pl^2/EI}$				
			0.4	0.8	1.2	1.6	2.0	1.0	2.0	4.0	8.0	12.0
1e. Conc. load	0.25	$y_{l/2}$	1.0167	1.0702	1.1729	1.3546	1.6902	0.9069	0.7082	0.3751	0.1273	0.0596
		θ_A	1.0144	1.0605	1.1485	1.3031	1.5863	0.9193	0.7447	0.4376	0.1756	0.0889
		θ_B	1.0185	1.0779	1.1923	1.3958	1.7744	0.8972	0.6805	0.3311	0.0990	0.0444
		$M_{l/4}$	1.0101	1.0425	1.1039	1.2104	1.4025	0.9427	0.8158	0.5752	0.3272	0.2217
	0.50	$y_{l/2}$	1.0163	1.0684	1.1686	1.3455	1.6722	0.9092	0.7152	0.3885	0.1407	0.0694
		θ_A	1.0169	1.0713	1.1757	1.3604	1.7016	0.9054	0.7039	0.3671	0.1204	0.0553
		$M_{l/2}$	1.0136	1.0570	1.1402	1.2870	1.5574	0.9242	0.7616	0.4820	0.2498	0.1667
2e. Uniform load	0	$y_{l/2}$	1.0165	1.0696	1.1714	1.3515	1.6839	0.9077	0.7107	0.3797	0.1319	0.0630
		θ_A	1.0163	1.0684	1.1686	1.3455	1.6722	0.9092	0.7152	0.3885	0.1407	0.0694
		$M_{l/2}$	1.0169	1.0713	1.1757	1.3604	1.7016	0.9054	0.7039	0.3671	0.1204	0.0553
	0.50	$y_{l/2}$	1.0165	1.0695	1.1714	1.3515	1.6839	0.9077	0.7107	0.3797	0.1319	0.0630
		θ_A	1.0180	1.0759	1.1873	1.3851	1.7524	0.8997	0.6875	0.3418	0.1053	0.0475
		θ_B	1.0149	1.0625	1.1540	1.3147	1.6099	0.9166	0.7368	0.4248	0.1682	0.0865
		$M_{l/2}$	1.0169	1.0713	1.1757	1.3604	1.7016	0.9054	0.7039	0.3671	0.1204	0.0553
2e. Uniformly increasing	0	$y_{l/2}$	1.0165	1.0696	1.1714	1.3515	1.6839	0.9077	0.7107	0.3797	0.1319	0.0630
		θ_A	1.0172	1.0722	1.1781	1.3656	1.7127	0.9044	0.7011	0.3643	0.1214	0.0570
		θ_B	1.0155	1.0651	1.1603	1.3280	1.6368	0.9134	0.7276	0.4097	0.1575	0.0803
		$M_{l/2}$	1.0169	1.0713	1.1757	1.3604	1.7016	0.9054	0.7039	0.3671	0.1204	0.0553
	0.50	$y_{l/2}$	1.0167	1.0702	1.1729	1.3545	1.6899	0.9069	0.7084	0.3754	0.1278	0.0601
		θ_A	1.0184	1.0776	1.1915	1.3942	1.7710	0.8976	0.6816	0.3329	0.1002	0.0450
		θ_B	1.0140	1.0588	1.1445	1.2948	1.5702	0.9215	0.7516	0.4521	0.1936	0.1048
		$M_{l/2}$	1.0183	1.0771	1.1900	1.3901	1.7604	0.8980	0.6812	0.3243	0.0800	0.0270
3e. Conc. moment	0	$y_{l/2}$	1.0169	1.0713	1.1757	1.3604	1.7016	0.9054	0.7039	0.3671	0.1204	0.0553
		θ_A	1.0108	1.0454	1.1114	1.2266	1.4365	0.9391	0.8060	0.5630	0.3281	0.2292
		θ_B	1.0190	1.0801	1.1979	1.4078	1.7993	0.8945	0.6728	0.3200	0.0932	0.0417
	0.25	$y_{l/2}$	1.0161	1.0677	1.1668	1.3418	1.6646	0.9101	0.7180	0.3932	0.1437	0.0704
		θ_A	1.0202	1.0852	1.2102	1.4318	1.8424	0.8873	0.6485	0.2595	0.0113	−0.0244
		θ_B	1.0173	1.0728	1.1795	1.3582	1.7174	0.9035	0.6982	0.3571	0.1131	0.0512

TABLE 9c ***Reaction and deflection coefficients for beams under simultaneous axial and transverse loading: left end simply supported, right end fixed***

Case no. in Table 3	Load location a/l	Coefficient listed for	Axial compressive load, $kl = \sqrt{Pl^2/EI}$					Axial tensile load, $kl = \sqrt{Pl^2/EI}$				
			0.6	1.2	1.8	2.4	3.0	1.0	2.0	4.0	8.0	12.0
1c. Conc. load	0.25	$y_{l/2}$	1.0190	1.0804	1.2005	1.4195	1.8478	0.9507	0.8275	0.5417	0.2225	0.1108
		θ_A	1.0172	1.0726	1.1803	1.3753	1.7530	0.9553	0.8429	0.5762	0.2576	0.1338
		M_B	1.0172	1.0728	1.1818	1.3812	1.7729	0.9554	0.8443	0.5881	0.3018	0.1940
	0.50	$y_{l/2}$	1.0170	1.0719	1.1786	1.3718	1.7458	0.9557	0.8444	0.5802	0.2647	0.1416
		θ_A	1.0199	1.0842	1.2101	1.4406	1.8933	0.9485	0.8202	0.5255	0.2066	0.1005
		M_B	1.0137	1.0579	1.1432	1.2963	1.5890	0.9642	0.8733	0.6520	0.3670	0.2412
2c. Uniform load	0	$y_{l/2}$	1.0176	1.0742	1.1846	1.3848	1.7736	0.9543	0.8397	0.5694	0.2524	0.1323
		θ_A	1.0183	1.0776	1.1933	1.4042	1.8162	0.9524	0.8334	0.5561	0.2413	0.1263
		M_B	1.0122	1.0515	1.1273	1.2635	1.5243	0.9681	0.8874	0.6900	0.4287	0.3033
	0.50	$y_{l/2}$	1.0163	1.0689	1.1709	1.3549	1.7094	0.9575	0.8502	0.5932	0.2778	0.1505
		θ_A	1.0202	1.0856	1.2139	1.4496	1.9147	0.9477	0.8179	0.5224	0.2087	0.1048
		M_B	1.0091	1.0383	1.0940	1.1920	1.3744	0.9760	0.9141	0.7545	0.5126	0.3774
2c. Uniformly increasing	0	$y_{l/2}$	1.0170	1.0719	1.1785	1.3716	1.7453	0.9557	0.8444	0.5799	0.2637	0.1405
		θ_A	1.0192	1.0814	1.2030	1.4255	1.8619	0.9502	0.8259	0.5394	0.2237	0.1136
		M_B	1.0105	1.0440	1.1084	1.2230	1.4399	0.9726	0.9028	0.7277	0.4799	0.3504
	0.50	$y_{l/2}$	1.0160	1.0674	1.1669	1.3463	1.6911	0.9584	0.8533	0.6003	0.2860	0.1571
		θ_A	1.0202	1.0855	1.2138	1.4499	1.9165	0.9478	0.8183	0.5245	0.2141	0.1105
		M_B	1.0071	1.0298	1.0726	1.1473	1.2843	0.9813	0.9325	0.8029	0.5900	0.4573
2c. Uniformly decreasing	0	$y_{l/2}$	1.0180	1.0762	1.1895	1.3957	1.7968	0.9532	0.8359	0.5608	0.2431	0.1256
		θ_A	1.0177	1.0751	1.1868	1.3900	1.7857	0.9539	0.8383	0.5673	0.2531	0.1347
		M_B	1.0142	1.0600	1.1489	1.3098	1.6207	0.9630	0.8698	0.6470	0.3701	0.2495
	0.50	$y_{l/2}$	1.0165	1.0695	1.1725	1.3584	1.7169	0.9571	0.8490	0.5902	0.2743	0.1477
		θ_A	1.0202	1.0856	1.2139	1.4495	1.9140	0.9477	0.8177	0.5216	0.2066	0.1026
		M_B	1.0104	1.0439	1.1078	1.2208	1.4327	0.9726	0.9023	0.7232	0.4625	0.3257
3c. Conc. moment	0	$y_{l/2}$	1.0199	1.0842	1.2101	1.4406	1.8933	0.9485	0.8202	0.5255	0.2066	0.1005
		θ_A	1.0122	1.0515	1.1273	1.2635	1.5243	0.9681	0.8874	0.6900	0.4287	0.3030
		M_B	1.0183	1.0779	1.1949	1.4105	1.8379	0.9525	0.8348	0.5684	0.2842	0.1704
	0.50	$y_{l/2}$	1.0245	1.1041	1.2613	1.5528	2.1347	0.9368	0.7812	0.4387	0.1175	0.0390
		θ_A	1.0168	1.0707	1.1750	1.3618	1.7186	0.9562	0.8452	0.5760	0.2437	0.1176
		M_B	0.9861	0.9391	0.8392	0.6354	0.1828	1.0346	1.1098	1.1951	0.9753	0.7055

TABLE 9d *Reaction and deflection coefficients for beams under simultaneous axial and transverse loading: fixed ends*

Case no. in Table 3	Load location a/l	Coefficient listed for	Axial compressive load, $kl = \sqrt{Pl^2/EI}$					Axial tensile load, $kl = \sqrt{Pl^2/EI}$				
			0.8	1.6	2.4	3.2	4.0	1.0	2.0	4.0	8.0	12.0
1d. Conc. load	0.25	$y_{l/2}$	1.0163	1.0684	1.1686	1.3455	1.6722	0.9756	0.9092	0.7152	0.3885	0.2228
		R_A	1.0007	1.0027	1.0064	1.0121	1.0205	0.9990	0.9960	0.9859	0.9613	0.9423
		M_A	1.0088	1.0366	1.0885	1.1766	1.3298	0.9867	0.9499	0.8350	0.6008	0.4416
		M_B	1.0143	1.0603	1.1498	1.3117	1.6204	0.9787	0.9213	0.7583	0.4984	0.3645
	0.50	$y_{l/2}$	1.0163	1.0684	1.1686	1.3455	1.6722	0.9756	0.9092	0.7152	0.3885	0.2228
		M_A	1.0136	1.0570	1.1402	1.2870	1.5574	0.9797	0.9242	0.7616	0.4820	0.3317
		$M_{l/2}$	1.0136	1.0570	1.1402	1.2870	1.5574	0.9797	0.9242	0.7616	0.4820	0.3317
2d. Uniform load	0	$y_{l/2}$	1.0163	1.0684	1.1686	1.3455	1.6722	0.9756	0.9092	0.7152	0.3885	0.2228
		M_A	1.0108	1.0454	1.1114	1.2266	1.4365	0.9837	0.9391	0.8060	0.5630	0.4167
		$M_{l/2}$	1.0190	1.0801	1.1979	1.4078	1.7993	0.9716	0.8945	0.6728	0.3200	0.1617
	0.50	$y_{l/2}$	1.0146	1.0667	1.1667	1.3434	1.6696	0.9741	0.9077	0.7141	0.3879	0.2224
		R_A	0.9982	0.9927	0.9828	0.9677	0.9453	1.0027	1.0106	1.0375	1.1033	1.1551
		M_A	1.0141	1.0595	1.1473	1.3045	1.5999	0.9789	0.9217	0.7571	0.4868	0.3459
		M_B	1.0093	1.0390	1.0950	1.1913	1.3622	0.9859	0.9470	0.8282	0.5976	0.4488
2d. Uniformly increasing	0	$y_{l/2}$	1.0163	1.0684	1.1686	1.3455	1.6722	0.9756	0.9092	0.7152	0.3885	0.2228
		R_A	0.9995	0.9979	0.9951	0.9908	0.9845	1.0008	1.0030	1.0108	1.0303	1.0463
		M_A	1.0124	1.0521	1.1282	1.2627	1.5107	0.9814	0.9307	0.7818	0.5218	0.3750
		M_B	1.0098	1.0410	1.1001	1.2026	1.3870	0.9853	0.9447	0.8221	0.5904	0.4445
	0.50	$y_{l/2}$	1.0161	1.0679	1.1672	1.3427	1.6667	0.9758	0.9099	0.7174	0.3927	0.2274
		R_A	0.9969	0.9875	0.9707	0.9449	0.9070	1.0047	1.0182	1.0648	1.1815	1.2778
		M_A	1.0141	1.0595	1.1476	1.3063	1.6076	0.9790	0.9222	0.7602	0.4995	0.3647
		M_B	1.0075	1.0312	1.0755	1.1507	1.2819	0.9887	0.9573	0.8594	0.6582	0.5168
3d. Conc. moment	0.25	$y_{l/2}$	1.0169	1.0713	1.1757	1.3604	1.7016	0.9746	0.9054	0.7039	0.3671	0.2001
		R_A	0.9993	0.9972	0.9932	0.9867	0.9763	1.0010	1.0038	1.0122	1.0217	1.0134
		M_A	1.0291	1.1227	1.3025	1.6203	2.2055	0.9563	0.8376	0.4941	−0.0440	−0.2412
		M_B	1.0151	1.0635	1.1571	1.3244	1.5380	0.9775	0.9164	0.7404	0.4517	0.3035
	0.50	$\theta_{l/2}$	1.0054	1.0220	1.0515	1.0969	1.1641	0.9918	0.9681	0.8874	0.6900	0.5346
		R_A	1.0027	1.0110	1.0260	1.0492	1.0842	0.9959	0.9842	0.9449	0.8561	0.7960
		M_A	1.0081	1.0331	1.0779	1.1477	1.2525	0.9877	0.9525	0.8348	0.5684	0.3881

verse shear, bending moment, slope, and deflection shown in Tables 10 and 11.

Table 10 lists the general equations just mentioned as well as boundary values and selected maximum values for the case of axial compressive loading plus transverse loading. Since, in general, axial tension is a less critical condition, where deflections, slopes, and moments are usually reduced by the axial load, Table 11 is much more compact and gives only the general equations and the left-end boundary values.

Although the principle of superposition does not apply to the problem considered here, this modification of the principle can be used: The moment (or deflection) for a combination of transverse loads can be found by adding the moments (or deflections) for each transverse load combined with the entire axial load. Thus a beam supported at the ends and subjected to a uniform load, a center load, and an axial compression would have a maximum bending moment (or deflection) given by the sum of the maximum moments (or deflections) for Table 10, case 1e and 2e, the end load being included once for each transverse load.

A problem closely related to the beam under combined axial and lateral loading occurs when the ends of a beam are axially restrained from motion along the axis of the beam (held) and a lateral load is applied. A solution can be effected by equating the increase in length of the neutral surface of the beam Pl/AE to the decrease in length due to the curvature of the neutral surface $\frac{1}{2}\int_0^l \theta^2\,dx$ (Eq. 14). In general, solving the resulting equation for P is difficult owing to the presence of the hyperbolic functions and the several powers of the load P in the equation. If the beam is long, slender, and heavily loaded, this will be necessary for good accuracy; but if the deflections are small, the deflection curve can be approximated with a sine or cosine curve, obtaining the results given in Table 12. The following examples will illustrate the use of the formulas in Tables 9 to 12.

EXAMPLES

1. A 4-in, 7.7-lb steel I-beam 20 ft long is simply supported at both ends and simultaneously subjected to a transverse load of 50 lb/ft (including its own weight), a concentrated lateral load of 600 lb acting vertically downward at a position 8 ft from the left end, and an axial compression of 3000 lb. It is required to determine the maximum fiber stress and the deflection at midlength.

Solution. Here $P = 3000$; $l = 240$; $I = 6$; $I/c = 3$; $A = 2.21$; $w_a = w_l = \frac{50}{12} = 4.17$; and $a = 0$ for case 2e; $W = 600$ and $a = 96$ for case 1e; $k = \sqrt{P/EI} = 0.00408$; and $kl = 0.98$. The solution will be carried out (*a*) ignoring deflection, (*b*) using coefficients from Table 9, and (*c*) using precise formulas from Table 10.

(*a*) $R_A = 860$ lb, and max $M_{8'} = 860(8) - 8(50)(4) = 5280$ ft-lb:

$$\text{Max compressive stress} = -\frac{P}{A} - \frac{M}{I/c} = -\frac{3000}{2.21} - \frac{5280(12)}{3} = -22{,}475 \text{ lb/in}^2$$

For the uniform load (Table 3, case 2e):

$$y_{l/2} = \frac{-5}{384}\frac{w_a l^4}{EI} = \frac{-5(4.17)(240^4)}{384(30 \cdot 10^6)(6)} = -1.00 \text{ in}$$

For the concentrated load (Table 3, case 1e):

$$R_A = 360 \text{ lb}$$

$$\theta_A = \frac{-600(96)[2(240) - 96](240 - 96)}{6(30 \cdot 10^6)(6)(240)} = -0.123$$

$$y_{l/2} = -0.123(120) + \frac{360(120^3)}{6(30 \cdot 10^6)(6)} - \frac{600(120 - 96)^3}{6(30 \cdot 10^6)(6)} = -0.907 \text{ in}$$

Thus

$$\text{Total midlength deflection} = -1.907 \text{ in}$$

(*b*) From Table 9*b* (simply supported ends), coefficients are given for concentrated loads at $l/4$ and $l/2$. Plotting curves of the coefficients versus kl and using linear interpolation to correct for $a = 0.4l$ give a value of $C_P = 1.112$ for the midlength deflection and 1.083 for the moment under the load. Similarly, for a uniform load on the entire span ($a = 0$), the values of C_P are found to be 1.111 for the midlength deflection and 1.115 for the moment at midlength. If it is assumed that this last coefficient is also satisfactory for the moment at $x = 0.4l$, the following deflections and moments are calculated:

$$\text{Max } M_{8'} = 360(8)(1.083) + [500(8) - 8(50)(4)](1.115) = 3120 + 2680 = 5800 \text{ ft-lb}$$

$$\text{Max compressive stress} = -\frac{P}{A} - \frac{M}{I/c} = -\frac{3000}{2.21} - \frac{5800(12)}{3} = -24{,}560 \text{ lb/in}^2$$

$$\text{Midlength deflection} = -0.907(1.112) - 1.00(1.111) = -2.12 \text{ in}$$

(*c*) From Table 10, cases 1e and 2e, $R_A = 860$ lb and

$$\theta_A = \frac{-600}{3000}\left[\frac{\sin 0.00408(240 - 96)}{\sin 0.98} - \frac{240 - 96}{240}\right] + \frac{-4.17}{0.00408(3000)}\left[\tan\frac{0.98}{2} - \frac{0.98}{2}\right]$$

$$= \frac{-600}{3000}\left(\frac{0.5547}{0.8305} - 0.6\right) - 0.341(0.533 - 0.49) = -0.0283$$

$$\text{Max } M_{8'} = \frac{860}{0.00408}\sin 0.00408(96) - \frac{-0.0283(3000)}{0.00408}\sin 0.392 - \frac{4.17}{0.00408^2}(1 - \cos 0.392)$$

$$= 80{,}500 + 7950 - 19{,}000 = 69{,}450 \text{ in-lb}$$

$$\text{Max compressive stress} = -\frac{3000}{2.21} - \frac{69{,}450}{3} = -24{,}500 \text{ lb/in}^2$$

$$\text{Midlength deflection} = \frac{-0.0283}{0.00408}\sin 0.49 + \frac{860}{0.00408(3000)}(0.49 - \sin 0.49)$$

$$- \frac{600}{0.00408(3000)}[0.00408(120 - 96) - \sin 0.00408(120 - 96)]$$

$$- \frac{4.17}{0.00408^2(3000)}\left[\frac{0.00408^2(120^2)}{2} - 1 + \cos 0.00408(120)\right]$$

$$= -3.27 + 1.36 - 0.00785 - 0.192 = -2.11 \text{ in}$$

The ease with which the coefficients C_P can be obtained from Table 9*a* to 9*d* makes this a very desirable way to solve problems of axially loaded beams. Some caution must be observed, however, when interpolating for the position of the load. For example, the concentrated moment in Tables 9*c* and 9*d* shows a large variation in C_P for the end moments when the load position

TABLE 10 ***Shear, moment, slope, and deflection formulas for beams under simultaneous axial compression and transverse loading***

NOTATION: P = axial compressive load (pounds); all other notation is the same as that for Table 3.

The following constants and functions are hereby defined in order to permit condensing the tabulated formulas which follow. $k = (P/EI)^{\frac{1}{2}}$, (*Note:* See page 94 for a definition of $\langle x - a \rangle^n$.) The function $\sin k\langle x - a \rangle$ is also defined as having a value of zero if $x < a$

$F_1 = \cos kx$	$F_{a1} = \langle x - a \rangle^0 \cos k(x - a)$	$C_1 = \cos kl$	$C_{a1} = \cos k(l - a)$
$F_2 = \sin kx$	$F_{a2} = \sin k\langle x - a \rangle$	$C_2 = \sin kl$	$C_{a2} = \sin k(l - a)$
$F_3 = 1 - \cos kx$	$F_{a3} = \langle x - a \rangle^0[1 - \cos k(x - a)]$	$C_3 = 1 - \cos kl$	$C_{a3} = 1 - \cos k(l - a)$
$F_4 = kx - \sin kx$	$F_{a4} = k\langle x - a \rangle - \sin k\langle x - a \rangle$	$C_4 = kl - \sin kl$	$C_{a4} = k(l - a) - \sin k(l - a)$
	$F_{a5} = \dfrac{k^2}{2}\langle x - a \rangle^2 - F_{a3}$		$C_{a5} = \dfrac{k^2}{2}(l - a)^2 - C_{a3}$
	$F_{a6} = \dfrac{k^3}{6}\langle x - a \rangle^3 - F_{a4}$		$C_{a6} = \dfrac{k^3}{6}(l - a)^3 - C_{a4}$

(*Note:* M_A and R_A as well as M_B and R_B are reactions, not applied loads. They exist only when the necessary end restraints are provided.)

1. Axial compressive load plus concentrated intermediate lateral load

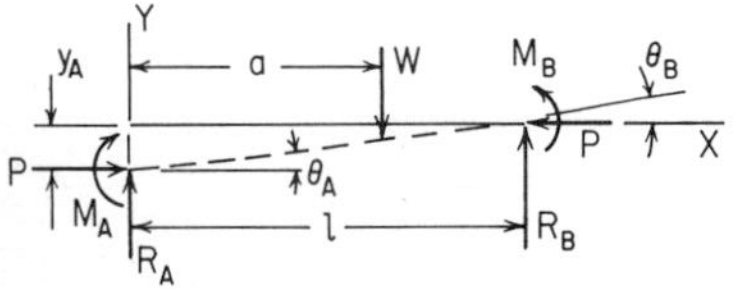

Transverse shear $= V = R_A F_1 - M_A k F_2 - \theta_A P F_1 - W F_{a1}$

Bending moment $= M = M_A F_1 + \dfrac{R_A}{k} F_2 - \dfrac{\theta_A P}{k} F_2 - \dfrac{W}{k} F_{a2}$

Slope $= \theta = \theta_A F_1 + \dfrac{M_A k}{P} F_2 + \dfrac{R_A}{P} F_3 - \dfrac{W}{P} F_{a3}$

Deflection $= y = y_A + \dfrac{\theta_A}{k} F_2 + \dfrac{M_A}{P} F_3 + \dfrac{R_A}{kP} F_4 - \dfrac{W}{kP} F_{a4}$

End restraints, reference no.	Boundary values	Selected maximum values of moments and deformations
1a. Left end free, right end fixed (cantilever) 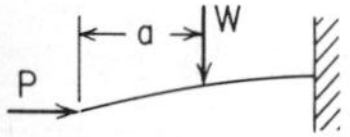 	$R_A = 0 \quad M_A = 0 \quad \theta_A = \dfrac{W}{P}\dfrac{C_{a3}}{C_1}$ $y_A = \dfrac{-W}{kP}\dfrac{C_2 C_{a3} - C_1 C_{a4}}{C_1}$ $R_B = W \quad \theta_B = 0 \quad y_B = 0$ $M_B = \dfrac{-W}{k}\dfrac{C_2 C_{a3} + C_1 C_{a2}}{C_1}$	Max $M = M_B$; max possible value $= \dfrac{-W}{k}\tan kl$ when $a = 0$ Max $\theta = \theta_A$; max possible value $= \dfrac{W}{P}\dfrac{1 - \cos kl}{\cos kl}$ when $a = 0$ Max $y = y_A$; max possible value $= \dfrac{-W}{kP}(\tan kl - kl)$ when $a = 0$
1b. Left end guided, right end fixed 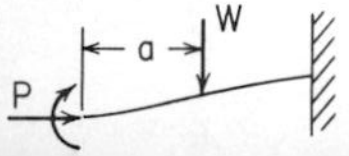 	$R_A = 0 \quad M_A = \dfrac{W}{k}\dfrac{C_{a3}}{C_2} \quad \theta_A = 0$ $y_A = \dfrac{-W}{kP}\dfrac{C_3 C_{a3} - C_2 C_{a4}}{C_2}$ $R_B = W \quad \theta_B = 0 \quad y_B = 0$ $-W \cos ka - \cos kl$	Max $+M = M_A$; max possible value $= \dfrac{W}{k}\tan\dfrac{kl}{2}$ when $a = 0$ Max $-M = M_B$; max possible value $= \dfrac{-W}{k}\tan\dfrac{kl}{2}$ when $a = 0$ Max $y = y_A$; max possible value $= \dfrac{-W}{kP}\left(2\tan\dfrac{kl}{2} - kl\right)$ when $a = 0$

right end fixed 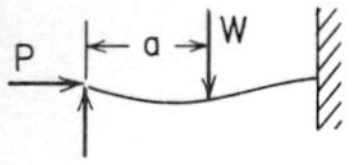 	$C_2C_3 - C_1C_4$ $\quad M_A = 0$ $\theta_A = \frac{-W}{P}\frac{C_4C_{a3} - C_3C_{a4}}{C_2C_3 - C_1C_4} \qquad y_A = 0$ $R_B = W - R_A \qquad \theta_B = 0 \qquad y_B = 0$ $M_B = \frac{-W}{k}\frac{kl\sin ka - ka\sin kl}{\sin kl - kl\cos kl}$	Max $-M = M_B$; max possible value occurs when $a = \frac{1}{k}\cos^{-1}\frac{\sin kl}{kl}$ If $a = l/2$ (transverse center load), then $R_A = W\frac{\sin kl - \sin\frac{kl}{2} - \frac{kl}{2}\cos kl}{\sin kl - kl\cos kl}$ $M_B = -Wl\frac{\sin\frac{kl}{2}\left(1 - \cos\frac{kl}{2}\right)}{\sin kl - kl\cos kl}$
1d. Left end fixed, right end fixed 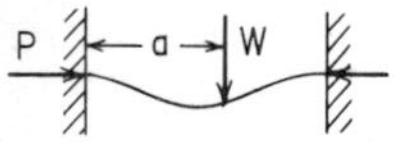 	$R_A = W\frac{C_3C_{a3} - C_2C_{a4}}{C_3^2 - C_2C_4}$ $M_A = \frac{-W}{k}\frac{C_4C_{a3} - C_3C_{a4}}{C_3^2 - C_2C_4}$ $\theta_A = 0 \qquad y_A = 0$ $R_B = W - R_A \qquad \theta_B = 0 \qquad y_B = 0$ $M_B = M_A + R_Al - W(l - a)$	Max $-M = M_A$ if $a < \frac{l}{2}$ If $a = \frac{l}{2}$ (transverse center load), then $R_A = R_B = \frac{W}{2} \qquad M_B = M_A = \frac{-W}{2k}\tan\frac{kl}{4}$ Max $+M = \frac{W}{2k}\tan\frac{kl}{4}$ at $x = \frac{l}{2}$ Max $y = \frac{-W}{kP}\left(\tan\frac{kl}{4} - \frac{kl}{4}\right)$ at $x = \frac{l}{2}$
1e. Left end simply supported, right end simply supported 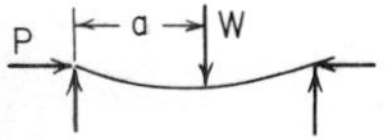 	$R_A = \frac{W}{l}(l - a) \qquad M_A = 0 \qquad y_A = 0$ $\theta_A = \frac{-W}{P}\left[\frac{\sin k(l - a)}{\sin kl} - \frac{l - a}{l}\right]$ $R_B = W\frac{a}{l} \qquad M_B = 0 \qquad y_B = 0$ $\theta_B = \frac{W}{P}\left(\frac{\sin ka}{\sin kl} - \frac{a}{l}\right)$	Max $M = \frac{W\sin k(l - a)}{k\sin kl}\sin ka$ at $x = a$ if $\frac{l}{2} < a < \frac{\pi}{2k}$ Max $M = \frac{W\sin k(l - a)}{k\sin kl}$ at $x = \frac{\pi}{2k}$ if $a > \frac{\pi}{2k}$ and $a > \frac{l}{2}$; max possible value of $M = \frac{W}{2k}\tan\frac{kl}{2}$ at $x = a$ when $a = \frac{l}{2}$ Max $\theta = \theta_B$ if $a > \frac{l}{2}$; max possible value occurs when $a = \frac{1}{k}\cos^{-1}\frac{\sin kl}{kl}$ Max y occurs at $x = \frac{1}{k}\cos^{-1}\frac{(l - a)\sin kl}{l\sin k(l - a)}$ if $a > \frac{l}{2}$; max possible value $= \frac{-W}{2kP}\left(\tan\frac{kl}{2} - \frac{kl}{2}\right)$ at $x = \frac{l}{2}$ when $a = \frac{l}{2}$

TABLE 10 *Shear, moment, slope, and deflection formulas for beams under simultaneous axial compression and transverse loading* (*Cont.*)

End restraints, reference no.	Boundary values	Selected maximum values of moments and deformations
1f. Left end guided, right end simply supported 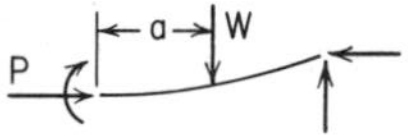	$R_A = 0 \qquad \theta_A = 0$ $M_A = \frac{W}{k}\frac{\sin k(l-a)}{\cos kl}$ $y_A = \frac{-W}{kP}\left[\frac{\sin k(l-a)}{\cos kl} - k(l-a)\right]$ $R_B = W \qquad M_B = 0 \qquad y_B = 0$ $\theta_B = \frac{W}{P}\left(\frac{\cos ka}{\cos kl} - 1\right)$	Max $M = M_A$; max possible value $= \frac{W}{k}\tan kl$ when $a = 0$ Max $\theta = \theta_B$; max possible value $= \frac{W}{P}\frac{1-\cos kl}{\cos kl}$ when $a = 0$ Max $y = y_A$; max possible value $= \frac{-W}{kP}(\tan kl - kl)$ when $a = 0$
2. Axial compressive load plus distributed lateral load 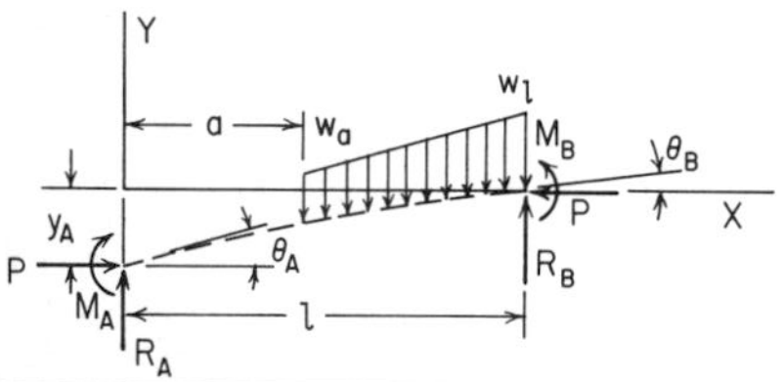		Transverse shear $= V = R_AF_1 - M_AkF_2 - \theta_APF_1 - \frac{w_a}{k}F_{a2} - \frac{w_l - w_a}{k^2(l-a)}F_{a3}$ Bending moment $= M = M_AF_1 + \frac{R_A}{k}F_2 - \frac{\theta_AP}{k}F_2 - \frac{w_a}{k^2}F_{a3} - \frac{w_l - w_a}{k^3(l-a)}F_{a4}$ Slope $= \theta = \theta_AF_1 + \frac{M_Ak}{P}F_2 + \frac{R_A}{P}F_3 - \frac{w_a}{kP}F_{a4} - \frac{w_l - w_a}{k^2P(l-a)}F_{a5}$ Deflection $= y = y_A + \frac{\theta_A}{k}F_2 + \frac{M_A}{P}F_3 + \frac{R_A}{kP}F_4 - \frac{w_a}{k^2P}F_{a5} - \frac{w_l - w_a}{k^3P(l-a)}F_{a6}$
End restraints, reference no.	Boundary values	Selected maximum values of moments and deformations

2a. Left end free, right end fixed (cantilever) 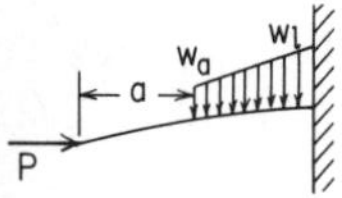	$R_A = 0 \qquad M_A = 0$
	$\theta_A = \frac{w_a}{kP}\frac{C_{a4}}{C_1} + \frac{w_l - w_a}{k^2P(l-a)}\frac{C_{a5}}{C_1}$
	$y_A = \frac{-w_a}{k^2P}\frac{C_2C_{a4} - C_1C_{a5}}{C_1} - \frac{w_l - w_a}{k^3P(l-a)}\frac{C_2C_{a5} - C_1C_{a6}}{C_1}$
	$R_B = \frac{w_a + w_l}{2}(l-a) \qquad \theta_B = 0$
	$M_B = \frac{-w_a}{k^2}\frac{C_2C_{a4} + C_1C_{a3}}{C_1} - \frac{w_l - w_a}{k^3(l-a)}\frac{C_2C_{a5} + C_1C_{a4}}{C_1}$
	$y_B = 0$
		If $a = 0$ and $w_a = w_l$ (uniform load on entire span), then
		Max $M = M_B = \frac{-w_a}{k^2}\left(1 + kl\tan kl - \frac{1}{\cos kl}\right)$
		Max $\theta = \theta_A = \frac{w_a}{kP}\left(\frac{kl}{\cos kl} - \tan kl\right)$
		Max $y = y_A = \frac{-w_a}{k^2P}\left(1 + kl\tan kl - \frac{k^2l^2}{2} - \frac{1}{\cos kl}\right)$
		If $a = 0$ and $w_a = 0$ (uniformly increasing load), then
		Max $M = M_B = \frac{-w_l}{k^2}\left(1 + \frac{kl}{2}\tan kl - \frac{\tan kl}{kl}\right)$
		Max $\theta = \theta_A = \frac{w_l}{kP}\left(\frac{1}{kl} + \frac{kl}{2\cos kl} - \frac{1}{kl\cos kl}\right)$
		$-w_l \quad kl \quad k^2l^2 \quad \tan kl$

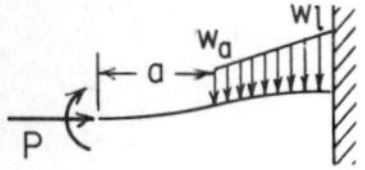	$M_A = \frac{w_a}{k^2}\frac{C_{a4}}{C_2} + \frac{w_l - w_a}{k^3(l-a)}\frac{C_{a5}}{C_2}$ $y_A = \frac{-w_a}{k^2P}\frac{C_3C_{a4} - C_2C_{a5}}{C_2} - \frac{w_l - w_a}{k^3P(l-a)}\frac{C_3C_{a5} - C_2C_{a6}}{C_2}$ $R_B = \frac{w_a + w_l}{2}(l-a) \qquad \theta_B = 0$ $M_B = \frac{-w_a}{k^2}\frac{C_2C_{a3} - C_1C_{a4}}{C_2} - \frac{(w_l - w_a)}{k^3(l-a)}\frac{C_2C_{a4} - C_1C_{a5}}{C_2}$ $y_B = 0$	$\text{Max } +M = M_A = \frac{w_a}{k^2}\left(\frac{kl}{\sin kl} - 1\right)$ $\text{Max } -M = M_B = \frac{-w_a}{k^2}\left(1 - \frac{kl}{\tan kl}\right)$ $\text{Max } y = y_A = \frac{-w_a l}{kP}\left(\tan\frac{kl}{2} - \frac{kl}{2}\right)$ If $a = 0$ and $w_a = 0$ (uniformly increasing load), then $\text{Max } +M = M_A = \frac{w_l}{k^2}\left(\frac{kl}{2\sin kl} - \frac{\tan(kl/2)}{kl}\right)$ $\text{Max } -M = M_B = \frac{-w_l}{k^2}\left(1 - \frac{kl}{2\tan kl} - \frac{1 - \cos kl}{kl \sin kl}\right)$ $\text{Max } y = y_A = \frac{-w_l}{k^2P}\left[\left(\frac{kl}{2} - \frac{2}{kl}\right)\tan\frac{kl}{2} - \frac{k^2l^2}{6} + 1\right]$

2c. Left end simply supported, right end fixed
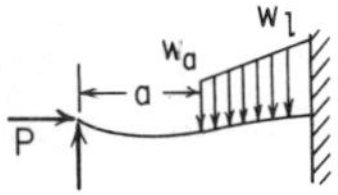
	$M_A = 0 \qquad y_A = 0$
	$R_A = \frac{w_a}{k}\frac{C_2C_{a4} - C_1C_{a5}}{C_2C_3 - C_1C_4} + \frac{w_l - w_a}{k^2(l-a)}\frac{C_2C_{a5} - C_1C_{a6}}{C_2C_3 - C_1C_4}$
	$\theta_A = \frac{-w_a}{kP}\left[\frac{C_4C_{a4} - C_3C_{a5}}{C_2C_3 - C_1C_4}\right] - \frac{w_l - w_a}{k^2P(l-a)}\frac{C_4C_{a5} - C_3C_{a6}}{C_2C_3 - C_1C_4}$
	$R_B = \frac{w_a + w_l}{2}(l-a) - R_A$
	$M_B = \frac{-w_a}{k^2}\left(\frac{C_2C_{a5} - klC_2C_{a4}}{C_2C_3 - C_1C_4} + C_{a3}\right) - \frac{(w_l - w_a)}{k^3(l-a)}\left(\frac{C_2C_{a6} - klC_2C_{a5}}{C_2C_3 - C_1C_4} + C_{a4}\right)$
	$\theta_B = 0 \qquad y_B = 0$
		If $a = 0$ and $w_a = w_l$ (uniform load on entire span), then
		$\text{Max } \theta = \theta_A = \frac{-w_a}{kP}\frac{4 - 2kl\sin kl - (2 - k^2l^2/2)(1 + \cos kl)}{\sin kl - kl\cos kl}$
		$\text{Max } -M = M_B = \frac{-w_a l}{k}\frac{\tan kl[\tan(kl/2) - kl/2]}{\tan kl - kl}$
		$R_A = \frac{w_a}{k}\frac{kl\sin kl - 1 + (1 - k^2l^2/2)\cos kl}{\sin kl - kl\cos kl}$
		If $a = 0$ and $w_a = 0$ (uniformly increasing load), then
		$\text{Max } \theta = \theta_A = \frac{-w_l l}{6P}\frac{2kl + kl\cos kl - 3\sin kl}{\sin kl - kl\cos kl}$
		$\text{Max } -M = M_B = \frac{-w_l}{k^2}\frac{(1 - k^2l^2/3)\tan kl - kl}{\tan kl - kl}$
		$R_A = \frac{w_l l}{6}\left(\frac{2\tan kl}{\tan kl - kl} - \frac{6}{k^2l^2} + 1\right)$

TABLE 10 ***Shear, moment, slope, and deflection formulas for beams under simultaneous axial compression and transverse loading*** ***(Cont.)***

End restraints, reference no.	Boundary values	Selected maximum values of moments and deformations

2d. Left end fixed, right end fixed
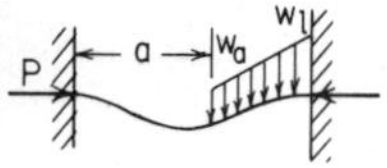
 	
	$\theta_A = 0 \qquad y_A = 0$
	$R_A = \frac{w_a}{k}\frac{C_3C_{a4} - C_2C_{a5}}{C_3^2 - C_2C_4} + \frac{w_l - w_a}{k^2(l-a)}\frac{C_3C_{a5} - C_2C_{a6}}{C_3^2 - C_2C_4}$
	$M_A = \frac{-w_a}{k^2}\frac{C_4C_{a4} - C_3C_{a5}}{C_3^2 - C_2C_4} - \frac{w_l - w_a}{k^3(l-a)}\frac{C_4C_{a5} - C_3C_{a6}}{C_3^2 - C_2C_4}$
	$R_B = \frac{w_a + w_l}{2}(l-a) - R_A$
	$M_B = M_A + R_A l - \frac{w_a}{2}(l-a)^2 - \frac{w_l - w_a}{6}(l-a)^2$
	$\theta_B = 0 \qquad y_B = 0$
		If $a = 0$ and $w_a = w_l$ (uniform load on entire span), then
		$\text{Max } -M = M_A = M_B = \frac{-w_a}{k^2}\left[1 - \frac{kl/2}{\tan(kl/2)}\right]$
		$\text{Max } +M = \frac{w_a}{k^2}\left[\frac{kl/2}{\sin(kl/2)} - 1\right] \quad \text{at } x = \frac{l}{2}$
		$\text{Max } y = \frac{-w_a l}{2kP}\left(\tan\frac{kl}{4} - \frac{kl}{4}\right) \quad \text{at } x = \frac{l}{2}$
		$R_A = R_B = \frac{w_a l}{2}$
		If $a = 0$ and $w_a = 0$ (uniformly increasing load), then
		$\text{Max } -M = M_B = \frac{-w_l}{k^2}\left(1 - \frac{kl/2 \sin kl - k^2l^2/6 - k^3l^3/3 \cos kl}{2 - 2\cos kl - kl \sin kl}\right)$
		$M_A = \frac{w_l l}{6k}\frac{3\sin kl - kl(2 + \cos kl)}{2 - 2\cos kl - kl\sin kl}$
		$R_A = \frac{w_l}{k^2 l}\left(\frac{k^2l^2}{6}\frac{3 - 3\cos kl - kl\sin kl}{2 - 2\cos kl - kl\sin kl} - 1\right)$

2e. Left end simply supported, right end simply supported
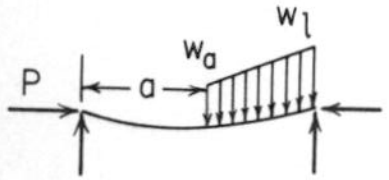
 	
	$M_A = 0 \qquad y_A = 0$
	$R_a = \frac{w_a}{2l}(l-a)^2 + \frac{w_l - w_a}{6l}(l-a)^2$
	$\theta_A = \frac{-w_a}{kP}\left[\frac{1 - \cos k(l-a)}{\sin kl} - \frac{k}{2l}(l-a)^2\right]$
	$- \frac{w_l - w_a}{kP}\left[\frac{k(l-a) - \sin k(l-a)}{k(l-a)\sin kl} - \frac{k}{6l}(l-a)^2\right]$
	$R_B = \frac{w_a + w_l}{2}(l-a) - R_A$
	$M_B = 0 \qquad y_B = 0$
	$\theta_B = \frac{w_a}{kP}\left[\frac{\cos ka - \cos kl}{\sin kl} - \frac{k(l^2 - a^2)}{2l}\right]$
	$+ \frac{w_l - w_a}{k^2P(l-a)}\left[\frac{k^2}{6l}(3al^2 - 2l^3 - a^3) + 1\right.$
		If $a = 0$ and $w_a = w_l$ (uniform load on entire span), then
		$\text{Max } +M = \frac{w_a}{k^2}\left[\frac{1}{\cos(kl/2)} - 1\right] \quad \text{at } x = \frac{l}{2}$
		$\text{Max } \theta = \theta_B = -\theta_A = \frac{w_a}{kP}\left(\tan\frac{kl}{2} - \frac{kl}{2}\right)$
		$\text{Max } y = \frac{-w_a}{k^2P}\left[\frac{1}{\cos(kl/2)} - \frac{k^2l^2}{8} - 1\right] \quad \text{at } x = \frac{l}{2}$
		If $a = 0$ and $w_a = 0$ (uniformly increasing load), then
		$M = \frac{w_l}{k^2}\left(\frac{\sin kx}{\sin kl} - \frac{x}{l}\right)$; max M occurs at $x = \frac{1}{k}\cos^{-1}\frac{\sin kl}{kl}$
		$\theta_A = \frac{-w_l}{kP}\left(\frac{1}{\sin kl} - \frac{1}{kl} - \frac{kl}{6}\right)$
		$\text{Max } \theta = \theta_B = \frac{w_l}{kP}\left(\frac{1}{kl} - \frac{kl}{3} - \frac{1}{\tan kl}\right)$

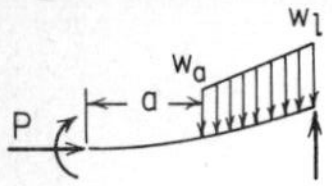

	$M_A = \frac{-w_a}{k^2}\frac{C_{a3}}{C_1} + \frac{w_l - w_a}{k^3(l-a)}\frac{C_{a4}}{C_1}$ $y_A = \frac{-w_a}{k^2P}\left[\frac{C_{a3}}{C_1} - \frac{k^2}{2}(l-a)^2\right] - \frac{w_l - w_a}{k^3P(l-a)}\left[\frac{C_{a4}}{C_1} - \frac{k^3}{6}(l-a)^3\right]$ $R_B = \frac{w_a + w_l}{2}(l-a) \qquad M_B = 0$ $\theta_B = \frac{w_a}{kP}\left[\frac{\sin kl - \sin ka}{\cos kl} - k(l-a)\right] + \frac{w_l - w_a}{k^2P(l-a)}\left[\frac{k(l-a)\sin kl - \cos ka}{\cos kl} - \frac{k^2(l-a)^2}{2} + 1\right]$ $y_B = 0$	$\text{Max } M = M_A = \frac{w_a}{k^2}\left(\frac{1}{\cos kl} - 1\right)$ $\text{Max } \theta = \theta_B = \frac{w_a}{kP}(\tan kl - kl)$ $\text{Max } y = y_A = \frac{-w_a}{k^2P}\left(\frac{1}{\cos kl} - 1 - \frac{k^2l^2}{2}\right)$ If $a = 0$ and $w_a = 0$ (uniformly increasing load), then $\text{Max } M = M_A = \frac{w_l}{k^3l}\frac{kl - \sin kl}{\cos kl}$ $\text{Max } \theta = \theta_B = \frac{w_l}{k^2Pl}\left(1 - \frac{k^2l^2}{2} - \frac{1 - kl\sin kl}{\cos kl}\right)$ $\text{Max } y = y_A = \frac{-w_l}{k^2P}\left(\frac{kl - \sin kl}{kl\cos kl} - \frac{k^2l^2}{6}\right)$

3. Axial compressive load plus concentrated intermediate moment

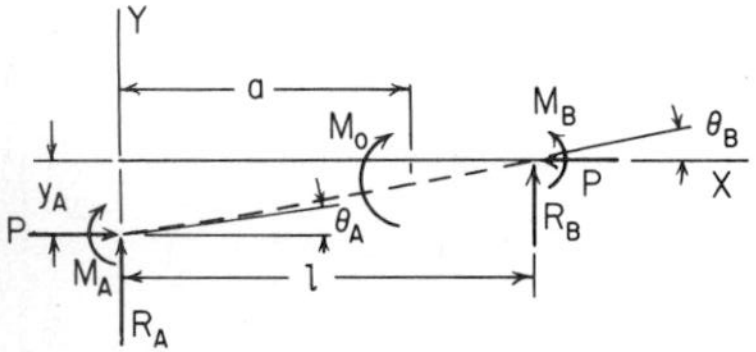

Transverse shear $= V = R_AF_1 - M_AkF_2 - \theta_APF_1 - M_okF_{a2}$

Bending moment $= M = M_AF_1 + \frac{R_A}{k}F_2 - \frac{\theta_AP}{k}F_2 + M_oF_{a1}$

Slope $= \theta = \theta_AF_1 + \frac{M_Ak}{P}F_2 + \frac{R_A}{P}F_3 + \frac{M_ok}{P}F_{a2}$

Deflection $= y = y_A + \frac{\theta_A}{k}F_2 + \frac{M_A}{P}F_3 + \frac{R_A}{kP}F_4 + \frac{M_o}{P}F_{a3}$

End restraints, reference no.	Boundary values	Selected maximum values of moments and deformations
3a. Left end free, right end fixed (cantilever) 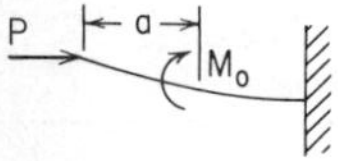 	$R_A = 0 \qquad M_A = 0$ $\theta_A = \frac{-M_ok}{P}\frac{\sin k(l-a)}{\cos kl}$ $y_A = \frac{M_o}{P}\left(\frac{\cos ka}{\cos kl} - 1\right)$ $R_B = 0 \qquad \theta_B = 0 \qquad y_B = 0$ $M_B = M_o\frac{\cos ka}{\cos kl}$	$\text{Max } M = M_B$; max possible value $= \frac{M_o}{\cos kl}$ when $a = 0$ $\text{Max } \theta = \theta_A$; max possible value $= \frac{-M_ok}{P}\tan kl$ when $a = 0$ $\text{Max } y = y_A$; max possible value $= \frac{M_o}{P}\left(\frac{1}{\cos kl} - 1\right)$ when $a = 0$

TABLE 10 ***Shear, moment, slope, and deflection formulas for beams under simultaneous axial compression and transverse loading*** **(Cont.)**

End restraints, reference no.	Boundary values	Selected maximum values of moments and deformations
3b. Left end guided, right end fixed 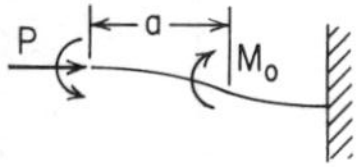	$R_A = 0 \qquad \theta_A = 0$ $M_A = -M_o \dfrac{\sin k(l-a)}{\sin kl}$ $y_A = \dfrac{M_o}{P}\left[\dfrac{\sin k(l-a) + \sin ka}{\sin kl} - 1\right]$ $R_B = 0 \qquad \theta_B = 0 \qquad y_B = 0$ $M_B = M_o \dfrac{\sin ka}{\sin kl}$	Max $+M = M_B$; max possible value $= M_o$ when $a = l$ Max $-M = M_A$; max possible value $= -M_o$ when $a = 0$ Max $\theta = \dfrac{-M_o k}{P} \dfrac{\sin k(l-a)}{\sin kl}$ at $x = a$; max possible value $= \dfrac{-M_o k}{2P} \tan \dfrac{kl}{2}$ when $a = \dfrac{l}{2}$ Max $y = y_A$; max possible value $= \dfrac{M_o}{P}\left[\dfrac{1}{\cos (kl/2)} - 1\right]$ when $a = \dfrac{l}{2}$
3c. Left end simply supported, right end fixed 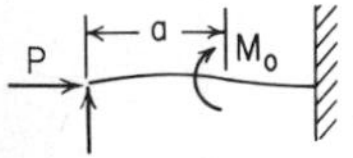	$M_A = 0 \qquad y_A = 0$ $R_A = -M_o k \dfrac{\cos ka - \cos kl}{\sin kl - kl \cos kl}$ $\theta_A = \dfrac{-M_o k}{P} \dfrac{C_3 C_{a3} - C_4 C_{a2}}{C_2 C_3 - C_1 C_4}$ $R_B = -R_A \qquad \theta_B = 0 \qquad y_B = 0$ $M_B = M_o \dfrac{\sin kl - kl \cos ka}{\sin kl - kl \cos kl}$	If $a = 0$ (concentrated end moment), then $R_A = -M_o k \dfrac{1 - \cos kl}{\sin kl - kl \cos kl}$ $\theta_A = \dfrac{-M_o k}{P} \dfrac{2 - 2\cos kl - kl \sin kl}{\sin kl - kl \cos kl}$ $M_B = -M_o \dfrac{kl - \sin kl}{\sin kl - kl \cos kl}$
3d. Left end fixed, right end fixed 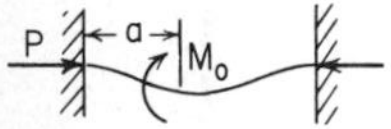	$\theta_A = 0 \qquad y_A = 0$ $R_A = -M_o k \dfrac{C_3 C_{a2} - C_2 C_{a3}}{C_3^2 - C_2 C_4}$ $M_A = -M_o \dfrac{C_3 C_{a3} - C_4 C_{a2}}{C_3^2 - C_2 C_4}$ $R_B = -R_A \qquad \theta_B = 0 \qquad y_B = 0$ $M_B = R_A l + M_A + M_o$	If $a = l/2$ (concentrated center moment), then $R_A = -M_o k \dfrac{[1/\cos (kl/2)] - 1}{2 \tan (kl/2) - kl}$ $M_A = -M_o \dfrac{1 - \cos kl - kl \sin (kl/2)}{2 - 2\cos kl - kl \sin kl}$ At the center, $y = 0$ and $\theta = \left(\dfrac{-M_o k}{2P}\right)\left(\dfrac{2 - 2\cos \frac{kl}{2} - \frac{kl}{2} \sin \frac{kl}{2}}{\sin \frac{kl}{2} - \frac{kl}{2} \cos \frac{kl}{2}}\right)$

3e. Left end simply supported, right end simply supported

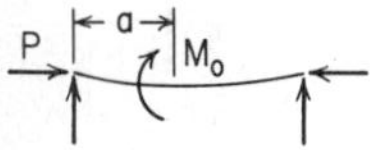

$M_A = 0 \qquad y_A = 0$

$R_A = \frac{-M_o}{l}$

$\theta_A = \frac{M_o}{Pl}\left[\frac{kl \cos k(l-a)}{\sin kl} - 1\right]$

$R_B = -R_A \qquad M_B = 0 \qquad y_B = 0$

$\theta_B = \frac{M_o}{Pl}\left(\frac{kl \cos ka}{\sin kl} - 1\right)$

If $a = 0$ (concentrated moment at the left end), then

$\theta_A = \frac{-M_o}{Pl}\left(1 - \frac{kl}{\tan kl}\right)$

$\theta_B = \frac{M_o}{Pl}\left(\frac{kl}{\sin kl} - 1\right)$

$M = M_o \cos kx\left(1 - \frac{\tan kx}{\tan kl}\right)$

If $a = l/2$ (concentrated moment at the center), then

$\theta_A = \theta_B = \frac{M_o}{Pl}\left[\frac{kl}{2 \sin (kl/2)} - 1\right]$ and $y = 0$ at the center

---

3f. Left end guided, right end simply supported

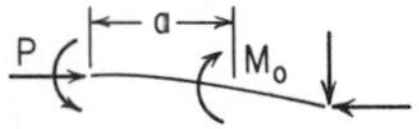

$R_A = 0 \qquad \theta_A = 0$

$M_A = -M_o \frac{\cos k(l-a)}{\cos kl}$

$y_A = \frac{M_o}{P}\left[\frac{\cos k(l-a)}{\cos kl} - 1\right]$

$R_B = 0 \qquad M_B = 0 \qquad y_B = 0$

$\theta_B = \frac{-M_o k}{P} \frac{\sin ka}{\cos kl}$

Max $M = M_A$; max possible value $= \frac{-M_o}{\cos kl}$ when $a = l$

Max $\theta = \theta_B$; max possible value $= \frac{-M_o k}{P} \tan kl$ when $a = l$

Max $y = y_A$; max possible value $= \frac{M_o}{P}\left(\frac{1}{\cos kl} - 1\right)$ when $a = l$

---

4. Axial compressive load plus externally created concentrated angular displacement

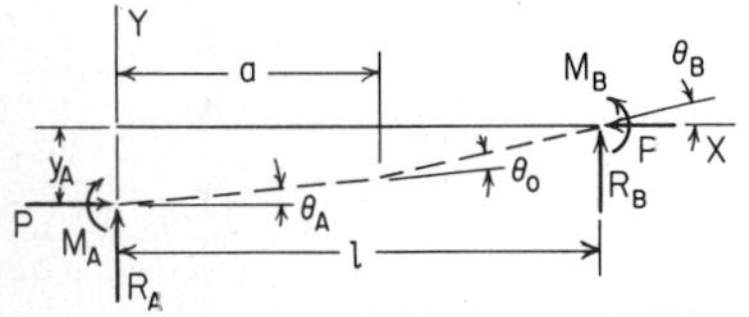

Transverse shear $= V = R_A F_1 - M_A k F_2 - \theta_A P F_1 - \theta_o P F_{a1}$

Bending moment $= M = M_A F_1 + \frac{R_A}{k} F_2 - \frac{\theta_A P}{k} F_2 - \frac{\theta_o P}{k} F_{a2}$

Slope $= \theta = \theta_A F_1 + \frac{M_A k}{P} F_2 + \frac{R_A}{P} F_3 + \theta_o F_{a1}$

Deflection $= y = y_A + \frac{\theta_A}{k} F_2 + \frac{M_A}{P} F_3 + \frac{R_A}{kP} F_4 + \frac{\theta_o}{k} F_{a2}$

TABLE 10 *Shear, moment, slope, and deflection formulas for beams under simultaneous axial compression and transverse loading* (Cont.)

End restraints, reference no.	Boundary values	Selected maximum values of moments and deformations
4a. Left end free, right end fixed 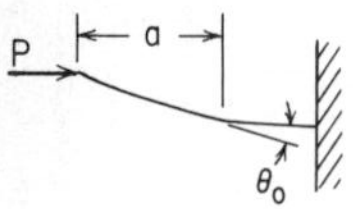	$R_A = 0 \qquad M_A = 0$ $\theta_A = -\theta_o \dfrac{\cos k(l-a)}{\cos kl}$ $y_A = \dfrac{\theta_o}{k}\dfrac{\sin ka}{\cos kl}$ $R_B = 0 \qquad \theta_B = 0 \qquad y_B = 0$ $M_B = \dfrac{\theta_o P}{k}\dfrac{\sin ka}{\cos kl}$	Max $M = M_B$; max possible value $= \dfrac{\theta_o P}{k}\tan kl$ when $a = l$ Max $\theta = \theta_A$; max possible value $= \dfrac{-\theta_o}{\cos kl}$ when $a = l$ Max $y = y_A$; max possible value $= \dfrac{\theta_o}{k}\tan kl$ when $a = l$
4b. Left end guided, right end fixed 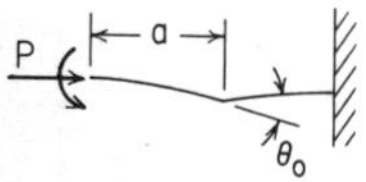	$R_A = 0 \qquad \theta_A = 0$ $M_A = \dfrac{-\theta_o P}{k}\dfrac{\cos k(l-a)}{\sin kl}$ $y_A = \dfrac{\theta_o}{k}\dfrac{\cos k(l-a) - \cos ka}{\sin kl}$ $R_B = 0 \qquad \theta_B = 0 \qquad y_B = 0$ $M_B = \dfrac{-\theta_o P}{k}\dfrac{\cos ka}{\sin kl}$	Max $-M = M_B$ if $a < \dfrac{l}{2}$; max possible value $= \dfrac{-\theta_o P}{k \sin kl}$ when $a = 0$ Max $-M = M_A$ if $a > \dfrac{l}{2}$; max possible value $= \dfrac{-\theta_o P}{k \sin kl}$ when $a = l$ Max $+y = y_A$; max possible value $= \dfrac{\theta_o}{k}\tan\dfrac{kl}{2}$ when $a = l$ Max $-y$ occurs at $x = a$; max possible value $= \dfrac{-\theta_o}{k}\tan\dfrac{kl}{2}$ at $x = 0$ when $a = 0$
4c. Left end simply supported, right end fixed 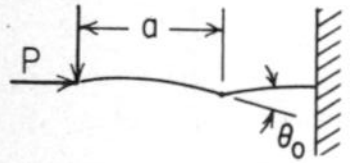	$M_A = 0 \qquad y_A = 0$ $R_A = -\theta_o P \dfrac{\sin ka}{\sin kl - kl\cos kl}$ $\theta_A = -\theta_o \dfrac{C_3 C_{a2} - C_4 C_{a1}}{C_2 C_3 - C_1 C_4}$ $R_B = -R_A \qquad \theta_B = 0 \qquad y_B = 0 \qquad M_B = R_A l$	
4d. Left end fixed, right end fixed 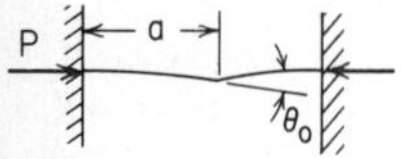	$\theta_A = 0 \qquad y_A = 0$ $R_A = -\theta_o P \dfrac{C_3 C_{a1} - C_2 C_{a2}}{C_3^2 - C_2 C_4}$ $M_A = \dfrac{-\theta_o P}{k}\dfrac{C_3 C_{a2} - C_4 C_{a1}}{C_3^2 - C_2 C_4}$ $R_B = -R_A \qquad \theta_B = 0 \qquad y_B = 0 \qquad M_B = M_A + R_A l$	

4e. Left end simply supported, right end simply supported 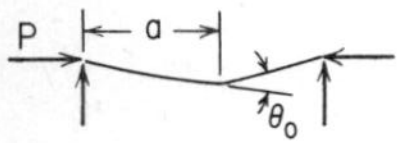	$M_A = 0 \quad y_A = 0 \quad R_A = 0$ $\theta_A = -\theta_o \dfrac{\sin k(l-a)}{\sin kl}$ $M_B = 0 \quad y_B = 0 \quad R_B = 0$ $\theta_B = \theta_o \dfrac{\sin ka}{\sin kl}$	Max $M = \dfrac{\theta_o P}{k} \dfrac{\sin k(l-a) \sin ka}{\sin kl}$ at $x = a$; max possible value $= \dfrac{\theta_o P}{k \cos (kl/2)}$ when $a = \dfrac{l}{2}$ Max $\theta = \theta_A$ if $a < l/2$; max possible value $= -\theta_o$ when $a = 0$ Max $y = \dfrac{-\theta_o}{k} \dfrac{\sin k(l-a) \sin ka}{\sin kl}$ at $x = a$; max possible value $= \dfrac{-\theta_o}{k \cos (kl/2)}$ when $a = \dfrac{l}{2}$
4f. Left end guided, right end simply supported 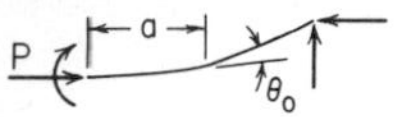	$R_A = 0 \quad \theta_A = 0$ $M_A = \dfrac{\theta_o P}{k} \dfrac{\sin k(l-a)}{\cos kl}$ $y_A = \dfrac{-\theta_o}{k} \dfrac{\sin k(l-a)}{\cos kl}$ $R_B = 0 \quad M_B = 0 \quad y_B = 0$ $\theta_B = \theta_o \dfrac{\cos ka}{\cos kl}$	Max $M = M_A$; max possible value $= \dfrac{\theta_o P}{k} \tan kl$ when $a = 0$ Max $\theta = \theta_B$; max possible value $= \dfrac{\theta_o}{\cos kl}$ when $a = 0$ Max $y = y_A$; max possible value $= \dfrac{-\theta_o}{k} \tan kl$ when $a = 0$

5. Axial compressive load plus externally created concentrated lateral displacement

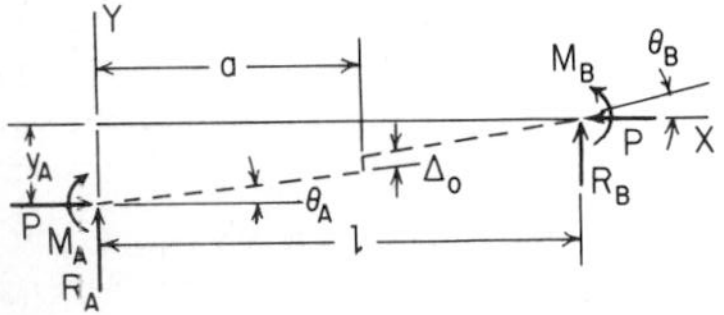

Transverse shear $= V = R_A F_1 - M_A k F_2 - \theta_A P F_1 + \Delta_o P k F_{a2}$

Bending moment $= M = M_A F_1 + \dfrac{R_A}{k} F_2 - \dfrac{\theta_A P}{k} F_2 - \Delta_o P F_{a1}$

Slope $= \theta = \theta_A F_1 + \dfrac{M_A k}{P} F_2 + \dfrac{R_A}{P} F_3 - \Delta_o k F_{a2}$

Deflection $= y = y_A + \dfrac{\theta_A}{k} F_2 + \dfrac{M_A}{P} F_3 + \dfrac{R_A}{kP} F_4 + \Delta_o F_{a1}$

End restraints, reference no.	Boundary values	Selected maximum values of moments and deformations
5a. Left end free, right end fixed (cantilever) 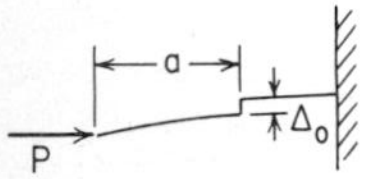	$R_A = 0 \quad M_A = 0$ $\theta_A = \Delta_o k \dfrac{\sin k(l-a)}{\cos kl}$ $y_A = -\Delta_o \dfrac{\cos ka}{\cos kl}$ $R_B = 0 \quad \theta_B = 0 \quad y_B = 0$ $M_B = -\Delta_o P \dfrac{\cos ka}{\cos kl}$	Max $M = M_B$; max possible value $= \dfrac{-\Delta_o P}{\cos kl}$ when $a = 0$ Max $\theta = \theta_A$; max possible value $= \Delta_o k \tan kl$ when $a = 0$ Max $y = y_A$; max possible value $= \dfrac{-\Delta_o}{\cos kl}$ when $a = 0$

TABLE 10 *Shear, moment, slope, and deflection formulas for beams under simultaneous axial compression and transverse loading* (*Cont.*)

End restraints, reference no.	Boundary values	Selected maximum values of moments and deformations
5b. Left end guided, right end fixed 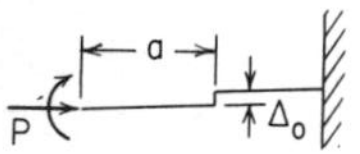	$R_A = 0 \quad \theta_A = 0$ $M_A = \Delta_o P \frac{\sin k(l-a)}{\sin kl}$ $y_A = -\Delta_o \frac{\sin k(l-a) + \sin ka}{\sin kl}$ $R_B = 0 \quad \theta_B = 0 \quad y_B = 0 \quad M_B = -\Delta_o P \frac{\sin ka}{\sin kl}$	Max $+M = M_A$; max possible value $= \Delta_o P$ when $a = 0$ Max $-M = M_B$; max possible value $= -\Delta_o P$ when $a = l$ Max $y = y_A$; max possible value $= \frac{-\Delta_o}{\cos(kl/2)}$ when $a = \frac{l}{2}$
5c. Left end simply supported, right end fixed 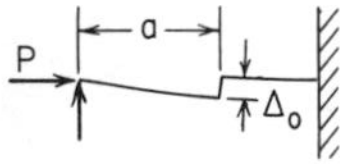	$M_A = 0 \quad y_A = 0$ $R_A = \Delta_o Pk \frac{\cos ka}{\sin kl - kl \cos kl}$ $\theta_A = -\Delta_o k \frac{C_3 C_{a1} + C_4 C_{a2}}{C_2 C_3 - C_1 C_4}$ $R_B = -R_A \quad \theta_B = 0 \quad y_B = 0 \quad M_B = R_A l$	
5d. Left end fixed, right end fixed 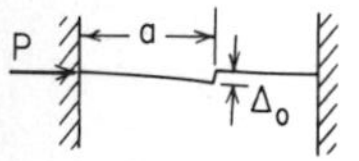	$\theta_A = 0 \quad y_A = 0$ $R_A = \Delta_o Pk \frac{C_3 C_{a2} + C_2 C_{a1}}{C_3^2 - C_2 C_4}$ $M_A = -\Delta_o P \frac{C_3 C_{a1} + C_4 C_{a2}}{C_3^2 - C_2 C_4}$ $R_B = -R_A \quad \theta_B = 0 \quad y_B = 0 \quad M_B = M_A + R_A l$	
5e. Left end simply supported, right end simply supported 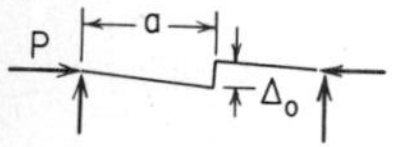	$R_A = 0 \quad M_A = 0 \quad y_A = 0$ $\theta_A = -\Delta_o k \frac{\cos k(l-a)}{\sin kl}$ $R_B = 0 \quad M_B = 0 \quad y_B = 0$ $\theta_B = -\Delta_o k \frac{\cos ka}{\sin kl}$	Max $+M = \Delta_o P \frac{\sin ka}{\sin kl} \cos k(l-a)$ at x just left of a; max possible value $= \Delta_o P$ when $a = l$ Max $-M = -\Delta_o P \frac{\cos ka}{\sin kl} \sin k(l-a)$ at x just right of a; max possible value $= -\Delta_o P$ when $a = 0$ Max $+y = \Delta_o \frac{\cos ka}{\sin kl} \sin k(l-a)$ at x just right of a; max possible value $= \Delta_o$ when $a = 0$ Max $-y = -\Delta_o \frac{\sin ka}{\sin kl} \cos k(l-a)$ at x just left of a; max possible value $= -\Delta_o$ when $a = l$

End restraints, reference no.	Boundary values	Selected maximum values of moments and deformations
5f. Left end guided, right end simply supported 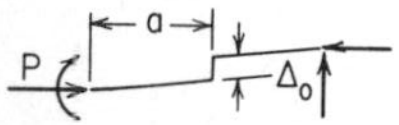	$R_A = 0 \quad \theta_A = 0$ $M_A = \Delta_o P \dfrac{\cos k(l-a)}{\cos kl}$ $y_A = -\Delta_o \dfrac{\cos k(l-a)}{\cos kl}$ $R_B = 0 \quad M_B = 0 \quad y_B = 0 \quad \theta_B = \Delta_o k \dfrac{\sin ka}{\cos kl}$	Max $M = M_A$; max possible value $= \dfrac{\Delta_o P}{\cos kl}$ when $a = l$ Max $\theta = \theta_B$; max possible value $= \Delta_o k \tan kl$ when $a = l$ Max $y = y_A$; max possible value $= \dfrac{-\Delta_o}{\cos kl}$ when $a = l$
6. Axial compressive load plus a uniform temperature variation from top to bottom in the portion from a to l; t is the thickness of the beam 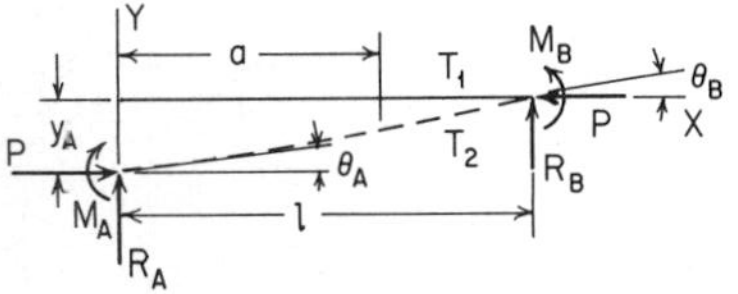	Transverse shear $= V = R_A F_1 - M_A k F_2 - \theta_A P F_1 - \dfrac{\gamma(T_2 - T_1)P}{kt} F_{a2}$ Bending moment $= M = M_A F_1 + \dfrac{R_A}{k} F_2 - \dfrac{\theta_A P}{k} F_2 - \dfrac{\gamma(T_2 - T_1)EI}{t} F_{a3}$ Slope $= \theta = \theta_A F_1 + \dfrac{M_A k}{P} F_2 + \dfrac{R_A}{P} F_3 + \dfrac{\gamma(T_2 - T_1)}{kt} F_{a2}$ Deflection $= y = y_A + \dfrac{\theta_A}{k} F_2 + \dfrac{M_A}{P} F_3 + \dfrac{R_A}{kP} F_4 + \dfrac{\gamma(T_2 - T_1)}{k^2 t} F_{a3}$	
6a. Left end free, right end fixed 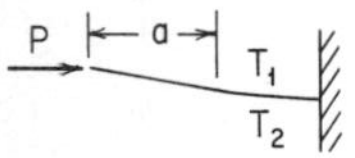	$R_A = 0 \quad M_A = 0$ $\theta_A = \dfrac{-\gamma(T_2 - T_1)}{kt} \dfrac{\sin k(l-a)}{\cos kl}$ $y_A = \dfrac{\gamma(T_2 - T_1)}{k^2 t}\left(\dfrac{\cos ka}{\cos kl} - 1\right)$ $R_B = 0 \quad \theta_B = 0 \quad y_B = 0 \quad M_B = P y_A$	Max $M = M_B$; max possible value $= \dfrac{\gamma(T_2 - T_1)EI}{t}\left(\dfrac{1}{\cos kl} - 1\right)$ when $a = 0$ Max $\theta = \theta_A$; max possible value $= \dfrac{-\gamma(T_2 - T_1)}{kt} \tan kl$ when $a = 0$ Max $y = y_A$; max possible value $= \dfrac{\gamma(T_2 - T_1)}{k^2 t}\left(\dfrac{1}{\cos kl} - 1\right)$ when $a = 0$
6b. Left end guided, right end fixed 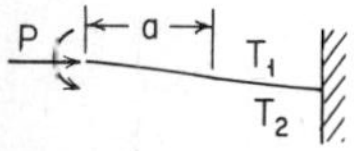	$R_A = 0 \quad \theta_A = 0$ $M_A = \dfrac{-\gamma(T_2 - T_1)EI}{t} \dfrac{\sin k(l-a)}{\sin kl}$ $y_A = \dfrac{\gamma(T_2 - T_1)}{k^2 t} \dfrac{C_3 C_{a2} - C_2 C_{a3}}{C_2}$ $R_B = 0 \quad \theta_B = 0 \quad y_B = 0$ $M_B = \dfrac{\gamma(T_2 - T_1)EI}{t}\left(\dfrac{\sin ka}{\sin kl} - 1\right)$	Max $-M = M_A$; max possible value $= \dfrac{-\gamma(T_2 - T_1)EI}{t}$ when $a = l$ (*Note:* There is no positive moment in the beam) Max $\theta = \dfrac{-\gamma(T_2 - T_1)}{kt} \dfrac{\sin ka}{\sin kl} \sin k(l-a)$ at $x = a$; max possible value $= \dfrac{-\gamma(T_2 - T_1)}{2kt} \tan \dfrac{kl}{2}$ when $a = \dfrac{l}{2}$ Max $y = y_A$; max possible value $= \dfrac{\gamma(T_2 - T_1)}{k^2 t}\left[\dfrac{1}{\cos (kl/2)} - 1\right]$ when $a = \dfrac{l}{2}$

TABLE 10 *Shear, moment, slope, and deflection formulas for beams under simultaneous axial compression and transverse loading* (*Cont.*)

End restraints, reference no.	Boundary values	Selected maximum values of moments and deformations
6c. Left end simply supported, right end fixed 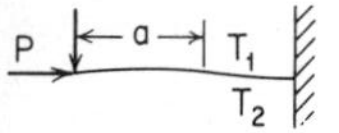 	$M_A = 0 \quad y_A = 0$ $R_A = \frac{-\gamma(T_2 - T_1)P}{kt} \frac{\cos ka - \cos kl}{\sin kl - kl \cos kl}$ $\theta_A = \frac{-\gamma(T_2 - T_1)}{kt} \frac{C_3 C_{a3} - C_4 C_{a2}}{C_2 C_3 - C_1 C_4}$ $R_B = -R_A \quad \theta_B = 0 \quad y_B = 0 \quad M_B = R_A l$	If $a = 0$ (temperature variation over entire span), then $\text{Max } -M = M_B = \frac{-\gamma(T_2 - T_1)P}{kt} \frac{1 - \cos kl}{\sin kl - kl \cos kl}$ $\text{Max } \theta = \theta_A = \frac{-\gamma(T_2 - T_1)}{kt} \frac{2 - 2\cos kl - kl \sin kl}{\sin kl - kl \cos kl}$
6d. Left end fixed, right end fixed 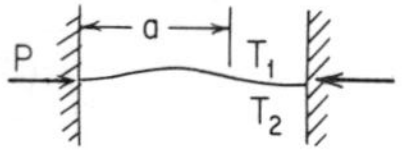 	$\theta_A = 0 \quad y_A = 0$ $R_A = \frac{-\gamma(T_2 - T_1)P}{kt} \frac{C_3 C_{a2} - C_2 C_{a3}}{C_3^2 - C_2 C_4}$ $M_A = \frac{-\gamma(T_2 - T_1)EI}{t} \frac{C_3 C_{a3} - C_4 C_{a2}}{C_3^2 - C_2 C_4}$ $R_B = -R_A \quad \theta_B = 0 \quad y_B = 0 \quad M_B = M_A + R_A l$	If $a = 0$ (temperature variation over entire span), then $R_A = R_B = 0$ $M = \frac{-\gamma(T_2 - T_1)EI}{t}$ everywhere in the span $\theta = 0$ and $y = 0$ everywhere in the span
6e. Left end simply supported, right end simply supported 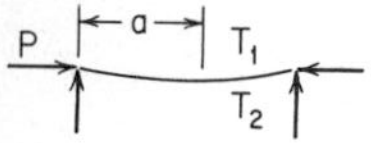 	$R_A = 0 \quad M_A = 0 \quad y_A = 0$ $\theta_A = \frac{-\gamma(T_2 - T_1)}{kt} \frac{1 - \cos k(l-a)}{\sin kl}$ $R_B = 0 \quad M_B = 0 \quad y_B = 0$ $\theta_B = \frac{\gamma(T_2 - T_1)}{kt} \frac{\cos ka - \cos kl}{\sin kl}$	Max y occurs at $x = \frac{1}{k} \tan^{-1} \frac{1 - \cos kl \cos ka}{\sin kl}$; max possible value = $\frac{-\gamma(T_2 - T_1)}{k^2 t} \left[\frac{1}{\cos (kl/2)} - 1 \right]$ at $x = \frac{l}{2}$ when $a = 0$ $\text{Max } M = P(\text{max } y)$ Max $\theta = \theta_B$; max possible value $= \frac{\gamma(T_2 - T_1)}{kt} \tan \frac{kl}{2}$ when $a = 0$
6f. Left end guided, right end simply supported 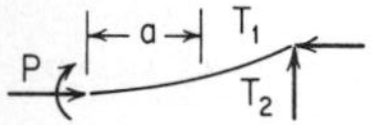 	$R_A = 0 \quad \theta_A = 0$ $M_A = \frac{\gamma(T_2 - T_1)EI}{t} \frac{1 - \cos k(l-a)}{\cos kl}$ $y_A = \frac{-\gamma(T_2 - T_1)}{k^2 t} \frac{1 - \cos k(l-a)}{\cos kl}$ $R_B = 0 \quad M_B = 0 \quad y_B = 0$ $\theta_B = \frac{\gamma(T_2 - T_1)}{kt} \frac{\sin kl - \sin ka}{\cos kl}$	Max $M = M_A$; max possible value $= \frac{\gamma(T_2 - T_1)EI}{t} \left(\frac{1}{\cos kl} - 1 \right)$ when $a = 0$ Max $y = y_A$; max possible value $= \frac{-\gamma(T_2 - T_1)}{k^2 t} \left(\frac{1}{\cos kl} - 1 \right)$ when $a = 0$ Max $\theta = \theta_B$; max possible value $= \frac{\gamma(T_2 - T_1)}{kt} \tan kl$ when $a = 0$

TABLE 11 ***Shear, moment, slope, and deflection formulas for beams under simultaneous axial tension and transverse loading***

NOTATION: P = axial tensile load (pounds); all other notation is the same as that for Table 3; see Table 10 for loading details

The following constants and functions are hereby defined in order to permit condensing the tabulated formulas which follow. $k = (P/EI)^{1/2}$. (*Note:* see page 94 for a definition of $\langle x - a\rangle^n$.) The function $\text{Sinh}\, k\langle x - a\rangle$ is also defined as having a value of zero if $x < a$

$F_1 = \text{Cosh}\, kx$
$F_2 = \text{Sinh}\, kx$
$F_3 = \text{Cosh}\, kx - 1$
$F_4 = \text{Sinh}\, kx - kx$

$F_{a1} = \langle x - a\rangle^0 \,\text{Cosh}\, k(x - a)$
$F_{a2} = \text{Sinh}\, k\langle x - a\rangle$
$F_{a3} = \langle x - a\rangle^0\, [\text{Cosh}\, k(x - a) - 1]$
$F_{a4} = \text{Sinh}\, k\langle x - a\rangle - k\langle x - a\rangle$

$$F_{a5} = F_{a3} - \frac{k^2}{2}\langle x - a\rangle^2$$

$$F_{a6} = F_{a4} - \frac{k^3}{6}\langle x - a\rangle^3$$

$C_1 = \text{Cosh}\, kl$
$C_2 = \text{Sinh}\, kl$
$C_3 = \text{Cosh}\, kl - 1$
$C_4 = \text{Sinh}\, kl - kl$

$C_{a1} = \text{Cosh}\, k(l - a)$
$C_{a2} = \text{Sinh}\, k(l - a)$
$C_{a3} = \text{Cosh}\, k(l - a) - 1$
$C_{a4} = \text{Sinh}\, k(l - a) - k(l - a)$

$$C_{a5} = C_{a3} - \frac{k^2}{2}(l - a)^2$$

$$C_{a6} = C_{a4} - \frac{k^3}{6}(l - a)^3$$

(*Note:* Load terms LT_V, LT_M, LT_θ, and LT_y are found at the end of the table for each of the several loadings.)

Axial tensile load plus lateral loading

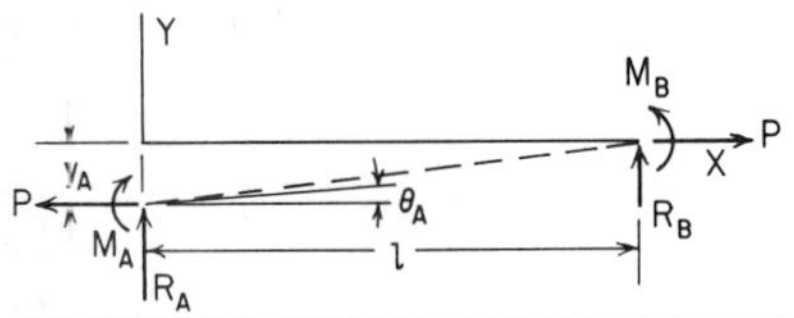

Transverse shear $= V = R_A F_1 + M_A k F_2 + \theta_A P F_1 + LT_V$

$$\text{Bending moment} = M = M_A F_1 + \frac{R_A}{k} F_2 + \frac{\theta_A P}{k} F_2 + LT_M$$

$$\text{Slope} = \theta = \theta_A F_1 + \frac{M_A k}{P} F_2 + \frac{R_A}{P} F_3 + LT_\theta$$

$$\text{Deflection} = y = y_A + \frac{\theta_A}{k} F_2 + \frac{M_A}{P} F_3 + \frac{R_A}{Pk} F_4 - LT_y$$

(*Note:* For each set of end restraints the two initial parameters not listed are zero. For example, with the left end free and the right end fixed, the values of R_A and M_A are zero.)

End restraints \ Lateral load		Case 1, Concentrated lateral load	Case 2, Distributed lateral load	Case 3, Concentrated moment	Case 4, Concentrated angular displacement	Case 5, Concentrated lateral displacement	Case 6, Uniform temperature variation
Left end free, right end fixed (a)	θ_A	$\frac{W}{P}\frac{C_{a3}}{C_1}$	$\frac{w_a}{kP}\frac{C_{a4}}{C_1} + \frac{(w_l - w_a)C_{a5}}{k^2P(l-a)C_1}$	$\frac{-M_o k}{P}\frac{C_{a2}}{C_1}$	$-\theta_o \frac{C_{a1}}{C_1}$	$-\Delta_o k \frac{C_{a2}}{C_1}$	$\frac{-\gamma(T_2 - T_1)}{kt}\frac{C_{a2}}{C_1}$
	y_A	$\frac{-W}{kP}\left(\frac{C_2C_{a3}}{C_1} - C_{a4}\right)$	$\frac{-w_a}{k^2P}\left(\frac{C_2C_{a4}}{C_1} - C_{a5}\right) + \frac{-(w_l - w_a)}{k^3P(l-a)}\left(\frac{C_2C_{a5}}{C_1} - C_{a6}\right)$	$\frac{M_o}{P}\left(\frac{C_2C_{a2}}{C_1} - C_{a3}\right)$	$\frac{\theta_o}{k}\left(\frac{C_2C_{a1}}{C_1} - C_{a2}\right)$	$\Delta_o\left(\frac{C_2C_{a2}}{C_1} - C_{a1}\right)$	$\frac{\gamma(T_2 - T_1)}{k^2t}\left(\frac{C_2C_{a2}}{C_1} - C_{a3}\right)$

Left end guided, right end fixed (b)	M_A	$\frac{W}{k}\frac{C_{a3}}{C_2}$	$\frac{w_a}{k^2}\frac{C_{a4}}{C_2}+\frac{w_l-w_a}{k^3(l-a)}\frac{C_{a5}}{C_2}$	$-M_o\frac{C_{a2}}{C_2}$	$\frac{-\theta_o P}{k}\frac{C_{a1}}{C_2}$	$-\Delta_o P\frac{C_{a2}}{C_2}$	$\frac{-\gamma(T_2-T_1)P}{k^2t}\frac{C_{a2}}{C_2}$
	y_A	$\frac{-W}{kP}\left(\frac{C_3C_{a3}}{C_2}-C_{a4}\right)$	$\frac{-w_a}{k^2P}\left(\frac{C_3C_{a4}}{C_2}-C_{a5}\right)+\frac{-(w_l-w_a)}{k^3P(l-a)}\left(\frac{C_3C_{a5}}{C_2}-C_{a6}\right)$	$\frac{M_o}{P}\left(\frac{C_3C_{a2}}{C_2}-C_{a3}\right)$	$\frac{\theta_o}{k}\left(\frac{C_3C_{a1}}{C_2}-C_{a2}\right)$	$\Delta_o\left(\frac{C_3C_{a2}}{C_2}-C_{a1}\right)$	$\frac{\gamma(T_2-T_1)}{k^2t}\left(\frac{C_3C_{a2}}{C_2}-C_{a3}\right)$
Left end simply supported, right end fixed (c)	R_A	$W\frac{C_2C_{a3}-C_1C_{a4}}{C_2C_3-C_1C_4}$	$\frac{w_a}{k}\frac{C_2C_{a4}-C_1C_{a5}}{C_2C_3-C_1C_4}+\frac{w_l-w_a}{k^2(l-a)}\frac{C_2C_{a5}-C_1C_{a6}}{C_2C_3-C_1C_4}$	$-M_ok\frac{C_2C_{a2}-C_1C_{a3}}{C_2C_3-C_1C_4}$	$-\theta_oP\frac{C_2C_{a1}-C_1C_{a2}}{C_2C_3-C_1C_4}$	$\Delta_okP\frac{C_1C_{a1}-C_2C_{a2}}{C_2C_3-C_1C_4}$	$\frac{-\gamma(T_2-T_1)P}{kt}\frac{C_2C_{a2}-C_1C_{a3}}{C_2C_3-C_1C_4}$
	θ_A	$\frac{-W}{P}\frac{C_4C_{a3}-C_3C_{a4}}{C_2C_3-C_1C_4}$	$\frac{-w_a}{kP}\frac{C_4C_{a4}-C_3C_{a5}}{C_2C_3-C_1C_4}+\frac{-(w_l-w_a)}{k^2P(l-a)}\frac{C_4C_{a5}-C_3C_{a6}}{C_2C_3-C_1C_4}$	$\frac{-M_ok}{P}\frac{C_3C_{a3}-C_4C_{a2}}{C_2C_3-C_1C_4}$	$-\theta_o\frac{C_3C_{a2}-C_4C_{a1}}{C_2C_3-C_1C_4}$	$\Delta_ok\frac{C_4C_{a2}-C_3C_{a1}}{C_2C_3-C_1C_4}$	$\frac{-\gamma(T_2-T_1)}{kt}\frac{C_3C_{a3}-C_4C_{a2}}{C_2C_3-C_1C_4}$
Left end fixed, right end fixed (d)	R_A	$W\frac{C_3C_{a3}-C_2C_{a4}}{C_3^2-C_2C_4}$	$\frac{w_a}{k}\frac{C_3C_{a4}-C_2C_{a5}}{C_3^2-C_2C_4}+\frac{w_l-w_a}{k^2(l-a)}\frac{C_3C_{a5}-C_2C_{a6}}{C_3^2-C_2C_4}$	$-M_ok\frac{C_3C_{a2}-C_2C_{a3}}{C_3^2-C_2C_4}$	$-\theta_oP\frac{C_3C_{a1}-C_2C_{a2}}{C_3^2-C_2C_4}$	$\Delta_oPk\frac{C_2C_{a1}-C_3C_{a2}}{C_3^2-C_2C_4}$	$\frac{-\gamma(T_2-T_1)P}{kt}\frac{C_3C_{a2}-C_2C_{a3}}{C_3^2-C_2C_4}$
	M_A	$\frac{-W}{k}\frac{C_4C_{a3}-C_3C_{a4}}{C_3^2-C_2C_4}$	$\frac{-w_a}{k^2}\frac{C_4C_{a4}-C_3C_{a5}}{C_3^2-C_2C_4}+\frac{-(w_l-w_a)}{k^3(l-a)}\frac{C_4C_{a5}-C_3C_{a6}}{C_3^2-C_2C_4}$	$-M_o\frac{C_3C_{a3}-C_4C_{a2}}{C_3^2-C_2C_4}$	$\frac{-\theta_oP}{k}\frac{C_3C_{a2}-C_4C_{a1}}{C_3^2-C_2C_4}$	$\Delta_oP\frac{C_4C_{a2}-C_3C_{a1}}{C_3^2-C_2C_4}$	$\frac{-\gamma(T_2-T_1)P}{k^2t}\frac{C_3C_{a3}-C_4C_{a2}}{C_3^2-C_2C_4}$
Left end simply supported, right end simply supported (e)	R_A	$\frac{W}{l}(l-a)$	$\frac{w_a}{2l}(l-a)^2+\frac{w_l-w_a}{6l}(l-a)^2$	$\frac{-M_o}{l}$	0	0	0
	θ_A	$\frac{-W}{Pkl}\left(\frac{C_4C_{a2}}{C_2}-C_{a4}\right)$	$\frac{-w_a}{Pk}\left[\frac{k(l-a)^2}{2l}-\frac{C_{a3}}{C_2}\right]+\frac{-(w_l-w_a)}{Pk^2(l-a)}\left[\frac{k^2(l-a)^3}{6l}-\frac{C_{a4}}{C_2}\right]$	$\frac{M_ok}{P}\left(\frac{1}{kl}-\frac{C_{a1}}{C_2}\right)$	$-\theta_o\frac{C_{a2}}{C_2}$	$-\Delta_ok\frac{C_{a1}}{C_2}$	$\frac{-\gamma(T_2-T_1)}{kt}\frac{C_{a3}}{C_2}$

TABLE 11 ***Shear, moment, slope, and deflection formulas for beams under simultaneous axial tension and transverse loading (Cont.)***

End restraints \ Lateral load		Case 1, Concentrated lateral load	Case 2, Distributed lateral load	Case 3, Concentrated moment	Case 4, Concentrated angular displacement	Case 5, Concentrated lateral displacement	Case 6, Uniform temperature variation
Left end guided, right end simply supported (f)	M_A	$\frac{W}{k}\frac{C_{a2}}{C_1}$	$\frac{w_a}{k^2}\frac{C_{a3}}{C_1}+\frac{w_l-w_a}{k^3(l-a)}\frac{C_{a4}}{C_1}$	$-M_o\frac{C_{a1}}{C_1}$	$\frac{-\theta_o P}{k}\frac{C_{a2}}{C_1}$	$-\Delta_o P\frac{C_{a1}}{C_1}$	$\frac{-\gamma(T_2-T_1)P}{k^2t}\frac{C_{a3}}{C_1}$
	y_A	$\frac{-W}{Pk}\left(\frac{C_3C_{a2}}{C_1}-C_{a4}\right)$	$\frac{-w_a}{k^2P}\left[\frac{k^2(l-a)^2}{2}-\frac{C_{a3}}{C_1}\right]+\frac{-(w_l-w_a)}{k^3P(l-a)}\left[\frac{k^3(l-a)^3}{6}-\frac{C_{a4}}{C_1}\right]$	$\frac{M_o}{P}\left(1-\frac{C_{a1}}{C_1}\right)$	$\frac{-\theta_o}{k}\frac{C_{a2}}{C_1}$	$-\Delta_o\frac{C_{a1}}{C_1}$	$\frac{-\gamma(T_2-T_1)}{k^2t}\frac{C_{a3}}{C_1}$
Load terms for all end restraints (a)–(f)	LT_V	$-WF_{a1}$	$\frac{-w_a}{k}F_{a2}-\frac{w_l-w_a}{k^2(l-a)}F_{a3}$	M_okF_{a2}	θ_oPF_{a1}	Δ_oPkF_{a2}	$\frac{\gamma(T_2-T_1)P}{kt}F_{a2}$
	LT_M	$\frac{-W}{k}F_{a2}$	$\frac{-w_a}{k^2}F_{a3}-\frac{w_l-w_a}{k^3(l-a)}F_{a4}$	M_oF_{a1}	$\frac{\theta_oP}{k}F_{a2}$	Δ_oPF_{a1}	$\frac{\gamma(T_2-T_1)P}{k^2t}F_{a3}$
	LT_θ	$\frac{-W}{P}F_{a3}$	$\frac{-w_a}{Pk}F_{a4}-\frac{(w_l-w_a)}{Pk^2(l-a)}F_{a5}$	$\frac{M_ok}{P}F_{a2}$	θ_oF_{a1}	Δ_okF_{a2}	$\frac{\gamma(T_2-T_1)}{kt}F_{a2}$
	LT_y	$\frac{-W}{Pk}F_{a4}$	$\frac{-w_a}{Pk^2}F_{a5}-\frac{w_l-w_a}{Pk^3(l-a)}F_{a6}$	$\frac{M_o}{P}F_{a3}$	$\frac{\theta_o}{k}F_{a2}$	Δ_oF_{a1}	$\frac{\gamma(T_2-T_1)}{k^2t}F_{a3}$

is changed from 0.25 to 0.50, especially under axial tension. Note that there are some cases in which C_P either changes sign or increases and then decreases when kl is increased; in these cases the loading produces both positive and negative moments and deflections in the span.

2. A solid round aluminum bar 1 in in diameter and 10 ft long is rigidly fixed at both ends by supports assumed to be sufficiently rigid to preclude any relative horizontal motion of the two supports. If this bar is subjected to a transverse center load of 300 lb at midlength, what is the center deflection and what is the maximum tensile stress in the bar?

Solution. Here P is an unknown tensile load; $W = 300$ and $a = 60$; $l = 120$; $A = 0.785$; $I = 0.0491$; and $E = 10 \cdot 10^6$. (This situation is described in Table 12, case 2.) The first equation is solved for y_{max}:

$$y_{max} + \frac{0.785}{16(0.0491)} y_{max}^3 = \frac{300(120^3)}{2(\pi^4)(10 \cdot 10^6)(0.0491)}$$

$$y_{max} + y_{max}^3 = 5.44$$

Therefore $y_{max} = 1.57$ in. The second equation is now solved for P:

$$P = \frac{\pi^2(10 \cdot 10^6)(0.785)}{4(120^2)} 1.57^2 = 3315 \text{ lb}$$

$$k = \sqrt{\frac{P}{EI}} = \left[\frac{3315}{(10 \cdot 10^6)(0.0491)}\right]^{\frac{1}{2}} = 0.0822$$

$$kl = 9.86$$

From Table 9, case 1d, the values of R_A and M_A can be calculated. (Note that θ_A and y_A are zero.) First evaluate the necessary constants:

$$C_2 = \text{Sinh } 9.86 = 9574.4$$

$$C_3 = \text{Cosh } 9.86 - 1 = 9574.4 - 1 = 9573.4$$

$$C_4 = \text{Sinh } 9.86 - 9.86 = 9564.5$$

$$C_{a3} = \text{Cosh} \frac{9.86}{2} - 1 = 69.193 - 1 = 68.193$$

$$C_{a4} = \text{Sinh } 4.93 - 4.93 = 69.186 - 4.93 = 64.256$$

$$R_A = W \frac{C_3 C_{a3} - C_2 C_{a4}}{C_3^2 - C_2 C_4} = 300 \frac{9573.4(68.193) - 9574.4(64.256)}{9573.4^2 - 9574.4(9564.5)} = 300(0.5) = 150 \text{ lb}$$

$$M_A = \frac{-W}{k} \frac{C_4 C_{a3} - C_3 C_{a4}}{C_3^2 - C_2 C_4} = \frac{-300}{0.0822} \frac{9564.5(68.193) - 9573.4(64.256)}{74{,}900}$$

$$= \frac{-300}{0.0822} 0.493 = -1800 \text{ in-lb}$$

$$\text{Max tensile stress} = \frac{P}{A} + \frac{Mc}{I} = \frac{3315}{0.785} + \frac{1800(0.5)}{0.0491} = 4220 + 18{,}330 = 22{,}550 \text{ lb/in}^2$$

$$\text{Midlength deflection} = \frac{-1800}{3315}\left(\text{Cosh} \frac{9.86}{2} - 1\right) + \frac{150}{3315(0.0822)}(\text{Sinh } 4.93 - 4.93)$$

$$= -37.0 + 35.4 = -1.6 \text{ in}$$

This compares favorably with the value $y_{max} = 1.57$ in obtained from the equation which was based on the assumption of a cosine curve for the deflection.

An alternative to working with the large numerical values of the hyperbolic sines and cosines as shown in the preceding calculations would be to simplify the equations for this case where the load is at the center by using the double-angle identities for hyperbolic functions. If this

TABLE 12 ***Beams restrained against horizontal displacement at the ends***

Case no., manner of loading and support	Formulas to solve for y_{max} and P
1. Ends pinned to rigid supports, concentrated center load W	$y_{max} + \frac{A}{4I} y_{max}^3 = \frac{2Wl^3}{\pi^4 EI}$ (Solve for y_{max}) $P = \frac{\pi^2 EA}{4l^2} y_{max}^2$ Use case 1e from Table 9*b* or Table 11 to determine maximum slopes and moments after solving for P
2. Ends fixed to rigid supports, concentrated center load W	$y_{max} + \frac{A}{16I} y_{max}^3 = \frac{Wl^3}{2\pi^4 EI}$ (Solve for y_{max}) $P = \frac{\pi^2 EA}{4l^2} y_{max}^2$ Use case 1d from Table 9*d* or Table 11 to determine maximum slopes and moments after solving for P
3. Ends pinned to rigid supports, uniformly distributed transverse load w on entire span	$y_{max} + \frac{A}{4I} y_{max}^3 = \frac{wl^4}{4\pi^4 EI}$ (Solve for y_{max}) $P = \frac{\pi^2 EA}{4l^2} y_{max}^2$ Use case 2*e* from Table 9*b* or Table 11 to determine maximum slopes and moments after solving for P
4. Ends fixed to rigid supports, uniformly distributed transverse load w on entire span	$y_{max} + \frac{A}{16I} y_{max}^3 = \frac{wl^4}{4\pi^4 EI}$ (Solve for y_{max}) $P = \frac{\pi^2 EA}{4l^2} y_{max}^2$ Use case 2d from Table 9*d* or Table 11 to determine maximum slopes and moments after solving for P
5. Same as case 1, except beam is perfectly flexible like a cable or chain and has an unstretched length l	$\tan\theta - \sin\theta = \frac{W}{2EA}$ or if $\theta < 12°$, $\theta = \left(\frac{W}{EA}\right)^{\frac{1}{3}}$ $P = \frac{W}{2\tan\theta}$ 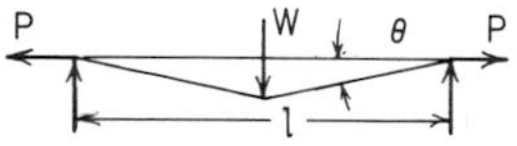
6. Same as case 3, except beam is perfectly flexible like a cable or chain and has an unstretched length l	$y_{max} = l\left(\frac{3wl}{64EA}\right)^{\frac{1}{3}}$ $P = \frac{wl^2}{8y_{max}}$ 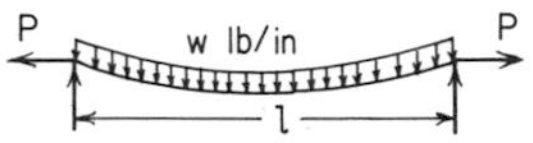

is done here, the expressions simplify to

$$R_A = \frac{W}{2} \qquad M_A = \frac{-W}{k} \operatorname{Tanh} \frac{kl}{4} \qquad y_{l/2} = \frac{-W}{kP}\left(\frac{kl}{4} - \operatorname{Tanh}\frac{kl}{4}\right)$$

Using these expressions gives $R_A = 150$, $M_A = -1800$, and $y_{l/2} = -1.63$. Table 8 for axial compression gives the formulas for these special cases, but when the lateral loads are not placed at midlength or any of the other common locations, a desk calculator or digital computer must be used. If tables of hyperbolic functions are employed, it should be kept in mind that adequate solutions can be made using values of kl close to the desired values if such values are given in the table and the desired ones are not. For example, if the values for the arguments 9.86 and 4.93 are not available but values for 10 and 5 are (note that it is necessary to maintain the correct ratio a/l), these values could be used with no noticeable change in the results.

7.8 *Beams of variable section*

Stress. For a beam, the cross section of which changes gradually, Eqs. 1, 4, and 10 to 12 (Art. 7.1) apply with sufficient accuracy; Eqs. 3 and 5 to 7 apply if I is treated as a variable, as in the examples that follow. All the formulas given in Table 3 for vertical shear and bending moments in *statically determinate* beams apply, but the formulas given for statically indeterminate beams and for deflection and slope are inapplicable to beams of nonuniform section unless the section varies in such a way that I is constant.

Accurate analysis (Ref. 3) shows that in an end-loaded cantilever beam of rectangular section which is symmetrically tapered in the plane of bending the maximum fiber stress is somewhat less than is indicated by Eq. 12, the error amounting to about 5 percent for a surface slope of 15° (wedge angle 30°) and about 10 percent for a surface slope of 20°. The maximum horizontal and vertical shear stress is shown to occur at the upper and lower surfaces instead of at the neutral axis and to be approximately three times as great as the average shear stress on the section for slopes up to 20°. It is very doubtful, however, if this shear stress is often critical even in wood beams, although it may possibly start failure in short, heavily reinforced concrete beams that are deepened or "haunched" at the ends. Such a failure, if observed, would probably be ascribed to compression since it would occur at a point of high compressive stress. It is also conceivable, of course, that this shear stress might be of importance in certain metal parts subject to repeated stress.

Abrupt changes in the section of a beam cause high local stresses, the effect of which is taken into account by using the proper factor of stress concentration (Art. 2.10, Table 37).

Deflection. Determining deflections or statically indeterminate reactions for beams of variable section can be considered in two categories: where the beam has a continuously varying cross section from one end to the other, and where the cross section varies in a stepwise fashion.

Considering the first category, where the section varies continuously, we sometimes find a variation where Eq. 5 (Art. 7.1) can be integrated directly, with the moment of inertia treated as a variable. In most instances, however, this is not easy, if possible, and a more productive approach is to integrate Eq. 6 (Art. 7.1) numerically using small incremental lengths Δx. This has been done for a limited number of cases, and the results are tabulated in Table 13*a* to *d*.

These tables give coefficients by which the stated reaction forces or moments or the stated deformations for uniform beams, as given in Table 3, must be multiplied to obtain the comparable reactions or deformations for the tapered beams. The coefficients are dependent upon the ratio of the moment of inertia at the right end of the beam I_B to the moment of inertia at the left end I_A, assuming that the uniform beam has a moment of inertia I_A. The coefficients are also dependent upon the manner of variation between the two end values. This variation is of the form $I_x = I_A(1 + Kx/l)^n$, where x is measured from the left end and $K = [(I_B/I_A)^{1/n} - 1]$. Thus if the beam is uniform, $n = 0$; if the width of a rectangular cross section varies linearly, $n = 1$; if the width of a rectangular cross section varies parabolically, $n = 2$; if the depth of a rectangular cross section varies linearly, $n = 3$; and if the lateral dimensions of any cross section vary linearly and proportionately, $n = 4$. Beams having similar variations in cross section can be analyzed approximately by comparing the given variations to those found in Table 13.

EXAMPLE

A tapered beam 30 in long with a depth varying linearly from 2 in at the left end to 4 in at the right end and with a constant width of 1.5 in is fixed on the right end and simply supported on the left end. A concentrated clockwise couple of 5000 in-lb is applied at midlength, and it is desired to know the maximum bending stress in the beam.

Solution. First determine the left-end reaction force for a uniform cross section. From Table 3, case 3c, the left reaction

$$R_A = \frac{-3M_o(l^2 - a^2)}{2l^3} = \frac{-3(5000)(30^2 - 15^2)}{2(30^3)} = -187.5 \text{ lb}$$

For the tapered beam

$$I_A = \frac{1.5(2^3)}{12} = 1 \text{ in}^4 \qquad I_B = \frac{1.5(4^3)}{12} = 8 \text{ in}^4$$

In Table 13*c* for $n = 3$, $I_B/I_A = 8$; and for case 3c with the loading at $l/2$, the coefficient is listed as 0.906. Therefore, the left-end reaction is $-187.5(0.906) = -170$ lb.

The maximum negative moment will occur just left of midlength and will equal $-170(15) = -2550$ in-lb. The maximum positive moment will occur just right of midlength and will equal $-2550 + 5000 = 2450$ in-lb. At midlength the moment of inertia $I = 1.5(3^3)/12 = 3.37$ in^4, and so the maximum stress is given by $\sigma = Mc/I = 2550(1.5)/3.37 = 1135$ lb/in^2 just left of midlength.

The second category of determining deflections, where the cross section varies in steps from one uniform section to another, can be solved in several ways. Equation 5 (Art. 7.1) can be integrated, matching slopes and deflections at the transition sections, or Eq. 6 can be integrated over the separate portions and summed to obtain the desired deflections. A third method utilizes the

TABLE 13a ***Reaction and deflection coefficients for tapered beams***

Moments of inertia vary as $(1 + Kx/l)^n$, where $n = 1.0$

Case no. in Table 3	Load location a/l	Multiplier listed for	I_B/I_A 0.25	0.50	2.0	4.0	8.0
1a	0	y_A	2.525	1.636	0.579	0.321	0.171
		θ_A	2.262	1.545	0.614	0.359	0.201
	0.50	y_A	2.898	1.755	0.543	0.284	0.146
		θ_A	2.811	1.731	0.548	0.289	0.149
1c	0.25	R_A	1.055	1.028	0.972	0.946	0.926
		θ_A	1.492	1.256	0.744	0.514	0.330
	0.50	R_A	1.148	1.073	0.936	0.887	0.852
		θ_A	1.740	1.365	0.682	0.435	0.261
1d	0.25	R_A	1.046	1.026	0.968	0.932	0.895
		M_A	1.137	1.077	0.905	0.797	0.686
	0.50	R_A	1.163	1.085	0.915	0.837	0.771
		M_A	1.326	1.171	0.829	0.674	0.542
1e	0.25	θ_A	1.396	1.220	0.760	0.531	0.342
		$y_{l/2}$	1.563	1.301	0.703	0.452	0.268
	0.50	θ_A	1.524	1.282	0.718	0.476	0.293
		$y_{l/2}$	1.665	1.349	0.674	0.416	0.239
2a. Uniform load	0	y_A	2.711	1.695	0.561	0.302	0.158
		θ_A	2.525	1.636	0.579	0.321	0.171
	0.50	y_A	3.091	1.806	0.532	0.275	0.140
		θ_A	3.029	1.790	0.535	0.278	0.142
2c. Uniform load	0	R_A	1.074	1.036	0.968	0.941	0.922
		θ_A	1.663	1.326	0.710	0.473	0.296
	0.50	R_A	1.224	1.104	0.917	0.858	0.818
		θ_A	1.942	1.438	0.653	0.403	0.237
2d. Uniform load	0	R_A	1.089	1.046	0.954	0.911	0.872
		M_A	1.267	1.137	0.863	0.733	0.615
	0.50	R_A	1.267	1.130	0.886	0.791	0.717
		M_A	1.481	1.234	0.794	0.625	0.491
2e. Uniform load	0	θ_A	1.508	1.271	0.729	0.492	0.309
		$y_{l/2}$	1.678	1.352	0.676	0.420	0.243
	0.50	θ_A	1.616	1.320	0.700	0.454	0.275
		$y_{l/2}$	1.765	1.389	0.658	0.398	0.225
2a. Uniformly increasing load	0	y_A	2.851	1.737	0.549	0.291	0.150
		θ_A	2.711	1.695	0.561	0.302	0.158
	0.50	y_A	3.220	1.839	0.525	0.270	0.137
		θ_A	3.172	1.827	0.527	0.272	0.138
2c. Uniformly increasing load	0	R_A	1.129	1.062	0.948	0.907	0.878
		θ_A	1.775	1.372	0.686	0.442	0.269
	0.50	R_A	1.275	1.124	0.907	0.842	0.799
		θ_A	2.063	1.479	0.639	0.388	0.225

TABLE 13a ***Reaction and deflection coefficients for tapered beams*** **(*Cont.*)**

Case no. in Table 3	Load location a/l	Multiplier listed for	I_B/I_A 0.25	0.50	2.0	4.0	8.0
2d. Uniformly increasing load	0	R_A	1.157	1.079	0.926	0.860	0.804
		M_A	1.353	1.177	0.833	0.685	0.559
	0.50	R_A	1.334	1.157	0.870	0.767	0.690
		M_A	1.573	1.269	0.777	0.601	0.468
2e. Uniformly increasing load	0	θ_A	1.561	1.295	0.714	0.472	0.291
		$y_{l/2}$	1.722	1.370	0.667	0.409	0.234
	0.50	θ_A	1.654	1.335	0.693	0.447	0.269
		$y_{l/2}$	1.806	1.404	0.651	0.392	0.221
3a	0	y_A	2.262	1.545	0.614	0.359	0.201
		θ_A	1.848	1.386	0.693	0.462	0.297
	0.50	y_A	2.566	1.658	0.566	0.305	0.159
		θ_A	2.443	1.622	0.575	0.313	0.164
3c	0	R_A	0.896	0.945	1.059	1.118	1.173
		θ_A	1.312	1.166	0.823	0.645	0.482
	0.50	R_A	1.016	1.014	0.977	0.952	0.929
		θ_A	1.148	1.125	0.794	0.565	0.365
3d	0.25	R_A	0.796	0.890	1.116	1.220	1.298
		M_A	1.614	1.331	0.653	0.340	0.106
	0.50	R_A	0.958	0.988	0.988	0.958	0.919
		M_A	0.875	0.965	0.965	0.875	0.758

TABLE 13b ***Reaction and deflection coefficients for tapered beams***

Moments of inertia vary as $(1 + Kx/l)^n$, where $n = 2.0$

Case no. in Table 3	Load location a/l	Multiplier listed for	I_B/I_A				
			0.25	0.50	2.0	4.0	8.0
1a	0	y_A	2.729	1.667	0.589	0.341	0.194
		θ_A	2.455	1.577	0.626	0.386	0.235
	0.50	y_A	3.105	1.783	0.549	0.296	0.157
		θ_A	3.025	1.761	0.555	0.301	0.161
1c	0.25	R_A	1.052	1.028	0.970	0.938	0.905
		θ_A	1.588	1.278	0.759	0.559	0.398
	0.50	R_A	1.138	1.070	0.932	0.867	0.807
		θ_A	1.867	1.390	0.695	0.468	0.306
1d	0.25	R_A	1.049	1.027	0.969	0.934	0.895
		M_A	1.155	1.082	0.909	0.813	0.713
	0.50	R_A	1.169	1.086	0.914	0.831	0.753
		M_A	1.358	1.177	0.833	0.681	0.548
1e	0.25	θ_A	1.509	1.246	0.778	0.586	0.428
		$y_{l/2}$	1.716	1.334	0.721	0.501	0.334
	0.50	θ_A	1.668	1.313	0.737	0.525	0.363
		$y_{l/2}$	1.840	1.385	0.692	0.460	0.294
2a. Uniform load	0	y_A	2.916	1.724	0.569	0.318	0.174
		θ_A	2.729	1.667	0.589	0.341	0.194
	0.50	y_A	3.282	1.830	0.537	0.283	0.148
		θ_A	3.226	1.816	0.540	0.287	0.150
2c. Uniform load	0	R_A	1.068	1.035	0.965	0.932	0.899
		θ_A	1.774	1.349	0.723	0.510	0.351
	0.50	R_A	1.203	1.098	0.910	0.831	0.761
		θ_A	2.076	1.463	0.664	0.430	0.271
2d. Uniform load	0	R_A	1.091	1.046	0.954	0.909	0.865
		M_A	1.290	1.142	0.866	0.741	0.628
	0.50	R_A	1.267	1.129	0.883	0.779	0.689
		M_A	1.509	1.239	0.795	0.625	0.486
2e. Uniform load	0	θ_A	1.645	1.301	0.747	0.542	0.382
		$y_{l/2}$	1.853	1.387	0.694	0.463	0.298
	0.50	θ_A	1.774	1.352	0.718	0.500	0.339
		$y_{l/2}$	1.955	1.426	0.675	0.438	0.274
2a. Uniformly increasing load	0	y_A	3.052	1.765	0.556	0.304	0.163
		θ_A	2.916	1.724	0.569	0.318	0.174
	0.50	y_A	3.395	1.860	0.529	0.276	0.143
		θ_A	3.354	1.849	0.532	0.279	0.144
2c. Uniformly increasing load	0	R_A	1.119	1.059	0.944	0.890	0.841
		θ_A	1.896	1.396	0.698	0.475	0.315
	0.50	R_A	1.244	1.116	0.898	0.810	0.736
		θ_A	2.196	1.503	0.649	0.411	0.255

TABLE 13b ***Reaction and deflection coefficients for tapered beams*** **(Cont.)**

Case no. in Table 3	Load location a/l	Multiplier listed for	I_B/I_A				
			0.25	0.50	2.0	4.0	8.0
2d. Uniformly increasing load	0	R_A	1.159	1.079	0.925	0.854	0.789
		M_A	1.379	1.182	0.836	0.691	0.565
	0.50	R_A	1.328	1.154	0.866	0.752	0.656
		M_A	1.596	1.272	0.777	0.598	0.457
2e. Uniformly increasing load	0	θ_A	1.708	1.326	0.732	0.521	0.360
		$y_{l/2}$	1.904	1.407	0.684	0.451	0.286
	0.50	θ_A	1.817	1.368	0.711	0.491	0.331
		$y_{l/2}$	2.001	1.442	0.668	0.430	0.268
3a	0	y_A	2.455	1.577	0.626	0.386	0.235
		θ_A	2.000	1.414	0.707	0.500	0.354
	0.50	y_A	2.786	1.691	0.575	0.323	0.177
		θ_A	2.667	1.657	0.586	0.333	0.185
3c	0	R_A	0.900	0.946	1.062	1.132	1.212
		θ_A	1.375	1.181	0.835	0.688	0.558
	0.50	R_A	1.021	1.015	0.977	0.946	0.911
		θ_A	1.223	1.148	0.814	0.622	0.451
3d	0.25	R_A	0.785	0.888	1.117	1.230	1.333
		M_A	1.682	1.347	0.660	0.348	0.083
	0.50	R_A	0.966	0.991	0.991	0.966	0.928
		M_A	0.890	0.972	0.974	0.905	0.807

TABLE 13c *Reaction and deflection coefficients for tapered beams*

Moments of inertia vary as $(1 + Kx/l)^n$, where $n = 3.0$

Case no. in Table 3	Load location a/l	Multiplier listed for	I_B/I_A 0.25	0.50	2.0	4.0	8.0
1a	0	y_A	2.796	1.677	0.593	0.349	0.204
		θ_A	2.520	1.587	0.630	0.397	0.250
	0.50	y_A	3.169	1.791	0.551	0.300	0.162
		θ_A	3.092	1.770	0.558	0.307	0.167
1c	0.25	R_A	1.051	1.027	0.969	0.936	0.899
		θ_A	1.626	1.286	0.764	0.573	0.422
	0.50	R_A	1.134	1.068	0.930	0.860	0.791
		θ_A	1.916	1.399	0.700	0.480	0.322
1d	0.25	R_A	1.050	1.027	0.969	0.934	0.895
		M_A	1.161	1.084	0.911	0.818	0.724
	0.50	R_A	1.171	1.086	0.914	0.829	0.748
		M_A	1.368	1.179	0.834	0.684	0.553
1e	0.25	θ_A	1.554	1.256	0.784	0.605	0.460
		$y_{l/2}$	1.774	1.346	0.728	0.519	0.362
	0.50	θ_A	1.723	1.324	0.743	0.543	0.391
		$y_{l/2}$	1.907	1.397	0.699	0.477	0.318
2a. Uniform load	0	y_A	2.981	1.734	0.572	0.324	0.182
		θ_A	2.796	1.677	0.593	0.349	0.204
	0.50	y_A	3.338	1.837	0.538	0.287	0.151
		θ_A	3.285	1.823	0.542	0.291	0.154
2c. Uniform load	0	R_A	1.066	1.034	0.965	0.928	0.891
		θ_A	1.817	1.357	0.727	0.522	0.370
	0.50	R_A	1.194	1.096	0.908	0.821	0.741
		θ_A	2.125	1.471	0.668	0.439	0.284
2d. Uniform load	0	R_A	1.092	1.046	0.954	0.908	0.863
		M_A	1.297	1.144	0.867	0.745	0.635
	0.50	R_A	1.266	1.128	0.882	0.776	0.680
		M_A	1.517	1.240	0.796	0.626	0.487
2e. Uniform load	0	θ_A	1.697	1.311	0.753	0.560	0.411
		$y_{l/2}$	1.919	1.400	0.700	0.480	0.322
	0.50	θ_A	1.833	1.363	0.724	0.517	0.365
		$y_{l/2}$	2.025	1.438	0.680	0.453	0.296
2a. Uniformly increasing load	0	y_A	3.115	1.773	0.559	0.309	0.169
		θ_A	2.981	1.734	0.572	0.324	0.182
	0.50	y_A	3.446	1.866	0.531	0.279	0.146
		θ_A	3.407	1.856	0.533	0.282	0.148
2c. Uniformly increasing load	0	R_A	1.114	1.058	0.942	0.885	0.829
		θ_A	1.942	1.404	0.702	0.486	0.332
	0.50	R_A	1.233	1.113	0.895	0.800	0.713
		θ_A	2.244	1.511	0.652	0.419	0.266

TABLE 13c ***Reaction and deflection coefficients for tapered beams (Cont.)***

Case no. in Table 3	Load location a/l	Multiplier listed for	I_B/I_A				
			0.25	0.50	2.0	4.0	8.0
2d. Uniformly increasing load	0	R_A	1.159	1.078	0.925	0.853	0.785
		M_A	1.386	1.183	0.837	0.694	0.569
	0.50	R_A	1.325	1.153	0.865	0.747	0.645
		M_A	1.602	1.273	0.777	0.598	0.456
2e. Uniformly increasing load	0	θ_A	1.764	1.337	0.738	0.538	0.387
		$y_{l/2}$	1.972	1.419	0.690	0.466	0.309
	0.50	θ_A	1.878	1.379	0.717	0.508	0.356
		$y_{l/2}$	2.072	1.454	0.674	0.445	0.288
3a	0	y_A	2.520	1.587	0.630	0.397	0.250
		θ_A	2.054	1.424	0.712	0.513	0.375
	0.50	y_A	2.858	1.702	0.579	0.330	0.185
		θ_A	2.741	1.668	0.590	0.342	0.194
3c	0	R_A	0.901	0.947	1.063	1.136	1.223
		θ_A	1.401	1.186	0.839	0.701	0.583
	0.50	R_A	1.022	1.015	0.977	0.945	0.906
		θ_A	1.257	1.157	0.820	0.642	0.483
3d	0.25	R_A	0.781	0.887	1.117	1.233	1.343
		M_A	1.705	1.352	0.663	0.355	0.088
	0.50	R_A	0.969	0.992	0.992	0.969	0.932
		M_A	0.897	0.975	0.977	0.916	0.828

TABLE 13d ***Reaction and deflection coefficients for tapered beams***

Moments of inertia vary as $(1 + Kx/l)^n$, where $n = 4.0$

Case no. in Table 3	Load location a/l	Multiplier listed for	I_B/I_A 0.25	0.50	2.0	4.0	8.0
1a	0	y_A	2.828	1.682	0.595	0.354	0.210
		θ_A	2.552	1.593	0.632	0.402	0.258
	0.50	y_A	3.200	1.796	0.553	0.303	0.165
		θ_A	3.124	1.774	0.559	0.310	0.170
1c	0.25	R_A	1.051	1.027	0.969	0.935	0.896
		θ_A	1.646	1.290	0.767	0.581	0.434
	0.50	R_A	1.131	1.068	0.929	0.857	0.784
		θ_A	1.941	1.404	0.702	0.485	0.331
1d	0.25	R_A	1.051	1.027	0.969	0.935	0.896
		M_A	1.164	1.085	0.912	0.821	0.730
	0.50	R_A	1.172	1.086	0.914	0.828	0.746
		M_A	1.373	1.180	0.835	0.686	0.556
1e	0.25	θ_A	1.578	1.260	0.787	0.615	0.476
		$y_{l/2}$	1.805	1.351	0.731	0.528	0.376
	0.50	θ_A	1.752	1.329	0.746	0.552	0.406
		$y_{l/2}$	1.941	1.404	0.702	0.485	0.331
2a. Uniform load	0	y_A	3.013	1.738	0.573	0.328	0.187
		θ_A	2.828	1.682	0.595	0.354	0.210
	0.50	y_A	3.365	1.841	0.539	0.289	0.154
		θ_A	3.314	1.827	0.543	0.293	0.157
2c. Uniform load	0	R_A	1.065	1.034	0.964	0.927	0.888
		θ_A	1.839	1.361	0.729	0.528	0.380
	0.50	R_A	1.190	1.095	0.907	0.817	0.731
		θ_A	2.151	1.476	0.670	0.443	0.290
2d. Uniform load	0	R_A	1.092	1.046	0.954	0.908	0.862
		M_A	1.301	1.145	0.867	0.747	0.639
	0.50	R_A	1.266	1.128	0.882	0.774	0.676
		M_A	1.521	1.241	0.796	0.627	0.488
2e. Uniform load	0	θ_A	1.724	1.316	0.756	0.569	0.426
		$y_{l/2}$	1.953	1.406	0.703	0.488	0.335
	0.50	θ_A	1.864	1.369	0.727	0.526	0.379
		$y_{l/2}$	2.061	1.445	0.683	0.461	0.307
2a. Uniformly increasing load	0	y_A	3.145	1.778	0.560	0.312	0.173
		θ_A	3.013	1.738	0.573	0.328	0.187
	0.50	y_A	3.470	1.869	0.532	0.281	0.147
		θ_A	3.432	1.859	0.534	0.284	0.150
2c. Uniformly increasing load	0	R_A	1.112	1.057	0.942	0.882	0.823
		θ_A	1.966	1.408	0.704	0.492	0.340
	0.50	R_A	1.227	1.111	0.894	0.794	0.701
		θ_A	2.269	1.515	0.653	0.423	0.271

TABLE 13d ***Reaction and deflection coefficients for tapered beams*** **(*Cont.*)**

Case no. in Table 3	Load location a/l	Multiplier listed for	I_B/I_A				
			0.25	0.50	2.0	4.0	8.0
2d. Uniformly increasing load	0	R_A	1.159	1.078	0.924	0.852	0.783
		M_A	1.390	1.184	0.837	0.695	0.572
	0.50	R_A	1.323	1.153	0.864	0.744	0.639
		M_A	1.605	1.274	0.777	0.598	0.456
2e. Uniformly increasing load	0	θ_A	1.793	1.343	0.741	0.547	0.402
		$y_{l/2}$	2.007	1.425	0.693	0.475	0.321
	0.50	θ_A	1.909	1.385	0.719	0.516	0.369
		$y_{l/2}$	2.108	1.461	0.677	0.453	0.299
3a	0	y_A	2.552	1.593	0.632	0.402	0.258
		θ_A	2.081	1.428	0.714	0.520	0.386
	0.50	y_A	2.893	1.707	0.581	0.334	0.190
		θ_A	2.778	1.674	0.592	0.346	0.200
3c	0	R_A	0.902	0.947	1.063	1.138	1.227
		θ_A	1.414	1.189	0.841	0.707	0.595
	0.50	R_A	1.023	1.015	0.976	0.944	0.904
		θ_A	1.275	1.161	0.823	0.652	0.499
3d	0.25	R_A	0.780	0.887	1.117	1.234	1.347
		M_A	1.716	1.354	0.665	0.359	0.092
	0.50	R_A	0.971	0.993	0.993	0.971	0.935
		M_A	0.902	0.976	0.979	0.922	0.839

advantages of the step function and its application to beam deflections as given in Table 3. In a given portion of the span where the cross section is uniform it is apparent that the shape of the elastic curve will remain the same if the internal bending moments and the moments of inertia are increased or decreased in proportion. By this means, a modified moment diagram can be constructed which could be applied to a beam with a single constant cross section and thereby produce an elastic curve identical to the one produced by the actual moments and the several moments of inertia present in the actual span. It is also apparent that this modified moment diagram could be produced by adding appropriate loads to the beam. (See Refs. 29 and 65.) In summary, then, a new loading is constructed which will produce the required elastic curve, and the solution for this loading is carried out by using the formulas in Table 3. This procedure will be illustrated by the following example.

EXAMPLE

The beam shown in Fig. 7.12 has a constant depth of 4 in and a step increase in the width from 2 in to 5 in at a point 5 ft from the left end. The left end is simply supported, and the right end is fixed; the loading is a uniform 200 lb/ft from $x = 3$ ft to the right end. Find the value of the reaction at the left end and the maximum stress.

Solution. For the left 5 ft, $I_1 = 2(4^3)/12 = 10.67\text{ in}^4$. For the right 5 ft, $I_2 = 5(4^3)/12 = 26.67\text{ in}^4$, or 2.5 times I_1.

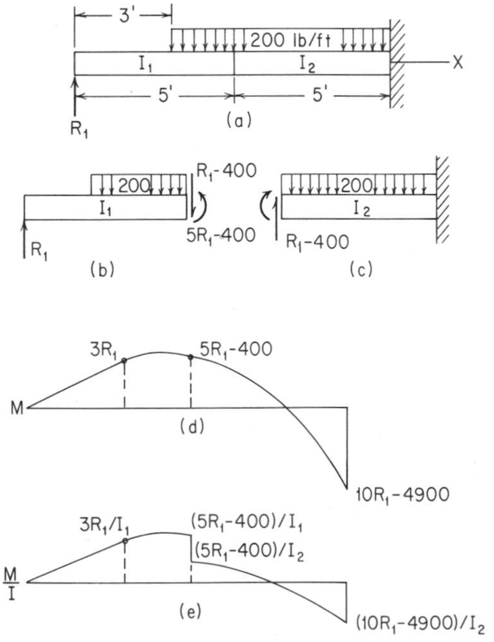

Fig. 7.12

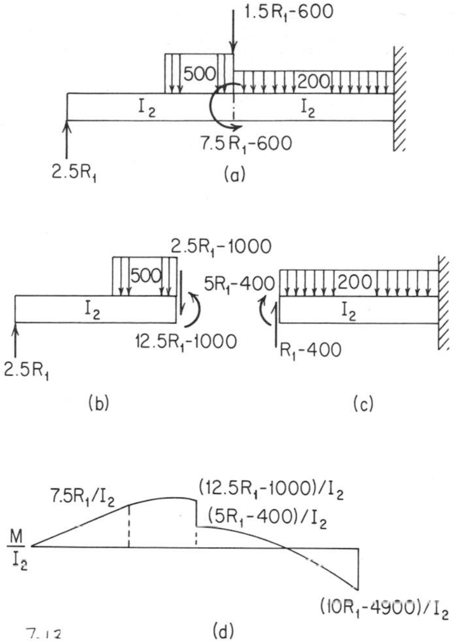

Fig. 7.13

The same M/I diagram shown in Fig. 7.12*e* can be produced by the loading shown in Fig. 7.13 acting upon a beam having a constant moment of inertia I_2. Note that all loads on the left portion simply have been increased by a factor of 2.5, while added loads at the 5-ft position reduce the effects of these added loads to those previously present on the right portion.

To find the left-end reaction for the beam loaded as shown in Fig. 7.13*a*, use Table 3, case 1c, where $W = 1.5R_1 - 600$ and $a = 5$; case 2c, where $w_a = w_l = 500$ and $a = 3$; case 2c, again, where $w_a = w_l = -300$ and $a = 5$; and finally case 3c, where $M_o = -(7.5R_1 - 600)$ and $a = 5$. Summing the expressions for R_A from these cases in the order given above, we obtain

$$R_A = 2.5R_1 = \frac{(1.5R_1 - 600)(10 - 5)^2}{2(10^3)}[2(10) + 5] + \frac{500(10 - 3)^3}{8(10^3)}[3(10) + 3]$$

$$+ \frac{(-300)(10 - 5)^3}{8(10^3)}[3(10) + 5] - \frac{3[-(7.5R_1 - 600)]}{2(10^3)}(10^2 - 5^2)$$

which gives $R_1 = 244$ lb.

From Fig. 7.12*a* we can observe that the maximum positive bending moment will occur at $x = 4.22$ ft, where the transverse shear will be zero. The maximum moments are therefore

$$\text{Max } +M = 244(4.22) - \frac{200}{2}(1.22^2) = 881 \text{ ft-lb}$$

$$\text{Max } -M = 244(10) - 4900 = -2460 \text{ ft-lb} \qquad \text{at the right end}$$

The maximum stresses are $\sigma = 881(12)(2)/10.67 = 1982$ lb/in² at $x = 4.22$ ft and $\sigma = 2460(12)(2)/26.67 = 2215$ lb/in² at $x = 10$ ft.

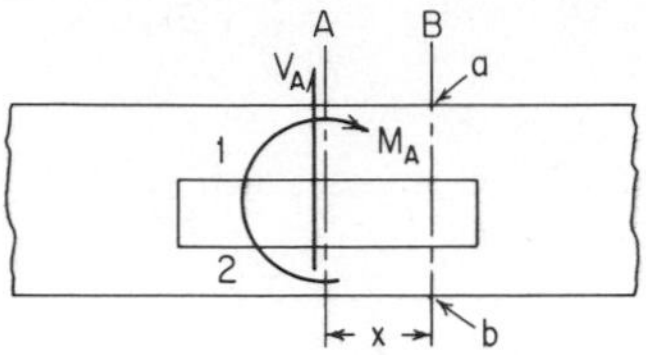

Fig. 7.14

7.9 *Slotted beams*

If the web of a beam is pierced by a hole or slot (Fig. 7.14), the stresses in the extreme fibers a and b at any section B are given by

$$\sigma_a = -\frac{M_A}{I/c} - \frac{V_A x I_1/(I_1 + I_2)}{(I/c)_1} \qquad \text{(compression)}$$

$$\sigma_b = \frac{M_A}{I/c} + \frac{V_A x I_2/(I_1 + I_2)}{(I/c)_2} \qquad \text{(tension)}$$

Here M_A is the bending moment at A (midlength of the slot), V_A is the vertical shear at A; I/c is the section modulus of the net beam section at B; I_1 and I_2 are the moments of inertia, and $(I/c)_1$ and $(I/c)_2$ are the section moduli of the cross sections of parts 1 and 2 about their own central axes. M and V are positive or negative according to the usual convention, and x is positive when measured to the right.

The preceding formulas are derived by replacing all forces acting on the beam to the left of A by an equivalent couple M_A and shear V_A acting at A. The couple produces a bending stress given by the first term of the formula. The shear divides between parts 1 and 2 in proportion to their respective I's and produces in each part an additional bending stress given by the second term of the formula. The stress at any other point in the cross section can be found similarly by adding the stresses due to M_A and those due to this secondary bending caused by the shear. (At the ends of the slot there is a stress concentration at the corners which is not taken into account here.)

The above analysis applies also to a beam with multiple slots of equal length; all that is necessary is to modify the term in brackets so that the numerator is the I of the part in question and the denominator is the sum of the I's of all the parts 1, 2, 3, etc. The formulas can also be used for a rigid frame consisting of beams of equal length joined at their ends by rigid members; thus in Fig. 7.14 parts 1 and 2 might equally well be two separate beams joined at their ends by rigid crosspieces.

7.10 *Beams of relatively great depth*

In beams of small span/depth ratio, the shear stresses are likely to be high and the resulting deflection due to shear may not be negligible. For span/

depth ratios of 3 or more, the deflection y_s due to shear is found by the method of unit loads to be

$$y_s = F \int \frac{Vv}{AG}\, dx \qquad (19)$$

or by Castigliano's first theorem to be

$$y_s = \frac{\partial U_s}{\partial P} \qquad (20)$$

In Eq. 19, V is the vertical shear due to the actual loads, v is the vertical shear due to a load of 1 lb acting at the section where the deflection is desired, A is the area of the section, G is the modulus of rigidity, F is a factor depending on the form of the cross section, and the integration extends over the entire length of the beam, with due regard to the sign of V and v. For a rectangular section, $F = \frac{6}{5}$; for a solid circular section, $F = \frac{10}{9}$; for a thin-walled hollow circular section, $F = 2$; for an I or box section having flanges and web of uniform thickness,

$$F = \left[1 + \frac{3(D_2^2 - D_1^2)D_1}{2D_2^3}\left(\frac{t_2}{t_1} - 1\right)\right]\frac{4D_2^2}{10r^2}$$

where D_1 = distance from neutral axis to the nearest surface of the flange
D_2 = distance from neutral axis to extreme fiber
t_1 = thickness of web (or webs in box beams)
t_2 = width of flange
r = radius of gyration of section with respect to the neutral axis

If the I- or box beam has flanges of nonuniform thickness, it may be replaced by an "equivalent" section whose flanges, of uniform thickness, have the same width and area as those of the actual section (Ref. 19). Approximate results may be obtained for I-beams using $F = 1$ and taking for A the area of the web.

Application of Eq. 19 to several common cases of loading yields the following results:

End support, center load P	$y_s = \frac{1}{4} F \frac{Pl}{AG}$
End support, uniform load W	$y_s = \frac{1}{8} F \frac{Wl}{AG}$
Cantilever, end load P	$y_s = F \frac{Pl}{AG}$
Cantilever, uniform load W	$y_s = \frac{1}{2} F \frac{Wl}{AG}$

In Eq. 20, $U_s = F \int (V^2/2AG)\, dx$, P is a vertical load, real or imaginary, applied at the section where y_s is to be found, and the other terms have the same meaning as in Eq. 19.

The deflection due to shear will usually be negligible in metal beams unless the span/depth ratio is extremely small; in wood beams, because of the small value of G compared with E, deflection due to shear is much more important. In computing deflections it may be allowed for by using for E a value obtained from bending tests (shear deflection ignored) on beams of similar proportions or a value about 10 percent less than that found by testing in direct compression if the span/depth ratio is between 12 and 24. For larger ratios the effect of shear is negligible, and for lower ratios it should be calculated by the preceding method.

For extremely short deep beams, the assumption of linear stress distribution, on which the simple theory of flexure is based, is no longer valid. Equation 1 gives sufficiently accurate results for span/depth ratios down to about 3; for still smaller ratios it was believed formerly that the actual stresses were smaller than the formula indicates (Refs. 1 and 2), but more recent analyses by numerical methods (Refs. 43 and 44) indicate that the contrary is true. These analyses show that at s/d between 1.5 and 1, depending on the manner of loading and support, the stress distribution changes radically and the ratio of maximum stress to Mc/I becomes greater than 1 and increases rapidly as s/d becomes still smaller. In the following table, the influence of s/d on both maximum fiber stress and maximum horizontal shear stress is shown in accordance with the solution given in Ref. 43. Reference 44 gives comparable results, and both strain-gage measurements (Ref. 45) and photoelastic studies (Ref. 46) support the conclusions reached in these analyses.

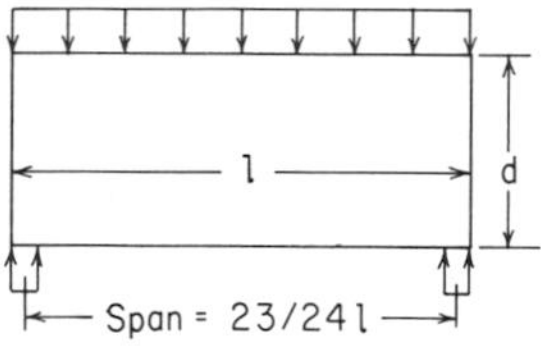

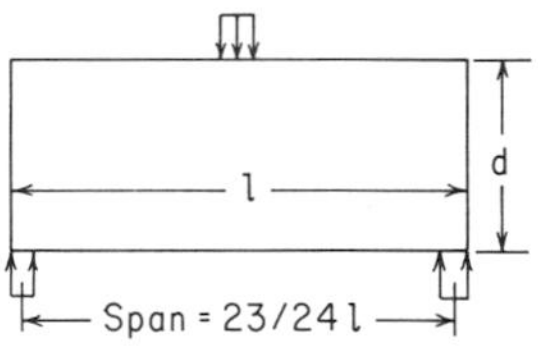

Ratio l/d	Ratio span/d	Uniform load over entire l			Uniform load over middle $\frac{1}{12}l$		
		$\frac{\max \sigma_t}{Mc/I}$	$\frac{\max \sigma_c}{Mc/I}$	$\frac{\max \tau}{V/A}$	$\frac{\max \sigma_t}{Mc/I}$	$\frac{\max \sigma_c}{Mc/I}$	$\frac{\max \tau}{V/A}$
3	2.875	1.025	1.030	1.58	0.970	1.655	1.57
2.5	2.395	1.046	1.035	1.60	0.960	1.965	1.60
2.0	1.915	1.116	1.022	1.64	0.962	2.525	1.70
1.5	1.4375	1.401	0.879	1.80	1.038	3.585	1.92
1	0.958	2.725	0.600	2.43	1.513	6.140	2.39
0.5	0.479	10.95	2.365	4.53	5.460	15.73	3.78
$\frac{1}{3}$	0.3193	24.70	5.160	6.05	12.35	25.55	7.23

These established facts concerning elastic stresses in short beams seem incompatible with the contrary influence of s/d on modulus of rupture, discussed in Art. 7.15, unless it is assumed that there is a very radical redistribution of stress as soon as plastic action sets in.

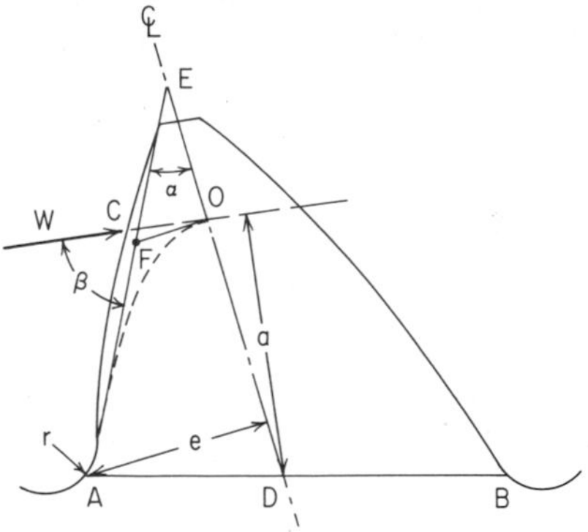

Fig. 7.15

The stress produced by a concentrated load acting on a very short *cantilever* beam or projection (gear tooth, sawtooth, screw thread) can be found by the following formula, due to Heywood (Chap. 2, Ref. 49) and modified by Kelley and Pedersen (Ref. 59). As given here, the formula follows this modification, with some changes in notation. Figure 7.15 represents the profile of the beam, assumed to be of uniform thickness t. ED is the axis or center line of the beam; it bisects the angle between the sides if these are straight; otherwise it is drawn through the centers of two unequal inscribed circles. W represents the load; its line of action, or load line, intersects the beam profile at C and the beam axis at O. The inscribed parabola, with vertex at O, is tangent to the fillet on the tension side of the beam at A, which is the point of maximum tensile stress. (A can be located by making AF equal to FE by trial, F being the intersection of a perpendicular to the axis at O and a trial tangent to the fillet.) B is the corresponding point on the compression side, and D is the intersection of the beam axis with section AB. The dimensions a and e are perpendicular, respectively, to the load line and to the beam axis; r is the fillet radius; and b is the straight-line distance from A to C. The tensile stress at A is given by

$$\sigma = \frac{W}{t}\left[1 + 0.26\left(\frac{e}{r}\right)^{0.7}\right]\left[\frac{1.5a}{e^2} + \frac{\cos\beta}{2e} + \frac{0.45}{(be)^{\frac{1}{2}}}\right]$$

Here the quantity in the first pair of brackets is the factor of stress concentration for the fillet. In the second pair of brackets, the first term represents the bending moment divided by the section modulus; the second term represents the effect of the component of the load along the tangent line, positive when tensile; and the third term represents what Heywood calls the *proximity effect,* which may be regarded as an adjustment for the very small span/depth ratio.

Kelley and Pedersen have suggested a further refinement in locating the point of maximum stress, putting it at an angular distance equal to $25° - \frac{1}{2}\alpha$, positive toward the root of the fillet. Heywood suggests locating this point at 30° from the outer end of the fillet, reducing this to 12° as the ratio of b to e increases; also, Heywood locates the moment of W about a point halfway between A and B instead of about D. For most cases the slightly different procedures seem to give comparable results and agree well with photoelastic analysis. However, more recent experimental studies (1963), including fatigue tests, indicate that actual stresses may considerably exceed those computed by the formula (Ref. 63).

7.11 *Beams of relatively great width*

Because of prevention of the lateral deformation that would normally accompany the fiber stresses, wide beams, such as thin metallic strips, are more rigid than the formulas of Art. 7.1 indicate. This stiffening effect is taken into account by using $E/(1 - \nu^2)$ instead of E in the formulas for deflection and curvature (Ref. 21).

In very short wide beams, such as the concrete slabs used as highway-bridge flooring, the deflection and fiber-stress distribution cannot be regarded as uniform across the width. In calculating the strength of such a slab, it is convenient to make use of the concept of *effective width*, i.e., the width of a spanwise strip which, acting as a beam with uniform extreme fiber stress equal to the maximum stress in the slab, develops the same resisting moment as does the slab. The effective width depends on the manner of support, manner of loading, and ratio of breadth to span b/a. It has been determined by Holl (Ref. 22) for a number of assumed conditions, and the results are given in the following table for a slab that is freely supported at each of two opposite edges (Fig. 7.16). Two kinds of loading are considered, viz., uniform load over the entire slab and load uniformly distributed over a central circular area of radius c. The ratio of the effective width e to the span a is given for each of a number of ratios of c to slab thickness h and each of a number of b/a values.

Loading	Values of e/a for				
	$b/a = 1$	$b/a = 1.2$	$b/a = 1.6$	$b/a = 2$	$b/a = \infty$
Uniform	0.960	1.145	1.519	1.900	
Central, $c = 0$	0.568	0.599	0.633	0.648	0.656
Central, $c = 0.125h$	0.581	0.614	0.649	0.665	0.673
Central, $c = 0.250h$	0.599	0.634	0.672	0.689	0.697
Central, $c = 0.500h$	0.652	0.694	0.740	0.761	0.770

For the same case (a slab that is supported at opposite edges and loaded on a central circular area) Westergaard (Ref. 23) gives $e = 0.58a + 4c$ as an approximate expression for effective width. Morris (Ref. 24) gives $e = \frac{1}{2}e_c + d$

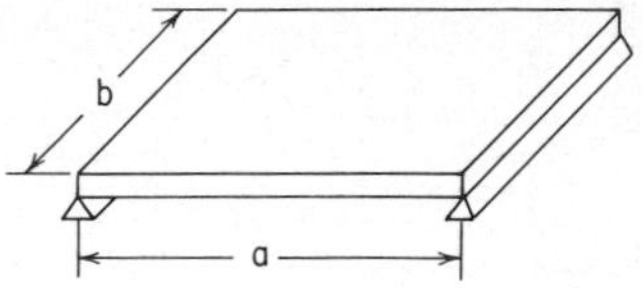

Fig. 7.16

as an approximate expression for the effective width for midspan *off-center* loading, where e_c is the effective width for central loading and d is the distance from the load to the nearer unsupported edge.

For a slab that is *fixed* at two opposite edges and uniformly loaded, the stresses and deflections may be calculated with sufficient accuracy by the ordinary beam formulas, replacing E by $E/(1 - \nu^2)$. For a slab thus supported and loaded at the center, the maximum stresses occur under the load, except for relatively large values of c, where they occur at the midpoints of the fixed edges. The effective widths are approximately as given in the following table (values from the curves of Ref. 22). Here b/a and c have the same meaning as in the preceding table, but it should be noted that values of e/b are given instead of e/a.

Values of c	Values of e/b for				Max stress at
	$b/a = 1$	$b/a = 1.2$	$b/a = 1.6$	$b/a = 2.0$	
0	0.51	0.52	0.53	0.53	Load
$0.01a$	0.52	0.54	0.55	0.55	Load
$0.03a$	0.58	0.59	0.60	0.60	Load
$0.10a$	0.69	0.73	0.81	0.86	Fixed edges

Holl (Ref. 22) discusses the deflections of a wide beam with two edges supported and the distribution of pressure under the supported edges. The problem of determining the effective width in concrete slabs and tests made for that purpose are discussed by Kelley (Ref. 25), who also gives a brief bibliography on the subject.

The case of a very wide *cantilever* slab under a concentrated load is discussed by MacGregor (Ref. 26), Holl (Ref. 27), Jaramillo (Ref. 47), Wellauer and Seireg (Ref. 48), Little (Ref. 49), Small (Ref. 50), and others. For the conditions represented in Fig. 7.17, a cantilever plate of infinite length with a concentrated load, the bending stress σ at any point can be expressed by $\sigma = K_m(6P/t^2)$, and the deflection y at any point by $y = K_y(Pa^2/\pi D)$, where K_m and K_y are dimensionless coefficients that depend upon the location of the load and the point, and D is as defined in Table 24. For the load at $x = c$, $z = 0$, the stress at any point on the fixed edge $x = 0$, $z = z$, and the deflection at any point on the free edge $x = a$, $z = z$, can be found by using the following values of K_m and K_y:

c/a \ z/a		0	0.25	0.50	1.0	1.5	2	∞
1.0	K_m	0.509	0.474	0.390	0.205	0.091	0.037	0
	K_y	0.524	0.470	0.380	0.215	0.108	0.049	0
0.75	K_m	0.428	0.387	0.284	0.140	0.059	0.023	0
	K_y	0.318	0.294	0.243	0.138	0.069	0.031	0
0.50	K_m	0.370	0.302	0.196	0.076	0.029	0.011	0
0.25	K_m	0.332	0.172	0.073	0.022	0.007	0.003	0

These values are based on the analysis of Jaramillo (Ref. 47), who assumes an infinite length for the plate, and are in good agreement, so far as comparable, with coefficients given by MacGregor (Ref. 26). They differ only slightly from results obtained by Holl (Ref. 27) for a length/span ratio of 4 and by Little (Ref. 49) for a length/span ratio of 5, and are in good agreement with available test data.

Wellauer and Seireg (Ref. 48) discuss the results of tests on beams of various proportions and explain and illustrate an empirical method by which the K_m values obtained by Jaramillo (Ref. 47) for the infinite plate under concentrated loading can be used to determine approximately the stress in a finite plate under any arbitrary transverse loading.

The stresses corresponding to the tabulated values of K_m are *spanwise* stresses; the maximum *crosswise* stress occurs under the load when the load is applied at the midpoint of the free edge and is approximately equal to the maximum spanwise stress for that loading.

Although the previous formulas are based on the assumption of infinite width of a slab, tests (Ref. 26) on a plate with a width of $8\frac{1}{2}$ in and span a of $1\frac{1}{4}$ in showed close agreement between calculated and measured deflections; and Holl's analysis (Ref. 27), based on the assumption of a plate width

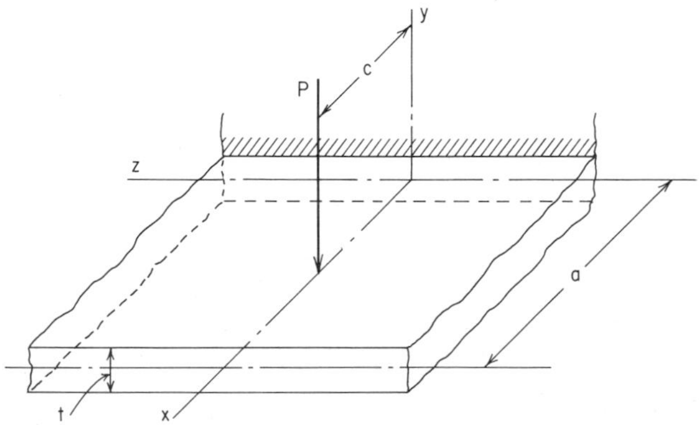

Fig. 7.17

four times the span, gives results that differ only slightly from MacGregor's (Ref. 26). The formulas given should therefore be applicable to slabs of breadth as small as four times the span.

7.12 *Beams with wide flanges; shear lag*

In thin metal construction, box, T-, or I-beams with very wide thin cover plates or flanges are sometimes used, and when a thin plate is stiffened by an attached member, a portion of the plate may be considered as a flange, acting integrally with the attached member which forms the web; examples are to be found in ship hulls, floors, tanks, and aircraft. In either type of construction the question arises as to what width of flange or plate would be considered effective; i.e., what width, uniformly stressed to the maximum stress that actually occurs, would provide a resisting moment equal to that of the actual stresses, which are greatest near the web and diminish as the distance from it increases.

This problem has been considered by several investigators; the results tabulated on page 192 are due to Hildebrand and Reissner (Ref. 38), Winter (Ref. 39), and Miller (Ref. 28).

Let b = actual net width of the flange (or clear distance between webs in continuous plate and stiffener construction); let l = span; and let b' = effective width of the flange at the section of maximum bending moment. Then the approximate value of b'/b, which varies with the loading and with the ratio l/b, can be found for beams of uniform section in the table on p. 192. (In this table the case numbers refer to the manner of loading and support represented in Table 3.)

Some of the more important conclusions stated in Ref. 38 can be summarized as follows:

The amount of shear lag depends not only on the method of loading and support and the ratio of span to flange width, but also on the ratio of G to E and on the ratio $m = (3I_w + I_s)/(I_w + I_s)$, where I_w and I_s are the moments of inertia about the neutral axis of the beam of the side plates and cover plates, respectively. (The values tabulated from Ref. 38 are for $G/E = 0.375$ and $m = 2$.) The value of b'/b increases with increasing m, but for values of m between 1.5 and 2.5 the variation is small enough to be disregarded. Shear lag at the critical section does not seem to be affected appreciably by the taper of the beam in width, but the taper in cover-plate thickness may have an important effect. In beams with fixed ends the effect of shear lag at the end sections is the same as for a cantilever of span equal to the distance from the point of inflection to the adjacent end.

In Ref. 39 it is stated that for a given l/b ratio the effect of shear lag is practically the same for box beams, I-beams, T-beams, and U-beams.

Flange in compression. The preceding discussion and tabulated factors apply to any case in which the flange is subjected to tension or to compression

Ratio effective width to total width b'/b for wide flanges

Case no. and load positions (from Table 3)	Reference no.	l/b													
		1	1.25	1.50	1.75	2	2.5	3	4	5	6	8	10	15	20
1a. $a = 0$	38	0.571	0.638	0.690	0.730	0.757	0.801	0.830	0.870	0.895	0.913	0.934	0.946		
2a. $w_a = w_l, a = 0$	38			0.550	0.600	0.632	0.685	0.724	0.780	0.815	0.842	0.876	0.899		
2a. $w_a = 0, a = 0$	38						0.609	0.650	0.710	0.751	0.784	0.826	0.858		
1e. $a = l/2$	38				0.530	0.571	0.638	0.686	0.757	0.801	0.830	0.870	0.895	0.936	0.946
1e. $a = l/2$	39							0.550	0.670	0.732	0.779	0.850	0.894	0.945	
1e. $a = l/2$	28						0.525			0.750					
1e. $a = l/4$	38				0.455	0.495	0.560	0.610	0.686	0.740	0.788	0.826	0.855	0.910	0.930
2e. $w_a = w_l, a = 0$	38				0.640	0.690	0.772	0.830	0.897	0.936	0.957	0.977	0.985	0.991	0.995
2e. $a = 0$	39							0.850	0.896	0.928	0.950	0.974	0.984	0.995	
2e. $a = 0$	28						0.725			0.875					

less than that required to produce elastic instability (see Chap. 14). When a thin flange or sheet is on the compression side, however, it may be stressed beyond the stability limit. For this condition, the effective width decreases with the actual stress. A formula for effective width used in aircraft design is

$$b' = Kt\sqrt{\frac{E}{s}}$$

where s is the maximum compressive stress (adjacent to the supporting web or webs) and K is a coefficient which may be conservatively taken as 0.85 for box beams and 0.60 for a T- or I-beam having flanges with unsupported outer edges.

A theoretical analysis that takes into account both compressive buckling and shear lag is described in Ref. 40. Problems involving shear lag and buckling are most frequently encountered in design with thin-gage metal; good guides to such design are the books "Specification for the Design of Cold-Formed Steel Structural Members" in 5 parts including commentary, published in 1968 by the American Iron and Steel Institute, and "Aluminum Construction Manual," published in 1959 by the Aluminum Association.

7.13 *Beams with very thin webs*

In beams with extremely thin webs, such as are used in airplane construction, buckling due to shear will occur at stresses well below the elastic limit. This can be prevented if the web is made *shear resistant* by the addition of stiffeners such as those used in plate girders, but the number of these required may be excessive. Instead of making the web shear resistant, it may be permitted to buckle elastically without damage, the shear being carried wholly in *diagonal tension.* This tension tends to pull the upper and lower flanges together, and to prevent this, vertical struts are provided which carry the vertical component of the diagonal web tension. A girder so designed is, in effect, a Pratt truss, the web replacing the diagonal-tension members and the vertical struts constituting the compression members. In appearance, these struts resemble the stiffeners of an ordinary plate girder, but their function is obviously quite different.

A beam of this kind is called a *diagonal-tension field beam,* or *Wagner beam,* after Professor Herbert Wagner of Danzig, who is largely responsible for developing the theory. Because of its rather limited field of application, only one example of the Wagner beam will be considered here, viz., a cantilever under end load.

Let $P =$ end load; $h =$ depth of the beam; $t =$ thickness of the web; $d =$ spacing of the vertical struts; $x =$ distance from the loaded end to the section in question; H_t and $H_c =$ total stresses in the tension and compression flanges, respectively, at the given section; C = total compression on a vertical strut; and f = unit diagonal tensile stress in the web. Then

$$H_t = \frac{Px}{h} - \frac{1}{2}P \qquad H_c = \frac{Px}{h} + \frac{1}{2}P \qquad C = \frac{Pd}{h} \qquad f = \frac{2P}{ht}$$

The vertical component of the web tension constitutes a beam loading on each individual flange between struts; the maximum value of the resulting bending moment occurs at the struts and is given by $M_f = \frac{1}{12}Pd^2/h$. The flexural stresses due to M_f must be added to the stresses due to H_t or H_c, which may be found simply by dividing H_t or H_c by the area of the corresponding flange.

The horizontal component of the web tension causes a bending moment $M = \frac{1}{8}Ph$ in the vertical strut at the end of the beam unless bending there is prevented by some system of bracing. This end strut must also distribute the load to the web, and should be designed to carry the load as a pin-ended column of length $\frac{1}{2}h$ as well as to resist the moment imposed by the web tension.

The intermediate struts are designed as pin-ended columns with lengths somewhat less than h. An adjacent portion of the web is included in the area of the column, the width of the strip considered effective being $30t$ in the case of aluminum and $60t$ in the case of steel.

Obviously the preceding formulas will apply also to a beam with end supports and center load if P is replaced by the reaction $\frac{1}{2}P$. Because of various simplifying assumptions made in the analysis, these formulas are conservative; in particular the formula for stress in the vertical struts or stiffeners gives results much larger than actual stresses that have been discovered experimentally. A more accurate analysis, together with experimental data from various sources, will be found in Refs. 30, 34, and 35.

7.14 *Beams not loaded in plane of symmetry; flexural center*

The formulas for stress and deflection given in Art. 7.1 are valid if the beam is loaded in a plane of symmetry; they are also valid if the applied loads are parallel to either principal central axis of the beam section, but unless the loads also pass through the *elastic axis,* the beam will be subjected to torsion as well as bending.

For the general case of a beam of any section loaded by a transverse load P in any plane, therefore, the solution comprises the following steps: (1) The load P is resolved into an equal and parallel force P' passing through the flexural center Q of the section, and a twisting couple T equal to the moment of P about Q; (2) P' is resolved at Q into rectangular components P'_u and P'_v, each parallel to a principal central axis of the section; (3) the flexural stresses and deflections due to P'_u and P'_v are calculated independently by the formulas of Art. 7.1 and superimposed to find the effect of P'; and (4) the stresses due to T are computed independently and superimposed on the stresses due to P', giving the stresses due to the actual loading. (It is to be noted that T may cause longitudinal fiber stresses as well as shear stresses. See Art. 9.3 and the example at the end of this article.) If there are several loads, the effect of each is calculated separately and these effects added. For a distributed load the same procedure is followed as for a concentrated load.

The above procedure requires the determination of the position of the *flexural center* Q. For any section having two or more axes of symmetry (rectangle, I-beam, etc.) and for any section having a point of symmetry (equilateral triangle, Z bar, etc.), Q is at the centroid. For any section having only one axis of symmetry, Q is on that axis but in general not at the centroid. For such sections and for unsymmetrical sections in general, the position of Q must be determined by calculation, direct experiment, or the soap-film method (Art. 4.4).

Table 14 gives the position of the flexural center for each of a number of sections.

Neutral Axis. When a beam is bent by one or more loads that lie in a plane not parallel to either principal central axis of the section, the neutral axis passes through the centroid but is not perpendicular to the plane of the loads. Let axes 1 and 2 be the principal central axes of the section, and let I_1 and I_2 represent the corresponding moments of inertia. Then, if the plane of the loads makes with axis 1 an angle α, the neutral axis makes with axis 2 an angle β such that $\tan \beta = (I_2/I_1) \tan \alpha$. It can be seen from this equation that the neutral axis tends to approach the principal central axis about which the moment of inertia is least.

EXAMPLE

Figure 7.18*a* represents a cantilever beam of channel section under a diagonal end load applied at one corner. It is required to determine the maximum resulting fiber stress.

Solution. For the section (Fig. 7.18*b*): $I_u = 5.61$; $I_v = 19.9$; $b = 3.875$; $h = 5.75$; and $t = \frac{1}{4}$. By the formula from Table 14, $e = b^2h^2t/4I_v = 1.55$ in; therefore the flexural center is at Q, as shown. When the load is resolved into vertical and horizontal components at Q and a couple, the results are as shown in Fig. 7.18*b*. (Vertical and horizontal components are used because the *principal central axes* U and V are vertical and horizontal.)

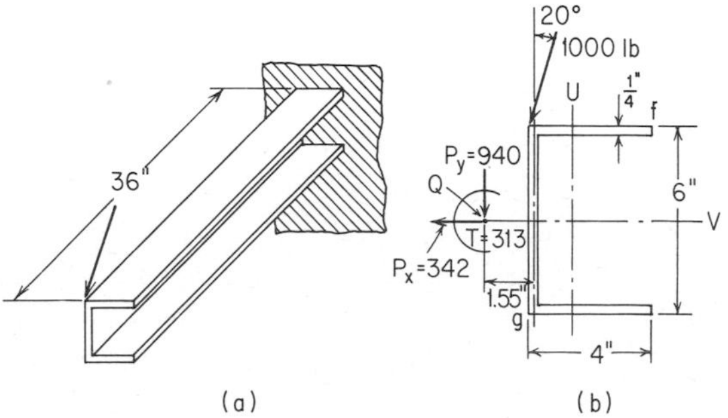

Fig. 7.18

TABLE 14 ***Position of flexural center Q for different sections***

Form of section	Position of Q
1. Any narrow section symmetrical about the x axis; centroid at $x = 0, y = 0$ 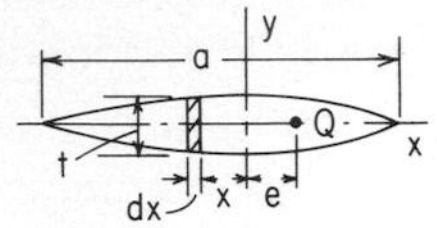	$e = \dfrac{1 + 3\nu}{1 + \nu} \dfrac{\int xt^3\,dx}{\int t^3\,dx}$ For narrow triangle (with $\nu = 0.25$), $e = 0.187a$ For any equilateral triangle, $e = 0$ (Refs. 32 and 52)
2. Beam composed of n elements of any form, connected or separate, with common neutral axis (e.g., multiple-spar airplane wing) 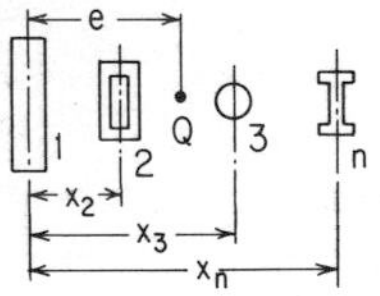	$e = \dfrac{E_2I_2x_2 + E_3I_3x_3 + \cdots + E_nI_nx_n}{E_1I_1 + E_2I_2 + E_3I_3 + \cdots + E_nI_n}$ where I_1, I_2, etc., are moments of inertia of the several elements about the X axis (that is, Q is at the centroid of the products EI for the several elements)
3. Semicircular area 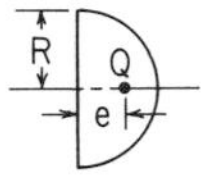	$e = \dfrac{8}{15\pi} \dfrac{3 + 4\nu}{1 + \nu} R$ (Q is to right of centroid) (Refs. 1 and 64) For sector of solid or hollow circular area, see Ref. 32
4. Angle 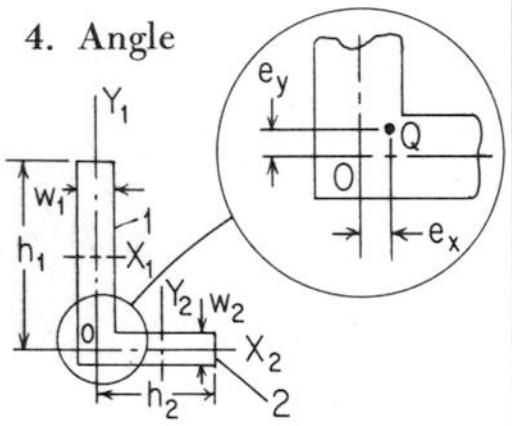	Leg 1 = rectangle w_1h_1; leg 2 = rectangle w_2h_2 I_1 = moment of inertia of leg 1 about Y_1 (central axis) I_2 = moment of inertia of leg 2 about Y_2 (central axis) $e_y = \dfrac{h_1}{2} \dfrac{I_1}{I_1 + I_2}$ (for e_x use X_1 and X_2 central axes) (Ref. 31) If w_1 and w_2 are small, $e_x = e_y = 0$ (practically) and Q is at 0
5. Channel 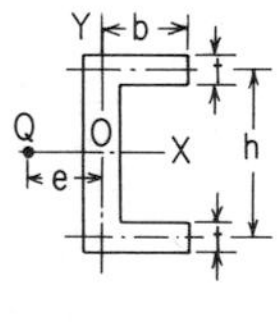	$e = h\dfrac{H_{xy}}{I_x}$ where H_{xy} = product of inertia of the half section (above X) with respect to axes X and Y, and I_x = moment of inertia of whole section with respect to axis X If t is uniform, $e = b^2h^2t/4I_x$
6. T 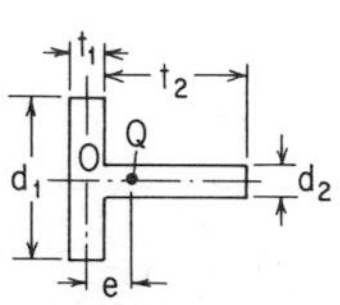	$e = \dfrac{1}{2}(t_1 + t_2)\dfrac{1}{1 + d_1^3t_1/d_2^3t_2}$ For a T-beam of ordinary proportions, Q may be assumed to be at 0

TABLE 14 ***Position of flexural center Q for different sections (Cont.)***

7. I with unequal flanges and thin web

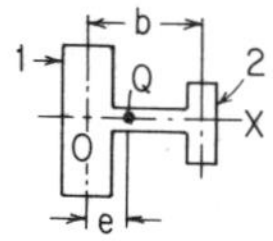

$$e = b\frac{I_2}{I_1 + I_2}$$

where I_1 and I_2, respectively, denote moments of inertia about X axis of flanges 1 and 2

8. D section (A = enclosed area)

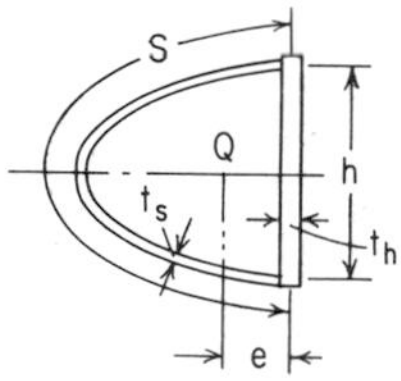

Values of eh/A

t_1/t_s \ S/h	1	1.5	2	3	4	5	6	7
0.5			1.0	0.800	0.665	0.570	0.500	0.445
0.6			0.910	0.712	0.588	0.498	0.434	0.386
0.7		0.980	0.831	0.641	0.525	0.443	0.384	0.338
0.8		0.910	0.770	0.590	0.475	0.400	0.345	0.305
0.9		0.850	0.710	0.540	0.430	0.360	0.310	0.275
1.0	1.0	0.800	0.662	0.500	0.400	0.330	0.285	0.250
1.2	0.905	0.715	0.525	0.380	0.304	0.285	0.244	0.215
1.6	0.765	0.588	0.475	0.345	0.270	0.221	0.190	0.165
2.0	0.660	0.497	0.400	0.285	0.220	0.181	0.155	0.135
3.0	0.500	0.364	0.285	0.200	0.155	0.125	0.106	0.091

(Ref. 31)

9. For thin-walled sections, such as lipped channels, hat sections, and sectors of circular tubes, see Table 21. The positions of the flexural centers and shear centers coincide

The maximum fiber stress will occur at the corner where the stresses due to the vertical and horizontal bending moments are of the same kind; at the upper-right corner f both stresses are tensile, and since f is farther from the U axis than the lower-left corner g where both stresses are compressive, it will sustain the greater stress. This stress will be simply the sum of the stresses due to the vertical and horizontal components of the load, or

$$\sigma = \frac{940(36)(3)}{19.9} + \frac{342(36)(2.765)}{5.61} = 5100 + 6070 = 11{,}200 \text{ lb/in}^2$$

The effect of the couple depends on the way in which the inner end of the beam is supported. If it is simply constrained against rotation in the planes of bending and twisting, the twisting moment will be resisted wholly by shear stress on the cross section, and these stresses can be found by the appropriate torsion formula of Table 20. If, however, the beam is built in so that the flanges are fixed in the *horizontal* plane, then part of the torque is resisted by the bending rigidity of the flanges and the corresponding moment causes a further fiber stress. This can be found by using the formulas of Art. 9.3.

For the channel section, K is given with sufficient accuracy by the formula $K = (t^3/3)(h + 2b)$ (Table 21, case 1), which gives $K = 0.073$. Taking $G = 12{,}000{,}000$ and $E = 30{,}000{,}000$, and the formula for C_w as

$$C_w = \frac{h^2 b^3 t}{12}\,\frac{2h + 3b}{h + 6b} = 38.4 \text{ in}^6$$

the value for β can be found. From Table 22 the formula for β is given as

$$\beta = \left(\frac{KG}{C_w E}\right)^{1/2} = \left[\frac{0.073(12)}{38.4(30)}\right]^{1/2} = 0.0276$$

From Table 22, case 1b, the value of θ'' at the wall is given as

$$\theta'' = \frac{T_o}{C_w E\beta} \operatorname{Tanh} \beta l = \frac{313}{38.4(30 \cdot 10^6)(0.0276)} \operatorname{Tanh} 0.0276(36) = 7.47(10^{-6})$$

Therefore the longitudinal compressive stress at f can be found from the expression for σ_x in Table 21, case 1, as

$$\sigma_x = \frac{hb}{2}\frac{h + 3b}{h + 6b} E\theta'' = \frac{6(4)}{2}\left[\frac{6 + 3(4)}{6 + 6(4)}\right](30)(10^6)(7.47)(10^{-6}) = 1610 \text{ lb/in}^2$$

The resultant fiber stress at f is 11,200 − 1610 = 9590 lb/in^2.

7.15 *Straight uniform beams (common case); ultimate strength*

When a beam is stressed beyond the elastic limit, plane sections remain plane or nearly so but unit stresses are no longer proportional to strains and hence no longer proportional to distance from the neutral surface. If the material has similar stress-strain curves in tension and compression, the stress distribution above and below the neutral surface will be similar and the neutral axis of any section which is symmetric about a horizontal axis will still pass through the centroid; if the material has different properties in tension and compression, then the neutral axis will shift away from the side on which the fibers yield the most; this shift causes an additional departure from the stress distribution assumed by the theory outlined in Art. 7.1.

Failure in bending. The strength of a beam of ordinary proportions is determined by the maximum bending moment it can sustain. For beams of nonductile material (cast iron, concrete, or seasoned wood) this moment may be calculated by the formula $M_m = \sigma'(I/c)$ if σ', the *modulus of rupture,* is known. The modulus of rupture depends on the material and other factors (see Art. 2.11), and attempts have been made to calculate it for a given material and section from the form of the complete stress-strain diagram. Thus for cast iron an approximate value of σ' may be found by the formula $\sigma' = K\sqrt{c/z'}\,\sigma_t$, where c is the distance to the extreme fiber, z' is the distance from the neutral axis to the centroid of the tensile part of the section, and K is an experimental coefficient equal to $\frac{6}{5}$ for sections that are flat at the top and bottom (rectangle, I, T, etc.) and $\frac{4}{3}$ for sections that are pointed or convex at the top and bottom (circle, diamond, etc.) (Ref. 4). Some tests indicate that this method of calculating the breaking strength of cast iron is sometimes inaccurate but generally errs on the side of safety (Ref. 5).

In general, the breaking strength of a beam can be predicted best from experimentally determined values of the rupture factor and tensile strength or the form factor and modulus of rupture. In Table 15 are given the rupture factors and form factors for a number of sections and materials. The rupture factors are based on the ultimate tensile strength for all materials except wood, for which it is based on ultimate compressive strength. The form factors are based on a rectangular section; for wood this section is 2 × 2 in, and for beams of greater depth a scale correction must be made by the formula given. For structural steel, wrought aluminum, and other ductile metals, where beams do not actually break, the modulus of rupture means the computed fiber stress at the maximum bending moment (Refs. 6 to 9).

The values given in Table 15 are for span/depth ratios of about 15 to 20. The effect of the span/depth ratio on the strength of cast-iron and plaster (brittle material) beams is shown in the following table, which gives the modulus of rupture (determined by center loading) for various span/depth ratios s/d in percentages of the modulus of rupture for $s/d = 20$ (Refs. 7, 10).

Material and section	s/d = Span/depth ratio								
	30	20	15	10	5	4	3	2	1
Cast iron, circular	97	100	102	105	110	112	114	118	123
Cast iron, rectangular	99	100	101	102	111				
Plaster, rectangular	100	100	101	102	110				

The table shows that the modulus of rupture increases as the span/depth ratio decreases, but the variation is slight except for relatively small values of s/d.

When the maximum bending moment occurs at but one section, as for a single concentrated load, the modulus of rupture is higher than when the maximum moment extends over a considerable part of the span. For instance, the modulus of rupture of short beams of brittle material is about 20 percent higher when determined by center loading than when determined by third-point loading. The disparity decreases as the s/d ratio increases.

Beams of ductile material (structural steel or aluminum) do not ordinarily fracture under static loading but fail through excessive deflection. For such beams, if they are of relatively thick section so as to preclude local buckling, the maximum bending moment is that which corresponds to plastic yielding throughout the section. This maximum moment, or "plastic" moment, is usually denoted by M_p and can be calculated by the formula $M_p = \sigma_y Z$, where σ_y is the lower yield point of the material and Z, called the *plastic section modulus*, is the arithmetical sum of the statical moments about the neutral axis of the parts of the cross section above and below that axis. Thus, for a rectangular section of depth d and width b,

$$Z = (\tfrac{1}{2}bd)(\tfrac{1}{4}d) + (\tfrac{1}{2}bd)(\tfrac{1}{4}d) = \tfrac{1}{4}bd^2.$$

This method of calculating the maximum resisting moment of a ductile-material beam is widely used in "plastic design," and is discussed further

TABLE 15 ***Form factors for beams***

Form of section (tensile side down)	Form factors F and rupture factors R			
	Cast iron	Cast aluminum alloy 195	Plaster	Wood
1. 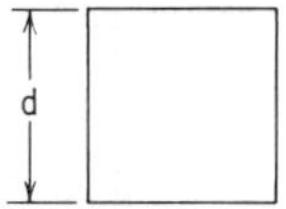	$F = 1$ $R = 2.3 - 0.02S$ where S = tensile strength, thousands of lb/in^2	$F = 1$ $R = 1.75$	$F = 1$ $R = 1.60$	For $d = 2$ in, $F = 1$ For $d > 2$ in, $F = 1 - 0.07\left(\sqrt{\frac{d}{2}} - 1\right)$ $R = 1.84$
2. 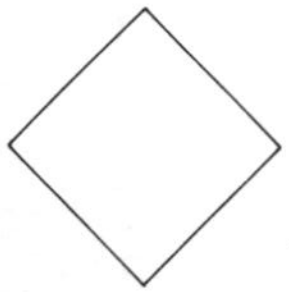	$F = 1.35$	$F = 1.26$	$F = 1.23$	$F = 1.414$
3. 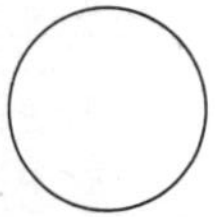	$F = 1.20$ $R = 2.80 - 0.025S$	$F = 1.12$	$F = 1.15$	$F = 1.18$
4. 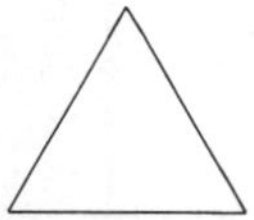	$F = 0.98$	$F = 0.88$	$F = 1.00$	
5. 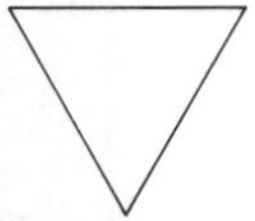	$F = 1.27$	$F = 1.43$	$F = 1.27$	

6. 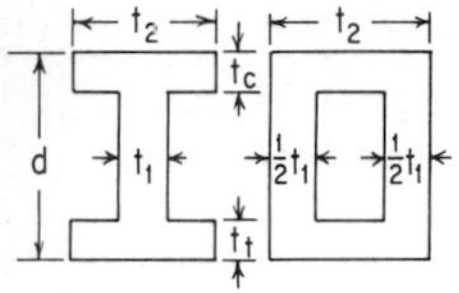	$F = 0.5 + 0.5\left[\frac{k(t_2 - t_1)}{t_2} + \frac{t_1}{t_2}\right]$ where $k = 3.6t_t/d$ $F = 0.84$ (for $d = 2.36$, $t_1 = 0.41$, $t_2 = 2.36$, and $t_c = t_t = 0.4$)	$F = 0.75$ (for $d = t_2 = 6t_1 = 6t_c = 6t_t$)	$F = 0.91$ (for $d = t_2 = 6t_1 = 6t_c = 6t_t$)	$F = 0.5 + 0.5\left[\frac{k(t_2 - t_1)}{t_2} + \frac{t_1}{t_2}\right]$ where $k = 1.6\frac{t_c}{d} - 0.07$, up to $\frac{t_c}{d} = 0.6$
7. 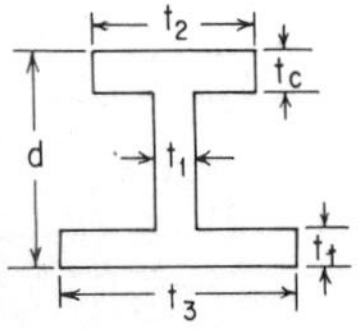	$F = 0.88$ (for $d = 2.36$, $t_1 = 0.33$, $t_2 = 1.18$, $t_3 = 2.44$, $t_c = 0.39$, and $t_t = 0.55$)			

in Art. 7.16. It is important to note that when the plastic moment has been developed, the neutral axis divides the cross-section area into halves and so is not always a centroidal axis. It is also important to note that the plastic moment is always greater than the moment required to just stress the extreme fiber to the lower yield point. This moment, which may be denoted by M_y, is equal to $\sigma_y I/c$, and so

$$\frac{M_p}{M_y} = \frac{Z}{I/c}.$$

This ratio $Z(I/c)$, called the *shape factor,* depends on the form of the cross section. For a solid rectangle it would be $\frac{1}{4}bd^2/\frac{1}{6}bd^2$, or 1.5; for an I section it is usually about 1.15.

In tubes and beams of thin open section, local buckling or crippling will sometimes occur before the full plastic resisting moment is realized, and the length of the member will have an influence. Tubes of steel or aluminum alloy generally will develop a modulus of rupture exceeding the ultimate tensile strength when the ratio of diameter to wall thickness is less than 50 for steel or 35 for aluminum. Wide-flanged steel beams will develop the full plastic resisting moment when the outstanding width/thickness ratio is less than 8.7 for $\sigma_y = 33{,}000$ or 8.3 for $\sigma_y = 36{,}000$. Charts giving the effective modulus of rupture of steel, aluminum, and magnesium tubes of various proportions may be found in Ref. 55.

Failure in shear. Failure by an actual shear fracture is likely to occur only in wood beams, where the shear strength parallel to the grain is, of course, small.

In I-beams and similar thin-webbed sections, the diagonal compression that accompanies shear (Art. 6.5) may lead to a buckling failure (see the discussion of *web buckling* that follows), and in beams of cast iron and concrete the diagonal tension that similarly accompanies shear may cause rupture. The formula for shear stress (Eq. 2) may be considered valid as long as the *fiber* stresses do not exceed the proportional limit, and therefore it may be used to calculate the vertical shear necessary to produce failure in any case where the ultimate shearing strength of the beam is reached while the fiber stresses, at the section of maximum shear, are still within the proportional limit.

Web buckling; local failure. An I-beam or similar thin-webbed member may fail by buckling of the web owing to diagonal compression when the shear stress reaches a certain value. Ketchum and Draffin (Ref. 11) and Wendt and Withey (Ref. 12) found that in light I-beams this type of buckling occurs when the shear stress, calculated by $\tau = 1.25V/\text{web area}$ (Ref. 11) or $\tau = V/\text{web area}$ (Ref. 12), reaches a value equal to the unit load that can be carried by a vertical strip of the beam as a round-ended column. For the thin webs of the beams tested, such a thin strip would be computed as a Euler column; for heavier beams an appropriate parabolic or other formula should be used (Chap. 11).

In plate girders, web buckling may be prevented by vertical or diagonal stiffeners, usually consisting of double angles that are riveted or welded, one on each side of the web. Steel-construction specifications (Ref. 13) require that such stiffeners be provided when h/t exceeds 70 and v exceeds $64{,}000{,}000/(h/t)^2$. Such stiffeners should have a moment of inertia (figured for an axis at the center line of the web) equal to at least $0.00000016H^4$ and should be spaced so that the clear distance between successive stiffeners is not more than $11{,}000t/\sqrt{v}$ or 84 in, whichever is least. Here h is the clear depth of the web between flanges, t is the web thickness, v is the shear stress V/ht, and H is the total depth of the web. In light-metal airplane construction, the stiffeners are sometimes designed to have a moment of inertia about an axis parallel to the web given by $I = (2.29d/t)(Vh/33E)^{\frac{4}{3}}$, where $V =$ the (total) vertical shear and $d =$ the stiffener spacing center-to-center (Ref. 14).

Buckling failure may occur also as a result of vertical compression at a support or concentrated load, which is caused by either column-type buckling of the web (Refs. 11 and 12) or crippling of the web at the toe of the fillet (Ref. 15). To guard against this latter type of failure, present specifications provide that for interior loads $R/t(N + 2k) \leq 24{,}000$ and for end reactions $R/t(N + k) \leq 24{,}000$, where R is the concentrated load or end reaction, t the web thickness, N the length of bearing, and k the distance from the outer face of the flange to the web toe of the fillet. Here R is in pounds and all linear dimensions are in inches.

Wood beams will crush locally if the supports are too narrow or if a load is applied over too small a bearing area. The unit bearing stress in either case is calculated by dividing the force by the nominal bearing area, no allowance being made for the nonuniform distribution of pressure consequent upon bending (Ref. 9). Metal beams also may be subjected to high local pressure stresses; these are discussed in Chap. 13.

Lateral buckling. The compression flange of an I-beam or similar member may fail as a column as a result of lateral buckling if it is unsupported. Such buckling may be *elastic* or *plastic;* that is, it may occur at a maximum fiber stress below or above the elastic limit. In the first case the buckling is an example of elastic instability, for which relevant formulas are given in Table 34 of Chap. 14. For buckling above the elastic range analytical solutions are difficult to obtain, and empirical expressions based on experiment are used (as will be shown to be true also of the columns discussed in Chap. 11).

Moore (Ref. 16) found that standard I-beams fail by lateral buckling when

$$s' = 40{,}000 - 60\,\frac{ml}{r}$$

where s' is the compressive stress in the extreme fiber (computed by Eq. 1), l is the span (in inches), r is the radius of gyration (in inches) of the beam section about a central axis parallel to the web, and m is a coefficient which depends on the manner of loading and support and has the following values:

Loading and Support	*Value of m*
End supports, uniform load	0.667
End supports, midpoint load	0.500
End supports, single load at any point	0.500
End supports, loads at third points	0.667
End supports, loads at quarter points	0.750
End supports, loads at sixth points	0.833
Cantilever beam, uniform load	0.667
Cantilever beam, end load	1.000
Fixed-ended beam, uniform load	0.281
Fixed-ended beam, midpoint load	0.250

For very light I-beams, Ketchum and Draffin (Ref. 11) found that the lower limit of test results is given by

$$s' = 24{,}000 - 40\frac{ml}{r}$$

where the terms have the same meaning and m the same values as given previously.

The beams tested by Moore generally failed at stresses below but very close to the yield point and so probably could be regarded as representing plastic buckling. The lighter beams tested by Ketchum and Draffin, however, failed at stresses below the limit of proportionality and are examples of elastic buckling.

In Ref. 13 rules are given for the reduction in allowable compressive stress according to the unbraced length of the compression flange. A review of the literature on this subject of the lateral buckling of structural members and a bibliography through 1959 are to be found in Ref. 58.

Narrow rectangular beams may fail also as a result of buckling of the compression edge. When this buckling occurs below the elastic limit, the strength is determined by elastic stability; formulas for this case are given in Table 34. For buckling at stresses beyond the elastic limit, no simple formula for the critical stress can be given, but methods for calculating this critical stress are given for aluminum beams by Dumont and Hill (Ref. 17) and for wood beams by Trayer and March (Ref. 18).

7.16 *Plastic, or ultimate strength, design*

The foregoing discussion of beams and frames is based for the most part on the assumption of purely elastic action and on the acceptance of maximum fiber stress as the primary criterion of safety. These constitute the basis of *elastic* analysis and design. An alternative and often preferred method of design, applicable to rigid frames and statically indeterminate beams made of materials capable of plastic action, is the method of *plastic,* or *ultimate strength,* design. It is based on the fact that such a frame or beam cannot deflect indefinitely or collapse until the full plastic moment M_p (see Art. 7.15) has been developed at each of several critical sections. If it is assumed that the plastic moment—a determinable couple—does indeed act at each such section, then the problem becomes a statically determinate one and the load corresponding to the collapse condition can be readily calculated.

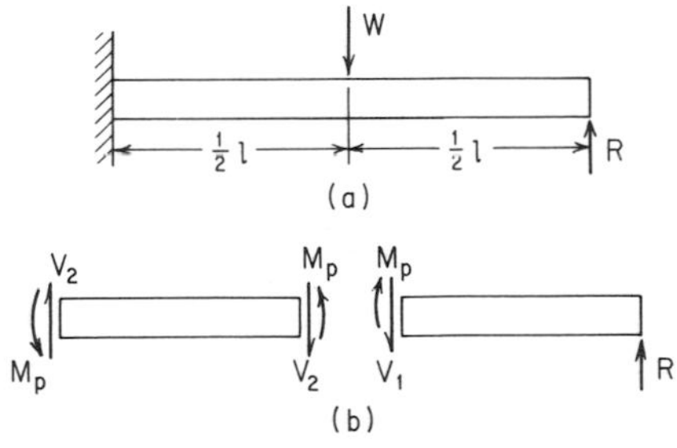

Fig. 7.19

A simple illustration of the procedure is afforded by the beam of Fig. 7.19*a*, corresponding to case 1c of Table 3. Suppose it is desired to determine the maximum value of the load W that the beam can support. It is shown by elastic analysis, and is indeed apparent from inspection, that the maximum bending moments occur at the load and at the left end of the beam. The maximum possible value of each such moment is M_p. It is evident that the beam cannot collapse until the moment at each of these points reaches this value. Therefore, when W has reached its maximum value and collapse is imminent, the beam is acted on by the force system represented in Fig. 7.19*b*; there is a *plastic hinge* and a known couple M_p at each of the critical sections, and the problem is statically determinate. For equilibrium of the right half, $R = M_p/(l/2)$ and $V_1 = R$; and for equilibrium of the left half, $V_2 = W - R$ and $[W - M_p/(l/2)]\frac{l}{2} = 2M_p$, or $W = 6M_p/l$.

In attempting to predict the collapse load on the basis of elastic analysis, it is easy to fall into the error of equating the maximum elastic moment $\frac{3}{16}Wl$ at the wall (Table 3) to M_p, thus obtaining $W = \frac{16}{3}M_p/l$. This erroneous procedure fails to take into account the fact that as W increases and yielding commences and progresses at the wall section, there is a redistribution of moments; the moment at the wall becomes less than $\frac{3}{16}Wl$, and the moment at the load becomes greater than $\frac{5}{32}Wl$ until finally each moment becomes equal to M_p. An important point to note is that although the elastic moments are affected by even a very slight departure from the assumed conditions—perfect fixity at one end and rigid support at the other—the collapse load is not thus affected. So long as the constraints are rigid enough to develop the plastic hinges as indicated, the ultimate load will be the same. Similarly, the method does not require that the beam be uniform in section, although a local reduction in section leading to the formation of a hinge at some point other than those assumed, of course, would alter the solution.

Because of its simplicity, this example may give an exaggerated impression of the ease of plastic analysis; but it does indicate that for any indeterminate structure with strength that is determined primarily by resistance to bending, the method is well suited to the determination of ultimate load and—through the use of a suitable factor of safety—to design. Its accuracy has been proved by good agreement between computed and experimental ultimate loads for a variety of frames. An extended discussion of plastic analysis is not appropriate here, but the interested reader will find an extensive literature on the subject (Refs. 60, 61).

REFERENCES

1. Timoshenko, S.: "Theory of Elasticity," Engineering Societies Monograph, McGraw-Hill Book Company, 1934.
2. Frocht, M. M.: A Photoelastic Investigation of Shear and Bending Stresses in Centrally Loaded Simple Beams, *Eng. Bull., Carnegie Inst. Technol.*, 1937.
3. Timoshenko, S.: "Strength of Materials," D. Van Nostrand Company, Inc., 1930.
4. Bach, C.: Zur Beigungsfestigkeit des Gusseisens, *Z. Vereines Dtsch. Ing.*, vol. 32, p. 1089, 1888.
5. Schlick, W. J., and B. A. Moore: Strength and Elastic Properties of Cast Iron, *Iowa Eng. Exp. Sta., Iowa State College, Bull.* 127, 1930.
6. Symposium on Cast Iron, *Proc. ASTM,* vol. 33, part II, p. 115, 1933.
7. Roark, R. J., R. S. Hartenberg, and R. Z. Williams: The Effect of Form and Scale on Strength, *Eng. Exp. Sta., Univ. Wis., Bull.* 82, 1938.
8. Newlin, J. A., and G. W. Trayer: Form Factors of Beams Subjected to Transverse Loading Only, *Natl. Adv. Comm. Aeron., Rept.* 181, 1924.
9. "Wood Handbook," Forest Products Laboratory, U.S. Dept. of Agriculture, 1974.
10. MacKenzie, J. T., and C. K. Donoho: A Study of the Effect of Span on the Transverse Test Results of Cast Iron, *Proc. ASTM,* vol. 37, part II, 1937.
11. Ketchum, M. S., and J. O. Draffin: Strength of Light I-beams, *Eng. Exp. Sta., Univ. Ill., Bull.* 241, 1932.
12. Wendt, K. F., and M. O. Withey: The Strength of Light Steel Joists, *Eng. Exp. Sta., Univ. Wis., Bull.* 79, 1934.
13. American Institute of Steel Construction: "Specifications for the Design, Fabrication and Erection of Structural Steel for Buildings," 1969.
14. Younger, J. E.: "Structural Design of Metal Airplanes," McGraw-Hill Book Company, 1935.
15. Lyse, I., and H. J. Godfrey: Investigation of Web Buckling in Steel Beams, *Trans. Am. Soc. Civil Eng.*, vol. 100, p. 675, 1935.
16. Moore, H. F.: The Strength of I-beams in Flexure, *Eng. Exp. Sta., Univ. Ill., Bull.* 68, 1913.
17. Dumont, C., and H. N. Hill: The Lateral Instability of Deep Rectangular Beams, *Nat. Adv. Comm. Aeron., Tech. Note* 601, 1937.
18. Trayer, G. W., and H. W. March: Elastic Instability of Members having Sections Common in Aircraft Construction, *Natl. Adv. Comm. Aeron., Rept.* 382, 1931.
19. Newlin, J. A., and G. W. Trayer: Deflection of Beams with Special Reference to Shear Deformation, *Natl. Adv. Comm. Aeron., Rept.* 180, 1924.
20. Pearson, K.: On the Flexure of Heavy Beams Subjected to Continuous Systems of Load, *Q. J. Pure Appl. Math.*, vol. 24, p. 63, 1890.

21. Timoshenko, S.: Mathematical Determination of the Modulus of Elasticity, *Mech. Eng.*, vol. 45, p. 259, 1923.
22. Holl, D. L.: Analysis of Thin Rectangular Plates Supported on Opposite Edges, *Iowa Eng. Exp. Sta., Iowa State College, Bull.* 129, 1936.
23. Westergaard, H. M.: Computation of Stress Due to Wheel Loads, *Public Roads,* U.S. Dept. of Agriculture, Bureau of Public Roads, vol. 11, p. 9, March, 1930.
24. Morris, C. T.: Concentrated Loads on Slabs, *Ohio State Univ. Eng. Exp. Sta. Bull.* 80, 1933.
25. Kelley, E. F.: Effective Width of Concrete Bridge Slabs Supporting Concentrated Loads, *Public Roads,* U.S. Dept. of Agriculture, Bureau of Public Roads, vol. 7, no. 1, 1926.
26. MacGregor, C. W.: Deflection of Long Helical Gear Tooth, *Mech. Eng.*, vol. 57, p. 225, 1935.
27. Holl, D. L.: Cantilever Plate with Concentrated Edge Load, *ASME Paper* A-8, *J. Appl. Mech.*, vol. 4, no. 1, 1937.
28. Miller, A. B.: Die mittragende Breite, and Über die mittragende Breite, *Luftfahrtforsch.*, vol. 4, no. 1, 1929.
29. Hetényi, M.: Application of Maclaurin Series to the Analysis of Beams in Bending, *J. Franklin Inst.*, vol. 254, 1952.
30. Kuhn, P., J. P. Peterson, and L. R. Levin: A Summary of Diagonal Tension, Parts I and II, *Natl. Adv. Comm. Aeron., Tech. Notes* 2661 and 2662, 1952.
31. Schwalbe, W. L. S.: The Center of Torsion for Angle and Channel Sections, *Trans. ASME,* vol. 54, no. 11, p. 125, 1932.
32. Young, A. W., E. M. Elderton, and K. Pearson: "On the Torsion Resulting from Flexure in Prisms with Cross-sections of Uniaxial Symmetry," Drapers' Co. Research Memoirs, tech. ser. 7, 1918.
33. Maurer, E. R., and M. O. Withey: "Strength of Materials," John Wiley & Sons, Inc., 1935.
34. Peery, D. J.: "Aircraft Structures," McGraw-Hill Book Company, 1950.
35. Sechler, E. E., and L. G. Dunn: "Airplane Structural Analysis and Design," John Wiley & Sons, Inc., 1942.
36. Griffel, W.: "Handbook of Formulas for Stress and Strain," Frederick Ungar Publishing Co., 1966.
37. Reissner, E.: Least Work Solutions of Shear Lag Problems, *J. Aeron. Sci.*, vol. 8, no. 7, p. 284, 1941.
38. Hildebrand, F. B., and E. Reissner: Least-work Analysis of the Problem of Shear Lag in Box Beams, *Natl. Adv. Comm. Aeron., Tech. Note* 893, 1943.
39. Winter, G.: Stress Distribution in and Equivalent Width of Flanges of Wide, Thin-wall Steel Beams, *Natl. Adv. Comm. Aeron., Tech. Note* 784, 1940.
40. Tate, M. B.: Shear Lag in Tension Panels and Box Beams, *Iowa Eng. Exp. Sta. Iowa State College, Eng. Rept.* 3, 1950.
41. Vlasov, V. Z., and U. N. Leontév: "Beams, Plates and Shells on Elastic Foundations," transl. from Russian, Israel Program for Scientific Translations, Jerusalem, NASA TT F-357, U.S., 1966.
42. Kameswara Rao, N.S.V., Y. C. Das, and M. Anandakrishnan: Variational Approach to Beams on Elastic Foundations, *Proc. Am. Soc. Civil Eng., J. Eng. Mech. Div.*, vol. 97, no. 2, April 1971.
43. White, Richard N.: Rectangular Plates Subjected to Partial Edge Loads: Their Elastic Stability and Stress Distribution, doctoral dissertation, University of Wisconsin, 1961.
44. Chow, L., Harry D. Conway, and George Winter: Stresses in Deep Beams, *Trans. Am. Soc. Civil Eng.*, vol. 118, p. 686, 1953.
45. Kaar, P. H.: Stress in Centrally Loaded Deep Beams, *Proc. Soc. Exp. Stress Anal.*, vol. 15, no. 1, p. 77, 1957.
46. Saad, S., and A. W. Hendry: Stresses in a Deep Beam with a Central Concentrated Load, *Exp. Mech., J. Soc. Exp. Stress Anal.*, vol. 18, no. 1, p. 192, June 1961.

47. Jaramillo, T. J.: Deflections and Moments due to a Concentrated Load on a Cantilever Plate of Infinite Length, *ASME J. Appl. Mech.*, vol. 17, no. 1, March 1950.
48. Wellauer, E. J., and A. Seireg: Bending Strength of Gear Teeth by Cantilever-plate Theory, *ASME J. Eng. Ind.*, vol. 82, August 1960.
49. Little, Robert W.: Bending of a Cantilever Plate, master's thesis, University of Wisconsin, 1959.
50. Small, N. C.: Bending of a Cantilever Plate Supported from an Elastic Half Space, *ASME J. Appl. Mech.*, vol. 28, no. 3, September 1961.
51. Hetényi, M.: Series Solutions for Beams on Elastic Foundations, *ASME J. Appl. Mech.*, vol. 38, no. 2, June 1971.
52. Duncan, W. J.: The Flexural Center or Center of Shear, *J. R. Aeron. Soc.*, vol. 57, September 1953.
53. Hetényi, Miklos: "Beams on Elastic Foundation," The University of Michigan Press, 1946.
54. O'Donnell, W. J.: The Additional Deflection of a Cantilever Due to the Elasticity of the Support, *ASME J. Appl. Mech.*, vol. 27, no. 3, September 1960.
55. "ANC Mil-Hdbk-5, Strength of Metal Aircraft Elements," Armed Forces Supply Support Center, March 1959.
56. Kleinlogel, A.: "Rigid Frame Formulas," Frederick Ungar Publishing Co., 1958.
57. Leontovich, Valerian: "Frames and Arches," McGraw-Hill Book Company, 1959.
58. Lee, G. C.: A Survey of Literature on the Lateral Instability of Beams, *Bull.* 63 *Weld. Res. Counc.*, August 1960.
59. Kelley, B. W., and R. Pedersen: The Beam Strength of Modern Gear Tooth Design, *Trans. SAE,* vol. 66, 1950.
60. Beedle, Lyman S.: "Plastic Design of Steel Frames," John Wiley & Sons, Inc., 1958.
61. "The Steel Skeleton," vol. II, "Plastic Behaviour and Design," Cambridge University Press, 1956.
62. Hiltscher, R.: Stress Distribution around Tunnel Openings of Rectangular Basic Profile with Circular Roof, *Der Bauingenieur,* vol. 8, 1957.
63. Weigle, R. E., R. R. Lasselle, and J. P. Purtell: Experimental Investigation of the Fatigue Behavior of Thread-type Projections, *Exp. Mech.*, vol. 3, no. 5, May 1963.
64. Leko, T.: On the Bending Problem of Prismatical Beam by Terminal Transverse Load, *ASME J. Appl. Mech.*, vol. 32, no. 1, March 1965.
65. Thomson, W. T.: Deflection of Beams by the Operational Method, *J. Franklin Inst.*, vol. 247, no. 6, June 1949.

EIGHT

Curved Beams

8.1 *Bending in the plane of the curve*

When a curved beam is bent in the plane of initial curvature, plane sections remain plane, but because of the different lengths of fibers on the inner and outer sides of the beam, the distribution of strain and stress is not linear; the neutral axis therefore does not pass through the centroid of the section and Eqs. 1 and 2 of Art. 7.1 do not apply. The error involved in their use is slight as long as the radius of curvature is more than about ten times the depth of the beam, but it becomes large for sharp curvatures. In Table 16 are given formulas for the position of the neutral axis and for the maximum fiber stresses in curved beams having different degrees of curvature and various forms of cross section. [In large part the formulas and tabulated coefficients are taken from the University of Illinois Circular by Wilson and Quereau (Ref. 1) with modifications suggested by Neugebauer (Ref. 28).]

Shear stress. Although Eq. 2 of Art. 7.1 does not apply to curved beams, Eq. 13, used as for a straight beam, gives the *maximum* shear stress with sufficient accuracy in most instances. A solution for the shear stress in a curved beam of narrow rectangular section is given by Case (Ref. 2); his formula shows that even for very sharp curvature (radius of beam axis equal to depth) the maximum shear stress is only about 10 percent in excess of the value $\frac{3}{2} V/A$ given by Eq. 13. It should be noted that this maximum shear stress occurs at the neutral axis of the curved beam, not at the central axis.

Radial stress. Owing to the radial components of the fiber stresses, radial stresses are set up in a curved beam; these are tensile when the bending

TABLE 16 ***Formulas for curved beams subjected to bending in the plane of the curve***

NOTATION: R = radius of curvature measured to centroid of section; c = distance from centroidal axis to extreme fiber on concave side of beam; A = area of section; h = distance from centroidal axis to neutral axis measured toward center of curvature; I = moment of inertia of cross section about centroidal axis perpendicular to plane of curvature; and $k_i = \sigma_i/\sigma$ and $k_o = \sigma_o/\sigma$ where σ_i = actual stress in extreme fiber on concave side, σ_o = actual stress in extreme fiber on convex side, and σ = fictitious unit stress in corresponding fiber as computed by ordinary flexure formula for a straight beam

Form and dimensions of cross section, reference no.	Formulas	Values of $\frac{h}{c}$, k_i, and k_o for various values of $\frac{R}{c}$										
1. Solid rectangular section 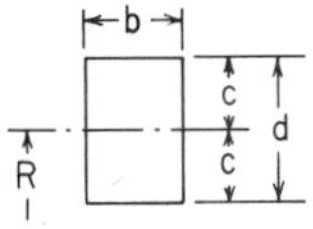	$\frac{h}{c} = \frac{R}{c} - \frac{2}{\ln\left(\frac{R/c+1}{R/c-1}\right)}$ (*Note:* h/c, k_i, and k_o are independent of the width b)	$\frac{R}{c}$ =	1.20	1.40	1.60	1.80	2.00	3.00	4.00	6.00	8.00	10.00
	$k_i = \frac{1}{3h/c}\,\frac{1-h/c}{R/c-1}$	$\frac{h}{c}$ =	0.366	0.284	0.236	0.204	0.180	0.115	0.085	0.056	0.042	0.033
	$k_o = \frac{1}{3h/c}\,\frac{1+h/c}{R/c+1}$	k_i =	2.888	2.103	1.798	1.631	1.523	1.288	1.200	1.124	1.090	1.071
		k_o =	0.566	0.628	0.671	0.704	0.730	0.810	0.853	0.898	0.922	0.937
2. Solid circular or elliptical section 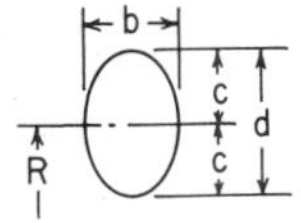	$\frac{h}{c} = \frac{1}{2}\left[\frac{R}{c} - \sqrt{(R/c)^2 - 1}\right]$	$\frac{R}{c}$ =	1.20	1.40	1.60	1.80	2.00	3.00	4.00	6.00	8.00	10.00
	$k_i = \frac{1}{4h/c}\,\frac{1-h/c}{R/c-1}$ (*Note:* h/c, k_i, and k_o are independent of the width b)	$\frac{h}{c}$ =	0.268	0.210	0.176	0.152	0.134	0.086	0.064	0.042	0.031	0.025
	$k_o = \frac{1}{4h/c}\,\frac{1+h/c}{R/c+1}$	k_i =	3.408	2.350	1.957	1.748	1.616	1.332	1.229	1.142	1.103	1.080
		k_o =	0.537	0.600	0.644	0.678	0.705	0.791	0.837	0.887	0.913	0.929
3. Hollow circular section 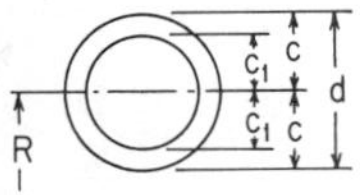	$\frac{h}{c} = \frac{1}{2}\left[\frac{2R}{c} - \sqrt{(R/c)^2-1} - \sqrt{(R/c)^2 - (c_1/c)^2}\right]$	(When $c_1/c = \frac{1}{2}$)										
	$k_i = \frac{1}{4h/c}\,\frac{1-h/c}{R/c-1}\left[1 + \left(\frac{c_1}{c}\right)^2\right]$	$\frac{R}{c}$ =	1.20	1.40	1.60	1.80	2.00	3.00	4.00	6.00	8.00	10.00
	$k_o = \frac{1}{4h/c}\,\frac{1+h/c}{R/c+1}\left[1 + \left(\frac{c_1}{c}\right)^2\right]$	$\frac{h}{c}$ =	0.323	0.256	0.216	0.187	0.166	0.107	0.079	0.052	0.039	0.031
	(*Note:* For thin-walled tubes the discussion on page 215 should be considered)	k_i =	3.276	2.267	1.895	1.697	1.573	1.307	1.211	1.130	1.094	1.074
		k_o =	0.582	0.638	0.678	0.708	0.733	0.810	0.852	0.897	0.921	0.936
4. Hollow elliptical section 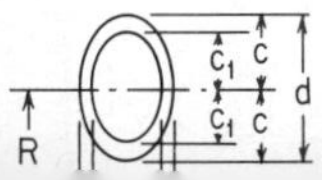	$\frac{h}{c} = \frac{R}{c} - \frac{\frac{1}{2}[1 - (b_1/b)(c_1/c)]}{\frac{R}{c} - \sqrt{(R/c)^2-1} - \frac{b_1/b}{c_1/c}\left[\frac{R}{c} - \sqrt{(R/c)^2 - c_1k)^2}\right]}$	(When $b_1/b = \frac{3}{5}$, $c_1/c = \frac{4}{5}$)										
	$k_i = \frac{1}{4h/c}\,\frac{1-h/c}{R/c-1}\,\frac{1-(b_1/b)(c_1/c)^3}{1-(b_1/b)(c_1/c)}$	$\frac{R}{c}$ =	1.20	1.40	1.60	1.80	2.00	3.00	4.00	6.00	8.00	10.00
	$\frac{1}{\quad}\,\frac{1+h/c}{\quad}\,\frac{1-(b_1/b)(c_1/c)^3}{\quad}$	$\frac{h}{c}$ =	0.354	0.279	0.233	0.202	0.178	0.114	0.085	0.056	0.042	0.034
		k_i =	3.033	2.154	1.825	1.648	1.535	1.291	1.202	1.125	1.083	1.063
		k_o =	0.579	0.637	0.677	0.709	0.734	0.812	0.854	0.899	0.916	0.930

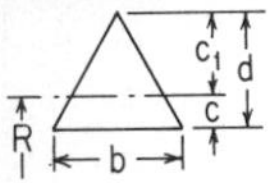

$$\left(\frac{R}{c} + 2\right) \ln\left(\frac{\cdots}{R/c - 1}\right) - 3$$

$$k_i = \frac{1}{2h/c}\,\frac{1 - h/c}{R/c - 1}$$

$$k_o = \frac{1}{4h/c}\,\frac{2 + h/c}{R/c + 2}$$

(*Note:* h/c, k_i, and k_o are independent of the width b)

R/c									
h/c = 0.434	0.348	0.296	0.259	0.232	0.155	0.117	0.079	0.060	0.048
k_i = 3.265	2.345	1.984	1.784	1.656	1.368	1.258	1.163	1.120	1.095
k_o = 0.438	0.497	0.539	0.573	0.601	0.697	0.754	0.821	0.859	0.883

6. Trapezoidal section

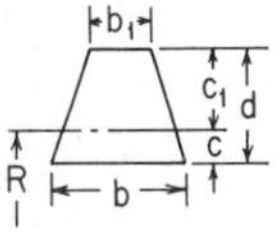

$$\frac{d}{c} = \frac{3(1 + b_1/b)}{1 + 2b_1/b} \qquad \frac{c_1}{c} = \frac{d}{c} - 1$$

$$\frac{h}{c} = \frac{R}{c} - \frac{\frac{1}{2}(1 - b_1/b)(d/c)^2}{\left[\frac{R}{c} + \frac{c_1}{c} - \frac{b_1}{b}\left(\frac{R}{c} - 1\right)\right] \ln\left(\frac{R/c + c_1/c}{R/c - 1}\right) - \left(1 - \frac{b_1}{b}\right)\frac{d}{c}}$$

$$k_i = \frac{1}{2h/c}\,\frac{1 - h/c}{R/c - 1}\,\frac{1 + 4b_1/b + (b_1/b)^2}{(1 + 2b_1/b)^2}$$

$$k_o = \frac{c_1/c}{2h/c}\,\frac{c_1/c + h/c}{R/c + c_1/c}\,\frac{1 + 4b_1/b + (b_1/b)^2}{(2 + b_1/b)^2}$$

(*Note:* While h/c, k_i, and k_o depend upon the width ratio b_1/b, they are independent of the width b)

(When $b_1/b = \frac{1}{2}$)

R/c = 1.20	1.40	1.60	1.80	2.00	3.00	4.00	6.00	8.00	10.00
h/c = 0.403	0.318	0.267	0.232	0.206	0.134	0.100	0.067	0.050	0.040
k_i = 3.011	2.183	1.859	1.681	1.567	1.314	1.219	1.137	1.100	1.078
k_o = 0.544	0.605	0.648	0.681	0.707	0.790	0.836	0.885	0.911	0.927

7. T-beam or channel section

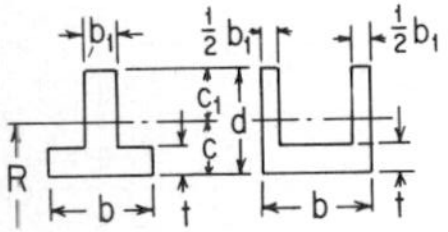

$$\frac{d}{c} = \frac{2[b_1/b + (1 - b_1/b)(t/d)]}{b_1/b + (1 - b_1/b)(t/d)^2} \qquad \frac{c_1}{c} = \frac{d}{c} - 1$$

$$\frac{h}{c} = \frac{R}{c} - \frac{(d/c)[b_1/b + (1 - b_1/b)(t/d)]}{\left(\frac{b_1}{b}\right) \ln\left[\frac{d/c + R/c - 1}{(d/c)(t/d) + R/c - 1}\right] + \ln\left[\frac{(d/c)(t/d) + R/c - 1}{R/c - 1}\right]}$$

$$k_i = \frac{I_c}{Ac^2(R/c - 1)}\,\frac{1 - h/c}{h/c} \qquad \text{where}$$

$$\frac{I_c}{Ac^2} = \frac{1}{3}\left(\frac{d}{c}\right)^2 \left[\frac{b_1/b + (1 - b_1/b)(t/d)^3}{b_1/b + (1 - b_1/b)(t/d)}\right] - 1$$

$$k_o = \frac{I_c}{Ac^2(h/c)}\,\frac{d/c + h/c - 1}{R/c + d/c - 1}\,\frac{1}{d/c - 1}$$

(*Note:* While h/c, k_i, and k_o depend upon the width ratio b_1/b, they are independent of the width b)

(When $b_1/b = \frac{1}{4}$, $t/d = \frac{1}{4}$)

R/c = 1.200	1.400	1.600	1.800	2.000	3.000	4.000	6.000	8.000	10.000
h/c = 0.502	0.419	0.366	0.328	0.297	0.207	0.160	0.111	0.085	0.069
k_i = 3.633	2.538	2.112	1.879	1.731	1.403	1.281	1.176	1.128	1.101
k_o = 0.583	0.634	0.670	0.697	0.719	0.791	0.832	0.879	0.905	0.922

TABLE 16 *Formulas for curved beams* (Cont.)

Form and dimensions of cross section, reference no.	Precise formula for h and approximate formula for k_i	Values of k_i, k_o, and $\dfrac{h}{R}$ for various values of $\dfrac{R}{c}$

8. Symmetrical I-beam or hollow rectangular section

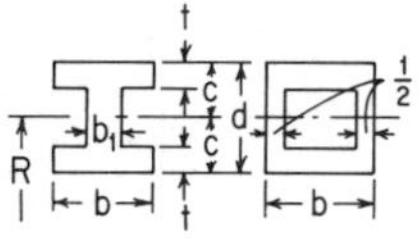

$$\frac{h}{c} = \frac{R}{c} - \frac{2[t/c + (1 - t/c)(b_1/b)]}{\ln\left[\dfrac{(R/c)^2 + (R/c + 1)(t/c) - 1}{(R/c)^2 - (R/c - 1)(t/c) - 1}\right] + \dfrac{b_1}{b}\ln\left[\dfrac{R/c - t/c + 1}{R/c + t/c - 1}\right]}$$

where

$$k_i = \frac{I_c}{Ac^2(R/c - 1)}\frac{1 - h/c}{h/c}$$

$$\frac{I_c}{Ac^2} = \frac{1}{3}\left[\frac{1 - (1 - b_1/b)(1 - t/c)^3}{1 - (1 - b_1/b)(1 - t/c)}\right]$$

$$k_o = \frac{I_c}{Ac^2(R/c + 1)}\frac{1 + h/c}{h/c}$$

(*Note:* While h/c, k_i, and k_o depend upon the width ratio b_1/b, they are independent of the width b)

(When $b_1/b = \tfrac{1}{3}$, $t/d = \tfrac{1}{6}$)

$\dfrac{R}{c} =$	1.20	1.40	1.60	1.80	2.00	3.00	4.00	6.00	8.00	10.00
$\dfrac{h}{c} =$	0.489	0.391	0.330	0.287	0.254	0.164	0.122	0.081	0.060	0.048
$k_i =$	2.516	1.876	1.630	1.496	1.411	1.225	1.156	1.097	1.071	1.055
$k_o =$	0.666	0.714	0.747	0.771	0.791	0.853	0.886	0.921	0.940	0.951

9. Unsymmetrical I-beam section

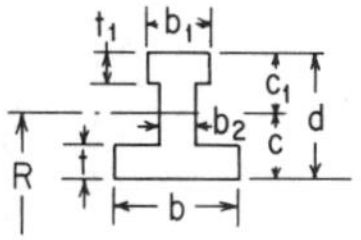

$$A = bd[b_1/b + (1 - b_2/b)(t/d) - (b_1/b - b_2/b)(1 - t_1/d)]$$

$$\frac{d}{c} = \frac{2A/bd}{(b_1/b - b_2/b)(2 - t_1/d)(t_1/d) + (1 - b_2/b)(t/d)^2 + b_2/b}$$

$$\frac{h}{c} = \frac{R}{c} - \frac{(A/bd)(d/c)}{\ln\left[\dfrac{R/c + t/c - 1}{R/c - 1}\right] + \dfrac{b_2}{b}\ln\left[\dfrac{R/c + c_1/c - t_1/c}{R/c + t/c - 1}\right] + \dfrac{b_1}{b}\ln\left[\dfrac{R/c + c_1/c}{R/c + c_1/c - t_1/c}\right]}$$

where

$$k_i = \frac{I_c}{Ac^2(R/c - 1)}\frac{1 - h/c}{h/c}$$

$$\frac{I_c}{Ac^2} = \frac{1}{3}\left(\frac{d}{c}\right)^2 \left[\frac{b_1/b + (1 - b_2/b)(t/d)^3 - (b_1/b - b_2/b)(1 - t_1/d)^3}{b_1/b + (1 - b_2/b)(t/d) - (b_1/b - b_2/b)(1 - t_1/d)}\right] - 1$$

$$k_o = \frac{I_c}{Ac^2(h/c)}\frac{d/c + h/c - 1}{R/c + d/c - 1}\frac{1}{d/c - 1}$$

(When $b_1/b = \tfrac{2}{3}$, $b_2/b = \tfrac{1}{6}$, $t_1/d = \tfrac{1}{6}$; $t/d = \tfrac{1}{3}$)

$\dfrac{R}{c} =$	1.20	1.40	1.60	1.80	2.00	3.00	4.00	6.00	8.00	10.00
$\dfrac{h}{c} =$	0.491	0.409	0.356	0.318	0.288	0.200	0.154	0.106	0.081	0.066
$k_i =$	3.589	2.504	2.083	1.853	1.706	1.385	1.266	1.165	1.120	1.094
$k_o =$	0.671	0.721	0.754	0.779	0.798	0.856	0.887	0.921	0.938	0.950

moment tends to straighten the beam and compressive under the reverse condition. Case (Ref. 2) gives a formula for this radial stress at any point in a beam of narrow rectangular section; the formula shows that the radial stress does not become equal to even the smaller of the extreme fiber stresses until the curvature is very sharp (the ratio of outer to inner radius is 7).

It can thus be seen that radial stresses are unlikely to be important in curved beams of rectangular section. In beams of I or T section, however, the radial stress may become critical when the web thickness is small. Seely and James (Ref. 3) discuss this question; they explain a method of calculating the stress and give experimental results secured with plaster models. Their conclusion is that the radial web stress is probably less important than the fiber stress for all values of R/c (the radius of curvature divided by the distance from the centroid of the section to the extreme fiber on the inner side of the beam) greater than about 2, provided that the web thickness is not less than about one-fifth the flange breadth and that stress concentration at the junction of flange and web is negligible. (This is with reference to tensile stresses in a curved beam under a *straightening* moment; in a beam with very thin web, the web might buckle under the radial compression produced by a moment that tended to increase the curvature.)

Curved beams with wide flanges. In reinforcing rings for large pipes, airplane fuselages, and ship hulls, the combination of a curved sheet and attached web or stiffener forms a curved beam with wide flanges. Formulas for the effective width of a flange in such a curved beam are given in Ref. 9 and are as follows.

When the flange is indefinitely wide (e.g., the inner flange of a pipe-stiffener ring), the effective width is

$$b' = 1.56\sqrt{Rt}$$

where b' is the total width assumed effective, R is the mean radius of curvature of the flange, and t is the thickness of the flange, all dimensions being in inches.

When the flange has a definite unsupported width b (gross width less web thickness), the ratio of effective to actual width b'/b is a function of qb, where

$$q = \sqrt[4]{\frac{3(1-\nu^2)}{R^2t^2}}$$

Corresponding values of qb and b'/b are as follows:

qb	1	2	3	4	5	6	7	8	9	10	11
b'/b	0.980	0.850	0.610	0.470	0.380	0.328	0.273	0.244	0.217	0.200	0.192

For the curved beam each flange should be considered as replaced by one of corresponding effective width b', and all calculations for direct, bending, and shear stresses, including corrections for curvature, should be based on this transformed section.

Bleich (Ref. 10) has shown that the radial components of the longitudinal flange stresses bend both flanges toward the web. The maximum transverse stress σ' due to this bending occurs at the junction of flange and web and is given by $\sigma' = \beta\sigma_m$; here σ_m is the longitudinal bending stress at the middle surface of the flange, and β is a coefficient that depends on the ratio c^2/Rt, where c is the actual unsupported projecting width of the flange to either side of the web, and R and t have the same meaning as before. Values of β may be found from the following table; they were taken from Ref. 10, where values of b' are similarly tabulated.

$c^2/Rt = 0$	0.1	0.2	0.3	0.4	0.5	0.6	0.8
$\beta = 0$	0.297	0.580	0.836	1.056	1.238	1.382	1.577
$c^2/Rt = 1$	1.2	1.4	1.5	2	3	4	5
$\beta = 1.677$	1.721	1.732	1.732	1.707	1.671	1.680	1.700

U-shaped members. A U-shaped member having a semicircular inner boundary and a rectangular outer boundary is sometimes used as a punch or riveter frame. Such a member can usually be analyzed as a curved beam having a concentric outer boundary, but when the back thickness is large, a more accurate analysis may be necessary. In Ref. 11 are presented the results of a photoelastic stress analysis of such members in which the effects of variations in the several dimensions were determined.

Deflections. Sharply curved beams are in general only a small portion of a larger structure. The contribution to deflection made by the curved portion can best be calculated by using the stresses at the inner and outer surfaces to calculate strains and the strains, in turn, to determine the rotations of the plane sections.

8.2 *Deflection of curved beams of large radius*

If for a curved beam the radius of curvature is large enough such that Eqs. 1 and 2 of Art. 7.1 are acceptable, i.e., the radius of curvature is greater than 10 times the depth, then Eq. 8 in Art. 7.1 can be used to find the change in slope. The deflection can be found most easily by Castigliano's first theorem using Eqs. 3 and 7. The following example shows how this is done.

EXAMPLE

Figure 8.1 represents a slender uniform bar curved to form the quadrant of a circle; it is fixed at the lower end and at the upper end is loaded by a vertical force V, a horizontal force H, and a couple M_0. It is desired to find the vertical deflection D_y, the horizontal deflection D_x, and the rotation θ of the upper end, denoted here by D_y, D_x, and θ, respectively.

Solution. According to Castigliano's first theorem, $D_y = \partial U/\partial V$, $D_x = \partial U/\partial H$, and $\theta = \partial U/\partial M_0$. Denoting the angular position of any section by x, it is evident that the moment there is $M = VR \sin x + HR(1 - \cos x) + M_0$. Disregarding shear and axial stress, and

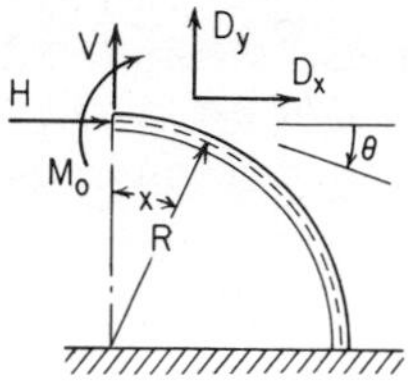

Fig. 8.1

replacing ds by $R\,dx$, we have (Eq. 3, Art. 7.1)

$$U = U_f = \int_0^{\pi/2} \frac{[VR \sin x + HR(1 - \cos x) + M_0]^2 R\,dx}{2EI}$$

Instead of integrating this and then carrying out the partial differentiations, we differentiate first and then integrate, and for convenience suppress the constant term EI until all computations are completed. Thus

$$D_y = \frac{\partial U}{\partial V}$$

$$= \int_0^{\pi/2} [VR \sin x + HR(1 - \cos x) + M_0](R \sin x)R\,dx$$

$$= VR^3(\tfrac{1}{2}x - \tfrac{1}{2}\sin x \cos x) - HR^3(\cos x + \tfrac{1}{2}\sin^2 x) - M_0R^2 \cos x \Big|_0^{\pi/2}$$

$$= \frac{(\pi/4)VR^3 + \frac{1}{2}HR^3 + M_0R^2}{EI}$$

$$D_x = \frac{\partial U}{\partial H}$$

$$= \int_0^{\pi/2} [VR \sin x + HR(1 - \cos x) + M_0]R(1 - \cos x)R\,dx$$

$$= VR^3(-\cos x - \tfrac{1}{2}\sin^2 x) + HR^3(\tfrac{3}{2}x - 2\sin x + \tfrac{1}{2}\sin x \cos x) + M_0R^2(x - \sin x)\Big|_0^{\pi/2}$$

$$= \frac{\frac{1}{2}VR^3 + (\frac{3}{4}\pi - 2)\,HR^3 + (\pi/2 - 1)\,M_0R^2}{EI}$$

$$\theta = \frac{\partial U}{\partial M_0}$$

$$= \int_0^{\pi/2} [VR \sin x + HR(1 - \cos x) + M_0]R\,dx$$

$$= -VR^2 \cos x + HR^2(x - \sin x) + M_0RX \Big|_0^{\pi/2}$$

$$= \frac{VR^2 + (\pi/2 - 1)\,HR^2 + (\pi/2)M_0R}{EI}$$

The deflection produced by any one load or any combination of two loads is found by setting the other load or loads equal to zero; thus, V alone would produce $D_x = \frac{1}{2}VR^3/EI$, and M alone would produce $D_y = M_0R^2/EI$. In this example all results are positive, indicating that D_x is in the direction of H, D_y in the direction of V, and θ in the direction of M_0.

Distortion of tubular sections. In curved beams of thin tubular section, the distortion of the cross section produced by the radial components of the fiber stresses reduces both the strength and stiffness. If the beam curvature is not so sharp as to make Eqs. 1 and 4 of Art. 7.1 inapplicable, the effect of this distortion of the section can be taken into account as follows.

In calculating deflection of curved beams of hollow circular section, replace I by KI, where

$$K = 1 - \frac{9}{[10 + 12(tR/a^2)^2]}$$

(Here R = the radius of curvature of the beam axis; a = the outer radius of tube section; and t = the thickness of tube wall.) In calculating the maximum bending stress in curved beams of hollow circular section, use the formulas

$$\sigma_{max} = \frac{Ma}{I} \frac{2}{3K\sqrt{3\beta}} \quad \text{at } y = \frac{a}{\sqrt{3\beta}} \text{ if } \frac{tR}{a^2} < 1.472$$

or

$$\sigma_{max} = \frac{Ma}{I} \frac{1 - \beta}{K} \quad \text{at } y = a \text{ if } \frac{tR}{a^2} > 1.472$$

where $\beta = \dfrac{6}{5 + 6(tR/a^2)^2}$ and y is measured from the neutral axis.

Torsional stresses and deflections are unchanged.

In calculating deflection or stress in curved beams of hollow square section and uniform wall thickness, replace I by

$$\frac{1 + 0.0270n}{1 + 0.0656n} I$$

where $n = b^4/R^2t^2$. (Here R = the radius of curvature of the beam axis; b = the length of the side of the square section; and t = the thickness of the section wall.)

The preceding formulas for circular sections are from von Kármán (Ref. 4); the formulas for square sections are from Timoshenko (Ref. 5), who also gives formulas for rectangular sections. Utecht (Ref. 8) gives curves of stress-intensification factors for curved, thin-walled circular tubes for combinations of *in-plane* and *out-of-plane* bending moments.

8.3 *Circular rings and arches*

In large pipelines, tanks, aircraft, and submarines the circular ring is an important structural element, and for correct design it is often necessary to calculate the stresses and deflections produced in such a ring under various conditions of loading and support. The circular arch of uniform section is often employed in light building construction and has other applications also.

Rings. A closed circular ring may be regarded as a *statically indeterminate beam* and analyzed as such by the use of Castigliano's first theorem. In Table 17 are given formulas thus derived for the bending moments, tensions, shears, horizontal and vertical deflections, and rotation of the load point in the plane of the ring for various loads and supports. By superposition, these formulas can be combined so as to cover almost any condition of loading and support likely to occur.

The ring formulas are based on the following assumptions: (1) The ring is of uniform cross section. (2) It is of such large radius in comparison with its radial thickness that the deflection theory for straight beams is applicable. (3) It is nowhere stressed beyond the elastic limit. (4) It is not so severely deformed as to lose its essentially circular shape. (5) Its deflection is due primarily to bending, but if it is desired, the deflections due to deformations caused by axial tension or compression in the ring and/or by transverse shear stresses in the ring may be included. To include these effects, we can evaluate first the coefficients α and β, the *axial stress deformation factor,* and the *transverse shear deformation factor,* and then the constants k_1 to k_5. Such corrections are most often necessary when composite or sandwich construction is employed. If no axial or shear stress corrections are desired, α and β are set equal to zero and all values of k are set equal to unity. (6) In the case of pipes acting as beams between widely spaced supports, the distribution of shear stress across the section of the pipe is in accordance with Art. 7.1, Eq. 2, and the direction of the resultant shear stress at any point of the cross section is tangential.

Note carefully the deformations given regarding the point or points of loading as compared with the deformations of the horizontal and vertical diameters. For many of the cases listed, the numerical values of load and deflection coefficients have been given for several positions of the loading. These coefficients do not include the effect of axial and shear deformation.

EXAMPLES

1. A pipe with a diameter of 13 ft and thickness of $\frac{1}{2}$ in is supported at intervals of 44 ft by rings, each ring being supported at the extremities of its horizontal diameter by vertical reactions acting at the centroids of the ring sections. It is required to determine the bending moments in a ring at the bottom, sides, and top, and the maximum bending moment when the pipe is filled with water.

Solution. We use the formulas for cases 4 and 18. Taking the weight of water as 62.4 lb/ft^3 and the weight of the shell as 20.4 lb/ft^2, the total weight W of 44 ft of pipe carried by one ring is found to be 401,100 lb. Therefore, for case 18, $W = 401{,}100$; and for case 4, $W = 200{,}550$ and $\theta = \pi/2$.

(At bottom)

$$M = M_C = 0.2387(401{,}100)(6.5)(12) - 0.50(200{,}550)(78)$$

or
$$M = 7.468(10^6) - 7.822(10^6) = -354{,}000 \text{ in-lb}$$

(At top)

$$M = M_A = 0.0796(401{,}100)(78) - 0.1366(200{,}550)(78) = 354{,}000 \text{ in-lb}$$

$$T = T_A = 0.2387(401{,}100) - 0.3183(200{,}550) = 31{,}900 \text{ lb}$$

$$V = V_A = 0$$

(At sides)

$$M = M_A - T_A R(1 - u) + V_A R z + LT_M$$

where for $x = \pi/2$, $u = 0$, $z = 1$, and $LT_M = (WR/\pi)(1 - u - xz/2) = [401{,}100(78)/\pi]$ $(1 - \pi/4) = 2.137(10^6)$ for case 18, and $LT_M = 0$ for case 4 since $(z - s) = 0$. Therefore $M = 354{,}000 - 31{,}900(78)(1 - 0) + 0 + 2.137(10^6) = 2800$ in-lb

The value of 2800 is due to the small differences in large numbers used in the superposition. An exact solution would give zero for this value. It is apparent that at least four digits must be carried.

To determine the location of maximum bending moment let $0 < x < \pi/2$ and examine the expression for M:

$$M = M_A - T_A R(1 - \cos x) + \frac{WR}{\pi}\left(1 - \cos x - \frac{x \sin x}{2}\right)$$

$$\frac{dM}{dx} = -T_A R \sin x + \frac{WR}{\pi} \sin x - \frac{WR}{2\pi} \sin x - \frac{WRx}{2\pi} \cos x$$

$$= 31{,}900\, R \sin x - 63{,}800\, Rx \cos x$$

At $x = x_1$, let $dM/dx = 0$ or $\sin x_1 = 2x_1 \cos x_1$, which yields $x_1 = 66.8°(1.166\text{ rad})$. At $x = x_1 = 66.8°$,

$$M = 354{,}000 - 31{,}900(78)(1 - 0.394) + \frac{401{,}100(78)}{\pi}\left[1 - 0.394 - \frac{1.166(0.919)}{2}\right]$$

or $M = -455{,}000$ in-lb (max negative moment)

Similarly, at $x = 113.2°$, $M = 455{,}000$ in-lb (max positive moment).

By applying the supporting reactions outside the center line of the ring at a distance a from the centroid of the section, side couples that are each equal to $Wa/2$ would be introduced. The effect of these, found by the formulas for case 3, would be to reduce the maximum moments, and it can be shown that the optimum condition obtains when $a = 0.04R$.

2. The pipe of Example 1 rests on soft ground, with which it is in contact over 150° of its circumference at the bottom. The supporting pressure of the soil may be assumed to be radial and uniform. It is required to determine the bending moment at the top and bottom and at the surface of the soil. Also the bending stresses at these locations and the change in the horizontal diameter must be determined.

Solution. A section of pipe 1 in long is considered. The loading may be considered as a combination of cases 12 to 14. Owing to the weight of the pipe (case 13, $w = 0.1416$),

$$M_A = \frac{0.1416(78^2)}{2} = 430 \text{ in-lb}$$

$$T_A = \frac{0.1416(78)}{2} = 5.52 \text{ lb}$$

$$V_A = 0$$

and at $x = 180 - \frac{150}{2} = 105° = 1.833$ rad

$$LT_M = -0.1416(78^2)[1.833(0.966) - 0.259 - 1] = -440 \text{ in-lb}$$

Therefore,

$$M_{105°} = 430 - 5.52(78)(1 + 0.259) - 440 = -552 \text{ in-lb}$$

$$M_C = 1.5(0.1416)(78)$$

$$= 1292 \text{ in-lb}$$

Owing to the weight of contained water (case 14, $\rho = 0.0361$),

$$M_A = \frac{0.0361(78^3)}{4} = 4283 \text{ in-lb}$$

$$T_A = \frac{0.0361(78^2)(3)}{4} = 164.7 \text{ lb}$$

$$V_A = 0$$

and at $x = 105°$

$$LT_M = 0.0361(78^3)\left[1 + 0.259 - \frac{1.833(0.966)}{2}\right] = 6400 \text{ in-lb}$$

Therefore

$$M_{105°} = 4283 - 164.7(78)(1 + 0.259) + 6400 = -5490 \text{ in-lb}$$

$$M_C = \frac{0.0361(78^3)(3)}{4} = 12{,}850 \text{ in-lb}$$

Owing to earth pressure and the reversed reaction (case 12, $\theta = 105°$),

$$2wR \sin\theta = 2\pi R(0.1416) + 0.0361\pi R^2 = 759 \text{ lb} \qquad (w = 5.04 \text{ lb/in})$$

$$M_A = -5.04(78^2)\left[\frac{1.833 + 2(0.966) - 1.833(-0.259)}{\pi} - 1 - 0.259\right]$$

$$= -2777 \text{ in-lb}$$

$$T_A = -5.04(78)\left[\frac{0.966 + 1.833(0.259)}{\pi} - 0.259\right] = -78.5 \text{ lb}$$

$$V_A = 0$$

$$LT_M = 0$$

$$M_{105°} = -2777 + 88.5(78)(1.259) = 4930 \text{ in-lb}$$

$$M_C = -5.04(78^2)\frac{1.833(1 - 0.259)}{\pi} = -13{,}260 \text{ in-lb}$$

Therefore,

$$M_A = 430 + 4283 - 2777 = 1936 \text{ in-lb}$$

$$\sigma_A = \frac{6M_A}{t^2} = 46{,}500 \text{ lb/in}^2$$

$$M_{105°} = -552 - 5490 + 4930 = -1112 \text{ in-lb}$$

$$\sigma_{105°} = 26{,}700 \text{ lb/in}^2$$

$$M_C = 1292 + 12{,}850 - 13{,}260 = 882 \text{ in-lb}$$

$$\sigma_C = 21{,}200 \text{ lb/in}^2$$

The change in the horizontal diameter is found similarly by superimposing the three cases. For E use $30(10^6)/(1 - 0.285^2) = 32.65(10^6)$ since a plate is being bent instead of a narrow beam (see page 188). For I use the moment of inertia of a 1-in-wide piece, 0.5 in thick:

$$I = \tfrac{1}{12}(1)(0.5^3) = 0.0104 \text{ in}^4 \qquad EI = 340{,}000$$

TABLE 17 *Formulas for circular rings*

NOTATION: W = load (pounds); w and v = unit loads (pounds per linear inch); ρ = weight of contained liquid (pounds per cubic inch); M_o = applied couple (inch-pounds); M_A and M are internal moments at A and x, respectively, positive as shown. T_A, T, V_A, and V are internal forces, positive as shown. E = modulus of elasticity (pounds per square inch); I = area moment of inertia of ring cross section (inches to the fourth). [Note that for a pipe or cylinder a representative 1-in segment may be used by replacing EI by $Et^3/12(1-\nu^2)$.] θ, x, and ϕ are angles (radians); $s = \sin\theta$, $c = \cos\theta$, $z = \sin x$, $u = \cos x$, $n = \sin\phi$, and $e = \cos\phi$. D_V and D_H are changes in the vertical and horizontal diameters, respectively, and an increase is positive. ΔR is the change in the lower half of the vertical diameter or the vertical motion relative to point C of a line connecting points B and D on the ring. Similarly ΔR_W is the vertical motion relative to point C of a horizontal line connecting the load points on the ring. D_{WH} is the change in length of a horizontal line connecting the load points on the ring. $\Delta\psi$ is the angular rotation (radians) of the load point in the plane of the ring and is positive in the direction of a positive moment at that point

The hoop stress deformation factor is $\alpha = I/AR^2$, where A is the cross-sectional area and R is the radius to the centroid of the cross section. The transverse (radial) shear deformation factor is $\beta = FEI/GAR^2$, where G is the shear modulus of elasticity and F is a shape factor for the cross section (see page 185). The following constants are hereby defined in order to simplify the expressions which follow. Note that all of these constants are unity if no correction for hoop stress or shear stress is necessary or desired. $k_1 = 1 + \alpha + \beta$, $k_2 = 1 - \alpha + \beta$, $k_3 = 1 + \alpha - \beta$, $k_4 = k_2/k_1$, $k_5 = k_2^2/k_1$

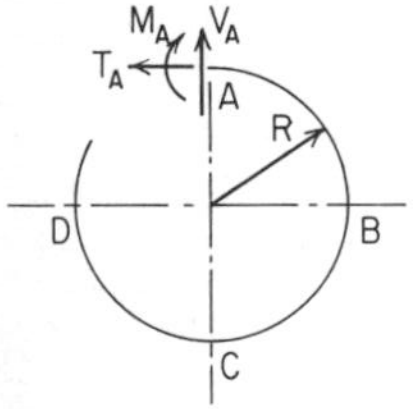

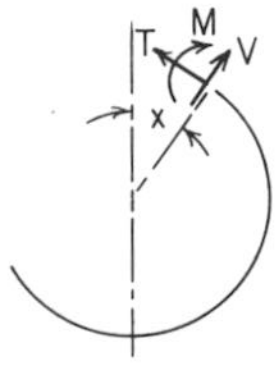

General formulas for moment, hoop load, and radial shear

$$M = M_A - T_A R(1-u) + V_A R z + LT_M$$
$$T = T_A u + V_A z + LT_T$$
$$V = -T_A z + V_A u + LT_V$$

where LT_M, LT_T, and LT_V are load terms given below for several types of load

Note: Due to symmetry in most of the cases presented, the loads beyond 180° are not included in the load terms. Only for cases 16, 17, and 19 should the equations for M, T, and V be used beyond 180°.

Note: The use of the bracket $\langle x-\theta\rangle^0$ is explained on page 94 and has a value of zero unless $x > \theta$

Reference no., loading, and load terms	Formulas for moments, loads, and deformations and some selected numerical values	
1. 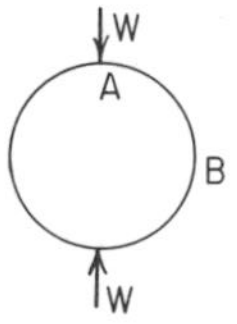 $LT_M = \frac{-WRz}{2}$ $LT_T = \frac{-Wz}{2}$ $LT_V = \frac{-Wu}{2}$	$M_A = \frac{WR}{\pi}$ $T_A = 0$ $V_A = 0$ $D_H = \frac{WR^3}{EI}\left(\frac{2}{\pi} - \frac{k_3}{2}\right)$ $D_V = \frac{-WR^3}{EI}\left(\frac{\pi k_1}{4} - \frac{2}{\pi}\right)$	Max $+M = M_A = 0.3183WR$ Max $-M = M_B = -0.1817WR$ If $\alpha = \beta = 0$, $D_H = 0.137\frac{WR^3}{EI}$ and $D_V = -0.149\frac{WR^3}{EI}$ For greater accuracy when the ring is relatively thick, multiply D_H by k_H and D_V by k_V, where k_H and k_V depend upon the ratio of outer radius R_o to inner radius R_i and have the following values: (see table below) (Ref. 19)

R_o/R_i	1.3	1.4	1.5	1.6	1.7	1.8	1.9
k_H	1.05	1.115	1.175	1.225	1.275	1.325	1.360
k_V	1.03	1.055	1.090	1.114	1.155	1.180	1.225

2.

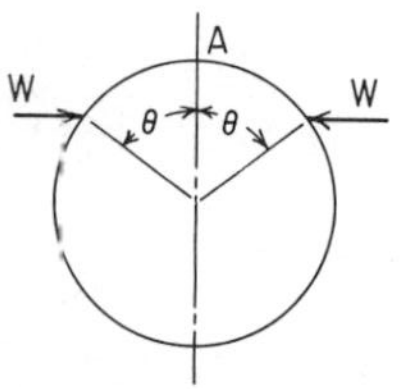

$LT_M = -WR(c-u)\langle x-\theta\rangle^0$

$LT_T = Wu\langle x-\theta\rangle^0$

$LT_V = -Wz\langle x-\theta\rangle^0$

$$M_A = -WR\left[\left(1-\frac{\theta}{\pi}\right)(1-c) - \frac{s}{\pi}(1-ck_4)\right]$$

$$T_A = -W\left(1-\frac{\theta}{\pi}+\frac{sck_4}{\pi}\right)$$

$$V_A = 0$$

$$D_H = \frac{-WR^3}{EI}\left[\frac{\theta}{2}k_1 - \frac{sc}{2}k_2 - \frac{2}{\pi}(s-\theta c)\right] \qquad \text{if } \theta \le \frac{\pi}{2}$$

$$D_V = \frac{WR^3}{EI}\left[\frac{2}{\pi}(s-\theta c) + c - 1 + \frac{s^2}{2}k_2\right]$$

$$\Delta R = \frac{WR^3}{EI}\left[\frac{k_2}{2\pi}(\theta - sck_4) - 0.1817(s-\theta c)\right] \qquad \text{if } \theta \le \frac{\pi}{2}$$

$$\Delta R_W = \frac{WR^3}{EI}\left\{\frac{\theta}{\pi}\left[sc(\pi-\theta) - 1 - c + \frac{s^2}{2}(4+k_2)\right] + \frac{s}{\pi}(1+c) - s^2 - \frac{s^3c}{2\pi}k_5\right\}$$

$$D_W = \frac{-WR^3}{EI}\left\{\frac{\theta}{\pi}[(\pi-\theta)(2c^2+k_1) + 2sc(2+k_2)] - sc(2+k_2) - \frac{s^2}{\pi}(2+c^2k_5)\right\}$$

$$\Delta\psi = \frac{WR^2}{EI}\left\{s - \frac{\theta}{\pi}[2s+(\pi-\theta)c] + \frac{s^2c}{\pi}k_4\right\}$$

$$\text{Max } +M = \frac{WRs}{\pi}(1-c^2k_4) \qquad \text{at } x = \theta$$

$$\text{Max } -M = M_A \qquad \text{if } \theta \le \frac{\pi}{2}$$

If $\alpha = \beta = 0$, $M = K_M WR$, $T = K_T W$, $D = K_D WR^3/EI$, $\Delta\psi = K_{\Delta\psi} WR^2/EI$, etc.

θ	30°	45°	60°
K_{M_A}	−0.0903	−0.1538	−0.1955
K_{M_θ}	0.0398	0.1125	0.2068
K_{T_A}	−0.9712	−0.9092	−0.8045
K_{D_H}	−0.0157	−0.0461	−0.0891
K_{D_V}	0.0207	0.0537	0.0930
$K_{\Delta R}$	0.0060	0.0179	0.0355
$K_{\Delta R_W}$	0.0119	0.0247	0.0391
K_{D_W}	−0.0060	−0.0302	−0.0770
$K_{\Delta\psi}$	0.0244	0.0496	0.0590

TABLE 17 *Formulas for circular rings* (*Cont.*)

Reference no., loading, and load terms	Formulas for moments, loads, and deformations and some selected numerical values
3. 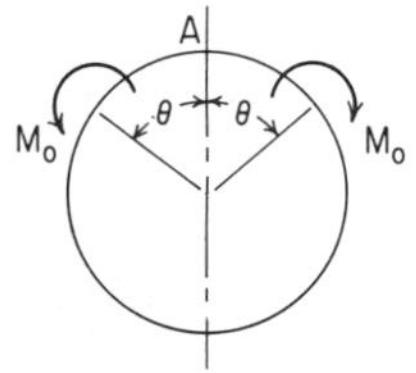 $LT_M = M_o\langle x-\theta\rangle^0$ $LT_T = 0$ $LT_V = 0$	$M_A = -M_o\left(1 - \frac{\theta}{\pi} - \frac{2s}{\pi k_1}\right)$ $T_A = \frac{M_o}{R}\frac{2s}{\pi k_1}$ $V_A = 0$ $D_H = \frac{M_o R^2}{EI}\left(\frac{2\theta}{\pi} - s\right) \quad \text{if } \theta \le \frac{\pi}{2}$ $D_V = \frac{M_o R^2}{EI}\left(\frac{2\theta}{\pi} - 1 + c\right)$ $\Delta R = \frac{M_o R^2}{EI}\left[\frac{1}{\pi}(\theta + sk_4) - \frac{\theta}{2}\right] \quad \text{if } \theta \le \frac{\pi}{2}$ $\Delta R_W = \frac{M_o R^2}{EI}\left\{\frac{\theta}{\pi}[1 + c - (\pi - \theta)s] + \frac{s^3}{\pi}k_4\right\}$ $D_{WH} = \frac{-M_o R^2}{EI}\left\{2s - \frac{2\theta}{\pi}[(\pi-\theta)c + 2s] + \frac{2s^2c}{\pi}k_4\right\}$ $\Delta\psi = \frac{M_o R}{EI}\left[\theta\left(1 - \frac{\theta}{\pi}\right) - \frac{2s^2}{\pi k_1}\right]$ Max $+M = M_o\left(\frac{\theta}{\pi} + \frac{2sc}{\pi k_1}\right)$ at x just greater than θ Max $-M = -M_o\left(1 - \frac{\theta}{\pi} - \frac{2sc}{\pi k_1}\right)$ at x just less than θ If $\alpha = \beta = 0$, $M = K_M M_o$, $T = K_T M_o/R$, $D = K_D M_o R^2/EI$, $\Delta\psi = K_{\Delta\psi} M_o R/EI$, etc.

θ	30°	45°	60°	90°
K_{M_A}	−0.5150	−0.2998	−0.1153	0.1366
K_{T_A}	0.3183	0.4502	0.5513	0.6366
K_{M_θ}	−0.5577	−0.4317	−0.3910	−0.5000
K_{D_H}	−0.1667	−0.2071	−0.1994	0.0000
K_{D_V}	0.1994	0.2071	0.1667	0.0000
$K_{\Delta R}$	0.0640	0.0824	0.0854	0.0329
$K_{\Delta R_W}$	0.1326	0.1228	0.1022	0.0329
$K_{D_{WH}}$	−0.0488	−0.0992	−0.1180	0.0000
$K_{\Delta\psi}$	0.2772	0.2707	0.2207	0.1488

4.

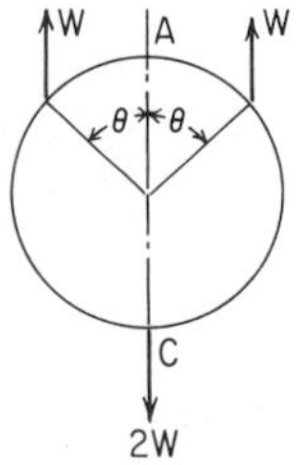

$LT_M = WR(z - s)\langle x - \theta \rangle^0$
$LT_T = Wz\langle x - \theta \rangle^0$
$LT_V = Wu\langle x - \theta \rangle^0$

$$M_A = -WR\left[\frac{1}{\pi}(1 + c + s^2k_4) - s\left(1 - \frac{\theta}{\pi}\right)\right]$$

$$T_A = -W\left(\frac{s^2}{\pi}k_4\right)$$

$$V_A = 0$$

$$D_H = \frac{-WR^3}{EI}\left[\frac{2}{\pi}(1 + c + \theta s) - \frac{s^2}{2}k_2 - k_3\right] \quad \text{if } \theta \le \frac{\pi}{2}$$

$$D_H = \frac{-WR^3}{EI}\left[\frac{2}{\pi}(1 + c + \theta s) + \frac{s^2}{2}k_2 - 2s\right] \quad \text{if } \theta \ge \frac{\pi}{2}$$

$$D_V = \frac{WR^3}{EI}\left[s - \frac{2}{\pi}(1 + c + \theta s) + \frac{\pi - \theta}{2}k_{-} - \frac{sc}{2}k_2\right]$$

$$\Delta R = \frac{WR^3}{EI}\left[0.1817(1 + c + \theta s) - 1 + \frac{\pi}{4}k_1 - \frac{s^2}{2\pi}k_5\right] \quad \text{if } \theta \le \frac{\pi}{2}$$

$$\Delta R = \frac{WR^3}{EI}\left[\frac{\pi - \theta}{2}(s + k_1) - \frac{1}{\pi}\left(1 + c + \theta s + \frac{s^2}{2}k_5\right) - \frac{1}{2}(1 + c + sck_2)\right] \quad \text{if } \theta \ge \frac{\pi}{2}$$

$$\Delta R_W = \frac{WR^3}{EI}\left\{\frac{1}{\pi}\left[\theta(\pi - \theta)s^2 - 2\theta s(1 + c) - (1 + c)^2 - \frac{s^4}{2}k_5\right] + \frac{sc}{2}k_3 + \frac{\pi - \theta}{2}k_1\right\}$$

$$D_{WH} = \frac{-WR^3}{EI}\left\{\frac{\theta}{\pi}[2(\pi - \theta)sc - 2 - 2c + s^2(4 + k_2)] + \frac{2s}{\pi}(1 + c) - 2s^2 - \frac{s^3c}{\pi}k_5\right\}$$

$$\Delta\psi = \frac{WR^2}{EI}\left\{\frac{\theta}{\pi}[1 + c - (\pi - \theta)s] + \frac{s^3}{\pi}k_4\right\}$$

$$\text{Max } -M = M_C = -WR\left[\frac{1}{\pi}(1 + c - s^2k_4) + \frac{s\theta}{\pi}\right]$$

Max $+M$ occurs at an angular position $x_1 = \arctan\frac{-\pi}{s^2k_4}$ if $\theta < x_1$

Max $+M$ occurs at the load if $\theta \ge x_1$

If $\alpha = \beta = 0$, $M = K_MWR$, $T = K_TW$, $D = K_DWR^3/EI$, $\Delta\psi = K_{\Delta\psi}WR^2/EI$, etc.

θ	30°	60°	90°	120°	150°
K_{M_A}	−0.2569	−0.1389	−0.1366	−0.1092	−0.0389
K_{T_A}	−0.0796	−0.2387	−0.3183	−0.2387	−0.0796
K_{M_C}	−0.5977	−0.5274	−0.5000	−0.4978	−0.3797
K_{M_θ}	−0.2462	−0.0195	0.1817	0.2489	0.1096
K_{D_H}	−0.2296	−0.1573	−0.1366	−0.1160	−0.0436
K_{D_V}	0.2379	0.1644	0.1488	0.1331	0.0597
$K_{\Delta R}$	0.1322	0.1033	0.0933	0.0877	0.0431
$K_{\Delta R_W}$	0.2053	0.1156	0.0933	0.0842	0.0271
$K_{D_{WH}}$	−0.0237	−0.0782	−0.1366	−0.1078	−0.0176
$K_{\Delta\psi}$	0.1326	0.1022	0.0329	0.0645	0.0667

TABLE 17 *Formulas for circular rings* (*Cont.*)

Reference no., loading, and load terms | Formulas for moments, loads, and deformations and some selected numerical values

5.

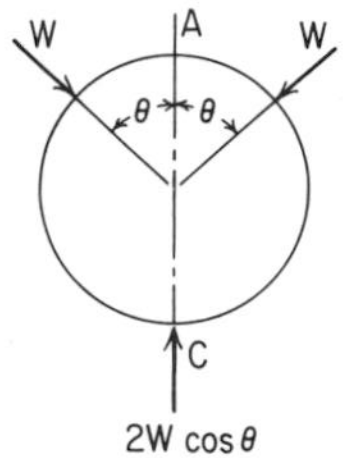

$LT_M = -WR \sin(x-\theta)\langle x-\theta\rangle^0$

$LT_T = -W \sin(x-\theta)\langle x-\theta\rangle^0$

$LT_V = -W \cos(x-\theta)\langle x-\theta\rangle^0$

$$M_A = -WR\left[s\left(1-\frac{\theta}{\pi}\right)-\frac{1}{\pi}(1+c)\right]$$

$$T_A = -W\left[s\left(1-\frac{\theta}{\pi}\right)\right]$$

$$V_A = 0$$

$$D_H = \frac{WR^3}{EI}\left[\frac{2}{\pi}(1+c)-\frac{\theta s}{2}k_1-ck_3\right] \quad \text{if } \theta \le \frac{\pi}{2}$$

$$D_H = \frac{WR^3}{EI}\left[\frac{2}{\pi}(1+c)-\frac{\pi-\theta}{2}sk_1\right] \quad \text{if } \theta \ge \frac{\pi}{2}$$

$$D_V = \frac{WR^3}{EI}\left[\frac{2}{\pi}(1+c)-\frac{s}{2}(2-k_2)-\frac{\pi-\theta}{2}ck_1\right]$$

$$\Delta R = \frac{WR^3}{EI}\left[\frac{1}{\pi}\left(1+c+\frac{\theta s}{2}k_2\right)-\frac{1}{2}(1-c)-\frac{\pi}{4}ck_1\right] \quad \text{if } \theta \le \frac{\pi}{2}$$

$$\Delta R = \frac{WR^3}{EI}\left[0.8183(1+c)-s+\frac{\theta s}{2\pi}k_2-\frac{\pi-\theta}{2}ck_1\right] \quad \text{if } \theta \ge \frac{\pi}{2}$$

$$\Delta R_W = \frac{WR^3}{EI}\left\{\frac{1}{\pi}\left[(1+c)(1+c+\theta s)+\frac{\theta s^3}{2}k_2\right]-s^3 - \frac{sc^2}{2}k_3-\frac{\pi-\theta}{2}ck_1\right\}$$

$$D_{WH} = \frac{-WR^3}{EI}\left\{\frac{\theta}{\pi}[(\pi-\theta)sk_1+c(2+2c+s^2k_2)]-s^2ck_2 - \frac{2s}{\pi}(1+c)\right\}$$

$$\Delta\psi = \frac{-WR^2}{EI}\left[\frac{\theta}{\pi}(1+c+s^2)-s^2\right]$$

$\text{Max } +M = M_C = \frac{WR}{\pi}(1+c-\theta s)$ if $\theta \le 60°$

$\text{Max } +M = M_\theta = WR\left[\frac{1}{\pi}(1+c+\theta sc)-sc\right]$ if $\theta \ge 60°$

$\text{Max } -M = M_C$ if $\theta \ge 90°$

$\text{Max } -M = M_A$ if $60° \le \theta \le 90°$

Max $-M$ occurs at an angular position $x_1 = \arctan\frac{-\pi c}{\theta s}$ if $\theta \le 60°$

If $\alpha = \beta = 0$, $M = K_M WR$, $T = K_T W$, $D = K_D WR^3/EI$, $\Delta\psi = K_{\Delta\psi} WR^2/EI$, etc.

θ	30°	60°	90°	120°	150°
K_{M_A}	0.1773	−0.0999	−0.1817	−0.1295	−0.0407
K_{T_A}	−0.4167	−0.5774	−0.5000	−0.2887	−0.0833
K_{M_C}	0.5106	0.1888	−0.1817	−0.4182	−0.3740
K_{M_θ}	0.2331	0.1888	0.3183	0.3035	0.1148
K_{D_H}	0.1910	0.0015	−0.1488	−0.1351	−0.0456
K_{D_V}	−0.1957	−0.0017	0.1366	0.1471	0.0620
$K_{\Delta R}$	−0.1115	−0.0209	0.0683	0.0936	0.0447
$K_{\Delta R_W}$	−0.1718	−0.0239	0.0683	0.0888	0.0278
$K_{D_{WH}}$	0.0176	−0.0276	−0.1488	−0.1206	−0.0182
$K_{\Delta\psi}$	−0.1027	0.0000	0.0000	0.0833	−0.0700

5.

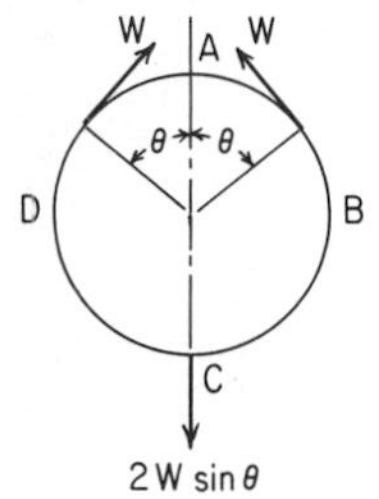

$\mathcal{L}T_M = -WR[1 - \cos(x-\theta)]\langle x-\theta\rangle^0$

$\mathcal{L}T_T = W\cos(x-\theta)\langle x-\theta\rangle^0$

$\mathcal{L}T_V = -W\sin(x-\theta)\langle x-\theta\rangle^0$

$$M_A = -WR\left[\frac{s}{\pi}(1+k_4) - \left(1-\frac{\theta}{\pi}\right)(1-c)\right]$$

$$T_A = -W\left[\frac{s}{\pi}k_4 + \left(1-\frac{\theta}{\pi}\right)c\right]$$

$$V_A = 0$$

$$D_H = \frac{WR^3}{EI}\left[\frac{s}{2}(2+k_3) - \frac{2}{\pi}(\theta+s) - \frac{\theta c}{2}k_1\right] \quad \text{if } \theta \le \frac{\pi}{2}$$

$$D_H = \frac{WR^3}{EI}\left[2 - \frac{2}{\pi}(\theta+s) - \frac{\pi-\theta}{2}ck_1 - \frac{s}{2}k_2\right] \quad \text{if } \theta \ge \frac{\pi}{2}$$

$$D_V = \frac{WR^3}{EI}\left[1 - c - \frac{2}{\pi}(\theta+s) - \frac{\pi-\theta}{2}sk_1\right]$$

$$\Delta R = \frac{WR^3}{EI}\left[\frac{1}{2}(\theta-s) - \frac{1}{\pi}\left(\theta + s - \frac{\theta c}{2}k_2 + \frac{s}{2}k_5\right) + \frac{\pi s}{4}k_1\right] \quad \text{if } \theta \le \frac{\pi}{2}$$

$$\Delta R = \frac{WR^3}{EI}\left[\frac{1}{2}(\pi - \theta - ck_2 - s - 2c) - \frac{\theta}{\pi}\left(1 - \frac{c}{2}k_2\right) - \frac{s}{2\pi}(2+k_5) + \frac{\pi-\theta}{2}sk_1\right] \quad \text{if } \theta \ge \frac{\pi}{2}$$

$$\Delta R_W = \frac{WR^3}{EI}\left\{\frac{\theta}{\pi}\left[(\pi-\theta)s - 1 - c - s^2\left(1-\frac{c}{2}k_2\right)\right] - \frac{s}{\pi}\left(1 + c + \frac{s^2}{2}k_5\right) - s\left[\frac{sc}{2}(2-k_3) - \frac{\pi-\theta}{2}k_1\right]\right\}$$

$$D_{WH} = \frac{-WR^3}{EI}\left\{\frac{1}{\pi}[\theta(\pi-\theta)(2+k_1)c - 2\theta sc + \theta sk_2 + 2s^2 - s^2ck_5] - s(2+c^2k_2)\left(1-\frac{\theta}{\pi}\right)\right\}$$

$$\Delta\psi = \frac{-WR^2}{EI}\left[\frac{\theta}{\pi}(\pi - \theta - s + sc) - sc - \frac{s^2}{\pi}k_4\right]$$

$$\text{Max } -M = M_C = -WR\left[\frac{s}{\pi}(1-k_4) + \frac{\theta}{\pi}(1+c)\right]$$

Max $+M$ occurs at an angular position $x_1 = \arctan\dfrac{-\pi s}{sk_4 - \theta c}$

(x_1 is always greater than θ and also greater than 90°)

If $\alpha = \beta = 0$, $M = K_M WR$, $T = K_T W$, $D = K_D WR^3/EI$, $\Delta\psi = K_{\Delta\psi} WR^2/EI$, etc.

θ	30°	60°	90°	120°	150°
K_{M_A}	−0.2067	−0.2180	−0.1366	−0.0513	−0.0073
K_{T_A}	−0.8808	−0.6090	−0.3183	−0.1090	−0.0148
K_{M_C}	−0.3110	−0.5000	−0.5000	−0.3333	−0.1117
K_{D_H}	−0.1284	−0.1808	−0.1366	−0.0559	−0.0083
K_{D_V}	0.1368	0.1889	0.1488	0.0688	0.0120
$K_{\Delta R}$	0.0713	0.1073	0.0933	0.0472	0.0088
$K_{\Delta R_W}$	0.1129	0.1196	0.0933	0.0460	0.0059
$K_{D_{WH}}$	−0.0170	−0.1063	−0.1366	−0.0548	−0.0036
$K_{\Delta\psi}$	0.0874	0.1180	0.0329	−0.0264	−0.0123

TABLE 17 Formulas for circular rings (Cont.)

Reference no., loading, and load terms	Formulas for moments, loads, and deformations and some selected numerical values

7. Ring under any number of equal radial forces equally spaced

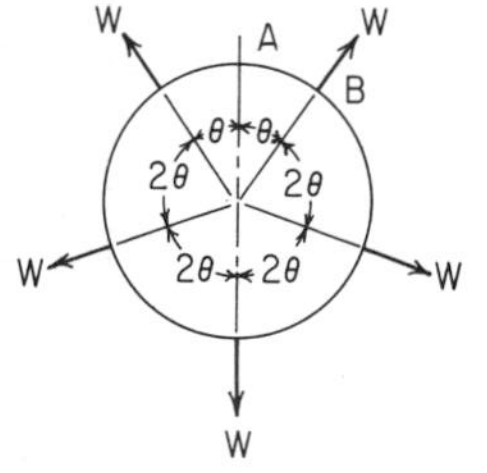

For $0 < x < \theta$ $\quad M = \frac{WR}{2}\left(\frac{u}{s} - \frac{1}{\theta}\right) \quad T = \frac{Wu}{2s} \quad V = \frac{-Wz}{2s}$

$\text{Max } +M = M_A = \frac{WR}{2}\left(\frac{1}{s} - \frac{1}{\theta}\right) \quad \text{Max } -M = \frac{-WR}{2}\left(\frac{1}{\theta} - \frac{c}{s}\right)$ at each load position

Radial displacement at each load point $= \frac{WR^3}{EI}\left[\frac{1}{4s^2}(\theta k_1 + sck_3) - \frac{1}{2\theta}\right] = R_B$

Radial displacement at $x = 0, 2\theta$, and so on $= \frac{-WR^3}{EI}\left[\frac{1}{2\theta} - \frac{1}{4s^2}(sk_3 + \theta ck_1)\right] = R_A$

If $\alpha = \beta = 0$, $M = K_M WR$, $R = K_R WR^3/EI$,

θ	15°	30°	45°	60°	90°
K_{M_A}	0.02199	0.04507	0.07049	0.09989	0.18169
K_{M_B}	−0.04383	−0.08890	−0.13662	−0.18879	−0.31831
K_{R_B}	0.00020	0.00168	0.00608	0.01594	0.07439
K_{R_A}	−0.00018	−0.00148	−0.00539	−0.01426	−0.06831

8.

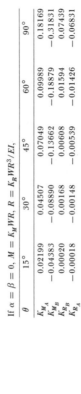

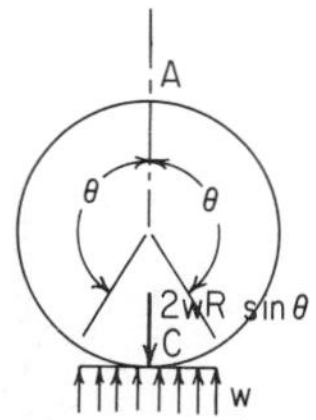

$\left(\textit{Note: } \theta \geq \frac{\pi}{2}\right)$

$LT_M = \frac{-wR^2}{2}(z - s)^2\langle x - \theta\rangle^0$

$LT_T = -wRz(z - s)\langle x - \theta\rangle^0$

$LT_V = -wRu(z - s)\langle x - \theta\rangle^0$

$$M_A = wR^2\left[\frac{1}{4} + \frac{s^2}{2} - \frac{1}{\pi}\left(s + \frac{3}{4}sc + \frac{s^3}{3}k_4 + \frac{\theta}{4} + \frac{\theta s^2}{2}\right)\right]$$

$$T_A = -wR\left(\frac{s^3}{3\pi}k_4\right)$$

$$V_A = 0$$

$$D_H = \frac{-wR^4}{EI}\left[\frac{1}{\pi}\left(2s + \frac{3sc}{2} + \theta s^2 + \frac{\theta}{2}\right) - \frac{1}{2} - s^2 + \frac{s^3}{6}k_2\right]$$

$$D_V = \frac{wR^4}{EI}\left[\frac{k_1}{2}(\pi s - \theta s - 1 - c) + \frac{k_2}{6}(1 + c^3) + \frac{1}{2}(1 + s^2) - \frac{1}{\pi}\left(2s + \frac{3sc}{2} + \theta s^2 + \frac{\theta}{2}\right)\right]$$

$$\Delta R = \frac{-wR^4}{EI}\left[\frac{1}{\pi}\left(s + \frac{3sc}{4} + \frac{\theta s^2}{2} + \frac{k_5 s^3}{6} + \frac{\theta}{4}\right) - \frac{1}{4} + \frac{s}{2} + \frac{3sc}{8} - \frac{\pi - \theta}{8}(1 + 2s^2) - \frac{k_2}{6}(1 + c^3) - \frac{k_1}{2}(\pi s - \theta s - 1 - c)\right]$$

$$\text{Max } -M = M_C = -wR^2\left[-\frac{1}{4} + \frac{1}{\pi}\left(s + \frac{3sc}{4} - \frac{s^3}{3}k_4 + \frac{\theta}{4} + \frac{\theta s^2}{2}\right)\right]$$

If $\alpha = \beta = 0$, $M = K_M wR^2$, $T = K_T wR$, $D = K_D wR^4/EI$, etc.

θ	90°	120°	135°	150°
K_{M_A}	−0.0494	−0.0329	−0.0182	−0.0065
K_{T_A}	−0.1061	−0.0689	−0.0375	−0.0133
K_{M_C}	−0.3372	−0.2700	−0.1932	−0.1050
K_{D_H}	−0.0533	−0.0362	−0.0204	−0.0074
K_{D_V}	0.0655	0.0464	0.0276	0.0108
$K_{\Delta R}$	0.0448	0.0325	0.0198	0.0080

9.

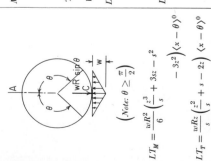

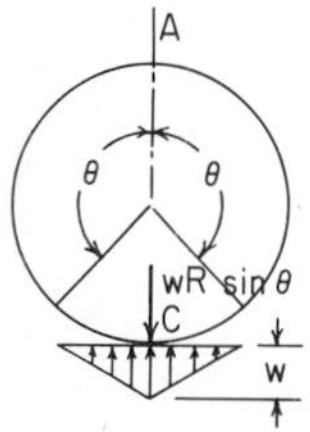

$\left(\textit{Note: } \theta \geq \frac{\pi}{2}\right)$

$$LT_M = \frac{wR^2}{6}\left(\frac{z^3}{s} + 3sz - s^2 - 3z^2\right)\langle x - \theta\rangle^0$$

$$LT_T = \frac{wRz}{s}\left(\frac{z^2}{s} + s - 2z\right)\langle x - \theta\rangle^0$$

$$LT_V = \frac{wRu}{2}\left(\frac{z^2}{s} + s - 2z\right)\langle x - \theta\rangle^0$$

$$M_A = wR^2\left\{\left(1 - \frac{\theta}{\pi}\right)\left(\frac{1}{4} + \frac{s^2}{6}\right) - \frac{1}{\pi}\left[\frac{11}{36}sc + \frac{1+c}{9s} + \frac{s}{2}\right] - \frac{s^3}{12\pi}k_4\right\}$$

$$T_A = -wR\left(\frac{s^3}{12\pi}k_4\right)$$

$$V_A = 0$$

$$D_H = \frac{wR^4}{EI}\left\{\frac{1}{2} + \frac{s^2}{3} - \frac{s^3}{24}k_2 - \frac{1}{\pi}\left[s + \frac{\theta}{2} + \frac{\theta}{3}s^2 + \frac{5sc}{6} + \frac{2}{9s}(1 + c^3)\right]\right\}$$

$$D_V = \frac{wR^4}{EI}\left\{\frac{1}{2} + \frac{s^2}{6} - \frac{k_2}{16}\left(\frac{\pi - \theta}{s} - s^2c - \frac{5c^3}{3} - \frac{8}{3}\right) - \frac{1}{\pi}\left[s + \frac{2}{9s}(1 + c^3) + \frac{5sc}{6} + \frac{\theta s^2}{3} + \frac{\theta}{2}\right] + \frac{k_1}{8}\left[(\pi - \theta)\left(\frac{1}{s} + 2s\right) - 3c - 4\right]\right\}$$

$$\Delta R = \frac{wR^4}{EI}\left\{(\pi - \theta)\left(\frac{1}{8} + \frac{s^2}{12}\right) + \frac{1}{4} - \frac{s}{4} - \frac{3sc}{8} - \frac{1 - 3c + 4c^3}{18s} - \frac{1}{\pi}\left(\frac{\theta}{4} + \frac{s}{2} + \frac{\theta s^2}{6} + \frac{3sc}{4} + \frac{1 - 3c + 4c^3}{9s}\right) + \frac{k_1}{8}\left[(\pi - \theta)\left(\frac{1}{s} + 2s\right) - 4 - 3c\right] + \frac{k_2}{48}\left[8 + 5c - 2s^2c - \frac{3}{s}(\pi - \theta)\right] - \frac{s^3}{24\pi}k_5\right\}$$

$$\text{Max } -M = M_C = -wR^2\left[-\frac{1}{4} + \frac{1}{\pi}\left(\frac{11}{36}sc + \frac{s}{2} + \frac{1+c}{9s} - \frac{s^3}{12}k4 + \frac{\theta}{4} + \frac{\theta s^2}{6}\right)\right]$$

If $\alpha = \beta = 0$, $M = K_M wR^2$, $T = K_T wR$, $D = K_D wR^4/EI$, etc.

θ	90°	120°	135°	150°
K_{M_A}	−0.0127	−0.0084	−0.0046	−0.0016
K_{T_A}	−0.0265	−0.0172	−0.0094	−0.0033
K_{M_C}	−0.1263	−0.0989	−0.0692	−0.0367
K_{D_H}	−0.0141	−0.0093	−0.0052	−0.0019
K_{D_V}	0.0185	0.0127	0.0074	0.0028
$K_{\Delta R}$	0.0131	0.0092	0.0054	0.0021

TABLE 17 *Formulas for circular rings* (*Cont.*)

Reference no., loading, and load terms	Formulas for moments, loads, and deformations and some selected numerical values
10. 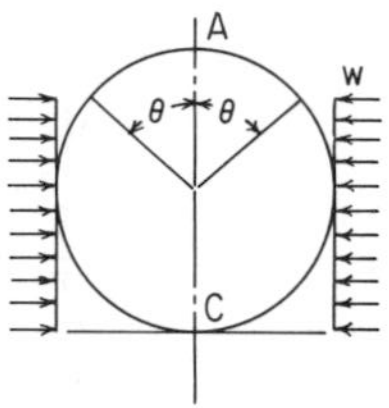 $LT_M = \frac{-wR^2}{2}(1 + u^2 - s^2 - 2uc)\langle x - \theta \rangle^0$ $LT_T = wRu(c - u)\langle x - \theta \rangle^0$ $LT_V = wRz(c - u)\langle x - \theta \rangle^0$	$M_A = -wR^2\left\{c + \frac{s^2}{2} - \frac{3}{4} + \frac{1}{\pi}\left[s - \theta c + \frac{3}{4}(\theta - sc) - \frac{\theta s^2}{2} - \frac{s^3}{3}k_4\right]\right\}$ $T_A = -wR\left[c + \frac{1}{\pi}\left(s - \theta c - \frac{s^3}{3}k_4\right)\right]$ $V_A = 0$ $D_H = \frac{-wR^4}{EI}\left[\frac{k_1}{2}(2 - s + \theta c) - \frac{k_2}{6}(2 - s^3) - \frac{1}{2} - \frac{1}{2\pi}(3sc - 3\theta + 2\theta s^2)\right]$ if $\theta \le \frac{\pi}{2}$ $D_H = \frac{-wR^4}{EI}\left[\frac{k_1}{2}(\pi c - \theta c + s) - \frac{s^3}{6}k_2 - \frac{1}{2} - c^2 + \frac{1}{2\pi}(2\theta c^2 + \theta - 3sc)\right]$ if $\theta \ge \frac{\pi}{2}$ $D_V = \frac{wR^4}{EI}\left[\frac{c^2}{2} - c + \frac{k_2}{6}(2 - c^3 + 3c) + \frac{1}{2\pi}(2\theta s^2 + 3sc - 3\theta)\right]$ $\Delta R = \frac{wR^4}{EI}\left[\frac{3}{8}(\theta - sc) - \frac{1}{4}(1 + \theta s^2) + k_2\left(\frac{1}{3} + \frac{\theta c}{2\pi} - \frac{s}{2\pi}\right) + \frac{s^3}{6\pi}k_5 + \frac{1}{4\pi}(3sc - 3\theta + 2\theta s^2)\right]$ if $\theta \le \frac{\pi}{2}$ $\Delta R = \frac{wR^4}{EI}\left[\frac{3}{8}(\pi - \theta + sc) - \frac{1}{4}(1 + \pi s^2 - \theta s^2) - c + \frac{s^3}{6\pi}k_5 + k_2\left(\frac{1}{3} + \frac{\theta c}{2\pi} - \frac{s}{2\pi} - \frac{c^3}{6}\right) + \frac{1}{4\pi}(3sc - 3\theta + 2\theta s^2)\right]$ if $\theta \ge \frac{\pi}{2}$ Max $-M = M_C = -wR^2\left\{\frac{1}{4} + \frac{1}{\pi}\left[\frac{3}{4}(\theta - sc) - s - \frac{\theta s^2}{2} + \theta c + \frac{s^3}{3}k_4\right]\right\}$ If $\alpha = \beta = 0$, $M = K_M wR^2$, $T = K_T wR$, $D = K_D wR^4/EI$, etc.

θ	0°	30°	45°	60°	90°	120°	135°	150°
K_{M_A}	−0.2500	−0.2434	−0.2235	−0.1867	−0.0872	−0.0185	−0.0052	−0.00076
K_{T_A}	−1.0000	−0.8676	−0.7179	−0.5401	−0.2122	−0.0401	−0.0108	−0.00155
K_{M_C}	−0.2500	−0.2492	−0.2448	−0.2315	−0.1628	−0.0633	−0.0265	−0.00663
K_{D_H}	−0.1667	−0.1658	−0.1610	−0.1470	−0.0833	−0.0197	−0.0057	−0.00086
K_{D_V}	0.1667	0.1655	0.1596	0.1443	0.0833	0.0224	0.0071	0.00118
$K_{\Delta R}$	0.0833	0.0830	0.0812	0.0756	0.0486	0.0147	0.0049	0.00086

11.

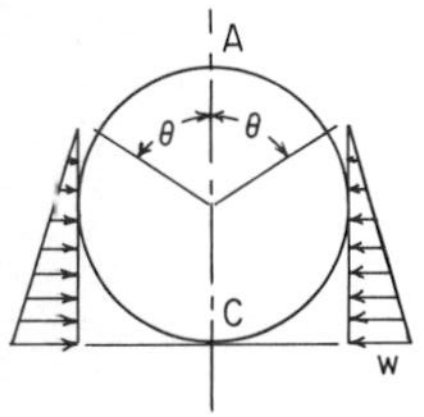

$$LT_M = \frac{wR^2}{1+c}\left(\frac{u^3}{6} - \frac{c^3}{6} + \frac{uc^2}{2} - \frac{u^2c}{2}\right)\langle x-\theta\rangle^0$$

$$LT_T = \frac{wRu}{2(1+c)}(c-u)^2\langle x-\theta\rangle^0$$

$$LT_V = \frac{wRz}{2(1+c)}(c-u)^2\langle x-\theta\rangle^0$$

$$M_A = \frac{wR^2}{1+c}\left[\left(1-\frac{\theta}{\pi}\right)\left(\frac{3c}{4} - \frac{c^2}{2} - \frac{c^3}{3} - \frac{s^2c}{2} - \frac{1}{4}\right) + \frac{1}{\pi}\left(\frac{s}{9} - \frac{3sc}{4} + \frac{11}{36}sc^2\right) + \frac{k_4}{8\pi}\left(\pi - \theta + sc + \frac{2s^3c}{3}\right)\right]$$

$$T_A = \frac{wR}{1+c}\left[\frac{1}{\pi}\left(\frac{\theta}{4} + \frac{\theta c^2}{2} - \frac{3sc}{4}\right) - \frac{1}{4} - \frac{c^2}{2} + \frac{k_4}{8\pi}\left(\pi - \theta + sc + \frac{2s^3c}{3}\right)\right]$$

$$V_A = 0$$

$$D_H = \frac{-wR^4}{EI(1+c)}\left[\frac{1}{\pi}\left(\frac{5\theta c}{6} - \frac{\theta s^2c}{3} - \frac{11}{18}sc^2 - \frac{2s}{9}\right) - \frac{c}{2} + \frac{k_1}{8}(\theta + 8c + 2\theta c^2 - 3sc) + \frac{k_2}{16}\left(sc + \frac{2s^3c}{3} - \frac{16c}{3} - \theta\right)\right] \quad \text{if } \theta \le \frac{\pi}{2}$$

$$D_H = \frac{-wR^4}{EI(1+c)}\left\{\frac{1}{\pi}\left(\frac{\theta c}{2} + \frac{\theta c^3}{3} - \frac{5sc^2}{6} - \frac{2s^3}{9}\right) - \frac{c^3}{3} - \frac{c}{2} + \frac{k_3}{8}[(\pi-\theta)(1+2c^2) + 3sc] - \frac{k_2}{16}\left(\pi - \theta + sc + \frac{2s^3c}{3}\right)\right\} \quad \text{if } \theta \ge \frac{\pi}{2}$$

$$D_V = \frac{wR^4}{EI(1+c)}\left\{\frac{1}{\pi}\left(\frac{2s^3}{9} + \frac{5sc^2}{6} - \frac{5\theta c}{6} + \frac{\theta s^2c}{3}\right) - \frac{1}{6} - \frac{c^2}{2} + \frac{c^3}{6} + \frac{k_2}{6}\left[(1+c)^2 - \frac{s^4}{4}\right]\right\}$$

$$\Delta R = \frac{wR^4}{EI(1+c)}\left[\frac{1}{\pi}\left(\frac{s}{9} + \frac{11}{36}sc^2 - \frac{5\theta c}{12} + \frac{\theta s^2c}{6}\right) - \frac{1}{18} - \frac{s}{18} - \frac{c}{4} + \frac{5\theta c}{24} - \frac{\theta s^2c}{12} - \frac{11}{72}sc^2 + \frac{k_2}{8\pi}\left(\theta + 2\theta c^2 - 3sc + \frac{8\pi c}{3}\right) + \frac{k_5}{16\pi}\left(\pi - \theta + sc + \frac{2s^3c}{3}\right)\right] \quad \text{if } \theta \le \frac{\pi}{2}$$

$$\Delta R = \frac{wR^4}{EI(1+c)}\left[\frac{1}{\pi}\left(\frac{s}{9} + \frac{11}{36}sc^2 - \frac{\theta c}{4} - \frac{\theta c^3}{6}\right) + (\pi - \theta)\left(\frac{c}{8} + \frac{c^3}{12}\right) - \frac{1}{6} + \frac{s}{18} - \frac{c}{4} - \frac{c^2}{2} + \frac{11}{72}sc^2 + \frac{k_2}{8\pi}\left(\frac{8\pi c}{3} + \theta + 2\theta c^2 - 3sc - \frac{\pi c^4}{3}\right) + \frac{k_5}{16\pi}\left(\pi - \theta + sc + \frac{2s^3c}{3}\right)\right] \quad \text{if } \theta \ge \frac{\pi}{2}$$

$$Max\ -M = M_C = \frac{-wR^2}{1+c}\left[\frac{c}{4} - \frac{1}{12} + \frac{\theta}{\pi}\left(\frac{5c}{12} + \frac{c^2}{2} - \frac{s^2c}{6} + \frac{1}{4}\right) - \frac{s}{\pi}\left(\frac{1}{9} + \frac{3c}{4} + \frac{11}{36}c^2\right) + \frac{k4}{8\pi}\left(\pi - \theta + sc + \frac{2s^3c}{3}\right)\right]$$

If $\alpha = \beta = 0$, $M = K_M wR^2$, $T = K_T wR$, $D = K_D wR^4/EI$, etc.

θ	0°	30°	45°	60°	90°	120°	135°	150°
K_{M_A}	−0.1042	−0.0939	−0.0808	−0.0635	−0.0271	−0.0055	−0.0015	−0.00022
K_{T_A}	−0.3125	−0.2679	−0.2191	−0.1628	−0.0625	−0.0116	−0.0031	−0.00045
K_{M_C}	−0.1458	−0.1384	−0.1282	−0.1129	−0.0688	−0.0239	−0.0096	−0.00232
K_{D_H}	−0.0833	−0.0774	−0.0693	−0.0575	−0.0274	−0.0059	−0.0017	−0.00025
K_{D_V}	0.0833	0.0074	0.0694	0.0579	0.0291	0.0071	0.0022	0.00035
$K_{\Delta R}$	0.0451	0.0424	0.0387	0.0332	0.0180	0.0048	0.0015	0.00026

TABLE 17 *Formulas for circular rings* (*Cont.*)

Reference no., loading, and load terms	Formulas for moments, loads, and deformations and some selected numerical values

12.

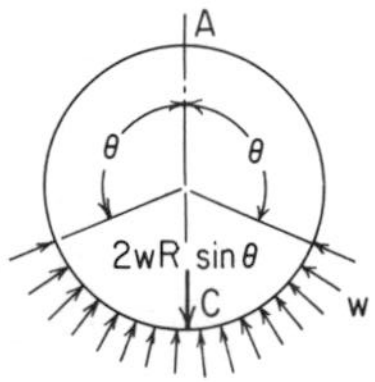

$LT_M = -wR^2[1 - \cos(x - \theta)]\langle x - \theta\rangle^0$

$LT_T = -wR[1 - \cos(x - \theta)]\langle x - \theta\rangle^0$

$LT_V = -wR\sin(x - \theta)\langle x - \theta\rangle^0$

$$M_A = -wR^2\left[\frac{1}{\pi}(\theta + 2s - \theta c) - 1 + c\right]$$

$$T_A = -wR\left[\frac{1}{\pi}(s - \theta c) + c\right]$$

$$V_A = 0$$

$$D_H = \frac{-wR^4}{EI}\left[\frac{2}{\pi}(\theta + s) - 2 + \frac{k_1}{2}(2 - s + \theta c) + k_3(1 - s)\right] \quad \text{if } \theta \le \frac{\pi}{2}$$

$$D_H = \frac{-wR^4}{EI}\left[\frac{k_1}{2}(\pi c - \theta c + s) - \frac{2}{\pi}(\pi - \theta - s)\right] \quad \text{if } \theta \ge \frac{\pi}{2}$$

$$D_V = \frac{wR^4}{EI}\left[1 - c - \frac{2}{\pi}(\theta + s) + \frac{s}{2}(\pi - \theta)k_1 - \alpha(1 + c)\right]$$

$$\Delta R = \frac{wR^4}{EI}\left[\frac{1}{2}(\theta - s) - \frac{1}{\pi}(\theta + s) - \frac{k_1}{4}(2 - \pi s) + \frac{k_2}{2\pi}(\pi - s + \theta c)\right] \quad \text{if } \theta \le \frac{\pi}{2}$$

$$\Delta R = \frac{wR^4}{EI}\left[0.8183(\pi - \theta - s) - 1 - c + \frac{k_2}{2\pi}(\pi - s + \theta c) - \frac{k_1}{2}(1 + c - \pi s + \theta s)\right] \quad \text{if } \theta \ge \frac{\pi}{2}$$

$$\text{Max } -M = M_C = -wR^2\left[\frac{\theta}{\pi}(1 + c)\right]$$

Max $+M$ occurs at an angular position $x_1 = \arctan\dfrac{-s\pi}{s - \theta c}$

If $\alpha = \beta = 0$, $M = K_M wR^2$, $T = K_T wR$, $D = K_D wR^4/EI$, etc.

θ	30°	60°	90°	120°	150°
K_{M_A}	−0.2067	−0.2180	−0.1366	−0.0513	−0.0073
K_{T_A}	−0.8808	−0.6090	−0.3183	−0.1090	−0.0148
K_{M_C}	−0.3110	−0.5000	−0.5000	−0.3333	−0.1117
K_{D_H}	−0.1284	−0.1808	−0.1366	−0.0559	−0.0083
K_{D_V}	0.1368	0.1889	0.1488	0.0688	0.0120
$K_{\Delta R}$	0.0713	0.1073	0.0933	0.0472	0.0088

13. Ring supported at base and loaded by own weight of w lb per linear inch 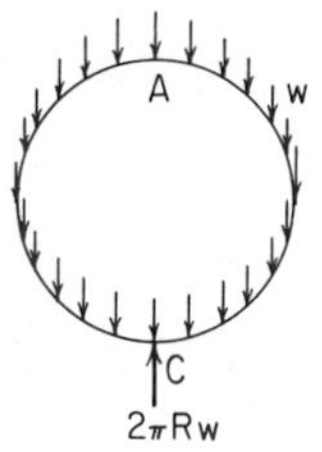 $LT_M = -wR^2(xz + u - 1)$ $LT_T = -wRxz$ $LT_V = -wRxu$	$M_A = \frac{wR^2}{2}k_4$ $T_A = \frac{wR}{2}k_4$ $V_A = 0$ $D_H = \frac{wR^4}{EI}\left(2 - \frac{\pi}{2}k_3\right)$ $D_V = \frac{-wR^4}{EI}\left(\frac{\pi^2}{4}k_1 - 2\right)$ $\Delta R = \frac{-wR^4}{EI}\left(\frac{3\pi^2}{16}k_1 - \frac{\pi}{2} + \frac{\alpha}{2}k_4\right)$	Max $+M = M_C = wR^2\left(2 - \frac{k_4}{2}\right)$ Max $-M$ occurs at an angular position x_1 where $\frac{x_1}{\tan x_1} = -\frac{k_4}{2}$ If $\alpha = \beta = 0$, $M_A = \frac{wR^2}{2}$ $T_A = \frac{wR}{2}$ $D_H = 0.4292\frac{wR^4}{EI}$ $D_V = -0.4674\frac{wR^4}{EI}$ $\Delta R = -0.2798\frac{wR^4}{EI}$ Max $+M = \frac{3}{2}wR^2$ at C Max $-M = -0.644wR^2$ at $x_1 = 1.84$ rad $(105.2°)$

TABLE 17 *Formulas for circular rings* (*Cont.*)

Reference no., loading, and load terms	Formulas for moments, loads, and deformations and some selected numerical values	
14. 1-in segment of pipe filled with liquid of specific weight ρ lb/in^3 and supported at the base 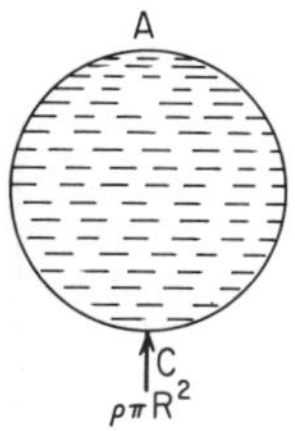 $LT_M = \rho R^3\left(1 - u - \frac{xz}{2}\right)$ $LT_T = \rho R^2\left(1 - u - \frac{xz}{2}\right)$ $LT_V = \rho R^2\left(\frac{z}{2} - \frac{xu}{2}\right)$	$M_A = \frac{\rho R^3}{4}$ $T_A = \frac{3\rho R^2}{4}$ $V_A = 0$ $D_H = \frac{\rho R^5 12(1-\nu^2)}{Et^3}\left[k_1 + k_3\left(1 - \frac{\pi}{4}\right) - 1\right]$ $D_V = \frac{\rho R^5 12(1-\nu^2)}{Et^3}\left(1 + 2\alpha - \frac{\pi^2}{8}k_1\right)$ $\Delta R = \frac{\rho R^5 12(1-\nu^2)}{Et^3}\left(\frac{\pi}{4} - 0.3003k_1 - \frac{5}{8}k_2\right)$ *Note:* For this case and case 15, $\alpha = \frac{t^2}{12R^2(1-\nu^2)}$ $\beta = \frac{t^2}{6R^2(1-\nu)}$	Max $+M = M_C = \frac{3\rho R^3}{4}$ Max $-M = -0.322\rho R^3$ at $x = 105.2°$ (1.84 rad) If $\alpha = \beta = 0$, $D_H = 0.2146\frac{\rho R^5 12(1-\nu^2)}{Et^3}$ $D_V = -0.2337\frac{\rho R^5 12(1-\nu^2)}{Et^3}$ $\Delta R = -0.1399\frac{\rho R^5 12(1-\nu^2)}{Et^3}$

15. 1-in segment of pipe partly filled with liquid of specific weight ρ lb/in^3 and supported at the base

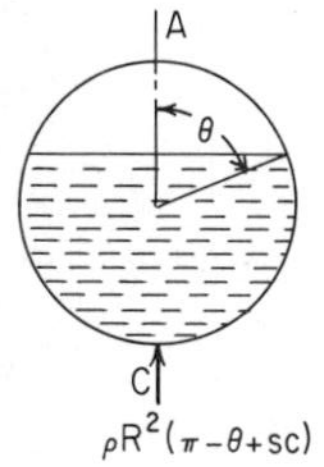

$$LT_M = \frac{\rho R^3}{2}[2c - z(x - \theta + sc) - u(1 + c^2)]\langle x - \theta\rangle^0$$

$$LT_T = \frac{\rho R^2}{2}[2c - z(x - \theta + sc) - u(1 + c^2)]\langle x - \theta\rangle^0$$

$$LT_V = \frac{\rho R^2}{2}[zc^2 - u(x - \theta + sc)]\langle x - \theta\rangle^0$$

Note: See case 14 for expressions for α and β

$$M_A = \rho R^3\left[\left(1 - \frac{\theta}{\pi}\right)\left(\frac{3}{4} + \frac{c^2}{2} - c\right) + \frac{5sc}{4\pi} - \frac{s}{\pi}\right]$$

$$T_A = \rho R^2\left[\left(1 - \frac{\theta}{\pi}\right)\left(\frac{1}{4} + \frac{c^2}{2}\right) - \frac{3sc}{4\pi}\right]$$

$$V_A = 0$$

$$D_H = \frac{\rho R^5 12(1 - \nu^2)}{Et^3}\left[1 - 2c + \frac{1}{\pi}(sc + 2\theta c - 2s - \theta) + \frac{k_1}{8}(\theta + 8c + 2\theta c^2 - 3sc) - \frac{k_3}{4}(\pi - 2\theta + 2sc - 4c)\right] \quad \text{if } \theta < \frac{\pi}{2}$$

$$D_H = \frac{\rho R^5 12(1 - \nu^2)}{Et^3}\left\{1 - 2c + \frac{1}{\pi}(sc + 2\theta c - 2s - \theta) + \frac{k_1}{8}[(\pi - \theta)(1 + 2c^2) + 3sc]\right\} \quad \text{if } \theta > \frac{\pi}{2}$$

$$D_V = \frac{\rho R^5 12(1 - \nu^2)}{Et^3}\left\{2 - \frac{s^2}{2} - c + \frac{1}{\pi}(sc + 2\theta c - 2s - \theta) - \frac{k_1}{8}[(\pi - \theta)(\pi - \theta + 2sc) + s^2] + \frac{\alpha}{2}(1 + c)^2\right\}$$

$$\Delta R = \frac{\rho R^5 12(1 - \nu^2)}{Et^3}\left[0.1817(s - \theta c) + 0.4092(sc - \theta) + \frac{\pi}{4} + \frac{k_1}{8}\left(1 + \pi\theta - \pi sc + 4c - \frac{3\pi^2}{4}\right) + \frac{k_2}{8\pi}(3sc - \pi - \theta - 2\theta c^2 - 4\pi c)\right] \quad \text{if } \theta < \frac{\pi}{2}$$

$$\Delta R = \frac{\rho R^5 12(1 - \nu^2)}{Et^3}\left\{0.4092(\pi - \theta - 2s + sc + 2\theta c) + \frac{1}{2}(1 + c^2 - \pi c) + \frac{k_1}{8}[1 + 3c^2 + 4c - (\pi - \theta)(\pi - \theta + 2sc)] - \frac{k_2}{8\pi}(\pi + \theta - 3sc + 2\theta c^2 + 4\pi c)\right\} \quad \text{if } \theta > \frac{\pi}{2}$$

$$\text{Max } +M = M_C = \rho R^3\left[\frac{3}{4} + \frac{\theta}{\pi}\left(c + \frac{c^2}{2} - \frac{1}{4}\right) - \frac{s}{\pi}\left(1 + \frac{c}{4}\right)\right]$$

If $\alpha = \beta = 0$, $M = K_M \rho R^3$, $T = K_T \rho R^2$, $D = K_D \rho R^5 12(1 - \nu^2)/Et^3$, etc.

θ	0°	30°	45°	60°	90°	120°	135°	150°
K_{M_A}	0.2500	0.2290	0.1935	0.1466	0.0567	0.0104	0.0027	0.00039
K_{T_A}	0.7500	0.6242	0.4944	0.3534	0.1250	0.0216	0.0056	0.00079
K_{M_C}	0.7500	0.7216	0.6619	0.5649	0.3067	0.0921	0.0344	0.00778
K_{D_H}	0.2146	0.2027	0.1787	0.1422	0.0597	0.0115	0.0031	0.00044
K_{D_V}	−0.2337	−0.2209	−0.1955	−0.1573	−0.0700	−0.0150	−0.0043	−0.00066
$K_{\Delta R}$	−0.1399	−0.1333	−0.1198	−0.0986	−0.0465	−0.0106	−0.0031	−0.00050

TABLE 17 *Formulas for circular rings* (*Cont.*)

Reference no., loading, and load terms	Formulas for moments, loads, and deformations and some selected numerical values

16.

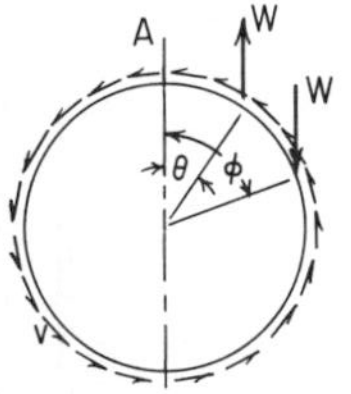

$$v = \frac{W}{2\pi R}(\sin\phi - \sin\theta)\ \text{lb/in}$$

$$LT_M = \frac{-WR}{2\pi}(n-s)(x-z) + WR(z-s)\langle x-\theta\rangle^0 - WR(z-n)\langle x-\phi\rangle^0$$

$$LT_T = \frac{W}{2\pi}(n-s)z + Wz\langle x-\theta\rangle^0 - Wz\langle x-\phi\rangle^0$$

$$LT_V = \frac{-W}{2\pi}(n-s)(1-u) + Wu\langle x-\theta\rangle^0 - Wu\langle x-\phi\rangle^0$$

$$M_A = \frac{-WR}{2\pi}[c - e - (\pi-\theta)s + (\pi-\phi)n + k_4(s^2-n^2)]$$

$$T_A = \frac{-W}{2\pi}[k_4(s^2-n^2)]$$

$$V_A = \frac{W}{2\pi}[\theta - \phi + s - n + k_4(sc - ne)]$$

If $\alpha = \beta = 0$, $M = K_M WR$, $T = K_T W$, $V = K_V W$,

θ	ϕ	0°	30°	60°	90°	120°	150°	180°
0°	K_{M_A}	0.0000	−0.1899	−0.2489	−0.2500	−0.2637	−0.2989	−0.3183
	K_{T_A}	0.0000	0.0398	0.1194	0.1592	0.1194	0.0398	0.0000
	K_{V_A}	0.0000	−0.2318	−0.3734	−0.4092	−0.4023	−0.4273	−0.5000
30°	K_{M_A}	0.1899	0.0000	−0.0590	−0.0601	−0.0738	−0.1090	−0.1284
	K_{T_A}	−0.0398	0.0000	0.0796	0.1194	0.0796	0.0000	−0.0398
	K_{V_A}	0.2318	0.0000	−0.1416	−0.1773	−0.1704	−0.1955	−0.2682
45°	K_{M_A}	0.2322	0.0423	−0.0167	−0.0178	−0.0315	−0.0667	−0.0861
	K_{T_A}	−0.0796	−0.0398	0.0398	0.0796	0.0398	−0.0398	−0.0796
	K_{V_A}	0.3171	0.0853	−0.0563	−0.0920	−0.0851	−0.1102	−0.1829
60°	K_{M_A}	0.2489	0.0590	0.0000	−0.0011	−0.0148	−0.0500	−0.0694
	K_{T_A}	−0.1194	−0.0796	0.0000	0.0398	0.0000	−0.0796	−0.1194
	K_{V_A}	0.3734	0.1416	0.0000	−0.0357	−0.0288	−0.0539	−0.1266
90°	K_{M_A}	0.2500	0.0601	0.0011	0.0000	−0.0137	−0.0489	−0.0683
	K_{T_A}	−0.1592	−0.1194	−0.0398	0.0000	−0.0398	−0.1194	−0.1592
	K_{V_A}	0.4092	0.1773	0.0357	0.0000	0.0069	−0.0182	−0.0909

17.

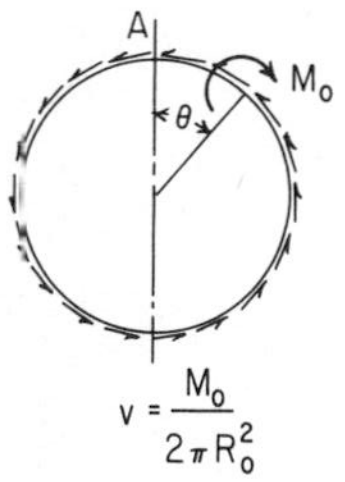

$$LT_M = \frac{-M_o}{2\pi}(x - z) + M_o\langle x - \theta\rangle^0$$

$$LT_T = \frac{M_o z}{2\pi R}$$

$$LT_V = \frac{-M_o}{2\pi R}(1 - u)$$

$$M_A = \frac{-M_o}{2\pi}(\pi - \theta - 2k_4 s)$$

$$T_A = \frac{M_o}{\pi R}k_4 s$$

$$V_A = \frac{-M_o}{2\pi R}(1 + 2k_4 c)$$

If $\alpha = \beta = 0$, $M = K_M M_o$, $T = K_T M_o/R$, $V = K_V M_o/R$,

θ	0°	30°	45°	60°	90°	120°	135°	150°	180°
K_{M_A}	−0.5000	−0.2575	−0.1499	−0.0577	0.0683	0.1090	0.1001	0.0758	0.0000
K_{T_A}	0.0000	0.1592	0.2251	0.2757	0.3183	0.2757	0.2250	0.1592	0.0000
K_{V_A}	−0.4775	−0.4348	−0.3842	−0.3183	−0.1592	0.0000	0.0659	0.1165	0.1592

TABLE 17 *Formulas for circular rings* (*Cont.*)

Reference no., loading, and load terms	Formulas for moments, loads, and deformations and some selected numerical values	
18. Bulkhead or supporting ring in pipe, supported at bottom and carrying total load W transferred by tangential shear of v lb/linear in distributed as shown 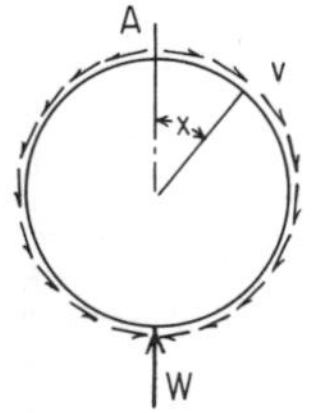 $v = \frac{W \sin x}{\pi R}$ $LT_M = \frac{WR}{\pi}\left(1 - u - \frac{xz}{2}\right)$ $LT_T = \frac{-W}{2\pi} xz$ $LT_V = \frac{W}{2\pi}(z - xu)$	$M_A = \frac{WR}{2\pi}\left(k_4 - \frac{1}{2}\right)$ $T_A = \frac{W}{2\pi}\left(k_4 + \frac{1}{2}\right)$ $V_A = 0$ $D_H = \frac{WR^3}{EI}\left(\frac{1}{\pi} - \frac{k_3}{4}\right)$ $D_V = \frac{-WR^3}{EI}\left(\frac{\pi}{8}k_1 - \frac{1}{\pi}\right)$ $\Delta R = \frac{-WR^3}{EI}\left[\frac{1}{4\pi}(1 - k_5 + \beta) + \frac{3\pi k_1}{32} - \frac{1}{4}\right]$	If $\alpha = \beta = 0$, $M_A = 0.0796WR$ $T_A = 0.2387W$ $V_A = 0$ $D_H = 0.0683\frac{WR^3}{EI}$ $D_V = -0.0744\frac{WR^3}{EI}$ $\Delta R = -0.0445\frac{WR^3}{EI}$ Max $+M = 0.2387WR$ at $x = \pi$ Max $-M = -0.1028WR$ at $x = 1.84$ rad $(105.2°)$

19. Ring loaded by own weight of w lb/linear in and supported by tangential shear of v lb/linear in distributed as shown 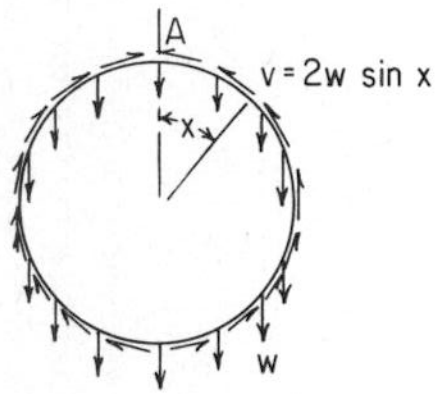 $LT_M = -wR^2(1 - u)$ $LT_T = 0$ $LT_V = -wRz$	$M_A = \frac{wR^2}{2}(1 - k_4)$ $T_A = \frac{-wR}{2}(1 + k_4)$ $V_A = 0$ $D_H = 0$ $D_V = 0$ $\Delta R = \frac{wR^4}{EI}\frac{\alpha(1 + \beta)}{k_1}$	If $\alpha = \beta = 0$, All moments $= 0$ All shears $= 0$ $T_A = -wR$ $T = -wRu$ $\Delta R = 0$

From case 12:

$$D_H = \frac{-5.04(78^4)}{340{,}000}\left[\frac{(\pi - 1.833)(-0.259) + 0.966}{2} - \frac{2}{\pi}(\pi - 1.833 - 0.966)\right]$$

$$= \frac{-5.04(78^4)}{340{,}000}(0.0954) = -52.37 \text{ in}$$

From case 13:

$$D_H = \frac{0.4292(0.1416)78^4}{340{,}000} = 6.616 \text{ in}$$

From case 14:

$$D_H = \frac{0.2146(0.0361)78^5}{340{,}000} = 65.79 \text{ in}$$

The total change in the horizontal diameter is 20 in. It must be understood at this point that the answers are somewhat in error since this large a deflection does violate the assumption that the loaded ring is very nearly circular. This was expected when the stresses were found to be so large in such a thin pipe.

Arches. Statically indeterminate arches can be solved by Castigliano's first theorem using Eqs. 3 and 7 in Art. 7.1 if the radius of curvature is not less than 10 times the depth of the cross section. If corrections for axial load and transverse shear are desired, the strain energy terms for these stresses may also be included. Such a technique applies to arches of variable cross section as well as variable radius of curvature. In such cases numerical summations would be used, having first subdivided the arch into a reasonable number of segments and average values used for T, V, M, E, and I for each segment.

Table 18 gives formulas for end reactions and end deformations for circular arches of constant radius and constant cross section under 14 different loadings and with 14 combinations of end conditions. The corrections for axial stress and transverse shear are accomplished as they were in Table 17 by the use of the constants α and β. Once the indeterminate reactions are known, the bending moments, axial loads, and transverse shear forces can be found by equilibrium equations. If deformations are desired for points away from the ends, the unit-load method (Eq. 6 in Art. 7.1) can be used to find them. The following example will illustrate the use of the formulas. Note that in many instances the answer depends upon the difference of similar large terms, and so appropriate attention to accuracy must be given.

EXAMPLE

A $WT4 \times 6.5$ structural steel T-beam is formed in the plane of its web into a circular arch of 50-in radius spanning a total angle of 120°. The right end is fixed, and the left end has a pin which is constrained to follow a horizontal slot in the support. The load is applied through a vertical bar welded to the beam, as shown in Fig. 8.2. Calculate the movement of the pin at the left end, the maximum bending stress, and the rotation of the bar at the point of attachment to the arch.

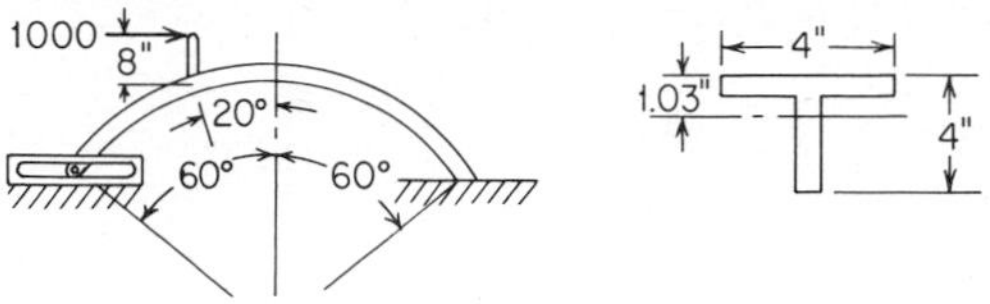

Fig. 8.2

Solution. The following material and cross-sectional properties may be used for this beam. $E = 30(10^6)$ lb/in^2, $G = 12(10^6)$ lb/in^2, $I_x = 2.90$ in^4, $A = 1.92$ in^2, flange thickness $= 0.254$ in, and web thickness $= 0.230$ in. The loading on the arch can be replaced by a concentrated moment of 8000 in-lb and a horizontal force of 1000 lb at a position indicated by $\phi = 20°(0.349$ rad). $R = 50$ in and $\theta = 60°$ (1.047 rad). For these loads and boundary conditions, cases 9b and 9j can be used.

Since the radius of 50 in is only a little more than 10 times the depth of 4 in, corrections for axial load and shear will be considered. The axial stress deformation factor $\alpha = I/AR^2 = 2.9/1.92(50^2) = 0.0006$. The transverse shear deformation factor $\beta = FEI/GAR^2$ where F will be approximated here by using $F = 1$ and $A =$ web area $= 4(0.23) = 0.92$. This gives $\beta = 1(30)(10^6)(2.90)/12(10^6)(0.92)(50^2) = 0.003$. The small values of α and β indicate that bending governs the deformations, and so the effect of axial load and transverse shear will be neglected. Note that $s = \sin 60°$, $c = \cos 60°$, $n = \sin 20°$, and $e = \cos 20°$.

For case 9b,

$$LF_H = 1000\left[\frac{1.0472 + 0.3491}{2}(1 + 2\cos 20° \cos 60°) - \frac{\sin 60° \cos 60°}{2} - \frac{\sin 20° \cos 20°}{2} - \cos 20° \sin 60° - \sin 20° \cos 60°\right]$$

$$= 1000(-0.00785) = -7.85$$

Similarly,

$$LF_V = 1000(-0.1867) = -186.7 \quad \text{and} \quad LF_M = 1000(-0.1040) = -104.0$$

For case 9j,

$$LF_H = \frac{8000}{50}(-0.5099) = -81.59$$

$$LF_V = \frac{8000}{50}(-1.6489) = -263.8$$

$$LF_M = \frac{8000}{50}(-1.396) = -223.4$$

Also,

$$B_{VV} = 1.0472 + 2(1.0472)\sin^2 60° - \sin 60° \cos 60° = 2.1850$$
$$B_{HV} = 0.5931$$
$$B_{MV} = 1.8138$$

Therefore,

$$V_A = -\frac{186.7}{2.1850} - \frac{263.8}{2.1850} = -85.47 - 120.74 = -206.2 \text{ lb}$$

$$\delta_{HA} = \frac{50^3}{30(10^6)(2.9)}[0.5931(-206.2) + 7.85 + 81.59] = -0.0472 \text{ in}$$

TABLE 18 *Reaction and deformation formulas for circular arches*

NOTATION: W = load (pounds); w = unit load (pounds per linear inch); M_o = applied couple (inch-pounds); θ_o = externally created concentrated angular displacement (radians); Δ_o = externally created concentrated lateral displacement (inches); $T - T_o$ = uniform temperature rise (degrees); T_1 and T_2 = temperatures on outside and inside, respectively (degrees). H_A and H_B are the horizontal end reactions at the left and right, respectively, and are positive to the left (pounds); V_A and V_B are the vertical end reactions at the left and right, respectively, and are positive upward (pounds); M_A and M_B are the reaction moments at the left and right, respectively, and are positive clockwise (inch-pounds). I is the area moment of inertia for bending in the plane of the arch

The axial stress deformation factor is $\alpha = I/AR^2$, where A is the cross-sectional area. The transverse shear deformation factor is $\beta = FEI/GAR^2$, where G is the shear modulus of elasticity and F is a shape factor for the cross section (see page 185). R = arch radius; $s = \sin\theta$; $c = \cos\theta$; $n = \sin\phi$; and $e = \cos\phi$. γ = temperature coefficient of expansion (inches per inch per degree)

General reaction and deformation expressions for cases 1 to 4; right end pinned in all four cases, no vertical motion at the left end

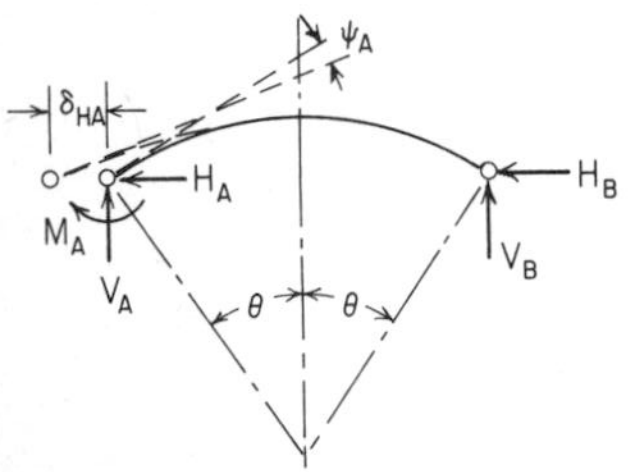

Deformation equations:

$$\text{Horizontal deflection at } A = \delta_{HA} = \frac{R^3}{EI}\left(A_{HH}H_A + A_{HM}\frac{M_A}{R} - LP_H\right) \quad (1)$$

$$\text{Angular rotation at } A = \psi_A = \frac{R^2}{EI}\left(A_{MH}H_A + A_{MM}\frac{M_A}{R} - LP_M\right) \quad (2)$$

where $A_{HH} = \theta + 2\theta c^2 - 3sc + \alpha(\theta + sc) + \beta(\theta - sc)$

$A_{MH} = A_{HM} = s - \theta c$

$A_{MM} = \dfrac{1}{4s^2}[\theta + 2\theta s^2 - sc + \alpha(\theta - sc) + \beta(\theta + sc)]$

and where LP_H and LP_M are loading terms given below for several types of load.

(*Note:* If desired, V_A, V_B, and H_B can be evaluated from equilibrium equations after calculating H_A and M_A

1. Left end pinned, right end pinned

Since $\delta_{HA} = 0$ and $M_A = 0$,

$$H_A = \frac{LP_H}{A_{HH}} \quad \text{and} \quad \psi_A = \frac{R^2}{EI}(A_{MH}H_A - LP_M)$$

The loading terms are given below.

Reference no., loading	Loading terms
1a. Concentrated vertical load 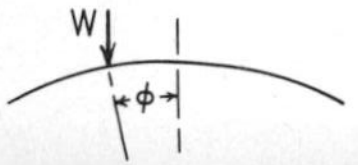	$LP_H = W\left[\frac{3}{2}c^2 - \frac{e^2}{2} - ec + \theta sc - \phi nc + (\alpha - \beta)\left(\frac{e^2 - c^2}{2}\right)\right]$ $LP_M = \frac{W}{4s^2}[2s^2(e - c - \theta s + \phi n) + \theta n + nsc - sne - \phi s + (\alpha + \beta)(\theta n - \phi s) + (\alpha - \beta)sn(e - c)]$

1b. Concentrated horizontal load 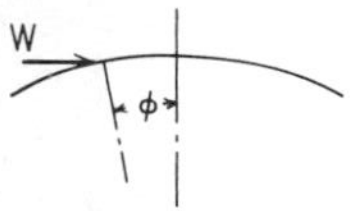	$LP_H = W\left[\frac{\theta}{2} + \theta c^2 - \frac{3}{2}sc + \phi ec + \frac{\phi}{2} - \frac{ne}{2} - nc + (\alpha + \beta)\left(\frac{\theta + \phi}{2}\right) + (\alpha - \beta)\left(\frac{sc + ne}{2}\right)\right]$ $LP_M = \frac{W}{4s^2}[\theta(c - e) + 2s^2(s + n - \theta c - \phi e) - sce + se^2 + (\alpha + \beta)\theta(c - e) + (\alpha - \beta)se(c - e)]$
1c. Concentrated radial load 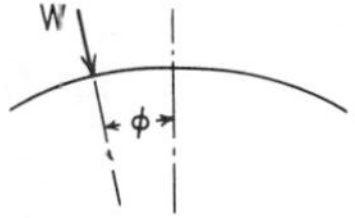	$LP_H = W\left[\frac{3}{2}c^2e + \theta\left(sce + \frac{n}{2} + c^2n\right) - \frac{e}{2} - c - \frac{3}{2}scn + \frac{\phi n}{2} + (\alpha + \beta)\frac{n}{2}(\theta + \phi) + (\alpha - \beta)\frac{s}{2}(se + cn)\right]$ $LP_M = \frac{W}{4s^2}[2s^2(1 + sn - ec) - 2\theta s^2(se + nc) - \phi se + \theta nc + (\alpha + \beta)(\theta nc - \phi se)]$
1d. Concentrated tangential load 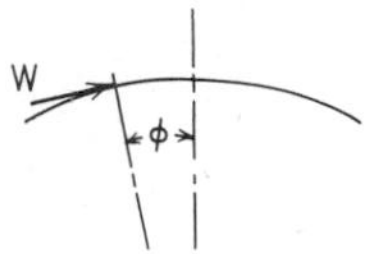	$LP_H = W\left[\phi c + \frac{e}{2}(\theta + \phi) - \frac{3}{2}c(nc + es) - \theta c(ns - ec) + (\alpha + \beta)\frac{e}{2}(\theta + \phi) + (\alpha - \beta)\frac{c}{2}(nc + se)\right]$ $LP_M = \frac{W}{4s^2}[2s^2(cn + se) + 2\theta s^2(sn - ce) + \phi s(n - 2s) - \theta(1 - ce) + s(e - c) + (\alpha + \beta)(\phi ns + \theta ec - \theta) + (\alpha - \beta)s(c - e)]$
1e. Uniform vertical load 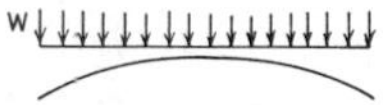	$LP_H = wR\left[\frac{sc^2}{2} - \frac{2s^3}{3} + \frac{\theta c}{2} - \theta c^3 + (\alpha - \beta)\frac{2s^3}{3}\right]$ $LP_M = wR\left(\frac{\theta c^2}{4} - \frac{\theta s^2}{4} - \frac{sc}{4}\right)$
1f. Uniform horizontal load on left side only 	$LP_H = wR\left[\frac{\theta}{2}\left(1 - \frac{5}{2}c + 2c^2 - 3c^3\right) - \frac{3}{2}sc + \frac{11}{4}sc^2 + \frac{s^3}{3} + (\alpha + \beta)\frac{1}{2}(\theta - \theta c + s) + (\alpha - \beta)\frac{s}{6}(1 + 3c - 4c^2)\right]$ $LP_M = \frac{wR}{4s^2}\left[\theta c(1 - 2s^2 - c + 3s^2c) + 2s^3(1 - 2c) + \frac{s}{3}(1 - c^3) - (\alpha + \beta)\frac{\theta}{2}(1 - c)^2 + (\alpha - \beta)\frac{s^3c}{6}\right]$
1g. Uniform horizontal load on right side only 	$LP_H = wR\left[-\frac{\theta}{2}\left(1 + \frac{c}{2} + 2c^2 - c^3\right) + \frac{3}{2}sc - \frac{s}{4} + \frac{7}{12}s^3 + (\alpha + \beta)\frac{1}{2}(s - \theta c - \theta) + (\alpha - \beta)\frac{s}{2}\left(1 - c - \frac{2}{3}s^2\right)\right]$ $LP_M = \frac{wR}{4s^2}\left[\theta(1 - s^2c^2 - c + 2s^2c) - s^3(2 - c) + \frac{s}{3}(c^3 - 1) + (\alpha + \beta)\frac{\theta}{2}(1 - c)^2 - (\alpha - \beta)\frac{s^3c}{6}\right]$

TABLE 18 ***Reaction and deformation formulas for circular arches (Cont.)***

Reference no., loading	Loading terms
1h. Uniform radial load 	$LP_H = wR[3sc^2 - \theta c - 2\theta c^3 + (\alpha + \beta)(s - \theta c) + (\alpha - \beta)s^3]$ $LP_M = wR(\theta c^2 - sc)$
1i. Uniform tangential load 	$LP_H = wR[\theta s + 2\theta sc^2 - 3s^2c + (\alpha + \beta)\theta s + (\alpha - \beta)s^2c]$ $LP_M = \frac{wR}{4s^2}[4s^2(1 + s^2 - \theta sc) - 2\theta(\theta + sc) + (\alpha + \beta)(2s^2 - 2\theta^2) + (\alpha - \beta)2s(\theta c - s)]$
1j. Concentrated couple 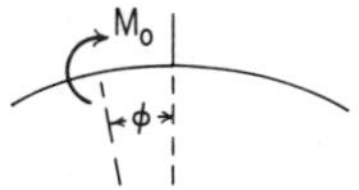	$LP_H = \frac{M_o}{R}(\phi c - n)$ $LP_M = \frac{M_o}{4s^2R}(2se - 2\phi s^2 - \theta - sc)$
1k. Concentrated angular displacement 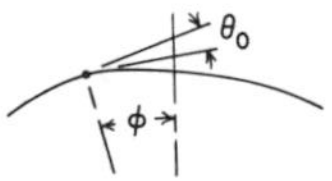	$LP_H = \frac{\theta_o EI}{R^2}(e - c)$ $LP_M = \frac{\theta_o EI}{R^2}\left(\frac{1}{2} + \frac{n}{2s}\right)$
1l. Concentrated radial displacement 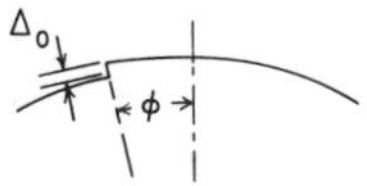	$LP_H = \frac{\Delta_o EI}{R^3}n$ $LP_M = \frac{\Delta_o EI}{R^3}\left(-\frac{e}{2s}\right)$
1m. Uniform temperature rise 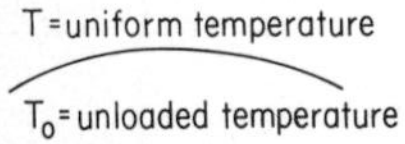	$LP_H = (T - T_o)\frac{\gamma EI}{R^2}(-2s)$ $LP_M = 0$
1n. Uniform temperature differential from outside to inside	$LP_H = (T_1 - T_2)\frac{\gamma EI}{Rt}(2s - 2\theta c)$ γEI ... (The average temperature is assumed to be T_o)

right end pinned

$$M_A = \frac{LP_M}{A_{MM}} R \qquad \text{and} \qquad \delta_{HA} = \frac{R^3}{EI}\left(A_{HM}\frac{M_A}{R} - LP_H\right)$$

Use load terms given above for cases 1a to 1n

---

3. Left end roller supported in vertical direction only, right end pinned

Since both M_A and H_A are zero, this is a statically determinate case and Eqs. 1 and 2 can be used to evaluate the deflections:

$$\delta_{HA} = \frac{-R^3}{EI} LP_H \qquad \text{and} \qquad \psi_A = \frac{-R^2}{EI} LP_M$$

Use load terms given above for cases 1a to 1n

---

4. Left end fixed, right end pinned

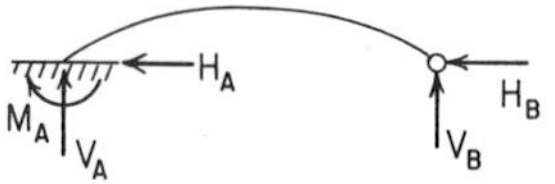

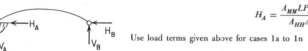

Since $\delta_{HA} = 0$ and $\psi_A = 0$,

$$H_A = \frac{A_{MM}LP_H - A_{HM}LP_M}{A_{HH}A_{MM} - A_{HM}^2} \qquad \text{and} \qquad \frac{M_A}{R} = \frac{A_{HH}LP_M - A_{HM}LP_H}{A_{HH}A_{MM} - A_{HM}^2}$$

Use load terms given above for cases 1a to 1n

General reaction and deformation expressions for cases 5 to 14, right end fixed in all 10 cases.

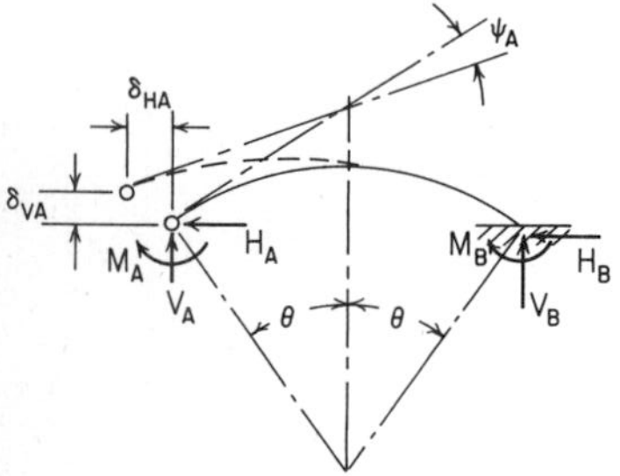

Deformation equations:

$$\text{Horizontal deflection at } A = \delta_{HA} = \frac{R^3}{EI}\left(B_{HH}H_A + B_{HV}V_A + B_{HM}\frac{M_A}{R} - LF_H\right) \tag{3}$$

$$\text{Vertical deflection at } A = \delta_{VA} = \frac{R^3}{EI}\left(B_{VH}H_A + B_{VV}V_A + B_{VM}\frac{M_A}{R} - LF_V\right) \tag{4}$$

$$\text{Angular rotation at } A = \psi_A = \frac{R^2}{EI}\left(B_{MH}H_A + B_{MV}V_A + B_{MM}\frac{M_A}{R} - LF_M\right) \tag{5}$$

where $B_{HH} = \theta + 2\theta c^2 - 3sc + \alpha(\theta + sc) + \beta(\theta - sc)$
$B_{HV} = B_{VH} = 2s^2 - 2\theta sc$
$B_{HM} = B_{MH} = 2s - 2\theta c$
$B_{VV} = \theta + 2\theta s^2 - sc + \alpha(\theta - sc) + \beta(\theta + sc)$
$B_{VM} = B_{MV} = 2\theta s$
$B_{MM} = 2\theta$

and where LF_H, LF_V, and LF_M are loading terms given below for several types of load

(*Note:* If desired, H_B, V_B, and M_B can be evaluated from equilibrium equations after calculating H_A, V_A, and M_A

TABLE 18 ***Reaction and deformation formulas for circular arches (Cont.)***

5. Left end fixed, right end fixed

Since $\delta_{HA} = 0$, $\delta_{VA} = 0$, and $\psi_A = 0$, Eqs. 3 to 5 must be solved simultaneously for H_A, V_A, and M_A/R
The loading terms are given below

Reference no., loading	Loading terms
5a. Concentrated vertical load 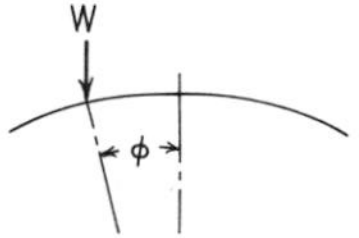	$LF_H = W\left[1 + \frac{c^2}{2} - \frac{e^2}{2} + ns - ec - (\theta + \phi)nc + (\alpha - \beta)\frac{e^2 - c^2}{2}\right]$ $LF_V = W\left[\left(\frac{\theta + \phi}{2}\right)(1 + 2ns) - nc + es - \frac{3}{2}sc + \frac{en}{2} + (\alpha + \beta)\frac{\theta + \phi}{2} - (\alpha - \beta)\frac{sc + en}{2}\right]$ $LF_M = W[e - c + (\theta + \phi)n]$
5b. Concentrated horizontal load 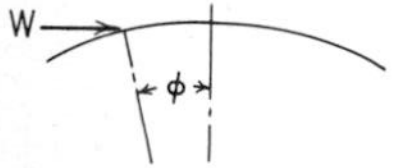	$LF_H = W\left[\left(\frac{\theta + \phi}{2}\right)(1 + 2ec) - \frac{sc}{2} - \frac{en}{2} - es - nc + (\alpha + \beta)\frac{\theta + \phi}{2} + (\alpha - \beta)\frac{en + sc}{2}\right]$ $LF_V = W\left[1 - \frac{3}{2}c^2 + sn + ec - \frac{e^2}{2} - (\theta + \phi)es + (\alpha - \beta)\frac{e^2 - c^2}{2}\right]$ $LF_M = W[s + n - (\theta + \phi)e]$
5c. Concentrated radial load 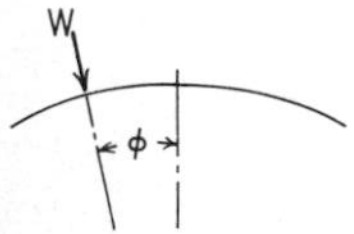	$LF_H = W\left[\frac{e}{2}(1 + c^2) - c - \frac{scn}{2} + \frac{n}{2}(\theta + \phi) + (\alpha + \beta)\frac{n}{2}(\theta + \phi) + (\alpha - \beta)\frac{s}{2}(se + cn)\right]$ $LF_V = W\left[\frac{e}{2}(\theta + \phi) + s - \frac{3c}{2}(se + cn) + n + (\alpha + \beta)\frac{e}{2}(\theta + \phi) - (\alpha - \beta)\frac{c}{2}(se + cn)\right]$ $LF_M = W(1 - ce + sn)$
5d. Concentrated tangential load 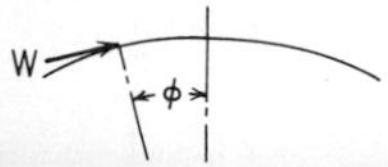	$LF_H = W\left[\left(c + \frac{e}{2}\right)(\theta + \phi) - s - \frac{sce}{2} - n - \frac{c^2n}{2} + (\alpha + \beta)\frac{e}{2}(\theta + \phi) + (\alpha - \beta)\frac{c}{2}(se + cn)\right]$ $LF_V = W\left[c - \left(s + \frac{n}{2}\right)(\theta + \phi) + \frac{3c}{2}(sn - ce) + \frac{e}{2} - (\alpha + \beta)\frac{n}{2}(\theta + \phi) + (\alpha - \beta)\frac{s}{2}(se + cn)\right]$ $LF_M = W(cn + se - \theta - \phi)$

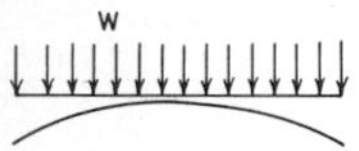	$LF_H = wR\left[\frac{s}{2} + \frac{5}{6}s^3 - \frac{3}{2}\theta c + \theta c^3 + (\alpha - \beta)\frac{2}{3}s^3\right]$ $LF_V = wR\left[2\theta s - \frac{3}{2}s^2c + \frac{\theta s}{2}(s^2 - c^2) + (\alpha + \beta)\theta s - (\alpha - \beta)s^2c\right]$ $LF_M = wR\left(\frac{\theta}{2} + \theta s^2 - \frac{sc}{2}\right)$
5f. Uniform horizontal load on left side only 	$LF_H = wR\left[\frac{\theta}{2} - \frac{3}{4}\theta c - \theta c^3 + \frac{7s}{4} - \frac{23}{12}s^3 - \frac{sc}{2} + (\alpha + \beta)\left(\frac{\theta}{2} - \theta c + \frac{s}{2}\right) + (\alpha - \beta)\left(\frac{sc}{2} - \frac{sc^2}{2} + \frac{s^3}{6}\right)\right]$ $LF_V = wR\left[\frac{5}{6} - \frac{3}{2}c^2 - \frac{5}{4}c + \frac{3}{4}\theta s - \theta s^3 + \frac{23}{12}c^3 + (\alpha - \beta)\left(\frac{1}{6} - \frac{c^2}{2} + \frac{c^3}{3}\right)\right]$ $LF_M = wR\left(s - \frac{7}{4}sc + \frac{3}{4}\theta - \theta s^2\right)$
5g. Uniform horizontal load on right side only 	$LF_H = wR\left[\frac{s^3}{12} + \frac{3}{4}s + \frac{sc}{2} - \frac{\theta}{2} - \frac{3}{4}\theta c + (\alpha + \beta)\left(\frac{s - \theta}{2}\right) + (\alpha - \beta)\left(\frac{sc^2}{2} - \frac{sc}{2} + \frac{s^3}{6}\right)\right]$ $LF_V = wR\left[\frac{3}{2}c^2 - \frac{5}{6} - \frac{c}{4} - \frac{5}{12}c^3 + \frac{3}{4}\theta s + (\alpha - \beta)\left(\frac{c^2}{2} - \frac{c^3}{3} - \frac{1}{6}\right)\right]$ $LF_M = wR\left(\frac{sc}{4} - s + \frac{3}{4}\theta\right)$
5h. Uniform radial load 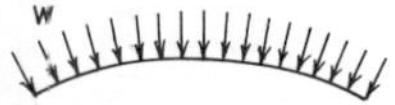	$LF_H = wR[3s - s^3 - 3\theta c + (\alpha + \beta)(s - \theta c) + (\alpha - \beta)s^3]$ $LF_V = wR[3\theta s - 3s^2c + (\alpha + \beta)\theta s - (\alpha - \beta)s^2c]$ $LF_M = wR(2\theta - 2sc)$
5i. Uniform tangential load 	$LF_H = wR[2c\theta^2 - \theta s - s^2c + (\alpha + \beta)\theta s + (\alpha - \beta)s^2c]$ $LF_V = wR[2s + \theta c - 3c^2s - 2s\theta^2 - 2\pi s\theta + (\alpha + \beta)(\theta c - s) + (\alpha - \beta)s^3]$ $LF_M = wR(2s^2 - 2\theta^2)$
5j. Concentrated couple 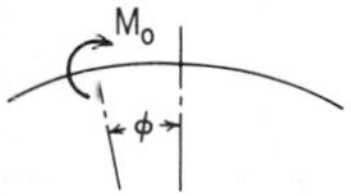	$LF_H = \frac{M_o}{R}[-n - s + (\theta + \phi)c]$ $LF_V = \frac{M_o}{R}[c - e - (\theta + \phi)s]$ $LF_M = \frac{M_o}{R}[-\theta - \phi]$

TABLE 18 *Reaction and deformation formulas for circular arches* (*Cont.*)

Reference no., loading	Loading terms
5k. Concentrated angular displacement 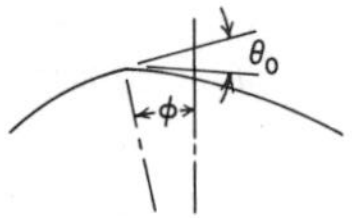	$LF_H = \frac{\theta_o EI}{R^2}(e - c)$ $LF_V = \frac{\theta_o EI}{R^2}(s - n)$ $LF_M = \frac{\theta_o EI}{R^2}(1)$
5l. Concentrated radial displacement 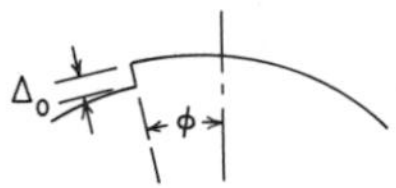	$LF_H = \frac{\Delta_o EI}{R^3}(n)$ $LF_V = \frac{\Delta_o EI}{R^3}(e)$ $LF_M = 0$
5m. Uniform temperature rise 	$LF_H = (T - T_o)\frac{\gamma EI}{R^2}(-2s)$ $LF_V = 0$ $LF_M = 0$
5n. Uniform temperature differential from outside to inside 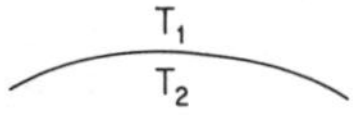	$LF_H = (T_1 - T_2)\frac{\gamma EI}{Rt}(2s - 2\theta c)$ where t = thickness from inside to outside. (The average temperature is assumed to be T_o) $LF_V = (T_1 - T_2)\frac{\gamma EI}{Rt}2\theta s$ $LF_M = (T_1 - T_2)\frac{\gamma EI}{Rt}2\theta$
6. Left end pinned, right end fixed 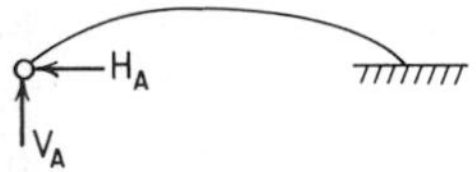	Since $\delta_{HA} = 0$, $\delta_{VA} = 0$, and $M_A = 0$, $H_A = \frac{B_{VV}LF_H - B_{HV}LF_V}{B_{HH}B_{VV} - B_{HV}^2}$ $V_A = \frac{B_{HH}LF_V - B_{HV}LF_H}{B_{HH}B_{VV} - B_{HV}^2}$ $\psi_A = \frac{R^2}{EI}(B_{MH}H_A + B_{MV}V_A - LF_M)$ Use load terms given above for cases 5a to 5n

7. Left end guided in horizontal direction, right end fixed

Since $\delta_{VA} = 0$, $\psi_A = 0$, and $H_A = 0$,

$$V_A = \frac{B_{MM}LF_V - B_{MV}LF_M}{B_{VV}B_{MM} - B_{MV}^2} \qquad \frac{M_A}{R} = \frac{B_{VV}LF_M - B_{MV}LF_V}{B_{VV}B_{MM} - B_{MV}^2}$$

$$\delta_{HA} = \frac{R^3}{EI}\left(B_{HV}V_A + B_{HM}\frac{M_A}{R} - LF_H\right)$$

Use load terms given above for cases 5a to 5n

---

8. Left end guided in vertical direction, right end fixed

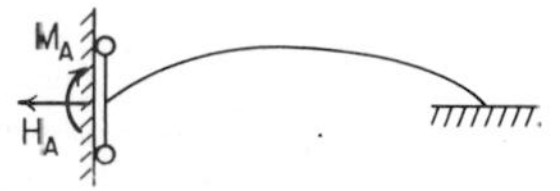

Since $\delta_{HA} = 0$, $\psi_A = 0$, and $V_A = 0$,

$$H_A = \frac{B_{MM}LF_H - B_{HM}LF_M}{B_{HH}B_{MM} - B_{HM}^2} \qquad \frac{M_A}{R} = \frac{B_{HH}LF_M - B_{HM}LF_H}{B_{HH}B_{MM} - B_{HM}^2}$$

$$\delta_{VA} = \frac{R^3}{EI}\left(B_{VH}H_A + B_{VM}\frac{M_A}{R} - LF_V\right)$$

Use load terms given above for cases 5a to 5n

---

9. Left end roller supported in vertical direction only, right end fixed

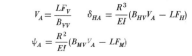

Since $\delta_{VA} = 0$, $H_A = 0$, and $M_A = 0$,

$$V_A = \frac{LF_V}{B_{VV}} \qquad \delta_{HA} = \frac{R^3}{EI}(B_{HV}V_A - LF_H)$$

$$\psi_A = \frac{R^2}{EI}(B_{MV}V_A - LF_M)$$

Use load terms given above for cases 5a to 5n

---

10. Left end roller supported in horizontal direction only, right end fixed

Since $\delta_{HA} = 0$, $V_A = 0$, and $M_A = 0$,

$$H_A = \frac{LF_H}{B_{HH}} \qquad \delta_{VA} = \frac{R^3}{EI}(B_{VH}H_A - LF_V)$$

$$\psi_A = \frac{R^2}{EI}(B_{MH}H_A - LF_M)$$

Use load terms given above for cases 5a to 5n

---

11. Left end restrained against rotation only, right end fixed

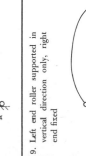

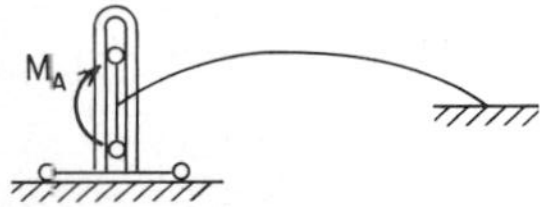

Since $\psi_A = 0$, $H_A = 0$, and $V_A = 0$,

$$\frac{M_A}{R} = \frac{LF_M}{B_{MM}} \qquad \delta_{HA} = \frac{R^3}{EI}\left(B_{HM}\frac{M_A}{R} - LF_H\right)$$

$$\delta_{VA} = \frac{R^3}{EI}\left(B_{VM}\frac{M_A}{R} - LF_V\right)$$

Use load terms given above for cases 5a to 5n

TABLE 18 *Reaction and deformation formulas for circular arches (Cont.)*

12. Left end free, right end fixed

Since $H_A = 0$, $V_A = 0$, and $M_A = 0$, this is a statically determinate problem. The deflections at the free end are given by

$$\delta_{HA} = \frac{-R^3}{EI} LF_H \qquad \delta_{VA} = \frac{-R^3}{EI} LF_V$$

$$\psi_A = \frac{-R^2}{EI} LF_M$$

Use load terms given above for cases 5a to 5n

---

13. Left end guided along an inclined surface at angle ζ, right end fixed

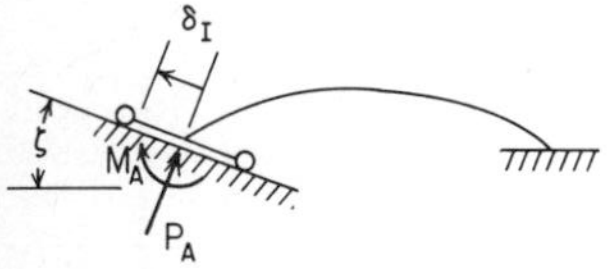

Since there is no deflection perpendicular to the incline, the following three equations must be solved for M_A, P_A, and δ_I:

$$\delta_I \frac{EI \cos\zeta}{R^3} = P_A(B_{HV} \cos\zeta - B_{HH} \sin\zeta) + B_{HM} \frac{M_A}{R} - LF_H$$

$$\delta_I \frac{EI \sin\zeta}{R^3} = P_A(B_{VV} \cos\zeta - B_{VH} \sin\zeta) + B_{VM} \frac{M_A}{R} - LF_V$$

$$0 = P_A(B_{MV} \cos\zeta - B_{MH} \sin\zeta) + B_{MM} \frac{M_A}{R} - LF_M$$

Use load terms given above for cases 5a to 5n

---

14. Left end roller supported along an inclined surface at angle ζ, right end fixed

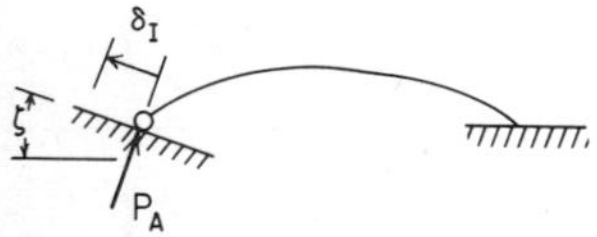

Since there is no deflection perpendicular to the incline and $M_A = 0$, the following equations give P_A, δ_I, and ψ_A:

$$P_A = \frac{LF_V \cos\zeta - LF_H \sin\zeta}{B_{HH} \sin^2\zeta - 2B_{HV} \sin\zeta \cos\zeta + B_{VV} \cos^2\zeta}$$

$$\delta_I = \frac{R^3}{EI}\{P_A[B_{HV}(\cos^2\zeta - \sin^2\zeta) + (B_{VV} - B_{HH}) \sin\zeta \cos\zeta] - LF_H \cos\zeta - LF_V \sin\zeta\}$$

$$\psi_A = \frac{R^2}{EI}[P_A(B_{MV} \cos\zeta - B_{MH} \sin\zeta) - LF_M]$$

Use load terms given above for cases 5a to 5n

The expression for the bending moment can now be obtained by an equilibrium equation for a position located by an angle x measured from the left end:

$$M_x = V_A R[\sin\theta - \sin(\theta - x)] + 8000\,\langle x - (\theta - \phi)\rangle^0$$
$$- 1000R[\cos(\theta - x) - \cos\phi]\langle x - (\theta - \phi)\rangle^0$$

At $x = 40°-$ $\quad M_x = -206.2(50)[\sin 60° - \sin(60° - 40°)] = -5403$ in-lb

At $x = 40°+$ $\quad M_x = -5403 + 8000 = 2597$ in-lb

At $x = 60°$ $\quad M_x = -206.2(50)(0.866) + 8000 - 1000(50)(1 - 0.940) = -3944$ in-lb

At $x = 120°$ $\quad M_x = 12{,}130$ in-lb

The maximum bending stress is therefore

$$\sigma = \frac{12{,}130(4 - 1.03)}{2.9} = 12{,}420 \text{ lb/in}^2$$

To obtain the rotation of the arch at the point of attachment of the bar, we first calculate the loads on the portion to the right of the loading and then establish an equivalent symmetric arch (see Fig. 8-3). Now from cases 12a, 12b, and 12j, where $\theta = \phi = 40°(0.698$ rad), we can

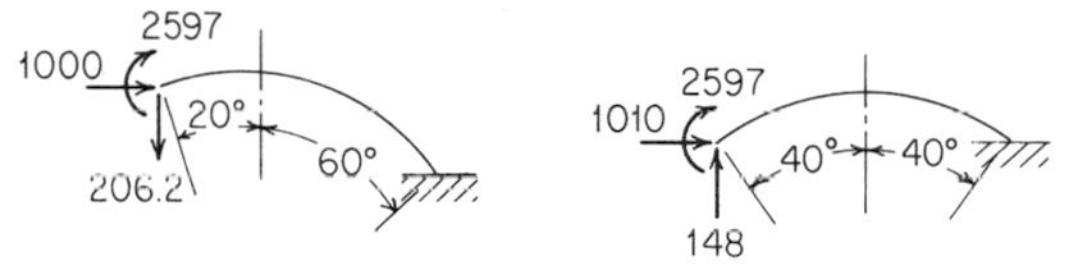

Fig. 8.3

determine the load terms:

For case 12a $\quad LF_M = -148[2(0.698)(0.643)] = -133$

For case 12b $\quad LF_M = 1010[0.643 + 0.643 - 2(0.698)(0.766)] = 218$

For case 12j $\quad LF_M = \dfrac{2597}{50}(-0.698 - 0.698) = -72.5$

Therefore, the rotation at the load is

$$\psi_A = \frac{-50^2}{30(10^6)(2.9)}(-133 + 218 - 72.5) = -0.00036 \text{ rad}$$

We would not expect the rotation to be in the opposite direction to the applied moment, but a careful examination of the problem shows that the point on the arch where the bar is fastened moves to the right 0.0128 in. Therefore, the net motion in the direction of the 1000-lb load on the end of the 8-in bar is 0.0099 in, and so the applied load does indeed do positive work on the system.

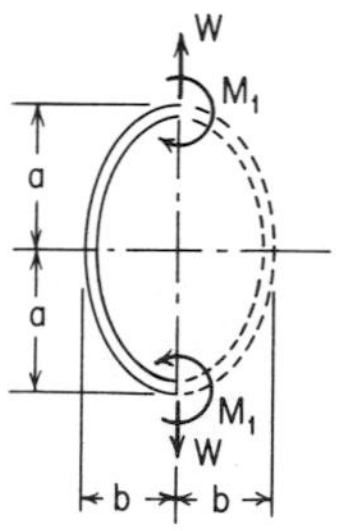

Fig. 8.4

8.4 *Elliptical rings*

For an elliptical ring of semiaxes a and b, under equal and opposite forces W (Fig. 8.4) the bending moment M_1 at the extremities of the major axis is given by $M_1 = K_1 Wa$, and for equal and opposite outward forces applied at the ends of the minor axis, the moment M_1 at the ends of the major axis is given by $M_1 = -K_2 Wa$, where K_1 and K_2 are coefficients which depend on the ratio a/b and have the following values:

a/b	1	1.1	1.2	1.3	1.4	1.5	1.6	1.7
K_1	0.318	0.295	0.274	0.255	0.240	0.227	0.216	0.205
K_2	0.182	0.186	0.191	0.195	0.199	0.203	0.206	0.208

a/b	1.8	1.9	2.0	2.1	2.2	2.3	2.4	2.5
K_1	0.195	0.185	0.175	0.167	0.161	0.155	0.150	0.145
K_2	0.211	0.213	0.215	0.217	0.219	0.220	0.222	0.223

Burke (Ref. 6) gives charts by which the moments and tensions in elliptical rings under various conditions of concentrated loading can be found; the preceding values of K were taken from these charts.

Timoshenko (Ref. 13) gives an analysis of an elliptical ring (or other ring with two axes of symmetry) under the action of a uniform outward pressure, which would apply to a tube of elliptical section under internal pressure. For this case $M = Kpa^2$, where M is the bending moment at a section a distance x along the ring from the end of the minor axis, p is the outward normal pressure per linear inch, and K is a coefficient that depends on the ratios b/a and x/S, where S is one-quarter of the perimeter of the ring. Values of K are given in the following table; M is positive when it produces tension at the inner surface of the ring:

x/S \ b/a	0.3	0.5	0.6	0.7	0.8	0.9
0	−0.172	−0.156	−0.140	−0.115	−0.085	−0.045
0.1	−0.167	−0.152	−0.135	−0.112	−0.082	−0.044
0.2	−0.150	−0.136	−0.120	−0.098	−0.070	−0.038
0.4	−0.085	−0.073	−0.060	−0.046	−0.030	−0.015
0.6	0.020	0.030	0.030	0.028	0.022	0.015
0.7	0.086	0.090	0.082	0.068	0.050	0.022
0.8	0.160	0.150	0.130	0.105	0.075	0.038
0.9	0.240	0.198	0.167	0.130	0.090	0.046
1.0	0.282	0.218	0.180	0.140	0.095	0.050

Values of M calculated by the preceding coefficients are correct only for a ring of uniform moment of inertia I; if I is not uniform, then a correction ΔM must be added. This correction is given by

$$\Delta M = \frac{-\int_0^x \left(\frac{M}{I}\right) dx}{\int_0^x \frac{dx}{I}}$$

The integrals can be evaluated graphically. Reference 12 gives charts for the calculation of moments in elliptical rings under uniform radial loading; the preceding values of K were taken from these charts.

8.5 *Curved beams loaded normal to plane of curvature*

This type of beam usually presents a statically indeterminate problem, the degree of indeterminacy depending upon the manner of loading and support. Both bending and twisting occur, and it is necessary to distinguish between an analysis that is applicable to compact or flangeless sections (circular, rectangular, etc.) in which torsion does not produce secondary bending and one that is applicable to flanged sections (I-beams, channels, etc.) in which torsion may be accompanied by such secondary bending (see Art. 9.3). It is also necessary to distinguish among three types of constraints that may or may not occur at the supports, namely: (1) the beam is prevented from *sloping,* its horizontal axis held horizontal by a bending couple; (2) the beam is prevented from *rolling,* its vertical axis held vertical by a twisting couple; and (3) in the case of a flanged section, the flanges are prevented from turning about their vertical axes by horizontal secondary bending couples. These types of constraints will be designated here as (1) fixed as to slope, (2) fixed as to roll, and (3) flanges fixed.

Compact sections. Table 19 treats the curved beam of uniform cross section under concentrated and distributed loads normal to the plane of curvature, out-of-plane concentrated bending moments, and concentrated and distributed torques. Expressions are given for transverse shear, bending moment, twisting moment, deflection, bending slope, and roll slope for 10 combinations of end conditions. To keep the presentation to a reasonable size, use is made of the singularity functions discussed previously (page 94) and an extensive list of constants and functions is given. In previous tables the representative functional values have been given, but in Table 19 the value of β depends upon both bending and torsional properties and so a useful set of tabular values would be too large to present. The curved beam or ring of circular cross section is so common, however, that numerical coefficients are given in the table for $\beta = 1.3$ which will apply to a solid or hollow circular cross section of material for which Poisson's ratio is 0.3.

Levy (Ref. 14) has treated the closed circular ring of arbitrary compact cross section for six loading cases. These cases have been chosen to permit appropriate superposition in order to solve a large number of problems, and both isolated and distributed out-of-plane loads are discussed. Hogan (Ref. 18) presents similar loadings and supports. In a similar way the information in Table 19 can be used by appropriate superposition to solve most out-of-plane loading problems on closed rings of compact cross section if strict attention is given to the symmetry and boundary conditions involved. Several simple examples of this reasoning are described in the following three cases.

TABLE 19 ***Formulas for curved beams loaded normal to the plane of curvature***

NOTATION: W = applied load normal to the plane of curvature (pounds); M_o = applied bending moment in a plane tangent to the curved axis of the beam (inch-pounds); T_o = applied twisting moment in a plane normal to the curved axis of the beam (inch-pounds); w = distributed load (pounds per linear inch); t_o = distributed twisting moment (inch-pounds per linear inch); V_A = reaction force, M_A = reaction bending moment, T_A = reaction twisting moment, y_A = deflection normal to the plane of curvature, Θ_A = slope of the beam axis in the plane of the moment M_A, and ψ_A = roll of the beam cross section in the plane of the twisting moment T_A, all at the left end of the beam. Similarly, V_B, M_B, T_B, y_B, Θ_B, and ψ_B are the reactions and displacements at the right end: V, M, T, y, Θ, and ψ are internal shear forces, moments, and displacements at an angular position x rad from the left end. All loads and reactions are positive as shown in the diagram; y is positive upward; Θ is positive when y increases as x increases; and ψ is positive in the direction of T

R = radius of curvature of the beam axis (inches); E = modulus of elasticity (pounds per square inch); I = area moment of inertia about the bending axis (inches to the fourth power) (note that this must be a principal axis of the beam cross section); G = modulus of rigidity (pounds per square inch); ν = Poisson's ratio; K = torsional stiffness constant of the cross section (inches to the fourth power) (see page 288); θ = angle in radians from the left end to the position of the loading; ϕ = angle (radians) subtended by the entire span of the curved beam. See page 94 for a definition of the term $\langle x - \theta \rangle^n$

The following constants and functions are hereby defined in order to permit condensing the tabulated formulas which follow. $\beta = EI/GK$

$$F_1 = \frac{1+\beta}{2} x \sin x - \beta(1 - \cos x) \qquad C_1 = \frac{1+\beta}{2} \phi \sin \phi - \beta(1 - \cos \phi)$$

$$F_2 = \frac{1+\beta}{2}(x \cos x - \sin x) \qquad C_2 = \frac{1+\beta}{2}(\phi \cos \phi - \sin \phi)$$

$$F_3 = -\beta(x - \sin x) - \frac{1+\beta}{2}(x \cos x - \sin x) \qquad C_3 = -\beta(\phi - \sin \phi) - \frac{1+\beta}{2}(\phi \cos \phi - \sin \phi)$$

$$F_4 = \frac{1+\beta}{2} x \cos x + \frac{1-\beta}{2} \sin x \qquad C_4 = \frac{1+\beta}{2} \phi \cos \phi + \frac{1-\beta}{2} \sin \phi$$

$$F_5 = -\frac{1+\beta}{2} x \sin x \qquad C_5 = -\frac{1+\beta}{2} \phi \sin \phi$$

$$F_6 = F_1 \qquad C_6 = C_1$$

$$F_7 = F_5 \qquad C_7 = C_5$$

$$F_8 = \frac{1-\beta}{2} \sin x - \frac{1+\beta}{2} x \cos x \qquad C_8 = \frac{1-\beta}{2} \sin \phi - \frac{1+\beta}{2} \phi \cos \phi$$

$$F_9 = F_2 \qquad C_9 = C_2$$

$$F_{a1} = \left\{\frac{1+\beta}{2}(x-\theta) \sin (x-\theta) - \beta[1 - \cos (x-\theta)]\right\} \langle x - \theta \rangle^0 \qquad C_{a1} = \frac{1+\beta}{2}(\phi - \theta) \sin (\phi - \theta) - \beta[1 - \cos (\phi - \theta)]$$

$$F_{a2} = \frac{1+\beta}{2}[(x-\theta) \cos (x-\theta) - \sin (x-\theta)]\langle x - \theta \rangle^0 \qquad C_{a2} = \frac{1+\beta}{2}[(\phi - \theta) \cos (\phi - \theta) - \sin (\phi - \theta)]$$

$F_{a3} = \{\beta[x - \theta - \sin(x - \theta)] - F_{a2}\}\langle x - \theta\rangle^0$

$F_{a4} = \left[\frac{1+\beta}{2}(x-\theta)\cos(x-\theta) + \frac{1-\beta}{2}\sin(x-\theta)\right]\langle x-\theta\rangle^0$

$F_{a5} = -\frac{1+\beta}{2}(x-\theta)\sin(x-\theta)\langle x-\theta\rangle^0$

$F_{a6} = F_{a1}$

$F_{a7} = F_{a5}$

$F_{a8} = \left[\frac{1-\beta}{2}\sin(x-\theta) - \frac{1+\beta}{2}(x-\theta)\cos(x-\theta)\right]\langle x-\theta\rangle^0$

$F_{a9} = F_{a2}$

$F_{a12} = \frac{1+\beta}{2}[(x-\theta)\sin(x-\theta) - 2 + 2\cos(x-\theta)]\langle x-\theta\rangle^0$

$F_{a13} = \left[\beta\left\{1 - \cos(x-\theta) - \frac{(x-\theta)^2}{2}\right\} - F_{a12}\right]\langle x-\theta\rangle^0$

$F_{a15} = -\frac{1+\beta}{2}[\sin(x-\theta) - (x-\theta)\cos(x-\theta)]\langle x-\theta\rangle^0$

$F_{a16} = [-\beta\{x - \theta - \sin(x-\theta)\} - F_{a15}]\langle x-\theta\rangle^0$

$F_{a18} = \left[1 - \cos(x-\theta) - \frac{1+\beta}{2}(x-\theta)\sin(x-\theta)\right]\langle x-\theta\rangle^0$

$F_{a19} = F_{a12}$

$C_{a3} = -\beta[\phi - \theta - \sin(\phi-\theta)] - C_{a2}$

$C_{a4} = \frac{1+\beta}{2}(\phi-\theta)\cos(\phi-\theta) + \frac{1-\beta}{2}\sin(\phi-\theta)$

$C_{a5} = -\frac{1+\beta}{2}(\phi-\theta)\sin(\phi-\theta)$

$C_{a6} = C_{a1}$

$C_{a7} = C_{a5}$

$C_{a8} = \frac{1-\beta}{2}\sin(\phi-\theta) - \frac{1+\beta}{2}(\phi-\theta)\cos(\phi-\theta)$

$C_{a9} = C_{a2}$

$C_{a12} = \frac{1+\beta}{2}[(\phi-\theta)\sin(\phi-\theta) - 2 + 2\cos(\phi-\theta)]$

$C_{a13} = \beta\left[1 - \cos(\phi-\theta) - \frac{(\phi-\theta)^2}{2}\right] - C_{a12}$

$C_{a15} = -\frac{1+\beta}{2}[\sin(\phi-\theta) - (\phi-\theta)\cos(\phi-\theta)]$

$C_{a16} = -\beta[\phi - \theta - \sin(\phi-\theta)] - C_{a15}$

$C_{a18} = 1 - \cos(\phi-\theta) - \frac{1+\beta}{2}(\phi-\theta)\sin(\phi-\theta)$

$C_{a19} = C_{a12}$

1. Concentrated intermediate lateral load

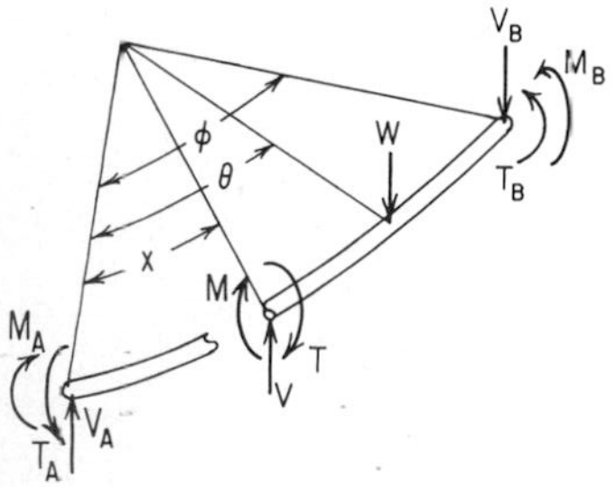

Transverse shear $= V = V_A - W\langle x-\theta\rangle^0$

Bending moment $= M = V_A R \sin x + M_A \cos x - T_A \sin x - WR\sin(x-\theta)\langle x-\theta\rangle^0$

Twisting moment $= T = V_A R(1-\cos x) + M_A \sin x + T_A \cos x - WR[1 - \cos(x-\theta)]\langle x-\theta\rangle^0$

Deflection $= y = y_A + \Theta_A R \sin x + \psi_A R(1-\cos x) + \frac{M_A R^2}{EI}F_1 + \frac{T_A R^2}{EI}F_2 + \frac{V_A R^3}{EI}F_3 - \frac{WR^3}{EI}F_{a3}$

Bending slope $= \Theta = \Theta_A \cos x + \psi_A \sin x + \frac{M_A R}{EI}F_4 + \frac{T_A R}{EI}F_5 + \frac{V_A R^2}{EI}F_6 - \frac{WR^2}{EI}F_{a6}$

Roll slope $= \psi = \psi_A \cos x - \Theta_A \sin x + \frac{M_A R}{EI}F_7 + \frac{T_A R}{EI}F_8 + \frac{V_A R^2}{EI}F_9 - \frac{WR^2}{EI}F_{a9}$

For tabulated values: $V = K_V W \quad M = K_M WR \quad T = K_T WR \quad y = K_y \frac{WR^3}{EI} \quad \Theta = K_\Theta \frac{WR^2}{EI} \quad \psi = K_\psi \frac{WR^2}{EI}$

TABLE 19 *Formulas for curved beams loaded normal to the plane of curvature (Cont.)*

End restraints, reference no.	Formulas for boundary values and selected numerical values

1a. Right end fixed, left end free

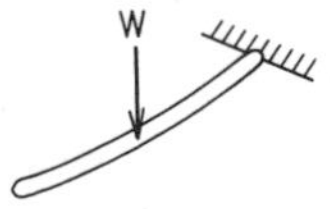

$V_A = 0 \qquad M_A = 0 \qquad T_A = 0$

$y_B = 0 \qquad \Theta_B = 0 \qquad \psi_B = 0$

$$y_A = \frac{-WR^3}{EI}[C_{a6} \sin \phi - C_{a9}(1 - \cos \phi) - C_{a3}] \qquad \Theta_A = \frac{WR^2}{EI}(C_{a6} \cos \phi - C_{a9} \sin \phi)$$

$$\psi_A = \frac{WR^2}{EI}(C_{a9} \cos \phi + C_{a6} \sin \phi)$$

$$V_B = -W$$

$$M_B = -WR \sin (\phi - \theta)$$

$$T_B = -WR[1 - \cos (\phi - \theta)]$$

If $\beta = 1.3$ (solid or hollow round cross section, $\nu = 0.3$),

ϕ	45°	90°			180°		
θ	0°	0°	30°	60°	0°	60°	120°
K_{yA}	−0.1607	−1.2485	−0.6285	−0.1576	−7.6969	−3.7971	−0.6293
$K_{\Theta A}$	0.3058	1.1500	0.3938	0.0535	2.6000	−0.1359	−0.3929
$K_{\psi A}$	0.0590	0.5064	0.3929	0.1269	3.6128	2.2002	0.3938
K_{VB}	−1.0000	−1.0000	−1.0000	−1.0000	−1.0000	−1.0000	−1.0000
K_{MB}	−0.7071	−1.0000	−0.8660	−0.5000	−0.0000	−0.8660	−0.8660
K_{TB}	−0.2929	−1.0000	−0.5000	−0.1340	−2.0000	−1.5000	−0.5000

1b. Right end fixed, left end simply supported

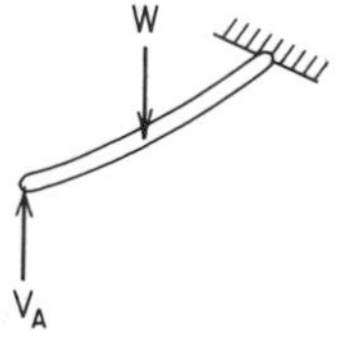

$M_A = 0 \qquad T_A = 0 \qquad y_A = 0$

$y_B = 0 \qquad \Theta_B = 0 \qquad \psi_B = 0$

$$V_A = W\frac{C_{a9}(1 - \cos \phi) - C_{a6} \sin \phi + C_{a3}}{C_9(1 - \cos \phi) - C_6 \sin \phi + C_3}$$

$$\Theta_A = \frac{WR^2}{EI}\,\frac{(C_{a3}C_9 - C_{a9}C_3) \sin \phi + (C_{a9}C_6 - C_{a6}C_9)(1 - \cos \phi) + (C_{a6}C_3 - C_{a3}C_6) \cos \phi}{C_9(1 - \cos \phi) - C_6 \sin \phi + C_3}$$

$$\psi_A = \frac{WR^2}{EI}\,\frac{[C_{a6}(C_3 + C_9) - C_6(C_{a3} + C_{a9})] \sin \phi + (C_{a9}C_3 - C_{a3}C_9) \cos \phi}{C_9(1 - \cos \phi) - C_6 \sin \phi + C_3}$$

$$V_B = V_A - W$$

$$M_B = V_A R \sin \phi - WR \sin (\phi - \theta)$$

$$T_B = V_A R(1 - \cos \phi) - WR[1 - \cos (\phi - \theta)]$$

If $\beta = 1.3$ (solid or hollow round cross section, $\nu = 0.3$),

ϕ	45°		90°		180°	
θ	15°	30°	30°	60°	60°	120°
K_{VA}	0.5136	0.1420	0.5034	0.1262	0.4933	0.0818
$K_{\Theta A}$	−0.0294	−0.0148	−0.1851	−0.0916	−1.4185	−0.6055
$K_{\psi A}$	0.0216	0.0106	0.1380	0.0630	0.4179	0.0984
K_{MB}	−0.1368	−0.1584	−0.3626	−0.3738	−0.8660	−0.8660
K_{TB}	0.0165	0.0075	0.0034	−0.0078	−0.5133	−0.3365
$K_{M\theta}$	0.1329	0.0710	0.2517	0.1093	0.4272	0.0708

1c. Right end fixed, left end supported and slope guided

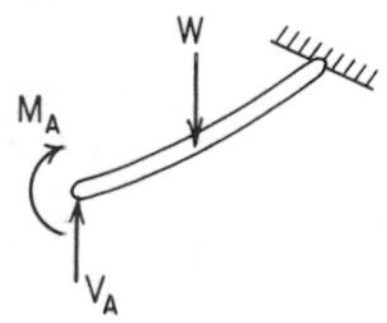

$T_A = 0 \quad y_A = 0 \quad \Theta_A = 0$
$y_B = 0 \quad \Theta_B = 0 \quad \psi_B = 0$

$$V_A = W\frac{(C_{a9}C_4 - C_{a6}C_7)(1-\cos\phi) + (C_{a6}C_1 - C_{a3}C_4)\cos\phi + (C_{a3}C_7 - C_{a9}C_1)\sin\phi}{(C_4C_9 - C_6C_7)(1-\cos\phi) + (C_1C_6 - C_3C_4)\cos\phi + (C_3C_7 - C_1C_9)\sin\phi}$$

$$M_A = WR\frac{(C_{a6}C_9 - C_{a9}C_6)(1-\cos\phi) + (C_{a3}C_6 - C_{a6}C_3)\cos\phi + (C_{a9}C_3 - C_{a3}C_9)\sin\phi}{(C_4C_9 - C_6C_7)(1-\cos\phi) + (C_1C_6 - C_3C_4)\cos\phi + (C_3C_7 - C_1C_9)\sin\phi}$$

$$\psi_A = \frac{WR^2}{EI}\,\frac{C_{a3}(C_4C_9 - C_6C_7) + C_{a6}(C_3C_7 - C_1C_9) + C_{a9}(C_1C_6 - C_3C_4)}{(C_4C_9 - C_6C_7)(1-\cos\phi) + (C_1C_6 - C_3C_4)\cos\phi + (C_3C_7 - C_1C_9)\sin\phi}$$

$$V_B = V_A - W$$

$$M_B = V_AR\sin\phi + M_A\cos\phi - WR\sin(\phi-\theta)$$

$$T_B = V_AR(1-\cos\phi) + M_A\sin\phi - WR[1-\cos(\phi-\theta)]$$

If $\beta = 1.3$ (solid or hollow round cross section, $\nu = 0.3$)

ϕ	45°		90°		180°	
θ	15°	30°	30°	60°	60°	120°
K_{VA}	0.7407	0.2561	0.7316	0.2392	0.6686	0.1566
K_{MA}	−0.1194	−0.0600	−0.2478	−0.1226	−0.5187	−0.2214
$K_{\psi A}$	−0.0008	−0.0007	−0.0147	−0.0126	−0.2152	−0.1718
K_{MB}	−0.0607	−0.1201	−0.1344	−0.2608	−0.3473	−0.6446
K_{TB}	−0.0015	−0.0015	−0.0161	−0.0174	−0.1629	−0.1869
$K_{M\theta}$	0.0764	0.0761	0.1512	0.1458	0.3196	0.2463

1d. Right end fixed, left end supported and roll guided

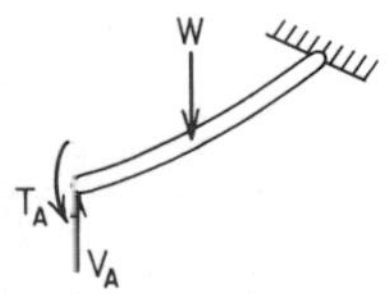

$M_A = 0 \quad y_A = 0 \quad \psi_A = 0$
$y_B = 0 \quad \Theta_B = 0 \quad \psi_B = 0$

$$V_A = W\frac{[(C_{a3} + C_{a9})C_5 - C_{a6}(C_2 + C_8)]\sin\phi + (C_{a3}C_8 - C_{a9}C_2)\cos\phi}{C_5(C_3 + C_9) - C_6(C_2 + C_8)\sin\phi + (C_3C_8 - C_2C_9)\cos\phi}$$

$$T_A = WR\frac{[C_{a6}(C_3 + C_9) - C_6(C_{a3} + C_{a9})]\sin\phi + (C_{a9}C_3 - C_{a3}C_9)\cos\phi}{C_5(C_3 + C_9) - C_6(C_2 + C_8)\sin\phi + (C_3C_8 - C_2C_9)\cos\phi}$$

$$\Theta_A = \frac{WR^2}{EI}\,\frac{C_{a3}(C_5C_9 - C_6C_8) + C_{a6}(C_3C_8 - C_2C_9) + C_{a9}(C_2C_6 - C_3C_5)}{C_5(C_3 + C_9) - C_6(C_2 + C_8)\sin\phi + (C_3C_8 - C_2C_9)\cos\phi}$$

$$V_B = V_A - W$$

$$M_B = V_AR\sin\phi - T_A\sin\phi - WR\sin(\phi-\theta)$$

$$T_B = V_AR(1-\cos\phi) + T_A\cos\phi - WR[1-\cos(\phi-\theta)]$$

If $\beta = 1.3$ (solid or hollow round cross section, $\nu = 0.3$)

ϕ	45°		90°		180°	
θ	15°	30°	30°	60°	60°	120°
K_{VA}	0.5053	0.1379	0.4684	0.1103	0.3910	0.0577
K_{TA}	−0.0226	−0.0111	−0.0862	−0.0393	−0.2180	−0.0513
$K_{\Theta A}$	−0.0252	−0.0127	−0.1320	−0.0674	−1.1525	−0.5429
K_{MB}	−0.1267	−0.1535	−0.3114	−0.3504	−0.8660	−0.8660
K_{TB}	−0.0019	−0.0015	−0.0316	−0.0237	−0.5000	−0.3333
$K_{M\theta}$	0.1366	0.0745	0.2773	0.1296	0.5274	0.0944

TABLE 19 *Formulas for curved beams loaded normal to the plane of curvature (Cont.)*

End restraints, reference no.	Formulas for boundary values and selected numerical values

1e. Right end fixed, left end fixed

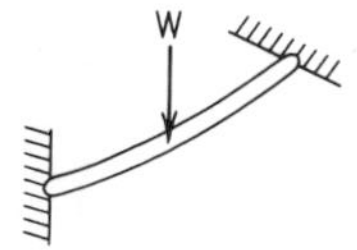

$y_A = 0 \quad \Theta_A = 0 \quad \psi_A = 0$

$y_B = 0 \quad \Theta_B = 0 \quad \psi_B = 0$

$$V_A = W\frac{C_{a3}(C_4C_8 - C_5C_7) + C_{a6}(C_2C_7 - C_1C_8) + C_{a9}(C_1C_5 - C_2C_4)}{C_1(C_5C_9 - C_6C_8) + C_4(C_3C_8 - C_2C_9) + C_7(C_2C_6 - C_3C_5)}$$

$$M_A = WR\frac{C_{a3}(C_5C_9 - C_6C_8) + C_{a6}(C_3C_8 - C_2C_9) + C_{a9}(C_2C_6 - C_3C_5)}{C_1(C_5C_9 - C_6C_8) + C_4(C_3C_8 - C_2C_9) + C_7(C_2C_6 - C_3C_5)}$$

$$T_A = WR\frac{C_{a3}(C_6C_7 - C_4C_9) + C_{a6}(C_1C_9 - C_3C_7) + C_{a9}(C_3C_4 - C_1C_6)}{C_1(C_5C_9 - C_6C_8) + C_4(C_3C_8 - C_2C_9) + C_7(C_2C_6 - C_3C_5)}$$

$$V_B = V_A - W$$

$$M_B = V_AR \sin\phi + M_A \cos\phi - T_A \sin\phi - WR \sin(\phi - \theta)$$

$$T_B = V_AR(1 - \cos\phi) + M_A \sin\phi + T_A \cos\phi - WR[1 - \cos(\phi - \theta)]$$

If $\beta = 1.3$ (solid or hollow round cross section, $\nu = 0.3$)

ϕ	45°	90°	180°	270°	360°	
θ	15°	30°	60°	90°	90°	180°
K_{VA}	0.7424	0.7473	0.7658	0.7902	0.9092	0.5000
K_{MA}	−0.1201	−0.2589	−0.5887	−0.8488	−0.9299	−0.3598
K_{TA}	0.0009	0.0135	0.1568	0.5235	0.7500	1.0000
K_{MB}	−0.0606	−0.1322	−0.2773	−0.2667	0.0701	−0.3598
K_{TB}	−0.0008	−0.0116	−0.1252	−0.3610	−0.2500	−1.0000
$K_{M\theta}$	0.0759	0.1427	0.2331	0.2667	0.1592	0.3598

1f. Right end supported and slope-guided, left end supported and slope-guided

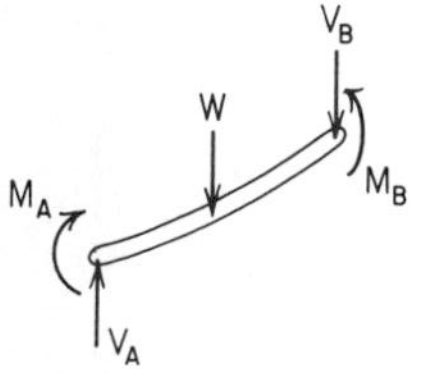

$T_A = 0 \quad y_A = 0 \quad \Theta_A = 0$

$T_B = 0 \quad y_B = 0 \quad \Theta_B = 0$

$$V_A = W\frac{[-C_1 \sin\phi + C_4(1 - \cos\phi)][1 - \cos(\phi - \theta)] + C_{a3} \sin^2\phi - C_{a6} \sin\phi(1 - \cos\phi)}{C_4(1 - \cos\phi)^2 + C_3 \sin^2\phi - (C_1 + C_6)(1 - \cos\phi)\sin\phi}$$

$$M_A = +WR\frac{C_{a6}(1 - \cos\phi)^2 - C_{a3}(1 - \cos\phi)\sin\phi + [C_3 \sin\phi - C_6(1 - \cos\phi)][1 - \cos(\phi - \theta)]}{C_4(1 - \cos\phi)^2 + C_3 \sin^2\phi - (C_1 + C_6)(1 - \cos\phi)\sin\phi}$$

$$\psi_A = \frac{WR^2}{EI}\,\frac{(C_{a3}C_4 - C_{a6}C_1)(1 - \cos\phi) - (C_{a3}C_6 - C_{a6}C_3)\sin\phi - (C_3C_4 - C_1C_6)[1 - \cos(\phi - \theta)]}{C_4(1 - \cos\phi)^2 + C_3 \sin^2\phi - (C_1 + C_6)(1 - \cos\phi)\sin\phi}$$

$$V_B = V_A - W$$

$$M_B = V_AR \sin\phi + M_A \cos\phi - WR \sin(\phi - \theta)$$

$$\psi_B = \psi_A \cos\phi + \frac{M_AR}{EI}C_7 + \frac{V_AR^2}{EI}C_9 - \frac{WR^2}{EI}C_{a9}$$

If $\beta = 1.3$ (solid or hollow round cross section, $\nu = 0.3$)

ϕ	45°	90°	180°	270°
θ	15°	30°	60°	90°
K_{VA}	0.7423	0.7457	0.7500	0.7414
K_{MA}	−0.1180	−0.2457	−0.5774	−1.2586
$K_{\psi A}$	−0.0024	−0.0215	−0.2722	−2.5702
K_{MB}	−0.0586	−0.1204	−0.2887	−0.7414
$K_{\psi B}$	−0.0023	−0.0200	−0.2372	−2.3554
$K_{M\theta}$	0.0781	0.1601	0.3608	0.7414

1g. Right end supported and slope-guided, left end supported and roll-guided

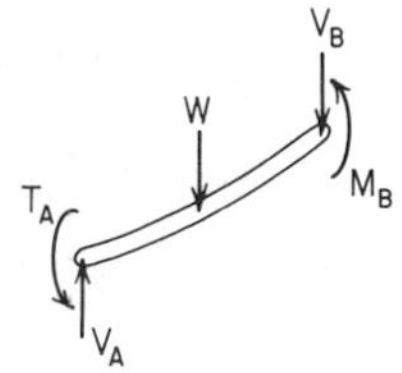

$M_A = 0 \quad y_A = 0 \quad \psi_A = 0$
$T_B = 0 \quad y_B = 0 \quad \Theta_B = 0$

$$V_A = W\frac{(C_5 \sin\phi - C_2 \cos\phi)[1 - \cos(\phi - \theta)] + C_{a3}\cos^2\phi - C_{a6}\sin\phi\cos\phi}{(C_5 \sin\phi - C_2 \cos\phi)(1 - \cos\phi) + C_3 \cos^2\phi - C_6 \sin\phi\cos\phi}$$

$$T_A = WR\frac{(C_3 \cos\phi - C_6 \sin\phi)[1 - \cos(\phi - \theta)] - (C_{a3}\cos\phi - C_{a6}\sin\phi)(1 - \cos\phi)}{(C_5 \sin\phi - C_2 \cos\phi)(1 - \cos\phi) + C_3 \cos^2\phi - C_6 \sin\phi\cos\phi}$$

$$\Theta_A = \frac{WR^2}{EI}\frac{(C_2C_6 - C_3C_5)[1 - \cos(\phi - \theta)] + (C_{a3}C_5 - C_{a6}C_2)(1 - \cos\phi) + (C_{a6}C_3 - C_{a3}C_6)\cos\phi}{(C_5 \sin\phi - C_2 \cos\phi)(1 - \cos\phi) + C_3 \cos^2\phi - C_6 \sin\phi\cos\phi}$$

$$V_B = V_A - W$$

$$M_B = V_A R \sin\phi - T_A \sin\phi - WR\sin(\phi - \theta)$$

$$\psi_B = -\Theta_A \sin\phi + \frac{T_A R}{EI}C_8 + \frac{V_A R^2}{EI}C_9 - \frac{WR^2}{EI}C_{a9}$$

If $\beta = 1.3$ (solid or hollow round cross section, $\nu = 0.3$)

ϕ	45°		90°		180°	
θ	15°	30°	30°	60°	60°	120°
K_{VA}	0.5087	0.1405	0.5000	0.1340	0.6257	0.2141
K_{TA}	−0.0212	−0.0100	−0.0774	−0.0327	−0.2486	−0.0717
$K_{\Theta A}$	−0.0252	−0.0127	−0.1347	−0.0694	−1.7627	−0.9497
K_{MB}	−0.1253	−0.1524	−0.2887	−0.3333	−0.8660	−0.8660
$K_{\psi B}$	−0.0016	−0.0012	−0.0349	−0.0262	−0.9585	−0.6390
$K_{M\theta}$	0.1372	0.0753	0.2887	0.1443	0.7572	0.2476

1h. Right end supported and slope-guided, left end simply supported

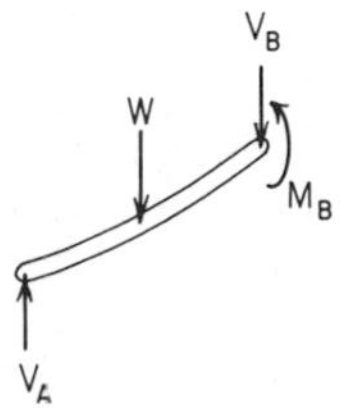

$M_A = 0 \quad T_A = 0 \quad y_A = 0$
$T_B = 0 \quad y_B = 0 \quad \Theta_B = 0$

$$V_A = W\frac{1 - \cos(\phi - \theta)}{1 - \cos\phi}$$

$$\Theta_A = \frac{WR^2}{EI}\left\{\frac{C_{a3}\sin\phi + C_6[1 - \cos(\phi - \theta)]}{1 - \cos\phi} - \frac{C_3 \sin\phi\,[1 - \cos(\phi - \theta)]}{(1 - \cos\phi)^2} - C_{a6}\right\}$$

$$\psi_A = \frac{WR^2}{EI}\left[\frac{C_{a6}\sin\phi - C_{a3}\cos\phi}{1 - \cos\phi} - (C_6 \sin\phi - C_3 \cos\phi)\frac{1 - \cos(\phi - \theta)}{(1 - \cos\phi)^2}\right]$$

$$V_B = V_A - W$$

$$M_B = V_A R \sin\phi - WR\sin(\phi - \theta)$$

$$\psi_B = \psi_A \cos\phi - \Theta_A \sin\phi + \frac{V_A R^2}{EI}C_9 - \frac{WR^2}{EI}C_{a9}$$

If $\beta = 1.3$ (solid or hollow round cross section, $\nu = 0.3$)

ϕ	45°		90°		180°	
θ	15°	30°	30°	60°	60°	120°
K_{VA}	0.4574	0.1163	0.5000	0.1340	0.7500	0.2500
$K_{\Theta A}$	−0.0341	−0.0169	−0.1854	−0.0909	−2.0859	−1.0429
$K_{\psi A}$	0.0467	0.0220	0.1397	0.0591	0.4784	0.1380
K_{MB}	−0.1766	−0.1766	−0.3660	−0.3660	−0.8660	−0.8660
$K_{\psi B}$	0.0308	0.0141	0.0042	−0.0097	−0.9878	−0.6475
$K_{M\theta}$	0.1184	0.0582	0.2500	0.1160	0.6495	0.2165

TABLE 19 ***Formulas for curved beams loaded normal to the plane of curvature*** ***(Cont.)***

End restraints, reference no.	Formulas for boundary values and selected numerical values

1i. Right end supported and roll-guided, left end supported and roll-guided

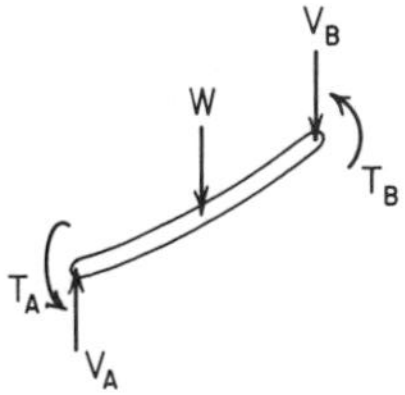

$M_A = 0 \quad y_A = 0 \quad \psi_A = 0$

$M_B = 0 \quad y_B = 0 \quad \psi_B = 0$

$$V_A = W\frac{(C_{a3} + C_{a9}) \sin \phi + (C_2 + C_8) \sin (\phi - \theta)}{(C_2 + C_3 + C_8 + C_9) \sin \phi}$$

$$T_A = WR\frac{(C_{a3} + C_{a9}) \sin \phi - (C_3 + C_9) \sin (\phi - \theta)}{(C_2 + C_3 + C_8 + C_9) \sin \phi}$$

$$\Theta_A = \frac{WR^2}{EI}\frac{[C_{a3}(C_8 + C_9) - C_{a9}(C_2 + C_3)] \sin \phi + (C_2C_9 - C_3C_8) \sin (\phi - \theta)}{(C_2 + C_3 + C_8 + C_9) \sin \phi}$$

$$V_B = V_A - W$$

$$T_B = V_AR(1 - \cos \phi) + T_A \cos \phi - WR[1 - \cos (\phi - \theta)]$$

$$\Theta_B = \Theta_A \cos \phi + \frac{V_AR^2}{EI}C_6 + \frac{T_AR}{EI}C_5 - \frac{WR^2}{EI}C_{a6}$$

If $\beta = 1.3$ (solid or hollow round cross section, $\nu = 0.3$)

ϕ	45°	90°	270°
θ	15°	30°	90°
K_{VA}	0.6667	0.6667	0.6667
K_{TA}	−0.0404	−0.1994	0.0667
$K_{\Theta A}$	−0.0462	−0.3430	−4.4795
K_{TB}	0.0327	0.1667	−1.3333
$K_{\Theta B}$	0.0382	0.3048	1.7333
$K_{M\theta}$	0.1830	0.4330	0.0000

1j. Right end supported and roll-guided, left end simply supported

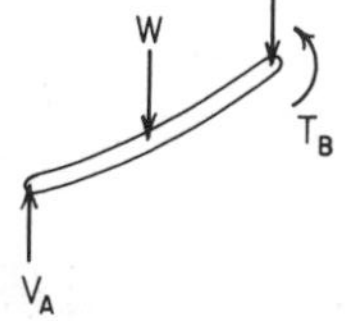

$M_A = 0 \quad T_A = 0 \quad y_A = 0$

$M_B = 0 \quad y_B = 0 \quad \psi_B = 0$

$$V_A = W\frac{\sin (\phi - \theta)}{\sin \phi}$$

$$\Theta_A = \frac{WR^2}{EI}\left\{\frac{C_{a3} \cos \phi - C_{a9}(1 - \cos \phi)}{\sin \phi} - [C_3 \cos \phi - C_9(1 - \cos \phi)]\frac{\sin (\phi - \theta)}{\sin^2 \phi}\right\}$$

$$\psi_A = \frac{WR^2}{EI}\left[C_{a3} + C_{a9} - (C_3 + C_9)\frac{\sin (\phi - \theta)}{\sin \phi}\right]$$

$$V_B = V_A - W$$

$$T_B = V_AR(1 - \cos \phi) - WR[1 - \cos (\phi - \theta)]$$

$$\Theta_B = \Theta_A \cos \phi + \psi_A \sin \phi + \frac{V_AR^2}{EI}C_6 - \frac{WR^2}{EI}C_{a6}$$

If $\beta = 1.3$ (solid or hollow round cross section, $\nu = 0.3$)

ϕ	45°		90°		270°	
θ	15°	30°	30°	60°	90°	180°
K_{VA}	0.7071	0.3660	0.8660	0.5000	0.0000	−1.0000
$K_{\Theta A}$	−0.0575	−0.0473	−0.6021	−0.5215	−3.6128	0.0000
$K_{\psi A}$	0.0413	0.0334	0.4071	0.3403	−4.0841	−8.1681
K_{TB}	0.0731	0.0731	0.3660	0.3660	−2.0000	−2.0000
$K_{\Theta B}$	0.0440	0.0509	0.4527	0.4666	6.6841	14.3810
$K_{M\theta}$	0.1830	0.1830	0.4330	0.4330	0.0000	0.0000

2. Concentrated intermediate bending moment

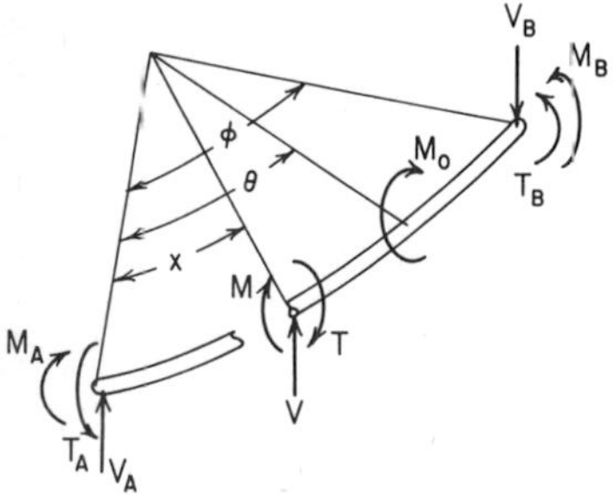

Transverse shear $= V = V_A$

Bending moment $= M = V_A R \sin x + M_A \cos x - T_A \sin x + M_o \cos(x-\theta)\langle x-\theta\rangle^0$

Twisting moment $= T = V_A R(1-\cos x) + M_A \sin x + T_A \cos x + M_c \sin(x-\theta)\langle x-\theta\rangle^0$

Vertical deflection $= y = y_A + \Theta_A R \sin x + \psi_A R(1-\cos x) + \frac{M_A R^2}{EI}F_1 + \frac{T_A R^2}{EI}F_2 + \frac{V_A R^3}{EI}F_3 + \frac{M_o R^2}{EI}F_{a1}$

Bending slope $= \Theta = \Theta_A \cos x + \psi_A \sin x + \frac{M_A R}{EI}F_4 + \frac{T_A R}{EI}F_5 + \frac{V_A R^2}{EI}F_6 + \frac{M_o R}{EI}F_{a4}$

Roll slope $= \psi = \psi_A \cos x - \Theta_A \sin x + \frac{M_A R}{EI}F_7 + \frac{T_A R}{EI}F_8 + \frac{V_A R^2}{EI}F_9 + \frac{M_o R}{EI}F_{a7}$

For tabulated values: $V = K_V \frac{M_o}{R}$ $\quad M = K_M M_o$ $\quad T = K_T M_o$ $\quad y = K_y \frac{M_o R^2}{EI}$ $\quad \Theta = K_\Theta \frac{M_o R}{EI}$ $\quad \psi = K_\psi \frac{M_o R}{EI}$

End restraints, reference no.	Formulas for boundary values and selected numerical values
2a. Right end fixed, left end free	$y_A = \frac{M_o R^2}{EI}[C_{a4}\sin\phi - C_{a7}(1-\cos\phi) - C_{a1}]$
	$\Theta_A = \frac{M_o R}{EI}(C_{a7}\sin\phi - C_{a4}\cos\phi)$
	$\psi_A = -\frac{M_o R}{EI}(C_{a4}\sin\phi + C_{a7}\cos\phi)$
$V_A = 0 \quad M_A = 0 \quad T_A = 0$	$V_B = 0 \qquad M_B = M_o \cos(\phi - \theta)$
$y_B = 0 \quad \Theta_B = 0 \quad \psi_B = 0$	$T_B = M_o \sin(\phi - \theta)$

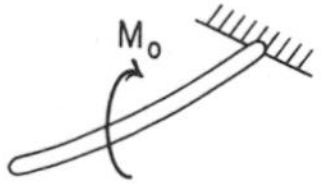

If $\beta = 1.3$ (solid or hollow round cross section, $\nu = 0.3$)

ϕ	45°	90°			180°		
θ	0°	0°	30°	60°	0°	60°	120°
K_{yA}	0.3058	1.1500	1.1222	0.6206	2.6000	4.0359	1.6929
$K_{\Theta A}$	−0.8282	−1.8064	−1.0429	−0.3011	−3.6128	−1.3342	0.4722
$K_{\psi A}$	0.0750	0.1500	−0.4722	−0.4465	0.0000	−2.0859	−1.0429
K_{MB}	0.7071	0.0000	0.5000	0.8660	−1.0000	−0.5000	0.5000
K_{TB}	0.7071	1.0000	0.8660	0.5000	0.0000	0.8660	0.8660

TABLE 19 ***Formulas for curved beams loaded normal to the plane of curvature (Cont.)***

End restraints, reference no.	Formulas for boundary values and selected numerical values

2b. Right end fixed, left end simply supported

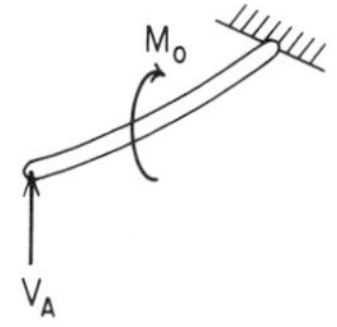

$M_A = 0 \quad T_A = 0 \quad y_A = 0$

$y_B = 0 \quad \Theta_B = 0 \quad \psi_B = 0$

$$V_A = \frac{-M_o}{R}\,\frac{C_{a7}(1-\cos\phi) - C_{a4}\sin\phi + C_{a1}}{C_9(1-\cos\phi) - C_6\sin\phi + C_3}$$

$$\Theta_A = -\frac{M_oR}{EI}\,\frac{(C_{a1}C_9 - C_{a7}C_3)\sin\phi + (C_{a7}C_6 - C_{a4}C_9)(1-\cos\phi) + (C_{a4}C_3 - C_{a1}C_6)\cos\phi}{C_9(1-\cos\phi) - C_6\sin\phi + C_3}$$

$$\psi_A = -\frac{M_oR}{EI}\,\frac{[C_{a4}(C_9 + C_3) - C_6(C_{a1} + C_{a7})]\sin\phi + (C_{a7}C_3 - C_{a1}C_9)\cos\phi}{C_9(1-\cos\phi) - C_6\sin\phi + C_3}$$

$V_B = V_A$

$M_B = V_AR\sin\phi + M_o\cos(\phi - \theta)$

$T_B = V_AR(1-\cos\phi) + M_o\sin(\phi-\theta)$

If $\beta = 1.3$ (solid or hollow round cross section, $\nu = 0.3$)

ϕ	45°	90°			180°		
θ	0°	0°	30°	60°	0°	60°	120°
K_{VA}	−1.9021	−0.9211	−0.8989	−0.4971	−0.3378	−0.5244	−0.2200
$K_{\Theta A}$	−0.2466	−0.7471	−0.0092	0.2706	−2.7346	0.0291	1.0441
$K_{\psi A}$	0.1872	0.6165	−0.0170	−0.1947	1.2204	−0.1915	−0.2483
K_{MB}	−0.6379	−0.9211	−0.3989	0.3689	−1.0000	−0.5000	0.5000
K_{TB}	0.1500	0.0789	−0.0329	0.0029	−0.6756	−0.1827	0.4261

2c. Right end fixed, left end supported and slope-guided

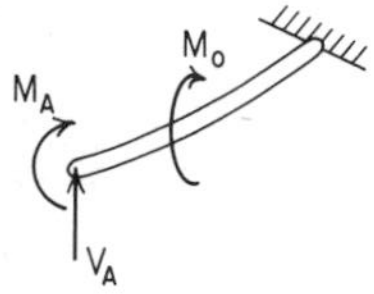

$T_A = 0 \quad y_A = 0 \quad \Theta_A = 0$

$y_B = 0 \quad \Theta_B = 0 \quad \psi_B = 0$

$$V_A = -\frac{M_o}{R}\,\frac{(C_{a7}C_4 - C_{a4}C_7)(1-\cos\phi) + (C_{a4}C_1 - C_{a1}C_4)\cos\phi + (C_{a1}C_7 - C_{a7}C_1)\sin\phi}{(C_4C_9 - C_6C_7)(1-\cos\phi) + (C_1C_6 - C_3C_4)\cos\phi + (C_3C_7 - C_1C_9)\sin\phi}$$

$$M_A = -M_o\,\frac{(C_{a4}C_9 - C_{a7}C_6)(1-\cos\phi) + (C_{a1}C_6 - C_{a4}C_3)\cos\phi + (C_{a7}C_3 - C_{a1}C_9)\sin\phi}{(C_4C_9 - C_6C_7)(1-\cos\phi) + (C_1C_6 - C_3C_4)\cos\phi + (C_3C_7 - C_1C_9)\sin\phi}$$

$$\psi_A = -\frac{M_oR}{EI}\,\frac{C_{a1}(C_4C_9 - C_6C_7) + C_{a4}(C_3C_7 - C_1C_9) + C_{a7}(C_1C_6 - C_3C_4)}{(C_4C_9 - C_6C_7)(1-\cos\phi) + (C_1C_6 - C_3C_4)\cos\phi + (C_3C_7 - C_1C_9)\sin\phi}$$

$V_B = V_A$

$M_B = V_AR\sin\phi + M_A\cos\phi + M_o\cos(\phi-\theta)$

$T_B = V_AR(1-\cos\phi) + M_A\sin\phi + M_o\sin(\phi-\theta)$

If $\beta = 1.3$ (solid or hollow round cross section, $\nu = 0.3$)

ϕ	45°		90°		180°	
θ	15°	30°	30°	60°	60°	120°
K_{VA}	−1.7096	−1.6976	−0.8876	−0.8308	−0.5279	−0.3489
K_{MA}	−0.0071	0.3450	−0.0123	0.3622	0.0107	0.3818
$K_{\psi A}$	−0.0025	0.0029	−0.0246	0.0286	−0.1785	0.2177
K_{MB}	−0.3478	0.0095	−0.3876	0.0352	−0.5107	0.1182
K_{TB}	−0.0057	0.0056	−0.0338	0.0314	−0.1899	0.1682

2d. Right end fixed, left end supported and roll-guided

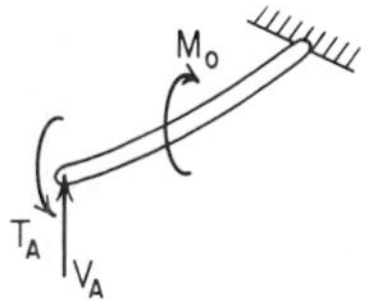

$M_A = 0 \quad y_A = 0 \quad \psi_A = 0$

$y_B = 0 \quad \Theta_B = 0 \quad \psi_B = 0$

$$V_A = -\frac{M_o}{R}\frac{[(C_{a1}+C_{a7})C_5 - C_{a4}(C_2+C_8)]\sin\phi + (C_{a1}C_8 - C_{a7}C_2)\cos\phi}{[C_5(C_3+C_9) - C_6(C_2+C_8)]\sin\phi + (C_3C_8 - C_2C_9)\cos\phi}$$

$$T_A = -M_o\frac{[C_{a4}(C_3+C_9) - (C_{a1}+C_{a7})C_6]\sin\phi + (C_{a7}C_3 - C_{a1}C_9)\cos\phi}{[C_5(C_3+C_9) - C_6(C_2+C_8)]\sin\phi + (C_3C_8 - C_2C_9)\cos\phi}$$

$$\Theta_A = -\frac{M_oR}{EI}\frac{C_{a1}(C_5C_9 - C_6C_8) + C_{a4}(C_3C_8 - C_2C_9) + C_{a7}(C_2C_3 - C_3C_5)}{[C_5(C_3+C_9) - C_6(C_2+C_8)]\sin\phi + (C_3C_8 - C_2C_9)\cos\phi}$$

$$V_B = V_A$$

$$M_B = V_AR\sin\phi - T_A\sin\phi + M_o\cos(\phi-\theta)$$

$$T_B = V_AR(1-\cos\phi) + T_A\cos\phi + M_o\sin(\phi-\theta)$$

If $\beta = 1.3$ (solid or hollow round cross section, $\nu = 0.3$)

ϕ	45°	90°			180°		
θ	0°	0°	30°	60°	0°	30°	60°
K_{VA}	−1.9739	−1.0773	−0.8946	−0.4478	−0.6366	−0.4775	−0.1592
K_{TA}	−0.1957	−0.3851	0.0106	0.1216	−0.6366	0.0999	0.1295
$K_{\Theta A}$	−0.2100	−0.5097	−0.0158	0.1956	−1.9576	−0.0928	0.8860
K_{MB}	−0.5503	−0.6923	−0.4052	0.2966	−1.000	−0.5000	0.5000
K_{TB}	−0.0094	−0.0773	−0.0286	0.0522	−0.6366	−0.1888	0.4182

2e. Right end fixed, left end fixed

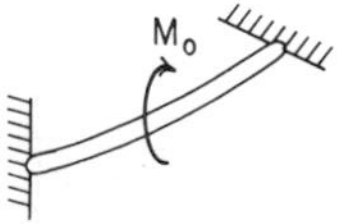

$y_A = 0 \quad \Theta_A = 0 \quad \psi_A = 0$

$y_B = 0 \quad \Theta_B = 0 \quad \psi_B = 0$

$$V_A = -\frac{M_o}{R}\frac{C_{a1}(C_4C_8 - C_5C_7) + C_{a4}(C_2C_7 - C_1C_8) + C_{a7}(C_1C_5 - C_2C_4)}{C_1(C_5C_9 - C_6C_8) + C_4(C_3C_8 - C_2C_9) + C_7(C_2C_6 - C_3C_5)}$$

$$M_A = -M_o\frac{C_{a1}(C_5C_9 - C_6C_8) + C_{a4}(C_3C_8 - C_2C_9) + C_{a7}(C_2C_6 - C_3C_5)}{C_1(C_5C_9 - C_6C_8) + C_4(C_3C_8 - C_2C_9) + C_7(C_2C_6 - C_3C_5)}$$

$$T_A = -M_o\frac{C_{a1}(C_6C_7 - C_4C_9) + C_{a4}(C_1C_9 - C_3C_7) + C_{a7}(C_3C_4 - C_1C_6)}{C_1(C_5C_9 - C_6C_8) + C_4(C_3C_8 - C_2C_9) + C_7(C_2C_6 - C_3C_5)}$$

$$V_B = V_A$$

$$M_B = V_AR\sin\phi + M_A\cos\phi - T_A\sin\phi + M_o\cos(\phi-\theta)$$

$$T_B = V_AR(1-\cos\phi) + M_A\sin\phi + T_A\cos\phi + M_o\sin(\phi-\theta)$$

If $\beta = 1.3$ (solid or hollow round cross section, $\nu = 0.3$)

ϕ	45°	90°	180°	270°	360°	
θ	15°	30°	60°	90°	90°	180°
K_{VA}	−1.7040	−0.8613	−0.4473	−0.3115	−0.1592	−0.3183
K_{MA}	−0.0094	−0.0309	−0.0474	0.0584	−0.0208	0.5000
K_{TA}	0.0031	0.0225	0.1301	0.2788	0.5908	−0.3183
K_{MB}	−0.3477	−0.3838	−0.4526	−0.4097	−0.0208	−0.5000
K_{TB}	−0.0036	−0.0262	−0.1586	−0.3699	−0.4092	−0.3183

TABLE 19 *Formulas for curved beams loaded normal to the plane of curvature (Cont.)*

End restraints, reference no.	Formulas for boundary values and selected numerical values
2f. Right end supported and slope-guided, left end supported and slope-guided	$V_A = +\frac{M_o}{R}\frac{[C_1 \sin\phi - C_4(1-\cos\phi)]\sin(\phi-\theta) - C_{a1}\sin^2\phi + C_{a4}\sin\phi(1-\cos\phi)}{C_4(1-\cos\phi)^2 + C_3\sin^2\phi - (C_1+C_6)(1-\cos\phi)\sin\phi}$
	$M_A = -M_o\frac{[C_3\sin\phi - C_6(1-\cos\phi)]\sin(\phi-\theta) - C_{a1}(1-\cos\phi)\sin\phi + C_{a4}(1-\cos\phi)^2}{C_4(1-\cos\phi)^2 + C_3\sin^2\phi - (C_1+C_6)(1-\cos\phi)\sin\phi}$
	$\psi_A = \frac{M_oR}{EI}\frac{(C_3C_4 - C_1C_6)\sin(\phi-\theta) + (C_{a1}C_6 - C_{a4}C_3)\sin\phi - (C_{a1}C_4 - C_{a4}C_1)(1-\cos\phi)}{C_4(1-\cos\phi)^2 + C_3\sin^2\phi - (C_1+C_6)(1-\cos\phi)\sin\phi}$
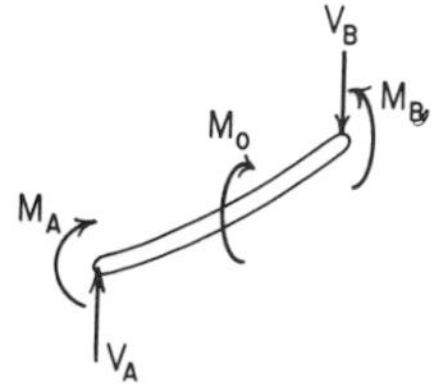	$V_B = V_A$
	$M_B = V_AR\sin\phi + M_A\cos\phi + M_o\cos(\phi-\theta)$
	$\psi_B = \psi_A\cos\phi + \frac{M_AR}{EI}C_7 + \frac{V_AR^2}{EI}C_9 + \frac{M_oR}{EI}C_{a7}$
$T_A = 0$ $y_A = 0$ $\Theta_A = 0$ $T_B = 0$ $y_B = 0$ $\Theta_B = 0$	

If $\beta = 1.3$ (solid or hollow round cross section, $\nu = 0.3$)

ϕ	45°	90°	180°	270°
θ	15°	30°	60°	90°
K_{VA}	−1.7035	−0.8582	−0.4330	−0.2842
K_{MA}	−0.0015	−0.0079	−0.0577	−0.2842
$M_{\psi A}$	−0.0090	−0.0388	−0.2449	−1.7462
K_{MB}	−0.3396	−0.3581	−0.4423	−0.7159
$K_{\psi B}$	−0.0092	−0.0418	−0.2765	−1.8667

End restraints, reference no.	Formulas for boundary values and selected numerical values
2g. Right end supported and slope-guided, left end supported and roll-guided	$V_A = -\frac{M_o}{R}\frac{C_{a1}\cos^2\phi - C_{a4}\sin\phi\cos\phi + (C_5\sin\phi - C_2\cos\phi)\sin(\phi-\theta)}{(C_5\sin\phi - C_2\cos\phi)(1-\cos\phi) + C_3\cos^2\phi - C_6\sin\phi\cos\phi}$
	$T_A = -M_o\frac{(C_{a4}\sin\phi - C_{a1}\cos\phi)(1-\cos\phi) + (C_3\cos\phi - C_6\sin\phi)\sin(\phi-\theta)}{(C_5\sin\phi - C_2\cos\phi)(1-\cos\phi) + C_3\cos^2\phi - C_6\sin\phi\cos\phi}$
	$\Theta_A = \frac{-M_oR}{EI}\frac{(C_{a1}C_5 - C_{a4}C_2)(1-\cos\phi) + (C_{a4}C_3 - C_{a1}C_6)\cos\phi + (C_2C_6 - C_3C_5)\sin(\phi-\theta)}{(C_5\sin\phi - C_2\cos\phi)(1-\cos\phi) + C_3\cos^2\phi - C_6\sin\phi\cos\phi}$
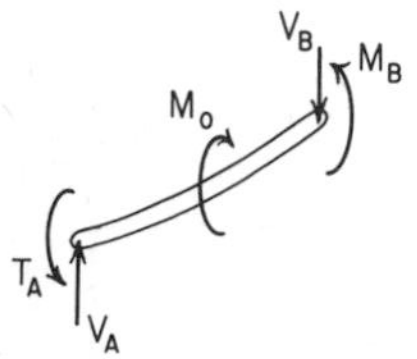	$V_B = V_A$
	$M_B = V_AR\sin\phi - T_A\sin\phi + M_o\cos(\phi-\theta)$
	$\psi_B = -\Theta_A\sin\phi + \frac{T_AR}{EI}C_8 + \frac{V_AR^2}{EI}C_9 + \frac{M_oR}{EI}C_{a7}$
$M_A = 0$ $y_A = 0$ $\psi_A = 0$ $T_B = 0$ $y_B = 0$ $\Theta_B = 0$	

If $\beta = 1.3$ (solid or hollow round cross section, $\nu = 0.3$)

ϕ	45°	90°			180°		
θ	0°	0°	30°	60°	0°	60°	120°
K_{VA}	−1.9576	−1.0000	−0.8660	−0.5000	−0.3378	−0.3888	−0.3555
K_{TA}	−0.1891	−0.3634	0.0186	0.1070	−0.6756	0.0883	0.1551
$K_{\Theta A}$	−0.2101	−0.5163	−0.0182	0.2001	−2.7346	−0.3232	1.3964
K_{MB}	−0.5434	−0.6366	−0.3847	0.2590	−1.0000	−0.5000	0.5000
$K_{\psi B}$	−0.0076	−0.0856	−0.0316	0.0578	−1.2204	−0.3619	0.8017

2h. Right end supported and slope-guided, left end simply supported

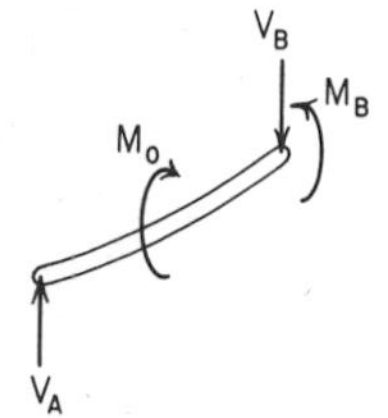

$M_A = 0$ $T_A = 0$ $y_A = 0$
$T_B = 0$ $y_B = 0$ $\Theta_B = 0$

$$V_A = -\frac{M_o}{R}\frac{\sin(\phi-\theta)}{1-\cos\phi}$$

$$\Theta_A = -\frac{M_oR}{EI}\left[\frac{C_{a1}\sin\phi + C_6\sin(\phi-\theta)}{1-\cos\phi} - \frac{C_3\sin\phi\sin(\phi-\theta)}{(1-\cos\phi)^2} - C_{a4}\right]$$

$$\psi_A = -\frac{M_oR}{EI}\left[\frac{C_{a4}\sin\phi - C_{a1}\cos\phi}{1-\cos\phi} + \frac{(C_3\cos\phi - C_6\sin\phi)\sin(\phi-\theta)}{(1-\cos\phi)^2}\right]$$

$$V_B = V_A$$
$$M_B = V_AR\sin\phi + M_o\cos(\phi-\theta)$$
$$\psi_B = \psi_A\cos\phi - \Theta_A\sin\phi + \frac{V_AR^2}{EI}C_9 + \frac{M_oR}{EI}C_{a7}$$

If $\beta = 1.3$ (solid or hollow round cross section, $\nu = 0.3$)

ϕ	45°	90°			180°		
θ	0°	0°	30°	60°	0°	60°	120°
K_{VA}	−2.4142	−1.0000	−0.8660	−0.5000	0.0000	−0.4330	−0.4330
$K_{\Theta A}$	−0.2888	−0.7549	−0.0060	0.2703	−3.6128	−0.2083	1.5981
$K_{\psi A}$	0.4161	0.6564	−0.0337	−0.1933	1.3000	−0.1700	−0.2985
K_{MB}	−1.0000	−1.0000	−0.3660	0.3660	−1.0000	−0.5000	0.5000
$K_{\psi B}$	0.2811	0.0985	−0.0410	0.0036	−1.3000	−0.3515	0.8200

2i. Right end supported and roll-guided, left end supported and roll-guided

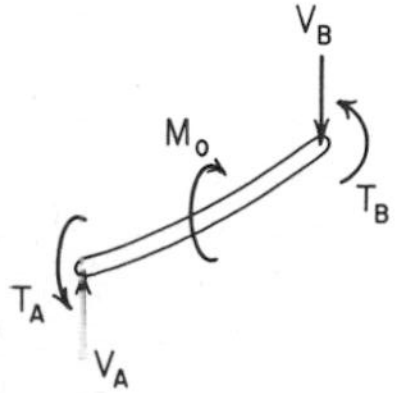

$M_A = 0$ $y_A = 0$ $\psi_A = 0$
$M_B = 0$ $y_B = 0$ $\psi_B = 0$

$$V_A = -\frac{M_o}{R}\frac{(C_{a1}+C_{a7})\sin^2\phi + (C_2+C_8)\cos(\phi-\theta)\sin\phi}{(C_2+C_3+C_8+C_9)\sin^2\phi}$$

$$T_A = -M_o\frac{(C_{a1}+C_{a7})\sin^2\phi - (C_3+C_9)\cos(\phi-\theta)\sin\phi}{(C_2+C_3+C_8+C_9)\sin^2\phi}$$

$$\Theta_A = -\frac{M_oR}{EI}\frac{[C_{a1}(C_8+C_9) - C_{a7}(C_2+C_3)]\sin\phi + (C_2C_9 - C_3C_8)\cos(\phi-\theta)}{(C_2+C_3+C_8+C_9)\sin^2\phi}$$

$$V_B = V_A$$
$$T_B = V_AR(1-\cos\phi) + T_A\cos\phi + M_o\sin(\phi-\theta)$$
$$\Theta_B = \Theta_A\cos\phi + \frac{T_AR}{EI}C_5 + \frac{V_AR^2}{EI}C_6 + \frac{M_oR}{EI}C_{a4}$$

If $\beta = 1.3$ (solid or hollow round cross section, $\nu = 0.3$)

ϕ	45°		90°		270°	
θ	0°	15°	0°	30°	0°	90°
K_{VA}	−1.2732	−1.2732	−0.6366	−0.6366	−0.2122	−0.2122
K_{TA}	−0.2732	−0.0485	−0.6366	−0.1366	−0.2122	0.7878
$K_{\Theta A}$	−0.3012	−0.0605	−0.9788	−0.2903	−5.1434	0.1259
K_{TB}	0.1410	0.0928	0.3634	0.2294	−1.2122	−0.2122
$K_{\Theta B}$	0.1658	0.1063	0.6776	0.3966	0.4259	2.0823

TABLE 19 *Formulas for curved beams loaded normal to the plane of curvature* (*Cont.*)

End restraints, reference no.	Formulas for boundary values and selected numerical values

2j. Right end supported and roll-guided, left end simply supported

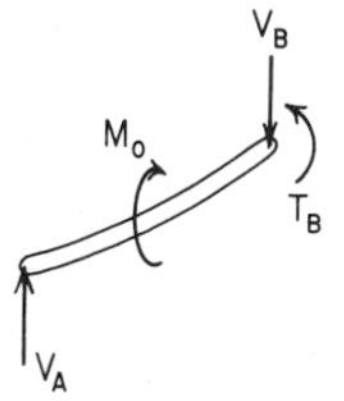

$M_A = 0 \quad T_A = 0 \quad y_A = 0$

$M_B = 0 \quad y_B = 0 \quad \psi_B = 0$

$$V_A = -\frac{M_o \cos(\phi - \theta)}{R \sin \phi}$$

$$\Theta_A = -\frac{M_o R}{EI}\left\{\frac{C_{a1}\cos\phi - C_{a7}(1-\cos\phi)}{\sin\phi} - \frac{[C_3\cos\phi - C_9(1-\cos\phi)]\cos(\phi-\theta)}{\sin^2\phi}\right\}$$

$$\psi_A = -\frac{M_o R}{EI}\left[C_{a1} + C_{a7} - \frac{(C_3 + C_9)\cos(\phi-\theta)}{\sin\phi}\right]$$

$$V_B = V_A$$

$$T_B = V_A R(1-\cos\phi) + M_o \sin(\phi - \theta)$$

$$\Theta_B = \Theta_A \cos\phi + \psi_A \sin\phi + \frac{V_A R^2}{EI}C_6 + \frac{M_o R}{EI}C_{a4}$$

If $\beta = 1.3$ (solid or hollow round cross section, $\nu = 0.3$)

ϕ	45°			90°		
θ	0°	15°	30°	0°	30°	60°
K_{VA}	−1.0000	−1.2247	−1.3660	0.0000	−0.5000	−0.8660
$K_{\Theta A}$	−0.3774	−0.0740	0.1322	−1.8064	−0.4679	0.6949
$K_{\psi A}$	0.2790	0.0495	−0.0947	1.3000	0.2790	−0.4684
K_{TB}	0.4142	0.1413	−0.1413	1.0000	0.3660	−0.3660
$K_{\Theta B}$	0.2051	0.1133	−0.0738	1.1500	0.4980	−0.4606

3. Concentrated intermediate twisting moment (torque)

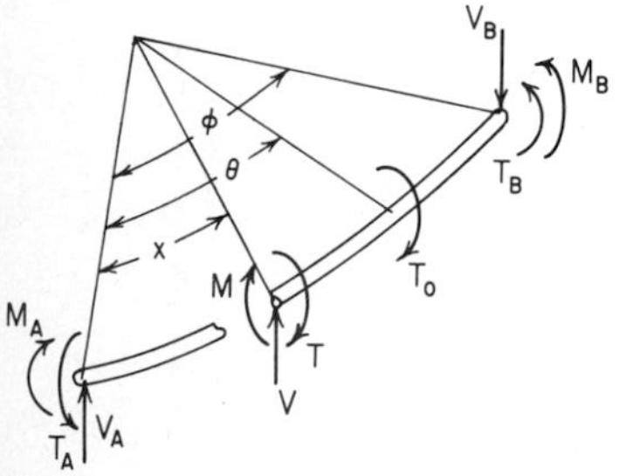

Transverse shear $= V = V_A$

Bending moment $= M = V_A R \sin x + M_A \cos x - T_A \sin x - T_o \sin(x-\theta)\langle x-\theta\rangle^0$

Twisting moment $= T = V_A R(1-\cos x) + M_A \sin x + T_A \cos x + T_o \cos(x-\theta)\langle x-\theta\rangle^0$

Vertical deflection $= y = y_A + \Theta_A R \sin x + \psi_A R(1-\cos x) + \frac{M_A R^2}{EI}F_1 + \frac{T_A R^2}{EI}F_2 + \frac{V_A R^3}{EI}F_3 + \frac{T_o R^2}{EI}F_{a2}$

Bending slope $= \Theta = \Theta_A \cos x + \psi_A \sin x + \frac{M_A R}{EI}F_4 + \frac{T_A R}{EI}F_5 + \frac{V_A R^2}{EI}F_6 + \frac{T_o R}{EI}F_{a5}$

Roll slope $= \psi = \psi_A \cos x - \Theta_A \sin x + \frac{M_A R}{EI}F_7 + \frac{T_A R}{EI}F_8 + \frac{V_A R^2}{EI}F_9 + \frac{T_o R}{EI}F_{a8}$

For tabulated values: $V = K_V \frac{T_o}{R} \quad M = K_M T_o \quad T = K_T T_o \quad y = K_y \frac{T_o R^2}{EI} \quad \Theta = K_\Theta \frac{T_o R}{EI} \quad \psi = K_\psi \frac{T_o R}{EI}$

3a. Right end fixed, left end free

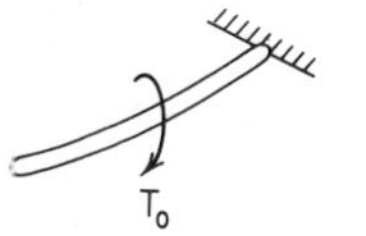

$V_A = 0 \quad M_A = 0 \quad T_A = 0$

$y_B = 0 \quad \Theta_B = 0 \quad \psi_B = 0$

$$y_A = \frac{T_o R^2}{EI}[C_{a5} \sin \phi - C_{a8}(1 - \cos \phi) - C_{a2}]$$

$$\Theta_A = -\frac{T_o R}{EI}(C_{a5} \cos \phi - C_{a8} \sin \phi)$$

$$\psi_A = -\frac{T_o R}{EI}(C_{a8} \cos \phi + C_{a5} \sin \phi)$$

$V_B = 0$

$M_B = -T_o \sin (\phi - \theta)$

$T_B = T_o \cos (\phi - \theta)$

If $\beta = 1.3$ (solid or hollow round cross section, $\nu = 0.3$)

ϕ	45°	90°			180°		
θ	0°	0°	30°	60°	0°	60°	120°
K_{yA}	−0.0590	−0.5064	0.0829	0.3489	−3.6128	0.0515	1.8579
$K_{\Theta A}$	−0.0750	−0.1500	−0.7320	−0.5965	0.0000	−2.0859	−1.0429
$K_{\psi A}$	0.9782	1.8064	1.0429	0.3011	3.6128	1.0744	−0.7320
K_{MB}	−0.7071	−1.0000	−0.8660	−0.5000	0.0000	−0.8660	−0.8660
K_{TB}	0.7071	0.0000	0.5000	0.8660	−1.0000	−0.5000	0.5000

3b. Right end fixed, left end simply supported

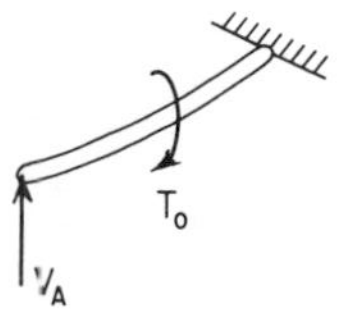

$M_A = 0 \quad T_A = 0 \quad y_A = 0$

$y_B = 0 \quad \Theta_B = 0 \quad \psi_B = 0$

$$V_A = -\frac{T_o}{R}\,\frac{C_{a8}(1 - \cos \phi) - C_{a5} \sin \phi + C_{a2}}{C_9(1 - \cos \phi) - C_6 \sin \phi + C_3}$$

$$\Theta_A = -\frac{T_o R}{EI}\,\frac{(C_{a2}C_9 - C_{a8}C_3) \sin \phi + (C_{a8}C_6 - C_{a5}C_9)(1 - \cos \phi) + (C_{a5}C_3 - C_{a2}C_6) \cos \phi}{C_9(1 - \cos \phi) - C_6 \sin \phi + C_3}$$

$$\psi_A = -\frac{T_o R}{EI}\,\frac{[C_{a5}(C_9 + C_3) - C_6(C_{a2} + C_{a8})] \sin \phi + (C_{a8}C_3 - C_{a2}C_9) \cos \phi}{C_9(1 - \cos \phi) - C_6 \sin \phi + C_3}$$

$V_B = V_A$

$M_B = V_A R \sin \phi - T_o \sin (\phi - \theta)$

$T_B = V_A R(1 - \cos \phi) + T_o \cos (\phi - \theta)$

If $\beta = 1.3$ (solid or hollow round cross section, $\nu = 0.3$)

ϕ	45°	90°			180°		
θ	0°	0°	30°	60°	0°	60°	120°
K_{VA}	0.3668	0.4056	−0.0664	−0.2795	0.4694	−0.0067	−0.2414
$K_{\Theta A}$	−0.1872	−0.6165	−0.6557	−0.2751	−1.2204	−2.0685	−0.4153
$K_{\psi A}$	0.9566	1.6010	1.0766	0.4426	1.9170	1.0985	0.1400
K_{MB}	−0.4477	−0.5944	−0.9324	−0.7795	0.0000	−0.8660	−0.8660
K_{TB}	0.8146	0.4056	0.4336	0.5865	−0.0612	−0.5134	0.0172

TABLE 19 ***Formulas for curved beams loaded normal to the plane of curvature (Cont.)***

End restraints, reference no.	Formulas for boundary values and selected numerical values

3c. Right end fixed, left end supported and slope-guided

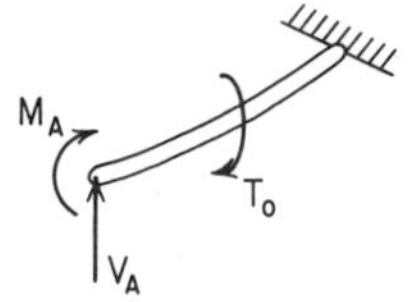

$T_A = 0 \quad y_A = 0 \quad \Theta_A = 0$

$y_B = 0 \quad \Theta_B = 0 \quad \psi_B = 0$

$$V_A = -\frac{T_o}{R}\,\frac{(C_{a8}C_4 - C_{a5}C_7)(1-\cos\phi) + (C_{a5}C_1 - C_{a2}C_4)\cos\phi + (C_{a2}C_7 - C_{a8}C_1)\sin\phi}{(C_4C_9 - C_6C_7)(1-\cos\phi) + (C_1C_6 - C_3C_4)\cos\phi + (C_3C_7 - C_1C_9)\sin\phi}$$

$$M_A = -T_o\,\frac{(C_{a5}C_9 - C_{a8}C_6)(1-\cos\phi) + (C_{a2}C_6 - C_{a5}C_3)\cos\phi + (C_{a8}C_3 - C_{a2}C_9)\sin\phi}{(C_4C_9 - C_6C_7)(1-\cos\phi) + (C_1C_6 - C_3C_4)\cos\phi + (C_3C_7 - C_1C_9)\sin\phi}$$

$$\psi_A = -\frac{T_oR}{EI}\,\frac{C_{a2}(C_4C_9 - C_6C_7) + C_{a5}(C_3C_7 - C_1C_9) + C_{a8}(C_1C_6 - C_3C_4)}{(C_4C_9 - C_6C_7)(1-\cos\phi) + (C_1C_6 - C_3C_4)\cos\phi + (C_3C_7 - C_1C_9)\sin\phi}$$

$V_B = V_A$

$M_B = V_AR\sin\phi + M_A\cos\phi - T_o\sin(\phi-\theta)$

$T_B = V_AR(1-\cos\phi) + M_A\sin\phi + T_o\cos(\phi-\theta)$

If $\beta = 1.3$ (solid or hollow round cross section, $\nu = 0.3$)

ϕ	45°	90°			180°		
θ	0°	0°	30°	60°	0°	60°	120°
K_{VA}	1.8104	1.1657	0.7420	0.0596	0.6201	0.2488	−0.1901
K_{MA}	−0.7589	−0.8252	−0.8776	−0.3682	−0.4463	−0.7564	−0.1519
$K_{\psi A}$	0.8145	1.0923	0.5355	0.2156	1.3724	0.1754	−0.0453
K_{MB}	0.0364	0.1657	−0.1240	−0.4404	0.4463	−0.1096	−0.7141
K_{TB}	0.7007	0.3406	0.3644	0.5575	0.2403	−0.0023	0.1199

3d. Right end fixed, left end supported and roll-guided

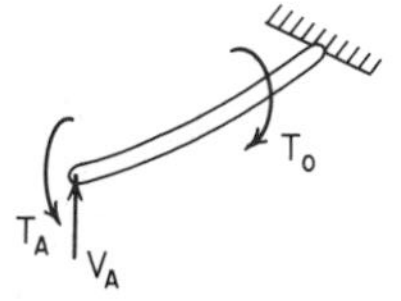

$M_A = 0 \quad y_A = 0 \quad \psi_A = 0$

$y_B = 0 \quad \Theta_B = 0 \quad \psi_B = 0$

$$V_A = -\frac{T_o}{R}\,\frac{[(C_{a2} + C_{a8})C_5 - C_{a5}(C_2 + C_8)]\sin\phi + (C_{a2}C_8 - C_{a8}C_2)\cos\phi}{[C_5(C_3 + C_9) - C_6(C_2 + C_8)]\sin\phi + (C_3C_8 - C_2C_9)\cos\phi}$$

$$T_A = -T_o\,\frac{[C_{a5}(C_3 + C_9) - C_6(C_{a2} + C_{a8})]\sin\phi + (C_{a8}C_3 - C_{a2}C_9)\cos\phi}{[C_5(C_3 + C_9) - C_6(C_2 + C_8)]\sin\phi + (C_3C_8 - C_2C_9)\cos\phi}$$

$$\Theta_A = -\frac{T_oR}{EI}\,\frac{C_{a2}(C_5C_9 - C_6C_8) + C_{a5}(C_3C_8 - C_2C_9) + C_{a8}(C_2C_6 - C_3C_5)}{[C_5(C_3 + C_9) - C_6(C_2 + C_8)]\sin\phi + (C_3C_8 - C_2C_9)\cos\phi}$$

$V_B = V_A$

$M_B = V_AR\sin\phi - T_A\sin\phi - T_o\sin(\phi-\theta)$

$T_B = V_AR(1-\cos\phi) + T_A\cos\phi + T_o\cos(\phi-\theta)$

If $\beta = 1.3$ (solid or hollow round cross section, $\nu = 0.3$)

ϕ	45°		90°		180°	
θ	15°	30°	30°	60°	60°	120°
K_{VA}	−0.3410	−0.4177	−0.3392	−0.3916	−0.2757	−0.2757
K_{TA}	−0.6694	−0.3198	−0.6724	−0.2765	−0.5730	−0.0730
$K_{\Theta A}$	−0.0544	−0.0263	−0.2411	−0.1046	−1.3691	−0.3262
K_{MB}	−0.2678	−0.3280	−0.5328	−0.6152	−0.8660	−0.8660
K_{TB}	0.2928	0.6175	0.1608	0.4744	−0.4783	0.0217

3e. Right end fixed, left end fixed

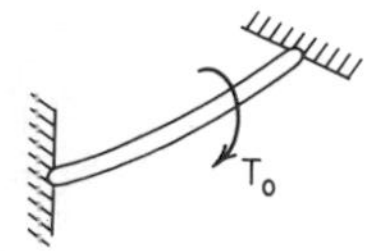

$y_A = 0 \quad \Theta_A = 0 \quad \psi_A = 0$
$y_B = 0 \quad \Theta_B = 0 \quad \psi_B = 0$

$$V_A = -\frac{T_o}{R}\frac{C_{a2}(C_4C_8 - C_5C_7) + C_{a5}(C_2C_7 - C_1C_8) + C_{a8}(C_1C_5 - C_2C_4)}{C_1(C_5C_9 - C_6C_8) + C_4(C_3C_8 - C_2C_9) + C_7(C_2C_6 - C_3C_5)}$$

$$M_A = -T_o\frac{C_{a2}(C_5C_9 - C_6C_8) + C_{a5}(C_3C_8 - C_2C_9) + C_{a8}(C_2C_6 - C_3C_5)}{C_1(C_5C_9 - C_6C_8) + C_4(C_3C_8 - C_2C_9) + C_7(C_2C_6 - C_3C_5)}$$

$$T_A = -T_o\frac{C_{a2}(C_6C_7 - C_4C_9) + C_{a5}(C_1C_9 - C_3C_7) + C_{a8}(C_3C_4 - C_1C_6)}{C_1(C_5C_9 - C_6C_8) + C_4(C_3C_8 - C_2C_9) + C_7(C_2C_6 - C_3C_5)}$$

$$V_B = V_A$$

$$M_B = V_A R \sin\phi + M_A\cos\phi - T_A\sin\phi - T_o\sin(\phi - \theta)$$

$$T_B = V_A R(1 - \cos\phi) + M_A\sin\phi + T_A\cos\phi + T_o\cos(\phi - \theta)$$

If $\beta = 1.3$ (solid or hollow round cross section, $\nu = 0.3$)

ϕ	45°	90°	180°	270°	360°	
θ	15°	30°	60°	90°	90°	180°
K_{VA}	0.1704	0.1705	0.1696	0.1625	0.1592	0.0000
K_{MA}	−0.2591	−0.4731	−0.6994	−0.7073	−0.7500	0.0000
K_{TA}	−0.6187	−0.4903	−0.1278	0.2211	0.1799	0.5000
K_{MB}	−0.1252	−0.2053	−0.1666	0.0586	0.2500	0.0000
K_{TB}	0.2953	0.1974	−0.0330	−0.1302	0.1799	−0.5000

3f. Right end supported and slope-guided, left end supported and slope-guided

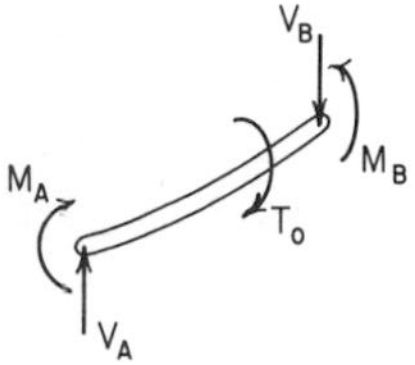

$T_A = 0 \quad y_A = 0 \quad \Theta_A = 0$
$T_B = 0 \quad y_B = 0 \quad \Theta_B = 0$

$$V_A = \frac{T_o}{R}\frac{[C_1\sin\phi - C_4(1 - \cos\phi)]\cos(\phi - \theta) - C_{a2}\sin^2\phi + C_{a5}(1 - \cos\phi)\sin\phi}{C_4(1 - \cos\phi)^2 + C_3\sin^2\phi - (C_1 + C_6)(1 - \cos\phi)\sin\phi}$$

$$M_A = -T_o\frac{[C_3\sin\phi - C_6(1 - \cos\phi)]\cos(\phi - \theta) - C_{a2}(1 - \cos\phi)\sin\phi + C_{a5}(1 - \cos\phi)^2}{C_4(1 - \cos\phi)^2 + C_3\sin^2\phi - (C_1 + C_6)(1 - \cos\phi)\sin\phi}$$

$$\psi_A = \frac{T_oR}{EI}\frac{(C_3C_4 - C_1C_6)\cos(\phi - \theta) + (C_{a2}C_6 - C_{a5}C_3)\sin\phi - (C_{a2}C_4 - C_{a5}C_1)(1 - \cos\phi)}{C_4(1 - \cos\phi)^2 + C_3\sin^2\phi - (C_1 + C_6)(1 - \cos\phi)\sin\phi}$$

$$V_B = V_A$$

$$M_B = V_A R\sin\phi + M_A\cos\phi - T_o\sin(\phi - \theta)$$

$$\psi_B = \psi_A\cos\phi + \frac{M_AR}{EI}C_7 + \frac{V_AR^2}{EI}C_9 + \frac{T_oR}{EI}C_{a8}$$

If $\beta = 1.3$ (solid or hollow round cross section, $\nu = 0.3$)

ϕ	45°		90°		180°	
θ	0°	15°	0°	30°	0°	60°
K_{VA}	1.0645	0.5147	0.8696	0.4252	0.5000	0.2500
K_{MA}	−1.4409	−1.4379	−0.8696	−0.9252	−0.3598	−0.7573
$K_{\psi A}$	1.6003	1.3211	1.2356	0.6889	1.4564	0.1746
K_{MB}	−0.9733	−1.1528	−0.1304	−0.4409	0.3598	−0.1088
$K_{\psi B}$	1.1213	1.1662	0.4208	0.4502	0.3500	−0.0034

TABLE 19 *Formulas for curved beams loaded normal to the plane of curvature* (*Cont.*)

End restraints, reference no.	Formulas for boundary values and selected numerical values
3g. Right end supported and slope-guided, left end supported and roll-guided. 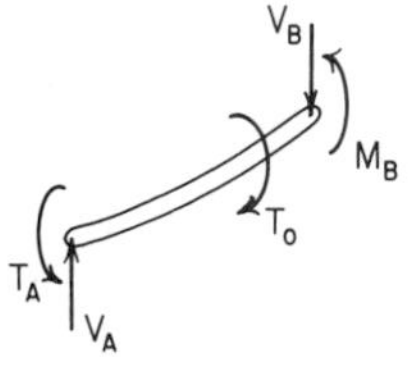 $M_A = 0 \quad y_A = 0 \quad \psi_A = 0$ $T_B = 0 \quad y_B = 0 \quad \Theta_B = 0$	$V_A = -\frac{T_o}{R}\frac{C_{a2}\cos^2\phi - C_{a5}\sin\phi\cos\phi + (C_5\sin\phi - C_2\cos\phi)\cos(\phi-\theta)}{(C_5\sin\phi - C_2\cos\phi)(1-\cos\phi) + C_3\cos^2\phi - C_6\sin\phi\cos\phi}$ $T_A = -T_o\frac{(C_{a5}\sin\phi - C_{a2}\cos\phi)(1-\cos\phi) + (C_3\cos\phi - C_6\sin\phi)\cos(\phi-\theta)}{(C_5\sin\phi - C_2\cos\phi)(1-\cos\phi) + C_3\cos^2\phi - C_6\sin\phi\cos\phi}$ $\Theta_A = -\frac{T_oR}{EI}\frac{(C_{a2}C_5 - C_{a5}C_2)(1-\cos\phi) + (C_{a5}C_3 - C_{a2}C_6)\cos\phi + (C_2C_6 - C_3C_5)\cos(\phi-\theta)}{(C_5\sin\phi - C_2\cos\phi)(1-\cos\phi) + C_3\cos^2\phi - C_6\sin\phi\cos\phi}$ $V_B = V_A$ $M_B = V_AR\sin\phi - T_A\sin\phi - T_o\sin(\phi-\theta)$ $\psi_B = -\Theta_A\sin\phi + \frac{T_AR}{EI}C_8 + \frac{V_AR^2}{EI}C_9 + \frac{T_oR}{EI}C_{a8}$
3h. Right end supported and slope-guided, left end simply supported 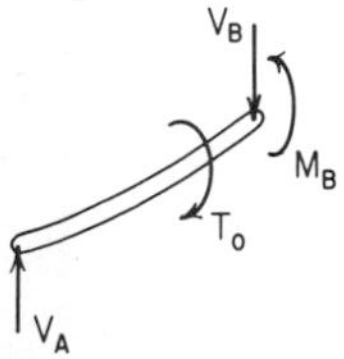 $M_A = 0 \quad T_A = 0 \quad y_A = 0$ $T_B = 0 \quad y_B = 0 \quad \Theta_B = 0$	$V_A = -\frac{T_o\cos(\phi-\theta)}{R(1-\cos\phi)}$ $\Theta_A = -\frac{T_oR}{EI}\left[\frac{C_{a2}\sin\phi + C_6\cos(\phi-\theta)}{1-\cos\phi} - \frac{C_3\sin\phi\cos(\phi-\theta)}{(1-\cos\phi)^2} - C_{a5}\right]$ $\psi_A = -\frac{T_oR}{EI}\left[\frac{C_{a5}\sin\phi - C_{a2}\cos\phi}{1-\cos\phi} + (C_3\cos\phi - C_6\sin\phi)\frac{\cos(\phi-\theta)}{(1-\cos\phi)^2}\right]$ $V_B = V_A$ $M_B = V_AR\sin\phi - T_o\sin(\phi-\theta)$ $\psi_B = \psi_A\cos\phi - \Theta_A\sin\phi + \frac{V_AR^2}{EI}C_9 + \frac{T_oR}{EI}C_{a8}$

3g. If $\beta = 1.3$ (solid or hollow round cross section, $\nu = 0.3$)

ϕ	45°		90°		180°	
θ	15°	30°	30°	60°	60°	120°
K_{VA}	−0.8503	−1.4915	−0.5000	−0.8660	−0.0512	−0.2859
K_{TA}	−0.8725	−0.7482	−0.7175	−0.4095	−0.6023	−0.0717
$K_{\Theta A}$	−0.0522	−0.0216	−0.2274	−0.0640	−1.9528	−0.2997
K_{MB}	−0.4843	−0.7844	−0.6485	−0.9566	−0.8660	−0.8660
$K_{\psi B}$	0.2386	0.5031	0.1780	0.5249	−0.9169	0.0416

3h. If $\beta = 1.3$ (solid or hollow round cross section, $\nu = 0.3$)

ϕ	45°	90°			180°		
θ	0°	0°	30°	60°	0°	60°	120°
K_{VA}	−2.4142	0.0000	−0.5000	−0.8660	0.5000	0.2500	−0.2500
$K_{\Theta A}$	−0.4161	−0.6564	−0.6984	−0.3328	−1.3000	−2.7359	−0.3929
$K_{\psi A}$	2.1998	1.8064	1.2961	0.7396	1.9242	1.1590	0.1380
K_{MB}	−2.4142	−1.0000	−1.3660	−1.3660	0.0000	−0.8660	−0.8660
$K_{\psi B}$	1.5263	0.5064	0.5413	0.7323	−0.1178	−0.9878	0.0332

3i. Right end supported and roll-guided, left end supported and roll-guided.

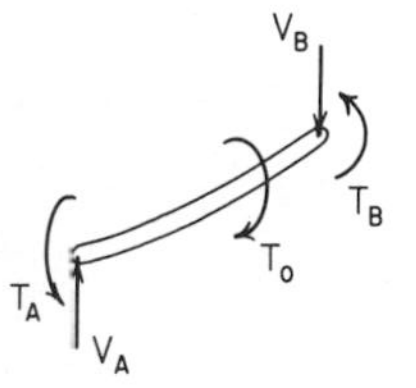

$M_A = 0 \quad y_A = 0 \quad \psi_A = 0$
$M_B = 0 \quad y_B = 0 \quad \psi_B = 0$

$$V_A = 0$$

$$T_A = -T_o \frac{(C_{a2} + C_{a8}) \sin^2 \phi - (C_3 + C_9) \sin (\phi - \theta) \sin \phi}{(C_2 + C_3 + C_8 + C_9) \sin^2 \phi}$$

$$\Theta_A = -\frac{T_o R}{EI} \frac{[C_{a2}(C_8 + C_9) - C_{a8}(C_2 + C_3)] \sin \phi - (C_2 C_9 - C_3 C_8) \sin (\phi - \theta)}{(C_2 + C_3 + C_8 + C_9) \sin^2 \phi}$$

$$V_B = 0$$

$$T_B = V_A R(1 - \cos \phi) + T_A \cos \phi + T_o \cos (\phi - \theta)$$

$$\Theta_B = \Theta_A \cos \phi + \frac{T_A R}{EI} C_5 + \frac{V_A R^2}{EI} C_6 + \frac{T_o R}{EI} C_{a5}$$

If $\beta = 1.3$ (solid or hollow round cross section, $\nu = 0.3$)

ϕ	45°	90°	270°
θ	15°	30°	90°
K_{VA}	0.0000	0.0000	0.0000
K_{TA}	−0.7071	−0.8660	0.0000
$K_{\Theta A}$	−0.0988	−0.6021	−3.6128
K_{TB}	0.3660	0.5000	−1.0000
$K_{\Theta B}$	0.0807	0.5215	0.0000

3j. Right end supported and roll-guided, left end simply supported

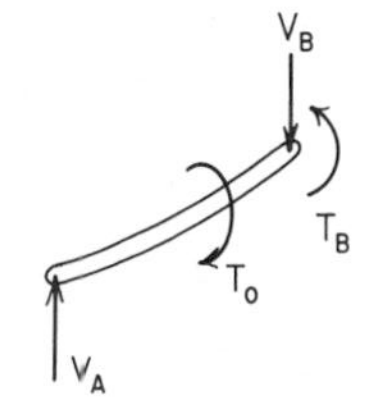

$M_A = 0 \quad T_A = 0 \quad y_A = 0$
$M_B = 0 \quad y_B = 0 \quad \psi_B = 0$

$$V_A = \frac{T_o \sin (\phi - \theta)}{R \sin \phi}$$

$$\Theta_A = -\frac{T_o R}{EI} \left\{ \frac{C_{a2} \cos \phi - C_{a8}(1 - \cos \phi)}{\sin \phi} + [C_3 \cos \phi - C_9(1 - \cos \phi)] \frac{\sin (\phi - \theta)}{\sin^2 \phi} \right\}$$

$$\psi_A = -\frac{T_o R}{EI} \left[C_{a2} + C_{a8} + (C_3 + C_9) \frac{\sin (\phi - \theta)}{\sin \phi} \right]$$

$$V_B = V_A$$

$$T_B = V_A R(1 - \cos \phi) + T_o \cos (\phi - \theta)$$

$$\Theta_B = \Theta_A \cos \phi + \psi_A \sin \phi + \frac{V_A R^2}{EI} C_6 + \frac{T_o R}{EI} C_{a5}$$

If $\beta = 1.3$ (solid or hollow round cross section, $\nu = 0.3$)

ϕ	45°			90°		
θ	0°	15°	30°	0°	30°	60°
K_{VA}	1.0000	0.7071	0.3660	1.0000	0.8660	0.5000
$K_{\Theta A}$	−0.2790	−0.2961	−0.1828	−1.3000	−1.7280	−1.1715
$K_{\psi A}$	1.0210	0.7220	0.3737	2.0420	1.7685	1.0210
K_{TB}	1.0000	1.0731	1.0731	1.0000	1.3660	1.3660
$K_{\Theta B}$	0.1439	0.1825	0.1515	0.7420	1.1641	0.9732

TABLE 19 *Formulas for curved beams loaded normal to the plane of curvature* (*Cont.*)

4. Uniformly distributed lateral load

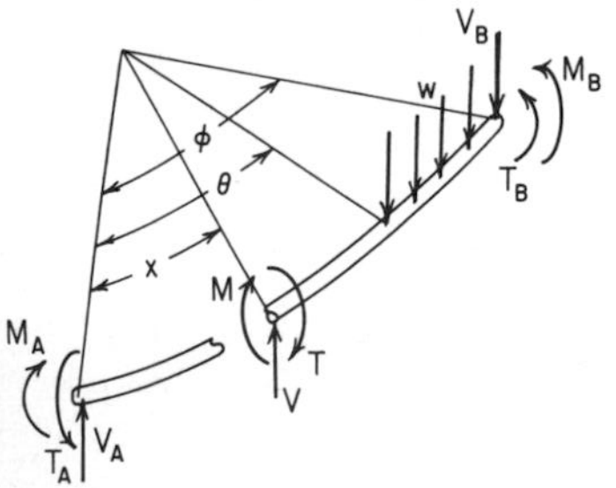

Transverse shear $= V = V_A - wR\langle x-\theta\rangle^1$

Bending moment $= M = V_A R \sin x + M_A \cos x - T_A \sin x - wR^2[1-\cos(x-\theta)]\langle x-\theta\rangle^0$

Twisting moment $= T = V_A R(1-\cos x) + M_A \sin x + T_A \cos x - wR^2[x-\theta-\sin(x-\theta)]\langle x-\theta\rangle^0$

Vertical deflection $= y = y_A + \Theta_A R \sin x + \psi_A R(1-\cos x) + \dfrac{M_A R^2}{EI}F_1 + \dfrac{T_A R^2}{EI}F_2 + \dfrac{V_A R^3}{EI}F_3 - \dfrac{wR^4}{EI}F_{a13}$

Bending slope $= \Theta = \Theta_A \cos x + \psi_A \sin x + \dfrac{M_A R}{EI}F_4 + \dfrac{T_A R}{EI}F_5 + \dfrac{V_A R^2}{EI}F_6 - \dfrac{wR^3}{EI}F_{a16}$

Roll slope $= \psi = \psi_A \cos x - \Theta_A \sin x + \dfrac{M_A R}{EI}F_7 + \dfrac{T_A R}{EI}F_8 + \dfrac{V_A R^2}{EI}F_9 - \dfrac{wR^3}{EI}F_{a19}$

For tabulated values: $V = K_V wR \qquad M = K_M wR^2 \qquad T = K_T wR^2 \qquad y = K_y \dfrac{wR^4}{EI} \qquad \Theta = K_\Theta \dfrac{wR^3}{EI} \qquad \psi = K_\psi \dfrac{wR^3}{EI}$

End restraints, reference no.	Formulas for boundary values and selected numerical values

4a. Right end fixed, left end free

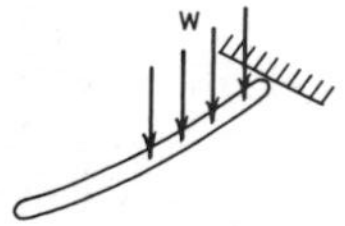

$V_A = 0 \qquad M_A = 0 \qquad T_A = 0$

$y_B = 0 \qquad \Theta_B = 0 \qquad \psi_B = 0$

$$y_A = -\frac{wR^4}{EI}[C_{a16}\sin\phi - C_{a19}(1-\cos\phi) - C_{a13}]$$

$$\Theta_A = \frac{wR^3}{EI}(C_{a16}\cos\phi - C_{a19}\sin\phi)$$

$$\psi_A = \frac{wR^3}{EI}(C_{a19}\cos\phi + C_{a16}\sin\phi)$$

$$V_B = -wR(\phi-\theta)$$

$$M_B = -wR^2[1-\cos(\phi-\theta)]$$

$$T_B = -wR^2[\phi-\theta-\sin(\phi-\theta)]$$

If $\beta = 1.3$ (solid or hollow round cross section, $\nu = 0.3$)

ϕ	45°	90°			180°		
θ	0°	0°	30°	60°	0°	60°	120°
K_{yA}	−0.0469	−0.7118	−0.2211	−0.0269	−8.4152	−2.2654	−0.1699
$K_{\Theta A}$	0.0762	0.4936	0.1071	0.0071	0.4712	−0.6033	−0.1583
$K_{\psi A}$	0.0267	0.4080	0.1583	0.0229	4.6000	1.3641	0.1071
K_{MB}	−0.2929	−1.0000	−0.5000	−0.1340	−2.0000	−1.5000	−0.5000
K_{TB}	−0.0783	−0.5708	−0.1812	−0.0236	−3.1416	−1.2284	−0.1812

4b. Right end fixed, left end simply supported

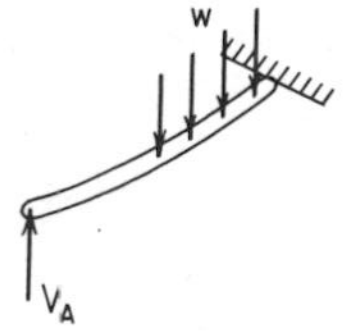

$M_A = 0 \quad T_A = 0 \quad y_A = 0$

$y_B = 0 \quad \Theta_B = 0 \quad \psi_B = 0$

$$V_A = wR\frac{C_{a19}(1-\cos\phi) - C_{a16}\sin\phi + C_{a13}}{C_9(1-\cos\phi) - C_6\sin\phi + C_3}$$

$$\Theta_A = \frac{wR^3}{EI}\,\frac{(C_{a13}C_9 - C_{a19}C_3)\sin\phi + (C_{a19}C_6 - C_{a16}C_9)(1-\cos\phi) + (C_{a16}C_3 - C_{a13}C_6)\cos\phi}{C_9(1-\cos\phi) - C_6\sin\phi + C_3}$$

$$\psi_A = \frac{wR^3}{EI}\,\frac{[C_{a16}(C_3+C_9) - C_6(C_{a13}+C_{a19})]\sin\phi + (C_{a19}C_3 - C_{a13}C_9)\cos\phi}{C_9(1-\cos\phi) - C_6\sin\phi + C_3}$$

$$V_B = V_A - wR(\phi - \theta)$$

$$M_B = V_A R\sin\phi - wR^2[1-\cos(\phi-\theta)]$$

$$T_B = V_A R(1-\cos\phi) - wR^2[\phi - \theta - \sin(\phi-\theta)]$$

If $\beta = 1.3$ (solid or hollow round cross section, $\nu = 0.3$)

ϕ	45°	90°			180°		
θ	0°	0°	30°	60°	0°	60°	120°
K_{VA}	0.2916	0.5701	0.1771	0.0215	1.0933	0.2943	0.0221
$K_{\Theta A}$	−0.1300	−0.1621	−0.0966	−0.0177	−2.3714	−1.3686	−0.2156
$K_{\psi A}$	0.0095	0.1192	0.0686	0.0119	0.6500	0.3008	0.0273
K_{MB}	−0.0867	−0.4299	−0.3229	−0.1124	−2.0000	−1.5000	−0.5000
K_{TB}	0.0071	−0.0007	−0.0041	−0.0021	−0.9549	−0.6397	−0.1370

4c. Right end fixed, left end supported and slope-guided

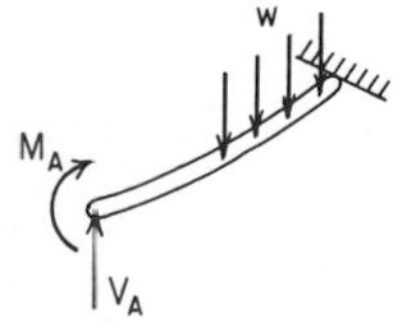

$T_A = 0 \quad y_A = 0 \quad \Theta_A = 0$

$y_B = 0 \quad \Theta_B = 0 \quad \psi_B = 0$

$$V_A = wR\frac{(C_{a19}C_4 - C_{a16}C_7)(1-\cos\phi) + (C_{a16}C_1 - C_{a13}C_4)\cos\phi + (C_{a13}C_7 - C_{a19}C_1)\sin\phi}{(C_4C_9 - C_6C_7)(1-\cos\phi) + (C_1C_6 - C_3C_4)\cos\phi + (C_3C_7 - C_1C_9)\sin\phi}$$

$$M_A = wR^2\frac{(C_{a16}C_9 - C_{a19}C_6)(1-\cos\phi) + (C_{a13}C_6 - C_{a16}C_3)\cos\phi + (C_{a19}C_3 - C_{a13}C_9)\sin\phi}{(C_4C_9 - C_6C_7)(1-\cos\phi) + (C_1C_6 - C_3C_4)\cos\phi + (C_3C_7 - C_1C_9)\sin\phi}$$

$$\psi_A = \frac{wR^3}{EI}\,\frac{C_{a13}(C_4C_9 - C_6C_7) + C_{a16}(C_3C_7 - C_1C_9) + C_{a19}(C_1C_6 - C_3C_4)}{(C_4C_9 - C_6C_7)(1-\cos\phi) + (C_1C_6 - C_3C_4)\cos\phi + (C_3C_7 - C_1C_9)\sin\phi}$$

$$V_B = V_A - wR(\phi - \theta)$$

$$M_B = V_A R\sin\phi + M_A\cos\phi - wR^2[1-\cos(\phi-\theta)]$$

$$T_B = V_A R(1-\cos\phi) + M_A\sin\phi - wR^2[\phi - \theta - \sin(\phi-\theta)]$$

If $\beta = 1.3$ (solid or hollow round cross section, $\nu = 0.3$)

ϕ	45°	90°			180°		
θ	0°	0°	30°	60°	0°	60°	120°
K_{VA}	0.3919	0.7700	0.2961	0.0434	1.3863	0.4634	0.0487
K_{MA}	−0.0527	−0.2169	−0.1293	−0.0237	−0.8672	−0.5005	−0.0789
$K_{\psi A}$	−0.0004	−0.0145	−0.0111	−0.0027	−0.4084	−0.3100	−0.0689
K_{MB}	−0.0531	−0.2301	−0.2039	−0.0906	−1.1328	−0.9995	−0.4211
K_{TB}	−0.0008	−0.0178	−0.0143	−0.0039	−0.3691	−0.3016	−0.0838

TABLE 19 *Formulas for curved beams loaded normal to the plane of curvature* (Cont.)

End restraints, reference no.	Formulas for boundary values and selected numerical values

4d. Right end fixed, left end supported and roll-guided

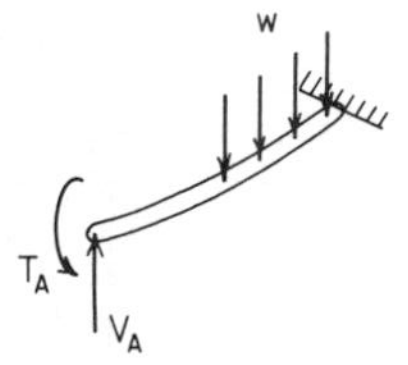

$M_A = 0 \quad y_A = 0 \quad \psi_A = 0$

$y_B = 0 \quad \Theta_B = 0 \quad \psi_B = 0$

$$V_A = wR\frac{[(C_{a13} + C_{a19})C_5 - C_{a16}(C_2 + C_8)] \sin\phi + (C_{a13}C_8 - C_{a19}C_2)\cos\phi}{[C_5(C_3 + C_9) - C_6(C_2 + C_8)]\sin\phi + (C_3C_8 - C_2C_9)\cos\phi}$$

$$T_A = wR^2\frac{[C_{a16}(C_3 + C_9) - C_6(C_{a13} + C_{a19})]\sin\phi + (C_{a19}C_3 - C_{a13}C_9)\cos\phi}{[C_5(C_3 + C_9) - C_6(C_2 + C_8)]\sin\phi + (C_3C_8 - C_2C_9)\cos\phi}$$

$$\Theta_A = \frac{wR^3}{EI}\,\frac{C_{a13}(C_5C_9 - C_6C_8) + C_{a16}(C_3C_8 - C_2C_9) + C_{a19}(C_2C_6 - C_3C_5)}{[C_5(C_3 + C_9) - C_6(C_2 + C_8)]\sin\phi + (C_3C_8 - C_2C_9)\cos\phi}$$

$V_B = V_A - wR(\phi - \theta)$

$M_B = V_AR\sin\phi - T_A\sin\phi - wR^2[1 - \cos(\phi - \theta)]$

$T_B = V_AR(1 - \cos\phi) + T_A\cos\phi - wR^2[\phi - \theta - \sin(\phi - \theta)]$

If $\beta = 1.3$ (solid or hollow round cross section, $\nu = 0.3$)

ϕ	45°	90°			180°		
θ	0°	0°	30°	60°	0°	60°	120°
K_{VA}	0.2880	0.5399	0.1597	0.0185	0.9342	0.2207	0.0154
K_{TA}	−0.0099	−0.0745	−0.0428	−0.0075	−0.3391	−0.1569	−0.0143
$K_{\Theta A}$	−0.0111	−0.1161	−0.0702	−0.0131	−1.9576	−1.1771	−0.1983
K_{MB}	−0.0822	−0.3856	−0.2975	−0.1080	−2.0000	−1.5000	−0.5000
K_{TB}	−0.0010	−0.0309	−0.0215	−0.0051	−0.9342	−0.6301	−0.1362

4e. Right end fixed, left end fixed

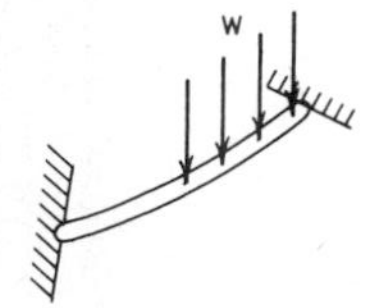

$y_A = 0 \quad \Theta_A = 0 \quad \psi_A = 0$

$y_B = 0 \quad \Theta_B = 0 \quad \psi_B = 0$

$$V_A = wR\frac{C_{a13}(C_4C_8 - C_5C_7) + C_{a16}(C_2C_7 - C_1C_8) + C_{a19}(C_1C_5 - C_2C_4)}{C_1(C_5C_9 - C_6C_8) + C_4(C_3C_8 - C_2C_9) + C_7(C_2C_6 - C_3C_5)}$$

$$M_A = wR^2\frac{C_{a13}(C_5C_9 - C_6C_8) + C_{a16}(C_3C_8 - C_2C_9) + C_{a19}(C_2C_6 - C_3C_5)}{C_1(C_5C_9 - C_6C_8) + C_4(C_3C_8 - C_2C_9) + C_7(C_2C_6 - C_3C_5)}$$

$$T_A = wR^2\frac{C_{a13}(C_6C_7 - C_4C_9) + C_{a16}(C_1C_9 - C_3C_7) + C_{a19}(C_3C_4 - C_1C_6)}{C_1(C_5C_9 - C_6C_8) + C_4(C_3C_8 - C_2C_9) + C_7(C_2C_6 - C_3C_5)}$$

$V_B = V_A - wR(\phi - \theta)$

$M_B = V_AR\sin\phi + M_A\cos\phi - T_A\sin\phi - wR^2[1 - \cos(\phi - \theta)]$

$T_B = V_AR(1 - \cos\phi) + M_A\sin\phi + T_A\cos\phi - wR^2[\phi - \theta - \sin(\phi - \theta)]$

If $\beta = 1.3$ (solid or hollow round cross section, $\nu = 0.3$)

ϕ	45°		90°		180°		360°
θ	0°	15°	0°	30°	0°	60°	0°
K_{VA}	0.3927	0.1548	0.7854	0.3080	1.5708	0.6034	3.1416
K_{MA}	−0.0531	−0.0316	−0.2279	−0.1376	−1.0000	−0.6013	−2.1304
K_{TA}	0.0005	0.0004	0.0133	0.0102	0.2976	0.2259	3.1416
K_{MB}	−0.0531	−0.0471	−0.2279	−0.2022	−1.0000	−0.8987	−2.1304
K_{TB}	−0.0005	−0.0004	−0.0133	−0.0108	−0.2976	−0.2473	−3.1416

4f. Right end supported and slope-guided, left end supported and slope-guided

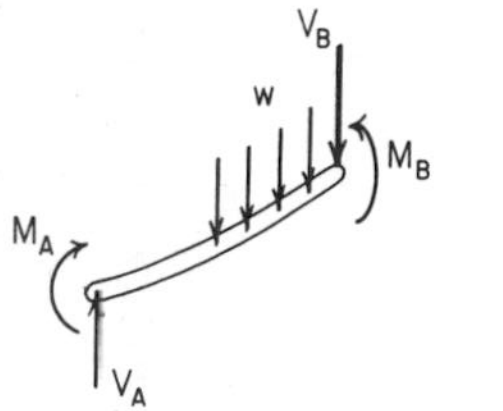

$T_A = 0 \quad y_A = 0 \quad \Theta_A = 0$
$T_B = 0 \quad y_B = 0 \quad \Theta_B = 0$

$$V_A = wR\frac{[C_4(1-\cos\phi) - C_1\sin\phi][\phi-\theta-\sin(\phi-\theta)] + C_{a13}\sin^2\phi - C_{a16}\sin\phi(1-\cos\phi)}{C_4(1-\cos\phi)^2 + C_3\sin^2\phi - (C_1+C_6)(1-\cos\phi)\sin\phi}$$

$$M_A = wR^2\frac{[C_3\sin\phi - C_6(1-\cos\phi)][\phi-\theta-\sin(\phi-\theta)] + C_{a16}(1-\cos\phi)^2 - C_{a13}\sin\phi(1-\cos\phi)}{C_4(1-\cos\phi)^2 + C_3\sin^2\phi - (C_1+C_6)(1-\cos\phi)\sin\phi}$$

$$\psi_A = \frac{wR^3}{EI}\,\frac{(C_{a13}C_4 - C_{a16}C_1)(1-\cos\phi) - (C_{a13}C_6 - C_{a16}C_3)\sin\phi - (C_3C_4 - C_1C_6)[\phi-\theta-\sin(\phi-\theta)]}{C_4(1-\cos\phi)^2 + C_3\sin^2\phi - (C_1+C_6)(1-\cos\phi)\sin\phi}$$

$$V_B = V_A - wR(\phi-\theta)$$

$$M_B = V_AR\sin\phi + M_A\cos\phi - wR^2[1-\cos(\phi-\theta)]$$

$$\psi_B = \psi_A\cos\phi + \frac{M_AR}{EI}C_7 + \frac{V_AR^2}{EI}C_9 - \frac{wR^3}{EI}C_{a19}$$

If $\beta = 1.3$ (solid or hollow round cross section, $\nu = 0.3$)

ϕ	45°		90°		180°	
θ	0°	15°	0°	30°	0°	60°
K_{VA}	0.3927	0.1549	0.7854	0.3086	1.5708	0.6142
K_{MA}	−0.0519	−0.0308	−0.2146	−0.1274	−1.0000	−0.6090
$K_{\psi A}$	−0.0013	−0.0010	−0.0220	−0.0171	−0.5375	−0.4155
K_{MB}	−0.0519	−0.0462	−0.2146	−0.1914	−1.0000	−0.8910
$K_{\psi B}$	−0.0013	−0.0010	−0.0220	−0.0177	−0.5375	−0.4393

4g. Right end supported and slope-guided, left end supported and roll-guided

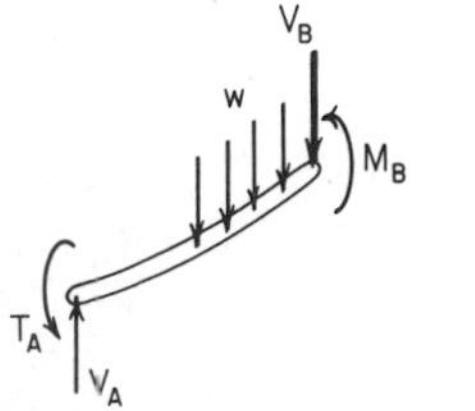

$M_A = 0 \quad y_A = 0 \quad \psi_A = 0$
$T_B = 0 \quad y_B = 0 \quad \Theta_B = 0$

$$V_A = wR\frac{(C_5\sin\phi - C_2\cos\phi)[\phi-\theta-\sin(\phi-\theta)] + C_{a13}\cos^2\phi - C_{a16}\sin\phi\cos\phi}{(C_5\sin\phi - C_2\cos\phi)(1-\cos\phi) + C_3\cos^2\phi - C_6\sin\phi\cos\phi}$$

$$T_A = wR\frac{(C_3\cos\phi - C_6\sin\phi)[\phi-\theta-\sin(\phi-\theta)] - (C_{a13}\cos\phi - C_{a16}\sin\phi)(1-\cos\phi)}{(C_5\sin\phi - C_2\cos\phi)(1-\cos\phi) + C_3\cos^2\phi - C_6\sin\phi\cos\phi}$$

$$\Theta_A = \frac{wR^3}{EI}\,\frac{(C_2C_6 - C_3C_5)[\phi-\theta-\sin(\phi-\theta)] + (C_{a13}C_5 - C_{a16}C_2)(1-\cos\phi) + (C_{a16}C_3 - C_{a13}C_6)\cos\phi}{(C_5\sin\phi - C_2\cos\phi)(1-\cos\phi) + C_3\cos^2\phi - C_6\sin\phi\cos\phi}$$

$$V_B = V_A - wR(\phi-\theta)$$

$$M_B = V_AR\sin\phi - T_A\sin\phi - wR^2[1-\cos(\phi-\theta)]$$

$$\psi_B = -\Theta_A\sin\phi + \frac{T_AR}{EI}C_8 + \frac{V_AR^2}{EI}C_9 - \frac{wR^3}{EI}C_{a19}$$

If $\beta = 1.3$ (solid or hollow round cross section, $\nu = 0.3$)

ϕ	45°	90°			180°		
θ	0°	0°	30°	60°	0°	60°	120°
K_{VA}	0.2896	0.5708	0.1812	0.0236	1.3727	0.5164	0.0793
K_{TA}	−0.0093	−0.0558	−0.0368	−0.0060	−0.3963	−0.1955	−0.0226
$K_{\Theta A}$	−0.0111	−0.1188	−0.0720	−0.0135	−3.0977	−1.9461	−0.3644
K_{MB}	−0.0815	−0.3634	−0.2820	−0.1043	−2.0000	−1.5000	−0.5000
$K_{\psi B}$	−0.0008	−0.0342	−0.0238	−0.0056	−1.7908	−1.2080	−0.2610

TABLE 19 *Formulas for curved beams loaded normal to the plane of curvature (Cont.)*

End restraints, reference no.	Formulas for boundary values and selected numerical values
4h. Right end supported and slope-guided, left end simply supported	

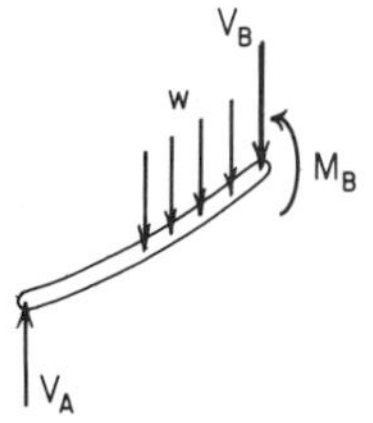

$M_A = 0 \quad T_A = 0 \quad y_A = 0$
$T_B = 0 \quad y_B = 0 \quad \Theta_B = 0$

$$V_A = wR\frac{\phi - \theta - \sin(\phi - \theta)}{1 - \cos\phi}$$

$$\Theta_A = \frac{wR^3}{EI}\left[\frac{C_{a13}\sin\phi + C_6[\phi - \theta - \sin(\phi - \theta)]}{1 - \cos\phi} - \frac{C_3\sin\phi[\phi - \theta - \sin(\phi - \theta)]}{(1 - \cos\phi)^2} - C_{a16}\right]$$

$$\psi_A = \frac{wR^3}{EI}\left\{\frac{C_{a16}\sin\phi - C_{a13}\cos\phi}{1 - \cos\phi} - (C_6\sin\phi - C_3\cos\phi)\frac{\phi - \theta - \sin(\phi - \theta)}{(1 - \cos\phi)^2}\right\}$$

$$V_B = V_A - wR(\phi - \theta)$$

$$M_B = V_AR\sin\phi - wR^2[1 - \cos(\phi - \theta)]$$

$$\psi_B = \psi_A\cos\phi - \Theta_A\sin\phi + \frac{V_AR^2}{EI}C_9 - \frac{wR^3}{EI}C_{a19}$$

If $\beta = 1.3$ (solid or hollow round cross section, $\nu = 0.3$)

ϕ	45°	90°			180°		
θ	0°	0°	30°	60°	0°	60°	120°
K_{VA}	0.2673	0.5708	0.1812	0.0236	1.5708	0.6142	0.0906
$K_{\Theta A}$	−0.0150	−0.1620	−0.0962	−0.0175	−3.6128	−2.2002	−0.3938
$K_{\psi A}$	0.0204	0.1189	0.0665	0.0109	0.7625	0.3762	0.0435
K_{MB}	−0.1039	−0.4292	−0.3188	−0.1104	−2.0000	−1.5000	−0.5000
$K_{\psi B}$	−0.0133	−0.0008	−0.0051	−0.0026	−1.8375	−1.2310	−0.2637

4i. Right end supported and roll-guided, left end supported and roll-guided

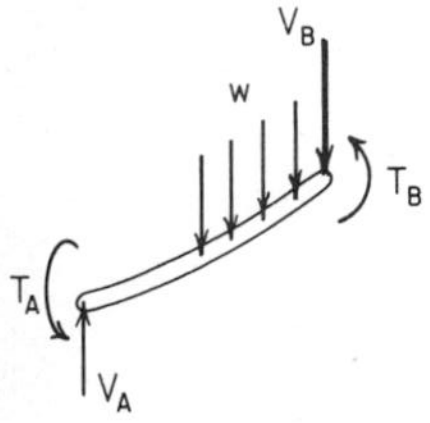

$M_A = 0 \quad y_A = 0 \quad \psi_A = 0$
$M_B = 0 \quad y_B = 0 \quad \psi_B = 0$

$$V_A = wR\frac{(C_{a13} + C_{a19})\sin\phi + (C_2 + C_8)[1 - \cos(\phi - \theta)]}{(C_2 + C_3 + C_8 + C_9)\sin\phi}$$

$$T_A = wR^2\frac{(C_{a13} + C_{a19})\sin\phi - (C_3 + C_9)[1 - \cos(\phi - \theta)]}{(C_2 + C_3 + C_8 + C_9)\sin\phi}$$

$$\Theta_A = \frac{wR^3}{EI}\frac{[C_{a13}(C_8 + C_9) - C_{a19}(C_2 + C_3)]\sin\phi + (C_2C_9 - C_3C_8)[1 - \cos(\phi - \theta)]}{(C_2 + C_3 + C_8 + C_9)\sin\phi}$$

$$V_B = V_A - wR(\phi - \theta)$$

$$T_B = V_AR(1 - \cos\phi) + T_A\cos\phi - wR^2[\phi - \theta - \sin(\phi - \theta)]$$

$$\Theta_B = \Theta_A\cos\phi + \frac{T_AR}{EI}C_5 + \frac{V_AR^2}{EI}C_6 - \frac{wR^3}{EI}C_{a16}$$

If $\beta = 1.3$ (solid or hollow round cross section, $\nu = 0.3$)

ϕ	45°		90°		270°	
θ	0°	15°	0°	30°	0°	90°
K_{VA}	0.3927	0.1745	0.7854	0.3491	2.3562	1.0472
K_{TA}	−0.0215	−0.0149	−0.2146	−0.1509	3.3562	3.0472
$K_{\Theta A}$	−0.0248	−0.0173	−0.3774	−0.2717	−10.9323	−6.2614
K_{TB}	0.0215	0.0170	0.2146	0.1679	−3.3562	−2.0944
$K_{\Theta B}$	0.0248	0.0194	0.3774	0.2912	10.9323	9.9484

4j. Right end supported and roll-guided, left end simply supported

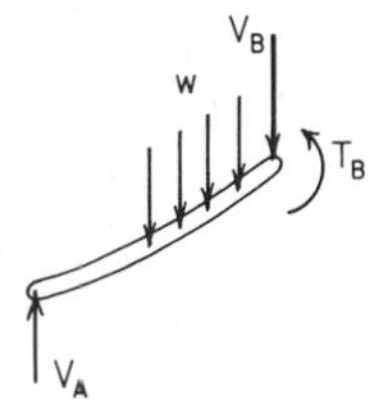

$M_A = 0 \quad T_A = 0 \quad y_A = 0$

$M_B = 0 \quad y_B = 0 \quad \psi_B = 0$

$$V_A = wR\frac{1-\cos(\phi-\theta)}{\sin\phi}$$

$$\Theta_A = \frac{wR^3}{EI}\left[\frac{C_{a13}\cos\phi - C_{a19}(1-\cos\phi)}{\sin\phi} - [C_3\cos\phi - C_9(1-\cos\phi)]\frac{1-\cos(\phi-\theta)}{\sin^2\phi}\right]$$

$$\psi_A = \frac{wR^3}{EI}\left[C_{a13} + C_{a19} - (C_3 + C_9)\frac{1-\cos(\phi-\theta)}{\sin\phi}\right]$$

$$V_B = V_A - wR(\phi-\theta)$$

$$T_B = V_AR(1-\cos\phi) - wR^2[\phi-\theta-\sin(\phi-\theta)]$$

$$\Theta_B = \Theta_A\cos\phi + \psi_A\sin\phi + \frac{V_AR^2}{EI}C_6 - \frac{wR^3}{EI}C_{a16}$$

If $\beta = 1.3$ (solid or hollow round cross section, $\nu = 0.3$)

ϕ	45°			90°		
θ	0°	15°	30°	0°	30°	60°
K_{VA}	0.4142	0.1895	0.0482	1.0000	0.5000	0.1340
$K_{\Theta A}$	−0.0308	−0.0215	−0.0066	−0.6564	−0.4679	−0.1470
$K_{\psi A}$	0.0220	0.0153	0.0047	0.4382	0.3082	0.0954
K_{TB}	0.0430	0.0319	0.0111	0.4292	0.3188	0.1104
$K_{\Theta B}$	0.0279	0.0216	0.0081	0.5367	0.4032	0.1404

5. Uniformly distributed torque

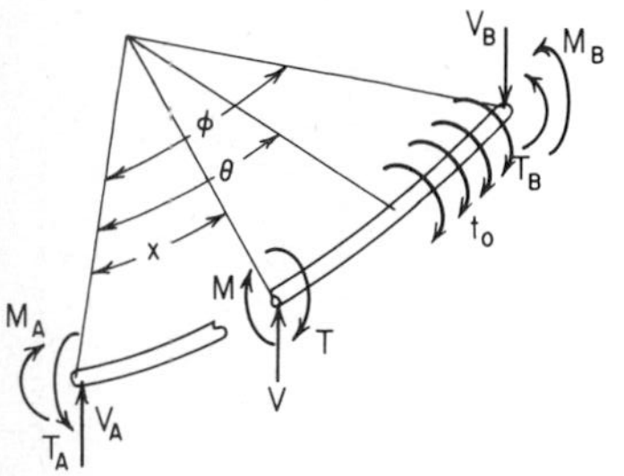

Transverse shear $= V = V_A$

Bending moment $= M = V_AR\sin x + M_A\cos x - T_A\sin x - t_oR[1-\cos(x-\theta)]\langle x-\theta\rangle^0$

Twisting moment $= T = V_AR(1-\cos x) + M_A\sin x + T_A\cos x + t_oR\sin(x-\theta)\langle x-\theta\rangle^0$

Vertical deflection $= y = y_A + \Theta_AR\sin x + \psi_AR(1-\cos x) + \frac{M_AR^2}{EI}F_1 + \frac{T_AR^2}{EI}F_2 + \frac{V_AR^3}{EI}F_3 + \frac{t_oR^3}{EI}F_{a12}$

Bending slope $= \Theta = \Theta_A\cos x + \psi_A\sin x + \frac{M_AR}{EI}F_4 + \frac{T_AR}{EI}F_5 + \frac{V_AR^2}{EI}F_6 + \frac{t_oR^2}{EI}F_{a15}$

Roll slope $= \psi = \psi_A\cos x - \Theta_A\sin x + \frac{M_AR}{EI}F_7 + \frac{T_AR}{EI}F_8 + \frac{V_AR^2}{EI}F_9 + \frac{t_oR^2}{EI}F_{a18}$

For tabulated values: $V = K_Vt_o \qquad M = K_Mt_oR \qquad T = K_Tt_oR \qquad y = K_y\frac{t_oR^3}{EI} \qquad \Theta = K_\Theta\frac{t_oR^2}{EI} \qquad \psi = K_\psi\frac{t_oR^2}{EI}$

TABLE 19 *Formulas for curved beams loaded normal to the plane of curvature* (*Cont.*)

End restraints, reference no.	Formulas for boundary values and selected numerical values
5a. Right end fixed, left end free 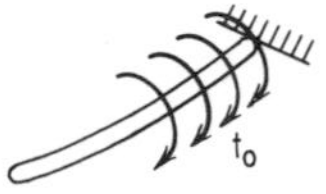 $V_A = 0 \quad M_A = 0 \quad T_A = 0$ $y_B = 0 \quad \Theta_B = 0 \quad \psi_B = 0$	$y_A = \frac{t_o R^3}{EI}[C_{a15} \sin \phi - C_{a18}(1 - \cos \phi) - C_{a12}]$ $\Theta_A = -\frac{t_o R^2}{EI}(C_{a15} \cos \phi - C_{a18} \sin \phi)$ $\psi_A = -\frac{t_o R^2}{EI}(C_{a18} \cos \phi + C_{a15} \sin \phi)$ $V_B = 0$ $M_B = -t_o R[1 - \cos(\phi - \theta)]$ $T_B = t_o R \sin(\phi - \theta)$
5b. Right end fixed, left end simply supported 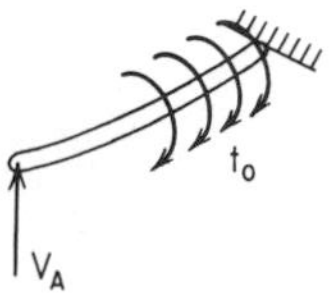 $M_A = 0 \quad T_A = 0 \quad y_A = 0$ $y_B = 0 \quad \Theta_B = 0 \quad \psi_B = 0$	$V_A = -t_o \frac{C_{a18}(1 - \cos \phi) - C_{a15} \sin \phi + C_{a12}}{C_9(1 - \cos \phi) - C_6 \sin \phi + C_3}$ $\Theta_A = -\frac{t_o R^2}{EI} \frac{(C_{a12}C_9 - C_{a18}C_3) \sin \phi + (C_{a18}C_6 - C_{a15}C_9)(1 - \cos \phi) + (C_{a15}C_3 - C_{a12}C_6) \cos \phi}{C_9(1 - \cos \phi) - C_6 \sin \phi + C_3}$ $\psi_A = -\frac{t_o R^2}{EI} \frac{[C_{a15}(C_9 + C_3) - C_6(C_{a12} + C_{a18})] \sin \phi + (C_{a18}C_3 - C_{a12}C_9) \cos \phi}{C_9(1 - \cos \phi) - C_6 \sin \phi + C_3}$ $V_B = V_A$ $M_B = V_A R \sin \phi - t_o R[1 - \cos(\phi - \theta)]$ $T_B = V_A R(1 - \cos \phi) + t_o R \sin(\phi - \theta)$

5a. If $\beta = 1.3$ (solid or hollow round cross section, $\nu = 0.3$)

ϕ	45°	90°			180°		
θ	0°	0°	30°	60°	0°	60°	120°
K_{yA}	0.0129	0.1500	0.2562	0.1206	0.6000	2.5359	1.1929
$K_{\Theta A}$	−0.1211	−0.8064	−0.5429	−0.1671	−3.6128	−2.2002	−0.3938
$K_{\psi A}$	0.3679	1.1500	0.3938	0.0535	2.0000	−0.5859	−0.5429
K_{MB}	−0.2929	−1.0000	−0.5000	−0.1340	−2.0000	−1.5000	−0.5000
K_{TB}	0.7071	1.0000	0.8660	0.5000	0.0000	0.8660	0.8660

5b. If $\beta = 1.3$ (solid or hollow round cross section, $\nu = 0.3$)

ϕ	45°	90°			180°		
θ	0°	0°	30°	60°	0°	60°	120°
K_{VA}	−0.0801	−0.1201	−0.2052	−0.0966	−0.0780	−0.3295	−0.1550
$K_{\Theta A}$	−0.0966	−0.6682	−0.3069	−0.0560	−3.4102	−1.3436	0.0092
$K_{\psi A}$	0.3726	1.2108	0.4977	0.1025	2.2816	0.6044	0.0170
K_{MB}	−0.3495	−1.1201	−0.7052	−0.2306	−2.0000	−1.5000	−0.5000
K_{TB}	0.6837	0.8799	0.6608	0.4034	−0.1559	0.2071	0.5560

5c. Right end fixed, left end supported and slope-guided

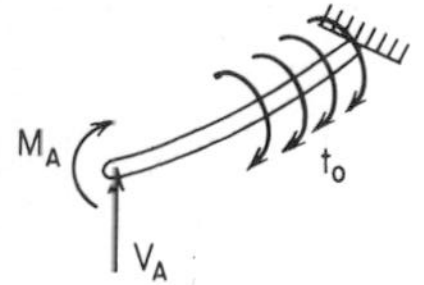

$T_A = 0 \quad y_A = 0 \quad \Theta_A = 0$

$y_B = 0 \quad \Theta_B = 0 \quad \psi_B = 0$

$$V_A = -t_o \frac{(C_{a18}C_4 - C_{a15}C_7)(1 - \cos\phi) + (C_{a15}C_1 - C_{a12}C_4)\cos\phi + (C_{a12}C_7 - C_{a18}C_1)\sin\phi}{(C_4C_9 - C_6C_7)(1 - \cos\phi) + (C_1C_6 - C_3C_4)\cos\phi + (C_3C_7 - C_1C_9)\sin\phi}$$

$$M_A = -t_oR \frac{(C_{a15}C_9 - C_{a18}C_6)(1 - \cos\phi) + (C_{a12}C_6 - C_{a15}C_3)\cos\phi + (C_{a18}C_3 - C_{a12}C_9)\sin\phi}{(C_4C_9 - C_6C_7)(1 - \cos\phi) + (C_1C_6 - C_3C_4)\cos\phi + (C_3C_7 - C_1C_9)\sin\phi}$$

$$\psi_A = -\frac{t_oR^2}{EI} \frac{C_{a12}(C_4C_9 - C_6C_7) + C_{a15}(C_3C_7 - C_1C_9) + C_{a18}(C_1C_6 - C_3C_4)}{(C_4C_9 - C_6C_7)(1 - \cos\phi) + (C_1C_6 - C_3C_4)\cos\phi + (C_3C_7 - C_1C_9)\sin\phi}$$

$V_B = V_A$

$M_B = V_AR\sin\phi + M_A\cos\phi - t_oR[1 - \cos(\phi - \theta)]$

$T_B = V_AR(1 - \cos\phi) + M_A\sin\phi + t_oR\sin(\phi - \theta)$

If $\beta = 1.3$ (solid or hollow round cross section, $\nu = 0.3$)

ϕ	45°	90°			180°		
θ	0°	0°	30°	60°	0°	60°	120°
K_{VA}	0.6652	0.7038	0.1732	−0.0276	0.3433	−0.1635	−0.1561
K_{MA}	−0.3918	−0.8944	−0.4108	−0.0749	−1.2471	−0.4913	0.0034
$K_{\psi A}$	0.2993	0.6594	0.2445	0.0563	0.7597	0.0048	0.0211
K_{MB}	−0.0996	−0.2962	−0.3268	−0.1616	−0.7529	−1.0087	−0.5034
K_{TB}	0.6249	0.8093	0.6284	0.3975	0.6866	0.5390	0.5538

5d. Right end fixed, left end supported and roll-guided

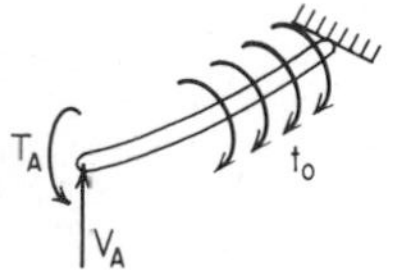

$M_A = 0 \quad y_A = 0 \quad \psi_A = 0$

$y_B = 0 \quad \Theta_B = 0 \quad \psi_B = 0$

$$V_A = -t_o \frac{[(C_{a12} + C_{a18})C_5 - C_{a15}(C_2 + C_8)]\sin\phi + (C_{a12}C_8 - C_{a18}C_2)\cos\phi}{[C_5(C_3 + C_9) - C_6(C_2 + C_8)]\sin\phi + (C_3C_8 - C_2C_9)\cos\phi}$$

$$T_A = -t_oR \frac{[C_{a15}(C_3 + C_9) - C_6(C_{a12} + C_{a18})]\sin\phi + (C_{a18}C_3 - C_{a12}C_9)\cos\phi}{[C_5(C_3 + C_9) - C_6(C_2 + C_8)]\sin\phi + (C_3C_8 - C_2C_9)\cos\phi}$$

$$\Theta_A = -\frac{t_oR^2}{EI} \frac{C_{a12}(C_5C_9 - C_6C_8) + C_{a15}(C_3C_8 - C_2C_9) + C_{a18}(C_2C_6 - C_3C_5)}{[C_5(C_3 + C_9) - C_6(C_2 + C_8)]\sin\phi + (C_3C_8 - C_2C_9)\cos\phi}$$

$V_B = V_A$

$M_B = V_AR\sin\phi - T_A\sin\phi - t_oR[1 - \cos(\phi - \theta)]$

$T_B = V_AR(1 - \cos\phi) + T_A\cos\phi + t_oR\sin(\phi - \theta)$

If $\beta = 1.3$ (solid or hollow round cross section, $\nu = 0.3$)

ϕ	45°	90°			180°		
θ	0°	0°	30°	60°	0°	60°	120°
K_{VA}	−0.2229	−0.4269	−0.3313	−0.1226	−0.6366	−0.4775	−0.1592
K_{TA}	−0.3895	−0.7563	−0.3109	−0.0640	−1.1902	−0.3153	−0.0089
$K_{\Theta A}$	−0.0237	−0.2020	−0.1153	−0.0165	−1.9576	−0.9588	0.0200
K_{MB}	−0.1751	−0.6706	−0.5204	−0.1926	−2.0000	−1.5000	−0.5000
K_{TB}	0.3664	0.5731	0.5347	0.3774	−0.0830	0.2264	0.5566

TABLE 19 ***Formulas for curved beams loaded normal to the plane of curvature (Cont.)***

End restraints, reference no.	Formulas for boundary values and selected numerical values

5e. Right end fixed, left end fixed

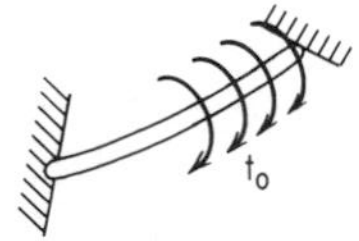

$y_A = 0 \quad \Theta_A = 0 \quad \psi_A = 0$

$y_B = 0 \quad \Theta_B = 0 \quad \psi_B = 0$

$$V_A = -t_o \frac{C_{a12}(C_4C_8 - C_5C_7) + C_{a15}(C_2C_7 - C_1C_8) + C_{a18}(C_1C_5 - C_2C_4)}{C_1(C_5C_9 - C_6C_8) + C_4(C_3C_8 - C_2C_9) + C_7(C_2C_6 - C_3C_5)}$$

$$M_A = -t_oR \frac{C_{a12}(C_5C_9 - C_6C_8) + C_{a15}(C_3C_8 - C_2C_9) + C_{a18}(C_2C_6 - C_3C_5)}{C_1(C_5C_9 - C_6C_8) + C_4(C_3C_8 - C_2C_9) + C_7(C_2C_6 - C_3C_5)}$$

$$T_A = -t_oR \frac{C_{a12}(C_6C_7 - C_4C_9) + C_{a15}(C_1C_9 - C_3C_7) + C_{a18}(C_3C_4 - C_1C_6)}{C_1(C_5C_9 - C_6C_8) + C_4(C_3C_8 - C_2C_9) + C_7(C_2C_6 - C_3C_5)}$$

$V_B = V_A$

$M_B = V_AR \sin\phi + M_A \cos\phi - T_A \sin\phi - t_oR[1 - \cos(\phi - \theta)]$

$T_B = V_AR(1 - \cos\phi) + M_A \sin\phi + T_A \cos\phi + t_oR \sin(\phi - \theta)$

If $\beta = 1.3$ (solid or hollow round cross section, $\nu = 0.3$)

ϕ	45°		90°		180°	
θ	0°	15°	0°	30°	0°	60°
K_{VA}	0.0000	−0.0444	0.0000	−0.0877	0.0000	−0.1657
K_{MA}	−0.1129	−0.0663	−0.3963	−0.2262	−1.0000	−0.4898
K_{TA}	−0.3674	−0.1571	−0.6037	−0.2238	−0.5536	−0.0035
K_{MB}	−0.1129	−0.1012	−0.3963	−0.3639	−1.0000	−1.0102
K_{TB}	0.3674	0.3290	0.6037	0.5522	0.5536	0.5382

5f. Right end supported and slope-guided, left end supported and slope-guided

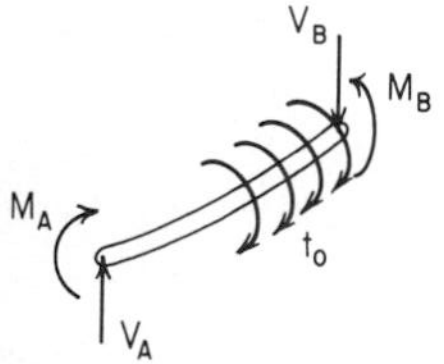

$T_A = 0 \quad y_A = 0 \quad \Theta_A = 0$

$T_B = 0 \quad y_B = 0 \quad \Theta_B = 0$

$$V_A = t_o \frac{[C_1 \sin\phi - C_4(1 - \cos\phi)] \sin(\phi - \theta) - C_{a12} \sin^2\phi + C_{a15}(1 - \cos\phi) \sin\phi}{C_4(1 - \cos\phi)^2 + C_3 \sin^2\phi - (C_1 + C_6)(1 - \cos\phi) \sin\phi}$$

$$M_A = -t_oR \frac{[C_3 \sin\phi - C_6(1 - \cos\phi)] \sin(\phi - \theta) - C_{a12}(1 - \cos\phi) \sin\phi + C_{a15}(1 - \cos\phi)^2}{C_4(1 - \cos\phi)^2 + C_3 \sin^2\phi - (C_1 + C_6)(1 - \cos\phi) \sin\phi}$$

$$\psi_A = \frac{t_oR^2}{EI} \frac{(C_3C_4 - C_1C_6) \sin(\phi - \theta) + (C_{a12}C_6 - C_{a15}C_3) \sin\phi - (C_{a12}C_4 - C_{a15}C_1)(1 - \cos\phi)}{C_4(1 - \cos\phi)^2 + C_3 \sin^2\phi - (C_1 + C_6)(1 - \cos\phi) \sin\phi}$$

$V_B = V_A$

$M_B = V_AR \sin\phi + M_A \cos\phi - t_oR[1 - \cos(\phi - \theta)]$

$$\psi_B = \psi_A \cos\phi + \frac{M_AR}{EI} C_7 + \frac{V_AR^2}{EI} C_9 + \frac{t_oR^2}{EI} C_{a18}$$

If $\beta = 1.3$ (solid or hollow round cross section, $\nu = 0.3$)

ϕ	45°		90°		180°	
θ	0°	15°	0°	30°	0°	60°
K_{VA}	0.0000	−0.2275	0.0000	−0.3732	0.0000	−0.4330
K_{MA}	−1.0000	−0.6129	−1.0000	−0.4928	−1.0000	−0.2974
$K_{\psi A}$	1.0000	0.6203	1.0000	0.5089	1.0000	0.1934
K_{MB}	−1.0000	−0.7282	−1.0000	−0.8732	−1.0000	−1.2026
$K_{\psi B}$	1.0000	0.7027	1.0000	0.7765	1.0000	0.7851

5g. Right end supported and slope-guided, left end supported and roll-guided

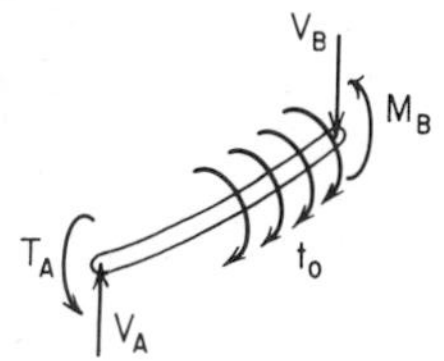

$M_A = 0 \quad y_A = 0 \quad \psi_A = 0$

$T_B = 0 \quad y_B = 0 \quad \Theta_B = 0$

$$V_A = -t_o \frac{C_{a12} \cos^2 \phi - C_{a15} \sin \phi \cos \phi + (C_5 \sin \phi - C_2 \cos \phi) \sin (\phi - \theta)}{(C_5 \sin \phi - C_2 \cos \phi)(1 - \cos \phi) + C_3 \cos^2 \phi - C_6 \sin \phi \cos \phi}$$

$$T_A = -t_o R \frac{(C_{a15} \sin \phi - C_{a12} \cos \phi)(1 - \cos \phi) + (C_3 \cos \phi - C_6 \sin \phi) \sin (\phi - \theta)}{(C_5 \sin \phi - C_2 \cos \phi)(1 - \cos \phi) + C_3 \cos^2 \phi - C_3 \sin \phi \cos \phi}$$

$$\Theta_A = -\frac{t_o R^2}{EI} \frac{(C_{a12} C_5 - C_{a15} C_2)(1 - \cos \phi) + (C_{a15} C_3 - C_{a12} C_6) \cos \phi - (C_2 C_6 - C_3 C_5) \sin (\phi - \theta)}{(C_5 \sin \phi - C_2 \cos \phi)(1 - \cos \phi) + C_3 \cos^2 \phi - C_6 \sin \phi \cos \phi}$$

$$V_B = V_A$$

$$M_B = V_A R \sin \phi - T_A \sin \phi - t_o R[1 - \cos (\phi - \theta)]$$

$$\psi_B = -\Theta_A \sin \phi + \frac{T_A R}{EI} C_8 + \frac{V_A R^2}{EI} C_9 + \frac{t_o R^2}{EI} C_{a18}$$

If $\beta = 1.3$ (solid or hollow round cross section, $\nu = 0.3$)

ϕ	45°	90°			180°		
θ	0°	0°	30°	60°	0°	60°	120°
K_{VA}	−0.8601	−1.0000	−0.8660	−0.5000	−0.5976	−0.5837	−0.4204
K_{TA}	−0.6437	−0.9170	−0.4608	−0.1698	−1.1953	−0.3014	0.0252
$K_{\Theta A}$	−0.0209	−0.1530	−0.0695	0.0158	−2.0590	−0.6825	0.6993
K_{MB}	−0.4459	−1.0830	−0.9052	−0.4642	−2.0000	−1.5000	−0.5000
$K_{\psi B}$	0.2985	0.6341	0.5916	0.4716	−0.1592	0.4340	1.0670

5h. Right end supported and slope-guided, left end simply supported

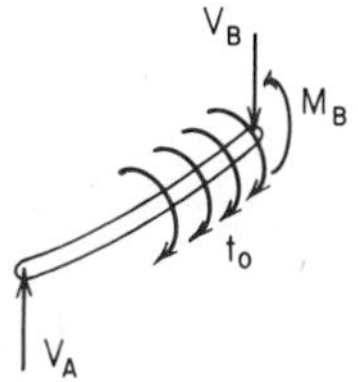

$M_A = 0 \quad T_A = 0 \quad y_A = 0$

$T_B = 0 \quad y_B = 0 \quad \Theta_B = 0$

$$V_A = -\frac{t_o \sin (\phi - \theta)}{1 - \cos \phi}$$

$$\Theta_A = -\frac{t_o R^2}{EI} \left[\frac{C_{a12} \sin \phi + C_6 \sin (\phi - \theta)}{1 - \cos \phi} - \frac{C_3 \sin \phi \sin (\phi - \theta)}{(1 - \cos \phi)^2} - C_{a15} \right]$$

$$\psi_A = -\frac{t_o R^2}{EI} \left[\frac{C_{a15} \sin \phi - C_{a12} \cos \phi}{1 - \cos \phi} + (C_3 \cos \phi - C_6 \sin \phi) \frac{\sin (\phi - \theta)}{(1 - \cos \phi)^2} \right]$$

$$V_B = V_A$$

$$M_B = V_A R \sin \phi - t_o R[1 - \cos (\phi - \theta)]$$

$$\psi_B = \psi_A \cos \phi - \Theta_A \sin \phi + \frac{V_A R^2}{EI} C_9 + \frac{t_o R^2}{EI} C_{a18}$$

If $\beta = 1.3$ (solid or hollow round cross section, $\nu = 0.3$)

ϕ	45°	90°			180°		
θ	0°	0°	30°	60°	0°	60°	120°
K_{VA}	−2.4142	−1.0000	−0.8660	−0.5000	0.0000	−0.4330	−0.4330
$K_{\Theta A}$	−0.2888	−0.7549	−0.3720	−0.0957	−3.6128	−1.0744	0.7320
$K_{\psi A}$	1.4161	1.6564	0.8324	0.3067	2.3000	0.5800	−0.0485
K_{MB}	−2.0000	−2.0000	−1.3660	−0.6340	−2.0000	−1.5000	−0.5000
$K_{\psi B}$	1.2811	1.0985	0.8250	0.5036	−0.3000	0.3985	1.0700

TABLE 19 *Formulas for curved beams loaded normal to the plane of curvature* (*Cont.*)

End restraints, reference no.	Formulas for boundary values and selected numerical values

5i. Right end supported and roll-guided, left end supported and roll-guided

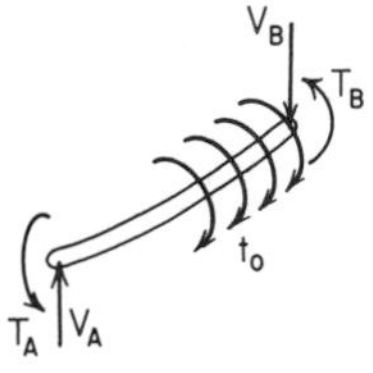

$M_A = 0 \quad y_A = 0 \quad \psi_A = 0$
$M_B = 0 \quad y_B = 0 \quad \psi_B = 0$

$$V_A = 0$$

$$T_A = -t_o R \frac{(C_{a12} + C_{a18}) \sin^2 \phi - (C_3 + C_9) \sin \phi \, [1 - \cos(\phi - \theta)]}{(C_2 + C_3 + C_8 + C_9) \sin^2 \phi}$$

$$\Theta_A = -\frac{t_o R^2}{EI} \frac{[C_{a12}(C_8 + C_9) - C_{a18}(C_2 + C_3)] \sin \phi - (C_2 C_9 - C_3 C_8)[1 - \cos(\phi - \theta)]}{(C_2 + C_3 + C_8 + C_9) \sin^2 \phi}$$

$$V_B = 0$$

$$T_B = T_A \cos \phi + t_o R \sin(\phi - \theta)$$

$$\Theta_B = \Theta_A \cos \phi + \frac{T_A R}{EI} C_5 + \frac{t_o R^2}{EI} C_{a15}$$

If $\beta = 1.3$ (solid or hollow round cross section, $\nu = 0.3$)

ϕ	45°		90°		270°	
θ	0°	15°	0°	30°	0°	90°
K_{VA}	0.0000	0.0000	0.0000	0.0000	0.0000	0.0000
K_{TA}	−0.4142	−0.1895	−1.0000	−0.5000	1.0000	2.0000
$K_{\Theta A}$	−0.0527	−0.0368	−0.6564	−0.4679	−6.5692	−2.3000
K_{TB}	0.4142	0.3660	1.0000	0.8660	−1.0000	0.0000
$K_{\Theta B}$	0.0527	0.0415	0.6564	0.5094	6.5692	7.2257

5j. Right end supported and roll-guided, left end simply supported

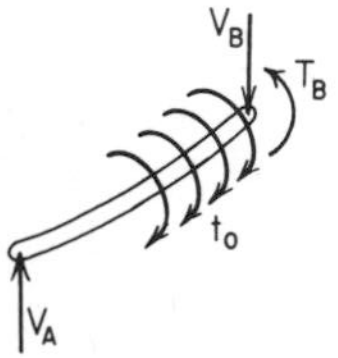

$M_A = 0 \quad T_A = 0 \quad y_A = 0$
$M_B = 0 \quad y_B = 0 \quad \psi_B = 0$

$$V_A = \frac{t_o[1 - \cos(\phi - \theta)]}{\sin \phi}$$

$$\Theta_A = -\frac{t_o R^2}{EI} \left\{ \frac{C_{a12} \cos \phi - C_{a18}(1 - \cos \phi)}{\sin \phi} + \frac{[C_3 \cos \phi - C_9(1 - \cos \phi)][1 - \cos(\phi - \theta)]}{\sin^2 \phi} \right\}$$

$$\psi_A = -\frac{t_o R^2}{EI} \left\{ C_{a12} + C_{a18} + \frac{(C_3 + C_9)[1 - \cos(\phi - \theta)]}{\sin \phi} \right\}$$

$$V_B = V_A$$

$$T_B = V_A R(1 - \cos \phi) + t_o R \sin(\phi - \theta)$$

$$\Theta_B = \Theta_A \cos \phi + \psi_A \sin \phi + \frac{V_A R^2}{EI} C_6 + \frac{t_o R^2}{EI} C_{a15}$$

If $\beta = 1.3$ (solid or hollow round cross section, $\nu = 0.3$)

ϕ	45°			90°		
θ	0°	15°	30°	0°	30°	60°
K_{VA}	0.4142	0.1895	0.0482	1.0000	0.5000	0.1340
$K_{\Theta A}$	−0.1683	−0.0896	−0.0247	−1.9564	−1.1179	−0.3212
$K_{\psi A}$	0.4229	0.1935	0.0492	2.0420	1.0210	0.2736
K_{TB}	0.8284	0.5555	0.2729	2.0000	1.3660	0.6340
$K_{\Theta B}$	0.1124	0.0688	0.0229	1.3985	0.8804	0.2878

1. If a closed circular ring is supported on any number of equally spaced simple supports (two or more) and if identical loading on each span is symmetrically placed relative to the center of the span, then each span can be treated by boundary condition *f* of Table 19. This boundary condition has both ends with no deflection or slope although they are free to roll as needed.

2. If a closed circular ring is supported on any even number of equally spaced simple supports and if the loading on any span is antisymmetrically placed relative to the center line of each span and symmetrically placed relative to each support, then boundary condition *f* can be applied to each full span. This problem can also be solved by applying boundary condition *g* to each half span. Boundary condition *g* has one end simply supported and slope-guided and the other end simply supported and roll-guided.

3. If a closed circular ring is supported on any even number of equally spaced simple supports (four or more) and if each span is symmetrically loaded relative to the center of the span with adjacent spans similarly loaded in opposite directions, then boundary condition *i* can be applied to each span. This boundary condition has both ends simply supported and roll-guided.

The following examples illustrate the applications of the formulas in Table 19 to both curved beams and closed rings with out-of-plane loads.

EXAMPLES

1. A piece of 8-in standard pipe is used to carry water across a passageway 40 ft wide. The pipe must come out of a wall normal to the surface and enter normal to a parallel wall at a position 16.56 ft down the passageway at the same elevation. To accomplish this a decision was made to bend the pipe into two opposite arcs of 28.28-ft radius with a total angle of 45° in each arc. If it is assumed that both ends are rigidly held by the walls, determine the maximum combined stress in the pipe due to its own weight and the weight of a full pipe of water.

Solution. An 8-in standard pipe has the following properties: $A = 8.4\ \text{in}^2$; $I = 72.5\ \text{in}^4$; $w = 2.38\ \text{lb/in}$; $E = 30(10^6)\ \text{lb/in}^2$; $\nu = 0.3$; $J = 145\ \text{in}^4$; OD = 8.625 in; ID = 7.981 in; and $t = 0.322$ in. The weight of water in a 1-in length of pipe is 1.81 lb. Owing to the symmetry of loading it is apparent that at the center of the span where the two arcs meet there is neither slope nor roll. An examination of Table 19 reveals that a curved beam that is fixed at the right end and roll- and slope-guided at the left end is not included among the 10 cases. Therefore, a solution will be carried out by considering a beam that is fixed at the right end and free at the left end with a uniformly distributed load over the entire span and both a concentrated moment and a concentrated torque on the left end. (These conditions are covered in cases 2a, 3a, and 4a.)

Since the pipe is round, $J = 2I$; and since $G = E/2(1+\nu)$, $\beta = 1.3$. Also note that for all three cases $\phi = 45°$ and $\theta = 0°$. For these conditions, numerical values of the coefficients are tabulated and the following expressions for the deformations and moments can be written directly from superposition of the three cases:

$$y_A = 0.3058\frac{M_oR^2}{EI} - 0.0590\frac{T_oR^2}{EI} - 0.0469\frac{(2.38 + 1.81)R^4}{EI}$$

$$\Theta_A = -0.8282\frac{M_oR}{EI} - 0.0750\frac{T_oR}{EI} + 0.0762\frac{4.19R^3}{EI}$$

$$\psi_A = 0.0750\frac{M_oR}{EI} + 0.9782\frac{T_oR}{EI} + 0.0267\frac{4.19R^3}{EI}$$

$$V_B = 0 + 0 - 4.19R(0.7854)$$

$$M_B = 0.7071M_o - 0.7071T_o - 0.2929(4.19)R^2$$

$$T_B = 0.7071M_o + 0.7071T_o - 0.0783(4.19)R^2$$

Since both Θ_A and ψ_A are zero and $R = 28.28(12) = 339.4$ in,

$$0 = -0.8282M_o - 0.0750T_o + 36{,}780$$

$$0 = 0.0750M_o + 0.9782T_o + 12{,}888$$

Solving these two equations gives $M_o = 45{,}920$ in-lb and $T_o = -16{,}700$ in-lb. Therefore,

$$y_A = -0.40 \text{ in} \qquad M_B = -97{,}100 \text{ in-lb}$$
$$T_B = -17{,}100 \text{ in-lb} \qquad V_B = -1120 \text{ lb}$$

The maximum combined stress would be at the top of the pipe at the wall where $\sigma = Mc/I = 97{,}100(4.3125)/72.5 = 5775$ lb/in² and $\tau = Tr/J = 17{,}100(4.3125)/145 = 509$ lb/in²:

$$\sigma_{max} = \frac{5775}{2} + \sqrt{\left(\frac{5775}{2}\right)^2 + 509^2} = 5819 \text{ lb/in}^2$$

2. A hollow steel rectangular beam 4 in wide, 8 in deep, and with 0.1-in wall thickness extends over a loading dock to be used as a crane rail. It is fixed to a warehouse wall at one end and is simply supported on a post at the other. The beam is curved in a horizontal plane with a radius of 15 ft and covers a total angular span of 60°. Calculate the torsional and bending stresses at the wall when a load of 3000 lb is 40° out from the wall. Neglect the weight of the beam.

Solution. The beam has the following properties: $R = 180$ in; $\phi = 60°(\pi/3 \text{ rad})$; $\theta = 40°$; $(\phi - \theta) = 20°(\pi/9 \text{ rad})$; $I = \frac{1}{12}[4(8^3) - 3.8(7.8^3)] = 20.39$ in⁴; $K = 2(0.1^2)(7.9^2)(3.9^2)/[8(0.1) + 4(0.1) - 2(0.1^2)] = 16.09$ in⁴ (see Table 20, case 16); $E = 30(10^6)$; $G = 12(10^6)$; and $\beta = 30(10^6)(20.39)/12(10^6)(16.09) = 3.168$. Equations for a curved beam that is fixed at one end and simply supported at the other with a concentrated load are found in Table 19, case 1b. To obtain the bending and twisting moments at the wall requires first the evaluation of the end reaction V_A, which, in turn, requires the following constants:

$$C_3 = -3.168\left(\frac{\pi}{3} - \sin 60°\right) - \frac{1 + 3.168}{2}\left(\frac{\pi}{3} - \cos 60° - \sin 60°\right) = 0.1397$$

$$C_{a3} = -3.168\left(\frac{\pi}{9} - \sin 20°\right) - C_{a2} = 0.006867$$

Similarly,

$$C_6 = C_1 = 0.3060 \qquad C_{a6} = C_{a1} = 0.05775$$
$$C_9 = C_2 = -0.7136 \qquad C_{a9} = C_{a2} = -0.02919$$

Therefore,

$$V_A = 3000\frac{-0.02919(1 - \cos 60°) - 0.05775 \sin 60° + 0.006867}{-0.7136(1 - \cos 60°) - 0.3060 \sin 60° + 0.1397} = 359.3 \text{ lb}$$

$$M_B = 359.3(180)(\sin 60°) - 3000(180)(\sin 20°) = -128{,}700 \text{ in-lb}$$

$$T_B = 359.3(180)(1 - \cos 60°) - 3000(180)(1 - \cos 20°) = -230 \text{ in-lb}$$

At the wall,

$$\sigma = \frac{Mc}{I} = \frac{128{,}700(4)}{20.39} = 25{,}240 \text{ lb/in}^2$$

$$\tau = \frac{VQ}{Ib} = \frac{(1000 - 359.3)[4(4)(2) - 3.9(3.8)(1.95)]}{20.39(0.2)} = 487 \text{ lb/in} \qquad \text{(due to transverse shear)}$$

$$\tau = \frac{T}{2t(a-t)(b-t)} = \frac{230}{2(0.1)(7.9)(3.9)} = 37.3 \text{ lb/in}^2 \qquad \text{(due to torsion)}$$

3. A solid round aluminum bar is in the form of a horizontal closed circular ring of 100-in radius resting on three equally spaced simple supports. A load of 1000 lb is placed midway between two supports, as shown in Fig. 8.5*a*. Calculate the deflection under this load if the bar is of such diameter as to make the maximum normal stress due to combined bending and torsion equal to 20,000 lb/in². Let $E = 10(10^6)$ lb/in² and $\nu = 0.3$.

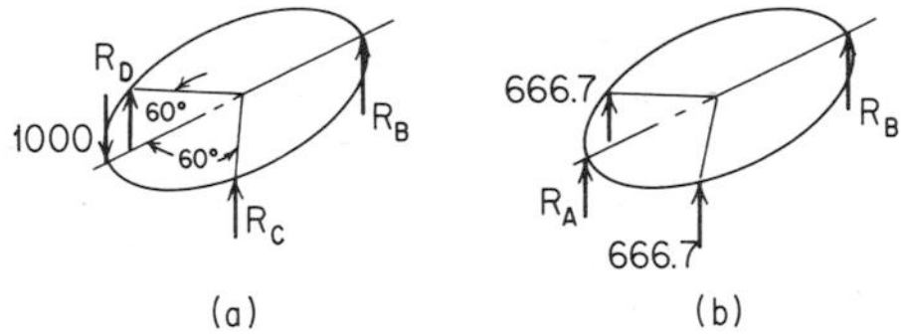

Fig. 8.5

Solution. The reactions R_B, R_C, and R_D are statically determinate, and a solution yields $R_B = -333.3$ lb and $R_C = R_D = 666.7$ lb. The internal bending and twisting moments are statically indeterminate, and so an energy solution would be appropriate. However, there are several ways that Table 19 can be used by superimposing various loadings. The method to be described here is probably the most straightforward.

Consider the equivalent loading shown in Fig. 8.5*b* where $R_B = -333.3$ lb and $R_A = -1000$ lb. The only difference is in the point of zero deflection. Owing to the symmetry of loading, one-half of the ring can be considered slope-guided at both ends, points A and B. Case 1f gives tabulated values of the necessary coefficients for $\phi = 180°$ and $\theta = 60°$. We can now solve for the following values:

$$V_A = -666.7(0.75) = -500 \text{ lb}$$

$$M_A = -666.7(100)(-0.5774) = 38{,}490 \text{ in-lb}$$

$$\psi_A = \frac{-666.7(100^2)}{EI}(-0.2722) = \frac{1.815(10^6)}{EI}$$

$$T_A = 0 \qquad y_A = 0 \qquad \Theta_A = 0$$

$$M_B = -666.7(100)(-0.2887) = 19{,}250 \text{ in-lb}$$

$$M_{60°} = -666.7(100)(0.3608) = -24{,}050 \text{ in-lb}$$

The equations for M and T can now be examined to determine the location of the maximum combined stress:

$$M_x = -50{,}000 \sin x + 38{,}490 \cos x + 66{,}667 \sin (x - 60°)\langle x - 60°\rangle^0$$
$$T_x = -50{,}000(1 - \cos x) + 38{,}490 \sin x + 66{,}667[1 - \cos(x - 60°)]\langle x - 60°\rangle^0$$

A careful examination of the expression for M shows no maximum values except at the ends and at the position of the load. The torque, however, has a maximum value of 13,100 in-lb at $x = 37.59°$ and a minimum value of -8790 in-lb at $x = 130.9°$. At these same locations the bending moments are zero. At the position of the load, the torque $T = 8330$ in-lb. Nowhere is the combined stress larger than the bending stress at point A. Therefore,

$$\sigma_A = 20{,}000 = \frac{M_A c}{I} = \frac{38{,}490\, d/2}{(\pi/64)d^4} = \frac{392{,}000}{d^3}$$

which gives

$$d = 2.70 \text{ in} \qquad \text{and} \qquad I = 2.609 \text{ in}^4$$

To obtain the deflection under the 1000-lb load in the original problem, first we must find the deflection at the position of the load of 666.7 lb in Fig. 8.5*b*. At $x = 60°$,

$$y_x = 0 + 0 + \frac{1.815(10^6)(100)}{10(10^6)2.609}(1 - \cos 60°) + \frac{38{,}490(100^2)}{10(10^6)(2.609)}F_1 + 0 + \frac{-500(100^3)}{10(10^6)(2.609)}F_3$$

where $F_1 = \frac{1 + 1.3}{2}\frac{\pi}{3}\sin 60° - 1.3(1 - \cos 60°) = 0.3929$ and $F_3 = 0.1583$. Therefore,

$$y_{60} = 3.478 + 5.796 - 3.033 = 6.24 \text{ in}$$

If the entire ring were now rotated as a rigid body about point B in order to lower points C and D by 6.24 in, point A would be lowered a distance of $6.24(2)/(1 + \cos 60°) = 8.32$ in, which is the downward deflection of the 1000-lb load.

Flanged sections. The formulas in Table 19 for flangeless or compact sections apply also to flanged sections when the ends are fixed as to slope only or when fixed as to slope and roll but not as to flange bending and if the loads are distributed or applied only at the ends. If the flanges are fixed or if concentrated loads are applied within the span, the additional torsional stiffness contributed by the bending resistance of the flanges [*warping restraint* (see Art. 9.3)] may appreciably affect the value and distribution of twisting and bending moments. The flange stresses caused by the secondary bending or warping may exceed the primary bending stresses. Refs. 15, 17, and 22 show methods of solution and give some numerical solutions for simple concentrated loads on curved I-beams with both ends fixed completely. Brookhart (Ref. 22) also includes results for additional boundary conditions and uniformly distributed loads. Results are compared with cases where the warping restraint was not considered.

Dabrowski (Ref. 23) gives a thorough presentation of the theory of curved thin-walled beams and works out many examples including multispan beams and beams with open cross sections, closed cross sections, and cross sections which contain both open and closed elements; an extensive bibliography is included. Vlasov (Ref. 27) also gives a very thorough derivation and discusses, among many other topics, vibrations, stability, laterally braced beams of open cross section, and thermal stresses. He also examines the corrections necessary to account for shear deformation in flanges being warped. Verden (Ref. 24) is primarily concerned with multispan curved beams and works out many examples. Sawko and Cope (Ref. 25) and Meyer (Ref. 26) apply finite element analysis to curved box girder bridges.

REFERENCES

1. Wilson, B. J., and J. F. Quereau: A Simple Method of Determining Stress in Curved Flexural Members, *Univ. Ill. Eng. Exp. Sta., Circ.* 16, 1927.
2. Case, J.: "Strength of Materials," Longmans, Green & Co., Ltd., 1925.

3. Seely, F. B., and R. V. James: The Plaster-model Method of Determining Stresses Applied to Curved Beams, *Univ. Ill. Eng. Exp. Sta. Bull.* 195, 1929.
4. von Kármán, Th: "Über die Formänderung dünnwandiger Rohre, insbesondere federnder Ausgleichrohre," *Z. Vereines Dtsch. Ing.*, vol. 55, p. 1889, 1911.
5. Timoshenko, S.: Bending Stresses in Curved Tubes of Rectangular Cross-section, *Trans. ASME,* vol. 45, p. 135, 1923.
6. Burke, W. F.: Working Charts for the Stress Analysis of Elliptic Rings, *Nat. Adv. Comm. Aeron., Tech. Note* 444, 1933.
7. Peery, D. J.: "Aircraft Structures," McGraw-Hill Book Company, 1950.
8. Utecht, E. A.: Stresses in Curved, Circular Thin-Wall Tubes, *ASME J. Appl. Mech.*, vol. 30, no. 1, March 1963.
9. Penstock Analysis and Stiffener Design, *U.S. Dept. of Agriculture, Bur. Reclamation, Boulder Canyon Proj. Final Repts., Pt. V, Bull. 5, 1940.*
10. Bleich, Hans: Stress Distribution in the Flanges of Curved T and I Beams, *U.S. Dept. of the Navy, David W. Taylor Model Basin, transl.* 228, 1950.
11. Mantle, J. B., and T. J. Dolan: A Photoelastic Study of Stresses in U-shaped Members, *Proc. Soc. Exp. Stress Anal.*, vol. 6, no. 1, 1948.
12. Stressed Skin Structures, Royal Aeronautical Society, data sheets.
13. Timoshenko, S.: "Strength of Materials," D. Van Nostrand Company, Inc., 1930.
14. Levy, Roy: Displacements of Circular Rings with Normal Loads, *Proc. Am. Soc. Civil Eng., J. Struct. Div.*, vol. 88, no. 1, February 1962.
15. Moorman, R. B. B.: Stresses in a Curved Beam under Loads Normal to the Plane of Its Axis, *Iowa Eng. Exp. Sta., Iowa State College, Bull.* 145, 1940.
16. Fisher, G. P.: Design Charts for Symmetrical Ring Girders, *ASME J. Appl. Mech.*, vol. 24, no. 1, March 1957.
17. Moorman, R. B. B.: Stresses in a Uniformly Loaded Circular-arc I-beam, *Univ. Missouri Bull., Eng. Ser.* 36, 1947.
18. Hogan, M. B.: *Utah Eng. Exp. Sta., Bulls.* 21, 27, and 31.
19. Blake, Alexander: Deflection of a Thick Ring in Diametral Compression, *ASME J. Appl. Mech.*, vol. 26, no. 2, June 1959.
20. Volterra, Enrico, and Tandall Chung: Constrained Circular Beam on Elastic Foundations, *Trans. Am. Soc. Civil Eng.,* vol. 120, 1955 (paper 2740).
21. Meck, H. R.: Three-Dimensional Deformation and Buckling of a Circular Ring of Arbitrary Section, *ASME J. Eng. Ind.*, vol. 91, no. 1, February 1969.
22. Brookhart, G. C.: Circular-Arc I-Type Girders, *Proc. Am. Soc. Civil Eng., J. Struct. Div.*, vol. 93, no. 6, December 1967.
23. Dabrowski, R.: "Curved Thin-Walled Girders, Theory and Analyses," Cement and Concrete Association, 1972.
24. Verden, Werner: "Curved Continuous Beams for Highway Bridges," Frederick Ungar Publishing Co., 1969 (English transl.)
25. Sawko, F., and R. J. Cope: Analysis of Multi-cell Bridges Without Transverse Diaphragms—A Finite Element Approach, *Struct. Eng.*, vol. 47, no. 11, November 1969.
26. Meyer, C.: Analysis and Design of Curved Box Girder Bridges, *Univ. of California, Berkeley, Struct. Eng. & Struct. Mech. Rept.* SESM-70-22, December 1970.
27. Vlasov, V. Z.: "Thin-Walled Elastic Beams," Clearing House for Federal Scientific and Technical Information, U.S. Dept. of Commerce, 1961.
28. Neugebauer, George H.: Private communication.

NINE

Torsion

9.1 *Straight bars of uniform circular section under pure torsion*

The formulas of this article are based on the following assumptions: (1) The bar is straight, of uniform circular section (solid or concentrically hollow), and of homogeneous isotropic material; (2) the bar is loaded only by equal and opposite twisting couples, which are applied at its ends in planes normal to its axis; and (3) the bar is not stressed beyond the elastic limit.

Behavior. The bar twists, each section rotating about the longitudinal axis. Plane sections remain plane, and radii remain straight. There is at any point a shear stress τ on the plane of the section; the magnitude of this stress is proportional to the distance from the center of the section, and its direction is perpendicular to the radius drawn through the point. Accompanying this shear stress there is an equal longitudinal shear stress on a radial plane and equal tensile and compressive stresses σ_t and σ_c at 45° (see Art. 6.6). The deformation and stresses described are represented in Fig. 9.1.

In addition to these deformations and stresses, there is some longitudinal strain and stress. It has been generally held that the longitudinal strain is a shortening, and that the longitudinal stress is a tension in the outer part

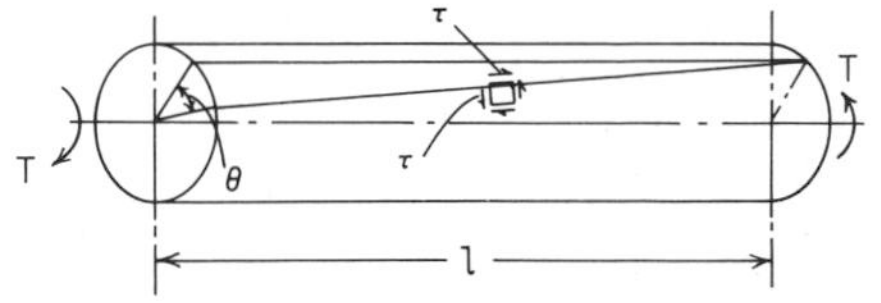

Fig. 9.1

and a balancing compression in the inner part (Ref. 5). Reiner (Ref. 25), on the basis of mathematical analysis, concluded that the longitudinal strain can be either a lengthening, a shortening, or zero, depending on certain physical characteristics of the material; he cites experiments carried out by Poynting (Ref. 26) in which steel and hard copper specimens showed elongation under torsion. In any event, for elastic loading, neither longitudinal deformation nor stress is likely to be large enough to have engineering significance in any case involving pure elastic torsion of a solid circular bar.

Formulas. Let T = twisting moment; l = length of the member; r = radius of the section; J = polar moment of inertia of the section; ρ = distance from the center of the section to any point q; τ = the shear stress; θ = angle of twist (radians); G = modulus of rigidity of the material; and U = strain energy. Then

$$\theta = \frac{Tl}{JG} \tag{1}$$

$$\tau = \frac{T\rho}{J} \qquad \text{(at } q\text{)} \tag{2}$$

$$\text{Max } \tau = \frac{Tr}{J} \qquad \text{(at surface)} \tag{3}$$

$$U = \frac{1}{2}\frac{T^2 l}{JG} \tag{4}$$

By substituting for J in Eqs. 1 and 3 its value $2I$ from Table 1, the following formulas are readily obtained: For solid section with radius r,

$$\theta = \frac{2Tl}{\pi r^4 G}$$

$$\text{Max } \tau = \frac{2T}{\pi r^3}$$

For hollow section with outer radius r_o and inner radius r_i,

$$\theta = \frac{2Tl}{\pi(r_o^4 - r_i^4)G}$$

$$\text{Max } \tau = \frac{2Tr_o}{\pi(r_o^4 - r_i^4)}$$

9.2 *Bars of noncircular uniform section under pure torsion*

The formulas of this article are based on the same assumptions as those of Art. 9.1, except that the cross section of the bar is not circular. It is important to note that the condition of loading implies that the end sections of the bar are free to warp, there being no constraining forces to hold them in their respective planes.

Behavior. The bar twists, each section rotating about its torsional center. Sections do not remain plane, but warp, and some radial lines through the torsional center do not remain straight. The distribution of shear stress on the section is not necessarily linear, and the direction of the shear stress is not necessarily normal to a radius.

Formulas. The torsional stiffness of the bar can be expressed by the general equation

$$T = \frac{\theta}{l} KG \qquad \text{or} \qquad \theta = \frac{Tl}{KG}$$

where K is a factor dependent on the form and dimensions of the cross section. For a *circular* section K is the polar moment of inertia J (Eq. 1); for other sections K is less than J and may be only a very small fraction of J. The maximum stress is a function of the twisting moment and of the form and dimensions of the cross section. In Table 20 formulas are given for K and for max τ for a variety of sections. The formulas for cases 1 to 3, 5, 10, and 12 are based on rigorous mathematical analysis. The equations for case 4 are given in a simplified form involving an approximation, with a resulting error not greater than 4 percent. The K formulas for cases 13 to 21 inclusive and the stress formulas for cases 13 to 18 inclusive are based on mathematical analysis but are approximate (Ref. 2); their accuracy depends upon how nearly the actual section conforms to the assumptions indicated as to form. The K formulas for the remaining cases and the stress formulas for cases 19 to 26 inclusive are based on the membrane analogy and are to be regarded as reasonably close approximations giving results that are rarely as much as 10 percent in error (Refs. 2 to 4 and 11).

It will be noted that formulas for K in cases 23 to 26 are based on the assumption of uniform flange thickness. For slightly tapering flanges, D should be taken as the diameter of the largest circle that can be inscribed in the actual section, and b as the average flange thickness. For sharply tapering flanges the method described by Griffith (Ref. 3) may be used. Charts relating especially to structural H and I sections are in Ref. 11.

The formulas of Table 20 make possible the calculation of the strength and stiffness of a bar of almost any form, but an understanding of the membrane analogy (Art. 4.4) makes it possible to draw certain conclusions as to the *comparative* torsional properties of different sections by simply visualizing the bubbles that would be formed over holes of corresponding size and shape. From the volume relationship, it can be seen that of two sections having the same area, the one more nearly circular is the stiffer, and that although any extension whatever of the section increases its torsional stiffness, narrow outstanding flanges and similar protrusions have little effect. It is also apparent that any member having a narrow section, such as a thin plate, has practically the same torsional stiffness when flat as when bent into the form of an open tube or into a channel or angle section.

From the slope relationship it can be seen that the greatest stresses (slopes) in a given section occur at the boundary adjacent to the thicker portions, and that the stresses are very low at the ends of outstanding flanges or protruding corners and very high at points where the boundary is sharply concave. Therefore a longitudinal slot or groove that is sharp at the bottom or narrow will cause high local stresses, and if it is deep will greatly reduce the torsional stiffness of the member. The direction of the shear stresses at any point is along the contour of the bubble surface at the corresponding point, and at points corresponding to high and low points of the bubble surface the shear stress becomes zero. Therefore there may be several points of zero shear stress in a section. Thus for an I section, there are high points of zero slope at the center of the largest inscribed circles (at the junction of web and flanges) and a low point of zero slope at the center of the web, and at these points in the section the shear stress is zero.

The preceding generalizations apply to solid sections, but it is possible to make somewhat similar generalizations concerning hollow or tubular sections from the formulas given for cases 10 to 16 inclusive. These formulas show that the strength and stiffness of a hollow section depend largely upon the area inclosed by the median boundary. For this reason a circular tube is stiffer and stronger than one of any other form, and the more nearly the form of any hollow section approaches the circular, the greater will be its strength and stiffness. It is also apparent from the formulas for strength that even a local reduction in the thickness of the wall of a tube, such as would be caused by a longitudinal groove, may greatly increase the maximum shear stress, though if the groove is narrow the effect on stiffness will be small.

The torsional strength and stiffness of thin-walled multicelled structures such as airplane wings and boat hulls can best be calculated by an arithmetical process of successive approximations (Ref. 12). The method of successive approximations can also be applied to solid sections (e.g., Table 20, cases 19 to 21), and any desired accuracy can be attained by sufficient repetition (Refs. 13 and 14).

EXAMPLES

1. It is required to compare the strength and stiffness of a circular steel tube, 4 in outside diameter and $\frac{5}{32}$ in thick, with the strength and stiffness of the same tube after it has been split by cutting full length along an element. No warping restraint is provided.

Solution. The strengths will be compared by comparing the twisting moments required to produce the same stress; the stiffnesses will be compared by comparing the values of K.

(*a*) For the tube (Table 20, case 10), $K = \frac{1}{2}\pi(r_o^4 - r_i^4) = \frac{1}{2}\pi[2^4 - (1\frac{27}{32})^4] = 6.98 \text{ in}^4$,

$$T = \tau \frac{\pi(r_o^4 - r_i^4)}{2r_o} = 3.49\tau \text{ in-lb}$$

(*b*) For the split tube (Table 20, case 17), $K = \frac{2}{3}\pi r t^3 = \frac{2}{3}\pi(1\frac{59}{64})(\frac{5}{32})^3 = 0.0154 \text{ in}^4$,

$$T = \tau \frac{4\pi^2 r^2 t^2}{6\pi r + 1.8t} = 0.097\tau \text{ in-lb}$$

The closed section is therefore more than 400 times as stiff as the open section, and more than 30 times as strong.

TABLE 20 ***Formulas for torsional deformation and stress***

GENERAL FORMULAS: $\theta = TL/KG$ and $\tau = T/Q$, where θ = angle of twist (radians); T = twisting moment (inch-pounds); L = length (inches); τ = unit shear stress (pounds per square inch); G = modulus of rigidity (pounds per square inch); K (inches to the fourth) and Q (inches cubed) are functions of the cross section

Form and dimensions of cross sections, other quantities involved, and case no.	Formula for K in $\theta = \frac{TL}{KG}$	Formula for shear stress
1. Solid circular section 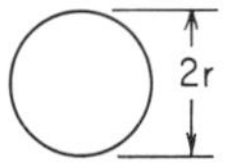	$K = \frac{1}{2}\pi r^4$	Max $\tau = \frac{2T}{\pi r^3}$ at boundary
2. Solid elliptical section 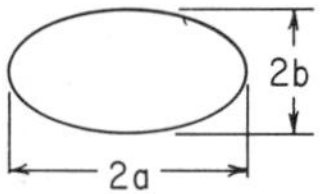	$K = \frac{\pi a^3 b^3}{a^2 + b^2}$	Max $\tau = \frac{2T}{\pi a b^2}$ at ends of minor axis
3. Solid square section 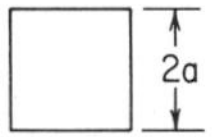	$K = 2.25a^4$	Max $\tau = \frac{0.6T}{a^3}$ at midpoint of each side
4. Solid rectangular section 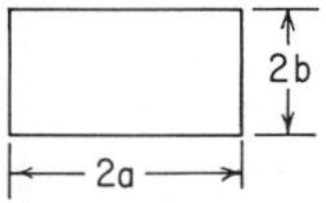	$K = ab^3\left[\frac{16}{3} - 3.36\frac{b}{a}\left(1 - \frac{b^4}{12a^4}\right)\right]$	Max $\tau = \frac{T(3a + 1.8b)}{8a^2b^2}$ at midpoint of each longer side

5. Solid triangular section (equilateral)

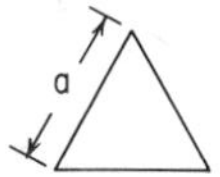

$$K = \frac{a^4\sqrt{3}}{80}$$

$\text{Max } \tau = \frac{20T}{a^3}$ at midpoint of each side

6. Isosceles triangle

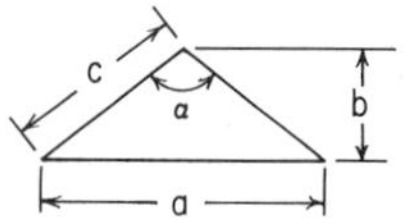

For $\frac{2}{3} < a/b < \sqrt{3}$ $(39° < \alpha < 82°)$

$K = \frac{a^3b^3}{15a^2 + 20b^2}$ approximate formula which is exact at $\alpha = 60°$ where $K = 0.02165c^4$

For $\sqrt{3} < a/b < 2\sqrt{3}$ $(82° < \alpha < 120°)$

$K = 0.0915b^4\left(\frac{a}{b} - 0.8592\right)$ approximate formula which is exact at $\alpha = 90°$ where $K = 0.0261c^4$ (errors < 4 percent) (Ref. 20)

For $39° < \alpha < 120°$

$Q = \frac{K}{b[0.200 + 0.309a/b - 0.0418(a/b)^2]}$ approximate formula which is exact at $\alpha = 60°$ and $\alpha = 90°$

For $\alpha = 60°$ $Q = 0.0768b^3 = 0.0500c^3$

For $\alpha = 90°$ $Q = 0.1604b^3 = 0.0567c^3$

Max τ at center of longest side

7. Circular segmental section

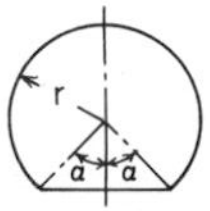

$K = Cr^4$ where C depends on α and has values as follows:

α	0°	30°	60°	80°	90°
C	$\frac{1}{2}\pi$	1.47	0.91	0.48	0.296

$\text{Max } \tau = \frac{T}{Q}$ at center of flat boundary. Here $Q = Cr^3$, where C depends on α and has values as follows:

α	0°	30°	60°	80°	90°
C	$\frac{1}{2}\pi$	1.25	0.80	0.49	0.35

(Ref. 16)

8. Circular sector

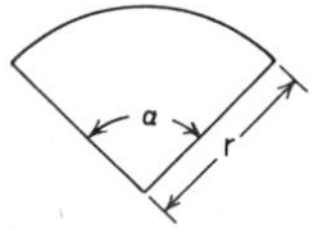

$K = Cr^4$ where C depends on α and has values as follows:

α	45°	60°	90°	120°
C	0.0181	0.0349	0.0825	0.148
α	180°	270°	300°	360°
C	0.296	0.528	0.686	0.878

$\text{Max } \tau = \frac{T}{Q}$ on radial boundary. Here $Q = Cr^3$, where C has values as follows:

α	60°	120°	180°
C	0.0712	0.227	0.35

(Ref. 17)

TABLE 20 *Formulas for torsional deformation and stress (Cont.)*

Form and dimensions of cross sections, other quantities involved, and case no.	Formula for K in $\theta = \dfrac{TL}{KG}$	Formula for shear stress
9. Circular shaft with opposite sides flattened 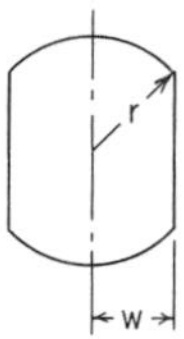	$K = cr^4$ where c depends on ratio w/r and has values as follows: w/r: $\frac{7}{8}$, $\frac{3}{4}$, $\frac{5}{8}$, $\frac{1}{2}$ c: 1.357, 1.076, 0.733, 0.438	$Q = cr^3$ where c depends on ratio w/r and has values as follows: w/r: $\frac{7}{8}$, $\frac{3}{4}$, $\frac{5}{8}$, $\frac{1}{2}$ c: 1.155, 0.912, 0.638, 0.471 (Ref. 21)
10. Hollow concentric circular section 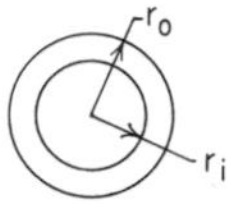	$K = \frac{1}{2}\pi(r_o^4 - r_i^4)$	Max $\tau = \dfrac{2Tr_o}{\pi(r_o^4 - r_i^4)}$ at outer boundary
11. Eccentric hollow circular section 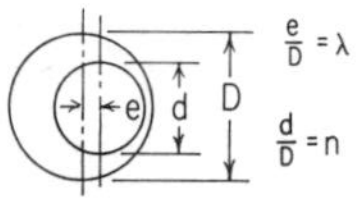	$K = \pi(D^4 - d^4)/32Q$ where $Q = 1 + \dfrac{16n^2}{(1-n^2)(1-n^4)}\lambda^2 + \dfrac{384n^4}{(1-n^2)^2(1-n^4)^4}\lambda^4$	Max $\tau = \dfrac{16TDF}{\pi(D^4 - d^4)}$ $F = 1 + \dfrac{4n^2}{1-n^2}\lambda + \dfrac{32n^2}{(1-n^2)(1-n^4)}\lambda^2 + \dfrac{48n^2(1+2n^2+3n^4+2n^6)}{(1-n^2)(1-n^4)(1-n^6)}\lambda^3 + \dfrac{64n^2(2+12n^2+19n^4+28n^6+18n^8+14n^{10}+3n^{12})}{(1-n^2)(1-n^4)(1-n^6)(1-n^8)}\lambda^4$ (Ref. 10)
12. Hollow elliptical section, outer and inner boundaries similar ellipses 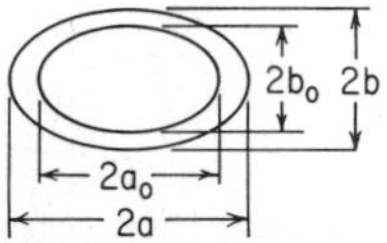	$K = \dfrac{\pi a^3 b^3}{a^2 + b^2}(1 - q^4)$ where $q = \dfrac{a_o}{a} = \dfrac{b_o}{b}$	Max $\tau = \dfrac{2T}{\pi a b^2(1 - q^4)}$ at ends of minor axis on outer surface

13. Hollow, thin-walled elliptical section of uniform thickness; U = length of median boundary, shown dotted: $U = \pi(a + b - t)\left[1 + 0.27\frac{(a-b)^2}{(a+b)^2}\right]$ (approximately) 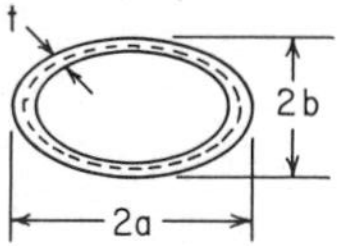	$K = \frac{4\pi^2 t[(a - \frac{1}{2}t)^2(b - \frac{1}{2}t)^2]}{U}$	Average $\tau = \frac{T}{2\pi t(a - \frac{1}{2}t)(b - \frac{1}{2}t)}$ (stress is nearly uniform if t is small)
14. Any thin tube of uniform thickness; U = length of median boundary; A = mean of areas enclosed by outer and inner boundaries, or (approximate) area within median boundary 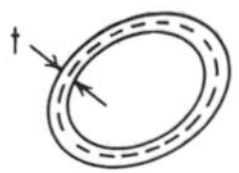	$K = \frac{4A^2 t}{U}$	Average $\tau = \frac{T}{2tA}$ Stress is nearly uniform if t is small)
15. Any thin tube. U and A as for case 14; t = thickness at any point 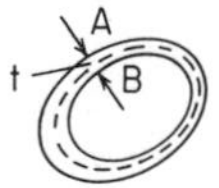	$K = \frac{4A^2}{\int dU/t}$	Average τ on any thickness $AB = \frac{T}{2tA}$ (max τ where t is a minimum)
16. Hollow rectangle 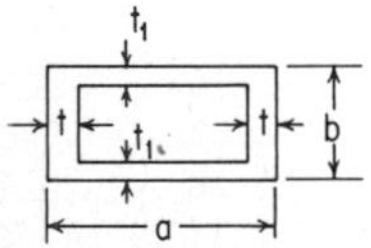	$K = \frac{2tt_1(a - t)^2(b - t_1)^2}{at + bt_1 - t^2 - t_1^2}$	Average $\tau = \frac{T}{2t(a - t)(b - t_1)}$ near midlength of short sides Average $\tau = \frac{T}{2t_1(a - t)(b - t_1)}$ near midlength of long sides (There will be higher stresses at inner corners unless fillets of fairly large radius are provided)

TABLE 20 ***Formulas for torsional deformation and stress (Cont.)***

Form and dimensions of cross sections, other quantities involved, and case no.	Formula for K in $\theta = \dfrac{TL}{KG}$	Formula for shear stress
17. Thin circular open tube of uniform thickness; r = mean radius 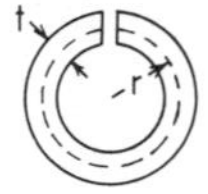	$K = \frac{2}{3}\pi r t^3$	Max $\tau = \dfrac{T(6\pi r + 1.8t)}{4\pi^2 r^2 t^2}$ along both edges remote from ends (this assumes t is small compared with mean radius; otherwise use the formulas given for cases 19 to 26)
18. Any thin open tube of uniform thickness; U = length of median line, shown dotted 	$K = \frac{1}{3}Ut^3$	Max $\tau = \dfrac{T(3U + 1.8t)}{U^2 t^2}$ along both edges remote from ends (this assumes t small compared with least radius of curvature of median line; otherwise use the formulas given for cases 19 to 26)
19. Any elongated section with axis of symmetry OX; U = length, A = area of section, I_x = moment of inertia about axis of symmetry 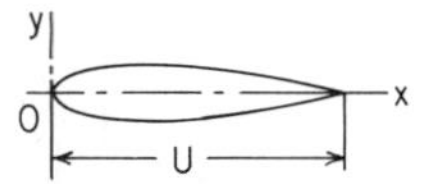	$K = \dfrac{4I_x}{1 + 16I_x/AU^2}$	
		For all solid sections of irregular form (cases 19 to 26 inclusive) the maximum shear stress occurs at or very near one of the points where the largest inscribed circle touches the boundary,* and of these, at the one where the curvature of the boundary is algebraically least. (Convexity represents positive and concavity negative curvature of the boundary.) At a point where the curvature is positive (boundary of section straight or convex) this maximum stress is given approximately by: $\tau = G\dfrac{\theta}{L}c$ or $\tau = \dfrac{T}{K}c$ where $c = \dfrac{D}{1 + \dfrac{\pi^2 D^4}{16A^2}}\left[1 + 0.15\left(\dfrac{\pi^2 D^4}{16A^2} - \dfrac{D}{2r}\right)\right]$ D = diameter of largest inscribed circle r = radius of curvature of boundary at the point (positive for this case) A = area of the section
20. Any elongated section or thin open tube; dU = elementary length along median line, t = thickness normal to median line, A = area of section 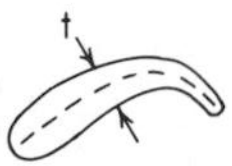	$K = \dfrac{\frac{1}{3}F}{1 + \frac{4}{3}F/AU^2}$ where $F = \int_0^U t^3\,dU$	

21. Any solid, fairly compact section without reentrant angles; J = polar moment of inertia about centroidal axis, A = area of section 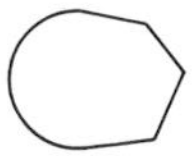	$K = \dfrac{A^4}{40J}$
22. Trapezoid 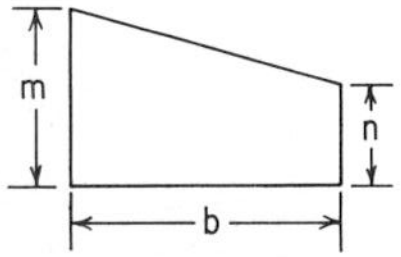	$K = \frac{1}{12}b(m+n)(m^2+n^2) - V_L m^4 - V_s n^4$ where $V_L = 0.10504 - 0.10s + 0.0848s^2 - 0.06746s^3 + 0.0515s^4$ $V_s = 0.10504 + 0.10s + 0.0848s^2 + 0.06746s^3 + 0.0515s^4$ $s = \dfrac{m-n}{b}$ (Ref. 11)
23. T section, flange thickness uniform; r, D, t, and t_1 as for case 26 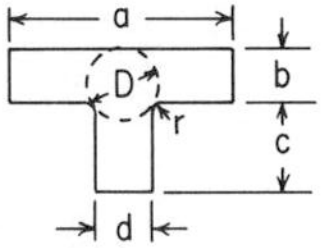	$K = K_1 + K_2 + \alpha D^4$ where $K_1 = ab^3\left[\frac{1}{3} - 0.21\frac{b}{a}\left(1 - \frac{b^4}{12a^4}\right)\right]$ $K_2 = cd^3\left[\frac{1}{3} - 0.105\frac{d}{c}\left(1 - \frac{d^4}{192c^4}\right)\right]$ $\alpha = \frac{t}{t_1}\left(0.15 + 0.10\frac{r}{b}\right)$
24. L section; r and D as for case 26; $b \geq d$ 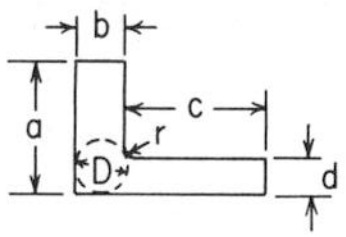	$K = K_1 + K_2 + \alpha D^4$ where $K_1 = ab^3\left[\frac{1}{3} - 0.21\frac{b}{a}\left(1 - \frac{b^4}{12a^4}\right)\right]$ $K_2 = cd^3\left[\frac{1}{3} - 0.105\frac{d}{c}\left(1 - \frac{d^4}{192c^4}\right)\right]$ $\alpha = \frac{d}{b}\left(0.07 + 0.076\frac{r}{b}\right)$

At a point where the curvature is negative (boundary of section concave or reentrant), this maximum stress is given approximately by:

$$\tau = G\frac{\theta}{L}c \qquad \text{or} \qquad \tau = \frac{T}{K}c$$

$$\text{where } c = \frac{D}{1 + \dfrac{\pi^2 D^4}{16A^2}}\left\{1 + \left[0.118\log_e\left(1 - \frac{D}{2r}\right) - 0.238\frac{D}{2r}\right]\text{Tanh}\,\frac{2\phi}{\pi}\right\}$$

and D, A, and r have the same meaning as before and ϕ = angle through which a tangent to the boundary rotates in turning or traveling around the reentrant portion, measured in radians (here r is *negative*). The preceding formulas should also be used for cases 17 and 18 when t is relatively large compared with radius of median line.)

*Unless at some other point on boundary there is a sharp reentrant angle, causing high local stress.

TABLE 20 *Formulas for torsional deformation and stress* (Cont.)

Form and dimensions of cross sections, other quantities involved, and case no.	Formula for K in $\theta = \frac{TL}{KG}$	Formula for shear stress
25. U or Z section 	K = sum of K's of constituent L sections computed as for case 24	
26. I section, flange thickness uniform; r = fillet radius, D = diameter largest inscribed circle, $t = b$ if $b < d$; $t = d$ if $d < b$; $t_1 = b$ if $b > d$; $t_1 = d$ if $d > b$ 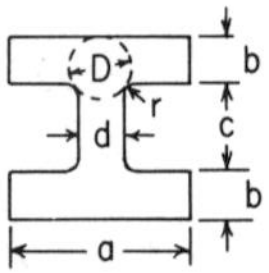	$K = 2K_1 + K_2 + 2\alpha D^4$ where $K_1 = ab^3\left[\frac{1}{3} - 0.21\frac{b}{a}\left(1 - \frac{b^4}{12a^4}\right)\right]$ $K_2 = \frac{1}{3}cd^3$ $\alpha = \frac{t}{t_1}\left(0.15 + 0.1\frac{r}{b}\right)$	

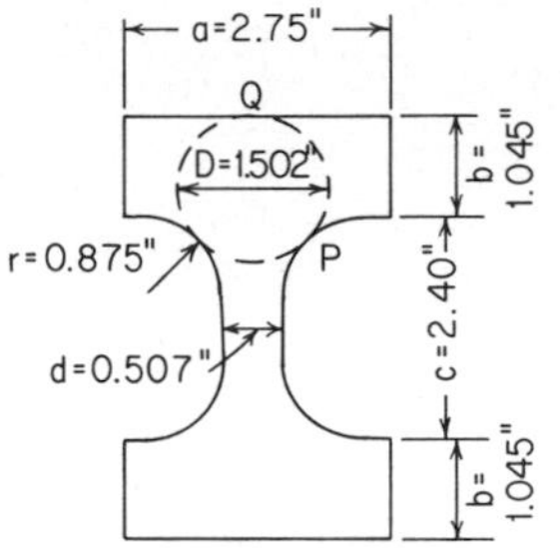

Fig. 9.2

2. It is required to determine the angle through which an airplane-wing spar of spruce, 8 ft long and having the section shown in Fig. 9.2, would be twisted by end torques of 500 in-lb, and to find the maximum resulting stress. For the material in question, $G = 100{,}000$ and $E = 1{,}500{,}000$.

Solution. All relevant dimensions are shown in Fig. 9.2, with notation corresponding to that used in the formulas. The first step is to compute K by the formula given for case 26 (Table 20), and we have

$$K = 2K_1 + K_2 + 2\alpha D^4$$

$$K_1 = 2.75(1.045^3)\left\{\frac{1}{3} - \frac{0.21(1.045)}{2.75}\left[1 - \frac{1.045^4}{12(2.75^4)}\right]\right\} = 0.796$$

$$K_2 = \tfrac{1}{3}(2.40)(0.507^3) = 0.104$$

$$\alpha = \frac{0.507}{1.045}\left[0.150 + \frac{0.1(0.875)}{1.045}\right] = 0.1133$$

$$K = 2(0.796) + 0.104 + 2(0.1133)(1.502^4) = 2.85 \text{ in}^4$$

Therefore

$$\theta = \frac{Tl}{KG} = \frac{500(96)}{2.85(100{,}000)} = 0.168 \text{ rad} = 9.64°$$

The maximum stress will probably be at P, the point where the largest inscribed circle touches the section boundary at a fillet. The formula is

$$\text{Max } \tau = \frac{T}{K}C$$

where

$$C = \frac{1.502}{1 + \dfrac{\pi^2(1.502^4)}{16(7.63^2)}}\left\{1 + \left[0.118 \log_e\left(1 - \frac{1.502}{2(-0.875)}\right) - 0.238\frac{1.502}{2(-0.875)}\right]\text{Tanh}\frac{2(\pi/2)}{\pi}\right\} = 1.73$$

Substituting the values of T, C, and K, it is found that

$$\text{Max } \tau = \frac{500}{2.85}(1.73) = 303 \text{ lb/in}^2$$

It will be of interest to compare this stress with that at Q, the other point where the maximum inscribed circle touches the boundary. Here the formula that applies is

$$\tau = \frac{T}{K}C$$

where $$C = \frac{1.502}{1 + \pi^2 \dfrac{1.502^4}{16(7.63^2)}} \left[1 + 0.15 \left(\frac{\pi^2(1.502^4)}{16(7.63^2)} - \frac{1.502}{\infty} \right) \right] = 1.437$$

(Here r = infinity because the boundary is straight.)

Substituting the values of T, C, and K as before, it is found that $\tau = 252$ lb/in^2.

9.3 *Effect of end constraint*

It was pointed out in Art. 9.2 that when noncircular bars are twisted, the sections do not remain plane but warp, and that the formulas of Table 20 are based on the assumption that this warping is not prevented. If one or both ends of a bar are so fixed that warping is prevented, or if the torque is applied to a section other than at the ends of a bar, the stresses and the angle of twist produced by the given torque are affected. In compact sections the effect is slight, but in the case of open thin-walled sections the effect may be considerable.

Behavior. To visualize the additional support created by warping restraint, consider a very thin rectangular cross section and an I-beam having the same thickness and the same total cross-sectional area as the rectangle. With no warping restraint the two sections will have essentially the same stiffness factor K (see Table 20, cases 4 and 26). With warping restraint provided at one end of the bar, the rectangle will be stiffened very little but the built-in flanges of the I-beam act as cantilever beams. The shear forces developed in the flanges as a result of the bending of these cantilevers will assist the torsional shear stresses in carrying the applied torque and greatly increase the stiffness of the bar unless the length is very great.

Formulas. Table 21 gives formulas for the warping stiffness factor C_w, the torsional stiffness factor K, the location of the shear center, the magnitudes and locations within the cross section of the maximum shear stresses due to simple torsion, the maximum shear stresses due to warping, and the maximum bending stresses due to warping. All the cross sections listed are assumed to have thin walls and the same thickness throughout the section unless otherwise indicated.

Table 22 provides the expressions necessary to evaluate the angle of rotation θ and the first three derivatives of θ along the span for a variety of loadings and boundary restraints. The formulas in this table are based on deformations from bending stresses in the thin-walled open cross sections due to warping restraint and consequently, since the transverse shear deformations of the beam action are neglected, are not applicable to cases where the torsion member is short or where the torsional loading is applied close to a support which provides warping restraint.

In a study of the effect on seven cross sections, all of which were approximately 4 in deep and had walls approximately 0.1 in thick, Schwabenlender (Ref. 28) tested them with one end fixed and the other end free to twist but not warp with the torsional loading applied to the latter end. He found that the effect of the transverse shear stress noticeably reduced the torsional stiffness of cross sections such as those shown in Table 21, cases 1 and 6 to 8, when the length was less than six times the depth; for sections such as those in cases 2 to 5 (Table 21), the effect became appreciable at even greater lengths. To establish an absolute maximum torsional stiffness constant we note that for any cross section, when the length approaches zero, the effective torsional stiffness constant K' cannot exceed J, the polar moment of inertia, where $J = I_x + I_y$ for axes through the centroid of the cross section. (Example 1 illustrates this last condition.)

Reference 19 gives formulas and graphs for the angle of rotation and the first three derivatives for 12 cases of torsional loading of open cross sections. Payne (Ref. 15) gives the solution for a box girder that is fixed at one end and has a torque applied at the other. (This solution was also presented in detail in the fourth edition of this book.) Chu (Ref. 29) and Vlasov (Ref. 30) discuss solutions for cross sections with both open and closed parts. Kollbrunner and Basler (Ref. 31) discuss the warping of continuous beams and consider the multicellular box section, among other cross sections.

EXAMPLES

1. A steel torsion member has a cross section in the form of a twin channel with flanges inward as dimensioned in Fig. 9.3. Both ends of this channel are rigidly welded to massive steel blocks to provide full warping restraint. A torsional load is to be applied to one end block while the other is fixed. Determine the angle of twist at the loaded end for an applied torque of 1000 in-lb for lengths of 100, 50, 25, and 10 in. Assume $E = 30(10^6)$ lb/in² and $\nu = 0.285$.

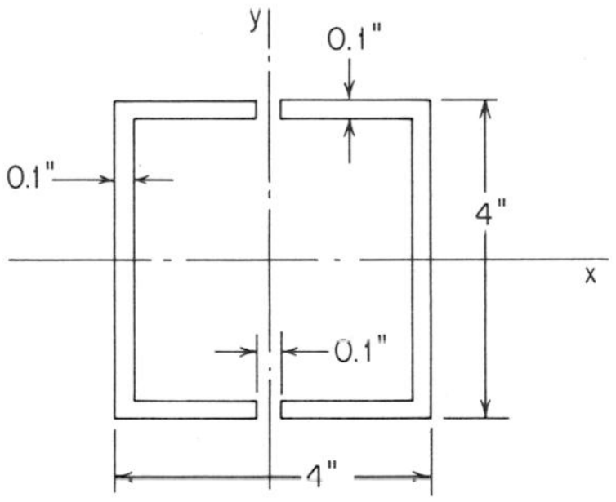

Fig. 9.3

Solution. First determine cross-sectional constants, noting that $b = 4 - 0.1 = 3.9$ in. $b_1 = 1.95 - 0.05 = 1.9$ in, $h = 4 - 0.1 = 3.9$ in, and $t = 0.1$ in. From Table 21, case 4,

$$K = \frac{t^3}{3}(2b + 4b_1) = \frac{0.1^3}{3}[2(3.9) + 4(1.9)] = 0.005133 \text{ in}^4$$

$$C_w = \frac{tb^2}{24}(8b_1^3 + 6h^2b_1 + h^2b + 12b_1^2h) = 28.93 \text{ in}^6$$

TABLE 21 ***Formulas for torsional properties and stresses in thin-walled open cross sections***

NOTATION: Point 0 indicates the shear center. e = distance from a reference to the shear center (inches); K = torsional stiffness constant (inches to the fourth power); C_w = warping constant (inches to the sixth power); τ_1 = shear stress due to torsional rigidity of the cross section (pounds per square inch); τ_2 = shear stress due to warping rigidity of the cross section (pounds per square inch); σ_x = bending stress due to warping rigidity of the cross section (pounds per square inch); E = modulus of elasticity of the material (pounds per square inch); and G = modulus of rigidity (shear modulus) of the material (pounds per square inch)

The appropriate values of θ', θ'', and θ''' are found in Table 22 for the loading and boundary restraints desired

Cross section, reference no.	Constants	Selected maximum values
1. Channel 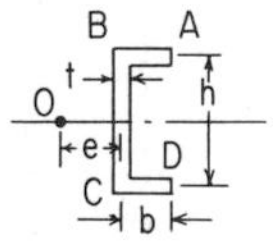	$e = \dfrac{3b^2}{h+6b}$ $K = \dfrac{t^3}{3}(h+2b)$ $C_w = \dfrac{h^2b^3t}{12}\dfrac{2h+3b}{h+6b}$	Max $\sigma_x = \dfrac{hb}{2}\dfrac{h+3b}{h+6b}E\theta''$ throughout the thickness at corners A and D Max $\tau_2 = \dfrac{hb^2}{4}\left[\dfrac{h+3b}{h+6b}\right]^2 E\theta'''$ throughout the thickness at a distance $b\dfrac{h+3b}{h+6b}$ from corners A and D Max $\tau_1 = tG\theta'$ at the surface everywhere
2. *C* section 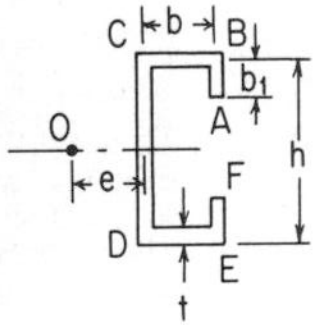	$e = b\dfrac{3h^2b+6h^2b_1-8b_1^3}{h^3+6h^2b+6h^2b_1+8b_1^3-12hb_1^2}$ $K = \dfrac{t^3}{3}(h+2b+2b_1)$ $C_w = t\left[\dfrac{h^2b^2}{2}\left(b_1+\dfrac{b}{3}-e-\dfrac{2eb_1}{b}+\dfrac{2b_1^2}{h}\right) + \dfrac{h^2e^2}{2}\left(b+b_1+\dfrac{h}{6}-\dfrac{2b_1^2}{h}\right)+\dfrac{2b_1^3}{3}(b+e)^2\right]$	Max $\sigma_x = \left[\dfrac{h}{2}(b-e)+b_1(b+e)\right]E\theta''$ throughout the thickness at corners A and F Max $\tau_2 = \left[\dfrac{h}{4}(b-e)(2b_1+b-e)+\dfrac{b_1^2}{2}(b+e)\right]E\theta'''$ throughout the thickness on the top and bottom flanges at a distance e from corners C and D Max $\tau_1 = tG\theta'$ at the surface everywhere
3. Hat section 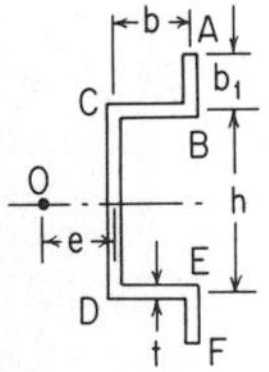	$e = b\dfrac{3h^2b+6h^2b_1-8b_1^3}{h^3+6h^2b+6h^2b_1+8b_1^3+12hb_1^2}$ $K = \dfrac{t^3}{3}(h+2b+2b_1)$ $C_w = t\left[\dfrac{h^2b^2}{2}\left(b_1+\dfrac{b}{3}-e-\dfrac{2eb_1}{b}-\dfrac{2b_1^2}{h}\right) + \dfrac{h^2e^2}{2}\left(b+b_1+\dfrac{h}{6}+\dfrac{2b_1^2}{h}\right)+\dfrac{2b_1^3}{3}(b+e)^2\right]$	$\sigma_x = \left[\dfrac{h}{2}(b-e)-b_1(b+e)\right]E\theta''$ throughout the thickness at corners A and F $\sigma_x = \dfrac{h}{2}(b-e)E\theta''$ throughout the thickness at corners B and E $\tau_2 = \left[\dfrac{h^2(b-e)^2}{8(b+e)}+\dfrac{b_1^2}{2}(b+e)-\dfrac{hb_1}{2}(b-e)\right]E\theta'''$ throughout the thickness at a distance $\dfrac{h(b-e)}{2(b+e)}$ from corner B toward corner A $\tau_2 = \left[\dfrac{b_1^2}{2}(b+e)-\dfrac{hb_1}{2}(b-e)-\dfrac{h}{4}(b-e)^2\right]E\theta'''$ throughout the thickness at a distance e from corner C toward corner B $\tau_1 = tG\theta'$ at the surface everywhere

4. Twin channel with flanges inward 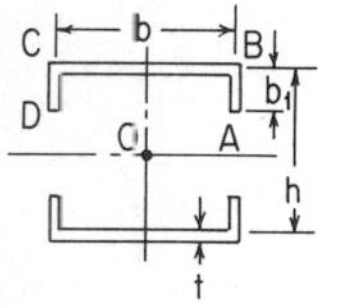	$K = \frac{t^3}{3}(2b + 4b_1)$ $C_w = \frac{tb^2}{24}(8b_1^3 + 6h^2b_1 + h^2b + 12b_1^2h)$	Max $\sigma_x = \frac{b}{2}\left(b_1 + \frac{h}{2}\right)E\theta''$ throughout the thickness at points A and D Max $\tau_2 = \frac{-b}{16}(4b_1^2 + 4b_1h + hb)E\theta'''$ throughout the thickness midway between corners B and C Max $\tau_1 = tG\theta'$ at the surface everywhere
5. Twin channel with flanges outward 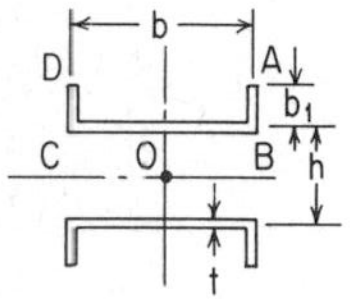	$K = \frac{t^3}{3}(2b + 4b_1)$ $C_w = \frac{tb^2}{24}(8b_1^3 + 6h^2b_1 + h^2b - 12b_1^2h)$	Max $\sigma_x = \frac{hb}{4}E\theta''$ throughout the thickness at points B and C if $h > b_1$ Max $\sigma_x = \left(\frac{hb}{4} - \frac{bb_1}{2}\right)E\theta''$ throughout the thickness at points A and D if $h < b_1$ Max $\tau_2 = \frac{b}{4}\left(\frac{h}{2} - b_1\right)^2 E\theta'''$ throughout the thickness at a distance $\frac{h}{2}$ from corner B toward point A if $b_1 > \frac{h}{2}\left(1 + \sqrt{\frac{1}{2} + \frac{b}{2h}}\right)$ Max $\tau_2 = \frac{b}{4}\left(b_1^2 - \frac{hb}{4} - hb_1\right)E\theta'''$ throughout the thickness at a point midway between corners B and C if $b_1 < \frac{h}{2}\left(1 + \sqrt{\frac{1}{2} + \frac{b}{2h}}\right)$ Max $\tau_1 = tG\theta'$ at the surface everywhere
6. Wide flanged beam with equal flanges 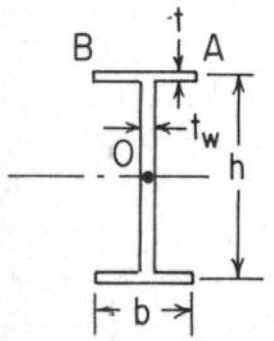	$K = \frac{1}{3}(2t^3b + t_w^3h)$ $C_w = \frac{h^2tb^3}{24}$	Max $\sigma_x = \frac{hb}{4}E\theta''$ throughout the thickness at points A and B Max $\tau_2 = -\frac{hb^2}{16}E\theta'''$ throughout the thickness at a point midway between A and B Max $\tau_1 = tG\theta'$ at the surface everywhere

TABLE 21 ***Formulas for torsional properties and stresses in thin-walled open cross sections (Cont.)***

Cross section, reference no.	Constants	Selected maximum values
7. Wide flanged beam with unequal flanges 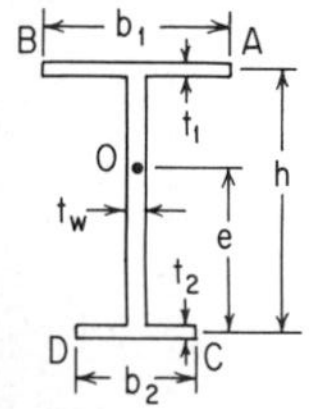	$e = \dfrac{t_1 b_1^3 h}{t_1 b_1^3 + t_2 b_2^3}$ $K = \frac{1}{3}(t_1^3 b_1 + t_2^3 b_2 + t_w^3 h)$ $C_w = \dfrac{h^2 t_1 t_2 b_1^3 b_2^3}{12(t_1 b_1^3 + t_2 b_2^3)}$	Max $\sigma_x = \dfrac{hb_1}{2} \dfrac{t_2 b_2^3}{t_1 b_1^3 + t_2 b_2^3} E\theta''$ throughout the thickness at points A and B if $t_2 b_2^2 > t_1 b_1^2$ Max $\sigma_x = \dfrac{hb_2}{2} \dfrac{t_1 b_1^3}{t_1 b_1^3 + t_2 b_2^3} E\theta''$ throughout the thickness at points C and D if $t_2 b_2^2 < t_1 b_1^2$ Max $\tau_2 = \dfrac{-1}{8} \dfrac{h t_2 b_2^3 b_1^2}{t_1 b_1^3 + t_2 b_2^3} E\theta'''$ throughout the thickness at a point midway between A and B if $t_2 b_2 > t_1 b_1$ Max $\tau_2 = \dfrac{-1}{8} \dfrac{h t_1 b_1^3 b_2^2}{t_1 b_1^3 + t_2 b_2^3} E\theta'''$ throughout the thickness at a point midway between C and D if $t_2 b_2 < t_1 b_1$ Max $\tau_1 = t_{max} G\theta'$ at the surface on the thickest portion
8. *Z* section 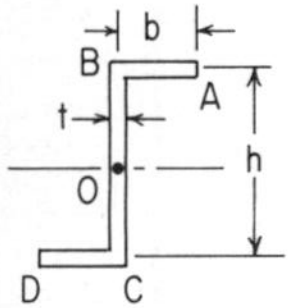	$K = \dfrac{t^3}{3}(2b + h)$ $C_w = \dfrac{th^2 b^3}{12}\left(\dfrac{b + 2h}{2b + h}\right)$	Max $\sigma_x = \dfrac{hb}{2} \dfrac{b + h}{2b + h} E\theta''$ throughout the thickness at points A and D Max $\tau_2 = \dfrac{-hb^2}{4}\left(\dfrac{b + h}{2b + h}\right)^2 E\theta'''$ throughout the thickness at a distance $\dfrac{b(b + h)}{2b + h}$ from point A Max $\tau_1 = tG\theta'$ at the surface everywhere
9. Segment of a circular tube 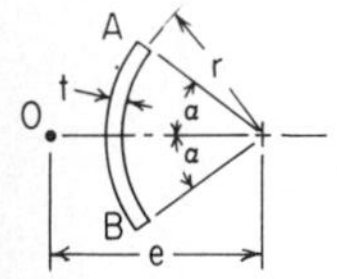	$e = 2r \dfrac{\sin\alpha - \alpha\cos\alpha}{\alpha - \sin\alpha\cos\alpha}$ $K = \frac{2}{3}t^3 r\alpha$ $C_w = \dfrac{2tr^5}{3}\left[\alpha^3 - 6\dfrac{(\sin\alpha - \alpha\cos\alpha)^2}{\alpha - \sin\alpha\cos\alpha}\right]$	Max $\sigma_x = (r^2\alpha - re\sin\alpha)E\theta''$ throughout the thickness at points A and B Max $\tau_2 = r^2\left[e(1 - \cos\alpha) - \dfrac{r\alpha^2}{2}\right]E\theta'''$ throughout the thickness at midlength Max $\tau_1 = tG\theta'$ at the surface everywhere

(*Note:* If t/r is small, α can be larger than π to evaluate constants for the case when the walls overlap)

10. 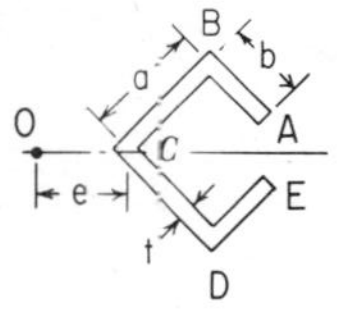	$e = 0.707ab^2 \dfrac{3a - 2b}{2a^3 - (a - b)^3}$ $K = \frac{2}{3}t^3(a + b)$ $C_w = \dfrac{ta^4b^3}{6} \dfrac{4a + 3b}{2a^3 - (a - b)^3}$	$\text{Max } \sigma_x = \dfrac{a^2b}{2} \dfrac{2a^2 + 3ab - b^2}{2a^3 - (a - b)^3} E\theta''$ throughout the thickness at points A and E $\tau_2 = \dfrac{-a^2b^2}{4} \dfrac{-a^2 - 2ab + b^2}{2a^3 - (a - b)^3} E\theta'''$ throughout the thickness at point C $\text{Max } \tau_1 = tG\theta'$ at the surface everywhere
11. 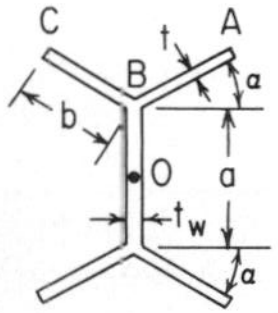	$K = \frac{1}{3}(4t^3b + t_w^3a)$ $C_w = \dfrac{a^2b^3t}{3} \cos^2 \alpha$ (*Note:* Expressions are equally valid for $+$ and $-\alpha$)	$\text{Max } \sigma_x = \dfrac{ab}{2} \cos \alpha \, E\theta''$ throughout the thickness at points A and C $\text{Max } \tau_2 = \dfrac{-ab^2}{4} \cos \alpha \, E\theta'''$ throughout the thickness at point B $\text{Max } \tau_1 = tG\theta'$ at the surface everywhere

TABLE 22 *Formulas for the angle of twist of uniform thin-walled elastic members under torsional loading*

NOTATION: T_o = applied torsional load (inch-pounds); t_o = applied distributed torsional load (inch-pounds per linear inch); T_A and T_B are the reaction end torques at the left and right ends, respectively. θ = angle of rotation at a distance x from the left end (radians). θ', θ'', and θ''' are the successive derivatives of θ with respect to the distance x. All rotations, applied torsional loads, and reaction end torques are positive as shown (*CCW* when viewed from the right end of the member). E is the modulus of elasticity of the material; C_w is the warping constant for the cross section; K is the torsional constant (see Table 21 for expressions for C_w and K); and G is the modulus of rigidity (shear modulus) of the material.

The following constants and functions are hereby defined in order to permit condensing the tabulated formulas which follow. See page 94 for a definition of $\langle x-a\rangle^n$. The function Sinh $\beta\langle x-a\rangle$ is also defined as zero if x is less than a. $\beta = (KG/C_wE)^{\frac{1}{2}}$

$F_1 = \text{Cosh}\,\beta x$	$F_{a1} = \langle x-a\rangle^0\,\text{Cosh}\,\beta(x-a)$	$C_1 = \text{Cosh}\,\beta l$	$C_{a1} = \text{Cosh}\,\beta(l-a)$	$A_1 = \text{Cosh}\,\beta a$
$F_2 = \text{Sinh}\,\beta x$	$F_{a2} = \text{Sinh}\,\beta\langle x-a\rangle$	$C_2 = \text{Sinh}\,\beta l$	$C_{a2} = \text{Sinh}\,\beta(l-a)$	$A_2 = \text{Sinh}\,\beta a$
$F_3 = \text{Cosh}\,\beta x - 1$	$F_{a3} = \langle x-a\rangle^0[\text{Cosh}\,\beta(x-a) - 1]$	$C_3 = \text{Cosh}\,\beta l - 1$	$C_{a3} = \text{Cosh}\,\beta(l-a) - 1$	
$F_4 = \text{Sinh}\,\beta x - \beta x$	$F_{a4} = \text{Sinh}\,\beta\langle x-a\rangle - \beta\langle x-a\rangle$	$C_4 = \text{Sinh}\,\beta l - \beta l$	$C_{a4} = \text{Sinh}\,\beta(l-a) - \beta(l-a)$	
	$F_{a5} = F_{a3} - \dfrac{\beta^2\langle x-a\rangle^2}{2}$		$C_{a5} = C_{a3} - \dfrac{\beta^2(l-a)^2}{2}$	
	$F_{a6} = F_{a4} - \dfrac{\beta^3\langle x-a\rangle^3}{6}$		$C_{a6} = C_{a4} - \dfrac{\beta^3(l-a)^3}{6}$	

1. Concentrated intermediate torque

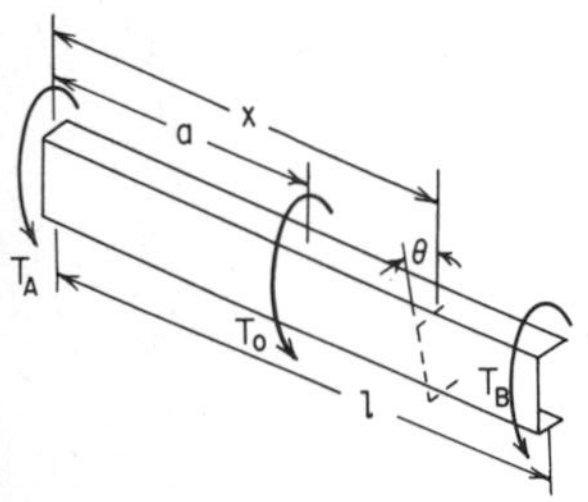

$$\theta = \theta_A + \frac{\theta'_A}{\beta}F_2 + \frac{\theta''_A}{\beta^2}F_3 + \frac{T_A}{C_wE\beta^3}F_4 + \frac{T_o}{C_wE\beta^3}F_{a4}$$

$$\theta' = \theta'_AF_1 + \frac{\theta''_A}{\beta}F_2 + \frac{T_A}{C_wE\beta^2}F_3 + \frac{T_o}{C_wE\beta^2}F_{a3}$$

$$\theta'' = \theta''_AF_1 + \theta'_A\beta F_2 + \frac{T_A}{C_wE\beta}F_2 + \frac{T_o}{C_wE\beta}F_{a2}$$

$$\theta''' = \theta'_A\beta^2F_1 + \theta''_A\beta F_2 + \frac{T_A}{C_wE}F_1 + \frac{T_o}{C_wE}F_{a1}$$

$$T = T_A + T_o\langle x-a\rangle^0$$

End restraints, reference no.	Boundary values	Selected special cases and maximum values
1a. Left end free to twist and warp, right end free to warp but not twist 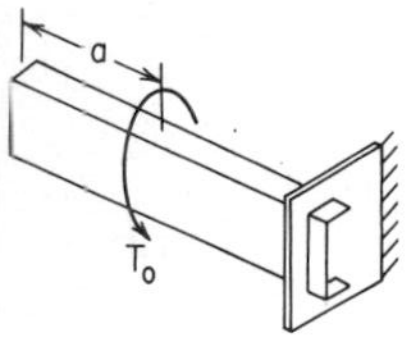	$T_A = 0 \quad \theta''_A = 0$ $\theta_A = \frac{T_o}{C_w E\beta^2}(l-a)$ $\theta'_A = \frac{-T_o}{C_w E\beta^2}\frac{C_{a2}}{C_2}$ $\theta_B = 0 \quad \theta''_B = 0 \quad T_B = -T_o$ $\theta'_B = \frac{-T_o}{C_w E\beta^2}\left(1 - \frac{A_2}{C_2}\right)$ $\theta'''_B = \frac{T_o}{C_w E}\frac{A_2}{C_2}$	Max $\theta = \theta_A$; max possible value $= \frac{T_o l}{C_w E\beta^2}$ when $a = 0$ Max $\theta' = \theta'_B$; max possible value $= \frac{-T_o}{C_w E\beta^2}$ when $a = 0$ Max $\theta'' = \frac{-T_o}{C_w E\beta}\frac{C_{a2}}{C_2}A_2$ at $x = a$; max possible value $= \frac{-T_o}{2C_w E\beta}\text{Tanh}\frac{\beta l}{2}$ when $a = l/2$ Max $-\theta''' = \frac{-T_o}{C_w E}\frac{C_{a2}}{C_2}A_1$ just left of $x = a$ Max $+\theta''' = \frac{T_o}{C_w E}\left(1 - \frac{C_{a2}}{C_2}A_1\right)$ just right of $x = a$; max possible value $= \frac{T_o}{C_w E}$ when a approaches l If $a = 0$ (torque applied at the left end), $\theta = \frac{T_o}{KG}(l-x) \quad \theta' = \frac{-T_o}{KG} \quad \theta'' = 0 \quad \theta''' = 0$
1b. Left end free to twist and warp, right end fixed (no twist or warp) 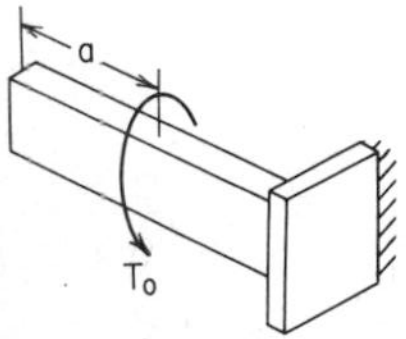	$T_A = 0 \quad \theta''_A = 0$ $\theta_A = \frac{T_o}{C_w E\beta^3}\left(\frac{C_2 C_{a3}}{C_1} - C_{a4}\right)$ $\theta'_A = \frac{-T_o}{C_w E\beta^2}\frac{C_{a3}}{C_1}$ $\theta_B = 0 \quad \theta'_B = 0 \quad T_B = -T_o$ $\theta''_B = \frac{-T_o}{C_w E\beta}\frac{A_2 - C_2}{C_1}$ $\theta'''_B = \frac{T_o}{C_w E}$	If $a = 0$ (torque applied at the left end), $\theta = \frac{T_o}{C_w E\beta^3}\left[\beta(l-x) - \text{Tanh}\,\beta l + \frac{\text{Sinh}\,\beta x}{\text{Cosh}\,\beta l}\right]$ $\theta' = \frac{-T_o}{C_w E\beta^2}\left(1 - \frac{\text{Cosh}\,\beta x}{\text{Cosh}\,\beta l}\right) \quad \theta'' = \frac{T_o}{C_w E\beta}\frac{\text{Sinh}\,\beta x}{\text{Cosh}\,\beta l}$ $\theta''' = \frac{T_o}{C_w E}\frac{\text{Cosh}\,\beta x}{\text{Cosh}\,\beta l}$ Max $\theta = \frac{T_o}{C_w E\beta^3}(\beta l - \text{Tanh}\,\beta l)$ at $x = 0$ Max $\theta' = \frac{-T_o}{C_w E\beta^2}\left(1 - \frac{1}{\text{Cosh}\,\beta l}\right)$ at $x = 0$ Max $\theta'' = \frac{T_o}{C_w E\beta}\text{Tanh}\,\beta l$ at $x = l$ Max $\theta''' = \frac{T_o}{C_w E}$ at $x = l$

TABLE 22 *Formulas for the angle of twist of uniform thin-walled elastic members under torsional loading* (Cont.)

End restraints, reference no.	Boundary values	Selected special cases and maximum values
1c. Left end free to twist but not warp, right end free to warp but not twist 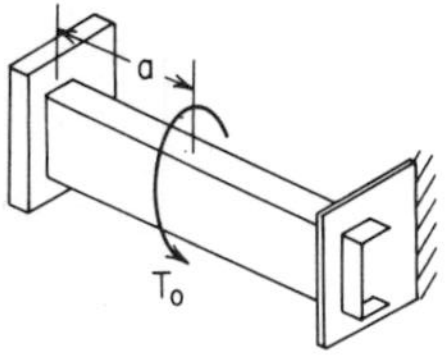	$T_A = 0 \quad \theta'_A = 0 \quad \theta'''_A = 0$ $\theta_A = \dfrac{T_o}{C_w E \beta^3}\left(\dfrac{C_3 C_{a2}}{C_1} - C_{a4}\right)$ $\theta''_A = \dfrac{-T_o}{C_w E \beta}\dfrac{C_{a2}}{C_1}$ $\theta_B = 0 \quad \theta''_B = 0 \quad T_B = -T_o$ $\theta'_B = \dfrac{-T_o}{C_w E \beta^2}\left(\dfrac{C_2 C_{a2}}{C_1} - C_{a3}\right)$ $\theta'''_B = \dfrac{-T_o}{C_w E}\left(\dfrac{C_2 C_{a2}}{C_1} - C_{a1}\right)$	If $a = 0$ (torque applied at the left end), $\theta = \dfrac{T_o}{C_w E \beta^3}[\text{Sinh}\,\beta x - \text{Tanh}\,\beta l\,\text{Cosh}\,\beta x + \beta(l - x)]$ $\theta' = \dfrac{T_o}{C_w E \beta^2}(\text{Cosh}\,\beta x - \text{Tanh}\,\beta l\,\text{Sinh}\,\beta x - 1)$ $\theta'' = \dfrac{T_o}{C_w E \beta}(\text{Sinh}\,\beta x - \text{Tanh}\,\beta l\,\text{Cosh}\,\beta x) \qquad \theta''' = \dfrac{T_o}{C_w E}(\text{Cosh}\,\beta x - \text{Tanh}\,\beta l\,\text{Sinh}\,\beta x)$ $\text{Max}\,\theta = \dfrac{T_o}{C_w E \beta^3}(\beta l - \text{Tanh}\,\beta l)$ at $x = 0$ $\text{Max}\,\theta' = \dfrac{-T_o}{C_w E \beta^2}\left(\dfrac{-1}{\text{Cosh}\,\beta l} + 1\right)$ at $x = l$ $\text{Max}\,\theta'' = \dfrac{-T_o}{C_w E \beta}\text{Tanh}\,\beta l$ at $x = 0$ $\text{Max}\,\theta''' = \dfrac{T_o}{C_w E}$ at $x = 0$
1d. Left end free to twist but not warp, right end fixed (no twist or warp) 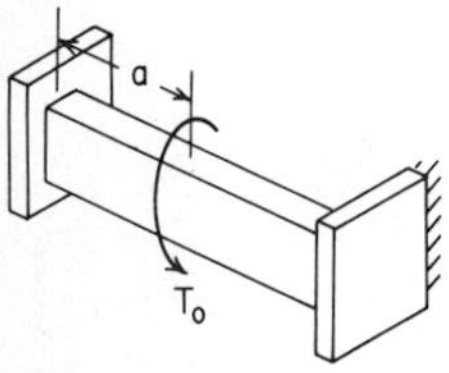	$T_A = 0 \quad \theta'_A = 0$ $\theta_A = \dfrac{T_o}{C_w E \beta^3}\left(\dfrac{C_3 C_{a3}}{C_2} - C_{a4}\right)$ $\theta''_A = \dfrac{-T_o}{C_w E \beta}\dfrac{C_{a3}}{C_2}$ $\theta_B = 0 \quad \theta'_B = 0 \quad T_B = -T_o$ $\theta''_B = \dfrac{-T_o}{C_w E \beta}\left(\dfrac{C_1 C_{a3}}{C_2} - C_{a2}\right)$ $\theta'''_B = \dfrac{T_o}{C_w E}$	If $a = 0$ (torque applied at the left end), $\theta = \dfrac{T_o}{C_w E \beta^3}\left[\text{Sinh}\,\beta x + \beta(l - x) - \text{Tanh}\,\dfrac{\beta l}{2}(1 + \text{Cosh}\,\beta x)\right]$ $\theta' = \dfrac{T_o}{C_w E \beta^2}\left(\text{Cosh}\,\beta x - 1 - \text{Tanh}\,\dfrac{\beta l}{2}\,\text{Sinh}\,\beta x\right)$ $\theta'' = \dfrac{T_o}{C_w E \beta}\left(\text{Sinh}\,\beta x - \text{Tanh}\,\dfrac{\beta l}{2}\,\text{Cosh}\,\beta x\right) \qquad \theta''' = \dfrac{T_o}{C_w E}\left(\text{Cosh}\,\beta x - \text{Tanh}\,\dfrac{\beta l}{2}\,\text{Sinh}\,\beta x\right)$ $\text{Max}\,\theta = \dfrac{T_o}{C_w E \beta^3}\left(\beta l - 2\,\text{Tanh}\,\dfrac{\beta l}{2}\right)$ at $x = 0$ $\text{Max}\,\theta' = \dfrac{T_o}{C_w E \beta^2}\left[\dfrac{1}{\text{Cosh}\,(\beta l/2)} - 1\right]$ at $x = \dfrac{l}{2}$ $\text{Max}\,\theta'' = \dfrac{\mp T_o}{C_w E \beta}\text{Tanh}\,\dfrac{\beta l}{2}$ at $x = 0$ and $x = l$, respectively $\text{Max}\,\theta''' = \dfrac{T_o}{C_w E}$ at $x = 0$ and $x = l$

1e. Both ends free to warp but not twist 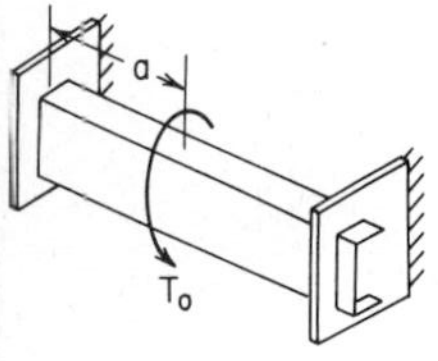 	$\theta_A = 0 \qquad \theta''_A = 0$ $T_A = -T_o\left(1 - \frac{a}{l}\right)$ $\theta'_A = \frac{T_o}{C_w E\beta^2}\left(1 - \frac{a}{l} - \frac{C_{a2}}{C_2}\right)$ $\theta_B = 0 \qquad \theta''_B = 0$ $\theta'_B = \frac{-T_o}{C_w E\beta^2}\left(\frac{a}{l} - \frac{A_2}{C_2}\right)$ $\theta'''_B = \frac{T_o}{C_w E}\frac{A_2}{C_2}$ $T_B = -T_o\frac{a}{l}$	If $a = l/2$ (torque applied at midlength), $\theta = \frac{T_o}{C_w E\beta^3}\left[\frac{\beta x}{2} - \frac{\text{Sinh}\,\beta x}{2\,\text{Cosh}\,(\beta l/2)} + \text{Sinh}\,\beta\left\langle x - \frac{l}{2}\right\rangle - \beta\left\langle x - \frac{l}{2}\right\rangle\right]$ $\theta' = \frac{T_o}{C_w E\beta^2}\left\{\frac{1}{2} - \frac{\text{Cosh}\,\beta x}{2\,\text{Cosh}\,(\beta l/2)} + \left\langle x - \frac{l}{2}\right\rangle^0\left[\text{Cosh}\,\beta\left(x - \frac{l}{2}\right) - 1\right]\right\}$ $\theta'' = \frac{-T_o}{C_w E\beta}\left[\frac{\text{Sinh}\,\beta x}{2\,\text{Cosh}\,(\beta l/2)} - \text{Sinh}\,\beta\left\langle x - \frac{l}{2}\right\rangle\right]$ $\theta''' = \frac{-T_o}{C_w E}\left[\frac{\text{Cosh}\,\beta x}{2\,\text{Cosh}\,(\beta l/2)} - \left\langle x - \frac{l}{2}\right\rangle^0 \text{Cosh}\,\beta\left(x - \frac{l}{2}\right)\right]$ $\text{Max}\,\theta = \frac{T_o}{2C_w E\beta^3}\left(\frac{\beta l}{2} - \text{Tanh}\,\frac{\beta l}{2}\right)$ at $x = \frac{l}{2}$ $\text{Max}\,\theta' = \frac{\pm T_o}{2C_w E\beta^2}\left[1 - \frac{1}{\text{Cosh}\,(\beta l/2)}\right]$ at $x = 0$ and $x = l$, respectively $\text{Max}\,\theta'' = \frac{-T_o}{2C_w E\beta}\,\text{Tanh}\,\frac{\beta l}{2}$ at $x = \frac{l}{2}$ $\text{Max}\,\theta''' = \frac{\mp T_o}{2C_w E}$ just left and just right of $x = \frac{l}{2}$, respectively
1f. Left end free to warp but not twist, right end fixed (no twist or warp) 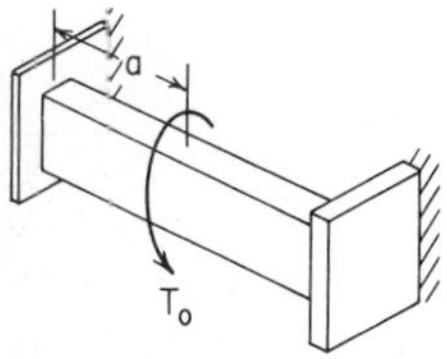 	$\theta_A = 0 \qquad \theta''_A = 0$ $T_A = -T_o\frac{C_1 C_{a4} - C_2 C_{a3}}{C_1 C_4 - C_2 C_3}$ $\theta'_A = \frac{T_o}{C_w E\beta^2}\frac{C_3 C_{a4} - C_4 C_{a3}}{C_1 C_4 - C_2 C_3}$ $\theta_B = 0 \qquad \theta'_B = 0$ $\theta''_B = \frac{T_o}{C_w E\beta}\frac{\beta l A_2 - \beta a C_2}{C_1 C_4 - C_2 C_3}$ $\theta'''_B = \frac{T_o}{C_w E}\frac{A_2 - \beta a C_1}{C_1 C_4 - C_2 C_3}$ $T_B = -T_o - T_A$	If $a = l/2$ (torque applied at midlength), $T_A = -T_o\frac{\text{Sinh}\,\beta l - (\beta l/2)\,\text{Cosh}\,\beta l - \text{Sinh}\,(\beta l/2)}{\text{Sinh}\,\beta l - \beta l\,\text{Cosh}\,\beta l}$ $\theta'_A = \frac{T_o}{C_w E\beta^2}\frac{2\,\text{Sinh}\,(\beta l/2) - \beta l\,\text{Cosh}\,(\beta l/2)}{\text{Sinh}\,\beta l - \beta l\,\text{Cosh}\,\beta l}\left(\text{Cosh}\,\frac{\beta l}{2} - 1\right)$

TABLE 22 ***Formulas for the angle of twist of uniform thin-walled elastic members under torsional loading (Cont.)***

End restraints, reference no.	Boundary values	Selected special cases and maximum values
1g. Both ends fixed (no twist or warp) 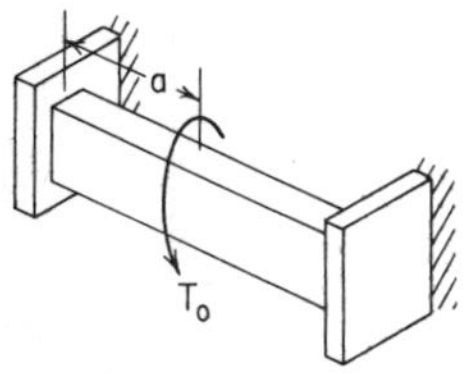 	$\theta_A = 0 \qquad \theta'_A = 0$ $\theta''_A = \frac{T_o}{C_w E\beta} \frac{C_3 C_{a4} - C_4 C_{a3}}{C_2 C_4 - C_3^2}$ $T_A = -T_o \frac{C_2 C_{a4} - C_3 C_{a3}}{C_2 C_4 - C_3^2}$ $\theta_B = 0 \qquad \theta'_B = 0$ $\theta''_B = \theta''_A C_1 + \frac{T_A}{C_w E\beta} C_2 + \frac{T_o}{C_w E\beta} C_{a2}$ $\theta'''_B = \theta''_A \beta C_2 + \frac{T_A}{C_w E} C_1 + \frac{T_o}{C_w E} C_{a1}$ $T_B = -T_o - T_A$	If $a = l/2$ (torque applied at midlength), $T_A = T_B = \frac{-T_o}{2}$ $\text{Max } \theta = \frac{T_o}{C_w E\beta^3}\left(\frac{\beta l}{4} - \text{Tanh}\,\frac{\beta l}{4}\right)$ at $x = \frac{l}{2}$ $\text{Max } \theta' = \frac{T_o}{2C_w E\beta^2}\left[1 - \frac{1}{\text{Cosh}\,(\beta l/4)}\right]$ at $x = \frac{l}{4}$ $\text{Max } \theta'' = \frac{\pm}{+} \frac{T_o}{2C_w E\beta} \text{Tanh}\,\frac{\beta l}{4}$ at $x = 0$, $x = \frac{l}{2}$, and $x = l$, respectively $\text{Max } -\theta''' = \frac{-T_o}{2C_w E}$ at $x = 0$ and just left of $x = \frac{l}{2}$ $\text{Max } +\theta''' = \frac{T_o}{2C_w E}$ at $x = l$ and just right of $x = \frac{l}{2}$
2. Uniformly distributed torque from a to l 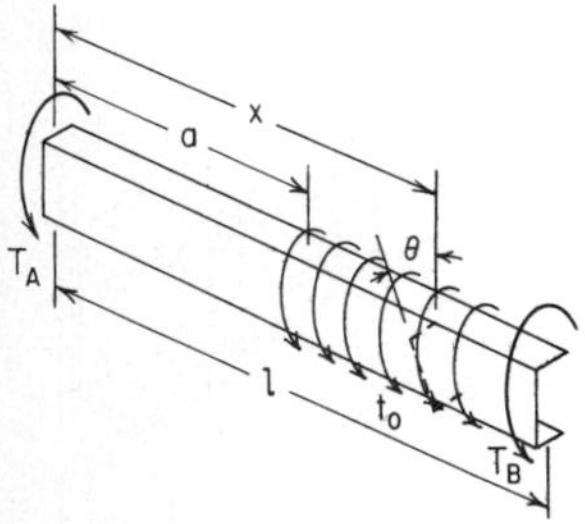 	$\theta = \theta_A + \frac{\theta'_A}{\beta} F_2 + \frac{\theta''_A}{\beta^2} F_3 + \frac{T_A}{C_w E\beta^3} F_4 + \frac{t_o}{C_w E\beta^4} F_{a5}$ $\theta' = \theta'_A F_1 + \frac{\theta''_A}{\beta} F_2 + \frac{T_A}{C_w E\beta^2} F_3 + \frac{t_o}{C_w E\beta^3} F_{a4}$ $\theta'' = \theta''_A F_1 + \theta'_A \beta F_2 + \frac{T_A}{C_w E\beta} F_2 + \frac{t_o}{C_w E\beta^2} F_{a3}$ $\theta''' = \theta'_A \beta^2 F_1 + \theta''_A \beta F_2 + \frac{T_A}{C_w E} F_1 + \frac{t_o}{C_w E\beta} F_{a2}$ $T = T_A + t_o\langle x - a\rangle$	

	Boundary values	Selected special cases and maximum values
2a. Left end free to twist and warp, right end free to warp but not twist 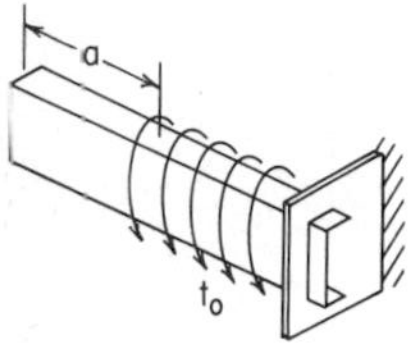 	$T_A = 0 \quad \theta''_A = 0$ $\theta_A = \frac{t_o}{2C_wE\beta^2}(l-a)^2$ $\theta'_A = \frac{-t_o}{C_wE\beta^3}\frac{C_{a3}}{C_2}$ $\theta_B = 0 \quad \theta''_B = 0$ $\theta'_B = \frac{-t_o}{C_wE\beta^3}\left[\frac{A_1 - C_1}{C_2} + \beta(l-a)\right]$ $\theta'''_B = \frac{-t_o}{C_wE\beta}\frac{A_1 - C_1}{C_2}$ $T_B = -t_o(l-a)$	If $a = 0$ (uniformly distributed torque over entire span), $\theta = \frac{t_o}{C_wE\beta^4}\left[\frac{\beta^2(l^2-x^2)}{2} + \frac{\text{Sinh}\,\beta(l-x) + \text{Sinh}\,\beta x}{\text{Sinh}\,\beta l} - 1\right]$ $\theta' = \frac{-t_o}{C_wE\beta^3}\left[\beta x + \frac{\text{Cosh}\,\beta(l-x) - \text{Cosh}\,\beta x}{\text{Sinh}\,\beta l}\right]$ $\theta'' = \frac{-t_o}{C_wE\beta^2}\left[1 - \frac{\text{Sinh}\,\beta(l-x) + \text{Sinh}\,\beta x}{\text{Sinh}\,\beta l}\right]$ $\theta''' = \frac{t_o}{C_wE\beta}\frac{\text{Cosh}\,\beta(l-x) - \text{Cosh}\,\beta x}{\text{Sinh}\,\beta l}$ $\text{Max}\,\theta = \frac{t_o l^2}{2C_wE\beta^2}$ at $x = 0$ $\text{Max}\,\theta' = \frac{-t_o}{C_wE\beta^3}\left(\beta l - \text{Tanh}\,\frac{\beta l}{2}\right)$ at $x = l$ $\text{Max}\,\theta'' = \frac{-t_o}{C_wE\beta^2}\left[1 - \frac{1}{\text{Cosh}\,(\beta l/2)}\right]$ at $x = \frac{l}{2}$ $\text{Max}\,\theta''' = \frac{\mp t_o}{C_wE\beta}\text{Tanh}\,\frac{\beta l}{2}$ at $x = 0$ and $x = l$, respectively
2b. Left end free to twist and warp, right end fixed (no twist or warp) 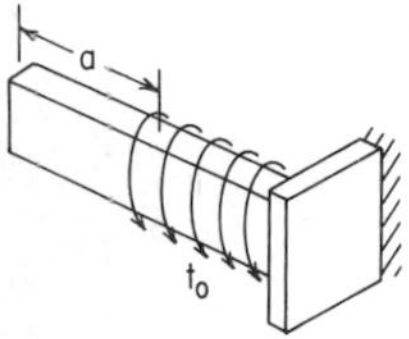 	$T_A = 0 \quad \theta''_A = 0$ $\theta_A = \frac{t_o}{C_wE\beta^4}\left(\frac{C_2C_{a4}}{C_1} - C_{a5}\right)$ $\theta'_A = \frac{-t_o}{C_wE\beta^3}\frac{C_{a4}}{C_1}$ $\theta_B = 0 \quad \theta'_B = 0$ $\theta''_B = \frac{-t_o}{C_wE\beta^2}\left(\frac{C_2C_{a4}}{C_1} - C_{a3}\right)$ $\theta'''_B = \frac{t_o}{C_wE}(l-a)$ $T_B = -t_o(l-a)$	If $a = 0$ (uniformly distributed torque over entire span) $\theta = \frac{-t_o}{C_wE\beta^4}\left[\frac{1 - \text{Cosh}\,\beta(l-x) + \beta l\,(\text{Sinh}\,\beta l - \text{Sinh}\,\beta x)}{\text{Cosh}\,\beta l} - \frac{\beta^2(l^2-x^2)}{2}\right]$ $\theta' = \frac{-t_o}{C_wE\beta^3}\left[\frac{\text{Sinh}\,\beta(l-x) - \beta l\,\text{Cosh}\,\beta x}{\text{Cosh}\,\beta l} + \beta x\right]$ $\theta'' = \frac{-t_o}{C_wE\beta^2}\left[1 - \frac{\text{Cosh}\,\beta(l-x) + \beta l\,\text{Sinh}\,\beta x}{\text{Cosh}\,\beta l}\right]$ $\theta''' = \frac{-t_o}{C_wE\beta}\frac{\text{Sinh}\,\beta(l-x) - \beta l\,\text{Cosh}\,\beta x}{\text{Cosh}\,\beta l}$ $\text{Max}\,\theta = \frac{t_o}{C_wE\beta^4}\left(1 + \frac{\beta^2 l^2}{2} - \frac{1 + \beta l\,\text{Sinh}\,\beta l}{\text{Cosh}\,\beta l}\right)$ at $x = 0$ The location of max θ' depends upon βl $\text{Max}\,\theta'' = \frac{t_o}{C_wE\beta^2}\left(\frac{1 + \beta l\,\text{Sinh}\,\beta l}{\text{Cosh}\,\beta l} - 1\right)$ at $x = l$ $\text{Max}\,\theta''' = \frac{t_o l}{C_wE}$ at $x = l$

TABLE 22 *Formulas for the angle of twist of uniform thin-walled elastic members under torsional loading* (*Cont.*)

End restraints, reference no.	Boundary values	Selected special cases and maximum values
2c. Left end free to twist but not warp, right end free to warp but not twist 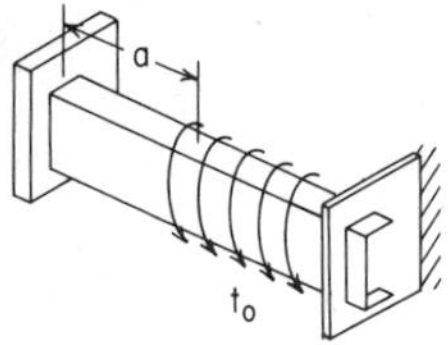 	$T_A = 0 \quad \theta'_A = 0 \quad \theta'''_A = 0$ $\theta_A = \frac{t_o}{C_w E \beta^4}\left(\frac{C_3 C_{a3}}{C_1} - C_{a5}\right)$ $\theta''_A = \frac{-t_o}{C_w E \beta^2}\frac{C_{a3}}{C_1}$ $\theta_B = 0 \quad \theta''_B = 0$ $\theta'_B = \frac{-t_o}{C_w E \beta^3}\left(\frac{C_2 C_{a3}}{C_1} - C_{a4}\right)$ $\theta'''_B = \frac{-t_o}{C_w E \beta}\left(\frac{C_2 C_{a3}}{C_1} - C_{a2}\right)$ $T_B = -t_o(l - a)$	If $a = 0$ (uniformly distributed torque over entire span), $\theta = \frac{t_o}{C_w E \beta^4}\left[\frac{\beta^2(l^2 - x^2)}{2} + \frac{\text{Cosh}\,\beta x}{\text{Cosh}\,\beta l} - 1\right]$ $\theta' = \frac{-t_o}{C_w E \beta^3}\left(\beta x - \frac{\text{Sinh}\,\beta x}{\text{Cosh}\,\beta l}\right)$ $\theta'' = \frac{-t_o}{C_w E \beta^2}\left(1 - \frac{\text{Cosh}\,\beta x}{\text{Cosh}\,\beta l}\right) \qquad \theta''' = \frac{t_o}{C_w E \beta}\frac{\text{Sinh}\,\beta x}{\text{Cosh}\,\beta l}$ $\text{Max}\,\theta = \frac{t_o}{C_w E \beta^4}\left(\frac{\beta^2 l^2}{2} + \frac{1}{\text{Cosh}\,\beta l} - 1\right)$ at $x = 0$ $\text{Max}\,\theta' = \frac{-t_o}{C_w E \beta^3}(\beta l - \text{Tanh}\,\beta l)$ at $x = l$ $\text{Max}\,\theta'' = \frac{-t_o}{C_w E \beta^2}\left(1 - \frac{1}{\text{Cosh}\,\beta l}\right)$ at $x = 0$ $\text{Max}\,\theta''' = \frac{t_o}{C_w E \beta}\text{Tanh}\,\beta l$ at $x = l$
2d. Left end free to twist but not warp, right end fixed (no twist or warp) 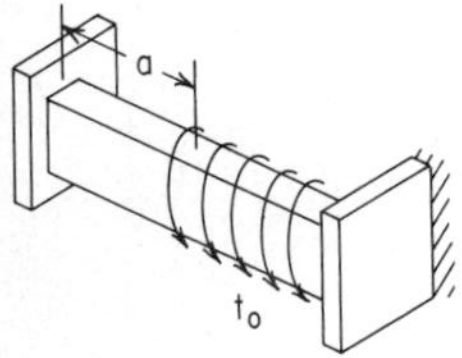 	$T_A = 0 \quad \theta'_A = 0$ $\theta_A = \frac{t_o}{C_w E \beta^4}\left(\frac{C_3 C_{a4}}{C_2} - C_{a5}\right)$ $\theta''_A = \frac{-t_o}{C_w E \beta^2}\frac{C_{a4}}{C_2}$ $\theta_B = 0 \quad \theta'_B = 0$ $\theta''_B = \frac{-t_o}{C_w E \beta^2}\left(\frac{C_1 C_{a4}}{C_2} - C_{a3}\right)$ $\theta'''_B = \frac{t_o}{C_w E}(l - a)$ $T_B = -t_o(l - a)$	If $a = 0$ (uniformly distributed torque over entire span), $\theta = \frac{t_o}{C_w E \beta^4}\left[\frac{\beta^2}{2}(l^2 - x^2) + \beta l \frac{\text{Cosh}\,\beta x - \text{Cosh}\,\beta l}{\text{Sinh}\,\beta l}\right]$ $\theta' = \frac{-t_o}{C_w E \beta^3}\left(\beta x - \frac{\beta l\,\text{Sinh}\,\beta x}{\text{Sinh}\,\beta l}\right)$ $\theta'' = \frac{-t_o}{C_w E \beta^2}\left(1 - \frac{\beta l\,\text{Cosh}\,\beta x}{\text{Sinh}\,\beta l}\right) \qquad \theta''' = \frac{t_o l}{C_w E}\frac{\text{Sinh}\,\beta x}{\text{Sinh}\,\beta l}$ $\text{Max}\,\theta = \frac{t_o l}{C_w E \beta^3}\left(\frac{\beta l}{2} - \text{Tanh}\,\frac{\beta l}{2}\right)$ at $x = 0$ $\text{Max}\,+\theta'' = \frac{t_o}{C_w E \beta^2}\left(\frac{\beta l}{\text{Tanh}\,\beta l} - 1\right)$ at $x = l$ $\text{Max}\,-\theta'' = \frac{-t_o}{C_w E \beta^2}\left(1 - \frac{\beta l}{\text{Sinh}\,\beta l}\right)$ at $x = 0$ $\text{Max}\,\theta''' = \frac{t_o l}{C_w E}$ at $x = l$

2e. Both ends free to warp but not twist

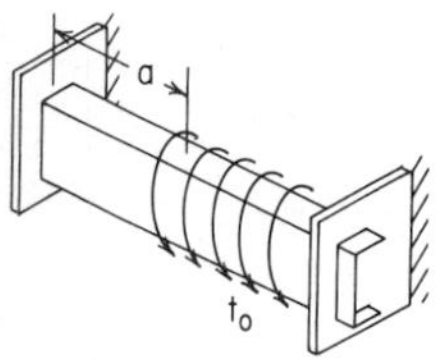

$$\theta_A = 0 \qquad \theta''_A = 0$$

$$T_A = \frac{-t_o}{2l}(l-a)^2$$

$$\theta'_A = \frac{t_o}{C_w E\beta^3}\left[\frac{\beta}{2l}(l-a)^2 - \frac{C_{a3}}{C_2}\right]$$

$$\theta_B = 0 \qquad \theta''_B = 0$$

$$\theta'_B = \frac{t_o}{C_w E\beta^3}\left[\frac{\beta}{2l}(l-a)^2 - \frac{C_1 C_{a3}}{C_2} + C_{a4}\right]$$

$$\theta'''_B = \frac{-t_o}{C_w E\beta}\,\frac{\text{Cosh}\,\beta a - \text{Cosh}\,\beta l}{\text{Sinh}\,\beta l}$$

$$T_B = \frac{-t_o}{2l}(l^2 - a^2)$$

If $a = 0$ (uniformly distributed torque over entire span),

$$\theta = \frac{t_o}{C_w E\beta^4}\left[\frac{\beta^2 x(l-x)}{2} + \frac{\text{Cosh}\,\beta(x - l/2)}{\text{Cosh}\,(\beta l/2)} - 1\right]$$

$$\theta' = \frac{t_o}{C_w E\beta^3}\left[\frac{\text{Sinh}\,\beta(x - l/2)}{\text{Cosh}\,(\beta l/2)} - \beta(x - l/2)\right]$$

$$\theta'' = \frac{t_o}{C_w E\beta^2}\left[\frac{\text{Cosh}\,\beta(x - l/2)}{\text{Cosh}\,(\beta l/2)} - 1\right]$$

$$\theta''' = \frac{t_o}{C_w E\beta}\,\frac{\text{Sinh}\,\beta(x - l/2)}{\text{Cosh}\,(\beta l/2)}$$

$$\text{Max}\,\theta = \frac{t_o}{C_w E\beta^4}\left[\frac{\beta^2 l^2}{8} + \frac{1}{\text{Cosh}\,(\beta l/2)} - 1\right] \qquad \text{at } x = \frac{l}{2}$$

$$\text{Max}\,\theta' = \frac{\pm t_o}{C_w E\beta^3}\left(\frac{\beta l}{2} - \text{Tanh}\,\frac{\beta l}{2}\right) \qquad \text{at } x = 0 \text{ and } x = l, \text{ respectively}$$

$$\text{Max}\,\theta'' = \frac{t_o}{C_w E\beta^2}\left[\frac{1}{\text{Cosh}\,(\beta l/2)} - 1\right] \qquad \text{at } x = \frac{l}{2}$$

$$\text{Max}\,\theta''' = \frac{\mp t_o}{C_w E\beta}\,\text{Tanh}\,\frac{\beta l}{2} \qquad \text{at } x = 0 \text{ and } x = l, \text{ respectively}$$

2f. Left end free to warp but not twist, right end fixed (no twist or warp)

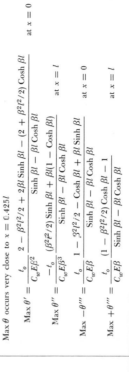

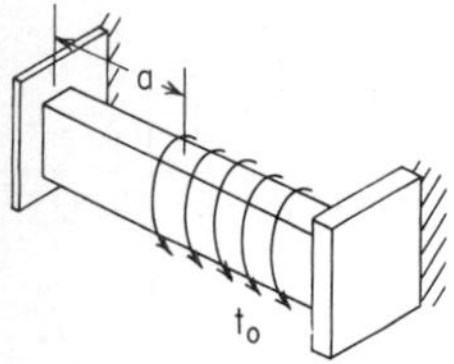

$$\theta_A = 0 \qquad \theta''_A = 0$$

$$T_A = \frac{-t_o}{\beta}\,\frac{C_1 C_{a5} - C_2 C_{a4}}{C_1 C_4 - C_2 C_3}$$

$$\theta'_A = \frac{t_o}{C_w E\beta^3}\,\frac{C_3 C_{a5} - C_4 C_{a4}}{C_1 C_4 - C_2 C_3}$$

If $a = 0$ (uniformly distributed torque over entire span),

Max θ occurs very close to $x = 0.425l$

$$\text{Max}\,\theta' = \frac{t_o}{C_w E\beta^2}\,\frac{2 - \beta^2 l^2/2 + 2\beta l\,\text{Sinh}\,\beta l - (2 + \beta^2 l^2/2)\,\text{Cosh}\,\beta l}{\text{Sinh}\,\beta l - \beta l\,\text{Cosh}\,\beta l} \qquad \text{at } x = 0$$

$$\text{Max}\,\theta'' = \frac{-t_o}{C_w E\beta^3}\,\frac{(\beta^2 l^2/2)\,\text{Sinh}\,\beta l + \beta l(1 - \text{Cosh}\,\beta l)}{\text{Sinh}\,\beta l - \beta l\,\text{Cosh}\,\beta l} \qquad \text{at } x = l$$

$$\text{Max}\,-\theta''' = \frac{t_o}{C_w E\beta}\,\frac{1 - \beta^2 l^2/2 - \text{Cosh}\,\beta l + \beta l\,\text{Sinh}\,\beta l}{\text{Sinh}\,\beta l - \beta l\,\text{Cosh}\,\beta l} \qquad \text{at } x = 0$$

$$\text{Max}\,+\theta''' = \frac{t_o}{C_w E\beta}\,\frac{(1 - \beta^2 l^2/2)\,\text{Cosh}\,\beta l - 1}{\text{Sinh}\,\beta l - \beta l\,\text{Cosh}\,\beta l} \qquad \text{at } x = l$$

TABLE 22 ***Formulas for the angle of twist of uniform thin-walled elastic members under torsional loading*** ***(Cont.)***

End restraints, reference no.	Boundary values	Selected special cases and maximum values
2g. Both ends fixed (no twist or warp) 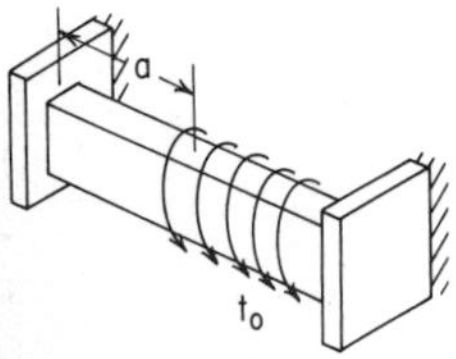	$\theta_A = 0 \qquad \theta'_A = 0$ $\theta''_A = \dfrac{t_o}{C_w E \beta^2} \dfrac{C_3 C_{a5} - C_4 C_{a4}}{C_2 C_4 - C_3^2}$ $T_A = \dfrac{-t_o}{\beta} \dfrac{C_2 C_{a5} - C_3 C_{a4}}{C_2 C_4 - C_3^2}$ $\theta'''_A = \dfrac{T_A}{C_w E}$	If $a = 0$ (uniformly distributed torque over entire span), $T_A = \dfrac{-t_o l}{2}$ $\theta = \dfrac{t_o l}{2 C_w E \beta^3} \left[\dfrac{\beta x}{l}(l - x) + \dfrac{\text{Cosh}\, \beta(x - l/2) - \text{Cosh}\,(\beta l/2)}{\text{Sinh}\,(\beta l/2)} \right]$ $\theta' = \dfrac{t_o l}{2 C_w E \beta^2} \left[1 - \dfrac{2x}{l} + \dfrac{\text{Sinh}\, \beta(x - l/2)}{\text{Sinh}\,(\beta l/2)} \right]$ $\theta'' = \dfrac{t_o}{C_w E \beta^2} \left[\dfrac{\beta l\, \text{Cosh}\, \beta(x - l/2)}{2\, \text{Sinh}\,(\beta l/2)} - 1 \right]$ $\theta''' = \dfrac{t_o}{C_w E \beta} \dfrac{\beta l\, \text{Sinh}\, \beta(x - l/2)}{2\, \text{Sinh}\,(\beta l/2)}$ $\text{Max}\ \theta = \dfrac{t_o l}{2 C_w E \beta^3} \left(\dfrac{\beta l}{4} - \text{Tanh}\, \dfrac{\beta l}{4} \right)$ at $x = \dfrac{l}{2}$ $\text{Max}\ -\theta'' = \dfrac{-t_o}{C_w E \beta^2} \left[1 - \dfrac{\beta l}{2\, \text{Sinh}\,(\beta l/2)} \right]$ at $x = \dfrac{l}{2}$ $\text{Max}\ +\theta'' = \dfrac{t_o}{C_w E \beta^2} \left[\dfrac{\beta l}{2\, \text{Tanh}\,(\beta l/2)} - 1 \right]$ at $x = 0$ and $x = l$ $\text{Max}\ \theta''' = \dfrac{\mp t_o l}{2 C_w E}$ at $x = 0$ and $x = l$, respectively

$$G = \frac{E}{2(1+\nu)} = \frac{30(10^6)}{2(1+0.285)} = 11.67(10^6)\ \text{lb/in}^2$$

$$\beta = \left(\frac{KG}{C_w E}\right)^{\frac{1}{2}} = \left[\frac{0.005133(11.67)(10^6)}{28.93(30)(10^6)}\right]^{\frac{1}{2}} = 0.00831$$

From Table 22, case 1d, when $a = 0$, the angular rotation at the loaded end is given as

$$\theta = \frac{T_o}{C_w E \beta^3}\left(\beta l - 2\ \text{Tanh}\,\frac{\beta l}{2}\right)$$

If we were to describe the total angle of twist at the loaded end in terms of an equivalent torsional stiffness constant K' in the expression

$\theta = T_o l / K'G$, then

$$K' = \frac{T_o l}{G\theta} \quad \text{or} \quad K' = K\frac{\beta l}{\beta l - 2\ \text{Tanh}\,(\beta l/2)}$$

The following table gives both K' and θ for the several lengths:

l	βl	$\text{Tanh}\,\frac{\beta l}{2}$	K'	θ
200	1.662	0.6810	0.0284	34.58°
100	0.831	0.3931	0.0954	5.15°
50	0.416	0.2048	0.3630	0.68°
25	0.208	0.1035	1.4333	0.09°
10	0.083	0.0415	8.926	0.006°

The stiffening effect of the fixed ends of the flanges is obvious even at a length of 200 in where $K' = 0.0284$ as compared with $K = 0.00513$. The warping restraint increases the stiffness more than five times. The large increases in K' at the shorter lengths of 25 and 10 in must be examined carefully. For this cross section $I_x = 3.88$ and $I_y = 3.96$, and so $J = 7.84\ \text{in}^4$. The calculated stiffness $K' = 8.926$ at $l = 10$ in is beyond the limiting value of 7.84, and so it is known to be in error because shear deformation was not included; therefore we would suspect the value of $K' = 1.433$ at $l = 25$ in as well. Indeed, Schwabenlender (Ref. 28) found that for a similar cross section the effect of shear deformation in the flanges reduced the stiffness by approximately 25 percent at a length of 25 in and by more than 60 percent at a length of 10 in.

2. A small cantilever crane rolls along a track welded from three pieces of 0.3-in-thick steel. The 20-ft-long track is solidly welded to a rigid foundation at the right end and simply supported 4 ft from the left end, which is free. The simple support also provides resistance to rotation about the beam axis but provides no restraint against warping or rotation in horizontal and

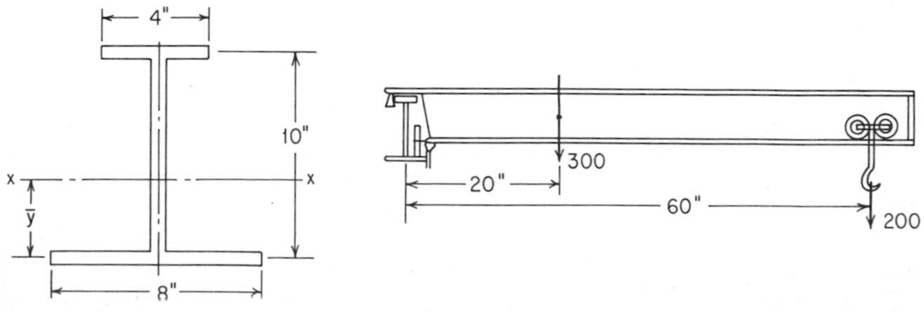

Fig. 9.4

vertical planes containing the axis. The crane weighs 300 lb and has a center of gravity which is 20 in out from the web of the track. A load of 200 lb is carried at the end of the crane 60 in from the web. It is desired to determine the maximum flexure stress and the maximum shear stress in the track and also the angle of inclination of the crane when it is located 8 ft from the welded end of the track.

Solution. The loading will be considered in two stages. First consider a vertical load of 500 lb acting 8 ft from the fixed end of a 16-ft beam fixed at one end and simply supported at the other. The following constants are needed:

$$\bar{y} = \frac{4(0.3)(10) + 9.7(0.3)(5)}{(4 + 9.7 + 8)(0.3)} = 4.08 \text{ in}$$

$$E = 30(10^6) \text{ lb/in}^2$$

$$I_x = \frac{4(0.3^3)}{12} + 4(0.3)(10 - 4.08)^2 + \frac{0.3(9.7^3)}{12} + 9.7(0.3)(5 - 4.08)^2$$

$$+ \frac{8(0.3^3)}{12} + 8(0.3)(4.08^2) = 107.3 \text{ in}^4$$

From Table 3, case 1c, $a = 8$ ft, $l = 16$ ft, $W = 500$ lb, $M_A = 0$, and $R_A = [500/2(16^3)](16 - 8)^2[2(16) + 8] = 156.25$ lb. Now construct the shear and moment diagrams.

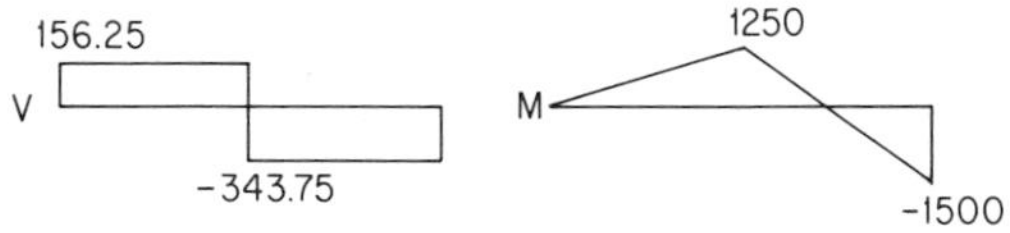

The second portion of the solution considers a beam fixed at the right end against both rotation and warping and free at the left end. It is loaded at a point 4 ft from the left end with an unknown torque T_c, which makes the angle of rotation zero at that point, and a known torque of $300(20) + 200(60) = 18{,}000$ in-lb at a point 12 ft from the left end. Again evaluate the following constants; assume $G = 12(10^6)$ lb/in²:

$$K = \tfrac{1}{3}(4 + 8 + 10)(0.3^3) = 0.198 \text{ in}^4 \qquad \text{(Table 21, case 7)}$$

$$C_w = \frac{(10^2)(0.3)(4^3)(8^3)}{(12)(4^3 + 8^3)} = 142.2 \text{ in}^6 \qquad \text{(Table 21, case 7)}$$

$$\beta = \left(\frac{KG}{C_w E}\right)^{\frac{1}{2}} = \left[\frac{(0.198)(12)(10^6)}{(142.2)(30)(10^6)}\right]^{\frac{1}{2}} = 0.0236 \text{ in}^{-1}$$

Therefore $\beta l = 0.0236(20)(12) = 5.664$, $\beta(l - a) = 0.0236(20 - 12)(12) = 2.2656$ for $a = 12$ ft; and $\beta(l - a) = 0.0236(20 - 4)(12) = 4.5312$ for $a = 4$ ft.

From Table 22, case 1b, consider two torsional loads: an unknown torque T_c at $a = 4$ ft and a torque of 18,000 in-lb at $a = 12$ ft. The following constants are needed:

$$C_1 = \text{Cosh } \beta l = \text{Cosh } 5.664 = 144.1515$$
$$C_2 = \text{Sinh } \beta l = \text{Sinh } 5.664 = 144.1480$$

For $a = 4$ ft,

$$C_{a3} = \text{Cosh } \beta(l - a) - 1 = \text{Cosh } 4.5312 - 1 = 45.4404$$
$$C_{a4} = \text{Sinh } \beta(l - a) - \beta(l - a) = \text{Sinh } 4.5312 - 4.5312 = 41.8984$$

For $a = 12$ ft,

$$C_{a3} = \text{Cosh } 2.2656 - 1 = 3.8703$$
$$C_{a4} = \text{Sinh } 2.2656 - 2.2656 = 2.5010$$

At the left end $T_A = 0$ and $\theta''_A = 0$:

$$\theta_A = \frac{18{,}000}{(142.2)(30)(10^6)(0.0236^3)}\left[\frac{(144.1480)(3.8703)}{144.1515} - 2.5010\right] + \frac{T_c}{(142.2)(30)(10^6)(0.0236^3)}\left[\frac{(144.1480)(45.4404)}{144.1515} - 41.8984\right]$$

$$= 0.43953 + 6.3148(10^{-5})T_c$$

Similarly, $\theta'_A = -0.0002034 - 1.327(10^{-7})\ T_c$.

To evaluate T_c the angle of rotation at $x = 4$ ft is set equal to zero:

$$\theta_c = 0 = \theta_A + \frac{\theta'_A}{\beta}F_{2(c)}$$

where

$$F_{2(c)} = \text{Sinh}\ [0.0236(48)] = \text{Sinh}\ 1.1328 = 1.3911$$

or

$$0 = 0.43953 + 6.3148(10^{-5})T_c - \frac{(0.0002034)(1.3911)}{0.0236} - \frac{(1.327)(10^{-7})(1.3911)}{0.0236}T_c$$

This gives $T_c = -7{,}728$ in-lb, $\theta_A = -0.04847$ rad, and $\theta'_A = 0.0008221$ rad/in.

To locate positions of maximum stress it is desirable to sketch curves of θ', θ'', and θ''' versus the position x.

$$\theta' = \theta'_A F_1 + \frac{T_c}{C_w E\beta^2}F_{a3(c)} + \frac{T_o}{C_w E\beta^2}F_{a3}$$

or

$$\theta' = 0.0008221\ \text{Cosh}\ \beta x - \frac{7{,}728}{(142.2)(30)(10^6)(0.0236^2)}[\text{Cosh}\ \beta(x-48) - 1]\langle x - 48\rangle^0 + \frac{18{,}000}{(142.2)(30)(10^6)(0.0236^2)}[\text{Cosh}\ \beta(x-144) - 1]\langle x - 144\rangle^0$$

This gives

$$\theta' = 0.0008221\ \text{Cosh}\ \beta x - 0.003253[\text{Cosh}\ \beta(x-48) - 1]\langle x-48\rangle^0 + 0.0007575[\text{Cosh}\ \beta(x-144) - 1]\langle x - 144\rangle^0$$

Similarly,

$$\theta'' = 0.00001940\ \text{Sinh}\ \beta x - 0.00007676\ \text{Sinh}\ \beta\langle x-48\rangle + 0.00001788\ \text{Sinh}\ \beta\langle x-144\rangle$$

$$\theta''' = 10^{-6}[0.458\ \text{Cosh}\ \beta x - 1.812\ \text{Cosh}\ \beta(x-48)\langle x-48\rangle^0 + 4.22\ \text{Cosh}\ \beta(x-144)\langle x-144\rangle^0]$$

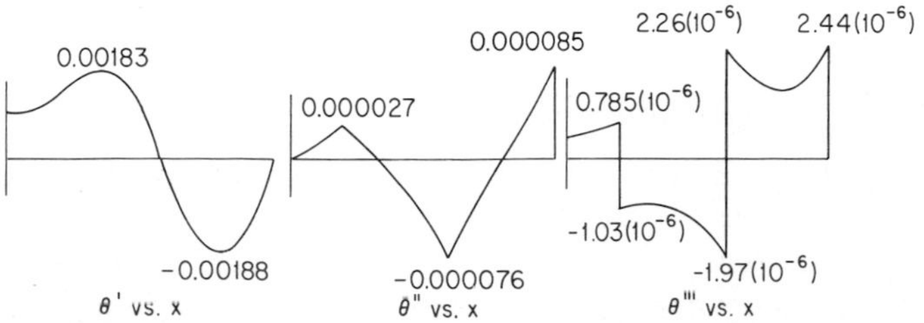

Maximum bending stresses are produced by the beam bending moments of +1250 ft-lb at $x = 12$ ft and −1500 ft-lb at $x = 20$ ft and by maximum values of θ'' of −0.000076 at $x = 12$ ft and +0.000085 at $x = 20$ ft. Since the largest magnitudes of both M_x and θ'' occur at $x = 20$ ft, the maximum bending stress will be at the wall. Therefore, at the wall,

$$\sigma_A = \frac{1500(12)(10 - 4.08 + 0.15)}{107.26} - \frac{10(4)}{2}\frac{0.3(8^3)}{0.3(4^3) + 0.3(8^3)}(30)(10^6)(0.000085)$$

$$= 970 - 45{,}300 = -44{,}300 \text{ lb/in}^2$$

$$\sigma_B = 970 + 45{,}300 = 46{,}300 \text{ lb/in}^2$$

$$\sigma_C = \frac{-1500(12)(4.08 + 0.15)}{107.26} + \frac{10(8)}{2}\frac{0.3(4^3)}{0.3(4^3) + 0.3(8^3)}(30)(10^6)(0.000085)$$

$$= -700 + 11{,}300 = 10{,}600 \text{ lb/in}^2$$

$$\sigma_D = -700 - 11{,}300 = -12{,}000 \text{ lb/in}^2$$

Maximum shear stresses are produced by θ', θ''', and beam shear V. The shear stress due to θ''' is maximum at the top of the web, that due to θ' is maximum on the surface anywhere, and that due to V is maximum at the neutral axis but is not much smaller at the top of the web. The largest shear stress in a given cross section is therefore found at the top of the web and is the sum of the absolute values of the three components at one of four possible locations at the top of the web. This gives

$$|\tau_{\max}| = \left|\frac{1}{8}\frac{10(0.3)(8^3)(4^2)}{0.3(4^3 + 8^3)}30(10^6)\theta'''\right| + |0.3(12)(10^6)\theta'| + \left|\frac{(2 - 0.15)(0.3)(10 - 4.08)}{107.26(0.3)}V\right|$$

$$= |533.3(10^6)\theta'''| + |3.6(10^6)\theta'| + |0.1021\,V|$$

The following maximum values of θ''', θ', V, and $\tau_{\max}$ are found at the given values of the position x:

x	θ'''	θ'	V	$\lvert\tau_{\max}\rvert$, lb/in²
48+	$-1.03(10^{-6})$	$1.41(10^{-3})$	156.3	5633
79.2	$-0.80(10^{-6})$	$1.83(10^{-3})$	156.3	7014
144−	$-1.97(10^{-6})$	$-0.28(10^{-3})$	156.3	2073
144+	$2.26(10^{-6})$	$-0.28(10^{-3})$	−343.7	2247
191.8	$1.37(10^{-6})$	$-1.88(10^{-3})$	−343.7	7522
240	$2.44(10^{-6})$	0	−343.7	1335

To obtain the rotation of the crane at the load, substitute $x = 144$ into

$$\theta = \theta_A + \frac{\theta'_A}{\beta}F_2 + \frac{T_c}{C_w E\beta^3}F_{a4(c)}$$

$$= -0.04847 + \frac{0.0008221}{0.0236}14.9414 - \frac{7728(4.7666 - 2.2656)}{142.2(30)(10^6)(0.0236^3)} = 0.1273 \text{ rad}$$

$$= 7.295°$$

9.4 *Effect of longitudinal stresses*

It was pointed out in Art. 9.1 that the elongation of the outer fibers consequent upon twist caused longitudinal stresses, but that in a bar of circular section these stresses were negligible. In a flexible bar, the section of which comprises one or more narrow rectangles, the stresses in the longitudinal fibers may become large; and since after twisting these fibers are inclined, the stresses in them have components, normal to the axis of twist, which contribute to the torsional resistance of the member.

The stress in the longitudinal fibers of a thin twisted strip and the effect of these stresses on torsional stiffness have been considered by Timoshenko (Ref. 5), Green (Ref. 6), and others. The following formulas apply to this case: Let $2a$ = width of strip; $2b$ = thickness of strip; τ, σ_t, and σ_c = maximum shear, maximum tensile, and maximum compressive stress due to twisting, respectively; T = applied twisting moment; and θ/l = angle of twist per unit length. Then

$$\sigma_t = \frac{E\tau^2}{12G^2}\left(\frac{a}{b}\right)^2 \tag{5}$$

$$\sigma_c = \tfrac{1}{2}\sigma_t \tag{6}$$

$$T = KG\frac{\theta}{l} + \frac{8}{45}E\left(\frac{\theta}{l}\right)^3 ba^5 \tag{7}$$

The first term on the right side of Eq. 7 ($KG\theta/l$) represents the part of the total applied torque T that is resisted by torsional shear; the second term represents the part that is resisted by the tensile stresses in the (helical) longitudinal fibers. It can be seen that this second part is small for small angles of twist but increases rapidly as θ/l increases.

To find the stresses produced by a given torque T, first the value of θ/l is found by Eq. 7, taking K as given for Table 20, case 4. Then τ is found by the stress formula for case 4, taking $KG\theta/l$ for the twisting moment. Finally σ_t and σ_c can be found by Eqs. 5 and 6.

This stiffening and strengthening effect of induced longitudinal stress will manifest itself in any bar having a section composed of narrow rectangles, such as an I, T, or channel, provided that the parts are so thin as to permit of a large unit twist without overstressing. At the same time the accompanying longitudinal compression (Eq. 6) may cause failure through elastic instability (see Table 34).

If a thin strip of width a and maximum thickness b is *initially* twisted (as by cold working) to an helical angle β, then there is an initial stiffening effect in torsion that can be expressed by the ratio of effective K to nominal K (as given in Table 22):

$$\frac{\text{Effective } K}{\text{Nominal } K} = 1 + C(1 + \nu)\beta^2\left(\frac{a}{b}\right)^2$$

where C is a numerical coefficient that depends on the shape of the cross section and is $\frac{2}{15}$ for a rectangle, $\frac{1}{8}$ for an ellipse, $\frac{1}{10}$ for a lenticular form, and $\frac{7}{60}$ for a double wedge (Ref. 22).

If a bar of any cross section is independently loaded in tension, then the corresponding longitudinal tensile stress σ_t similarly will provide a resisting torque that again depends on the angle of twist, and the total applied torque corresponding to any angle of twist θ is $T = (KG + \sigma_t J)\,\theta/l$, where J is the centroidal polar moment of inertia of the cross section. If the longitudinal loading causes a compressive stress σ_c, the equation becomes

$$T = (KG - \sigma_c J)\frac{\theta}{l}$$

Bending also influences the torsional stiffness of a rod unless the cross section has (1) two axes of symmetry, (2) point symmetry, or (3) one axis of symmetry that is normal to the plane of bending. (The influences of longitudinal loading and bending are discussed in Ref. 23.)

9.5 *Ultimate strength of bars in torsion*

When twisted to failure, bars of ductile material usually break in shear, the surface of fracture being normal to the axis and practically flat. Bars of brittle material usually break in tension, the surface of fracture being helicoidal.

Circular sections. The formulas of Art. 9.1 apply only when the maximum stress does not exceed the elastic limit. If Eq. 3 is used with T equal to the twisting moment at failure, a fictitious value of τ is obtained, which is called the *modulus of rupture in torsion* and which for convenience will be denoted here by τ'. For solid bars of steel, τ' slightly exceeds the ultimate tensile strength when the length is only about twice the diameter but drops to about 80 percent of the tensile strength when the length becomes 25 times the diameter. For solid bars of aluminum, τ' is about 90 percent of the tensile strength.

For tubes, the modulus of rupture decreases with the ratio of diameter D to wall thickness t. Younger (Ref. 7) gives the following approximate formula, applicable to tubes of steel and aluminum:

$$\tau' = \frac{1600\,\tau_0'}{(D/t - 2)^2 + 1600}$$

where τ' is the modulus of rupture in torsion of the tube and τ_0' is the modulus of rupture in torsion of a solid circular bar of the same material. (Curves giving τ' as a function of D/t for various steels and light alloys may be found in Ref. 18.)

For a solid bar of cast iron, the ratio of τ' to the tensile strength (rupture factor) seems to vary considerably, ranging from a little over 100 to 190 percent. The variation is probably due in part to differences in length and end conditions of the specimens tested.

Noncircular sections. The rupture factors for cast iron and plaster specimens of different sections in torsion are given in Table 38. These rupture factors are the ratios of the calculated maximum shear stress at rupture (calculated by the appropriate formula from Table 20) to the tensile strength of the material. As usual, it is seen that the rupture factor increases as the degree of stress localization increases.

Torsion combined with bending. In the case of thin tubular members that fail by buckling, it may be assumed that no failure will occur so long as $(\tau/\tau')^2 + (\sigma/\sigma')^2 < 1$. Here τ and σ are the maximum actual stresses, and τ' and σ' are the moduli of rupture in torsion and bending, respectively.

9.6 *Torsion of curved bars; helical springs*

The formulas of Arts. 9.1 and 9.2 can be applied to slightly curved bars without significant error, but for sharply curved bars, such as helical springs, account must be taken of the influence of curvature and slope. Among others, Wahl (Ref. 8) and Ancker and Goodier (Ref. 24) have discussed this problem, and the former presents charts which greatly facilitate the calculation of stress and deflection for springs of noncircular section. Of the following formulas cited, those for round wire were taken from Ref. 24, and those for square and rectangular wire from Ref. 8 (with some changes of notation).

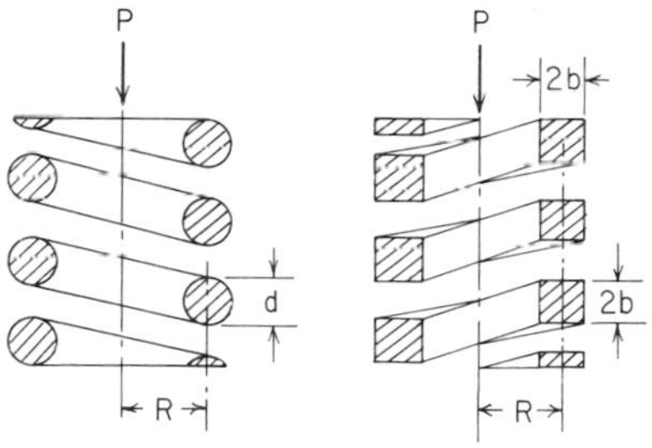

Fig. 9.5

Let R = radius of coil measured from spring axis to center of section (Fig. 9.5); d = diameter of circular section; $2b$ = thickness of square section; P = load (either tensile or compressive); n = number of active turns in spring; α = pitch angle of spring; f = total stretch or shortening of spring; and τ = maximum shear stress produced. Then for a spring of *circular* wire,

$$f = \frac{64PR^3n}{Gd^4}\left[1 - \frac{3}{64}\left(\frac{d}{R}\right)^2 + \frac{3+\nu}{2(1+\nu)}(\tan\alpha)^2\right]$$

$$\tau = \frac{16PR}{\pi d^3}\left[1 + \frac{5}{8}\frac{d}{R} + \frac{7}{32}\left(\frac{d}{R}\right)^2\right]$$

For a spring of *square* wire,

$$f = \frac{2.789PR^3n}{Gb^4} \quad \text{for } c > 3$$

$$\tau = \frac{4.8PR}{8b^3}\left(1 + \frac{1.2}{c} + \frac{0.56}{c^2} + \frac{0.5}{c^3}\right)$$

where $c = R/b$.

For a spring of *rectangular* wire, section $2a \times 2b$ where $a > b$,

$$f = \frac{3\pi PR^3 n}{8Gb^4} \frac{1}{a/b - 0.627[\text{Tanh}\,(\pi b/2a) + 0.004]}$$

for $c > 3$ if the long dimension $2a$ is parallel to the spring axis or for $c > 5$ if the long dimension $2a$ is perpendicular to the spring axis,

$$\tau = \frac{PR(3b + 1.8a)}{8b^2a^2}\left(1 + \frac{1.2}{c} + \frac{0.56}{c^2} + \frac{0.5}{c^3}\right)$$

It should be noted that in each of these cases the maximum stress is given by the ordinary formula for the section in question (from Table 20) multiplied by a corrective factor that takes account of curvature, and these corrective factors can be used for any curved bar of the corresponding cross section. Also, for compression springs with the end turns ground down for even bearing, n should be taken as the actual number of turns (including the tapered end turns) less 2. For tension springs n should be taken as the actual number of turns or slightly more.

Unless laterally supported, compression springs that are relatively long will buckle when compressed beyond a certain critical deflection. This critical deflection depends on the ratio of L, the free length, to D, the mean diameter, and is indicated approximately by the following tabulation, based on Ref. 27.

$L/D = 1$	2	3	4	5	6	7	8
Critical deflection/L = 0.72	0.71	0.68	0.63	0.53	0.39	0.27	0.17

Precise formula. For very accurate calculation of the extension of a spring, as is necessary in designing precision spring scales, account must be taken of the change in slope and radius of the coils caused by stretching. Sayre (Ref. 9) gives a formula which takes into account, not only the effect of this change in form, but also the deformation due to direct transverse shear and flexure. This formula can be written as

$$f = P\left\{\left[\frac{R_0^2 L}{GK} - \frac{R_0^2 H_0^2}{GKL}\left(1 - \frac{GK}{EI}\right) + \frac{FL}{AG}\right] - \left[\frac{R_0^2}{3GKL}\left(3 - \frac{2GK}{EI}\right)(H^2 + HH_0 - 2H_0^2)\right]\right\}$$

where f = stretch of the spring; P = load; R_0 = initial radius of the coil; H = variable length of the effective portion of the stretched spring; H_0 = initial value of H; L = actual developed length of the wire of which the spring is made; A = cross-sectional area of this wire; K = the torsional-stiffness factor for the wire section, as given in Table 20 ($K = \frac{1}{2}\pi r^4$ for a circle; $K = 2.25a^4$ for a square; etc.); F = the section factor for shear deformation (Art. 7.10, Eq. 19; $F = \frac{10}{9}$ for a circle or ellipse, $F = \frac{6}{5}$ for a square or rectangle); and I = moment of inertia of the wire section about a central axis parallel to the spring axis. The first term in brackets represents the initial rate of stretch, and the second term in brackets represents the change in this rate due to change in form consequent upon stretch. The final expression shows that f is not a linear function of P.

9.7 *Miscellaneous formulas for circular shafts*

The most common torsion member is the straight shaft of circular cross section, and a number of additional relations and formulas for this case will be given.

Shaft of varying diameter. If the diameter changes gradually, the stress at any section is given with sufficient accuracy by the formulas for uniform bars. If the change in section is abrupt, as at a shoulder with a small fillet, the maximum stress should be found by the use of a suitable factor of stress concentration k. Values of k are given in Table 37.

Shaft under combined torsion and bending. The bending stress σ and the torsional stress τ can be found separately, and the resultant shear and normal stresses can be found by the formulas of Art. 6.7. Or the equivalent bending moment M' may be used to find the maximum tensile and compressive stresses and the equivalent twisting moment T' to find the maximum shear stress. The following formulas apply to circular sections:

$$M' = \tfrac{1}{2}(M + \sqrt{M^2 + T^2})$$

$$T' = \sqrt{M^2 + T^2}$$

$$\text{Max } \sigma = \frac{M'r}{I}$$

$$\text{Max } \tau = \frac{T'r}{J}$$

where M = bending moment and T = twisting moment at the section in question. The use of the equivalent bending moment and the equivalent twisting moment is especially convenient in design since the radius required for a given allowable stress can be found directly.

Relation between twisting moment and power transmitted. A shaft rotating uniformly at n rpm and transmitting H hp is subjected to a steady twisting moment T (inch-pounds) given by

$$T = 63{,}024 \frac{H}{n}$$

For a solid circular shaft of radius r the resulting stress is

$$\tau = 40{,}100 \frac{H}{nr^3}$$

and the resulting unit twist (radians per inch) is

$$\frac{\theta}{l} = 40{,}100 \frac{H}{nr^4G}$$

REFERENCES

1. Prescott, J.: "Applied Elasticity," Longmans, Green & Co., Ltd., 1924.
2. Trayer, G. W., and H. W. March: The Torsion of Members having Sections Common in Aircraft Construction, *Natl. Adv. Comm. Aeron., Rept.* 334, 1930.
3. Griffith, A. A.: The Determination of the Torsional Stiffness and Strength of Cylindrical Bars of any Shape, *Repts. Memo.* 334, *Adv. Comm. Aeron.* (British), 1917.
4. Taylor, G. I., and A. A. Griffith: The Use of Soap Films in Solving Torsion Problems, *Reports and Memoranda* 333, *Adv. Comm. Aeron.* (British), 1917.
5. Timoshenko, S.: "Strength of Materials," pt. II, D. Van Nostrand Company, Inc., 1930.
6. Green, A. E.: The Equilibrium and Elastic Stability of a Thin Twisted Strip, *Proc. R. Soc. Lond., Ser. A*, vol. 154, 1936.
7. Younger, J. E.: "Structural Design of Metal Airplanes," McGraw-Hill Book Company, 1935.
8. Wahl, A. M.: "Mechanical Springs," 2d ed., McGraw-Hill Book Company, 1963. See also *ibid.*, Helical Compression and Tension Springs, *ASME Paper* A-38, *J. Appl. Mech.*, vol. 2, no. 1, 1935.
9. Sayre, M. F.: New Spring Formulas and New Materials for Precision Spring Scales, *Trans. ASME,* vol. 58, p. 379, 1936.
10. Wilson, T. S.: The Eccentric Circular Tube, *Aircr. Eng.*, vol. 14, no. 157, March 1942.
11. Lyse, I., and B. G. Johnston: Structural Beams in Torsion, *Inst. Res., Lehigh Univ., Circ.* 113, 1935.
12. Baron, F. M.: Torsion of Multi-connected Thin-walled Cylinders, *ASME Paper* A-72, *J. Appl. Mech.*, vol. 9, no. 2, June 1942.
13. Shortley, G. H., R. Wells, and B. Fried: *Eng. Exp. Sta., Ohio State Univ., Bull.* 107, September 1940.
14. Pletta, D. H., and F. J. Maher: The Torsional Properties of Round-edged Flat Bars, *Eng. Exp. Sta., Va. Polytech. Inst., Bull.* 50, 1942.
15. Payne, J. H.: Torsion in Box Beams, *Aircr. Eng.*, January 1942.
16. Weigand, A.: The Problem of Torsion in Prismatic Members of Circular Segmental Cross Section, *Natl. Adv. Comm. Aeron., Tech. Memo.* 1182, 1948.
17. Timoshenko, S. P., and J. N. Goodier: "Theory of Elasticity," 2d ed., McGraw-Hill Book Company, 1951.
18. "ANC Mil-Hdbk-5, Strength of Metal Aircraft Elements," Armed Forces Supply Support Center, March, 1959.
19. Bethlehem Steel Co.: Torsion Analysis of Rolled Steel Sections.
20. Nuttall, Henry: Torsion of Uniform Rods with Particular Reference to Rods of Triangular Cross Section, *ASME J. Appl. Mech.*, vol. 19, no. 4, December 1952.
21. Carter, W. J., and J. B. Oliphant: Torsion of a Circular Shaft with Diametrically Opposite Flat Sides, *ASME J. Appl. Mech.*, vol. 19, no. 3, September 1952.
22. Chu, Chen: The Effect of Initial Twist on the Torsional Rigidity of Thin Prismatical Bars and Tubular Members, *Proc. 1st U.S. Nat. Congr. Appl. Mech.,* p. 265, 1951.
23. Engel, H. L., and J. N. Goodier: Measurements of Torsional Stiffness Changes and Instability Due to Tension, Compression and Bending, *ASME J. Appl. Mech.*, vol. 20, no. 4, December 1953.
24. Ancker, C. J., Jr., and J. N. Goodier: Pitch and Curvature Correction for Helical Springs, *ASME J. Appl. Mech.*, vol. 25, no. 4, December 1958.
25. Reiner, M.: The Complete Elasticity Law for Some Metals According to Poynting's Observations, *Appl. Sci. Res., Sec.* A, vol. 5.
26. Poynting, J. H.: *Proc. R. Soc. Lond.*, A32, 1909; and A36, 1912.
27. "Mechanical Springs: Their Engineering and Design," 1st ed., William D. Gibson Co., Division of Associated Spring Corp., 1944.

28. Schwabenlender, C. W.: Torsion of Members of Open Cross Section, masters thesis, University of Wisconsin, 1965.
29. Chu, Kuang-Han, and A. Longinow: Torsion in Sections with Open and Closed Parts, *Proc. Am. Soc. Civil Eng., J. Struct. Div.*, vol. 93, no. 6, December 1967.
30. Vlasov, V. Z.: "Thin-Walled Elastic Beams," Clearing House for Federal Scientific and Technical Information, U.S. Depart. of Commerce, 1961.
31. Kollbrunner, C. F., and K. Basler: "Torsion in Structures," Springer-Verlag, 1969.

TEN

Flat Plates

10.1 *Common case*

The formulas of this article are based on the following assumptions: (1) The plate is flat, of uniform thickness, and of homogeneous isotropic material; (2) the thickness is not more than about one-quarter of the least transverse dimension, and the maximum deflection is not more than about one-half the thickness; (3) all forces—loads and reactions—are normal to the plane of the plate; and (4) the plate is nowhere stressed beyond the elastic limit. For convenience in discussion, it will be assumed further that the plane of the plate is horizontal.

Behavior. The plate deflects. The middle surface (halfway between top and bottom surfaces) remains unstressed; at other points there are biaxial stresses in the plane of the plate. Straight lines in the plate that were originally vertical remain straight but become inclined; therefore the intensity of either principal stress at points on any such line is proportional to the distance from the middle surface, and the maximum stresses occur at the surfaces of the plate.

Formulas. Unless otherwise indicated the formulas given in Tables 24 to 26 are based on very closely approximate mathematical analysis and may be accepted as sufficiently accurate so long as the assumptions stated hold true. Certain additional facts of importance in relation to these formulas are as follows.

Concentrated loading. It will be noted that all formulas for maximum stress due to a load applied over a small area give very high values when the radius of the loaded area approaches zero. Analysis by a more precise method (Ref. 12) shows that the actual maximum stress produced by a load concentrated on a very small area of radius r_o can be found by replacing the actual r_o by a so-called *equivalent radius* r'_o, which depends largely upon the thickness of the plate t and to a lesser degree on its least transverse dimension. Holl (Ref. 13) shows how b varies with the width of a flat plate. Westergaard (Ref. 14) gives an approximate expression for this equivalent radius:

$$r'_o = \sqrt{1.6r_o^2 + t^2} - 0.675t$$

This formula, which applies to a plate of any form, may be used for all values of r_o less than $0.5t$; for larger values the actual r_o may be used.

Use of the equivalent radius makes possible the calculation of the finite maximum stresses produced by a (nominal) point loading whereas the ordinary formula would indicate that these stresses were infinite.

Edge conditions. The formulas of Tables 24 to 26 are given for various combinations of edge support: free, guided (zero slope but free to move vertically), and simply supported or fixed. No exact edge condition is likely to be realized in ordinary construction, and a condition of true edge fixity is especially difficult to obtain. Even a small horizontal force at the line of contact may appreciably reduce the stress and deflection in a simply supported plate; however, a very slight yielding at nominally fixed edges will greatly relieve the stresses there while increasing the deflection and center stresses. For this reason it is usually advisable to design a fixed-edged plate that is to carry uniform load for somewhat higher center stresses than are indicated by theory.

10.2 *Bending of uniform-thickness plates with circular boundaries*

In Table 24 cases 1 to 15 consider annular and solid circular plates of constant thickness under *axisymmetric* loading for several combinations of boundary conditions. In addition to the formulas, tabulated values of deformation and moment coefficients are given for many common loading cases. The remaining cases include concentrated loading and plates with some circular and straight boundaries. Only the deflections due to bending strains are included; in Art. 10.3 the additional deflections due to shear strains are considered.

Formulas. For cases 1 to 15 (Table 24), expressions are given for deformations and reactions at the edges of the plates as well as general equations which allow the evaluation of *deflections, slopes, moments,* and *shears* at any point in the plate. The several axisymmetric loadings include uniform, uniformly increasing, and parabolically increasing normal pressure over a portion of the plate. This permits the approximation of any reasonable axisymmetric distributed loading by fitting an approximate second-order curve to the variation in loading and solving the problem by superposition. (See the Examples at the end of this article.)

In addition to the usual loadings, Table 24 also includes loading cases that may be described best as *externally applied conditions which force a lack of flatness into the plate.* For example, in cases 6 and 14 expressions are given for a manufactured concentrated change in slope in a plate, which could also be used if a plastic hinge were to develop in a plate and the change in slope at the plastic hinge is known or assumed. Similarly, case 7 treats a plate with a small step manufactured into the otherwise flat surface and gives the reactions which develop when this plate is forced to conform to the specified boundary conditions. These cases are also useful when considering known boundary rotations or lateral displacements. (References 46, 47, 57, and 58 present tables and graphs for many of the loadings given in these cases.)

The use of the constants C_1 to C_9 and the functions F_1 to F_9, L_1 to L_{19}, and G_1 to G_{19} in Table 24 appears to be a formidable task at first. However, when we consider the large number of cases it is possible to present in a limited space, the reason for this method of presentation becomes clear. With careful inspection, we find that the constants and functions with *like subscripts* are the same except for the change in variable. We also note the use of the *singularity function* $\langle r - r_o \rangle^0$, which is given a value of 0 for $r < r_o$ and a value of 1 for $r > r_o$. In Table 23 values are listed for all of the preceding functions for several values of the variables b/r, b/a, r_o/a, and r_o/r; also listed are five of the most used denominators for the several values of b/a. (Note that these values are for $\nu = 0.30$.)

EXAMPLES

1. A solid circular steel plate, 0.2 in thick and 20 in in diameter, is simply supported along the edge and loaded with a uniformly distributed load of 3 lb/in^2. It is required to determine the center deflection, the maximum stress, and the deflection equation. Given: $E = 30(10^6)$ lb/in^2 and $\nu = 0.285$.

Solution. This plate and loading is covered in Table 24, case 10a. The following constants are obtained:

$$D = \frac{30(10^6)(0.2^3)}{12(1 - 0.285^2)} = 21{,}800 \qquad q = 3 \qquad a = 10 \qquad r_o = 0$$

Since $r_o = 0$,

$$y_c = \frac{-qa^4}{64D}\frac{5 + \nu}{1 + \nu} = \frac{-3(10^4)(5.285)}{64(21{,}800)(1.285)} = -0.0883 \text{ in}$$

and

$$\text{Max } M = M_c = \frac{qa^2}{16}(3 + \nu) = \frac{3(10^2)(3.285)}{16} = 61.5 \text{ in-lb/in}$$

Therefore

$$\text{Max } \sigma = \frac{6M_c}{t^2} = \frac{6(61.5)}{0.2^2} = 9240 \text{ lb/in}^2$$

The general deflection equation for these several cases is

$$y = y_c + \frac{M_c r^2}{2D(1 + \nu)} + LT_y$$

where for this case $LT_y = (-qr^4/D)G_{11}$. For $r_o = 0$, $G_{11} = \frac{1}{64}$ (note that $r > r_o$ everywhere in the plate, so that $\langle r - r_o \rangle^0 = 1$); therefore,

$$y = -0.0883 + \frac{61.5r^2}{2(21{,}800)(1.285)} - \frac{3r^4}{21{,}800(64)}$$
$$= -0.0883 + 0.001098r^2 - 0.00000215r^4$$

As a check, the deflection at the outer edge can be evaluated as

$$y_a = -0.0883 + 0.001098(10^2) - 0.00000215(10^4)$$
$$= -0.0883 + 0.1098 - 0.0215 = 0$$

2. An annular aluminum plate with an outer radius of 20 in and an inner radius of 5 in is to be loaded with an annular line load of 40 lb/in at a radius of 10 in. Both the inner and outer edges are simply supported, and it is required to determine the maximum deflection and maximum stress as a function of the plate thickness. Given: $E = 10(10^6)$ lb/in^2 and $\nu = 0.30$.

Solution. The solution to this loading and support condition is found in Table 24, case 1c, where $b/a = 0.25$, $r_o/a = 0.50$, $a = 20$ in, and $w = 40$ lb/in. No numerical solutions are presented for this combination of b/a and r_o/a, and so either the equations for C_1, C_3, C_7, C_9, L_3, and L_9 must be evaluated or values for these coefficients must be found in Table 23. Since the values of C are found for the variable b/a, from Table 23, under the column headed 0.250, the following coefficients are determined.

$$C_1 = 0.881523 \qquad C_3 = 0.033465 \qquad C_7 = 1.70625$$
$$C_9 = 0.266288 \qquad C_1C_9 - C_3C_7 = 0.177640$$

The values of L are found for the variable r_o/a, and so from Table 23, under the column headed 0.500, the following coefficients are determined:

$$L_3 = 0.014554 \qquad \text{and} \qquad L_9 = 0.290898$$

Whether the numbers in Table 23 can be interpolated and used successfully depends upon the individual problem. In some instances, where lesser degrees of accuracy are required, interpolation can be used; in other instances, requiring greater degrees of accuracy, it would be better to solve the problem for values of b and r_o that do fit Table 23 and then interpolate between the values of the final deflections or stresses.

Using the preceding coefficients, the reaction force and slope can be determined at the inside edge and then the deflection equation developed (note that $y_b = 0$ and $M_{rb} = 0$):

$$\theta_b = \frac{-wa^2}{D}\frac{C_3L_9 - C_9L_3}{C_1C_9 - C_3C_7} = \frac{-40(20)^2}{D}\frac{0.033465(0.290898) - 0.266288(0.014554)}{0.177640}$$

$$\theta_b = \frac{-527.8}{D} \text{ rad}$$

$$Q_b = w\frac{C_1L_9 - C_7L_3}{C_1C_9 - C_3C_7} = 40\frac{0.881523(0.290898) - 1.70625(0.014554)}{0.177640} = 52.15 \text{ lb/in}$$

Therefore

$$y = 0 - \frac{527.8r}{D}F_1 + 0 + \frac{52.15r^3}{D}F_3 - \frac{40r^3}{D}G_3$$

Substituting the appropriate expressions for F_1, F_3, and G_3 would produce an equation for y as a function of r, but a reduction of this equation to simple form and an evaluation to determine the location and magnitude of maximum deflection would be extremely time-consuming. Table 23 can be used again to good advantage to evaluate y at specific values of r, and an excellent approximation to the maximum deflection can be obtained.

b/r	r	F_1	$-527.8rF_1$	F_3	$52.15r^3F_3$	r_o/r	G_3	$-40r^3G_3$	$y(D)$
1.00	5.000	0.000	0.0	0.000	0.0		0.000	0.0	0.0
0.90	5.555	0.09858	−289.0	0.000158	1.4		0.000	0.0	−287.6
0.80	6.250	0.194785	−642.5	0.001191	15.2		0.000	0.0	−627.3
0.70	7.143	0.289787	−1092.0	0.003753	71.3		0.000	0.0	−1020.7
0.60	8.333	0.385889	−1697.1	0.008208	247.7		0.000	0.0	−1449.4
0.50	10.000	0.487773	−2574.2	0.014554	759.0	1.00	0.000	0.0	−1815.2
0.40	12.500	0.605736	−3996.0	0.022290	2270.4	0.80	0.001191	−93.0	−1818.6
0.33	15.000	0.704699	−5578.6	0.027649	4866.4	0.67	0.005019	−677.6	−1389.8
0.30	16.667	0.765608	−6734.2	0.030175	7285.4	0.60	0.008208	−1520.0	−968.8
0.25	20.000	0.881523	−9304.5	0.033465	13961.7	0.50	0.014554	−4657.3	−0.1

An examination of the last column on the right shows a zero deflection at the outer edge and indicates that the maximum deflection is located at a radius near 11.25 in and has a value of approximately

$$\frac{-1900}{D} = \frac{-1900(12)(1-0.3^2)}{10(10^6)t^3} = \frac{-0.00207}{t^3} \text{ in}$$

The maximum bending moment will be either a tangential moment at the inside edge or a radial moment at the load line:

$$M_{tb} = \frac{\theta_b D(1-\nu^2)}{b} = \frac{-527.8(1-0.3^2)}{5} = -96.2 \text{ in-lb/in}$$

$$M_{r(r_o)} = \theta_b \frac{D}{r} F_{7(r_o)} + Q_b r F_{9(r_o)}$$

where at $r = r_o$, $b/r = 0.5$. Therefore

$$F_{7(r_o)} = 0.6825$$
$$F_{9(r_o)} = 0.290898$$

$$M_{r(r_o)} = \frac{-527.8}{10}(0.6825) + 52.15(10)(0.290898)$$
$$= -36.05 + 151.5 = 115.45 \text{ in-lb/in}$$

The maximum bending stress in the plate is

$$\sigma = \frac{6(115.45)}{t^2} = \frac{693}{t^2} \text{ lb/in}^2$$

3. A flat phosphor bronze disk with a thickness of 0.020 in and a diameter of 4 in is upset locally in a die to produce a sudden change in slope in the radial direction of 0.05 rad at a radius of $\frac{3}{4}$ in. It is then clamped between two flat dies as shown in Fig. 10.1. It is required to determine the maximum bending stress due to the clamping. Given: $E = 16(10^6)$ lb/in² and $\nu = 0.30$.

$$D = \frac{16(10^6)(0.02^3)}{12(1-0.30^2)} = 11.72$$

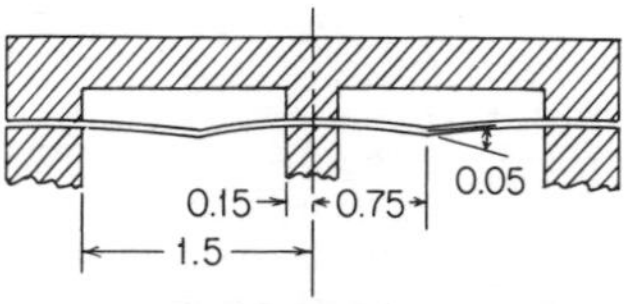

Fig. 10.1

Solution. This example of forcing a known change in slope into a plate clamped at both inner and outer edges is covered in Table 24, case 6h, where $\theta_o = 0.05$, $b/a = 0.10$, and $r_o/a = 0.50$. These dimensions were chosen to fit the tabulated data for a case where $\nu = 0.30$. For this case $M_{rb} = -2.054(0.05)(11.72)/1.5 = -0.803$ in-lb/in; $Q_b = -0.0915(0.05)(11.72)/1.5^2 = -0.0238$ lb/in; $y_b = 0$; and $\theta_b = 0$. The expression for M_r then becomes

$$M_r = -0.803F_8 - 0.0238rF_9 + \frac{0.05(11.72)}{r}G_7$$

An examination of the numerical values of F_8 and F_9 shows that F_8 decreases slightly less than F_9 increases as r increases, but the larger coefficient of the first term indicates that M_{rb} is indeed the maximum moment. The maximum stress is therefore $\sigma = 0.803(6)/0.02^2 = 12{,}050$ lb/in^2 in tension on the top surface at the inner edge. The maximum deflection is at $r_o = 0.75$ in and equals $-0.1071(0.05)(1.5) = -0.00803$ in.

4. A circular steel plate 2 in thick and 20 ft in diameter is simply supported at the outer edge and supported on a center support which can be considered to provide uniform pressure over a diameter of 1.8 in. The plate is loaded in an axisymmetric manner with a distributed load which increases linearly with radius from a value of 0 at $r = 4$ ft to a value of 2000 lb/ft^2 at the outer edge. Determine the maximum bending stress. Given: $E = 30(10^6)$ lb/in^2 and $\nu = 0.30$.

Solution. Table 24, case 11a, deals with this loading and a simply supported outer edge. For this problem $q = \frac{2000}{144} = 13.9$ lb/in^2, $a = 120$ in, and $r_o = 48$ in; and so $r_o/a = 0.4$. From the tabulated data for these quantities, $K_{y_c} = -0.01646$, $K_{\theta_a} = 0.02788$, and $K_{M_c} = 0.04494$. Therefore

$$y_c = \frac{-0.01646(13.9)(120^4)}{D} = \frac{-0.475(10^8)}{D} \text{ in}$$

$$M_c = 0.04494(13.9)(120^2) = 9000 \text{ in-lb/in}$$

Case 16 (Table 24) considers the center load over a small circular area. It is desired to determine W such that the max $y = 0.475(10^8)/D$. Therefore

$$-\frac{W120^2}{16\pi D}\frac{3+0.3}{1+0.3} = \frac{0.475(10^8)}{D}$$

which gives $W = -65{,}000$ lb. The maximum moment is at the center of the plate where

$$M_r = \frac{W}{4\pi}\left[(1+\nu)ln\frac{a}{b} + 1\right]$$

The equivalent radius b is given by

$$1.6r_o^2 + t^2 - 0.675t = 1.6(0.9^2) + 2^2 - 0.675(2) = 0.95 \text{ in.}$$

Therefore,

$$\text{Max } M = \frac{-65{,}000}{4\pi}\left(1.3ln\frac{120}{0.95} + 1\right) = -37{,}500 \text{ in-lb/in}$$

The maximum stress is at the center of the plate where

$$\sigma = \frac{6M}{t^2} = \frac{6(-37{,}500 + 9000)}{2^2} = -43{,}200 \text{ lb/in}^2 \qquad \text{(tension on the top surface)}$$

TABLE 23 ***Numerical values for functions used in Table 24***

Numerical values for the plate coefficients F, C, L, and G for values of b/r, b/a, r_o/a, and r_o/r, respectively, from 0.05 to 1.0. Poisson's ratio is 0.30. The table headings are given for G_1 to G_{19} for the various values of r_o/r. Also listed in the last five lines are values for the most used denominators for the ratios b/a

r_o/r	1.000	0.900	0.800	0.750	0.700	0.677	0.600	0.500
G_1	0.000	0.098580346	0.19478465	0.2423283	0.2897871	0.3215349	0.3858887	0.487773
G_2	0.000	0.004828991	0.01859406	0.0284644	0.0401146	0.0487855	0.0680514	0.100857
G_3	0.000	0.000158070	0.00119108	0.0022506	0.0037530	0.0050194	0.0082084	0.014554
G_4	1.000	0.973888889	0.95750000	0.9541667	0.9550000	0.9583333	0.9733333	1.025000
G_5	0.000	0.095000000	0.18000000	0.2187500	0.2550000	0.2777778	0.3200000	0.375000
G_6	0.000	0.004662232	0.01725742	0.0258495	0.0355862	0.0425624	0.0572477	0.079537
G_7	0.000	0.096055556	0.20475000	0.2654167	0.3315000	0.3791667	0.4853333	0.682500
G_8	1.000	0.933500000	0.87400000	0.8468750	0.8215000	0.8055556	0.7760000	0.737500
G_9	0.000	0.091560902	0.16643465	0.1976669	0.2247621	0.2405164	0.2664220	0.290898
G_{11}	0.000	0.000003996	0.00006104	0.0001453	0.0002935	0.0004391	0.0008752	0.001999
G_{12}	0.000	0.000000805	0.00001240	0.0000297	0.0000603	0.0000905	0.0001820	0.000422
G_{13}	0.000	0.000000270	0.00000418	0.0000100	0.0000205	0.0000308	0.0000623	0.000146
G_{14}	0.000	0.000158246	0.00119703	0.0022693	0.0038011	0.0051026	0.0084257	0.015272
G_{15}	0.000	0.000039985	0.00030618	0.0005844	0.0009861	0.0013307	0.0022227	0.004111
G_{16}	0.000	0.000016107	0.00012431	0.0002383	0.0004039	0.0005468	0.0009196	0.001721
G_{17}	0.000	0.004718219	0.01775614	0.0268759	0.0374539	0.0452137	0.0621534	0.090166
G_{18}	0.000	0.001596148	0.00610470	0.0093209	0.0131094	0.0159275	0.0221962	0.032948
G_{19}	0.000	0.000805106	0.00310827	0.0047694	0.0067426	0.0082212	0.0115422	0.017341
$C_1C_6 - C_3C_4$	0.000	0.000305662	0.00222102	0.0041166	0.0067283	0.0088751	0.0141017	0.023878
$C_1C_9 - C_3C_7$	0.000	0.009010922	0.03217504	0.0473029	0.0638890	0.0754312	0.0988254	0.131959
$C_2C_6 - C_3C_5$	0.000	0.000007497	0.00010649	0.0002435	0.0004705	0.0006822	0.0012691	0.002564
$C_2C_9 - C_3C_8$	0.000	0.000294588	0.00205369	0.0037205	0.0059332	0.0076903	0.0117606	0.018605
$C_4C_9 - C_6C_7$	0.000	0.088722311	0.15582772	0.1817463	0.2028510	0.2143566	0.2315332	0.243886

r_o/r	0.400	0.333	0.300	0.250	0.200	0.125	0.100	0.050
G_1	0.605736	0.704699	0.765608	0.881523	1.049227	1.547080	1.882168	3.588611
G_2	0.136697	0.161188	0.173321	0.191053	0.207811	0.229848	0.235987	0.245630
G_3	0.022290	0.027649	0.030175	0.033465	0.035691	0.035236	0.033390	0.025072
G_4	1.135000	1.266667	1.361667	1.562500	1.880000	2.881250	3.565000	7.032500
G_5	0.420000	0.444444	0.455000	0.468750	0.480000	0.492187	0.495000	0.498750
G_6	0.099258	0.109028	0.112346	0.114693	0.112944	0.099203	0.090379	0.062425
G_7	0.955500	1.213333	1.380167	1.706250	2.184000	3.583125	4.504500	9.077250
G_8	0.706000	0.688889	0.681500	0.671875	0.664000	0.655469	0.653500	0.650875
G_9	0.297036	0.289885	0.282550	0.266288	0.242827	0.190488	0.166993	0.106089
G_{11}	0.003833	0.005499	0.006463	0.008057	0.009792	0.012489	0.013350	0.014843
G_{12}	0.000827	0.001208	0.001435	0.001822	0.002266	0.003027	0.003302	0.003872
G_{13}	0.000289	0.000427	0.000510	0.000654	0.000822	0.001121	0.001233	0.001474
G_{14}	0.024248	0.031211	0.034904	0.040595	0.046306	0.054362	0.056737	0.060627
G_{15}	0.006691	0.008790	0.009945	0.011798	0.013777	0.016917	0.017991	0.020139
G_{16}	0.002840	0.003770	0.004290	0.005138	0.006065	0.007589	0.008130	0.009252
G_{17}	0.119723	0.139340	0.148888	0.162637	0.175397	0.191795	0.196271	0.203191
G_{18}	0.044939	0.053402	0.057723	0.064263	0.070816	0.080511	0.083666	0.089788
G_{19}	0.023971	0.028769	0.031261	0.035098	0.039031	0.045057	0.047086	0.051154
$C_1C_6 - C_3C_4$	0.034825	0.041810	0.044925	0.048816	0.051405	0.051951	0.051073	0.047702
$C_1C_9 - C_3C_7$	0.158627	0.170734	0.174676	0.177640	0.176832	0.168444	0.163902	0.153133
$C_2C_6 - C_3C_5$	0.004207	0.005285	0.005742	0.006226	0.006339	0.005459	0.004800	0.002829
$C_2C_9 - C_3C_8$	0.024867	0.027679	0.028408	0.028391	0.026763	0.020687	0.017588	0.009740
$C_4C_9 - C_6C_7$	0.242294	0.234900	0.229682	0.220381	0.209845	0.193385	0.188217	0.179431

TABLE 24 ***Formulas for flat circular plates of constant thickness***

NOTATION: W = total applied load (pounds); w = unit line load (pounds per inch of circumference); q = load per unit area (pounds per square inch); M_o = unit applied line moment loading (inch-pounds per inch of circumference); θ_o = externally applied change in angular displacement (radians); y_o = externally applied step in the normal displacement (inches); y = vertical deflection of plate (inches); θ = slope of plate measured from the horizontal (radians); M_r = unit radial bending moment (inch-pounds per inch of circumference); M_t = unit tangential bending moment (inch-pounds per inch of radius); Q = unit shear force (pounds per inch of circumference); E = modulus of elasticity (pounds per square inch); ν = Poisson's ratio; γ = temperature coefficient of expansion (inches per inch per degree); a = outer radius (inches); b = inner radius for annular plate (inches); t = plate thickness (inches); r = radial location of quantity being evaluated (inches); r_o = radial location of unit line loading or the start of a distributed load (inches). F_1 to F_9 and G_1 to G_{19} are the several functions of the radial location r. C_1 to C_9 are plate constants dependent upon the ratio a/b. L_1 to L_{19} are loading constants dependent upon the ratio a/r_o. When used as subscripts, r and t refer to radial and tangential directions, respectively. When used as subscripts, a and b refer to an evaluation of the quantity subscripted at the outer edge and inner edge, respectively. When used as a subscript, c refers to the subscripted quantity evaluated at the center of the plate

Positive signs are associated with the several quantities in the following manner: Deflections y and y_o are positive upward; slopes θ and θ_o are positive when the deflection y increases positively as r increases; moments M_r, M_t, and M_o are positive when creating compression on the top surface; and the shear force Q is positive when acting upward on the inner edge of a given annular section

Bending stresses can be found from the moments M_r and M_t by the expression $\sigma = 6M/t^2$. The plate constant $D = Et^3/12(1-\nu^2)$. The singularity function brackets $\langle \ \rangle$ indicate that the expression contained within the brackets must be equated to zero unless $r > r_o$, after which they are treated as any other brackets.

General Plate Functions and Constants for Solid and Annular Circular Plates

$$F_1 = \frac{1+\nu}{2}\frac{b}{r}\ln\frac{r}{b} + \frac{1-\nu}{4}\left(\frac{r}{b} - \frac{b}{r}\right) \qquad C_1 = \frac{1+\nu}{2}\frac{b}{a}\ln\frac{a}{b} + \frac{1-\nu}{4}\left(\frac{a}{b} - \frac{b}{a}\right)$$

$$F_2 = \frac{1}{4}\left[1 - \left(\frac{b}{r}\right)^2\left(1 + 2\ln\frac{r}{b}\right)\right] \qquad C_2 = \frac{1}{4}\left[1 - \left(\frac{b}{a}\right)^2\left(1 + 2\ln\frac{a}{b}\right)\right]$$

$$F_3 = \frac{b}{4r}\left\{\left[\left(\frac{b}{r}\right)^2 + 1\right]\ln\frac{r}{b} + \left(\frac{b}{r}\right)^2 - 1\right\} \qquad C_3 = \frac{b}{4a}\left\{\left[\left(\frac{b}{a}\right)^2 + 1\right]\ln\frac{a}{b} + \left(\frac{b}{a}\right)^2 - 1\right\}$$

$$F_4 = \frac{1}{2}\left[(1+\nu)\frac{b}{r} + (1-\nu)\frac{r}{b}\right] \qquad C_4 = \frac{1}{2}\left[(1+\nu)\frac{b}{a} + (1-\nu)\frac{a}{b}\right]$$

$$F_5 = \frac{1}{2}\left[1 - \left(\frac{b}{r}\right)^2\right] \qquad C_5 = \frac{1}{2}\left[1 - \left(\frac{b}{a}\right)^2\right]$$

$$F_6 = \frac{b}{4r}\left[\left(\frac{b}{r}\right)^2 - 1 + 2\ln\frac{r}{b}\right] \qquad C_6 = \frac{b}{4a}\left[\left(\frac{b}{a}\right)^2 - 1 + 2\ln\frac{a}{b}\right]$$

$$F_7 = \frac{1}{2}(1-\nu^2)\left(\frac{r}{b} - \frac{b}{r}\right) \qquad C_7 = \frac{1}{2}(1-\nu^2)\left(\frac{a}{b} - \frac{b}{a}\right)$$

$$F_8 = \frac{1}{2}\left[1 + \nu + (1-\nu)\left(\frac{b}{r}\right)^2\right] \qquad C_8 = \frac{1}{2}\left[1 + \nu + (1-\nu)\left(\frac{b}{a}\right)^2\right]$$

$F_9 =$ [illegible] $\qquad C_9 = \frac{b}{a}\left\{\frac{1+\nu}{2}\ln\frac{a}{b} + \frac{1-\nu}{4}\left[1 - \left(\frac{b}{a}\right)^2\right]\right\}$

$$\ldots \quad 2 \quad a \quad r_o \quad 4 \quad r_o \quad a$$

$$L_2 = \frac{1}{4}\left[1 - \left(\frac{r_o}{a}\right)^2\left(1 + 2\ln\frac{a}{r_o}\right)\right]$$

$$L_3 = \frac{r_o}{4a}\left\{\left[\left(\frac{r_o}{a}\right)^2 + 1\right]\ln\frac{a}{r_o} + \left(\frac{r_o}{a}\right)^2 - 1\right\}$$

$$L_4 = \frac{1}{2}\left[(1+\nu)\frac{r_o}{a} + (1-\nu)\frac{a}{r_o}\right]$$

$$L_5 = \frac{1}{2}\left[1 - \left(\frac{r_o}{a}\right)^2\right]$$

$$L_6 = \frac{r_o}{4a}\left[\left(\frac{r_o}{a}\right)^2 - 1 + 2\ln\frac{a}{r_o}\right]$$

$$L_7 = \frac{1}{2}(1-\nu^2)\left(\frac{a}{r_o} - \frac{r_o}{a}\right)$$

$$L_8 = \frac{1}{2}\left[1 + \nu + (1-\nu)\left(\frac{r_o}{a}\right)^2\right]$$

$$L_9 = \frac{r_o}{a}\left\{\frac{1+\nu}{2}\ln\frac{a}{r_o} + \frac{1-\nu}{4}\left[1 - \left(\frac{r_o}{a}\right)^2\right]\right\}$$

$$L_{11} = \frac{1}{64}\left\{1 + 4\left(\frac{r_o}{a}\right)^2 - 5\left(\frac{r_o}{a}\right)^4 - 4\left(\frac{r_o}{a}\right)^2\left[2 + \left(\frac{r_o}{a}\right)^2\right]\ln\frac{a}{r_o}\right\}$$

$$L_{12} = \frac{a}{14{,}400(a - r_o)}\left\{64 - 225\frac{r_o}{a} - 100\left(\frac{r_o}{a}\right)^3 + 261\left(\frac{r_o}{a}\right)^5 + 60\left(\frac{r_o}{a}\right)^3\left[3\left(\frac{r_o}{a}\right)^2 + 10\right]\ln\frac{a}{r_o}\right\}$$

$$L_{13} = \frac{a^2}{14{,}400(a - r_o)^2}\left\{25 - 128\frac{r_o}{a} + 225\left(\frac{r_o}{a}\right)^2 - 25\left(\frac{r_o}{a}\right)^4 - 97\left(\frac{r_o}{a}\right)^6 - 60\left(\frac{r_o}{a}\right)^4\left[5 + \left(\frac{r_o}{a}\right)^2\right]\ln\frac{a}{r_o}\right\}$$

$$L_{14} = \frac{1}{16}\left[1 - \left(\frac{r_o}{a}\right)^4 - 4\left(\frac{r_o}{a}\right)^2\ln\frac{a}{r_o}\right]$$

$$L_{15} = \frac{a}{720(a - r_o)}\left[16 - 45\frac{r_o}{a} + 9\left(\frac{r_o}{a}\right)^5 + 20\left(\frac{r_o}{a}\right)^3\left(1 + 3\ln\frac{a}{r_o}\right)\right]$$

$$L_{16} = \frac{a^2}{1440(a - r_o)^2}\left[15 - 64\frac{r_o}{a} + 90\left(\frac{r_o}{a}\right)^2 - 6\left(\frac{r_o}{a}\right)^6 - 5\left(\frac{r_o}{a}\right)^4\left(7 + 12\ln\frac{a}{r_o}\right)\right]$$

$$L_{17} = \frac{1}{4}\left\{1 - \frac{1-\nu}{4}\left[1 - \left(\frac{r_o}{a}\right)^4\right] - \left(\frac{r_o}{a}\right)^2\left[1 + (1+\nu)\ln\frac{a}{r_o}\right]\right\}$$

$$G_1 = \left[\frac{\quad}{2}\,\frac{\quad}{r}\ln\frac{\quad}{r_o} + \frac{\quad}{4}\left(\frac{\quad}{r_o} - \frac{\quad}{r}\right)\right]\langle r - r_o\rangle^0$$

$$G_2 = \frac{1}{4}\left[1 - \left(\frac{r_o}{r}\right)^2\left(1 + 2\ln\frac{r}{r_o}\right)\right]\langle r - r_o\rangle^0$$

$$G_3 = \frac{r_o}{4r}\left\{\left[\left(\frac{r_o}{r}\right)^2 + 1\right]\ln\frac{r}{r_o} + \left(\frac{r_o}{r}\right)^2 - 1\right\}\langle r - r_o\rangle^0$$

$$G_4 = \frac{1}{2}\left[(1+\nu)\frac{r_o}{r} + (1-\nu)\frac{r}{r_o}\right]\langle r - r_o\rangle^0$$

$$G_5 = \frac{1}{2}\left[1 - \left(\frac{r_o}{r}\right)^2\right]\langle r - r_o\rangle^0$$

$$G_6 = \frac{r_o}{4r}\left[\left(\frac{r_o}{r}\right)^2 - 1 + 2\ln\frac{r}{r_o}\right]\langle r - r_o\rangle^0$$

$$G_7 = \frac{1}{2}(1-\nu^2)\left(\frac{r}{r_o} - \frac{r_o}{r}\right)\langle r - r_o\rangle^0$$

$$G_8 = \frac{1}{2}\left[1 + \nu + (1-\nu)\left(\frac{r_o}{r}\right)^2\right]\langle r - r_o\rangle^0$$

$$G_9 = \frac{r_o}{r}\left\{\frac{1+\nu}{2}\ln\frac{r}{r_o} + \frac{1-\nu}{4}\left[1 - \left(\frac{r_o}{r}\right)^2\right]\right\}\langle r - r_o\rangle^0$$

$$G_{11} = \frac{1}{64}\left\{1 + 4\left(\frac{r_o}{r}\right)^2 - 5\left(\frac{r_o}{r}\right)^4 - 4\left(\frac{r_o}{r}\right)^2\left[2 + \left(\frac{r_o}{r}\right)^2\right]\ln\frac{r}{r_o}\right\}\langle r - r_o\rangle^0$$

$$G_{12} = \frac{r\langle r - r_o\rangle^0}{14{,}400(r - r_o)}\left\{64 - 225\frac{r_o}{r} - 100\left(\frac{r_o}{r}\right)^3 + 261\left(\frac{r_o}{r}\right)^5 + 60\left(\frac{r_o}{r}\right)^3\left[3\left(\frac{r_o}{r}\right)^2 + 10\right]\ln\frac{r}{r_o}\right\}$$

$$G_{13} = \frac{r^2\langle r - r_o\rangle^0}{14{,}400(r - r_o)^2}\left\{25 - 128\frac{r_o}{r} + 225\left(\frac{r_o}{r}\right)^2 - 25\left(\frac{r_o}{r}\right)^4 - 97\left(\frac{r_o}{r}\right)^6 - 60\left(\frac{r_o}{r}\right)^4\left[5 + \left(\frac{r_o}{r}\right)^2\right]\ln\frac{r}{r_o}\right\}$$

$$G_{14} = \frac{1}{16}\left[1 - \left(\frac{r_o}{r}\right)^4 - 4\left(\frac{r_o}{r}\right)^2\ln\frac{r}{r_o}\right]\langle r - r_o\rangle^0$$

$$G_{15} = \frac{r\langle r - r_o\rangle^0}{720(r - r_o)}\left[16 - 45\frac{r_o}{r} + 9\left(\frac{r_o}{r}\right)^5 + 20\left(\frac{r_o}{r}\right)^3\left(1 + 3\ln\frac{r}{r_o}\right)\right]$$

$$G_{16} = \frac{r^2\langle r - r_o\rangle^0}{1440(r - r_o)^2}\left[15 - 64\frac{r_o}{r} + 90\left(\frac{r_o}{r}\right)^2 - 6\left(\frac{r_o}{r}\right)^6 - 5\left(\frac{r_o}{r}\right)^4\left(7 + 12\ln\frac{r}{r_o}\right)\right]$$

$$G_{17} = \frac{1}{4}\left\{1 - \frac{1-\nu}{4}\left[1 - \left(\frac{r_o}{r}\right)^4\right] - \left(\frac{r_o}{r}\right)^2\left[1 + (1+\nu)\ln\frac{r}{r_o}\right]\right\}\langle r - r_o\rangle^0$$

TABLE 24 *Formulas for flat circular plates of constant thickness* (*Cont.*)

$$L_{18} = \frac{a}{720(a - r_o)}\left\{\left[20\left(\frac{r_o}{a}\right)^3 + 16\right](4 + \nu) - 45\frac{r_o}{a}(3 + \nu) - 9\left(\frac{r_o}{a}\right)^5(1 - \nu) + 60\left(\frac{r_o}{a}\right)^3(1 + \nu)\ln\frac{a}{r_o}\right\}$$

$$G_{18} = \frac{r\langle r - r_o\rangle^0}{720(r - r_o)}\left\{\left[20\left(\frac{r_o}{r}\right)^3 + 16\right](4 + \nu) - 45\frac{r_o}{r}(3 + \nu) - 9\left(\frac{r_o}{r}\right)^5(1 - \nu) + 60\left(\frac{r_o}{r}\right)^3(1 + \nu)\ln\frac{r}{r_o}\right\}$$

$$L_{19} = \frac{a^2}{1440(a - r_o)^2}\left[15(5 + \nu) - 64\frac{r_o}{a}(4 + \nu) + 90\left(\frac{r_o}{a}\right)^2(3 + \nu) - 5\left(\frac{r_o}{a}\right)^4(19 + 7\nu) + 6\left(\frac{r_o}{a}\right)^6(1 - \nu) - 60\left(\frac{r_o}{a}\right)^4(1 + \nu)\ln\frac{a}{r_o}\right]$$

$$G_{19} = \frac{r^2\langle r - r_o\rangle^0}{1440(r - r_o)^2}\left[15(5 + \nu) - 64\frac{r_o}{r}(4 + \nu) + 90\left(\frac{r_o}{r}\right)^2(3 + \nu) - 5\left(\frac{r_o}{r}\right)^4(19 + 7\nu) + 6\left(\frac{r_o}{r}\right)^6(1 - \nu) - 60\left(\frac{r_o}{r}\right)^4(1 + \nu)\ln\frac{r}{r_o}\right]$$

Case 1. Annular plate with a uniform annular line load of w lb/in at a radius r_o

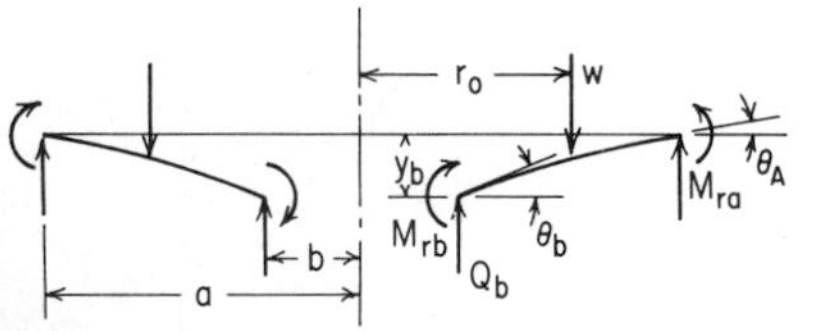

General expressions for deformations, moments, and shears:

$$y = y_b + \theta_b r F_1 + M_{rb}\frac{r^2}{D}F_2 + Q_b\frac{r^3}{D}F_3 - w\frac{r^3}{D}G_3$$

$$\theta = \theta_b F_4 + M_{rb}\frac{r}{D}F_5 + Q_b\frac{r^2}{D}F_6 - w\frac{r^2}{D}G_6$$

$$M_r = \theta_b\frac{D}{r}F_7 + M_{rb}F_8 + Q_b r F_9 - wrG_9$$

$$M_t = \frac{\theta D(1 - \nu^2)}{r} + \nu M_r$$

$$Q = Q_b\frac{b}{r} - w\frac{r_o}{r}\langle r - r_o\rangle^0$$

For the numerical data given below, $\nu = 0.3$

$$y = K_y\frac{wa^3}{D} \qquad \theta = K_\theta\frac{wa^2}{D} \qquad M = K_M wa \qquad Q = K_Q w$$

Case no., edge restraints	Boundary values	Special cases
1a. Outer edge simply supported, inner edge free 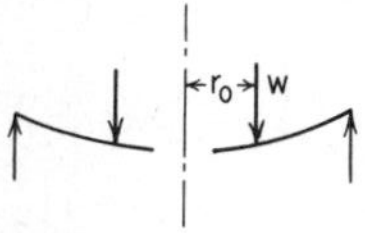	$M_{rb} = 0 \quad Q_b = 0 \quad y_a = 0 \quad M_{ra} = 0$ $y_b = \frac{-wa^3}{D}\left(\frac{C_1 L_9}{C_7} - L_3\right)$ $\theta_b = \frac{wa^2}{DC_7}L_9$ $\theta_a = \frac{wa^2}{D}\left(\frac{C_4 L_9}{C_7} - L_6\right)$ [illegible]	Max $y = y_b$ Max $M = M_{tb}$ If $r_o = b$ (load at inner edge),

b/a	0.1	0.3	0.5	0.7	0.9
K_{y_b}	−0.0364	−0.1266	−0.1934	−0.1927	−0.0938
K_{θ_b}	0.0371	0.2047	0.4262	0.6780	0.9532
K_{θ_a}	0.0418	0.1664	0.3573	0.6119	0.9237
$K_{M_{tb}}$	0.3374	0.6210	0.7757	0.8814	0.9638

Case	Formulas	Special cases
inner edge guided 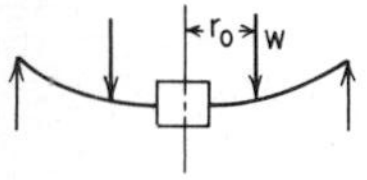	$y_b = \frac{-wa^3}{D}\left(\frac{C_2L_9}{C_8} - L_3\right)$ $M_{rb} = \frac{wa}{C_8}L_9$ $\theta_a = \frac{wa^2}{D}\left(\frac{C_5L_9}{C_8} - L_6\right)$ $Q_a = -w\frac{r_o}{a}$	Max $y = y_b$ Max $M = M_{rb}$ If $r_o = b$ (load at inner edge),
1c. Outer edge simply supported, inner edge simply supported 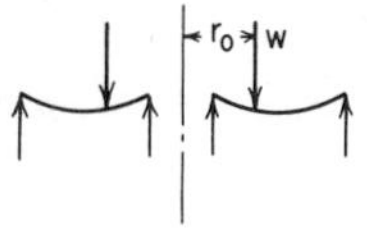	$y_b = 0 \quad M_{rb} = 0 \quad y_a = 0 \quad M_{ra} = 0$ $\theta_b = \frac{-wa^2}{D}\frac{C_3L_9 - C_9L_3}{C_1C_9 - C_3C_7}$ $Q_b = w\frac{C_1L_9 - C_7L_3}{C_1C_9 - C_3C_7}$ $\theta_a = \theta_bC_4 + Q_b\frac{a^2}{D}C_6 - \frac{wa^2}{D}L_6$ $Q_a = Q_b\frac{b}{a} - \frac{wr_o}{a}$	
1d. Outer edge simply supported, inner edge fixed 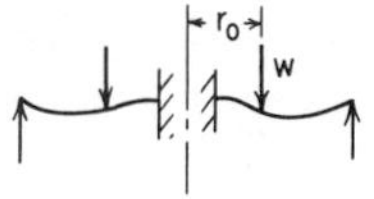	$y_b = 0 \quad \theta_b = 0 \quad y_a = 0 \quad M_{ra} = 0$ $M_{rb} = -wa\frac{C_3L_9 - C_9L_3}{C_2C_9 - C_3C_8}$ $Q_b = w\frac{C_2L_9 - C_8L_3}{C_2C_9 - C_3C_8}$ $\theta_a = M_{rb}\frac{a}{D}C_5 + Q_b\frac{a^2}{D}C_6 - \frac{wa^2}{D}L_6$ $Q_a = Q_b\frac{b}{a} - \frac{wr_o}{a}$	

Case 1b (inner edge guided), if $r_o = b$ (load at inner edge):

b/a	0.1	0.3	0.5	0.7	0.9
K_{y_b}	−0.0269	−0.0417	−0.0252	−0.0072	−0.0003
K_{θ_a}	0.0361	0.0763	0.0684	0.0342	0.0047
$K_{M_{rb}}$	0.2555	0.4146	0.3944	0.2736	0.0981

Case 1c:

b/a	0.1		0.5		0.7
r_o/a	0.5	0.7	0.7	0.9	0.9
$K_{y_{max}}$	−0.0102	−0.0113	−0.0023	−0.0017	−0.0005
K_{θ_a}	0.0278	0.0388	0.0120	0.0122	0.0055
K_{θ_b}	−0.0444	−0.0420	−0.0165	−0.0098	−0.0048
$K_{M_{tb}}$	−0.4043	−0.3819	−0.0301	−0.0178	−0.0063
$K_{M_{ro}}$	0.1629	0.1689	0.1161	0.0788	0.0662
K_{Q_b}	2.9405	2.4779	0.8114	0.3376	0.4145

Case 1d:

b/a	0.1		0.5		0.9
r_o/a	0.5	0.7	0.7	0.9	0.9
$K_{y_{max}}$	−0.0066	−0.0082	−0.0010	−0.0010	−0.0003
K_{θ_a}	0.0194	0.0308	0.0056	0.0084	0.0034
$K_{M_{rb}}$	−0.4141	−0.3911	−0.1172	−0.0692	−0.0519
K_{Q_b}	3.3624	2.8764	1.0696	0.4901	0.5972

TABLE 24 ***Formulas for flat circular plates of constant thickness*** **(Cont.)**

Case no., edge restraints	Boundary values	Special cases
1e. Outer edge fixed, inner edge free	$M_{rb} = 0 \quad Q_b = 0 \quad y_a = 0 \quad \theta_a = 0$	If $r_o = b$ (load at inner edge),

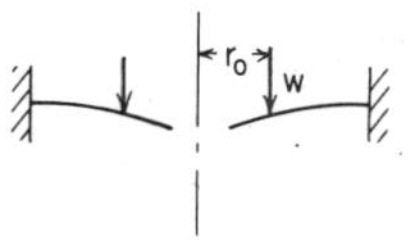

$$y_b = \frac{-wa^3}{D}\left(\frac{C_1 L_6}{C_4} - L_3\right)$$

$$\theta_b = \frac{wa^2}{DC_4} L_6$$

$$M_{ra} = -wa\left(L_9 - \frac{C_7 L_6}{C_4}\right)$$

$$Q_a = \frac{-wr_o}{a}$$

b/a	0.1	0.3	0.5	0.7	0.9
K_{y_b}	−0.0143	−0.0330	−0.0233	−0.0071	−0.0003
K_{θ_b}	0.0254	0.0825	0.0776	0.0373	0.0048
$K_{M_{ra}}$	−0.0528	−0.1687	−0.2379	−0.2124	−0.0911
$K_{M_{tb}}$	0.2307	0.2503	0.1412	0.0484	0.0048

(*Note:* $|M_{ra}| > |M_{tb}|$ if $b/a > 0.385$)

Case no., edge restraints	Boundary values	Special cases
1f. Outer edge fixed, inner edge guided	$\theta_b = 0 \quad Q_b = 0 \quad y_a = 0 \quad \theta_a = 0$	If $r_o = b$ (load at inner edge),

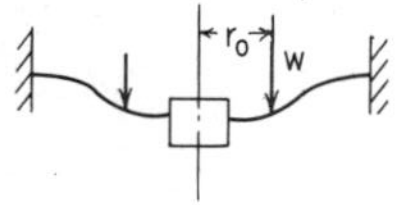

$$y_b = \frac{-wa^3}{D}\left(\frac{C_2 L_6}{C_5} - L_3\right)$$

$$M_{rb} = \frac{wa}{C_5} L_6$$

$$M_{ra} = -wa\left(L_9 - \frac{C_8 L_6}{C_5}\right)$$

$$Q_a = \frac{-wr_o}{a}$$

b/a	0.1	0.3	0.5	0.7	0.9
K_{y_b}	−0.0097	−0.0126	−0.0068	−0.0019	−0.0001
$K_{M_{rb}}$	0.1826	0.2469	0.2121	0.1396	0.0491
$K_{M_{ra}}$	−0.0477	−0.1143	−0.1345	−0.1101	−0.0458

Case no., edge restraints	Boundary values	Special cases
1g. Outer edge fixed, inner edge simply supported	$y_b = 0 \quad M_{rb} = 0 \quad y_a = 0 \quad \theta_a = 0$	

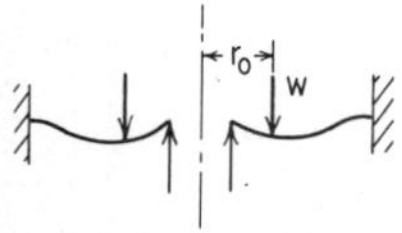

$$\theta_b = \frac{-wa^2}{D} \frac{C_3 L_6 - C_6 L_3}{C_1 C_6 - C_3 C_4}$$

$$Q_b = w \frac{C_1 L_6 - C_4 L_3}{C_1 C_6 - C_3 C_4}$$

$$M_{ra} = \theta_b \frac{D}{a} C_7 + Q_b a C_9 - waL_9$$

$$Q_a = Q_b \frac{b}{a} - \frac{wr_o}{a}$$

b/a	0.1		0.5		0.7
r_o/a	0.5	0.7	0.7	0.9	0.9
$K_{y_{max}}$	−0.0053	−0.0041	−0.0012	−0.0004	−0.0002
K_{θ_b}	−0.0262	−0.0166	−0.0092	−0.0023	−0.0018
$K_{M_{tb}}$	−0.2388	−0.1513	−0.0167	−0.0042	−0.0023
$K_{M_{ro}}$	0.1179	0.0766	0.0820	0.0208	0.0286
$K_{M_{ra}}$	−0.0893	−0.1244	−0.0664	−0.0674	−0.0521
K_{Q_b}	1.9152	1.0495	0.5658	0.0885	0.1784

1h. Outer edge fixed, inner edge fixed

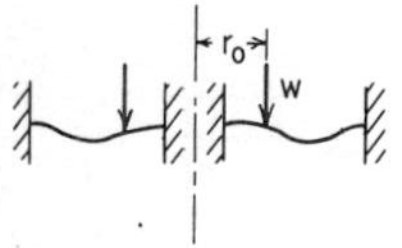

$y_b = 0 \qquad \theta_b = 0 \qquad y_a = 0 \qquad \theta_a = 0$

$$M_{rb} = -wa\frac{C_3L_6 - C_6L_3}{C_2C_6 - C_3C_5}$$

$$Q_b = w\frac{C_2L_6 - C_5L_3}{C_2C_6 - C_3C_5}$$

$$M_{ra} = M_{rb}C_8 + Q_baC_9 - waL_9$$

$$Q_a = Q_b\frac{b}{a} - \frac{wr_o}{a}$$

b/a	0.1		0.5		0.7
r_o/a	0.5	0.7	0.7	0.9	0.9
$K_{y_{max}}$	−0.0038	−0.0033	−0.0006	−0.0003	−0.0001
$K_{M_{rb}}$	−0.2792	−0.1769	−0.0856	−0.0216	−0.0252
$K_{M_{ra}}$	−0.0710	−0.1128	−0.0404	−0.0608	−0.0422
$K_{M_{ro}}$	0.1071	0.0795	0.0586	0.0240	0.0290
K_{Q_b}	2.4094	1.3625	0.8509	0.1603	0.3118

1i. Outer edge guided, inner edge simply supported

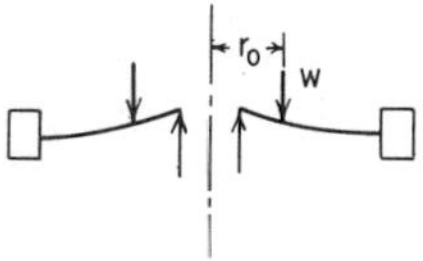

$y_b = 0 \qquad M_{rb} = 0 \qquad \theta_a = 0 \qquad Q_a = 0$

$$\theta_b = \frac{-wa^2}{DC_4}\left(\frac{r_oC_6}{b} - L_6\right)$$

$$Q_b = \frac{wr_o}{b}$$

$$y_a = \frac{-wa^3}{D}\left[\frac{C_1}{C_4}\left(\frac{r_oC_6}{b} - L_6\right) - \frac{r_oC_3}{b} + L_3\right]$$

$$M_{ra} = wa\left[\frac{C_7}{C_4}\left(L_6 - \frac{r_oC_6}{b}\right) + \frac{r_oC_9}{b} - L_9\right]$$

If $r_o = a$ (load at outer edge),

$$\text{Max } y = y_a = \frac{-wa^4}{bD}\left(\frac{C_1C_6}{C_4} - C_3\right)$$

$$\text{Max } M = M_{ra} = \frac{wa^2}{b}\left(C_9 - \frac{C_6C_7}{C_4}\right) \qquad \text{if } \frac{b}{a} > 0.385$$

$$\text{Max } M = M_{tb} = \frac{-wa^3}{b^2}(1 - \nu^2)\frac{C_6}{C_4} \qquad \text{if } \frac{b}{a} < 0.385$$

(For numerical values see case 1e after computing the loading at the inner edge)

1j. Outer edge guided, inner edge fixed

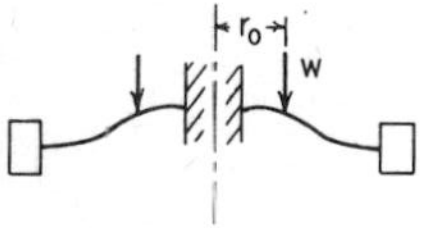

$y_b = 0 \qquad \theta_b = 0 \qquad \theta_a = 0 \qquad Q_a = 0$

$$M_{rb} = \frac{-wa}{C_5}\left(\frac{r_oC_6}{b} - L_6\right)$$

$$Q_b = \frac{wr_o}{b}$$

$$y_a = \frac{-wa^3}{D}\left[\frac{C_2}{C_5}\left(\frac{r_oC_6}{b} - L_6\right) - \frac{r_oC_3}{b} + L_3\right]$$

$$M_{ra} = wa\left[\frac{C_8}{C_5}\left(L_6 - \frac{r_oC_6}{b}\right) + \frac{r_oC_9}{b} - L_9\right]$$

If $r_o = a$ (load at outer edge),

$$\text{Max } y = y_a = \frac{-wa^4}{bD}\left(\frac{C_2C_6}{C_5} - C_3\right)$$

$$\text{Max } M = M_{rb} = \frac{-wa^2C_6}{bC_5}$$

(For numerical values see case 1f after computing the loading at the inner edge)

TABLE 24 ***Formulas for flat circular plates of constant thickness*** **(*Cont.*)**

Case no., edge restraints	Boundary values	Special cases
1k. Outer edge free, inner edge simply supported 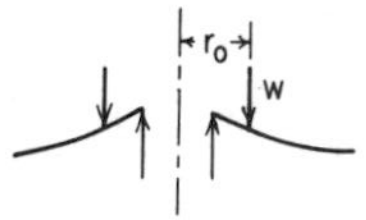	$y_b = 0 \quad M_{rb} = 0 \quad M_{ra} = 0 \quad Q_a = 0$ $\theta_b = \frac{-wa^2}{DC_7}\left(\frac{r_oC_9}{b} - L_9\right)$ $Q_b = \frac{wr_o}{b}$ $y_a = \frac{-wa^3}{D}\left[\frac{C_1}{C_7}\left(\frac{r_oC_9}{b} - L_9\right) - \frac{r_oC_3}{b} + L_3\right]$ $\theta_a = \frac{-wa^2}{D}\left[\frac{C_4}{C_7}\left(\frac{r_oC_9}{b} - L_9\right) - \frac{r_oC_6}{b} + L_6\right]$	If $r_o = a$ (load at outer edge), $\text{Max } y = y_a = \frac{-wa^4}{bD}\left(\frac{C_1C_9}{C_7} - C_3\right)$ $\text{Max } M = M_{tb} = \frac{-wa^3}{b^2}(1 - \nu^2)\frac{C_9}{C_7}$ (For numerical values see case 1b after computing the loading at the inner edge)
1l. Outer edge free, inner edge fixed 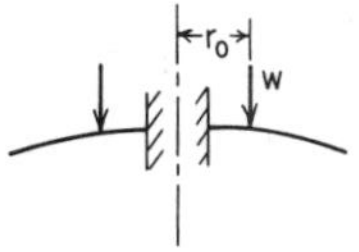	$y_b = 0 \quad \theta_b = 0 \quad M_{ra} = 0 \quad Q_a = 0$ $M_{rb} = \frac{-wa}{C_8}\left(\frac{r_oC_9}{b} - L_9\right)$ $Q_b = \frac{wr_o}{b}$ $y_a = \frac{-wa^3}{D}\left[\frac{C_2}{C_8}\left(\frac{r_oC_9}{b} - L_9\right) - \frac{r_oC_3}{b} + L_3\right]$ $\theta_a = \frac{-wa^2}{D}\left[\frac{C_5}{C_8}\left(\frac{r_oC_9}{b} - L_9\right) - \frac{r_oC_6}{b} + L_6\right]$	If $r_o = a$ (load at outer edge), $\text{Max } y = y_a = \frac{-wa^4}{bD}\left(\frac{C_2C_9}{C_8} - C_3\right)$ $\text{Max } M = M_{rb} = \frac{-wa^2}{b}\frac{C_9}{C_8}$ (For numerical values see case 1b after computing the loading at the inner edge)

Case 2. Annular plate with a uniformly distributed load of q lb/in^2 over the portion from r_o to a

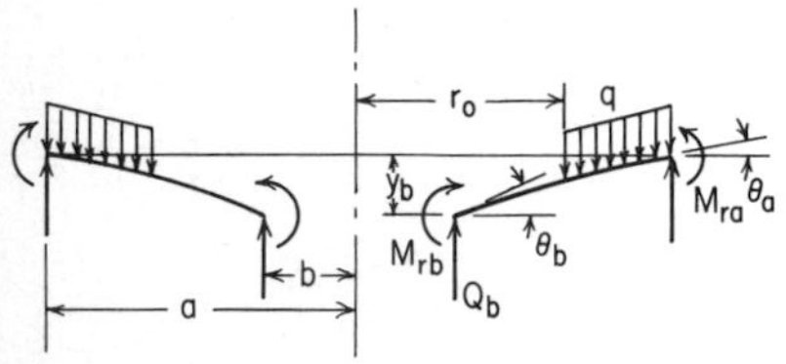

General expressions for deformations, moments, and shears:

$$y = y_b + \theta_b r F_1 + M_{rb}\frac{r^2}{D}F_2 + Q_b\frac{r^3}{D}F_3 - q\frac{r^4}{D}G_{11}$$

$$\theta = \theta_b F_4 + M_{rb}\frac{r}{D}F_5 + Q_b\frac{r^2}{D}F_6 - q\frac{r^3}{D}G_{14}$$

$$M_r = \theta_b\frac{D}{r}F_7 + M_{rb}F_8 + Q_b r F_9 - qr^2G_{17}$$

$$M_t = \frac{\theta D(1-\nu^2)}{r} + \nu M_r$$

$$Q = Q_b\frac{b}{r} - \frac{q}{2r}(r^2 - r_o^2)\langle r - r_o\rangle^0$$

For the numerical data given below, $\nu = 0.3$

$$y = K_y\frac{qa^4}{D} \qquad \theta = K_\theta\frac{qa^3}{D} \qquad M = K_Mqa^2 \qquad Q = K_Qqa$$

2a. Outer edge simply supported, inner edge free

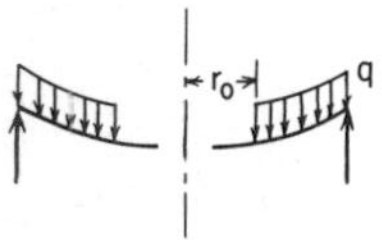

$M_{rb} = 0 \qquad Q_b = 0 \qquad y_a = 0 \qquad M_{ra} = 0$

$$y_b = \frac{-qa^4}{D}\left(\frac{C_1 L_{17}}{C_7} - L_{11}\right)$$

$$\theta_b = \frac{qa^3}{DC_7} L_{17}$$

$$\theta_a = \frac{qa^3}{D}\left(\frac{C_4 L_{17}}{C_7} - L_{14}\right)$$

$$Q_a = \frac{-q}{2a}(a^2 - r_o^2)$$

Special cases

Max $y = y_b$ Max $M = M_{tb}$

If $r_o = b$ (uniform load over entire plate),

b/a	0.1	0.3	0.5	0.7	0.9
K_{y_b}	−0.0687	−0.0761	−0.0624	−0.0325	−0.0048
K_{θ_a}	0.0986	0.1120	0.1201	0.1041	0.0477
K_{θ_b}	0.0436	0.1079	0.1321	0.1130	0.0491
$K_{M_{tb}}$	0.3965	0.3272	0.2404	0.1469	0.0497

2b. Outer edge simply supported, inner edge guided

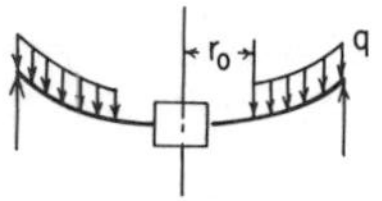

$\theta_b = 0 \qquad Q_b = 0 \qquad y_a = 0 \qquad M_{ra} = 0$

$$y_b = \frac{-qa^4}{D}\left(\frac{C_2 L_{17}}{C_8} - L_{11}\right)$$

$$M_{rb} = \frac{qa^2}{C_8} L_{17}$$

$$\theta_a = \frac{qa^3}{D}\left(\frac{C_5 L_{17}}{C_8} - L_{14}\right)$$

$$Q_a = \frac{-q}{2a}(a^2 - r_o^2)$$

Max $y = y_b$ Max $M = M_{rb}$

If $r_o = b$ (uniform load over entire plate),

b/a	0.1	0.3	0.5	0.7	0.9
K_{y_b}	−0.0575	−0.0314	−0.0103	−0.0015	−0.00002
K_{θ_a}	0.0919	0.0645	0.0306	0.0078	0.00032
$K_{M_{rt}}$	0.3003	0.2185	0.1223	0.0456	0.00505

2c. Outer edge simply supported, inner edge simply supported

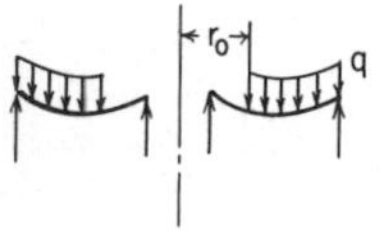

$y_b = 0 \qquad M_{rb} = 0 \qquad y_a = 0 \qquad M_{ra} = 0$

$$\theta_b = \frac{-qa^3}{D}\,\frac{C_3 L_{17} - C_9 L_{11}}{C_1 C_9 - C_3 C_7}$$

$$Q_b = qa\,\frac{C_1 L_{17} - C_7 L_{11}}{C_1 C_9 - C_3 C_7}$$

$$\theta_a = \theta_b C_4 + Q_b \frac{a^2}{D} C_6 - \frac{qa^3}{D} L_{14}$$

$$Q_a = Q_b \frac{b}{a} - \frac{q}{2a}(a^2 - r_o^2)$$

If $r_o = b$ (uniform load over entire plate),

b/a	0.1	0.3	0.5	0.7
$K_{y_{max}}$	−0.0060	−0.0029	−0.0008	−0.0001
K_{θ_b}	−0.0264	−0.0153	−0.0055	−0.0012
K_{θ_a}	0.0198	0.0119	0.0047	0.0011
$K_{M_{tb}}$	−0.2401	−0.0463	−0.0101	−0.0015
$K_{M_{rmax}}$	0.0708	0.0552	0.0300	0.0110
K_{Q_b}	1.8870	0.6015	0.3230	0.1684

TABLE 24 ***Formulas for flat circular plates of constant thickness*** **(*Cont.*)**

Columns: Case no., edge restraints | Boundary values | Special cases

2d. Outer edge simply supported, inner edge fixed

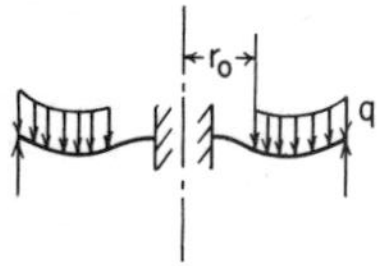

Boundary values:

$y_b = 0 \qquad \theta_b = 0 \qquad y_a = 0 \qquad M_{ra} = 0$

$$M_{rb} = -qa^2 \frac{C_3 L_{17} - C_9 L_{11}}{C_2 C_9 - C_3 C_8}$$

$$Q_b = qa \frac{C_2 L_{17} - C_8 L_{11}}{C_2 C_9 - C_3 C_8}$$

$$\theta_a = M_{rb} \frac{a}{D} C_5 + Q_b \frac{a^2}{D} C_6 - \frac{qa^3}{D} L_{14}$$

$$Q_a = Q_b \frac{b}{a} - \frac{q}{2a}(a^2 - r_o^2)$$

Special cases: If $r_o = b$ (uniform load over entire plate),

b/a	0.1	0.3	0.5	0.7
$K_{y_{max}}$	−0.0040	−0.0014	−0.0004	−0.00004
K_{θ_a}	0.0147	0.0070	0.0026	0.00056
$K_{M_{rb}}$	−0.2459	−0.0939	−0.0393	−0.01257
K_{Q_b}	2.1375	0.7533	0.4096	0.21259

2e. Outer edge fixed, inner edge free

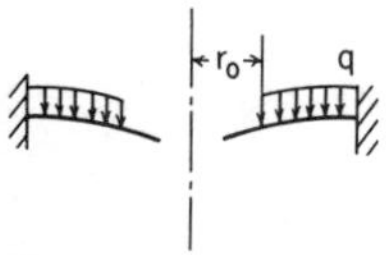

Boundary values:

$M_{rb} = 0 \qquad Q_b = 0 \qquad y_a = 0 \qquad \theta_a = 0$

$$y_b = \frac{-qa^4}{D}\left(\frac{C_1 L_{14}}{C_4} - L_{11}\right)$$

$$\theta_b = \frac{qa^3 L_{14}}{DC_4}$$

$$M_{ra} = -qa^2\left(L_{17} - \frac{C_7}{C_4} L_{14}\right)$$

$$Q_a = \frac{-q}{2a}\left(a^2 - r_o^2\right)$$

Special cases: If $r_o = b$ (uniform load over entire plate),

b/a	0.1	0.3	0.5	0.7	0.9
K_{y_b}	−0.0166	−0.0132	−0.0053	−0.0009	−0.00001
K_{θ_b}	0.0159	0.0256	0.0149	0.0040	0.00016
$K_{M_{ra}}$	−0.1246	−0.1135	−0.0800	−0.0361	−0.00470
$K_{M_{tb}}$	0.1448	0.0778	0.0271	0.0052	0.00016

2f. Outer edge fixed, inner edge guided

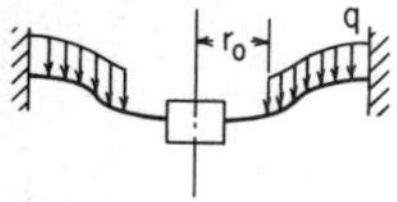

Boundary values:

$\theta_b = 0 \qquad Q_b = 0 \qquad y_a = 0 \qquad \theta_a = 0$

$$y_b = \frac{-qa^3}{D}\left(\frac{C_2 L_{14}}{C_5} - L_{11}\right)$$

$$M_{rb} = \frac{qa^2 L_{14}}{C_5}$$

$$M_{ra} = -qa^2\left(L_{17} - \frac{C_8}{C_5} L_{14}\right)$$

$$Q_a = \frac{-q}{2a}(a^2 - r_o^2)$$

Special cases: If $r_o = b$ (uniform load over entire plate),

b/a	0.1	0.3	0.5	0.7	0.9
K_{y_b}	−0.0137	−0.0068	−0.0021	−0.0003	
$K_{M_{rb}}$	0.1146	0.0767	0.0407	0.0149	0.00167
$K_{M_{ra}}$	−0.1214	−0.0966	−0.0601	−0.0252	−0.00316

2g. Outer edge fixed, inner edge simply supported

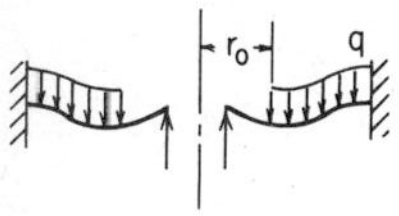

$y_b = 0 \qquad M_{rb} = 0 \qquad y_a = 0 \qquad \theta_a = 0$

$$\theta_b = \frac{-qa^3}{D}\frac{C_3L_{14} - C_6L_{11}}{C_1C_6 - C_3C_4}$$

$$Q_b = qa\frac{C_1L_{14} - C_4L_{11}}{C_1C_6 - C_3C_4}$$

$$M_{ra} = \theta_b\frac{D}{a}C_7 + Q_baC_9 - qa^2L_{17}$$

$$Q_a = Q_b\frac{b}{a} - \frac{q}{2a}(a^2 - r_o^2)$$

If $r_o = b$ (uniform load over entire plate),

b/a	0.1	0.3	0.5	0.7	0.9
$K_{y_{max}}$	−0.0025	−0.0012	−0.0003		
K_{θ_b}	−0.0135	−0.0073	−0.0027	−0.0006	
$K_{M_{tb}}$	0.1226	0.0221	0.0048	0.0007	
$K_{M_{ra}}$	−0.0634	−0.0462	−0.0262	−0.0102	−0.0012
K_{Q_b}	1.1591	0.3989	0.2262	0.1221	0.0383

2h. Outer edge fixed, inner edge fixed

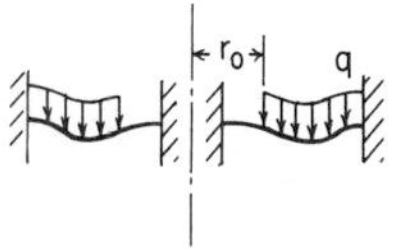

$y_b = 0 \qquad \theta_b = 0 \qquad y_a = 0 \qquad \theta_a = 0$

$$M_{rb} = -qa^2\frac{C_3L_{14} - C_6L_{11}}{C_2C_6 - C_3C_5}$$

$$Q_b = qa\frac{C_2L_{14} - C_5L_{11}}{C_2C_6 - C_3C_5}$$

$$M_{ra} = M_{rb}C_8 + Q_baC_9 - qa^2L_{17}$$

$$Q_a = Q_b\frac{b}{a} - \frac{q}{2a}(a^2 - r_o^2)$$

If $r_o = b$ (uniform load over entire plate),

b/a	0.1	0.3	0.5	0.7
$K_{y_{max}}$	−0.0018	−0.0006	−0.0002	
$K_{M_{rb}}$	−0.1433	−0.0570	−0.0247	−0.0081
$K_{M_{ra}}$	−0.0540	−0.0347	−0.0187	−0.0070
K_{Q_b}	1.4127	0.5414	0.3084	0.1650

2i. Outer edge guided, inner edge simply supported

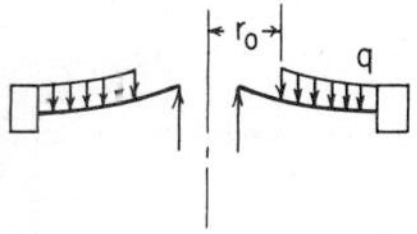

$y_b = 0 \qquad M_{rb} = 0 \qquad \theta_a = 0 \qquad Q_a = 0$

$$\theta_b = \frac{-qa^3}{DC_4}\left[\frac{C_6}{2ab}(a^2 - r_o^2) - L_{14}\right]$$

$$Q_b = \frac{q}{2b}(a^2 - r_o^2)$$

$$y_a = \theta_baC_1 + Q_b\frac{a^3}{D}C_3 - \frac{qa^4}{D}L_{11}$$

$$M_{ra} = \theta_b\frac{D}{a}C_7 + Q_baC_9 - qa^2L_{17}$$

If $r_o = b$ (uniform load over entire plate),

b/a	0.1	0.3	0.5	0.7	0.9
K_{y_a}	−0.0543	−0.0369	−0.0122	−0.0017	−0.00002
K_{θ_b}	−0.1096	−0.0995	−0.0433	−0.0096	−0.00034
$K_{M_{ra}}$	0.1368	0.1423	0.0985	0.0412	0.00491
$K_{M_{tb}}$	−0.9971	−0.3018	−0.0788	−0.0125	−0.00035

TABLE 24 *Formulas for flat circular plates of constant thickness* (Cont.)

<table>
<tr><th>Case no., edge restraints</th><th>Boundary values</th><th>Special cases</th></tr>
<tr><td>2j. Outer edge guided, inner edge fixed
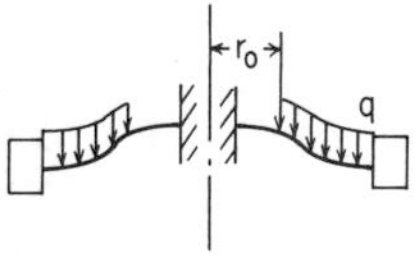</td>
<td>$y_b = 0 \quad \theta_b = 0 \quad \theta_a = 0 \quad Q_a = 0$
$M_{rb} = \frac{-qa^2}{C_5}\left[\frac{C_6}{2ab}(a^2 - r_o^2) - L_{14}\right]$
$Q_b = \frac{q}{2b}(a^2 - r_o^2)$
$y_a = M_{rb}\frac{a^2}{D}C_2 + Q_b\frac{a^3}{D}C_3 - \frac{qa^4}{D}L_{11}$
$M_{ra} = M_{rb}C_8 + Q_b aC_9 - qa^2L_{17}$</td>
<td>If $r_o = b$ (uniform load over entire plate),
<table>
<tr><th>b/a</th><th>0.1</th><th>0.3</th><th>0.5</th><th>0.7</th><th>0.9</th></tr>
<tr><td>K_{y_a}</td><td>−0.0343</td><td>−0.0123</td><td>−0.0030</td><td>−0.0004</td><td></td></tr>
<tr><td>$K_{M_{rb}}$</td><td>−0.7892</td><td>−0.2978</td><td>−0.1184</td><td>−0.0359</td><td>−0.00351</td></tr>
<tr><td>$K_{M_{ra}}$</td><td>0.1146</td><td>0.0767</td><td>0.0407</td><td>0.0149</td><td>0.00167</td></tr>
</table></td></tr>
<tr><td>2k. Outer edge free, inner edge simply supported
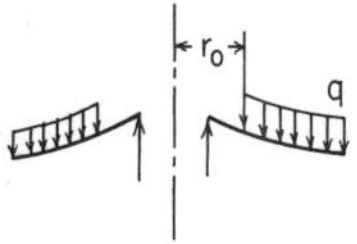</td>
<td>$y_b = 0 \quad M_{rb} = 0 \quad M_{ra} = 0 \quad Q_a = 0$
$\theta_b = \frac{-qa^3}{DC_7}\left[\frac{C_9}{2ab}(a^2 - r_o^2) - L_{17}\right]$
$Q_b = \frac{q}{2b}(a^2 - r_o^2)$
$y_a = \theta_b aC_1 + Q_b\frac{a^3}{D}C_3 - \frac{qa^4}{D}L_{11}$
$\theta_a = \theta_b C_4 + Q_b\frac{a^2}{D}C_6 - \frac{qa^3}{D}L_{14}$</td>
<td>If $r_o = b$ (uniform load over entire plate),
<table>
<tr><th>b/a</th><th>0.1</th><th>0.3</th><th>0.5</th><th>0.7</th><th>0.9</th></tr>
<tr><td>K_{y_a}</td><td>−0.1115</td><td>−0.1158</td><td>−0.0826</td><td>−0.0378</td><td>−0.0051</td></tr>
<tr><td>K_{θ_b}</td><td>−0.1400</td><td>−0.2026</td><td>−0.1876</td><td>−0.1340</td><td>−0.0515</td></tr>
<tr><td>K_{θ_a}</td><td>−0.1082</td><td>−0.1404</td><td>−0.1479</td><td>−0.1188</td><td>−0.0498</td></tr>
<tr><td>$K_{M_{tb}}$</td><td>−1.2734</td><td>−0.6146</td><td>−0.3414</td><td>−0.1742</td><td>−0.0521</td></tr>
</table></td></tr>
<tr><td>2l. Outer edge free, inner edge fixed
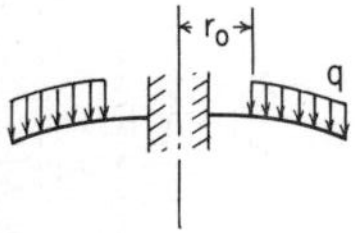</td>
<td>$y_b = 0 \quad \theta_b = 0 \quad M_{ra} = 0 \quad Q_a = 0$
$M_{rb} = \frac{-qa^2}{C_8}\left[\frac{C_9}{2ab}(a^2 - r_o^2) - L_{17}\right]$
$Q_b = \frac{q}{2b}(a^2 - r_o^2)$
$y_a = M_{rb}\frac{a^2}{D}C_2 + Q_b\frac{a^3}{D}C_3 - \frac{qa^4}{D}L_{11}$</td>
<td>If $r_o = b$ (uniform load over entire plate),
<table>
<tr><th>b/a</th><th>0.1</th><th>0.3</th><th>0.5</th><th>0.7</th><th>0.9</th></tr>
<tr><td>K_{y_a}</td><td>−0.0757</td><td>−0.0318</td><td>−0.0086</td><td>−0.0011</td><td></td></tr>
<tr><td>K_{θ_a}</td><td>−0.0868</td><td>−0.0512</td><td>−0.0207</td><td>−0.0046</td><td>−0.00017</td></tr>
<tr><td>$K_{M_{rb}}$</td><td>−0.9646</td><td>−0.4103</td><td>−0.1736</td><td>−0.0541</td><td>−0.00530</td></tr>
</table></td></tr>
</table>

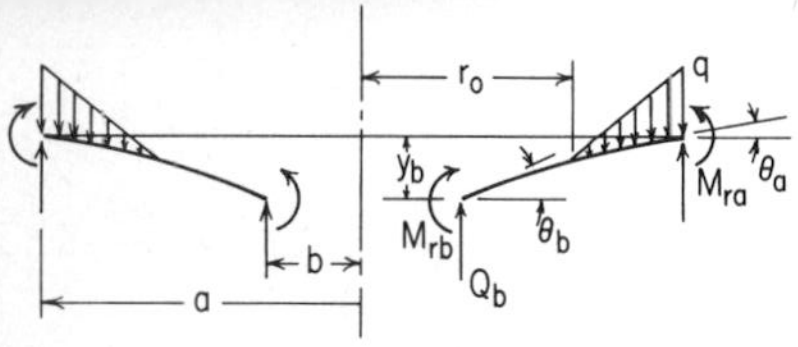

General expressions for deformations, moments, and shears:

$$y = y_b + \theta_b r F_1 + M_{rb}\frac{r^2}{D}F_2 + Q_b\frac{r^3}{D}F_3 - q\frac{r^4}{D}\frac{r-r_o}{a-r_o}G_{12}$$

$$\theta = \theta_b F_4 + M_{rb}\frac{r}{D}F_5 + Q_b\frac{r^2}{D}F_6 - q\frac{r^3}{D}\frac{r-r_o}{a-r_o}G_{15}$$

$$M_r = \theta_b\frac{D}{r}F_7 + M_{rb}F_8 + Q_b r F_9 - qr^2\frac{r-r_o}{a-r_o}G_{18}$$

$$M_t = \frac{\theta D(1-\nu^2)}{r} + \nu M_r$$

$$Q = Q_b\frac{b}{r} - \frac{q}{6r(a-r_o)}(2r^3 - 3r_o r^2 + r_o^3)\langle r - r_o\rangle^0$$

For the numerical data given below, $\nu = 0.3$

$$y = K_y\frac{qa^4}{D} \qquad \theta = K_\theta\frac{qa^3}{D} \qquad M = K_M qa^2 \qquad Q = K_Q qa$$

Case no., edge restraints	Boundary values	Special cases
3a. Outer edge simply supported, inner edge free 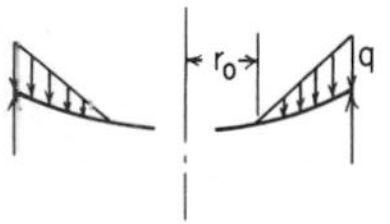	$M_{rb} = 0 \quad Q_b = 0 \quad y_a = 0 \quad M_{ra} = 0$ $y_b = \frac{-qa^4}{D}\left(\frac{C_1L_{18}}{C_7} - L_{12}\right)$ $\theta_b = \frac{qa^3}{DC_7}L_{18}$ $\theta_a = \frac{qa^3}{D}\left(\frac{C_4L_{18}}{C_7} - L_{15}\right)$ $Q_a = \frac{-q}{6a}(2a^2 - r_oa - r_o^2)$	Max $y = y_b$ Max $M = M_{tb}$ If $r_o = b$ (linearly increasing load from b to a), (see table below)
3b. Outer edge simply supported, inner edge guided 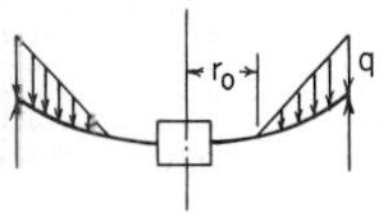	$\theta_b = 0 \quad Q_b = 0 \quad y_a = 0 \quad M_{ra} = 0$ $y_b = \frac{-qa^4}{D}\left(\frac{C_2L_{18}}{C_8} - L_{12}\right)$ $M_{rb} = \frac{qa^2L_{18}}{C_8}$ $\theta_a = \frac{qa^3}{D}\left(\frac{C_5L_{18}}{C_8} - L_{15}\right)$ $Q_a = \frac{-q}{6a}(2a^2 - r_oa - r_o^2)$	Max $y = y_b$ Max $M = M_{rb}$ If $r_o = b$ (linearly increasing load from b to a), (see table below)

Case 3a, $r_o = b$:

b/a	0.1	0.3	0.5	0.7	0.9
K_{y_b}	−0.0317	−0.0306	−0.0231	−0.0114	−0.0016
K_{θ_a}	0.0482	0.0470	0.0454	0.0368	0.0161
K_{θ_b}	0.0186	0.0418	0.0483	0.0396	0.0166
$K_{M_{tb}}$	0.1690	0.1269	0.0879	0.0514	0.0168

Case 3b, $r_o = b$:

b/a	0.1	0.3	0.5	0.7	0.9
K_{y_b}	−0.0269	−0.0133	−0.0041	−0.0006	−0.00001
K_{θ_a}	0.0454	0.0286	0.0126	0.0031	0.00012
$K_{M_{rb}}$	0.1280	0.0847	0.0447	0.0160	0.00171

TABLE 24 *Formulas for flat circular plates of constant thickness* (*Cont.*)

Case no., edge restraints	Boundary values	Special cases
3c. Outer edge simply supported, inner edge simply supported 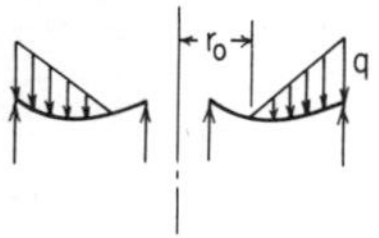	$y_b = 0 \quad M_{rb} = 0 \quad y_a = 0 \quad M_{ra} = 0$ $\theta_b = \frac{-qa^3}{D}\frac{C_3L_{18} - C_9L_{12}}{C_1C_9 - C_3C_7}$ $Q_b = qa\frac{C_1L_{18} - C_7L_{12}}{C_1C_9 - C_3C_7}$ $\theta_a = \theta_b C_4 + Q_b\frac{a^2}{D}C_6 - \frac{qa^3}{D}L_{15}$ $Q_a = Q_b\frac{b}{a} - \frac{q}{6a}(2a^2 - r_oa - r_o^2)$	If $r_o = b$ (linearly increasing load from b to a), (see table below)

Case 3c, special cases — if $r_o = b$ (linearly increasing load from b to a):

b/a	0.1	0.3	0.5	0.7
$K_{y_{max}}$	−0.0034	−0.0015	−0.0004	−0.0001
K_{θ_b}	−0.0137	−0.0077	−0.0027	−0.0006
K_{θ_a}	0.0119	0.0068	0.0026	0.0006
$K_{M_{tb}}$	−0.1245	−0.0232	−0.0049	−0.0007
$K_{M_{rmax}}$	0.0407	0.0296	0.0159	0.0057
K_{Q_b}	0.8700	0.2417	0.1196	0.0591

Case no., edge restraints	Boundary values	Special cases
3d. Outer edge simply supported, inner edge fixed 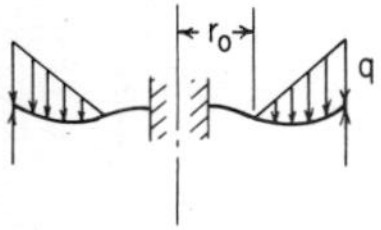	$y_b = 0 \quad \theta_b = 0 \quad y_a = 0 \quad M_{ra} = 0$ $M_{rb} = -qa^2\frac{C_3L_{18} - C_9L_{12}}{C_2C_9 - C_3C_8}$ $Q_b = qa\frac{C_2L_{18} - C_8L_{12}}{C_2C_9 - C_3C_8}$ $\theta_a = M_{rb}\frac{a}{D}C_5 + Q_b\frac{a^2}{D}C_6 - \frac{qa^3}{D}L_{15}$ $Q_a = Q_b\frac{b}{a} - \frac{q}{6a}(2a^2 - r_oa - r_o^2)$	If $r_o = b$ (linearly increasing load from b to a), (see table below)

Case 3d, special cases — if $r_o = b$ (linearly increasing load from b to a):

b/a	0.1	0.3	0.5	0.7
$K_{y_{max}}$	−0.0024	−0.0008	−0.0002	−0.00002
K_{θ_a}	0.0093	0.0044	0.0016	0.00034
$K_{M_{rb}}$	−0.1275	−0.0470	−0.0192	−0.00601
K_{Q_b}	0.9999	0.3178	0.1619	0.08029

Case no., edge restraints	Boundary values	Special cases
3e. Outer edge fixed, inner edge free 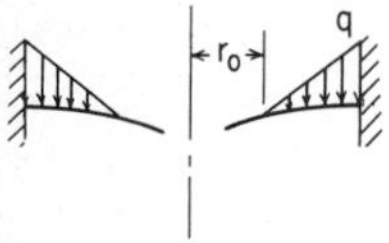	$M_{rb} = 0 \quad Q_b = 0 \quad y_a = 0 \quad \theta_a = 0$ $y_b = \frac{-qa^4}{D}\left(\frac{C_1L_{15}}{C_4} - L_{12}\right)$ $\theta_b = \frac{qa^3L_{15}}{DC_4}$ $M_{ra} = -qa^2\left(L_{18} - \frac{C_7}{C_4}L_{15}\right)$ $Q_a = \frac{-q}{6a}(2a^2 - r_oa - r_o^2)$	If $r_o = b$ (linearly increasing load from b to a), (see table below)

Case 3e, special cases — if $r_o = b$ (linearly increasing load from b to a):

b/a	0.1	0.3	0.5	0.7	0.9
K_{y_b}	−0.0062	−0.0042	−0.0015	−0.00024	
K_{θ_b}	0.0051	0.0073	0.0040	0.00103	0.00004
$K_{M_{ra}}$	−0.0609	−0.0476	−0.0302	−0.01277	−0.00159
$K_{M_{tb}}$	0.0459	0.0222	0.0073	0.00134	0.00004

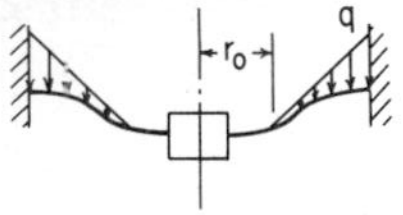	$y_b = \frac{-qa^4}{D}\left(\frac{C_2L_{15}}{C_5} - L_{12}\right)$ $M_{rb} = \frac{qa^2L_{15}}{C_5}$ $M_{ra} = -qa^2\left(L_{18} - \frac{C_8}{C_5}L_{15}\right)$ $Q_a = \frac{-q}{6a}(2a^2 - r_oa - r_o^2)$	see table below
3g. Outer edge fixed, inner edge simply supported 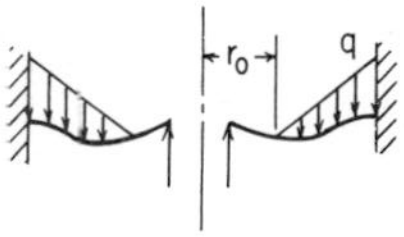	$y_b = 0 \quad M_{rb} = 0 \quad y_a = 0 \quad \theta_a = 0$ $\theta_b = \frac{-qa^3}{D}\frac{C_3L_{15} - C_6L_{12}}{C_1C_6 - C_3C_4}$ $Q_b = qa\frac{C_1L_{15} - C_4L_{12}}{C_1C_6 - C_3C_4}$ $M_{ra} = \theta_b\frac{D}{a}C_7 + Q_baC_9 - qa^2L_{18}$ $Q_a = Q_b\frac{b}{a} - \frac{q}{6a}(2a^2 - r_oa - r_o^2)$	If $r_o = b$ (linearly increasing load from b to a),
3h. Outer edge fixed, inner edge fixed 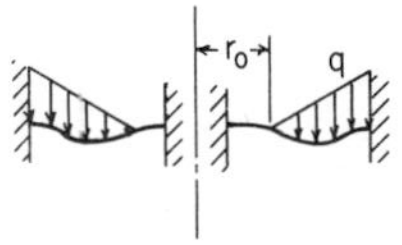	$y_b = 0 \quad \theta_b = 0 \quad y_a = 0 \quad \theta_a = 0$ $M_{rb} = -qa^2\frac{C_3L_{15} - C_6L_{12}}{C_2C_6 - C_3C_5}$ $Q_b = qa\frac{C_2L_{15} - C_5L_{12}}{C_2C_6 - C_3C_5}$ $M_{ra} = M_{rb}C_8 + Q_baC_9 - qa^2L_{18}$ $Q_a = Q_b\frac{b}{a} - \frac{q}{6a}(2a^2 - r_oa - r_o^2)$	If $r_o = b$ (linearly increasing load from b to a),
3i. Outer edge guided, inner edge simply supported 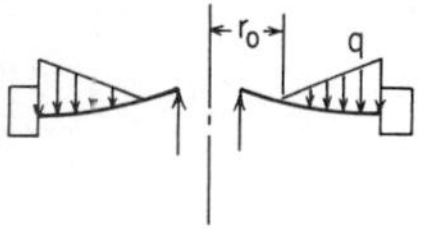	$y_b = 0 \quad M_{rb} = 0 \quad \theta_a = 0 \quad Q_a = 0$ $\theta_b = \frac{-qa^3}{DC_4}\left[\frac{C_6}{6ab}(2a^2 - r_oa - r_o^2) - L_{15}\right]$ $Q_b = \frac{q}{6b}(2a^2 - r_oa - r_o^2)$ $y_a = \theta_baC_1 + Q_b\frac{a^3}{D}C_3 - \frac{qa^4}{D}L_{12}$ $M_{ra} = \theta_b\frac{D}{a}C_7 + Q_baC_9 - qa^2L_{18}$	If $r_o = b$ (linearly increasing load from b to a),

b/a	0.1	0.3	0.5	0.7	0.9
K_{y_b}	−0.0053	−0.0024	−0.0007	−0.0001	
$K_{M_{rb}}$	0.0364	0.0219	0.0110	0.0039	0.00042
$K_{M_{ra}}$	−0.0599	−0.0428	−0.0249	−0.0099	−0.00120

3g. If $r_o = b$ (linearly increasing load from b to a),

b/a	0.1	0.3	0.5	0.7	0.9
$K_{y_{max}}$	−0.0013	−0.0005	−0.0002		
K_{θ_b}	−0.0059	−0.0031	−0.0011	−0.0002	
$K_{M_{tb}}$	0.0539	0.0094	0.0020	0.0003	
$K_{M_{ra}}$	−0.0381	−0.0264	−0.0145	−0.0056	−0.0006
K_{Q_b}	0.4326	0.1260	0.0658	0.0339	0.0104

3h. If $r_o = b$ (linearly increasing load from b to a),

b/a	0.1	0.3	0.5	0.7	0.9
$K_{y_{max}}$	−0.0009	−0.0003	−0.0001		
$K_{M_{rb}}$	−0.0630	−0.0242	−0.0102	−0.0033	−0.00035
$K_{M_{ra}}$	−0.0340	−0.0215	−0.0114	−0.0043	−0.00048
K_{Q_b}	0.5440	0.1865	0.0999	0.0514	0.01575

3i. If $r_o = b$ (linearly increasing load from b to a),

b/a	0.1	0.3	0.5	0.7	0.9
K_{y_a}	−0.0389	−0.0254	−0.0082	−0.0011	−0.00001
K_{θ_b}	−0.0748	−0.0665	−0.0283	−0.0062	−0.00022
$K_{M_{ra}}$	0.1054	0.1032	0.0689	0.0282	0.00330
$K_{M_{tb}}$	−0.6808	−0.2017	−0.0516	−0.0080	−0.00022

TABLE 24 *Formulas for flat circular plates of constant thickness* (*Cont.*)

Case no., edge restraints | Boundary values | Special cases

3j. Outer edge guided, inner edge fixed

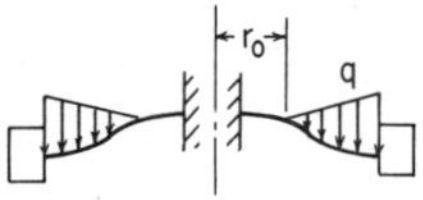

Boundary values:

$y_b = 0 \qquad \theta_b = 0 \qquad \theta_a = 0 \qquad Q_a = 0$

$$M_{rb} = \frac{-qa^2}{C_5}\left[\frac{C_6}{6ab}(2a^2 - r_o a - r_o^2) - L_{15}\right]$$

$$Q_b = \frac{q}{6b}(2a^2 - r_o a - r_o^2)$$

$$y_a = M_{rb}\frac{a^2}{D}C_2 + Q_b\frac{a^3}{D}C_3 - \frac{qa^4}{D}L_{12}$$

$$M_{ra} = M_{rb}C_8 + Q_b a C_9 - qa^2 L_{18}$$

Special cases: If $r_o = b$ (linearly increasing load from b to a),

b/a	0.1	0.3	0.5	0.7	0.9
K_{y_a}	−0.0253	−0.0089	−0.0022	−0.0003	
$K_{M_{rb}}$	−0.5388	−0.1990	−0.0774	−0.0231	−0.00221
$K_{M_{ra}}$	0.0903	0.0594	0.0312	0.0113	0.00125

3k. Outer edge free, inner edge simply supported

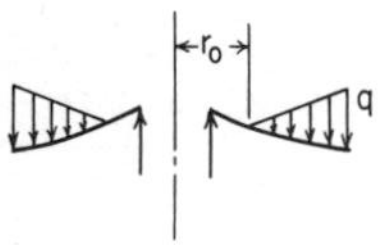

Boundary values:

$y_b = 0 \qquad M_{rb} = 0 \qquad M_{ra} = 0 \qquad Q_a = 0$

$$\theta_b = \frac{-qa^3}{DC_7}\left[\frac{C_9}{6ab}(2a^2 - r_o a - r_o^2) - L_{18}\right]$$

$$Q_b = \frac{q}{6b}(2a^2 - r_o a - r_o^2)$$

$$y_a = \theta_b a C_1 + Q_b\frac{a^3}{D}C_3 - \frac{qa^4}{D}L_{12}$$

$$\theta_a = \theta_b C_4 + Q_b\frac{a^2}{D}C_6 - \frac{qa^3}{D}L_{15}$$

Special cases: If $r_o = b$ (linearly increasing load from b to a),

b/a	0.1	0.3	0.5	0.7	0.9
K_{y_a}	−0.0830	−0.0826	−0.0574	−0.0258	−0.0034
K_{θ_b}	−0.0982	−0.1413	−0.1293	−0.0912	−0.0346
K_{θ_a}	−0.0834	−0.1019	−0.1035	−0.0812	−0.0335
$K_{M_{tb}}$	−0.8937	−0.4286	−0.2354	−0.1186	−0.0350

3l. Outer edge free, inner edge fixed

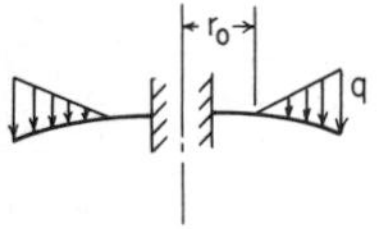

Boundary values:

$y_b = 0 \qquad \theta_b = 0 \qquad M_{ra} = 0 \qquad Q_a = 0$

$$M_{rb} = \frac{-qa^2}{C_8}\left[\frac{C_9}{6ab}(2a^2 - r_o a - r_o^2) - L_{18}\right]$$

$$Q_b = \frac{q}{6b}(2a^2 - r_o a - r_o^2)$$

$$y_a = M_{rb}\frac{a^2}{D}C_2 + Q_b\frac{a^3}{D}C_3 - \frac{qa^4}{D}L_{12}$$

$$\theta_a = M_{rb}\frac{a}{D}C_5 + Q_b\frac{a^2}{D}C_6 - \frac{qa^3}{D}L_{15}$$

Special cases: If $r_o = b$ (linearly increasing load from b to a),

b/a	0.1	0.3	0.5	0.7	0.9
K_{y_a}	−0.0579	−0.0240	−0.0064	−0.0008	
K_{θ_a}	−0.0684	−0.0397	−0.0159	−0.0035	−0.00013
$K_{M_{rb}}$	−0.6769	−0.2861	−0.1197	−0.0368	−0.00356

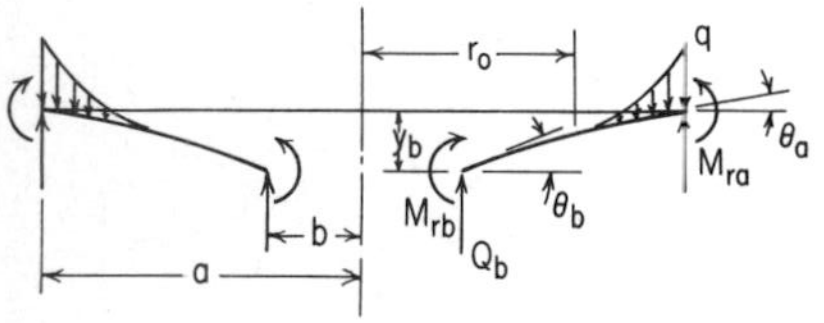

General expressions for deformations, moments, and shears:

$$y = y_b + \theta_b r F_1 + M_{rb}\frac{r^2}{D}F_2 + Q_b\frac{r^3}{D}F_3 - q\frac{r^4}{D}\left(\frac{r - r_o}{a - r_o}\right)^2 G_{13}$$

$$\theta = \theta_b F_4 + M_{rb}\frac{r}{D}F_5 + Q_b\frac{r^2}{D}F_6 - q\frac{r^3}{D}\left(\frac{r - r_o}{a - r_o}\right)^2 G_{16}$$

$$M_r = \theta_b\frac{D}{r}F_7 + M_{rb}F_8 + Q_b r F_9 - qr^2\left(\frac{r - r_o}{a - r_o}\right)^2 G_{19}$$

$$M_t = \frac{\theta D(1 - \nu^2)}{r} - \nu M_r$$

$$Q = Q_b\frac{b}{r} - \frac{q}{12r(a - r_o)^2}(3r^4 - 8r_o r^3 + 6r_o^2 r^2 - r_o^4)\langle r - r_o\rangle^0$$

For the numerical data given below, $\nu = 0.3$

$$y = K_y\frac{qa^4}{D} \qquad \theta = K_\theta\frac{qa^3}{D} \qquad M = K_M qa^2 \qquad Q = K_Q qa$$

Case no., edge restraints	Boundary values	Special cases
4a. Outer edge simply supported, inner edge free 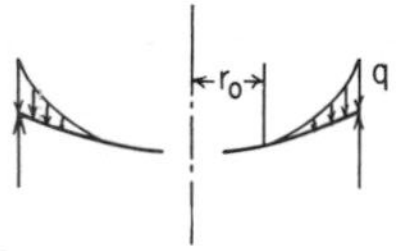	$M_{rb} = 0 \quad Q_b = 0 \quad y_a = 0 \quad M_{ra} = 0$ $y_b = \frac{-qa^4}{D}\left(\frac{C_1 L_{19}}{C_7} - L_{13}\right)$ $\theta_b = \frac{qa^3}{DC_7}L_{19}$ $\theta_a = \frac{qa^3}{D}\left(\frac{C_4 L_{19}}{C_7} - L_{16}\right)$ $Q_a = \frac{-q}{12a}(3a^2 - 2ar_o - r_o^2)$	Max $y = y_b$ Max $M = M_{tb}$ If $r_o = b$ (parabolically increasing load from b to a),
4b. Outer edge simply supported, inner edge guided 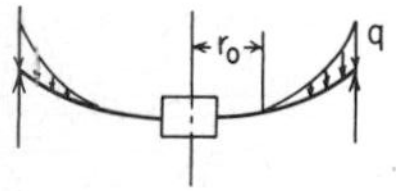	$\theta_b = 0 \quad Q_b = 0 \quad y_a = 0 \quad M_{ra} = 0$ $y_b = \frac{-qa^4}{D}\left(\frac{C_2 L_{19}}{C_8} - L_{13}\right)$ $M_{rb} = \frac{qa^2 L_{19}}{C_8}$ $\theta_a = \frac{qa^3}{D}\left(\frac{C_5 L_{19}}{C_8} - L_{16}\right)$ $Q_a = \frac{-q}{12a}(3a^2 - 2ar_o - r_o^2)$	Max $y = y_b$ Max $M = M_{rb}$ If $r_o = b$ (parabolically increasing load from b to a),

Case 4a special cases:

b/a	0.1	0.3	0.5	0.7	0.9
K_{y_b}	−0.0184	−0.0168	−0.0122	−0.0059	−0.0008
K_{θ_a}	0.0291	0.0266	0.0243	0.0190	0.0082
K_{θ_b}	0.0105	0.0227	0.0254	0.0203	0.0084
$K_{M_{tb}}$	0.0951	0.0687	0.0462	0.0264	0.0085

Case 4b special cases:

b/a	0.1	0.3	0.5	0.7	0.9
K_{y_b}	−0.0158	−0.0074	−0.0022	−0.0003	
K_{θ_a}	0.0275	0.0166	0.0071	0.0017	0.00007
$K_{M_{rb}}$	0.0721	0.0459	0.0235	0.0082	0.00086

TABLE 24 *Formulas for flat circular plates of constant thickness* (*Cont.*)

Case no., edge restraints	Boundary values	Special cases
4c. Outer edge simply supported, inner edge simply supported 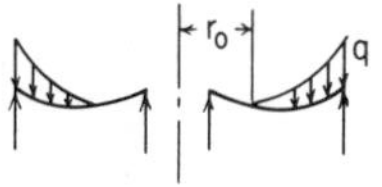	$y_b = 0 \quad M_{rb} = 0 \quad y_a = 0 \quad M_{ra} = 0$ $\theta_b = \frac{-qa^3}{D}\frac{C_3L_{19} - C_9L_{13}}{C_1C_9 - C_3C_7}$ $Q_b = qa\frac{C_1L_{19} - C_7L_{13}}{C_1C_9 - C_3C_7}$ $\theta_a = \theta_b C_4 + Q_b\frac{a^2}{D}C_6 - \frac{qa^3}{D}L_{16}$ $Q_a = Q_b\frac{b}{a} - \frac{q}{12a}(3a^2 - 2ar_o - r_o^2)$	If $r_o = b$ (parabolically increasing load from b to a), see table 4c below
4d. Outer edge simply supported, inner edge fixed 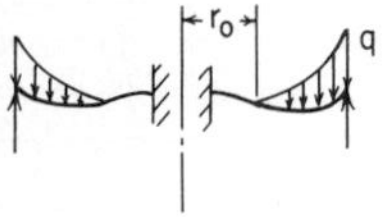	$y_b = 0 \quad \theta_b = 0 \quad y_a = 0 \quad M_{ra} = 0$ $M_{rb} = -qa^2\frac{C_3L_{19} - C_9L_{13}}{C_2C_9 - C_3C_8}$ $Q_b = qa\frac{C_2L_{19} - C_8L_{13}}{C_2C_9 - C_3C_8}$ $\theta_a = M_{rb}\frac{a}{D}C_5 + Q_b\frac{a^2}{D}C_6 - \frac{qa^3}{D}L_{16}$ $Q_a = Q_b\frac{b}{a} - \frac{q}{12a}(3a^2 - 2ar_o - r_o^2)$	If $r_o = b$ (parabolically increasing load from b to a), see table 4d below
4e. Outer edge fixed, inner edge free 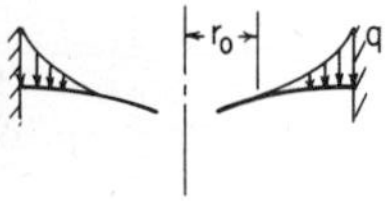	$M_{rb} = 0 \quad Q_b = 0 \quad y_a = 0 \quad \theta_a = 0$ $y_b = \frac{-qa^4}{D}\left(\frac{C_1L_{16}}{C_4} - L_{13}\right)$ $\theta_b = \frac{qa^3L_{16}}{DC_4}$ $M_{ra} = -qa^2\left(L_{19} - \frac{C_7}{C_4}L_{16}\right)$ $Q_a = \frac{-q}{12a}(3a^2 - 2ar_o - r_o^2)$	If $r_o = b$ (parabolically increasing load from b to a), see table 4e below

4c. If $r_o = b$ (parabolically increasing load from b to a),

b/a	0.1	0.3	0.5	0.7
$K_{y_{max}}$	−0.0022	−0.0009	−0.0003	
K_{θ_b}	−0.0083	−0.0046	−0.0016	−0.0003
K_{θ_a}	0.0080	0.0044	0.0017	0.0004
$K_{M_{tb}}$	−0.0759	−0.0139	−0.0029	−0.0004
$K_{M_{r\,max}}$	0.0267	0.0185	0.0098	0.0035
K_{Q_b}	0.5068	0.1330	0.0633	0.0305

4d. If $r_o = b$ (parabolically increasing load from b to a),

b/a	0.1	0.3	0.5	0.7
$K_{y_{max}}$	−0.0016	−0.0005	−0.0001	−0.00002
K_{θ_a}	0.0064	0.0030	0.0011	0.00023
$K_{M_{rb}}$	−0.0777	−0.0281	−0.0113	−0.00349
K_{Q_b}	0.5860	0.1785	0.0882	0.04276

4e. If $r_o = b$ (parabolically increasing load from b to a),

b/a	0.1	0.3	0.5	0.7	0.9
K_{y_b}	−0.0031	−0.0019	−0.0007	−0.0001	
K_{θ_b}	0.0023	0.0032	0.0017	0.0004	0.00002
$K_{M_{ra}}$	−0.0368	−0.0269	−0.0162	−0.0066	−0.00081
$K_{M_{tb}}$	0.0208	0.0096	0.0031	0.0006	0.00002

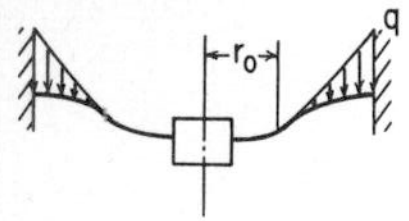

$y_b = \frac{?}{D}\left(\frac{? L_{16}}{C_5} - L_{13}\right)$

$M_{rb} = \frac{qa^2 L_{16}}{C_5} \qquad M_{ra} = -qa^2\left(L_{19} - \frac{C_8}{C_5}L_{16}\right)$

$Q_a = \frac{-q}{12a}(3a^2 - 2ar_o - r_o^2)$

b/a	0.1	0.3	0.5	0.7	0.9
K_{y_b}	−0.0026	−0.0011	−0.0003		
$K_{M_{rb}}$	0.0164	0.0094	0.0046	0.0016	0.00016
$K_{M_{ra}}$	−0.0364	−0.0248	−0.0140	−0.0054	−0.00066

4g. Outer edge fixed, inner edge simply supported

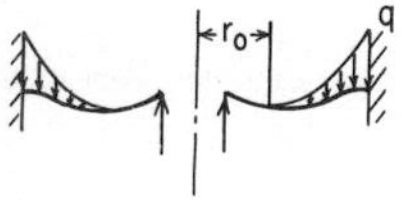

$y_b = 0 \qquad M_{rb} = 0 \qquad y_a = 0 \qquad \theta_a = 0$

$\theta_b = \frac{-qa^3}{D}\frac{C_3L_{16} - C_6L_{13}}{C_1C_6 - C_3C_4}$

$Q_b = qa\frac{C_1L_{16} - C_4L_{13}}{C_1C_6 - C_3C_4}$

$M_{ra} = \theta_b\frac{D}{a}C_7 + Q_b aC_9 - qa^2L_{19}$

$Q_a = Q_b\frac{b}{a} - \frac{q}{12a}(3a^2 - 2ar_o - r_o^2)$

If $r_o = b$ (parabolically increasing load from b to a),

b/a	0.1	0.3	0.5	0.7
$K_{y_{max}}$	−0.0007	−0.0003	−0.0001	
K_{θ_b}	−0.0031	−0.0016	−0.0006	−0.00012
$K_{M_{tb}}$	0.0285	0.0049	0.0010	0.00015
$K_{M_{ra}}$	−0.0255	−0.0172	−0.0093	−0.00352
K_{Q_b}	0.2136	0.0577	0.0289	0.01450

4h. Outer edge fixed, inner edge fixed

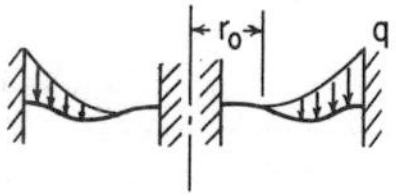

$y_b = 0 \qquad \theta_b = 0 \qquad y_a = 0 \qquad \theta_a = 0$

$M_{rb} = -qa^2\frac{C_3L_{16} - C_6L_{13}}{C_2C_6 - C_3C_5}$

$Q_b = qa\frac{C_2L_{16} - C_5L_{13}}{C_2C_6 - C_3C_5}$

$M_{ra} = M_{rb}C_8 + Q_b aC_9 - qa^2L_{19}$

$Q_a = Q_b\frac{b}{a} - \frac{q}{12a}(3a^2 - 2ar_o - r_o^2)$

If $r_o = b$ (parabolically increasing load from b to a),

b/a	0.1	0.3	0.5	0.7
$K_{y_{max}}$	−0.0005	−0.0002	−0.00005	
$K_{M_{rb}}$	−0.0333	−0.0126	−0.00524	−0.00168
$K_{M_{ra}}$	−0.0234	−0.0147	−0.00773	−0.00287
K_{Q_b}	0.2726	0.0891	0.04633	0.02335

4i. Outer edge guided, inner edge simply supported

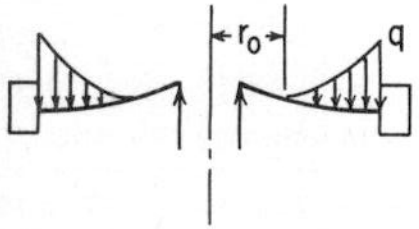

$y_b = 0 \qquad M_{rb} = 0 \qquad \theta_a = 0 \qquad Q_a = 0$

$\theta_b = \frac{-qa^3}{DC_4}\left[\frac{C_6}{12ab}(3a^2 - 2ar_o - r_o^2) - L_{16}\right]$

$Q_b = \frac{q}{12b}(3a^2 - 2ar_o - r_o^2)$

$y_a = \theta_b aC_1 + Q_b\frac{a^3}{D}C_3 - \frac{qa^4}{D}L_{13}$

$M_{ra} = \theta_b\frac{D}{a}C_7 + Q_b aC_9 - qa^2L_{19}$

If $r_o = b$ (parabolically increasing load from b to a),

b/a	0.1	0.3	0.5	0.7	0.9
K_{y_a}	−0.0302	−0.0193	−0.0061	−0.0008	−0.00001
K_{θ_b}	−0.0567	−0.0498	−0.0210	−0.0045	−0.00016
$K_{M_{ra}}$	0.0859	0.0813	0.0532	0.0215	0.00249
$K_{M_{tb}}$	−0.5156	−0.1510	−0.0381	−0.0059	−0.00016

TABLE 24 *Formulas for flat circular plates of constant thickness* (Cont.)

Columns: Case no., edge restraints | Boundary values | Special cases

4j. Outer edge guided, inner edge fixed

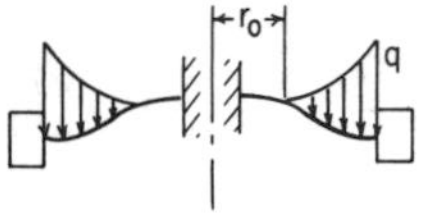

Boundary values:

$y_b = 0 \qquad \theta_b = 0 \qquad \theta_a = 0 \qquad Q_a = 0$

$$M_{rb} = \frac{-qa^2}{C_5}\left[\frac{C_6}{12ab}(3a^2 - 2ar_o - r_o^2) - L_{16}\right]$$

$$Q_b = \frac{q}{12b}(3a^2 - 2ar_o - r_o^2)$$

$$y_a = M_{rb}\frac{a^2}{D}C_2 + Q_b\frac{a^3}{D}C_3 - \frac{qa^4}{D}L_{13}$$

$$M_{ra} = M_{rb}C_8 + Q_b aC_9 - qa^2L_{19}$$

Special cases:

If $r_o = b$ (parabolically increasing load from b to a),

b/a	0.1	0.3	0.5	0.7	0.9
K_{y_a}	−0.0199	−0.0070	−0.0017	−0.0002	
$K_{M_{rb}}$	−0.4081	−0.1490	−0.0573	−0.0169	−0.00161
$K_{M_{ra}}$	0.0745	0.0485	0.0253	0.0091	0.00100

4k. Outer edge free, inner edge simply supported

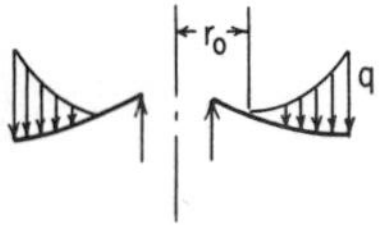

Boundary values:

$y_b = 0 \qquad M_{rb} = 0 \qquad M_{ra} = 0 \qquad Q_a = 0$

$$\theta_b = \frac{-qa^3}{DC_7}\left[\frac{C_9}{12ab}(3a^2 - 2ar_o - r_o^2) - L_{19}\right]$$

$$Q_b = \frac{q}{12b}(3a^2 - 2ar_o - r_o^2)$$

$$y_a = \theta_b aC_1 + Q_b\frac{a^3}{D}C_3 - \frac{qa^4}{D}L_{13}$$

$$\theta_a = \theta_b C_4 + Q_b\frac{a^2}{D}C_6 - \frac{qa^3}{D}L_{16}$$

Special cases:

If $r_o = b$ (parabolically increasing load from b to a),

b/a	0.1	0.3	0.5	0.7	0.9
K_{y_a}	−0.0662	−0.0644	−0.0441	−0.0196	−0.0026
K_{θ_b}	−0.0757	−0.1087	−0.0989	−0.0693	−0.0260
K_{θ_a}	−0.0680	−0.0802	−0.0799	−0.0618	−0.0252
$K_{M_{tb}}$	−0.6892	−0.3298	−0.1800	−0.0900	−0.0263

4l. Outer edge free, inner edge fixed

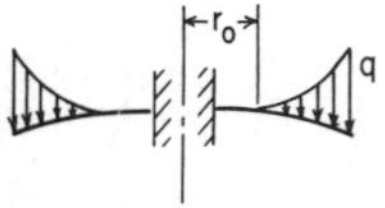

Boundary values:

$y_b = 0 \qquad \theta_b = 0 \qquad M_{ra} = 0 \qquad Q_a = 0$

$$M_{rb} = \frac{-qa^2}{C_8}\left[\frac{C_9}{12ab}(3a^2 - 2ar_o - r_o^2) - L_{19}\right]$$

$$Q_b = \frac{q}{12b}(3a^2 - 2ar_o - r_o^2)$$

$$y_a = M_{rb}\frac{a^2}{D}C_2 + Q_b\frac{a^3}{D}C_3 - \frac{qa^4}{D}L_{13}$$

$$\theta_a = M_{rb}\frac{a}{D}C_5 + Q_b\frac{a^2}{D}C_6 - \frac{qa^3}{D}L_{16}$$

Special cases:

If $r_o = b$ (parabolically increasing load from b to a),

b/a	0.1	0.3	0.5	0.7	0.9
K_{y_a}	−0.0468	−0.0193	−0.0051	−0.0006	−0.00001
K_{θ_a}	−0.0564	−0.0324	−0.0128	−0.0028	−0.00010
$K_{M_{rb}}$	−0.5221	−0.2202	−0.0915	−0.0279	−0.00268

Case 5. Annular plate with a uniform line moment of M_o in-lb/in at a radius r_o

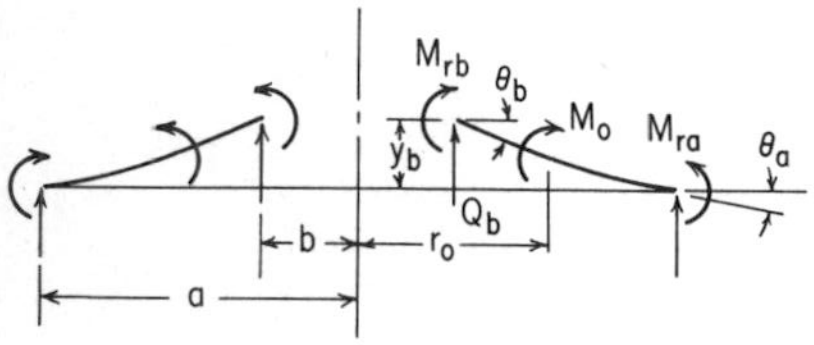

General expressions for deformations, moments, and shears:

$$y = y_b + \theta_b r F_1 + M_{rb}\frac{r^2}{D}F_2 + Q_b\frac{r^3}{D}F_3 + M_o\frac{r^2}{D}G_2$$

$$\theta = \theta_b F_4 + M_{rb}\frac{r}{D}F_5 + Q_b\frac{r^2}{D}F_6 + M_o\frac{r}{D}G_5$$

$$M_r = \theta_b\frac{D}{r}F_7 + M_{rb}F_8 + Q_b r F_9 + M_o G_8$$

$$M_t = \frac{\theta D(1-\nu^2)}{r} + \nu M_r$$

$$Q = Q_b\frac{b}{r}$$

For the numerical data given below, $\nu = 0.3$

$$y = K_y\frac{M_o a^2}{D} \qquad \theta = K_\theta\frac{M_o a}{D} \qquad M = K_M M_o \qquad Q = K_Q\frac{M_o}{a}$$

Case no., edge restraints | **Boundary values** | **Special cases**

5a. Outer edge simply supported, inner edge free

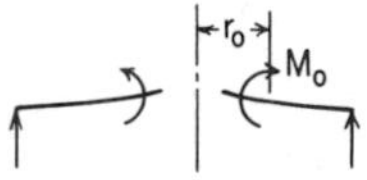

Boundary values:

$M_{rb} = 0 \qquad Q_b = 0 \qquad y_a = 0 \qquad M_{ra} = 0$

$$y_b = \frac{M_o a^2}{D}\left(\frac{C_1 L_8}{C_7} - L_2\right)$$

$$\theta_b = \frac{-M_o a}{D C_7}L_8$$

$$\theta_a = \frac{-M_o a}{D}\left(\frac{C_4 L_8}{C_7} - L_5\right)$$

$Q_a = 0$

Special cases:

Max $y = y_b$ Max $M = M_{tb}$

If $r_o = b$ (moment M_o at the inner edge),

b/a	0.1	0.3	0.5	0.7	0.9
K_{y_b}	0.0371	0.2047	0.4262	0.6780	0.9532
K_{θ_a}	−0.0222	−0.2174	−0.7326	−2.1116	−9.3696
K_{θ_b}	−0.1451	−0.4938	−1.0806	−2.4781	−9.7183
$K_{M_{tb}}$	−1.0202	−1.1978	−1.6667	−2.9216	−9.5263

If $r_o = a$ (moment M_o at the outer edge),

b/a	0.1	0.3	0.5	0.7	0.9
K_{y_b}	0.4178	0.5547	0.7147	0.8742	1.0263
K_{θ_a}	−0.7914	−0.9866	−1.5018	−2.8808	−10.1388
K_{θ_b}	−0.2220	−0.7246	−1.4652	−3.0166	−10.4107
$K_{M_{tb}}$	−2.0202	−2.1978	−2.6667	−3.9216	−10.5263

TABLE 24 Formulas for flat circular plates of constant thickness (Cont.)

Case no., edge restraints	Boundary values	Special cases

5b. Outer edge simply supported, inner edge guided

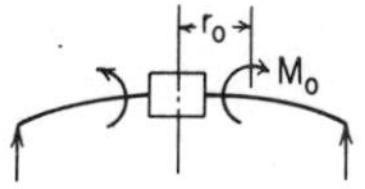

Boundary values:

$\theta_b = 0 \qquad Q_b = 0 \qquad y_a = 0 \qquad M_{ra} = 0$

$$y_b = \frac{M_o a^2}{D}\left(\frac{C_2 L_8}{C_8} - L_2\right)$$

$$M_{rb} = \frac{-M_o L_8}{C_8}$$

$$\theta_a = \frac{-M_o a}{D}\left(\frac{C_5 L_8}{C_8} - L_5\right)$$

$$Q_a = 0$$

Special cases:

Max $y = y_b$ Max $M = M_{rb}$

If $r_o = a$ (moment M_o at the outer edge),

b/a	0.1	0.3	0.5	0.7	0.9
K_{y_b}	0.3611	0.2543	0.1368	0.0488	0.0052
K_{θ_a}	−0.7575	−0.6676	−0.5085	−0.3104	−0.1018
$K_{M_{rb}}$	−1.5302	−1.4674	−1.3559	−1.2173	−1.0712

5c. Outer edge simply supported, inner edge simply supported

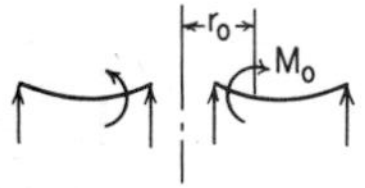

Boundary values:

$y_b = 0 \qquad M_{rb} = 0 \qquad y_a = 0 \qquad M_{ra} = 0$

$$\theta_b = \frac{M_o a}{D}\,\frac{C_3 L_8 - C_9 L_2}{C_1 C_9 - C_3 C_7}$$

$$Q_b = \frac{-M_o}{a}\,\frac{C_1 L_8 - C_7 L_2}{C_1 C_9 - C_3 C_7}$$

$$\theta_a = \theta_b C_4 + Q_b \frac{a^2}{D} C_6 + \frac{M_o a}{D} L_5$$

$$Q_a = Q_b \frac{b}{a}$$

Special cases:

If $r_o = b$ (moment M_o at the inner edge),

b/a	0.1	0.3	0.5	0.7	0.9
$K_{y_{max}}$	−0.0095	−0.0167	−0.0118	−0.0050	−0.0005
K_{θ_a}	0.0204	0.0518	0.0552	0.0411	0.0158
K_{θ_b}	−0.1073	−0.1626	−0.1410	−0.0929	−0.0327
$K_{M_{tb}}$	−0.6765	−0.1933	0.0434	0.1793	0.2669
K_{Q_b}	−1.0189	−1.6176	−2.2045	−3.5180	−10.1611

If $r_o = a$ (moment M_o at the outer edge),

b/a	0.1	0.3	0.5	0.7	0.9
$K_{y_{max}}$	0.0587	0.0390	0.0190	0.0063	0.0004
K_{θ_a}	−0.3116	−0.2572	−0.1810	−0.1053	−0.0339
K_{θ_b}	0.2037	0.1728	0.1103	0.0587	0.0175
$K_{M_{tb}}$	1.8539	0.5240	0.2007	0.0764	0.0177
K_{Q_b}	11.4835	4.3830	3.6964	4.5358	10.9401

5d. Outer edge simply supported, inner edge fixed

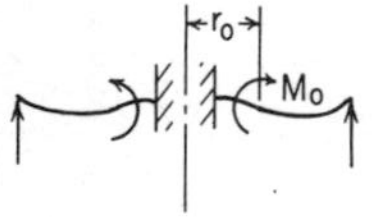

Boundary values:

$y_b = 0 \qquad \theta_b = 0 \qquad y_a = 0 \qquad M_{ra} = 0$

$$M_{rb} = M_o \frac{C_3 L_8 - C_9 L_2}{C_2 C_9 - C_3 C_8}$$

$$Q_b = \frac{-M_o}{a}\,\frac{C_2 L_8 - C_8 L_2}{C_2 C_9 - C_3 C_8}$$

$$\theta_a = M_{rb}\frac{a}{D} C_5 + Q_b \frac{a^2}{D} C_6 + \frac{M_o a}{D} L_5$$

$$Q_a = Q_b \frac{b}{a}$$

Special cases:

If $r_o = a$ (moment M_o at the outer edge),

b/a	0.1	0.3	0.5	0.7	0.9
$K_{y_{max}}$	0.0449	0.0245	0.0112	0.0038	0.0002
K_{θ_a}	−0.2729	−0.2021	−0.1378	−0.0793	−0.0255
$K_{M_{rb}}$	1.8985	1.0622	0.7823	0.6325	0.5366
K_{Q_b}	−13.4178	−6.1012	−5.4209	−6.7611	−16.3923

5e. Outer edge fixed, inner edge free

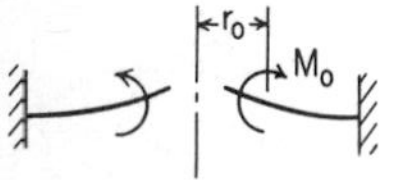

$M_{rb} = 0 \qquad Q_b = 0 \qquad y_a = 0 \qquad \theta_a = 0$

$$y_b = \frac{M_o a^2}{D}\left(\frac{C_1 L_5}{C_4} - L_2\right)$$

$$\theta_b = \frac{-M_o a}{DC_4} L_5$$

$$M_{ra} = M_o\left(L_8 - \frac{C_7}{C_4} L_5\right)$$

$$Q_a = 0$$

If $r_o = b$ (moment M_o at the inner edge),

b/a	0.1	0.3	0.5	0.7	0.9
K_{y_b}	0.0254	0.0825	0.0776	0.0373	0.0048
K_{θ_b}	−0.1389	−0.3342	−0.3659	−0.2670	−0.0976
$K_{M_{tb}}$	−0.9635	−0.7136	−0.3659	−0.0471	0.2014

5f. Outer edge fixed, inner edge guided

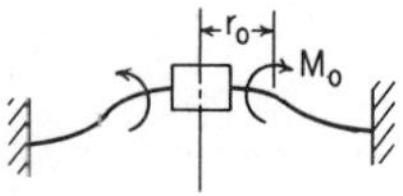

$\theta_b = 0 \qquad Q_b = 0 \qquad y_a = 0 \qquad \theta_a = 0$

$$y_b = \frac{M_o a^2}{D}\left(\frac{C_2 L_5}{C_5} - L_2\right)$$

$$M_{rb} = \frac{-M_o}{C_5} L_5$$

$$M_{ra} = M_o\left(L_8 - \frac{C_8}{C_5} L_5\right)$$

$$Q_a = 0$$

b/a	0.1		0.5		0.7
r_o/a	0.5	0.7	0.7	0.9	0.9
K_{y_b}	0.0779	0.0815	0.0285	0.0207	0.0101
$K_{M_{rb}}$	−0.7576	−0.5151	−0.6800	−0.2533	−0.3726
$K_{M_{ra}}$	0.2424	0.4849	0.3200	0.7467	0.6274

5g. Outer edge fixed, inner edge simply supported

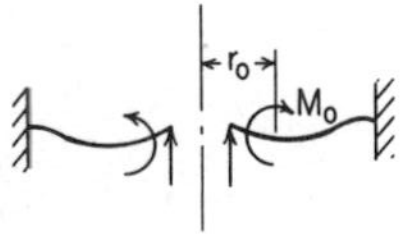

$y_b = 0 \qquad M_{rb} = 0 \qquad y_a = 0 \qquad \theta_a = 0$

$$\theta_b = \frac{M_o a}{D} \frac{C_3 L_5 - C_6 L_2}{C_1 C_6 - C_3 C_4}$$

$$Q_b = \frac{-M_o}{a} \frac{C_1 L_5 - C_4 L_2}{C_1 C_6 - C_3 C_4}$$

$$M_{ra} = \theta_b \frac{D}{a} C_7 + Q_b a C_9 + M_o L_8$$

$$Q_a = Q_b \frac{b}{a}$$

If $r_o = b$ (moment M_o at the inner edge),

b/a	0.1	0.3	0.5	0.7	0.9
$K_{y_{max}}$	−0.0067	−0.0102	−0.0066	−0.0029	−0.0002
K_{θ_b}	−0.0940	−0.1278	−0.1074	−0.0699	−0.0245
K_{Q_b}	−1.7696	−2.5007	−3.3310	−5.2890	−15.2529

5h. Outer edge fixed, inner edge fixed

$y_b = 0 \qquad \theta_b = 0 \qquad y_a = 0 \qquad \theta_a = 0$

$$M_{rb} = M_o \frac{C_3 L_5 - C_6 L_2}{C_2 C_6 - C_3 C_5}$$

$$Q_b = \frac{-M_o}{a} \frac{C_2 L_5 - C_5 L_2}{C_2 C_6 - C_3 C_5}$$

$$M_{ra} = M_{rb} C_8 + Q_b a C_9 + M_o L_8$$

$$Q_a = Q_b \frac{b}{a}$$

b/a	0.1		0.5		0.7
r_o/a	0.5	0.7	0.7	0.9	0.9
$K_{M_{rb}}$	0.7096	1.0185	0.2031	0.3895	0.3925
$K_{M_{ra}}$	−0.1407	0.0844	−0.2399	0.3391	0.0238
$K_{M_{ro}}$	−0.5045	−0.5371	−0.4655	−0.4671	−0.5540
$K_{M_{ro}}$	0.4955	0.4629	0.5345	0.5329	0.4460
K_{Q_b}	−8.0354	−8.3997	−4.1636	−3.0307	−5.4823

(*Note:* The two values of $K_{M_{ro}}$ are for positions just before and after the applied moment M_o)

TABLE 24 *Formulas for flat circular plates of constant thickness* (*Cont.*)

Case 6. Annular plate with an externally applied angular displacement θ_o on an annulus with a radius r_o

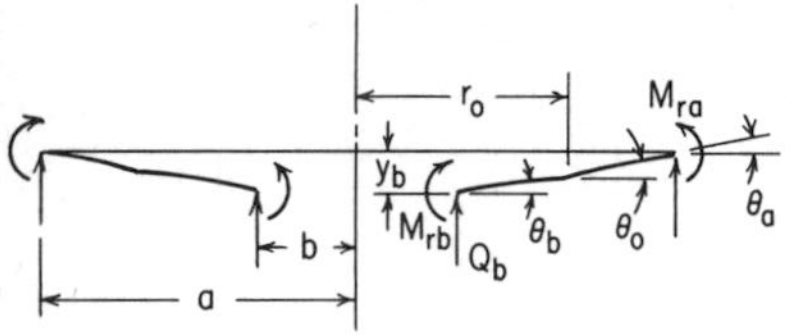

General expressions for deformations, moments, and shears:

$$y = y_b + \theta_b r F_1 + M_{rb}\frac{r^2}{D}F_2 + Q_b\frac{r^3}{D}F_3 + \theta_o r G_1$$

$$\theta = \theta_b F_4 + M_{rb}\frac{r}{D}F_5 + Q_b\frac{r^2}{D}F_6 + \theta_o G_4$$

$$M_r = \theta_b\frac{D}{r}F_7 + M_{rb}F_8 + Q_b r F_9 + \frac{\theta_o D}{r}G_7$$

$$M_t = \frac{\theta D(1-\nu^2)}{r} + \nu M_r$$

$$Q = Q_b\frac{b}{r}$$

For the numerical data given below, $\nu = 0.3$

$$y = K_y\theta_o a \qquad \theta = K_\theta\theta_o \qquad M = K_M\theta_o\frac{D}{a} \qquad Q = K_Q\theta_o\frac{D}{a^2}$$

Case no., edge restraints	Boundary values	Special cases

6a. Outer edge simply supported, inner edge free

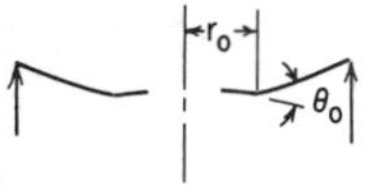

Boundary values:

$M_{rb} = 0 \qquad Q_b = 0 \qquad y_a = 0 \qquad M_{ra} = 0$

$$y_b = \theta_o a\left(\frac{C_1L_7}{C_7} - L_1\right)$$

$$\theta_b = \frac{-\theta_o}{C_7}L_7$$

$$\theta_a = -\theta_o\left(\frac{C_4L_7}{C_7} - L_4\right)$$

$Q_a = 0$

Special cases:

b/a	0.1		0.5		0.7
r_o/a	0.5	0.7	0.7	0.9	0.9
K_{y_b}	−0.2026	−0.1513	−0.0529	−0.0299	−0.0146
K_{y_o}	−0.2821	−0.2224	−0.1468	−0.0844	−0.0709
K_{θ_b}	−0.1515	−0.0736	−0.4857	−0.1407	−0.2898
$K_{M_{tb}}$	−1.3788	−0.6697	−0.8840	−0.2562	−0.3767
$K_{M_{to}}$	1.1030	0.9583	0.6325	0.8435	0.7088

6b. Outer edge simply supported, inner edge guided

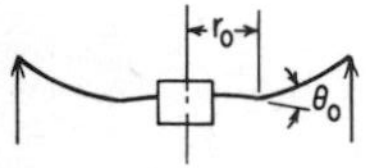

Boundary values:

$\theta_b = 0 \qquad Q_b = 0 \qquad y_a = 0 \qquad M_{ra} = 0$

$$y_b = \theta_o a\left(\frac{C_2L_7}{C_8} - L_1\right)$$

$$M_{rb} = \frac{-\theta_o D L_7}{a C_8}$$

$$\theta_a = -\theta_o\left(\frac{C_5L_7}{C_8} - L_4\right)$$

Special cases:

b/a	0.1		0.5		0.7
r_o/a	0.5	0.7	0.7	0.9	0.9
K_{y_b}	−0.2413	−0.1701	−0.2445	−0.0854	−0.0939
K_{θ_a}	0.5080	0.7039	0.7864	0.9251	0.9441
$K_{M_{rb}}$	−1.0444	−0.5073	−0.4495	−0.1302	−0.1169
$K_{M_{tb}}$	0.3133	0.1522	0.1349	0.0391	0.0351

inner edge simply supported

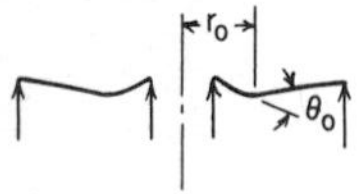

$y_b = 0 \qquad M_{rb} = 0 \qquad y_a = 0 \qquad M_{ra} = 0$

$$\theta_b = \theta_o \frac{C_3L_7 - C_9L_1}{C_1C_9 - C_3C_7}$$

$$Q_b = \frac{-\theta_o D}{a^2} \frac{C_1L_7 - C_7L_1}{C_1C_9 - C_3C_7}$$

$$\theta_a = \theta_b C_4 + Q_b \frac{a^2}{D} C_6 + \theta_o L_4$$

$$Q_a = Q_b \frac{b}{a}$$

b/a	0.1		0.5		0.7
r_o/a	0.5	0.7	0.7	0.9	0.9
K_{y_o}	−0.1629	−0.1689	−0.1161	−0.0788	−0.0662
K_{θ_b}	−0.3579	−0.2277	−0.6023	−0.2067	−0.3412
K_{θ_a}	0.2522	0.5189	0.3594	0.7743	0.6508
$K_{M_{tb}}$	−3.2572	−2.0722	−1.0961	−0.3762	−0.4435
$K_{M_{to}}$	0.6152	0.6973	0.4905	0.7851	0.6602
K_{Q_b}	5.5679	4.1574	0.2734	0.1548	0.0758

6d. Outer edge simply supported, inner edge fixed

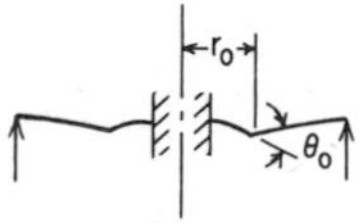

$y_b = 0 \qquad \theta_b = 0 \qquad y_a = 0 \qquad M_{ra} = 0$

$$M_{rb} = \frac{\theta_o D}{a} \frac{C_3L_7 - C_9L_1}{C_2C_9 - C_3C_8}$$

$$Q_b = \frac{-\theta_o D}{a^2} \frac{C_2L_7 - C_8L_1}{C_2C_9 - C_3C_8}$$

$$\theta_a = M_{rb} \frac{a}{D} C_5 + Q_b \frac{a^2}{D} C_6 + \theta_o L_4$$

$$Q_a = Q_b \frac{b}{a}$$

b/a	0.1		0.5		0.7
r_o/a	0.5	0.7	0.7	0.9	0.9
K_{y_o}	−0.1333	−0.1561	−0.0658	−0.0709	−0.0524
K_{θ_a}	0.1843	0.4757	0.1239	0.6935	0.4997
$K_{M_{rb}}$	−3.3356	−2.1221	−4.2716	−1.4662	−3.6737
K_{Q_b}	8.9664	6.3196	9.6900	3.3870	12.9999

6e. Outer edge fixed, inner edge free

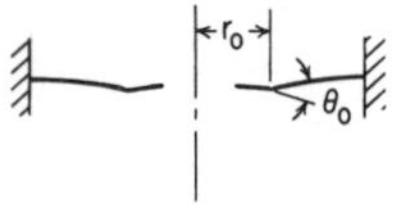

$M_{rb} = 0 \qquad Q_b = 0 \qquad y_a = 0 \qquad \theta_a = 0$

$$y_b = \theta_o a \left(\frac{C_1L_4}{C_4} - L_1 \right)$$

$$\theta_b = \frac{-\theta_o L_4}{C_4}$$

$$M_{ra} = \frac{\theta_o D}{a} \left(L_7 - \frac{C_7}{C_4} L_4 \right)$$

$$Q_a = 0$$

b/a	0.1		0.5		0.7
r_o/a	0.5	0.7	0.7	0.9	0.9
K_{y_b}	0.0534	0.2144	0.1647	0.3649	0.1969
K_{y_o}	−0.0975	−0.0445	−0.0155	−0.0029	−0.0013
K_{θ_b}	−0.2875	−0.2679	−0.9317	−0.9501	−1.0198
$K_{M_{tb}}$	−2.6164	−2.4377	−1.6957	−1.7293	−1.3257

TABLE 24 ***Formulas for flat circular plates of constant thickness*** *(Cont.)*

Case no., edge restraints	Boundary values	Special cases

6f. Outer edge fixed, inner edge guided

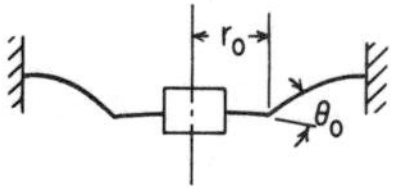

Boundary values:

$\theta_b = 0 \qquad Q_b = 0 \qquad y_a = 0 \qquad \theta_a = 0$

$$y_b = \theta_o a\left(\frac{C_2 L_4}{C_5} - L_1\right)$$

$$M_{rb} = \frac{-\theta_o D L_4}{a C_5}$$

$$M_{ra} = \frac{\theta_o D}{a}\left(L_7 - \frac{C_8}{C_5} L_4\right)$$

$$Q_a = 0$$

Special cases:

b/a	0.1		0.5		0.7
r_o/a	0.5	0.7	0.7	0.9	0.9
K_{y_b}	0.0009	0.1655	−0.0329	0.1634	0.0546
K_{y_o}	−0.1067	−0.0472	−0.0786	−0.0094	−0.0158
$K_{M_{rb}}$	−2.0707	−1.9293	−2.5467	−2.5970	−3.8192
$K_{M_{ra}}$	−0.6707	−0.9293	−1.5467	−1.8193	−3.0414

6g. Outer edge fixed, inner edge simply supported

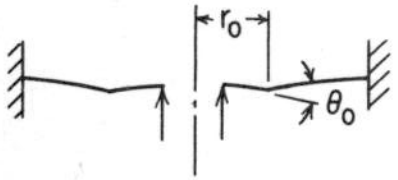

Boundary values:

$y_b = 0 \qquad M_{rb} = 0 \qquad y_a = 0 \qquad \theta_a = 0$

$$\theta_b = \theta_o \frac{C_3 L_4 - C_6 L_1}{C_1 C_6 - C_3 C_4}$$

$$Q_b = \frac{-\theta_o D}{a^2} \frac{C_1 L_4 - C_4 L_1}{C_1 C_6 - C_3 C_4}$$

$$M_{ra} = \theta_b \frac{D}{a} C_7 + Q_b a C_9 + \frac{\theta_o D}{a} L_7$$

$$Q_a = Q_b \frac{b}{a}$$

Special cases:

b/a	0.1		0.5		0.7
r_o/a	0.5	0.7	0.7	0.9	0.9
K_{y_o}	−0.1179	−0.0766	−0.0820	−0.0208	−0.0286
K_{θ_b}	−0.1931	0.1116	−0.3832	0.2653	0.0218
$K_{M_{ra}}$	−0.8094	−1.6653	−1.9864	−4.2792	−6.1794
$K_{M_{tb}}$	−1.7567	1.0151	−0.6974	0.4828	0.0284
K_{Q_b}	−3.7263	−14.9665	−7.0690	−15.6627	−27.9529

6h. Outer edge fixed, inner edge fixed

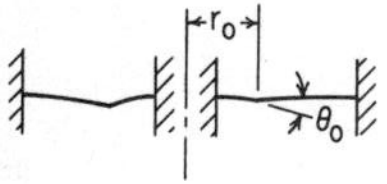

Boundary values:

$y_b = 0 \qquad \theta_b = 0 \qquad y_a = 0 \qquad \theta_a = 0$

$$M_{rb} = \frac{\theta_o D}{a} \frac{C_3 L_4 - C_6 L_1}{C_2 C_6 - C_3 C_5}$$

$$Q_b = \frac{-\theta_o D}{a^2} \frac{C_2 L_4 - C_5 L_1}{C_2 C_6 - C_3 C_5}$$

$$M_{ra} = M_{rb} C_8 + Q_b a C_9 + \theta_o \frac{D}{a} L_7$$

$$Q_a = Q_b \frac{b}{a}$$

Special cases:

b/a	0.1		0.5		0.7
r_o/a	0.5	0.7	0.7	0.9	0.9
K_{y_o}	−0.1071	−0.0795	−0.0586	−0.0240	−0.0290
$K_{M_{rb}}$	−2.0540	1.1868	−3.5685	2.4702	0.3122
$K_{M_{ra}}$	−0.6751	−1.7429	−0.8988	−5.0320	−6.3013
K_{Q_b}	−0.0915	−17.067	4.8176	−23.8910	−29.6041

Case 7. Annular plate with an externally applied vertical deformation y_o at a radius r_o

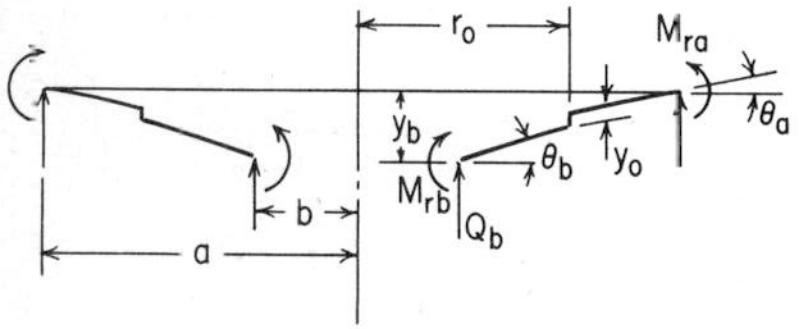

General expressions for deformations, moments, and shears:

$$y = y_b + \theta_b r F_1 + M_{rb}\frac{r^2}{D}F_2 + Q_b\frac{r^3}{D}F_3 + y_o\langle r - r_o\rangle^0$$

$$\theta = \theta_b F_4 + M_{rb}\frac{r}{D}F_5 + Q_b\frac{r^2}{D}F_6$$

$$M_r = \theta_b\frac{D}{r}F_7 + M_{rb}F_8 + Q_b r F_9$$

$$M_t = \frac{\theta D(1-\nu^2)}{r} + \nu M_r$$

$$Q = Q_b\frac{b}{r}$$

For the numerical data given below, $\nu = 0.3$

$$y = K_y y_o \qquad \theta = K_\theta\frac{y_o}{a} \qquad M = K_M y_o\frac{D}{a^2} \qquad Q = K_Q y_o\frac{D}{a^3}$$

Case no., edge restraints	Boundary values	Special cases
7e. Outer edge simply supported, inner edge simply supported 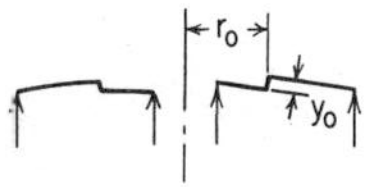	$y_b = 0 \quad M_{rb} = 0 \quad y_a = 0 \quad M_{ra} = 0$ $\theta_b = \dfrac{-y_o C_9}{a(C_1C_9 - C_3C_7)}$ $Q_b = \dfrac{y_o D C_7}{a^3(C_1C_9 - C_3C_7)}$ $\theta_a = \dfrac{y_o}{a}\dfrac{C_7C_6 - C_9C_4}{C_1C_9 - C_3C_7}$ $Q_a = Q_b\dfrac{b}{a}$	See below

b/a	0.1	0.3	0.5	0.7	0.9
K_{θ_b}	−1.0189	−1.6176	−2.2045	−3.5180	−10.1611
K_{θ_a}	−1.1484	−1.3149	−1.8482	−3.1751	−9.8461
$K_{M_{tb}}$	−9.2716	−4.9066	−4.0121	−4.5734	−10.2740
K_{Q_b}	27.4828	7.9013	5.1721	5.1887	10.6599

(*Note:* Constants given are valid for all values of $r_o > b$)

TABLE 24 *Formulas for flat circular plates of constant thickness* (Cont.)

Case no., edge restraints	Boundary values	Special cases
7d. Outer edge simply supported, inner edge fixed 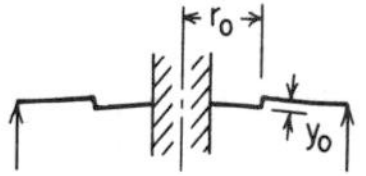	$y_b = 0 \quad \theta_b = 0 \quad y_a = 0 \quad M_{ra} = 0$ $M_{rb} = \dfrac{-y_o D C_9}{a^2(C_2C_9 - C_3C_8)}$ $Q_b = \dfrac{y_o D C_8}{a^3(C_2C_9 - C_3C_8)}$ $\theta_a = \dfrac{y_o}{a}\dfrac{C_6C_8 - C_5C_9}{C_2C_9 - C_3C_8}$ $Q_a = Q_b\dfrac{b}{a}$	see below
7g. Outer edge fixed, inner edge simply supported 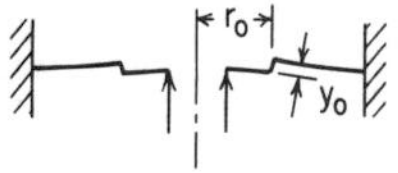	$y_b = 0 \quad M_{rb} = 0 \quad y_a = 0 \quad \theta_a = 0$ $\theta_b = \dfrac{-y_o C_6}{a(C_1C_6 - C_3C_4)}$ $Q_b = \dfrac{y_o D C_4}{a^3(C_1C_6 - C_3C_4)}$ $M_{ra} = \dfrac{y_o D}{a^2}\dfrac{C_4C_9 - C_6C_7}{C_1C_6 - C_3C_4}$ $Q_a = Q_b\dfrac{b}{a}$	see below
7h. Outer edge fixed, inner edge fixed 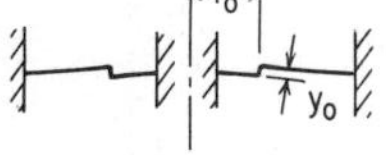	$y_b = 0 \quad \theta_b = 0 \quad y_a = 0 \quad \theta_a = 0$ $M_{rb} = \dfrac{-y_o D C_6}{a^2(C_2C_6 - C_3C_5)}$ $Q_b = \dfrac{y_o D C_5}{a^3(C_2C_6 - C_3C_5)}$ $M_{ra} = \dfrac{y_o D}{a^2}\dfrac{C_5C_9 - C_6C_8}{C_2C_6 - C_3C_5}$ $Q_a = Q_b\dfrac{b}{a}$	see below

Special cases, 7d:

b/a	0.1	0.3	0.5	0.7	0.9
K_{θ_a}	−1.3418	−1.8304	−2.7104	−4.7327	−14.7530
$K_{M_{rb}}$	−9.4949	−9.9462	−15.6353	−37.8822	−310.808
K_{Q_b}	37.1567	23.9899	39.6394	138.459	3168.83

(*Note:* Constants given are valid for all values of $r_o > b$)

Special cases, 7g:

b/a	0.1	0.3	0.5	0.7	0.9
K_{θ_b}	−1.7696	−2.5008	−3.3310	−5.2890	−15.2528
$K_{M_{ra}}$	3.6853	5.1126	10.2140	30.1487	290.2615
$K_{M_{tb}}$	−16.1036	−7.5856	−6.0624	−6.8757	−15.4223
K_{Q_b}	69.8026	30.3098	42.9269	141.937	3168.165

(*Note:* Constants given are valid for all values of $r_o > b$)

Special cases, 7h:

b/a	0.1	0.3	0.5	0.7	0.9
$K_{M_{rb}}$	−18.8284	−19.5643	−31.0210	−75.6312	−621.8586
$K_{M_{ra}}$	4.9162	9.0548	19.6681	59.6789	579.6755
K_{Q_b}	103.1218	79.2350	146.258	541.958	12671.35

(*Note:* Constants given are valid for all values of $r_o > b$)

Case 8. Annular plate with, from r_o to a, a uniform temperature differential ΔT between the bottom and the top surface (the midplane temperature is assumed to be unchanged, and so no in-plane forces develop)

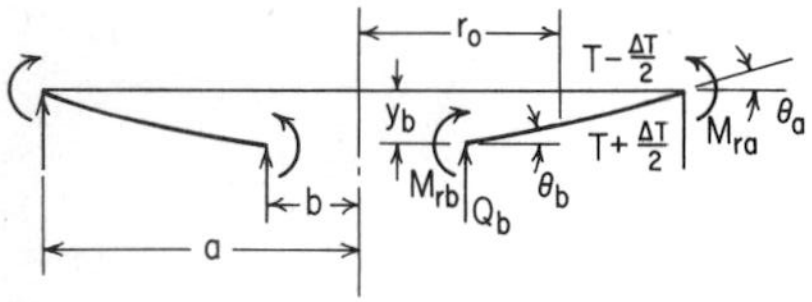

General expressions for deformations, moments, and shears:

$$y = y_b + \theta_b r F_1 + M_{rb}\frac{r^2}{D}F_2 + Q_b\frac{r^3}{D}F_3 + \frac{\gamma(1+\nu)\,\Delta T}{t}r^2 G_2$$

$$\theta = \theta_b F_4 + M_{rb}\frac{r}{D}F_5 + Q_b\frac{r^2}{D}F_6 + \frac{\gamma(1+\nu)\,\Delta T}{t}rG_5$$

$$M_r = \theta_b\frac{D}{r}F_7 + M_{rb}F_8 + Q_b r F_9 + \frac{\gamma(1+\nu)\,\Delta T}{t}D(G_8 - 1)$$

$$M_t = \frac{\theta D(1-\nu^2)}{r} + \nu M_r - \frac{\gamma(1-\nu^2)\,\Delta T D}{t}\langle r - r_o\rangle^0$$

$$Q = Q_b\frac{b}{r}$$

For the numerical data given below, $\nu = 0.3$

$$y = K_y\frac{\gamma\,\Delta T a^2}{t} \qquad \theta = K_\theta\frac{\gamma\,\Delta T a}{t} \qquad M = K_M\frac{\gamma\,\Delta T D}{t} \qquad Q = K_Q\frac{\gamma\,\Delta T D}{at}$$

Case no., edge restraints	Boundary values	Special cases
8a. Outer edge simply supported, inner edge free 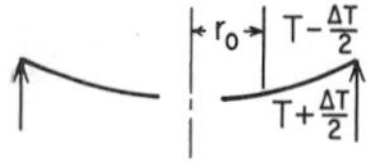 	$M_{rb} = 0 \quad Q_b = 0 \quad y_a = 0 \quad M_{ra} = 0$ $y_b = \frac{-\gamma(1+\nu)\,\Delta T a^2}{t}\left[L_2 + \frac{C_1}{C_7}(1 - L_8)\right]$ $\theta_b = \frac{\gamma(1+\nu)\,\Delta T a}{tC_7}(1 - L_8)$ $Q_a = 0$ $\theta_a = \frac{\gamma(1+\nu)\,\Delta T a}{t}\left[L_5 + \frac{C_4}{C_7}(1 - L_8)\right]$	If $r_o = b$ (ΔT over entire plate), see table below (*Note:* There are no moments in the plate)

b/a	0.1	0.3	0.5	0.7	0.9
K_{y_b}	−0.4950	−0.4550	−0.3750	−0.2550	−0.0950
K_{θ_a}	1.0000	1.0000	1.0000	1.0000	1.0000
K_{θ_b}	0.1000	0.3000	0.5000	0.7000	0.9000

TABLE 24 *Formulas for flat circular plates of constant thickness* (*Cont.*)

Case no., edge restraints	Boundary values	Special cases

8b. Outer edge simply supported, inner edge guided

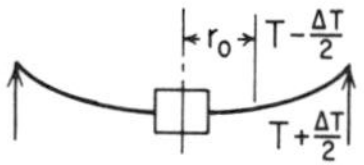

$\theta_b = 0 \qquad Q_b = 0 \qquad y_a = 0 \qquad M_{ra} = 0$

$$y_b = \frac{-\gamma(1+\nu)\,\Delta T a^2}{t}\left[L_2 + \frac{C_2}{C_8}(1 - L_8)\right]$$

$$M_{rb} = \frac{\gamma(1+\nu)\,\Delta T D}{t C_8}(1 - L_8)$$

$$\theta_a = \frac{\gamma(1+\nu)\,\Delta T a}{t}\left[L_5 + \frac{C_5}{C_8}(1 - L_8)\right]$$

$$Q_a = 0$$

If $r_o = b$ (ΔT over entire plate),

b/a	0.1	0.3	0.5	0.7	0.9
K_{y_b}	−0.4695	−0.3306	−0.1778	−0.0635	−0.0067
K_{θ_a}	0.9847	0.8679	0.6610	0.4035	0.1323
$K_{M_{rb}}$	0.6893	0.6076	0.4627	0.2825	0.0926
$K_{M_{tb}}$	−0.7032	−0.7277	−0.7712	−0.8253	−0.8822

8c. Outer edge simply supported, inner edge simply supported

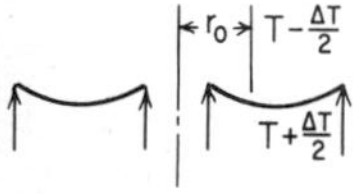

$y_b = 0 \qquad M_{rb} = 0 \qquad y_a = 0 \qquad M_{ra} = 0$

$$\theta_b = \frac{-\gamma(1+\nu)\,\Delta T a}{t}\,\frac{C_9 L_2 + C_3(1 - L_8)}{C_1 C_9 - C_3 C_7}$$

$$Q_b = \frac{\gamma(1+\nu)\,\Delta T D}{at}\,\frac{C_7 L_2 + C_1(1 - L_8)}{C_1 C_9 - C_3 C_7}$$

$$\theta_a = \theta_b C_4 + Q_b \frac{a^2}{D} C_6 + \frac{\gamma(1+\nu)\,\Delta T a}{t} L_5$$

$$Q_a = Q_b \frac{b}{a}$$

If $r_o = b$ (ΔT over entire plate),

b/a	0.1	0.3	0.5	0.7	0.9
$K_{y_{max}}$	−0.0865	−0.0701	−0.0388	−0.0142	
K_{θ_b}	−0.4043	−0.4360	−0.3267	−0.1971	−0.0653
K_{θ_a}	0.4316	0.4017	0.3069	0.1904	0.0646
$K_{M_{tb}}$	−4.5894	−2.2325	−1.5045	−1.1662	−0.9760
K_{Q_b}	13.6040	3.5951	1.9395	1.3231	1.0127

8d. Outer edge simply supported, inner edge fixed

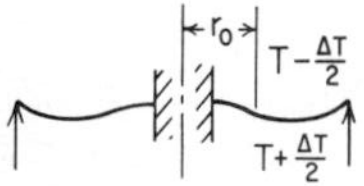

$y_b = 0 \qquad \theta_b = 0 \qquad y_a = 0 \qquad M_{ra} = 0$

$$M_{rb} = \frac{-\gamma(1+\nu)\,\Delta T D}{t}\,\frac{C_9 L_2 + C_3(1 - L_8)}{C_2 C_9 - C_3 C_8}$$

$$Q_b = \frac{\gamma(1+\nu)\,\Delta T D}{at}\,\frac{C_8 L_2 + C_2(1 - L_8)}{C_2 C_9 - C_3 C_8}$$

$$\theta_a = M_{rb}\frac{a}{D} C_5 + Q_b \frac{a^2}{D} C_6 + \frac{\gamma(1+\nu)\,\Delta T a}{t} L_5$$

$$Q_a = Q_b \frac{b}{a}$$

If $r_o = b$ (ΔT over entire plate),

b/a	0.1	0.3	0.5	0.7	0.9
$K_{y_{max}}$	−0.0583	−0.0318	−0.0147	−0.0049	
K_{θ_a}	0.3548	0.2628	0.1792	0.1031	0.0331
$K_{M_{rb}}$	−3.7681	−2.6809	−2.3170	−2.1223	−1.9975
K_{Q_b}	17.4431	7.9316	7.0471	8.7894	21.3100

8e. Outer edge fixed, inner edge free

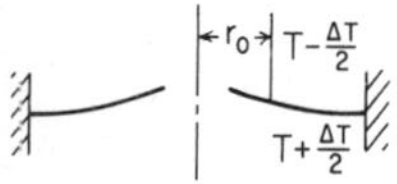

$M_{rb} = 0 \qquad Q_b = 0 \qquad y_a = 0 \qquad \theta_a = 0$

$$y_b = \frac{-\gamma(1+\nu)\,\Delta T a^2}{t}\left(L_2 - \frac{C_1}{C_4}L_5\right)$$

$$\theta_b = \frac{-\gamma(1+\nu)\,\Delta T a}{tC_4}L_5$$

$$M_{ra} = \frac{-\gamma(1+\nu)\,\Delta T D}{t}\left(\frac{C_7}{C_4}L_5 + 1 - L_8\right)$$

$Q_a = 0$

If $r_o = b$ (ΔT over entire plate),

b/a	0.1	0.3	0.5	0.7	0.9
K_{y_b}	0.0330	0.1073	0.1009	0.0484	0.0062
K_{θ_b}	−0.1805	−0.4344	−0.4756	−0.3471	−0.1268
$K_{M_{ra}}$	−1.2635	−1.0136	−0.6659	−0.3471	−0.0986
$K_{M_{tb}}$	−2.5526	−2.2277	−1.7756	−1.3613	−1.0382

8f. Outer edge fixed, inner edge guided

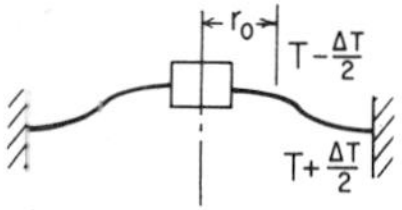

$\theta_b = 0 \qquad Q_b = 0 \qquad y_a = 0 \qquad \theta_a = 0$

$$y_b = \frac{-\gamma(1+\nu)\,\Delta T a^2}{t}\left(L_2 - \frac{C_2}{C_5}L_5\right)$$

$$M_{rb} = \frac{-\gamma(1+\nu)\,\Delta T D}{tC_5}L_5$$

$$M_{ra} = \frac{-\gamma(1+\nu)\,\Delta T D}{t}\left(\frac{C_8}{C_5}L_5 + 1 - L_8\right)$$

$Q_a = 0$

If $r_o = b$ (ΔT over entire plate),

(All deflections are zero and $K_{M_r} = K_{M_t} = -1.30$ everywhere in the plate)

b/a	0.1		0.5		0.7
r_o/a	0.5	0.7	0.7	0.9	0.9
K_{y_b}	0.1013	0.1059	0.0370	0.0269	0.0132
$K_{M_{rb}}$	−0.9849	−0.6697	−0.8840	−0.3293	−0.4843
$K_{M_{ra}}$	−0.9849	−0.6697	−0.8840	−0.3293	−0.4843
$K_{M_{to}}$	−1.5364	−1.3405	−1.3267	−1.0885	−1.1223

8g. Outer edge fixed, inner edge simply supported

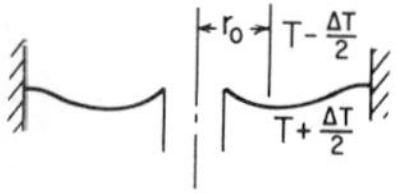

$y_b = 0 \qquad M_{rb} = 0 \qquad y_a = 0 \qquad \theta_a = 0$

$$\theta_b = \frac{-\gamma(1+\nu)\,\Delta T a}{t}\,\frac{C_6L_2 - C_3L_5}{C_1C_6 - C_3C_4}$$

$$Q_b = \frac{\gamma(1+\nu)\,\Delta T D}{at}\,\frac{C_4L_2 - C_1L_5}{C_1C_6 - C_3C_4}$$

$$M_{ra} = \theta_b\frac{D}{a}C_7 + Q_b a C_9 - \frac{\gamma(1+\nu)\Delta T D}{t}(1 - L_8)$$

$$Q_a = Q_b\frac{b}{a}$$

If $r_o = b$ (ΔT over entire plate),

b/a	0.1	0.3	0.5	0.7	0.9
$K_{y_{max}}$	−0.0088	−0.0133	−0.0091	−0.0039	
K_{θ_b}	−0.1222	−0.1662	−0.1396	−0.0909	−0.0319
$K_{M_{tb}}$	−2.0219	−1.4141	−1.1641	−1.0282	−0.9422
$K_{M_{ra}}$	−1.3850	−1.5620	−1.6962	−1.8076	−1.9050
K_{Q_b}	−2.3005	−3.2510	−4.3303	−6.8757	−19.8288

TABLE 24 *Formulas for flat circular plates of constant thickness* (Cont.)

Case no., edge restraints	Boundary values	Special cases
8h. Outer edge fixed, inner edge fixed 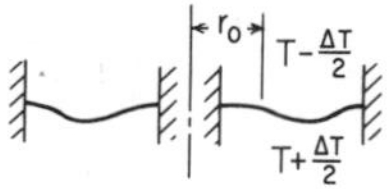	$y_b = 0 \quad \theta_b = 0 \quad y_a = 0 \quad \theta_a = 0$ $M_{rb} = \frac{-\gamma(1+\nu)\,\Delta TD}{t} \frac{C_6L_2 - C_3L_5}{C_2C_6 - C_3C_5}$ $Q_b = \frac{\gamma(1+\nu)\,\Delta TD}{at} \frac{C_5L_2 - C_2L_5}{C_2C_6 - C_3C_5}$ $M_{ra} = M_{rb}C_8 + Q_baC_9 - \frac{\gamma(1+\nu)\,\Delta TD}{t}(1 - L_8)$ $Q_a = Q_b\frac{b}{a}$	If $r_o = b$ (ΔT over entire plate), (All deflections are zero and $K_{M_r} = K_{M_t} = -1.30$ everywhere in the plate.)

b/a	0.1		0.5		0.7
r_o/a	0.5	0.7	0.7	0.9	0.9
$K_{M_{rb}}$	0.9224	1.3241	0.2640	0.5063	0.5103
$K_{M_{ra}}$	−1.4829	−1.1903	−1.6119	−0.8592	−1.2691
$K_{M_{ta}}$	−1.3549	−1.2671	−1.3936	−1.1677	−1.2907
K_{Q_b}	−10.4460	−10.9196	−5.4127	−3.9399	−7.1270

Cases 9 to 15. Solid circular plate under the several indicated loadings

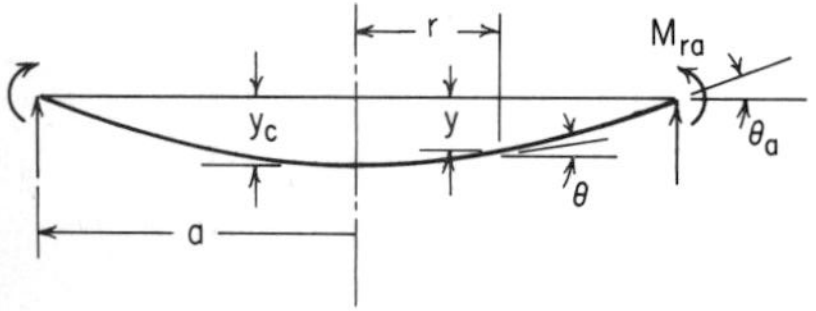

General expressions for deformations, moments, and shears:

$$y = y_c + \frac{M_cr^2}{2D(1+\nu)} + LT_y$$ (*Note:* y_c is the center deflection)

$$\theta = \frac{M_cr}{D(1+\nu)} + LT_\theta$$ (*Note:* M_c is the moment at the center)

$$M_r = M_c + LT_M$$

$$M_t = \frac{\theta D(1-\nu^2)}{r} + \nu M_r$$ (*Note:* For $r < r_o$, $M_t = M_r = M_c$)

$$Q_r = LT_Q$$

For the numerical data given below, $\nu = 0.3$ (*Note:* ln = natural logarithm)

Case no., loading, load terms	Edge restraint	Boundary values	Special cases
9. Uniform annular line load 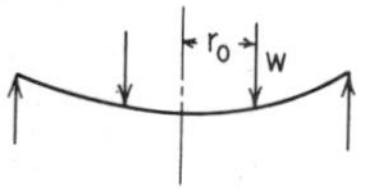 $LT_y = \frac{-wr^3}{D}G_3$	9a. Simply supported	$y_a = 0 \quad M_{ra} = 0$ $y_c = \frac{-wa^3}{2D}\left(\frac{L_9}{1+\nu} - 2L_3\right)$ $M_c = waL_9$ $Q_a = -w\frac{r_o}{a}$ $\theta_a = \frac{wr_o(a^2 - r_o^2)}{2D(1+\nu)a}$	$y = K_y\frac{wa^3}{D} \quad \theta = K_\theta\frac{wa^2}{D} \quad M = K_Mwa$

r_o/a	0.2	0.4	0.6	0.8
K_{y_c}	−0.05770	−0.09195	−0.09426	−0.06282
K_{θ_a}	0.07385	0.12923	0.14769	0.11077
K_{M_c}	0.24283	0.29704	0.26642	0.16643

Note: If r_o approaches 0, see case 16

$LT_\theta = \frac{-wr^2}{D} G_6$

$LT_M = -wrG_9$

$LT_Q = \frac{-wr_o}{r} \langle r - r_o \rangle^0$

9b. Fixed

$y_c = \frac{-wa^3}{2D}(L_6 - 2L_3)$

$M_c = wa(1 + \nu)L_6$

$M_{ra} = \frac{-wr_o}{2a^2}(a^2 - r_o^2)$

$y_a = 0 \qquad \theta_a = 0$

r_o/a	0.2	0.4	0.6	0.8
K_{y_c}	−0.02078	−0.02734	−0.02042	−0.00744
K_{M_c}	0.14683	0.12904	0.07442	0.02243
$K_{M_{ra}}$	−0.09600	−0.16800	−0.19200	−0.14400

(*Note:* If r_o approaches 0, see case 17)

10. Uniformly distributed load from r_o to a

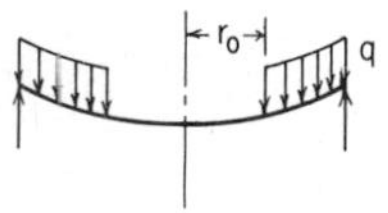

$LT_y = \frac{-qr^4}{D} G_{11}$

$LT_\theta = \frac{-qr^3}{D} G_{14}$

$LT_M = -qr^2 G_{17}$

$LT_Q = \frac{-q}{2r}(r^2 - r_o^2)\langle r - r_o \rangle^0$

10a. Simply supported

$y_a = 0 \qquad M_{ra} = 0$

$y_c = \frac{-qa^4}{2D}\left(\frac{L_{17}}{1+\nu} - 2L_{11}\right)$

$M_c = qa^2 L_{17}$

$\theta_a = \frac{q}{8Da(1+\nu)}(a^2 - r_o^2)^2$

$Q_a = \frac{-q}{2a}(a^2 - r_o^2)$

$y = K_y \frac{qa^4}{D} \qquad \theta = K_\theta \frac{qa^3}{D} \qquad M = K_M qa^2$

r_o/a	0.0	0.2	0.4	0.6	0.8
K_{y_c}	−0.06370	−0.05767	−0.04221	−0.02303	−0.00677
K_{θ_a}	0.09615	0.08862	0.06785	0.03939	0.01246
K_{M_c}	0.20625	0.17540	0.11972	0.06215	0.01776

Note: If $r_o = 0$, $\quad G_{11} = \frac{1}{64} \quad G_{14} = \frac{1}{16} \quad G_{17} = \frac{(3+\nu)}{16}$

$y_c = \frac{-qa^4(5+\nu)}{64D(1+\nu)} \qquad M_c = \frac{qa^2(3+\nu)}{16} \qquad \theta_a = \frac{qa^3}{8D(1+\nu)}$

10b. Fixed

$y_a = 0 \qquad \theta_a = 0$

$y_c = \frac{-qa^4}{2D}(L_{14} - 2L_{11})$

$M_c = qa^2(1+\nu)L_{14}$

$M_{ra} = \frac{-q}{8a^2}(a^2 - r_o^2)^2$

r_o/a	0.0	0.2	0.4	0.6	0.8
K_{y_c}	−0.01563	−0.01336	−0.00829	−0.00334	−0.00054
K_{M_c}	0.08125	0.06020	0.03152	0.01095	0.00156
$K_{M_{ra}}$	−0.12500	−0.11520	−0.08820	−0.05120	−0.01620

Note: If $r_o = 0$, $\quad G_{11} = \frac{1}{64} \quad G_{14} = \frac{1}{16} \quad G_{17} = \frac{(3+\nu)}{16}$

$y_c = \frac{-qa^4}{64D} \qquad M_c = \frac{qa^2(1+\nu)}{16} \qquad M_{ra} = \frac{-qa^2}{8}$

11. Linearly increasing load from r_o to a

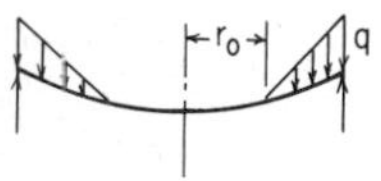

$LT_y = \frac{-qr^4}{D}\frac{r - r_o}{a - r_o} G_{12}$

$LT_\theta = \frac{-qr^3}{D}\frac{r - r_o}{a - r_o} G_{15}$

11a. Simply supported

$M_{ra} = 0 \qquad y_a = 0$

$y_c = \frac{-qa^4}{2D}\left(\frac{L_{18}}{1+\nu} - 2L_{12}\right)$

$M_c = qa^2 L_{18}$

$\theta_a = \frac{qa^3}{D}\left(\frac{L_{18}}{1+\nu} - L_{15}\right)$

$Q_a = \frac{-q}{6a}(2a^2 - r_o a - r_o^2)$

$y = K_y \frac{qa^4}{D} \qquad \theta = K_\theta \frac{qa^3}{D} \qquad M = K_M qa^2$

r_o/a	0.0	0.2	0.4	0.6	0.8
K_{y_c}	−0.03231	−0.02497	−0.01646	−0.00836	−0.00234
K_{θ_a}	0.05128	0.04070	0.02788	0.01485	0.00439
K_{M_c}	0.09555	0.07082	0.04494	0.02220	0.00610

Note: If $r_o = 0$, $\quad G_{12} = \frac{1}{225} \quad G_{15} = \frac{1}{45} \quad G_{18} = \frac{(4+\nu)}{45}$

$y_c = \frac{-qa^4(6+\nu)}{150D(1+\nu)} \qquad M_c = \frac{qa^2(4+\nu)}{45} \qquad \theta_a = \frac{qa^3}{15D(1+\nu)}$

TABLE 24 ***Formulas for flat circular plates of constant thickness*** **(*Cont.*)**

Case no., loading, load terms	Edge restraint	Boundary values	Special cases
$LT_M = -qr^2 \frac{r - r_o}{a - r_o} G_{18}$ $LT_Q = \frac{-q(2r^3 - 3r_o r + r_o^3)}{6r(a - r_o)}$ $\langle r - r_o \rangle^0$	11b. Fixed	$y_a = 0 \quad \theta_a = 0$ $y_c = \frac{-qa^4}{2D}(L_{15} - 2L_{12})$ $M_c = qa^2(1 + \nu)L_{15}$ $M_{ra} = -qa^2[L_{18} - (1 + \nu)L_{15}]$	r_o/a: 0.0, 0.2, 0.4, 0.6, 0.8 K_{y_c}: −0.00667, −0.00462, −0.00252, −0.00093, −0.00014 K_{M_c}: 0.02889, 0.01791, 0.00870, 0.00289, 0.00040 $K_{M_{ra}}$: −0.06667, −0.05291, −0.03624, −0.01931, −0.00571 *Note:* If $r_o = 0$, $G_{12} = \frac{1}{225}$ $G_{15} = \frac{1}{45}$ $G_{18} = \frac{(4 + \nu)}{45}$ $y_c = \frac{-qa^4}{150D}$ $M_c = \frac{qa^2(1 + \nu)}{45}$ $M_{ra} = \frac{-qa^2}{15}$
12. Parabolically increasing load from r_o to a 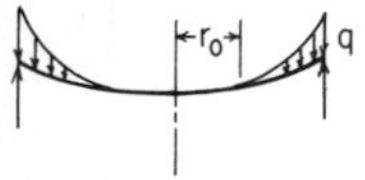 $LT_y = \frac{-qr^4}{D}\left(\frac{r - r_o}{a - r_o}\right)^2 G_{13}$ $LT_\theta = \frac{-qr^3}{D}\left(\frac{r - r_o}{a - r_o}\right)^2 G_{16}$ $LT_M = -qr^2\left(\frac{r - r_o}{a - r_o}\right)^2 G_{19}$ $LT_Q = \frac{-q(3r^4 - 8r_o r^3 + 6r_o^2 r^2 - r_o^4)}{12r(a - r_o)^2}$ $\langle r - r_o \rangle^0$	12a. Simply supported	$y_a = 0 \quad M_{ra} = 0$ $y_c = \frac{-qa^4}{2D}\left(\frac{L_{19}}{1 + \nu} - 2L_{13}\right)$ $M_c = qa^2 L_{19}$ $\theta_a = \frac{qa^3}{D}\left(\frac{L_{19}}{1 + \nu} - L_{16}\right)$ $Q_a = \frac{-q}{12a}(3a^2 - 12ar_o - r_o^2)$	$y = K_y \frac{qa^4}{D}$ $\theta = K_\theta \frac{qa^3}{D}$ $M = K_M qa^2$ r_o/a: 0.0, 0.2, 0.4, 0.6, 0.8 K_{y_c}: −0.01949, −0.01419, −0.00893, −0.00438, −0.00119 K_{θ_a}: 0.03205, 0.02396, 0.01560, 0.00796, 0.00227 K_{M_c}: 0.05521, 0.03903, 0.02397, 0.01154, 0.00311 *Note:* If $r_o = 0$, $G_{13} = \frac{1}{576}$ $G_{16} = \frac{1}{96}$ $G_{19} = \frac{(5 + \nu)}{96}$ $y_c = \frac{-qa^4(7 + \nu)}{288D(1 + \nu)}$ $M_c = \frac{qa^2(5 + \nu)}{96}$ $\theta_a = \frac{qa^3}{24D(1 + \nu)}$
	12b. Fixed	$y_a = 0 \quad \theta_a = 0$ $y_c = \frac{-qa^4}{2D}(L_{16} - 2L_{13})$ $M_c = qa^2(1 + \nu)L_{16}$ $M_{ra} = -qa^2[L_{19} - (1 + \nu)L_{16}]$	r_o/a: 0.0, 0.2, 0.4, 0.6, 0.8 K_{y_c}: −0.00347, −0.00221, −0.00113, −0.00040, −0.000058 K_{M_c}: 0.01354, 0.00788, 0.00369, 0.00120, 0.000162 $K_{M_{ra}}$: −0.04167, −0.03115, −0.02028, −0.01035, −0.002947 *Note:* If $r_o = 0$, $G_{13} = \frac{1}{576}$ $G_{16} = \frac{1}{96}$ $G_{19} = \frac{(5 + \nu)}{96}$ $y_c = \frac{-qa^4}{288D}$ $M_c = \frac{qa^2(1 + \nu)}{96}$ $M_{ra} = \frac{-qa^2}{24}$

13. Uniform line moment at r_o

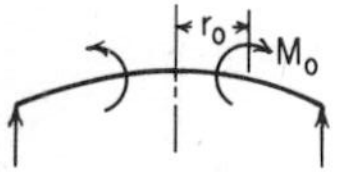

$LT_y = \frac{M_o r^2}{D} G_2$

$LT_\theta = \frac{M_o r}{D} G_5$

$LT_M = M_o G_8$

$LT_Q = 0$

13a. Simply supported

$y_a = 0 \qquad M_{ra} = 0 \qquad Q_a = 0$

$$y_c = \frac{M_o r_o^2}{2D}\left(\frac{1}{1+\nu} + \ln\frac{a}{r_o}\right)$$

$$M_c = -M_o L_8$$

$$\theta_a = \frac{-M_o r_o^2}{Da(1+\nu)}$$

$y = K_y \frac{M_o a^2}{D} \qquad \theta = K_\theta \frac{M_o a}{D} \qquad M = K_M M_o$

r_o/a	0.2	0.4	0.6	0.8	1.0
K_{y_c}	0.04757	0.13484	0.23041	0.31756	0.38462
K_{θ_a}	−0.03077	−0.12308	−0.27692	−0.49231	−0.76923
K_{M_c}	−0.66400	−0.70600	−0.77600	−0.87400	−1.00000

13b. Fixed

$y_a = 0 \qquad \theta_a = 0 \qquad Q_a = 0$

$$y_c = \frac{M_o r_o^2}{2D} \ln\frac{a}{r_o}$$

$$M_c = \frac{-M_o(1+\nu)}{2a^2}(a^2 - r_o^2)$$

$$M_{ra} = \frac{M_o r_o^2}{a^2}$$

r_o/a	0.2	0.4	0.6	0.8
K_{y_c}	0.03219	0.07330	0.09195	0.07141
K_{M_c}	−0.62400	−0.54600	−0.41600	−0.23400
$K_{M_{ra}}$	0.04000	0.16000	0.36000	0.64000

14. Externally applied angular displacement at a radius r_o

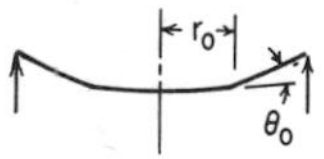

$LT_y = \theta_o r G_1$

$LT_\theta = \theta_o G_4$

$LT_M = \frac{\theta_o D}{r} G_7$

$LT_Q = 0$

14a. Simply supported

$y_a = 0 \qquad M_{ra} = 0 \qquad Q_a = 0$

$$y_c = \frac{-\theta_o r_o(1+\nu)}{2} \ln\frac{a}{r_o}$$

$$M_c = \frac{-\theta_o D(1-\nu^2)}{2r_o a^2}(a^2 - r_o^2)$$

$$\theta_a = \frac{\theta_o r_o}{a}$$

$y = K_y \theta_o a \qquad \theta = K_\theta \theta_o \qquad M = K_M \theta_o \frac{D}{a}$

r_o/a	0.2	0.4	0.6	0.8
K_{y_c}	−0.20923	−0.23824	−0.19922	−0.11603
K_{θ_a}	0.20000	0.40000	0.60000	0.80000
K_{M_c}	−2.18400	−0.95550	−0.48533	−0.20475
$K_{M_{to}}$	2.33600	1.31950	1.03133	0.93275

14b. Fixed

$y_a = 0 \qquad \theta_a = 0 \qquad Q_a = 0$

$$y_c = \frac{\theta_o r_o}{2}\left[1 - (1+\nu)\ln\frac{a}{r_o}\right]$$

$$M_c = \frac{-\theta_o D(1+\nu)}{a} L_4$$

$$M_{ra} = \frac{-\theta_o D r_o}{a^2}(1+\nu)$$

r_o/a	0.2	0.4	0.6	0.8	1.0
K_{y_c}	−0.10923	−0.03824	−0.10078	−0.28396	−0.50000
K_{M_c}	−2.44400	−1.47550	−1.26533	−1.24475	−1.30000
$K_{M_{ra}}$	−0.26000	−0.52000	−0.78000	−1.04000	−1.30000

TABLE 24 *Formulas for flat circular plates of constant thickness* (*Cont.*)

Case no., loading, load terms	Edge restraint	Boundary values	Special cases
15. Uniform temperature differential ΔT between the bottom and top surface from r_o to a 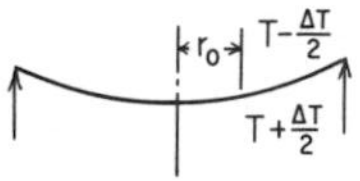	15a. Simply supported	$y_a = 0 \quad M_{ra} = 0 \quad Q_a = 0$ $y_c = \frac{-\gamma\,\Delta T}{2t}\left[a^2 - r_o^2 - r_o^2(1+\nu)\ln\frac{a}{r_o}\right]$ $M_c = \frac{\gamma D(1+\nu)\,\Delta T}{t}(1 - L_8)$ $\theta_a = \frac{\gamma\,\Delta T}{ta}(a^2 - r_o^2)$	$y = K_y\frac{\gamma\,\Delta T a^2}{t} \quad \theta = K_\theta\frac{\gamma\,\Delta T a}{t} \quad M = K_M\frac{\gamma\,\Delta T D}{t}$ (see table 15a below)
$LT_y = \frac{\gamma(1+\nu)\,\Delta T}{t}r^2G_2$ $LT_\theta = \frac{\gamma(1+\nu)\,\Delta T}{t}rG_5$ $LT_M = \frac{\gamma D(1+\nu)\,\Delta T}{t}(G_8 - 1)$ $LT_Q = 0$	15b. Fixed	$y_a = 0 \quad \theta_a = 0 \quad Q_a = 0$ $y_c = \frac{\gamma(1+\nu)\,\Delta T}{2t}r_o^2\ln\frac{a}{r_o}$ $M_c = \frac{-\gamma D(1+\nu)^2\,\Delta T}{2ta^2}(a^2 - r_o^2)$ $M_{ra} = \frac{-\gamma D(1+\nu)\,\Delta T}{ta^2}(a^2 - r_o^2)$	(see table 15b below)

Special cases, 15a:

r_o/a	0.0	0.2	0.4	0.6	0.8
K_{y_c}	−0.50000	−0.43815	−0.32470	−0.20047	−0.08717
K_{θ_a}	1.00000	0.96000	0.84000	0.64000	0.36000
K_{M_c}	0.00000	0.43680	0.38220	0.29120	0.16380
$K_{M_{to}}$	−0.00000	−0.47320	−0.52780	−0.61880	−0.74620

Special cases, 15b:

r_o/a	0.0	0.2	0.4	0.6	0.8
K_{y_c}	0.00000	0.04185	0.09530	0.11953	0.09283
K_{M_c}	−1.30000	−0.81120	−0.70980	−0.54080	−0.30420
$K_{M_{ra}}$	−1.30000	−1.24800	−1.09200	−0.83200	−0.46800
$K_{M_{to}}$	−1.30000	−1.72120	−1.61980	−1.45080	−1.21420

Note: The term $\frac{-\gamma(1-\nu^2)\,\Delta T D}{t}\langle r - r_o\rangle^0$ must be added to M_t for this case 15. Also, if $r_o = 0$, then $G_2 = \frac{1}{4}$, $G_5 = \frac{1}{2}$, $G_8 = (1+\nu)/2$, and $\langle r - r_o\rangle^0 = 1$ for all values of r.

Cases 16 to 31. The following cases include loadings on circular plates or plates bounded by some circular boundaries (each case is complete in itself) (*Note:* ln = natural logarithm)

Case no., loading, restraints	Formulas	Special cases
16. Uniform load over a very small central circular area of radius r_o; edge simply supported 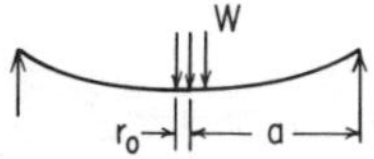 $W = q\pi r_o^2$	For $r > r_o$ $y = \frac{-W}{16\pi D}\left[\frac{3+\nu}{1+\nu}(a^2 - r^2) - 2r^2\ln\frac{a}{r}\right]$ $\theta = \frac{Wr}{4\pi D}\left(\frac{1}{1+\nu} + \ln\frac{a}{r}\right)$ $M_r = \frac{W}{16\pi}\left[4(1+\nu)\ln\frac{a}{r} + (1-\nu)\left(\frac{a^2 - r^2}{a^2}\right)\frac{r_o'^2}{r^2}\right]$ where $r_o' = \sqrt{1.6r_o^2 + t^2} - 0.675t$ if $r_o < 0.5t$ or $r_o' = r_o$ if $r_o > 0.5t$ $M_t = \frac{W}{16\pi}\left[4(1+\nu)\ln\frac{a}{r} + (1-\nu)\left(4 - \frac{r_o'^2}{r^2}\right)\right]$	$\text{Max } y = \frac{-Wa^2}{16\pi D}\frac{3+\nu}{1+\nu}$ at $r = 0$ $\text{Max } \theta = \frac{Wa}{4\pi D(1+\nu)}$ at $r = a$ $\text{Max } M_r = \frac{W}{4\pi}\left[(1+\nu)\ln\frac{a}{r_o'} + 1\right]$ at $r = 0$ $\text{Max } M_t = \text{max } M_r$ at $r = 0$

17. Uniform load over a very small central circular area of radius r_o; edge fixed

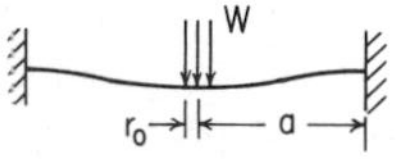

$W = q\pi r_o^2$

For $r > r_o$

$$y = \frac{-W}{16\pi D}\left[a^2 - r^2\left(1 + 2\ln\frac{a}{r}\right)\right]$$

$$\theta = \frac{Wr}{4\pi D}\ln\frac{a}{r}$$

$$M_r = \frac{W}{4\pi}\left[(1+\nu)\ln\frac{a}{r} - 1 + \frac{(1-\nu)r_o'^2}{4r^2}\right]$$

where $r_o' = \sqrt{1.6r_o^2 + t^2} - 0.675t$ if $r_o < 0.5t$

or $r_o' = r_o$ if $r_o > 0.5t$

$$M_t = \frac{W}{4\pi}\left[(1+\nu)\ln\frac{a}{r} - \nu + \frac{\nu(1-\nu)r_o'^2}{4r^2}\right]$$

Max $y = \dfrac{-Wa^2}{16\pi D}$ at $r = 0$

Max $\theta = 0.0293\dfrac{Wa}{D}$ at $r = 0.368a$

Max $+M_r = \dfrac{W}{4\pi}(1+\nu)\ln\dfrac{a}{r_o}$ at $r = 0$

Max $-M_r = \dfrac{-W}{4\pi}$ at $r = a$

Max $+M_t = \max +M_r$ at $r = 0$

Max $-M_t = \dfrac{-\nu W}{4\pi}$ at $r = a$

---

18. Uniform load over a small eccentric circular area of radius r_o; edge simply supported

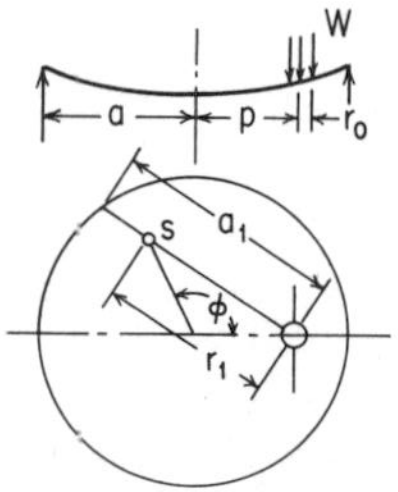

Note:

$r_o' = \sqrt{1.6r_o^2 + t^2} - 0.675t$ if $r_o < 0.5t$

or $r_o' = r_o$ if $r_o > 0.5t$

$$\text{Max } M_r = \text{max } M_t = \frac{W}{4\pi}\left[1 + (1+\nu)\ln\left(\frac{a-p}{r_o'}\right) - \frac{(1-\nu)r_o'^2}{4(a-p)^2}\right] \text{ at the load}$$

At any point s,

$$M_r = \text{max } M_r \frac{(1+\nu)\ln(a_1/r_1)}{1 + (1+\nu)\ln(a_1/r_o')} \qquad M_t = \text{max } M_t \frac{(1+\nu)\ln(a_1/r_1) + 1 - \nu}{1 + (1+\nu)\ln(a_1/r_o')}$$

$$y = K_o(r^3 - b_o a r^2 + c_o a^3) + K_1(r^4 - b_1 a r^3 + c_1 a^3 r)\cos\phi + K_2(r^4 - b_2 a r^3 + c_2 a^2 r^2)\cos 2\phi$$

where
$$K_o = \frac{W}{\pi D a^4}\frac{2(1+\nu)}{9(5+\nu)}(p^3 - b_o a p^2 + c_o a^3)$$

$$K_1 = \frac{W}{\pi D a^6}\frac{2(3+\nu)}{3(9+\nu)}(p^4 - b_1 a p^3 + c_1 a^3 p)$$

$$K_2 = \frac{W}{\pi D a^6}\frac{(4+\nu)^2}{(9+\nu)(5+\nu)}(p^4 - b_2 a p^3 + c_2 a^2 p^2)$$

where $b_o = \dfrac{3(2+\nu)}{2(1+\nu)}$ $\quad b_1 = \dfrac{3(4+\nu)}{2(3+\nu)}$ $\quad b_2 = \dfrac{2(5+\nu)}{4+\nu}$ $\quad c_o = \dfrac{4+\nu}{2(1+\nu)}$ $\quad c_1 = \dfrac{6+\nu}{2(3+\nu)}$ $\quad c_2 = \dfrac{6+\nu}{4+\nu}$

(See Ref. 1)

TABLE 24 *Formulas for flat circular plates of constant thickness* (*Cont.*)

Case no., loading, load terms	

19. Uniform load over a small eccentric circular area of radius r_o; edge fixed

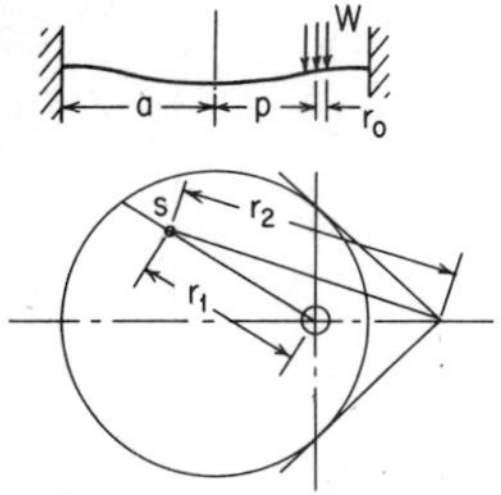

At any point s,

$$y = \frac{-W}{16\pi D}\left[\frac{p^2 r_2^2}{a^2} - r_1^2\left(1 + 2\ln\frac{p r_2}{a r_1}\right)\right]$$

At the load point,

$$y = \frac{-W}{16\pi D}\frac{(a^2 - p^2)^2}{a^2}$$

$$M_r = \frac{W(1+\nu)}{16\pi}\left[4\ln\left(\frac{a-p}{r'_o}\right) + \left(\frac{r'_o}{a-p}\right)^2\right] = \max M \text{ if } r_o < 0.6(a-p)$$

(*Note:* r'_o defined in case 18)

At the near edge,

$$M_r = \frac{-W}{8\pi}\left[2 - \left(\frac{r'_o}{a-p}\right)^2\right] = \max, M \text{ if } r_o > 0.6(a-p)$$

[Formulas due to Michell (Ref. 2). See Ref. 60 for modified boundary conditions]

20. Central couple on a simply supported plate (trunnion loading)

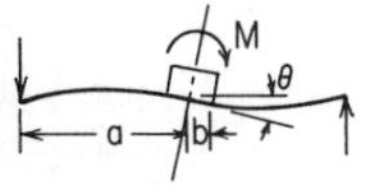

At the inner edge,

$$r = b \qquad \max \sigma_r = \frac{\beta M}{a t^2} \qquad \theta = \frac{\alpha M}{E t^3}$$

b/a	0.10	0.15	0.20	0.25	0.30	0.40	0.50	0.60	0.70	0.80
β	9.478	6.252	4.621	3.625	2.947	2.062	1.489	1.067	0.731	0.449
α	1.403	1.058	0.820	0.641	0.500	0.301	0.169	0.084	0.035	0.010

(Ref. 22)

21. Central couple on a plate with a fixed edge (trunnion loading)

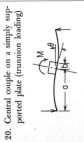

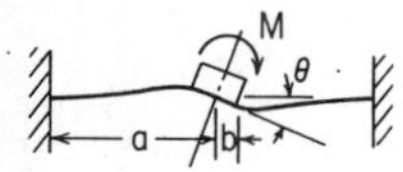

At the inner edge,

$$r = b \qquad \max \sigma_r = \frac{\beta M}{a t^2} \qquad \theta = \frac{\alpha M}{E t^3}$$

At the outer edge,

$$r = a \qquad \max \sigma_r = \frac{\beta r_o M}{a^2 t^2}$$

b/a	0.10	0.15	0.20	0.25	0.30	0.40	0.50	0.60	0.70	0.80
β	9.36	6.08	4.41	3.37	2.66	1.73	1.146	0.749	0.467	0.262
α	1.149	0.813	0.595	0.439	0.320	0.167	0.081	0.035	0.013	0.003

(Ref. 21)

22. Linearly distributed load symmetrical about a diameter; edge simply supported

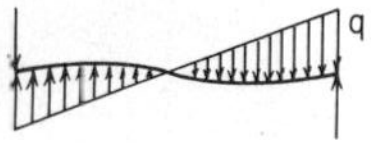

$$\text{Max } M_r = \frac{qa^2(5+\nu)}{72\sqrt{3}} \qquad \text{at } r = 0.577a$$

$$\text{Max } M_t = \frac{qa^2(5+\nu)(1+3\nu)}{72(3+\nu)} \qquad \text{at } r = 0.675a$$

$$\text{Max edge reaction per linear inch} = \frac{qa}{4}$$

$$\text{Max } y = 0.042\frac{qa^4}{Et^3} \qquad \text{at } r = 0.503a \text{ (for } \nu = 0.3)$$

(Refs. 20 and 21)

---

23. Central couple balanced by linearly distributed pressure (footing)

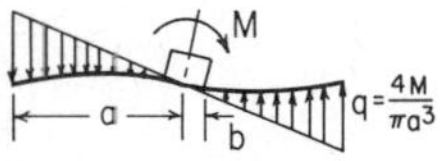

(At inner edge) max $\sigma_r = \beta \dfrac{M}{at^2}$ where β is given in the following table:

a/b	1.25	1.50	2.00	3.00	4.00	5.00
β	0.1625	0.4560	1.105	2.250	3.385	4.470

(Values for $\nu = 0.3$)

(Ref. 21)

---

24. Concentrated load applied at the outer edge of an annular plate with a fixed inner edge

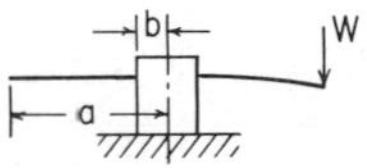

(At inner edge) max $\sigma_r = \beta \dfrac{W}{t^2}$ where β is given in the following table:

a/b	1.25	1.50	2.00	3.00	4.00	5.00
β	3.7	4.25	5.2	6.7	7.9	8.8

(Values for $\nu = 0.3$)

(Refs. 21 and 22)

(See Ref. 64 for this loading on a plate with radially varying thickness. See graphs in Ref. 59 for the load distributed over an arc at the edge. See Ref. 60 for the load W placed away from the edge)

---

25. Solid circular plate with a uniformly distributed load q over the shaded segment

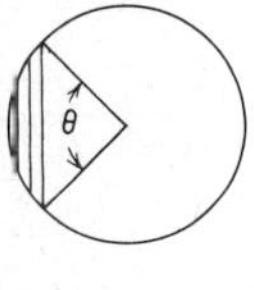

$$\text{Max } \sigma = \text{max } \sigma_r = \beta \frac{qa^2}{t^2}$$

$\text{Max } y = \alpha \dfrac{qa^4}{Et^3}$ on the symmetrical diameter at the value of r given in the table

Edge	Coefficient	θ: 90°	θ: 120°	θ: 180°
Supported	α	0.0244, $r = 0.39a$	0.0844, $r = 0.30a$	0.345, $r = 0.15a$
	β	0.306, $r = 0.60a$		
Fixed	α	0.00368, $r = 0.50a$	0.0173, $r = 0.4a$	0.0905, $r = 0.20a$
	β	0.285, $r = a$		

Values for $\nu = \frac{1}{3}$

(Ref. 39)

TABLE 24 *Formulas for flat circular plates of constant thickness* (*Cont.*)

Case no., loading, load terms

26. Solid circular plate, uniform load q over the shaded sector

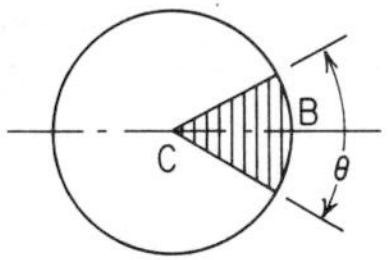

For simply supported edges:

Max $\sigma = \sigma_r$ near the center along the loaded radius of symmetry (values not given)

σ_r at the center $= \dfrac{\theta}{360}\sigma_r$ at the center of a fully loaded plate

Max $y = \alpha_1 \dfrac{qa^4}{Et^3}$ at approximately $\frac{1}{4}$ the radius from center along the radius of symmetry (α_1 given in table)

For fixed edges:

Max $\sigma = \sigma_r$ at point $B = \beta \dfrac{qa^2}{t^2}$

Max $y = \alpha_2 \dfrac{qa^4}{Et^3}$ at approximately $\frac{1}{4}$ the radius from center along the radius of symmetry (β and α_2 given in table)

Edge condition	Coefficient	θ 30°	60°	90°	120°	150°	180°
Simply supported	α_1	0.061	0.121	0.179	0.235	0.289	0.343
Fixed	α_2	0.017	0.034	0.050	0.064	0.077	0.089
	β	0.240	0.371	0.457	0.518	0.564	0.602

[*Note:* For either edge condition $y_c = (\theta/360)y_c$ for a fully loaded plate]

(Ref. 38)

27. Solid circular sector, uniformly distributed load q over the entire surface; edges simply supported

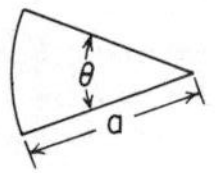

Max $\sigma_r = \beta \dfrac{qa^2}{t^2}$ max $\sigma_t = \beta_1 \dfrac{qa^2}{t^2}$ max $y = \alpha \dfrac{qa^4}{Et^3}$

θ	45°	60°	90°	180°
β	0.102	0.147	0.240	0.522
β_1	0.114	0.155	0.216	0.312
α	0.0054	0.0105	0.0250	0.0870

(Values for $\nu = 0.3$)

(Ref. 21)

28. Solid circular sector, uniformly distributed load q over the entire surface; straight edges simply supported, curved edge fixed

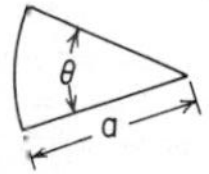

Max $\sigma = \sigma_r$ at curved edge $= \beta \dfrac{qa^2}{t^2}$ Max $y = \alpha \dfrac{qa^4}{Et^3}$

θ	45°	60°	90°	180°
β	0.1500	0.2040	0.2928	0.4536
α	0.0035	0.0065	0.0144	0.0380

(Values for $\nu = 0.3$)

(Ref. 21)

29. Solid circular sector of infinite radius, uniformly distributed load q over entire surface; straight edges fixed

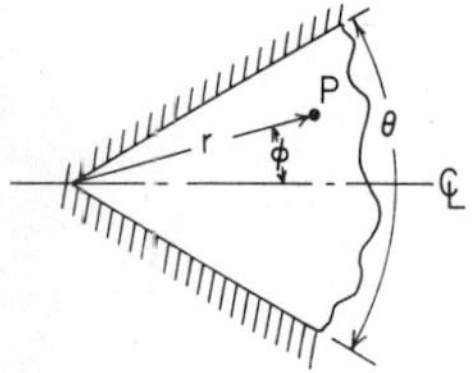

At point P:

$$\sigma_r = \frac{9qr^2}{8t^2}\left[\frac{3+\nu}{3} - \frac{4\cos\theta\cos 2\phi - (1-\nu)\cos 4\phi}{2\cos^2\theta + 1}\right]$$

$$\sigma_t = \frac{9qr^2}{8t^2}\left[\frac{1+3\nu}{3} - \frac{4\nu\cos\theta\cos 2\phi + (1-\nu)\cos 4\phi}{2\cos^2\theta + 1}\right]$$

$$y = \frac{-3(1-\nu^2)qr^4}{16Et^3}\left(1 + \frac{\cos 4\phi - 4\cos\theta\cos 2\phi}{2\cos^2\theta + 1}\right)$$

(*Note:* θ should not exceed 60°)

At the edge, $\phi = \pm\theta/2$:

$$\sigma_t = \frac{3qr^2}{2t^2}\,\frac{\sin^2\theta}{1 + 2\cos^2\theta}$$

$$\sigma_r = \nu\sigma_t$$

Along the center line, $\phi = 0$:

$$\sigma_r = \frac{3qr^2}{4t^2}\,\frac{3(1-\cos\theta)^2 - \nu\sin^2\theta}{1 + 2\cos^2\theta}$$

$$\sigma_t = \frac{3qr^2}{4t^2}\,\frac{3\nu(1-\cos\theta)^2 - \sin^2\theta}{1 + 2\cos^2\theta}$$

$$y = \frac{-3(1-\nu^2)qr^4}{8Et^3}\,\frac{(1-\cos\theta)^2}{1 + 2\cos^2\theta}$$

(Ref. 37)

30. Solid semicircular plate, uniformly distributed load q over the entire surface; all edges fixed

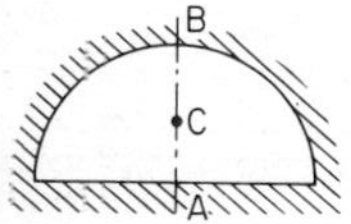

Max $\sigma = \sigma_r$ at $A = -0.42 \dfrac{qa^2}{t^2}$ (values for $\nu = 0.2$)

σ_r at $B = -0.36 \dfrac{qa^2}{t^2}$

σ_r at $C = 0.21 \dfrac{qa^2}{t^2}$

(Ref. 40)

TABLE 24 ***Formulas for flat circular plates of constant thickness*** ***(Cont.)***

Case no., loading, load terms	

31. Semicircular annular plate, uniformly loaded over entire surface; outer edge supported, all other edges free

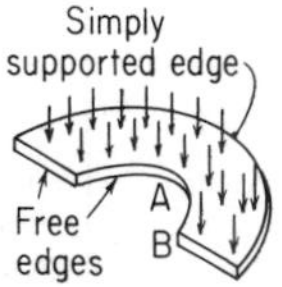

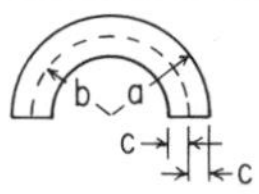

(At A) $$\sigma_t = \frac{6qcb}{t^2}\left(\frac{b}{c} - \frac{1}{3}\right)\left[c_1\left(1 - \gamma_1^2\frac{c}{b}\right) + c_2\left(1 - \gamma_2^2\frac{c}{b}\right) + \frac{c}{b}\right]K \quad (= \text{max stress})$$

(At B) $$y = \frac{24qc^2b^2}{Et^3}\left(\frac{b}{c} - \frac{1}{3}\right)\left[c_1 \operatorname{Cosh}\frac{\gamma_1\pi}{2} + c_2 \operatorname{Cosh}\frac{\gamma_2\pi}{2} + \frac{c}{b}\right] \quad (= \text{max deflection})$$

where $$c_1 = \frac{1}{\left(\frac{b}{c} - \gamma_1^2\right)(\lambda - 1)\operatorname{Cosh}\frac{\gamma_1\pi}{2}} \qquad c_2 = \frac{1}{\left(\frac{b}{c} - \gamma_2^2\right)\left(\frac{1}{\lambda} - 1\right)\operatorname{Cosh}\frac{\gamma_2\pi}{2}}$$

$$\gamma_1 = \frac{\gamma}{\sqrt{2}}\sqrt{1 + \sqrt{1 - \frac{4b^2}{c^2\gamma^4}}}, \qquad \gamma_2 = \frac{\gamma}{\sqrt{2}}\sqrt{1 - \sqrt{1 - \frac{4b^2}{c^2\gamma^4}}}$$

$$\gamma = \sqrt{\frac{2b}{c} + 4\left(1 - \frac{0.625t}{2c}\right)\frac{G}{E}\left(1 + \frac{b}{c}\right)^2} \qquad \lambda = \frac{\gamma_1\left(\frac{b}{c} - \gamma_1^2 + \lambda_1\right)\left(\frac{b}{c} - \gamma_2^2\right)\operatorname{Tanh}\frac{\gamma_1\pi}{2}}{\gamma_2\left(\frac{b}{c} - \gamma_2^2 + \lambda_1\right)\left(\frac{b}{c} - \gamma_1^2\right)\operatorname{Tanh}\frac{\gamma_2\pi}{2}} \qquad \lambda_1 = 4\left(1 - \frac{0.625t}{2c}\right)\frac{G}{E}\left(1 + \frac{b}{c}\right)^2$$

K = function of $(b - c)/(b + c)$ and has values as follows:

$(b - c)/(b + c)$	0.05	0.10	0.2	0.3	0.4	0.5	0.6	0.7	0.8	0.9	1.0
K	2.33	2.20	1.95	1.75	1.58	1.44	1.32	1.22	1.13	1.06	1.0

[Formulas due to Wahl (Ref. 10)]

32. Elliptical plate, uniformly distributed load q over entire surface 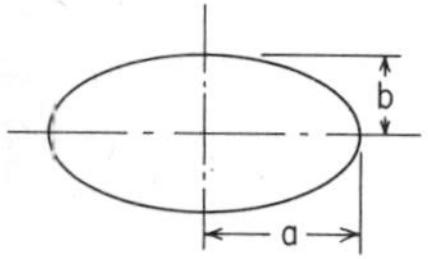 $\alpha = \frac{b}{a}$	32a. Simply supported	At the center: $\text{Max } \sigma = \sigma_b = -[2.816 + 1.581\nu - (1.691 + 1.206\nu)\alpha]\frac{qb^2}{t^2}$ $\text{Max } y = -[2.649 + 0.15\nu - (1.711 + 0.75\nu)\alpha]\frac{qb^4(1 - \nu^2)}{Et^3}$ [Approximate formulas for $0.2 < \alpha < 1.0$ (see numerical data in Refs. 21 and 56)]
	32b. Fixed	At the edge of span b: $\max \sigma = \sigma_b = \frac{6qb^2}{t^2(3 + 2\alpha^2 + 3\alpha^4)}$ At the edge of span a: $\sigma_a = \frac{6qb^2\alpha^2}{t^2(3 + 2\alpha^2 + 3\alpha^4)}$ At the center: $\sigma_b = \frac{-3qb^2(1 + \nu\alpha^2)}{t^2(3 + 2\alpha^2 + 3\alpha^4)} \qquad \sigma_a = \frac{-3qb^2(\alpha^2 + \nu)}{t^2(3 + 2\alpha^2 + 3\alpha^4)} \qquad \max y = \frac{-3qb^4(1 - \nu^2)}{2Et^3(3 + 2\alpha^2 + 3\alpha^4)}$ (Ref. 5)
33. Elliptical plate, uniform load over a small concentric circular area of radius r_o (note definition of r'_o in case 18) 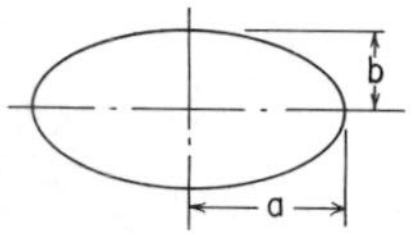 $\alpha = \frac{b}{a}$	33a. Simply supported	At the center: $\text{Max } \sigma = \sigma_b = \frac{-3W}{2\pi t^2}\left[(1 + \nu)\ln\frac{b}{r'_o} + \nu(6.57 - 2.57\alpha)\right]$ $\text{Max } y = \frac{-Wb^2}{Et^3}(0.76 - 0.18\alpha) \qquad \text{for } \nu = 0.25$ [Approximate formulas by interpolation between cases of circular plate and infinitely long narrow strip (Ref. 4)]
	33b. Fixed	At the center: $\sigma_b = \frac{-3W(1 + \nu)}{2\pi t^2}\left(\ln\frac{2b}{r'_o} - 0.317\alpha - 0.376\right)$ $\text{Max } y = \frac{-Wb^2}{Et^3}(0.326 - 0.104\alpha) \qquad \text{for } \nu = 0.25$ [Approximate formulas by interpolation between cases of circular plate and infinitely long narrow strip (Ref. 6).]

TABLE 25 ***Shear deflections for flat circular plates of constant thickness***

NOTATION: y_{sb}, y_{sa}, and y_{sr_o} are the deflections at b, a, and r_o, respectively, caused by transverse shear stresses. K_{sb}, K_{sa}, and K_{sr_o} are deflection coefficients defined by the relationships $y_s = K_s wa/tG$ for an annular line load and $y_s = K_s qa^2/tG$ for all distributed loadings (See Table 24 for all other notation and for the loading cases referenced)

Case no.	Shear deflection coefficients	Tabulated values for specific cases					
1a, 1b, 1e, 1f, 9	$K_{sr_o} = K_{sb} = -1.2\frac{r_o}{a}\ln\frac{a}{r_o}$ (*Note:* $r_o > 0$)	r_o/a	0.1	0.3	0.5	0.7	0.9
		K_{sb}	−0.2763	−0.4334	−0.4159	−0.2996	−0.1138
2a, 2b, 2e, 2f, 10	$K_{sr_o} = K_{sb} = -0.30\left[1 - \left(\frac{r_o}{a}\right)^2\left(1 + 2\ln\frac{a}{r_o}\right)\right]$	K_{sb}	−0.2832	−0.2080	−0.1210	−0.0481	−0.0058
3a, 3b, 3e, 3f, 11	$K_{sr_o} = K_{sb} = \frac{-a}{30(a - r_o)}\left[4 - 9\frac{r_o}{a} + \left(\frac{r_o}{a}\right)^3\left(5 + 6\ln\frac{a}{r_o}\right)\right]$	K_{sb}	−0.1155	−0.0776	−0.0430	−0.0166	−0.0019
4a, 4b, 4e, 4f, 12	$K_{sr_o} = K_{sb} = \frac{-a^2}{120(a - r_o)^2}\left[9 - 32\frac{r_o}{a} + 36\left(\frac{r_o}{a}\right)^2 - \left(\frac{r_o}{a}\right)^4\left(13 + 12\ln\frac{a}{r_o}\right)\right]$	K_{sb}	−0.0633	−0.0411	−0.0223	−0.0084	−0.00098
1i, 1j, 1k, 1l	$K_{sr_o} = K_{sa} = -1.2\frac{r_o}{a}\ln\frac{r_o}{b}$ (*Note:* $b > 0$)	r_o/a	0.2	0.4	0.6	0.8	1.0
		b/a					
		0.1	−0.1664	−0.6654	−1.2901	−1.9963	−2.7631
		0.3		−0.1381	−0.4991	−0.9416	−1.4448
		0.5			−0.1313	−0.4512	−0.8318
		0.7	Values of K_{sa}			−0.1282	−0.4280
		0.9					−0.1264
2i, 2j, 2k, 2l	$K_{sr_o} = -0.60\left[1 - \left(\frac{r_o}{a}\right)^2\right]\ln\frac{r_o}{b}$ (*Note:* $b > 0$)	r_o/a	0.1	0.3	0.5	0.7	0.9
		b/a					
		0.1	−0.0000	−0.5998	−0.7242	−0.5955	−0.2505
		0.3		−0.0000	−0.2299	−0.2593	−0.1252
		0.5			−0.0000	−0.1030	−0.0670
		0.7	Values of K_{sr_o}			−0.0000	−0.0287
		0.9					−0.0000
	$K_{sa} = -0.30\left[2\ln\frac{a}{b} - 1 + \left(\frac{r_o}{a}\right)^2\left(1 - 2\ln\frac{r_o}{b}\right)\right]$ (*Note:* $b > 0$)	0.1	−1.0846	−1.0493	−0.9151	−0.6565	−0.2567
		0.3		−0.4494	−0.4208	−0.3203	−0.1315
		0.5			−0.1909	−0.1640	−0.0732
		0.7	Values of K_{sa}			−0.0610	−0.0349
		0.9					−0.0062

3i, 3j, 3k, 3l

$$K_{sr_o} = -0.20\left[2 - \frac{r_o}{a} - \left(\frac{r_o}{a}\right)^2\right]\ln\frac{r_o}{b} \qquad (\textit{Note: } b > 0)$$

Values of K_{sr_o}

b/a \ r_o/a	0.1	0.3	0.5	0.7	0.9
0.1	−0.0000	−0.3538	−0.4024	−0.3152	−0.1274
0.3		−0.0000	−0.1277	−0.1373	−0.0637
0.5			−0.0000	−0.0545	−0.0341
0.7				−0.0000	−0.0146
0.9					−0.0000

$$K_{sa} = \frac{-a}{30(a - r_o)}\left[6\left(2 - 3\frac{r_o}{a}\right)\ln\frac{a}{b} - 4 + 9\frac{r_o}{a} - \left(\frac{r_o}{a}\right)^3\left(5 - 6\ln\frac{r_o}{b}\right)\right] \qquad (\textit{Note: } b > 0)$$

Values of K_{sa}

b/a \ r_o/a	0.1	0.3	0.5	0.7	0.9
0.1	−0.7549	−0.6638	−0.5327	−0.3565	−0.1316
0.3		−0.3101	−0.2580	−0.1785	−0.0679
0.5			−0.1303	−0.0957	−0.0383
0.7				−0.0412	−0.0187
0.9					−0.0042

4i, 4j, 4k, 4l

$$K_{sr_o} = -0.10\left[3 - 2\frac{r_o}{a} - \left(\frac{r_o}{a}\right)^2\right]\ln\frac{r_o}{b} \qquad (\textit{Note: } b > 0)$$

Values of K_{sr_o}

b/a \ r_o/a	0.1	0.3	0.5	0.7	0.9
0.1	−0.0000	−0.2538	−0.2817	−0.2160	−0.0857
0.3		−0.0000	−0.0894	−0.0941	−0.0428
0.5			−0.0000	−0.0373	−0.0229
0.7				−0.0000	−0.0098
0.9					−0.0000

$$K_{sa} = \frac{-a^2}{120(a - r_o)^2}\left\{12\left[3 - 8\frac{r_o}{a} + 6\left(\frac{r_o}{a}\right)^2\right]\ln\frac{a}{b} - 9 + 32\frac{r_o}{a} - 36\left(\frac{r_o}{a}\right)^2 + \left(\frac{r_o}{a}\right)^4\left(13 - 12\ln\frac{r_o}{b}\right)\right\} \qquad (\textit{Note: } b > 0)$$

Values of K_{sa}

b/a \ r_o/a	0.1	0.3	0.5	0.7	0.9
0.1	−0.5791	−0.4908	−0.3807	−0.2472	−0.0888
0.3		−0.2370	−0.1884	−0.1252	−0.0460
0.5			−0.0990	−0.0685	−0.0261
0.7				−0.0312	−0.0129
0.9					−0.0031

10.3 Circular-plate deflection due to shear

The formulas for deflection given in Table 24 take into account bending stresses only; there is, in every case, some additional deflection due to shear. Usually this is so slight as to be negligible, but in circular pierced plates with large openings the deflection due to shear may constitute a considerable portion of the total deflection. Wahl (Ref. 19) suggests that this is the case when the thickness is greater than one-third the difference in inner and outer diameters for plates with simply supported edges, or greater than one-sixth this difference for plates with one or both edges fixed.

Table 25 gives formulas for the additional deflection due to shear in which the form factor F has been taken equal to 1.2, as in Art. 7.10. All of the cases listed have shear forces which are statically determinate. For the *indeterminate* cases, the shear deflection, along with the bending deflection, must be considered in the determination of the reactions if shear deflection is significant.

Essenburg and Gulati (Ref. 61) discuss the problem in which two plates when loaded touch over a portion of the surface. They indicate that the consideration of shear deformation is essential in developing the necessary expressions. Two examples are worked out.

EXAMPLE

An annular plate with an inner radius of 1.4 in, an outer radius of 2 in, and a thickness of 0.50 in is simply supported at the inner edge and loaded with an annular line load of 800 lb/in at a radius of 1.8 in. The deflection of the free outer edge is desired. Given: $E = 18(10^6)$ lb/in^2 and $\nu = 0.30$.

Solution. In order to evaluate the deflection due to bending one can refer to Table 24, case 1k. Since $b/a = 0.7$, in Table 23, under the column headed 0.700, we obtain the following constants:

$$C_1 = 0.2898 \qquad C_3 = 0.003753 \qquad C_7 = 0.3315 \qquad C_9 = 0.2248$$

Similarly, $r_o/a = 0.9$, and again in Table 23, under the column headed 0.900, we obtain the additional constants $L_3 = 0.0001581$ and $L_9 = 0.09156$.

The plate constant $D = Et^3/12(1 - \nu^2) = 18(10^6)(0.5)^3/12(1 - 0.3^2) = 206{,}000$, and the shear modulus $G = E/2(1 + \nu) = 18(10^6)/2(1 + 0.3) = 6.92(10^6)$ lb/in^2. The bending deflection of the outer edge is given by

$$y_a = \frac{-wa^3}{D}\left[\frac{C_1}{C_7}\left(\frac{r_oC_9}{b} - L_9\right) - \frac{r_oC_3}{b} + L_3\right]$$

or

$$y_a = \frac{-800(2)^3}{206{,}000}\left\{\frac{0.2898}{0.3315}\left[\frac{1.8(0.2248)}{1.4} - 0.09156\right] - \frac{1.8(0.003753)}{1.4} + 0.0001581\right\}$$

$$= \frac{-800(2)^3}{206{,}000}(0.16774) = -0.00521 \text{ in}$$

For the deflection due to shear we refer to Table 25, case 1k, and obtain

$$y_{sa} = \frac{-wa}{tG}\left(1.2\frac{r_o}{a}\ln\frac{r_o}{b}\right) = \frac{800(2)}{0.5(6.92)(10^6)}\left[1.2(0.9)\ln\frac{1.8}{1.4}\right] = -0.000125 \text{ in}$$

Thus, the total deflection of the outer edge is $-0.00521 - 0.000125 = -0.00534$ in. Note that the thickness 0.50 is somewhat more than one-third the difference in inner and outer diameters 1.2, and the shear deflection is only 2.4 percent of the bending deflection.

10.4 *Bimetallic circular plates*

The expressions in Table 24 for cases of axisymmetric loading can be used for bimetallic plates with the following equivalent values of the plate stiffness constant D and Poisson's ratio ν:

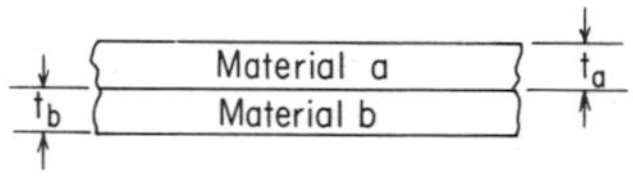

Fig. 10.2

$$\text{Equivalent } D_e = \frac{E_a t_a^3}{12(1-\nu_a^2)} K_{2p}$$

where

$$K_{2p} = 1 + \frac{E_b t_b^3 (1-\nu_a^2)}{E_a t_a^3 (1-\nu_b^2)} + \frac{3(1-\nu_a^2)(1+t_b/t_a)^2(1+E_a t_a/E_b t_b)}{(1+E_a t_a/E_b t_b)^2 - (\nu_a + \nu_b E_a t_a/E_b t_b)^2}$$

$$\text{Equivalent } \nu_e = \nu_a \frac{K_{3p}}{K_{2p}}$$

where

$$K_{3p} = 1 + \frac{\nu_b E_b t_b^3 (1-\nu_a^2)}{\nu_a E_a t_a^3 (1-\nu_a^2)} + \frac{3(1-\nu_a^2)(1+t_b/t_a)^2(1+\nu_b E_a t_a/\nu_a E_b t_b)}{(1+E_a t_a/E_b t_b)^2 - (\nu_a + \nu_b E_a t_a/E_b t_b)^2}$$

A bimetallic plate deforms laterally into a spherical surface when its uniform temperature differs from T_o, the temperature at which the plate is flat. Cases 8 and 15 (Table 24) can be used to solve for reaction moments and forces as well as the deformations of a bimetallic plate subjected to a uniform temperature T provided that any guided and/or fixed edges are not capable of developing in-plane resisting forces but instead allow the plate to expand or contract in its plane as necessary. To use these cases we need only to replace the term $\gamma(1+\nu)\Delta T/t$ by

$$\frac{6(\gamma_b - \gamma_a)(T - T_o)(t_a + t_b)(1+\nu_e)}{t_b^2 K_{1p}}$$

where

$$K_{1p} = 4 + 6\frac{t_a}{t_b} + 4\left(\frac{t_a}{t_b}\right)^2 + \frac{E_a t_a^3 (1-\nu_b)}{E_b t_b^3 (1-\nu_a)} + \frac{E_b t_b (1-\nu_a)}{E_a t_a (1-\nu_b)}$$

and replace D by the equivalent stiffness D_e given previously.

After the moments and deformations have been determined, the flexural stresses can be evaluated. The stresses due to the bending moments caused by restraints and any applied loads are given by the following expressions: In the top surface of material a, in the direction of any moment M

$$\sigma = \frac{-6M}{t_a^2 K_{2p}}\left[1 + \frac{(1-\nu_a^2)(1+t_b/t_a)(1+E_a t_a/E_b t_b)}{(1+E_a t_a/E_b t_b)^2 - (\nu_a + \nu_b E_a t_a/E_b t_b)^2}\right]$$

In the bottom surface of material b,

$$\sigma = \frac{6M}{t_a^2 K_{2p}}\left[\frac{E_b t_b(1-\nu_a^2)}{E_a t_a(1-\nu_b^2)} + \frac{t_a}{t_b}\,\frac{(1-\nu_a^2)(1+t_b/t_a)(1+E_a t_a/E_b t_b)}{(1+E_a t_a/E_b t_b)^2 - (\nu_a + \nu_b E_a t_a/E_b t_b)^2}\right]$$

Even when no restraints are imposed, the distortion of a bimetallic plate due to a temperature change is accompanied by flexural stresses in the two materials. This differs from the plate made of a single material, which deforms free of stress when subjected to a linear temperature variation through the thickness when there are no restraints. Therefore, the following stresses must be added algebraically to the preceding stresses due to bending moments, if any: In the top surface of material a, in all directions

$$\sigma = \frac{-(\gamma_b - \gamma_a)(T - T_o)E_a}{(1-\nu_a)K_{1p}}\left[3\frac{t_a}{t_b} + 2\left(\frac{t_a}{t_b}\right)^2 - \frac{E_b t_b(1-\nu_a)}{E_a t_a(1-\nu_b)}\right]$$

In the bottom surface of material b,

$$\sigma = \frac{(\gamma_b - \gamma_a)(T - T_o)E_b}{(1-\nu_b)K_{1p}}\left[3\frac{t_a}{t_b} + 2 - \frac{E_a t_a^3(1-\nu_b)}{E_b t_b^3(1-\nu_a)}\right]$$

EXAMPLE

An annular bimetallic plate has a 3-in outer diameter and a 2.1-in inner diameter. The top portion is 0.020-in-thick stainless steel, and the bottom is 0.030-in-thick titanium. For the stainless steel $E = 28(10^6)$ lb/in^2, $\nu = 0.3$, and $\gamma = 9.6(10^{-6})$ in/in/°F; and for the titanium $E = 17(10^6)$ lb/in^2, $\nu = 0.3$, and $\gamma = 5.7(10^{-6})$ in/in/°F. The outer edge is simply supported, and the inner edge is elastically supported by a spring which develops 500 lb of load for each inch of deflection. It is necessary to determine the center deflection and the maximum stress for a temperature rise of 50°F.

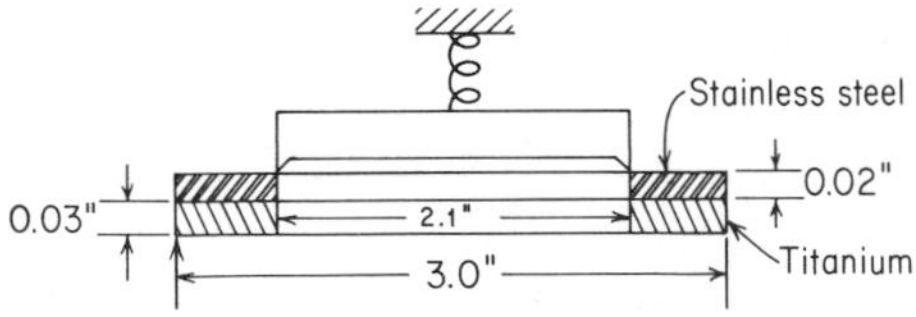

Fig. 10.3

Solution. First evaluate the constants K_{1p}, K_{2p}, and K_{3p}, the equivalent stiffness D_e, and the equivalent Poisson's ratio ν_e:

$$K_{1p} = 4 + 6\frac{0.02}{0.03} + 4\left(\frac{2}{3}\right)^2 + \frac{28}{17}\left(\frac{2}{3}\right)^3\left(\frac{1-0.3}{1-0.3}\right) + \frac{17}{28}\left(\frac{3}{2}\right)\left(\frac{1-0.3}{1-0.3}\right)$$

$$= 11.177$$

Since $\nu_a = \nu_b$ for this example, $K_{3p} = K_{2p} = 11.986$ and the equivalent Poisson's ratio $\nu_e = 0.3$. Therefore

$$D_e = \frac{28(10^6)(0.02^3)}{12(1-0.3^2)}(11.986) = 246 \text{ in-lb}$$

Table 24, case 8a, treats an annular plate with the inner edge free and the outer edge simply supported. The term $\gamma\Delta T/t$ must be replaced by

$$\frac{6(\gamma_b - \gamma_a)(T - T_o)(t_a + t_b)}{t_b^2 K_{1p}} = \frac{6(5.7 - 9.6)(10^{-6})(50)(0.02 + 0.03)}{(0.03^2)(11.177)} = -0.00582$$

Since $b/a = 1.05/1.5 = 0.7$ and $\nu_e = 0.3$, the tabulated data can be used and $K_{yb} = -0.255$ and $K_{\theta b} = 0.700$. Therefore, $y_b = -0.255(-0.00582)(1.5^2) = 0.00334$ in and $\theta_b = 0.7(-0.00582)(1.5) = -0.0061$ rad. There are no moments or edge loads in the plate, and so $M_{rb} = 0$, and $Q_b = 0$. Case 1a treats an annular plate with an annular line load. For $r_o = b$ and $b/a = 0.7$, $K_{yb} = -0.1927$ and $K_{\theta b} = 0.6780$. Therefore, $y_b = -0.1927w(1.5^3)/246 = -0.002645w$, $\theta_b = -0.678w(1.5^2)/246 = 0.0062w$ rad, $M_{rb} = 0$, and $Q_b = 0$.

Equating the deflection of the inner edge of the plate to the deflection of the elastic support gives $y_b = 0.00334 - 0.002645w = 2\pi(1.05)w/500 = 0.0132w$. Solving for w, we obtain $w = 0.211$ lb/in for a total center load of 1.39 lb. The deflection of the inner edge is $y_b = 0.0132(0.211) = 0.00279$ in. The maximum moment developed in the plate is the tangential moment at the inner edge: $M_{tb} = 0.8814(0.211)(1.5) = 0.279$ in-lb/in. The stresses can now be computed. On the top surface of the stainless steel

$$\sigma = \frac{-6(0.279)}{0.02^2(11.986)}\left\{1 + \frac{(1-0.3^2)(1+3/2)[1+28(2)/17(3)]}{[1+28(2)/17(3)]^2 - [0.3 + 0.3(28)(2)/17(3)]^2}\right\}$$
$$- \frac{(5.7-9.6)(10^{-6})(50)(28)(10^6)}{(1-0.3)(11.177)}\left[3\left(\frac{2}{3}\right) + 2\left(\frac{2}{3}\right)^2 - \frac{17}{28}\left(\frac{3}{2}\right)\right]$$

$$= -765 + 1381 = 616 \text{ lb/in}^2$$

Similarly, on the bottom surface of the titanium

$$\sigma = 595 - 1488 = -893 \text{ lb/in}^2$$

10.5 *Nonuniform loading of circular plates*

The case of a circular plate under a nonuniformly distributed loading symmetrical about the center can be solved by treating the load as a series of elementary ring loadings and summing the stresses and deflections produced by such loadings. The number of ring loadings into which the actual load should be resolved depends upon the rate at which the distributed load varies along the radius and the accuracy desired. In general, a division of the load into rings each having a width equal to one-fifth the loaded length of the radius should be sufficient.

If the nonuniformly distributed loading can be reasonably approximated by a second-order curve, the loadings in Table 24, cases 2 to 4 can be superimposed in the necessary proportions. (This technique is illustrated in Art. 10.6.) Heap (Ref. 48) gives tabular data for circular plates loaded with a lateral pressure varying inversely with the square of the radius.

Concentrated loads. In Refs. 60 and 75 to 79 similar numerical techniques are discussed for concentrated loads on either of two concentric annular plates in combination with edge beams in some cases. The numerical data presented are limited but are enough to enable the reader to approximate many other cases.

10.6 *Circular plates on elastic foundations*

Discussions of the theory of bending of circular plates on elastic foundations can be found in Refs. 21 and 46, and in Ref. 41 of Chap. 7. The complexity of these solutions prohibits their inclusion in this handbook, but a simple *iterative* approach to this problem is possible. The procedure consists in evaluating the deflection of the loaded plate without the elastic foundation and then superimposing a given fraction of the foundation reaction resulting from this deflection until finally the given fraction increases to 1 and the assumed and calculated foundation reactions are equal.

EXAMPLE

Given the same problem stated in Example 1 of Art. 10-2 (page 326), but in addition to the simply supported edge an elastic foundation with a modulus of 20 lb/in²/in is present under the entire plate.

Solution. An examination of the deflection equation resulting from the uniform load shows that the term involving r^4 is significant only near the outer edge where the effect of foundation pressure would not be very large. We must also account for the fact that the foundation reactions will reduce the plate deflections or the procedure described may not converge. Therefore, for a first trial let us assume that the foundation pressure is given by

$$q_f = 20(-0.0883 + 0.001098r^2)(0.50) = -0.883 + 0.01098r^2$$

The total loading on the plate then consists of a uniform load of $3 - 0.883 = 2.117$ lb/in² and a parabolically increasing load of 1.098 lb/in² maximum value. From Table 24, case 10a,

$$y_c = \frac{-qa^4(5+\nu)}{64D(1+\nu)} = \frac{-2.117(10^4)(5.285)}{64(21{,}800)(1.285)} = -0.063 \text{ in}$$

$$M_c = \frac{qa^2}{16}(3+\nu) = \frac{2.117(10^2)(3.285)}{16} = 43.5 \text{ in-lb/in}$$

$$LT_y = \frac{-qr^4}{D}G_{11} = \frac{-2.117r^4}{21{,}800}\left(\frac{1}{64}\right) = -1.517(10^{-6})r^4$$

From Table 24, case 12a,

$$y_c = \frac{-qa^4(7+\nu)}{288D(1+\nu)} = \frac{-1.098(10^4)(7.285)}{288(21{,}800)(1.285)} = -0.00992 \text{ in}$$

$$M_c = \frac{qa^2(5+\nu)}{96} = \frac{1.098(10^2)(5.285)}{96} = 6.05 \text{ in-lb/in}$$

$$LT_y = \frac{-qr^6}{Da^2}G_{13} = \frac{-1.098r^6}{21{,}800(10^2)}\left(\frac{25}{14{,}400}\right) = -8.75(10^{-10})r^6$$

Using these values, the deflection equation can be written

$$y = -0.0623 - 0.00992 + \frac{(43.5+6.05)r^2}{2(21{,}800)1.285} - 1.517(10^{-6})r^4 - 8.75(10^{-10})r^6$$

or $$y = -0.0722 + 0.000885r^2 - 1.517(10^{-6})r^4 - 8.75(10^{-10})r^6$$

This deflection would create a foundation reaction

$$q_f = 20(-0.0722 + 0.000885r^2) = -1.445 + 0.0177r^2$$

if the higher-order terms were neglected. Again applying a 50 percent factor to the difference between the assumed and calculated foundation pressure gives an improved loading from the foundation:

$$q_f = -1.164 + 0.01434r^2$$

Repeating the previous steps again, we obtain

$$y_c = -0.0623\left(\frac{3-1.164}{2.117}\right) - 0.00992\left(\frac{0.01434}{0.01098}\right) = -0.0671 \text{ in}$$

$$M_c = 43.5\left(\frac{3-1.164}{2.117}\right) + 6.05\left(\frac{0.01434}{0.01098}\right) = 45.61 \text{ in-lb/in}$$

$$y = -0.0671 + 0.000813r^2$$
$$q_f = -1.342 + 0.01626r^2$$

Successive repetitions of the previous steps give improved values for q_f:

$$q_f = -1.306 + 0.1584r^2 \qquad q_f = -1.296 + 0.1566r^2 \qquad q_f = -1.290 + 0.1566r^2$$

Using values from the last iteration, the final answers are

$$y_c = -0.0645 \text{ in} \qquad M_c = 43.8 \text{ in-lb/in} \qquad \text{and max } \sigma = 6580 \text{ psi}$$

An exact analysis using expressions from Ref. 46 gives

$$y_c = -0.0637 \text{ in} \qquad \text{and} \qquad M_c = 43.3 \text{ in-lb/in}$$

10.7 *Circular plates of variable thickness*

For any circular plate of variable thickness, loaded symmetrically with respect to the center, the stresses and deflections can be found as follows: The plate is divided into an arbitrary number of concentric rings, each of which is assumed to have a uniform thickness equal to its mean thickness. Each such ring is loaded by radial moments M_a and M_b at its outer and inner circumferences, respectively, by vertical shears at its inner and outer circumferences, and by whatever load is distributed over its surface. The shears are known, each being equal to the total load on the plate within the corresponding circumference. The problem is to determine the edge moments, and this is done by making use of the fact that the slope of each ring at its inner circumference is equal to the slope of the next inner ring at its outer circumference. This condition, together with the known slope (or moment) at the outer edge of the plate and the known slope (or moment) at the inside edge or center of the plate, enables as many equations to be written as there are unknown quantities M. Having found all the edge moments, stresses and deflections can be calculated for each ring by the appropriate formulas of Table 24 and the deflections added to find the deflection of the plate.

A more direct solution (Ref. 21) is available if the plate is of such form that the variation in thickness can be expressed fairly closely by the equation $t = t_o e^{-nx^2/6}$, where t is the thickness at any point a distance r from the center, t_o is the thickness at the center, e is the base for the napierian system of logarithms (2.718), x is the ratio r/a, and n is a number chosen so as to make the equation agree with the actual variation in thickness. The constant n is positive for a plate that decreases in thickness toward the edge and negative for a plate that increases in thickness toward the edge. For a plate of uniform thickness, $n = 0$; and for a plate twice as thick at the center as at the edge, $n = +4.16$. The maximum stress and deflection for a uniformly loaded circular plate are given by max $\sigma = \beta q a^2/t_o^2$ and max $y = \alpha q a^4/E t_0^3$, where β and α depend on n, and for values of n from 4 to -4 can be found by interpolation from the following table:

Edge conditions		n								
		+4	+3	+2	+1	0	−1	−2	−3	−4
Edges supported	β	1.63	1.55	1.45	1.39	1.24	1.16	1.04	0.945	0.855
Case 10a, $r_o = 0$	α	1.220	1.060	0.924	0.804	0.695	0.600	0.511	0.432	0.361
Edges fixed	β	2.14	1.63	1.31	0.985	0.75	0.55	0.43	0.32	0.26
Case 10b, $r_o = 0$	α	0.4375	0.3490	0.276	0.217	0.1707	0.1343	0.1048	0.0830	0.0653

For the loadings in the preceding table as well as for a simply supported plate with an edge moment, Ref. 46 gives graphs and tables which permit the evaluation of radial and tangential stresses throughout the plate. This same reference gives extensive tables of moment and shear coefficients for a variety of loadings and support conditions for plates in which the thickness varies as $t = t_a r/a^{-n/3}$, where t_a is the thickness at the outer edge: Values are tabulated for $n = 0$, 1, 1.5, and 2 and for $\nu = \frac{1}{6}$.

Stresses and deflections for plates with thicknesses varying linearly with radius are tabulated in Refs. 46 and 57. Annular plates with the outer edges fixed and the inner edges guided and with thicknesses increasing linearly with the radii from zero at the center are discussed in Ref. 36 and tabulated in previous editions of this handbook. A uniformly loaded circular plate with a fixed edge and a thickness varying linearly along a diameter is discussed by Strock and Yu (Ref 65). Conway (Ref. 66) considers the necessary proportions for a rib along the diameter of a uniformly loaded, clamped circular plate to affect a minimum weight design for a given maximum stress.

Perforated plates. Slot and O'Donnell (Ref. 62) present the relationship between the effective elastic constants for *thick perforated plates* under bending and *thin perforated plates* under in-plane loading. Numerical results are presented in the form of tables and graphs, and many references are listed.

10.8 *Disk springs*

The *conical disk,* or *Belleville spring* (Fig. 10.4), is not a flat plate, of course, but it may appropriately be considered in this chapter because it bears a

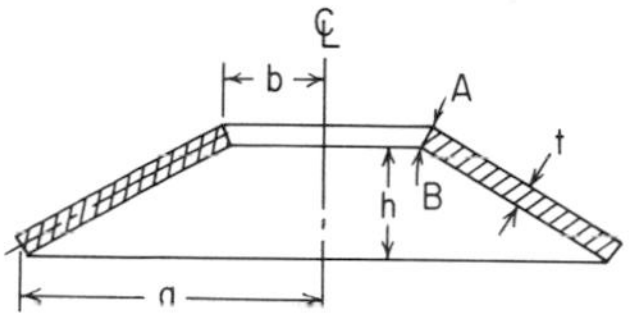

Fig. 10.4

superficial resemblance to a flat ring and is sometimes erroneously analyzed by the formulas for case 1a. The stress and deflection produced in a spring of this type are not proportional to the applied load because the change in form consequent upon deflection markedly changes the load-deflection and load-stress relationships. This is indeed the peculiar advantage of this form of spring because it makes it possible to secure almost any desired variation of "spring rate" and also possible to obtain a considerable range of deflection under almost constant load. The largest stresses occur at the inner edge.

Formulas for deflection and stress (Ref. 27) are

$$P = \frac{E\delta}{(1-\nu^2)Ma^2}\left[(h-\delta)\left(h-\frac{\delta}{2}\right)t + t^3\right]$$

$$\text{Stress at } A = \frac{-E\delta}{(1-\nu^2)Ma^2}\left[C_1\left(h-\frac{\delta}{2}\right) + C_2 t\right]$$

$$\text{Stress at } B = \frac{-E\delta}{(1-\nu^2)Ma^2}\left[C_1\left(h-\frac{\delta}{2}\right) - C_2 t\right]$$

where P = total applied load; E = modulus of elasticity; δ = deflection; h = cone height of either inner or outer surface; t = thickness; a and b are the outer and inner radii of the middle surface; and M, C_1, and C_2 are constants whose values are functions of a/b and are given in the following table:

a/b	M	C_1	C_2
1.0	0		
1.2	0.31	1.02	1.05
1.4	0.46	1.07	1.14
1.6	0.57	1.14	1.23
1.8	0.64	1.18	1.30
2.0	0.70	1.23	1.39
2.2	0.73	1.27	1.46
2.6	0.76	1.35	1.60
3.0	0.78	1.43	1.74
3.4	0.80	1.50	1.88
3.8	0.80	1.57	2.00
4.2	0.80	1.64	2.14
4.6	0.80	1.71	2.26
5.0	0.79	1.77	2.38

The formulas for stress may give either positive or negative results, depending upon δ; a negative result indicates compressive stress, and a positive result a tensile stress. It is to be noted that P also may become negative.

Wempner (Refs. 67 and 68) derives more exacting expressions for the conical spring. Unless the center hole is small or the cone angle is outside the range normally used for disk springs, however, the differences are slight. Reference 69 presents useful design curves based on Ref. 27.

10.9 *Narrow ring under distributed torque about its axis*

When the inner radius b is almost as great as the outer radius a, the loading for cases 1a, 1k, 2a, 2k, and so on, becomes almost equivalent to that shown in Fig. 10.5, which represents a ring subjected to a uniformly distributed

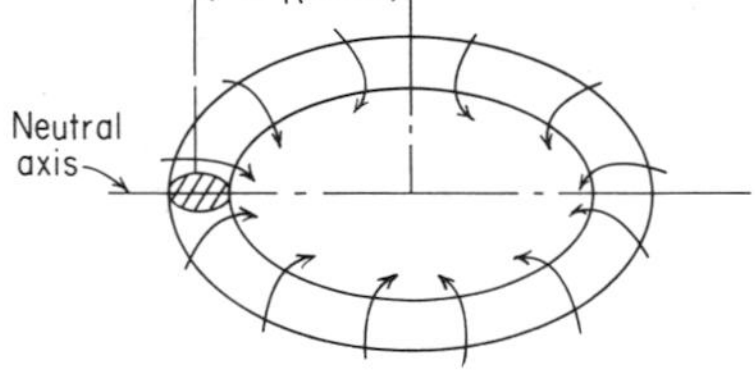

Fig. 10.5

moment of M in-lb/linear inch acting about its axis. An approximation to this type of loading also occurs in clamping, or "follower," rings used for joining pipe; here the bolt forces and the balancing gasket or flange pressure produce the distributed moment, which obviously tends to "roll" the ring, or turn it inside out, so to speak.

Under this loading the ring, whatever the shape of its cross section (as long as it is reasonably compact) is subjected to a bending moment at every section equal to *MR*, the neutral axis being the central axis of the cross section in the plane of the ring. The maximum resulting stress occurs at the extreme fiber and is given by Eq. 12 of Art. 7.1; that is,

$$\sigma = \frac{MR}{I/c}$$

The ring does not bend, and there is no twisting, but every section rotates in its own plane about its centroid through an angle

$$\theta = \frac{MR^2}{EI} = \frac{\sigma R}{Ec}$$

These formulas may be used to obtain approximate results for the cases of flat-plate loading listed previously when the difference between a and b is small, as well as for pipe flanges, etc. Paul (Ref. 70) discusses the collapse or inversion of rings due to plastic action.

10.10 *Bending of uniform-thickness plates with straight boundaries*

Formulas. No general expression for deflection as a function of position in a plate is given since solutions for plates with straight boundaries are generally obtained numerically for specific ratios of plate dimensions, load location, and boundary conditions. In a few instances Poisson's ratio is included in the expressions given, but in most cases a specific value of Poisson's ratio has been used in obtaining the tabulated numerical results and the value used is indicated. Reference 47 includes results obtained using several values of Poisson's ratio and shows the range of values that can be expected as this ratio is changed. Errors in deflection should not exceed 7 or 8 percent and in maximum stress 15 percent for values of Poisson's ratio in the range from 0.15 to 0.30. Since much of the data are obtained using finite-difference approximations for the plate differential equations and a limited number of elements have been used, it is not always possible to identify maximum values if they occur at points between the chosen grid points.

Table 26 presents maximum values where possible and the significant values otherwise for deflections normal to the plate surface, bending stresses, and in many cases the boundary reaction forces R. For rectangular plates with simply supported edges the maximum stresses are shown to be near the center of the plate. There are, however, stresses of similar magnitude near the corners if the corners are held down as has been assumed for all cases presented. Reference 21 discusses the increase in stress at the center of the plate when the corners are permitted to rise. For a uniformly loaded square plate this increase in stress is approximately 35 percent.

It is impractical to include plates of all possible shapes and loadings, but

TABLE 26 ***Formulas for flat plates with straight boundaries and constant thickness***

NOTATION: The notation for Table 24 applies with the following modifications: a and b refer to plate dimensions, and when used as subscripts for stress, they refer to the stresses in directions parallel to the sides a and b, respectively. σ is a bending stress which is positive when tensile on the bottom and compressive on the top if loadings are considered vertically downward. R is the reaction force normal to the plate surface exerted by the boundary support on the edge of the plate (pounds per inch). r'_o is the equivalent radius of contact for a load concentrated on a very small area and is given by $r'_o = \sqrt{1.6r_o^2 + t^2} - 0.675t$ if $r_o < 0.5t$ and $r'_o = r_o$ if $r_o > 0.5t$

Case no., shape, and supports	Case no., loading	Formulas and tabulated specific values
1. Rectangular plate; all edges simply supported 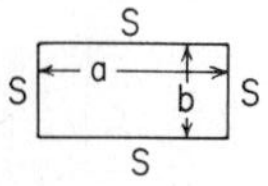	1a. Uniform over entire plate	(At center) $\text{Max } \sigma = \sigma_b = \dfrac{\beta q b^2}{t^2}$ and $\max y = \dfrac{-\alpha q b^4}{Et^3}$ (At center of long sides) $\text{Max } R = \gamma q b$ (see table 1a below) (Ref. 21 for $\nu = 0.3$)
	1b. Uniform over small concentric circle of radius r_o (note definition of r'_o)	(At center) $\text{Max } \sigma = \dfrac{3W}{2\pi t^2}\left[(1+\nu)\ln\dfrac{2b}{\pi r'_o} + \beta\right]$ $\text{Max } y = \dfrac{-\alpha W b^2}{Et^3}$ (see table 1b below) (Ref. 21 for $\nu = 0.3$)

Case 1a:

a/b	1.0	1.2	1.4	1.6	1.8	2.0	3.0	4.0	5.0	∞
β	0.2874	0.3762	0.4530	0.5172	0.5688	0.6102	0.7134	0.7410	0.7476	0.7500
α	0.0444	0.0616	0.0770	0.0906	0.1017	0.1110	0.1335	0.1400	0.1417	0.1421
γ	0.420	0.455	0.478	0.491	0.499	0.503	0.505	0.502	0.501	0.500

Case 1b:

a/b	1.0	1.2	1.4	1.6	1.8	2.0	∞
β	0.435	0.650	0.789	0.875	0.927	0.958	1.000
α	0.1267	0.1478	0.1621	0.1715	0.1770	0.1805	0.1851

1c. Uniform over central rectangular area

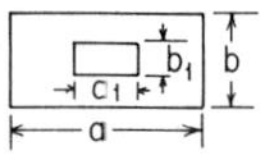

(At center) $\quad \text{Max } \sigma = \sigma_b = \dfrac{\beta W}{t^2} \quad$ where $W = qa_1b_1$

a_1/b → / b_1/b ↓	$a = b$: 0	0.2	0.4	0.6	0.8	1.0	$a = 1.4b$: 0	0.2	0.4	0.8	1.2	1.4	$a = 2b$: 0	0.4	0.8	1.2	1.6	2.0
0		1.82	1.38	1.12	0.93	0.76		2.0	1.55	1.12	0.84	0.75		1.64	1.20	0.97	0.78	0.64
0.2	1.82	1.28	1.08	0.90	0.76	0.63	1.78	1.43	1.23	0.95	0.74	0.64	1.73	1.31	1.03	0.84	0.68	0.57
0.4	1.39	1.07	0.84	0.72	0.62	0.52	1.59	1.13	1.00	0.80	0.62	0.55	1.32	1.08	0.88	0.74	0.60	0.50
0.6	1.12	0.90	0.72	0.60	0.52	0.43	1.10	0.91	0.82	0.68	0.53	0.47	1.04	0.90	0.76	0.64	0.54	0.44
0.8	0.92	0.76	0.62	0.51	0.42	0.36	0.90	0.76	0.68	0.57	0.45	0.40	0.87	0.76	0.63	0.54	0.44	0.38
1.0	0.76	0.63	0.52	0.42	0.35	0.30	0.75	0.62	0.57	0.47	0.38	0.33	0.71	0.61	0.53	0.45	0.38	0.30

(Values from charts of Ref. 8; $\nu = 0.3$.)

1d. Uniformly increasing along length

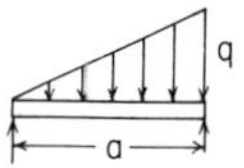

$$\text{Max } \sigma = \frac{\beta q b^2}{t^2} \quad \text{and} \quad \max y = \frac{-\alpha q b^4}{E t^3}$$

a/b	1	1.5	2.0	2.5	3.0	3.5	4.0
β	0.16	0.26	0.34	0.38	0.43	0.47	0.49
α	0.022	0.043	0.060	0.070	0.078	0.086	0.091

(Values from charts of Ref. 8; $\nu = 0.3$.)

1e. Uniformly increasing along width

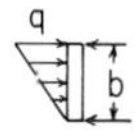

$$\text{Max } \sigma = \frac{\beta q b^2}{t^2} \quad \text{and} \quad \max y = \frac{-\alpha q b^4}{E t^3}$$

a/b	1	1.5	2.0	2.5	3.0	3.5	4.0
β	0.16	0.26	0.32	0.35	0.37	0.38	0.38
α	0.022	0.042	0.056	0.063	0.067	0.069	0.070

(Values from charts of Ref. 8; $\nu = 0.3$.)

TABLE 26 *Formulas for flat plates with straight boundaries and constant thickness* (*Cont.*)

Case no., shape, and supports	Case no., loading	Formulas and tabulated specific values
	1f. Uniform over entire plate plus uniform tension or compression P lb/linear in applied to short edges 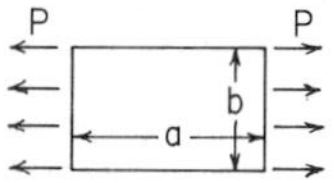	Max $y = \alpha \frac{qb^4}{Et^3}$ Max $\sigma_a = \beta_x \frac{qb^2}{t^2}$ Max $\sigma_b = \beta_y \frac{qb^2}{t^2}$. Here α, β_x, and β_y depend on ratios $\frac{a}{b}$ and $\frac{P}{P_E}$, where $P_E = \frac{\pi^2 E t^3}{3(1 - \nu^2)b^2}$, and have the following values:

Coef.	a/b \ P/P_E	0	0.15	0.25	0.50	0.75	1	2	3	4	5
					P, Tension						
α	1	0.044	0.039		0.030		0.023	0.015	0.011	0.008	0.0075
	$1\frac{1}{2}$	0.084	0.075		0.060		0.045	0.0305	0.024	0.019	0.0170
	2	0.110	0.100		0.084		0.067	0.0475	0.0375	0.030	0.0260
	3	0.133	0.125		0.1135		0.100	0.081	0.066	0.057	0.0490
	4	0.140	0.136		0.1280		0.118	0.102	0.089	0.080	0.072
β_x	1	0.287					0.135	0.096	0.072	0.054	0.045
	$1\frac{1}{2}$	0.300					0.150	0.105	0.078	0.066	0.048
	2	0.278					0.162	0.117	0.093	0.075	0.069
	3	0.246					0.180	0.150	0.126	0.105	0.093
	4	0.222					0.192	0.168	0.156	0.138	0.124
β_y	1	0.287					0.132	0.084	0.054	0.036	0.030
	$1\frac{1}{2}$	0.487					0.240	0.156	0.114	0.090	0.072
	2	0.610					0.360	0.258	0.198	0.162	0.138
	3	0.713					0.510	0.414	0.348	0.294	0.258
	4	0.741					0.624	0.540	0.480	0.420	0.372
					P, Compression						
α	1	0.044		0.060	0.094	0.180					
	$1\frac{1}{2}$	0.084		0.109	0.155	0.237					
	2	0.110		0.139	0.161	0.181					
	3	0.131		0.145	0.150	0.150					
	4	0.140		0.142	0.142	0.138					
β_x	1	0.287		0.372	0.606	1.236					
	$1\frac{1}{2}$	0.300		0.372	0.522	0.846					
	2	0.278		0.330	0.390	0.450					
	3	0.246		0.228	0.228	0.210					
	4	0.222		0.225	0.225	0.225					
β_y	1	0.287		0.420	0.600	1.260					
	$1\frac{1}{2}$	0.487		0.624	0.786	1.380					
	2	0.610		0.720	0.900	1.020					
	3	0.713		0.750	0.792	0.750					
	4	0.741		0.750	0.750	0.750					

In the above formulas σ_a and σ_b are stresses due to bending only; add direct stress P/t to σ_a (Ref. 41)

1g. Uniform over entire plate plus uniform tension P lb/linear in applied to all edges

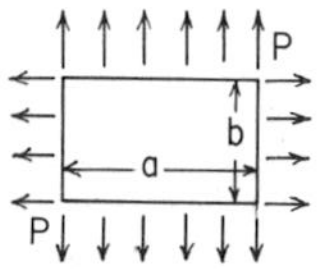

Max $y = \alpha \frac{qb^4}{Et^3}$ Max $\sigma_a = \beta_x \frac{qb^2}{t^2}$ Max $\sigma_b = \beta_y \frac{qb^2}{t^2}$. Here α, β_x, and β_y depend on ratios $\frac{a}{b}$ and $\frac{P}{P_E}$, where $P_E = \frac{\pi^2 Et^3}{3(1 = \nu^2)b^2}$, and have the following values:

Coef.	a/b \ P/P_E	0	0.15	0.5	1	2	3	4	5
α	1	0.044	0.035	0.022	0.015	0.008	0.006	0.004	0.003
	$1\frac{1}{2}$	0.084	0.060	0.035	0.022	0.012	0.008	0.006	0.005
	2	0.110	0.075	0.042	0.025	0.014	0.010	0.007	0.006
	3	0.133	0.085	0.045	0.026	0.016	0.011	0.008	0.007
	4	0.140	0.088	0.046	0.026	0.016	0.011	0.008	0.007
β_x	1	0.287	0.216	0.132	0.084	0.048	0.033	0.026	0.021
	$1\frac{1}{2}$	0.300	0.204	0.117	0.075	0.045	0.031	0.024	0.020
	2	0.278	0.189	0.111	0.072	0.044	0.031	0.024	0.020
	3	0.246	0.183	0.108	0.070	0.043	0.031	0.025	0.020
	4	0.222	0.183	0.108	0.074	0.047	0.032	0.027	0.024
β_y	1	0.287	0.222	0.138	0.090	0.051	0.036	0.030	0.024
	$1\frac{1}{2}$	0.487	0.342	0.186	0.108	0.066	0.042	0.036	0.030
	2	0.610	0.302	0.216	0.132	0.072	0.051	0.042	0.036
	3	0.713	0.444	0.234	0.141	0.078	0.054	0.042	0.036
	4	0.741	0.456	0.240	0.144	0.078	0.054	0.042	0.036

In the above formulas σ_a and σ_b are stresses due to bending only; add direct stresses P/t to both σ_a and σ_b. (Ref. 42)

2. Rectangular plate; three edges simply supported, one edge (b) free

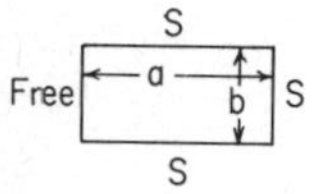

2a. Uniform over entire plate

Max $\sigma = \frac{\beta q b^2}{t^2}$ and max $y = \frac{-\alpha q b^4}{Et^3}$

a/b	0.50	0.667	1.0	1.5	2.0	4.0
β	0.36	0.45	0.67	0.77	0.79	0.80
α	0.080	0.106	0.140	0.160	0.165	0.167

(Ref. 8 for $\nu = 0.3$)

2d. Uniformly increasing along the a side

Max $\sigma = \frac{\beta q b^2}{t^2}$ and max $y = \frac{-\alpha q b^4}{Et^3}$

a/b	0.50	0.667	1.0	1.5	2.0	2.5	3.0	3.5	4.0
β	0.11	0.16	0.20	0.28	0.32	0.35	0.36	0.37	0.37
α	0.026	0.033	0.040	0.050	0.058	0.064	0.067	0.069	0.070

(Ref. 8 for $\nu = 0.3$)

TABLE 26 *Formulas for flat plates with straight boundaries and constant thickness* (*Cont.*)

Case no., shape, and supports	Case no., loading	Formulas and tabulated specific values
3. Rectangular plate; three edges simply supported, one short edge (b) fixed	3a. Uniform over entire plate	Max $\sigma = \dfrac{\beta q b^2}{t^2}$ and max $y = \dfrac{-\alpha q b^4}{E t^3}$

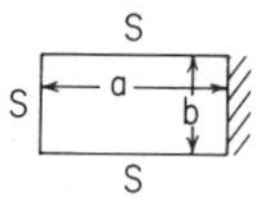

a/b	1	1.5	2.0	2.5	3.0	3.5	4.0
β	0.50	0.67	0.73	0.74	0.75	0.75	0.75
α	0.030	0.071	0.101	0.122	0.132	0.137	0.139

(Values from charts of Ref. 8; $\nu = 0.3$)

Case no., shape, and supports	Case no., loading	Formulas and tabulated specific values
4. Rectangular plate; three edges simply supported, one long edge (a) fixed	4a. Uniform over entire plate	Max $\sigma = \dfrac{\beta q b^2}{t^2}$ and max $y = \dfrac{-\alpha q b^4}{E t^3}$

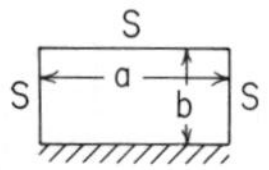

a/b	1	1.5	2.0	2.5	3.0	3.5	4.0
β	0.50	0.66	0.73	0.74	0.74	0.75	0.75
α	0.030	0.046	0.054	0.056	0.057	0.058	0.058

(Values from charts of Ref. 8; $\nu = 0.3$)

Case no., shape, and supports	Case no., loading	Formulas and tabulated specific values
5. Rectangular plate; two long edges simply supported, two short edges fixed	5a. Uniform over entire plate	(At center of short edges) Max $\sigma = \dfrac{-\beta q b^2}{t^2}$ (At center) Max $y = \dfrac{-\alpha q b^4}{E t^3}$

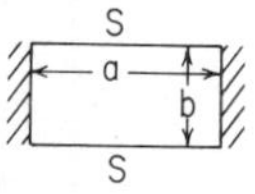

a/b	1	1.2	1.4	1.6	1.8	2	∞
β	0.4182	0.5208	0.5988	0.6540	0.6912	0.7146	0.750
α	0.0210	0.0349	0.0502	0.0658	0.0800	0.0922	

(Ref. 21)

Case no., shape, and supports	Case no., loading	Formulas and tabulated specific values
6. Rectangular plate, two long edges fixed, two short edges simply supported	6a. Uniform over entire plate	(At center of long edges) Max $\sigma = \dfrac{-\beta q b^2}{t^2}$ (At center) Max $y = \dfrac{-\alpha q b^4}{E t^3}$

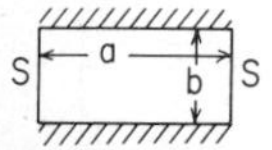

a/b	1	1.2	1.4	1.6	1.8	2	∞
β	0.4182	0.4626	0.4860	0.4968	0.4971	0.4973	0.500
α	0.0210	0.0243	0.0262	0.0273	0.0280	0.0283	0.0285

(Ref. 21)

7. Rectangular plate; one edge fixed, opposite edge free, remaining edges simply supported

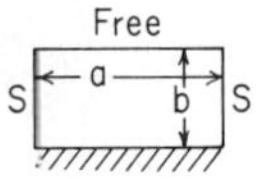

7a. Uniform over entire plate

(At center of fixed edge) $\sigma = \frac{-\beta_1 q b^2}{t^2}$ and $R = \gamma_1 q b$

(At center of free edge) $\sigma = \frac{\beta_2 q b^2}{t^2}$

(At end of free edge) $R = \gamma_2 q b$

a/b	0.25	0.50	0.75	1.0	1.5	2.0	3.0
β_1	0.044	0.176	0.380	0.665	1.282	1.804	2.450
β_2	0.048	0.190	0.386	0.565	0.730	0.688	0.434
γ_1	0.183	0.368	0.541	0.701	0.919	1.018	1.055
γ_2	0.131	0.295	0.526	0.832	1.491	1.979	2.401

(Ref. 49 for $\nu = 0.2$)

7aa. Uniform over $\frac{2}{3}$ of plate from fixed edge

(At center of fixed edge) $\sigma = \frac{-\beta q b^2}{t^2}$ and $R = \gamma q b$

a/b	0.25	0.50	0.75	1.0	1.5	2.0	3.0
β	0.044	0.161	0.298	0.454	0.730	0.932	1.158
γ	0.183	0.348	0.466	0.551	0.645	0.681	0.689

(Ref. 49 for $\nu = 0.2$)

7aaa. Uniform over $\frac{1}{3}$ of plate from fixed edge

(At center of fixed edge) $\sigma = \frac{-\beta q b^2}{t^2}$ and $R = \gamma q b$

a/b	0.25	0.50	0.75	1.0	1.5	2.0	3.0
β	0.040	0.106	0.150	0.190	0.244	0.277	0.310
γ	0.172	0.266	0.302	0.320	0.334	0.338	0.338

(Ref. 49 for $\nu = 0.2$)

7d. Uniformly decreasing from fixed edge to free edge

(At center of fixed edge) $\sigma = \frac{-\beta q b^2}{t^2}$ and $R = \gamma q b$

a/b	0.25	0.50	0.75	1.0	1.5	2.0	3.0
β	0.037	0.120	0.212	0.321	0.523	0.677	0.866
γ	0.159	0.275	0.354	0.413	0.482	0.509	0.517

(Ref. 49 for $\nu = 0.2$)

TABLE 26 *Formulas for flat plates with straight boundaries and constant thickness* (*Cont.*)

Case no., shape, and supports	Case no., loading	Formulas and tabulated specific values

7dd. Uniformly decreasing from fixed edge to zero at $\frac{2}{3}b$

(At center of fixed edge) $\sigma = \dfrac{-\beta q b^2}{t^2}$ and $R = \gamma q b$

a/b	0.25	0.50	0.75	1.0	1.5	2.0	3.0
β	0.033	0.094	0.146	0.200	0.272	0.339	0.400
γ	0.148	0.233	0.277	0.304	0.330	0.339	0.340

(Ref. 49 for $\nu = 0.2$)

7ddd. Uniformly decreasing from fixed edge to zero at $\frac{1}{3}b$

(At center of fixed edge) $\sigma = \dfrac{-\beta q b^2}{t^2}$ and $R = \gamma q b$

a/b	0.25	0.50	0.75	1.0	1.5	2.0	3.0
β	0.023	0.048	0.061	0.073	0.088	0.097	0.105
γ	0.115	0.149	0.159	0.164	0.167	0.168	0.168

(Ref. 49 for $\nu = 0.2$)

7f. Distributed line load w lb/in along free edge

(At center of fixed edge) $\sigma_b = \dfrac{-\beta_1 w b}{t^2}$ and $R = \gamma_1 w$

(At center of free edge) $\sigma_a = \dfrac{\beta_2 w b}{t^2}$

(At ends of free edge) $R = \gamma_2 w$

a/b	0.25	0.50	0.75	1.0	1.5	2.0	3.0
β_1	0.000	0.024	0.188	0.570	1.726	2.899	4.508
β_2	0.321	0.780	1.204	1.554	1.868	1.747	1.120
γ_1	0.000	0.028	0.160	0.371	0.774	1.004	1.119
γ_2	1.236	2.381	3.458	4.510	6.416	7.772	9.031

(Ref. 49 for $\nu = 0.2$)

8. Rectangular plate, all edges fixed

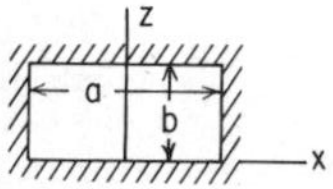

8a. Uniform over entire plate

(At center of long edge) $\text{Max } \sigma = \dfrac{-\beta_1 q b^2}{t^2}$

(At center) $\sigma = \dfrac{\beta_2 q b^2}{t^2}$ and $\max y = \dfrac{\alpha q b^4}{E t^3}$

a/b	1.0	1.2	1.4	1.6	1.8	2.0	∞
β_1	0.3078	0.3834	0.4356	0.4680	0.4872	0.4974	0.5000
β_2	0.1386	0.1794	0.2094	0.2286	0.2406	0.2472	0.2500
α	0.0138	0.0188	0.0226	0.0251	0.0267	0.0277	0.0284

(Ref. 21 for $\nu = 0.3$)

8b. Uniform over small concentric circle of radius r_o (note definition of r'_o)

(At center) $\sigma_b = \frac{3W}{2\pi t^2}\left[(1+\nu)\ln\frac{2b}{\pi r'_o} + \beta_1\right]$ and $\max y = \frac{\alpha W b^2}{Et^3}$

(At center of long edge) $\sigma_b = \frac{-\beta_2 W}{t^2}$

a/b	1.0	1.2	1.4	1.6	1.8	2.0	∞
β_1	−0.238	−0.078	0.011	0.053	0.068	0.067	0.067
β_2	0.7542	0.8940	0.9624	0.9906	1.0000	1.004	1.008
α	0.0611	0.0706	0.0754	0.0777	0.0785	0.0788	0.0791

(Ref. 21 for $\nu = 0.3$)

8d. Uniformly decreasing parallel to side b

(At $x = 0, z = 0$) $\text{Max } \sigma_b = \frac{-\beta_1 q b^2}{t^2}$

(At $x = 0, z = 0.4b$) $\sigma_b = \frac{\beta_2 q b^2}{t^2}$ and $\sigma_a = \frac{\beta_3 q b^2}{t^2}$

(At $x = 0, z = b$) $\sigma_b = \frac{-\beta_4 q b^2}{t^2}$

$\left(\text{At } x = \pm\frac{a}{2}, z = 0.45b\right)$ $\text{Max } \sigma_a = \frac{-\beta_5 q b^2}{t^2}$

$\text{Max } y = \frac{-\alpha q b^4}{Et^3}$

a/b	0.6	0.8	1.0	1.2	1.4	1.6	1.8	2.0
β_1	0.1132	0.1778	0.2365	0.2777	0.3004	0.3092	0.3100	0.3068
β_2	0.0410	0.0633	0.0869	0.1038	0.1128	0.1255	0.1157	0.1148
β_3	0.0637	0.0688	0.0762	0.0715	0.0610	0.0509	0.0415	0.0356
β_4	0.0206	0.0497	0.0898	0.1249	0.1482	0.1615	0.1680	0.1709
β_5	0.1304	0.1436	0.1686	0.1800	0.1845	0.1874	0.1902	0.1908
α	0.0016	0.0047	0.0074	0.0097	0.0113	0.0126	0.0133	0.0136

(Ref. 28 for $\nu = 0.3$)

TABLE 26 *Formulas for flat plates with straight boundaries and constant thickness* (*Cont.*)

Case no., shape, and supports	Case no., loading	Formulas and tabulated specific values
9. Rectangular plate, three edges fixed, one edge (a) simply supported 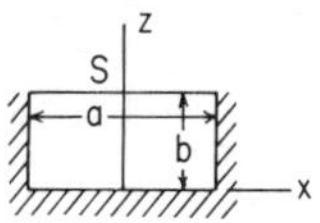	9a. Uniform over entire plate	(At $x = 0, z = 0$) Max $\sigma_b = \dfrac{-\beta_1 q b^2}{t^2}$ and $R = \gamma_1 q b$ (At $x = 0, z = 0.6b$) $\sigma_b = \dfrac{\beta_2 q b^2}{t^2}$ and $\sigma_a = \dfrac{\beta_3 q b^2}{t^2}$ (At $x = 0, z = b$) $R = \gamma_2 q b$ $\left(\text{At } x = \pm\dfrac{a}{2}, z = 0.6b\right)$ $\sigma_a = \dfrac{-\beta_4 q b^2}{t^2}$ and $R = \gamma_3 q b$

a/b	0.25	0.50	0.75	1.0	1.5	2.0	3.0
β_1	0.020	0.081	0.173	0.307	0.539	0.657	0.718
β_2	0.004	0.018	0.062	0.134	0.284	0.370	0.422
β_3	0.016	0.061	0.118	0.158	0.164	0.135	0.097
β_4	0.031	0.121	0.242	0.343	0.417	0.398	0.318
γ_1	0.115	0.230	0.343	0.453	0.584	0.622	0.625
γ_2	0.123	0.181	0.253	0.319	0.387	0.397	0.386
γ_3	0.125	0.256	0.382	0.471	0.547	0.549	0.530

(Ref. 49 for $\nu = 0.2$)

Case no., shape, and supports	Case no., loading	Formulas and tabulated specific values
	9aa. Uniform over $\frac{2}{3}$ of plate from fixed edge 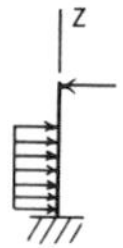	(At $x = 0, z = 0$) Max $\sigma_b = \dfrac{-\beta_1 q b^2}{t^2}$ and $R = \gamma_1 q b$ (At $x = 0, z = 0.6b$) $\sigma_b = \dfrac{\beta_2 q b^2}{t^2}$ and $\sigma_a = \dfrac{\beta_3 q b^2}{t^2}$ (At $x = 0, z = b$) $R = \gamma_2 q b$ $\left(\text{At } x = \pm\dfrac{a}{2}, z = 0.4b\right)$ $\sigma_a = \dfrac{-\beta_4 q b^2}{t^2}$ and $R = \gamma_3 q b$

a/b	0.25	0.50	0.75	1.0	1.5	2.0	3.0
β_1	0.020	0.080	0.164	0.274	0.445	0.525	0.566
β_2	0.003	0.016	0.044	0.093	0.193	0.252	0.286
β_3	0.012	0.043	0.081	0.108	0.112	0.091	0.066
β_4	0.031	0.111	0.197	0.255	0.284	0.263	0.204
γ_1	0.115	0.230	0.334	0.423	0.517	0.542	0.543
γ_2	0.002	0.015	0.048	0.088	0.132	0.139	0.131
γ_3	0.125	0.250	0.345	0.396	0.422	0.417	0.405

(Ref. 49 for $\nu = 0.2$)

9aaa. Uniform over $\frac{1}{3}$ of plate from fixed edge

(At $x = 0, z = 0$) Max $\sigma_b = \dfrac{-\beta_1 q b^2}{t^2}$ and $R = \gamma_1 q b$

(At $x = 0, z = 0.2b$) $\sigma_b = \dfrac{\beta_2 q b^2}{t^2}$ and $\sigma_a = \dfrac{\beta_3 q b^2}{t^2}$

(At $x = 0, z = b$) $R = \gamma_2 q b$

$\left(\text{At } x = \pm\dfrac{a}{2}, z = 0.2b\right)$ $\sigma_a = \dfrac{-\beta_4 q b^2}{t^2}$ and $R = \gamma_3 q b$

a/b	0.25	0.50	0.75	1.0	1.5	2.0	3.0
β_1	0.020	0.068	0.108	0.148	0.194	0.213	0.222
β_2	0.005	0.026	0.044	0.050	0.047	0.041	0.037
β_3	0.013	0.028	0.031	0.026	0.016	0.011	0.008
β_4	0.026	0.063	0.079	0.079	0.068	0.056	0.037
γ_1	0.114	0.210	0.261	0.290	0.312	0.316	0.316
γ_2	0.000	0.000	0.004	0.011	0.020	0.021	0.020
γ_3	0.111	0.170	0.190	0.185	0.176	0.175	0.190

(Ref. 49 for $\nu = 0.2$)

9d. Uniformly decreasing from fixed edge to simply supported edge

(At $x = 0, z = 0$) Max $\sigma_b = \dfrac{-\beta_1 q b^2}{t^2}$ and $R = \gamma_1 q b$

$\left(\text{At } x = \pm\dfrac{a}{2}, z = 0.4b\right)$ $\sigma_a = \dfrac{-\beta_2 q b^2}{t^2}$ and $R = \gamma_2 q b$

a/b	0.25	0.50	0.75	1.0	1.5	2.0	3.0
β_1	0.018	0.064	0.120	0.192	0.303	0.356	0.382
β_2	0.019	0.068	0.124	0.161	0.181	0.168	0.132
γ_1	0.106	0.195	0.265	0.323	0.383	0.399	0.400
γ_2	0.075	0.152	0.212	0.245	0.262	0.258	0.250

(Ref. 49 for $\nu = 0.2$)

TABLE 26 *Formulas for flat plates with straight boundaries and constant thickness* (*Cont.*)

Case no., shape, and supports	Case no., loading	Formulas and tabulated specific values
	9dd. Uniformly decreasing from fixed edge to zero at $\frac{2}{3}b$	(At $x = 0$, $z = 0$) Max $\sigma_b = \dfrac{-\beta_1 q b^2}{t^2}$ and $R = \gamma_1 q b$

$$\left(\text{At } x = \pm\frac{a}{2},\, z = 0.4b \text{ if } a \ge b \text{ or } z = 0.2b \text{ if } a < b\right)\quad \sigma_a = \frac{-\beta_2 q b^2}{t^2}\quad \text{and}\quad R = \gamma_2 q b$$

a/b	0.25	0.50	0.75	1.0	1.5	2.0	3.0
β_1	0.017	0.056	0.095	0.140	0.201	0.228	0.241
β_2	0.019	0.050	0.068	0.098	0.106	0.097	0.074
γ_1	0.101	0.177	0.227	0.262	0.294	0.301	0.301
γ_2	0.082	0.129	0.146	0.157	0.165	0.162	0.158

(Ref. 49 for $\nu = 0.2$)

| | 9ddd. Uniformly decreasing from fixed edge to zero at $\frac{1}{3}b$ | (At $x = 0$, $z = 0$)   Max $\sigma_b = \dfrac{-\beta_1 q b^2}{t^2}$ and $R = \gamma_1 q b$ |

$$\left(\text{At } x = \pm\frac{a}{2},\, z = 0.2b\right)\quad \sigma_a = \frac{-\beta_2 q b^2}{t^2}\quad \text{and}\quad R = \gamma_2 q b$$

a/b	0.25	0.50	0.75	1.0	1.5	2.0	3.0
β_1	0.014	0.035	0.047	0.061	0.075	0.080	0.082
β_2	0.010	0.024	0.031	0.030	0.025	0.020	0.013
γ_1	0.088	0.130	0.146	0.155	0.161	0.162	0.162
γ_2	0.046	0.069	0.079	0.077	0.074	0.074	0.082

(Ref. 49 for $\nu = 0.2$)

10. Rectangular plate; three edges fixed, one edge (*a*) free

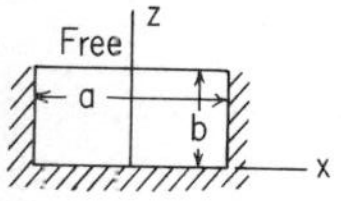

10a. Uniform over entire plate

(At $x = 0, z = 0$) Max $\sigma_b = \dfrac{-\beta_1 q b^2}{t^2}$ and $R = \gamma_1 q b$

(At $x = 0, z = b$) $\sigma_a = \dfrac{\beta_2 q b^2}{t^2}$

$\left(\text{At } x = \pm\dfrac{a}{2}, z = b\right)$ $\sigma_a = \dfrac{-\beta_3 q b^2}{t^2}$ and $R = \gamma_2 q b$

a/b	0.25	0.5	0.75	1.0	1.5	2.0	3.0
β_1	0.020	0.081	0.173	0.321	0.727	1.226	2.105
β_2	0.016	0.066	0.148	0.259	0.484	0.605	0.519
β_3	0.031	0.126	0.286	0.511	1.073	1.568	1.982
γ_1	0.114	0.230	0.341	0.457	0.673	0.845	1.012
γ_2	0.125	0.248	0.371	0.510	0.859	1.212	1.627

(Ref. 49 for $\nu = 0.2$)

10aa. Uniform over $\frac{2}{3}$ of plate from fixed edge

(At $x = 0, z = 0$) Max $\sigma_b = \dfrac{-\beta_1 q b^2}{t^2}$ and $R = \gamma_1 q b$

$\left(\text{At } x = \pm\dfrac{a}{2}, z = 0.6b \text{ for } a > b \text{ or } z = 0.4b \text{ for } a \leq b\right)$ $\sigma_a = \dfrac{-\beta_2 q b^2}{t^2}$ and $R = \gamma_2 q b$

a/b	0.25	0.50	0.75	1.0	1.5	2.0	3.0
β_1	0.020	0.080	0.164	0.277	0.501	0.710	1.031
β_2	0.031	0.110	0.198	0.260	0.370	0.433	0.455
γ_1	0.115	0.230	0.334	0.424	0.544	0.615	0.674
γ_2	0.125	0.250	0.344	0.394	0.399	0.409	0.393

(Ref. 49 for $\nu = 0.2$)

10aaa. Uniform over $\frac{1}{3}$ of plate from fixed edge

(At $x = 0, z = 0$) Max $\sigma_b = \dfrac{-\beta_1 q b^2}{t^2}$ and $R = \gamma_1 q b$

$\left(\text{At } x = \pm\dfrac{a}{2}, z = 0.2b\right)$ $\sigma_a = \dfrac{-\beta_2 q b^2}{t^2}$ and $R = \gamma_2 q b$

a/b	0.25	0.50	0.75	1.0	1.5	2.0	3.0
β_1	0.020	0.068	0.110	0.148	0.202	0.240	0.290
β_2	0.026	0.063	0.084	0.079	0.068	0.057	0.040
γ_1	0.115	0.210	0.257	0.291	0.316	0.327	0.335
γ_2	0.111	0.170	0.194	0.185	0.174	0.170	0.180

(Ref. 49 for $\nu = 0.2$)

TABLE 26 *Formulas for flat plates with straight boundaries and constant thickness* (Cont.)

Case no., shape, and supports	Case no., loading	Formulas and tabulated specific values
	10d. Uniformly decreasing from fixed edge to zero at free edge	(At $x = 0$, $z = 0$) Max $\sigma_b = \dfrac{-\beta_1 qb^2}{t^2}$ and $R = \gamma_1 qb$ $\left(\text{At } x = \pm\dfrac{a}{2},\ z = b \text{ if } a > b \text{ or } z = 0.4b \text{ if } a \leq b\right)$ $\sigma_a = \dfrac{-\beta_2 qb^2}{t^2}$ and $R = \gamma_2 qb$

a/b	0.25	0.50	0.75	1.0	1.5	2.0	3.0
β_1	0.018	0.064	0.120	0.195	0.351	0.507	0.758
β_2	0.019	0.068	0.125	0.166	0.244	0.387	0.514
γ_1	0.106	0.195	0.265	0.324	0.406	0.458	0.505
γ_2	0.075	0.151	0.211	0.242	0.106	0.199	0.313

(Ref. 49 for $\nu = 0.2$)

Case no., shape, and supports	Case no., loading	Formulas and tabulated specific values
	10dd. Uniformly decreasing from fixed edge to zero at $\frac{2}{3}b$	(At $x = 0$, $z = 0$) Max $\sigma_b = \dfrac{-\beta_1 qb^2}{t^2}$ and $R = \gamma_1 qb$ $\left(\text{At } x = \pm\dfrac{a}{2},\ z = 0.4b \text{ if } a \geq b \text{ or } z = 0.2b \text{ if } a < b\right)$ $\sigma_b = \dfrac{-\beta_2 qb^2}{t^2}$ and $R = \gamma_2 qb$

a/b	0.25	0.50	0.75	1.0	1.5	2.0	3.0
β_1	0.017	0.056	0.095	0.141	0.215	0.277	0.365
β_2	0.019	0.050	0.068	0.099	0.114	0.113	0.101
γ_1	0.102	0.177	0.227	0.263	0.301	0.320	0.336
γ_2	0.082	0.129	0.146	0.157	0.163	0.157	0.146

(Ref. 49 for $\nu = 0.2$)

10ddd. Uniformly decreasing from fixed edge to zero at $\frac{1}{3}b$

(At $x = 0, z = 0$) Max $\sigma_b = \frac{-\beta_1 q b^2}{t^2}$ and $R = \gamma_1 q b$

$\left(\text{At } x = \pm\frac{a}{2}, z = 0.2b\right)$ $\sigma_a = \frac{-\beta_2 q b^2}{t^2}$ and $R = \gamma_2 q b$

a/b	0.25	0.50	0.75	1.0	1.5	2.0	3.0
β_1	0.014	0.035	0.047	0.061	0.076	0.086	0.100
β_2	0.010	0.024	0.031	0.030	0.025	0.020	0.014
γ_1	0.088	0.130	0.146	0.156	0.162	0.165	0.167
γ_2	0.046	0.069	0.079	0.077	0.073	0.073	0.079

(Ref. 49 for $\nu = 0.2$)

11. Rectangular plate; two adjacent edges fixed, two remaining edges free

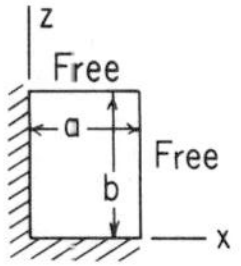

11a. Uniform over entire plate

(At $x = a, z = 0$) Max $\sigma_b = \frac{-\beta_1 q b^2}{t^2}$ and $R = \gamma_1 q b$

$\left(\text{At } x = 0, z = b \text{ if } a > \frac{b}{2} \text{ or } z = 0.8b \text{ if } a \leq \frac{b}{2}\right)$ $\sigma_c = \frac{-\beta_2 q b^2}{t^2}$ and $R = \gamma_2 q b$

a/b	0.125	0.25	0.375	0.50	0.75	1.0
β_1	0.050	0.182	0.353	0.631	1.246	1.769
β_2	0.047	0.188	0.398	0.632	1.186	1.769
γ_1	0.312	0.572	0.671	0.874	1.129	1.183
γ_2	0.127	0.264	0.413	0.557	0.829	1.183

(Ref. 49 for $\nu = 0.2$)

11aa. Uniform over plate from $z = 0$ to $z = \frac{2}{3}b$

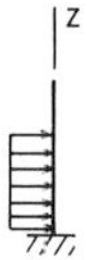

(At $x = a, z = 0$) Max $\sigma_b = \frac{-\beta_1 q b^2}{t^2}$ and $R = \gamma_1 q b$

$\left(\text{At } x = 0, z = 0.6b \text{ if } a > \frac{b}{2} \text{ or } z = 0.4b \text{ if } a \leq \frac{b}{2}\right)$ $\sigma_a = \frac{-\beta_2 q b^2}{t^2}$ and $R = \gamma_2 q b$

a/b	0.125	0.25	0.375	0.50	0.75	1.0
β_1	0.050	0.173	0.297	0.465	0.758	0.963
β_2	0.044	0.143	0.230	0.286	0.396	0.435
γ_1	0.311	0.543	0.563	0.654	0.741	0.748
γ_2	0.126	0.249	0.335	0.377	0.384	0.393

(Ref. 49 for $\nu = 0.2$)

TABLE 26 *Formulas for flat plates with straight boundaries and constant thickness* (*Cont.*)

Case no., shape, and supports	Case no., loading	Formulas and tabulated specific values

11aaa. Uniform over plate from $z = 0$ to $z = \frac{1}{3}b$

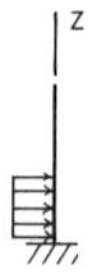

(At $x = a, z = 0$) $\quad \text{Max } \sigma_b = \dfrac{-\beta_1 q b^2}{t^2}$ and $R = \gamma_1 q b$

$\left(\text{At } x = 0, z = 0.4b \text{ if } a > \dfrac{b}{2} \text{ or } z = 0.2b \text{ if } a \le \dfrac{b}{2}\right) \quad \sigma_a = \dfrac{-\beta_2 q b^2}{t^2}$ and $R = \gamma_2 q b$

a/b	0.125	0.25	0.375	0.50	0.75	1.0
β_1	0.034	0.099	0.143	0.186	0.241	0.274
β_2	0.034	0.068	0.081	0.079	0.085	0.081
γ_1	0.222	0.311	0.335	0.343	0.349	0.347
γ_2	0.109	0.162	0.180	0.117	0.109	0.105

(Ref. 49 for $\nu = 0.2$)

11d. Uniformly decreasing from $z = 0$ to $z = b$

(At $x = a, z = 0$) $\quad \text{Max } \sigma_b = \dfrac{-\beta_1 q b^2}{t^2}$ and $R = \gamma_1 q b$

$\left(\text{At } x = 0, z = b \text{ if } a = b, \text{ or } z = 0.6b \text{ if } \dfrac{b}{2} < a < b, \text{ or } z = 0.4b \text{ if } a < \dfrac{b}{2}\right) \quad \sigma_a = \dfrac{-\beta_2 q b^2}{t^2}$ and $R = \gamma_2 q b$

a/b	0.125	0.25	0.375	0.50	0.75	1.0
β_1	0.043	0.133	0.212	0.328	0.537	0.695
β_2	0.028	0.090	0.148	0.200	0.276	0.397
γ_1	0.271	0.423	0.419	0.483	0.551	0.559
γ_2	0.076	0.151	0.205	0.195	0.230	0.192

(Ref. 49 for $\nu = 0.2$)

11dd. Uniformly decreasing from $z = 0$ to $z = \frac{2}{3}b$

(At $x = a, z = 0$) $\quad \text{Max } \sigma_b = \dfrac{-\beta_1 q b^2}{t^2}$ and $R = \gamma_1 q b$

(At $x = 0$, $z = 0.4b$ if $a \ge 0.375b$ or $z = 0.2b$ if $a < 0.375b$) $\quad \sigma_a = \dfrac{-\beta_2 q b^2}{t^2}$ and $R = \gamma_2 q b$

a/b	0.125	0.25	0.375	0.50	0.75	1.0
β_1	0.040	0.109	0.154	0.215	0.304	0.362
β_2	0.026	0.059	0.089	0.107	0.116	0.113
γ_1	0.250	0.354	0.316	0.338	0.357	0.357
γ_2	0.084	0.129	0.135	0.151	0.156	0.152

(Ref. 49 for $\nu = 0.2$)

	11ddd. Uniformly decreasing from $z = 0$ to $z = \frac{1}{3}b$	(At $x = a$, $z = 0$) $\text{Max } \sigma_b = \dfrac{-\beta_1 qb^2}{t^2}$ and $R = \gamma_1 qb$

(At $x = 0$, $z = 0.2b$) $\sigma_a = \dfrac{-\beta_2 qb^2}{t^2}$ and $R = \gamma_2 qb$

a/b	0.125	0.25	0.375	0.50	0.75	1.0
β_1	0.025	0.052	0.071	0.084	0.100	0.109
β_2	0.014	0.028	0.031	0.029	0.025	0.020
γ_1	0.193	0.217	0.170	0.171	0.171	0.171
γ_2	0.048	0.072	0.076	0.075	0.072	0.072

(Ref. 49 for $\nu = 0.2$)

---

12. Continuous plate; supported at equal intervals a on circular supports of radius r_0

12a. Uniform over entire surface

(At edge of support)

$$\sigma_a = \frac{0.15q}{t^2}\left(a - \frac{4}{3}\,r_0\right)^2\left(\frac{1}{n} + 4\right) \qquad \text{when } 0.15 < n < 0.30$$ (Ref. 9)

$$\text{or } \sigma_a = \frac{3qa^2}{2\pi t^2}\left[(1 + \nu)\ln\frac{a}{r_0} - 21(1-\nu)\frac{r_0^2}{a^2} - 0.55 - \frac{1.50}{\nu}\right] \qquad \text{when } n < 0.15$$ (Ref. 11)

$$\text{where } n = \frac{2r_0}{a}$$

---

13. Continuous plate; supported continuously on an elastic foundation of modulus k (lb/in²/in)

13b. Uniform over a small circle of radius r_0, remote from edges

(Under the load)

$$\text{Max } \sigma = \frac{3W(1+\nu)}{2\pi t^2}\left(\ln\frac{L_e}{r_0} + 0.6159\right) \qquad \text{where } L_e = \sqrt[4]{\frac{Et^3}{12(1-\nu^2)k}}$$ (Ref. 14)

$$\text{Max foundation pressure } q_0 = \frac{W}{8L_e^2}$$

$$\text{Max deflection } y = \frac{-W}{8kL_e^2}$$

TABLE 26 ***Formulas for flat plates with straight boundaries and constant thickness*** **(*Cont.*)**

<table>
<tr><th>Case no., shape, and supports</th><th>Case no., loading</th><th>Formulas and tabulated specific values</th></tr>
<tr><td></td><td>13bb. Uniform over a small circle of radius r_o, adjacent to edge but remote from corner</td><td>(Under the load)
$$\text{Max } \sigma = \frac{0.863W(1+\nu)}{t^2}\left(\ln \frac{L_e}{r_o} + 0.207\right)$$
$$\text{Max deflection } y = 0.408(1+0.4\nu)\frac{W}{kL_e^2}$$
(Ref. 14)</td></tr>
<tr><td></td><td>13bbb. Uniform over a small circle of radius r_o, adjacent to a corner</td><td>(At the corner) $\quad \text{Max deflection } y = \left(1.1 - 1.245\frac{r_o}{L_e}\right)\frac{W}{kL_e^2}$
(At a distance $= 2.38\sqrt{r_o L_e}$ from the corner along diagonal)
$$\text{Max } \sigma = \frac{3W}{t^2}\left[1 - 1.083\left(\frac{r_o}{L_e}\right)^{0.6}\right]$$
(Ref. 14)</td></tr>
<tr><td>14. Parallelogram plate (skew slab); all edges simply supported
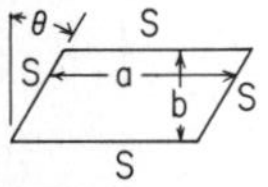
</td><td>14a. Uniform over entire plate</td><td>(At center of plate) $\quad \text{Max } \sigma = \frac{\beta q b^2}{t^2}$ and $\text{max } y = \frac{\alpha q b^4}{Et^3}$
For $a/b = 2.0$
<table><tr><th>θ</th><th>0°</th><th>30°</th><th>45°</th><th>60°</th><th>75°</th></tr><tr><td>β</td><td>0.585</td><td>0.570</td><td>0.539</td><td>0.463</td><td>0.201</td></tr><tr><td>α</td><td>0.119</td><td>0.118</td><td>0.108</td><td>0.092</td><td>0.011</td></tr></table>(Ref. 24 for $\nu = 0.2$)</td></tr>
<tr><td>15. Parallelogram plate (skew slab); shorter edges simply supported, longer edges free
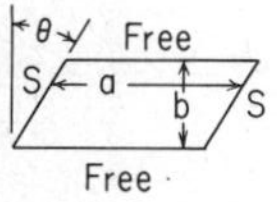
</td><td>15a. Uniform over entire plate</td><td>(Along free edge) $\quad \text{Max } \sigma = \frac{\beta_1 q b^2}{t^2}$ and $\text{max } y = \frac{\alpha_1 q b^4}{Et^3}$
(At center of plate) $\quad \text{Max } \sigma = \frac{\beta_2 q b^2}{t^2}$ and $\text{max } y = \frac{\alpha_2 q b^4}{Et^3}$
For $a/b = 2.0$
<table><tr><th>θ</th><th>0°</th><th>30°</th><th>45°</th><th>60°</th></tr><tr><td>β_1</td><td>3.05</td><td>2.20</td><td>1.78</td><td>0.91</td></tr><tr><td>β_2</td><td>2.97</td><td>2.19</td><td>1.75</td><td>1.00</td></tr><tr><td>α_1</td><td>2.58</td><td>1.50</td><td>1.00</td><td>0.46</td></tr><tr><td>α_2</td><td>2.47</td><td>1.36</td><td>0.82</td><td>0.21</td></tr></table>(Ref. 24 for $\nu = 0.2$)</td></tr>
</table>

16. Parallelogram plate (skew slab); all edges fixed

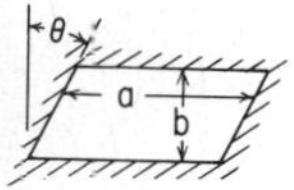

16a. Uniform over entire plate

(Along longer edge toward obtuse angle) $\text{Max } \sigma = \dfrac{\beta_1 q b^2}{t^2}$

(At center of plate) $\sigma = \dfrac{\beta_2 q b^2}{t^2}$ and $\max y = \dfrac{\alpha q b^4}{E t^3}$

θ	a/b	1.00	1.25	1.50	1.75	2.00	2.25	2.50	3.00
0°	β_1	0.308	0.400	0.454	0.481	0.497			
	β_2	0.138	0.187	0.220	0.239	0.247			
	α	0.0135	0.0195	0.0235	0.0258	0.0273			
15°	β_1	0.320	0.412	0.483	0.531	0.553			
	β_2	0.135	0.200	0.235	0.253	0.261			
	α	0.0127	0.0189	0.0232	0.0257	0.0273			
30°	β_1		0.400	0.495	0.547	0.568	0.580		
	β_2		0.198	0.221	0.235	0.245	0.252		
	α		0.0168	0.0218	0.0249	0.0268	0.0281		
45°	β_1			0.394	0.470	0.531	0.575	0.601	
	β_2			0.218	0.244	0.260	0.265	0.260	
	α			0.0165	0.0208	0.0242	0.0265	0.0284	
60°	β_1					0.310	0.450	0.538	0.613
	β_2					0.188	0.204	0.214	0.224
	α					0.0136	0.0171	0.0198	0.0245

(Ref. 53 for $\nu = \frac{1}{3}$)

17. Equilateral triangle; all edges simply supported

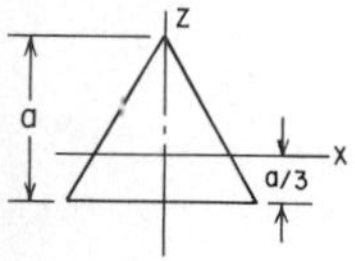

17a. Uniform over entire plate

(At $x = 0, z = -0.062a$) $\text{Max } \sigma_z = \dfrac{0.1488 q a^2}{t^2}$

(At $x = 0, z = 0.129a$) $\text{Max } \sigma_x = \dfrac{0.1554 q a^2}{t^2}$

(At $x = 0, z = 0$) $\text{Max } y = \dfrac{-q a^4 (1 - \nu^2)}{81 E t^3}$

(Refs. 21 and 23 for $\nu = 0.3$)

17b. Uniform over small circle of radius r_o at $x = 0, z = 0$

(At $x = 0, z = 0$) $\text{Max } \sigma = \dfrac{3W}{2\pi t^2}\left[(1 + \nu) \ln \dfrac{0.378a}{r'_o} + 1\right]$

$\text{Max } y = 0.069 W (1 - \nu^2) a^4 / E t^3$

TABLE 26 ***Formulas for flat plates with straight boundaries and constant thickness* (*Cont.*)**

Case no., shape, and supports	Case no., loading	Formulas and tabulated specific values
18. Right-angle isosceles triangle; all edges simply supported 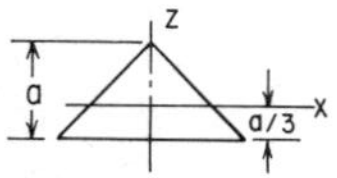	18a. Uniform over entire plate	$\text{Max } \sigma = \sigma_z = \frac{0.262qa^2}{t^2}$ $\text{Max } \sigma_x = \frac{0.225qa^2}{t^2}$ $\text{Max } y = \frac{0.038qa^4}{Et^3}$ (Ref. 21 for $\nu = 0.3$)
19. Regular polygonal plate; all edges simply supported  Number of sides $= n$	19a. Uniform over entire plate	(At center) $\sigma = \frac{\beta qa^2}{t^2}$ and $\max y = \frac{-\alpha qa^4}{Et^3}$ (At center of straight edge) $\text{Max slope} = \frac{\xi qa^3}{Et^3}$

n	3	4	5	6	7	8	9	10	15	∞
β	1.302	1.152	1.086	1.056	1.044	1.038	1.038	1.044	1.074	1.236
α	0.910	0.710	0.635	0.599	0.581	0.573	0.572	0.572	0.586	0.695
ξ	1.535	1.176	1.028	0.951	0.910	0.888	0.877	0.871	0.883	1.050

(Ref. 55 for $\nu = 0.3$)

Case no., shape, and supports	Case no., loading	Formulas and tabulated specific values
20. Regular polygonal plate; all edges fixed  Number of sides $= n$	20a. Uniform over entire plate	(At center) $\sigma = \frac{\beta_1 qa^2}{t^2}$ and $\max y = \frac{-\alpha qa^4}{Et^3}$ (At center of straight edge) $\text{Max } \sigma = \frac{-\beta_2 qa^2}{t^2}$

n	3	4	5	6	7	8	9	10	∞
β_1	0.589	0.550	0.530	0.518	0.511	0.506	0.503	0.500	0.4875
β_2	1.423	1.232	1.132	1.068	1.023	0.990	0.964	0.944	0.750
α	0.264	0.221	0.203	0.194	0.188	0.184	0.182	0.180	0.171

(Ref. 55 for $\nu = 0.3$)

many more cases can be found in the literature. Bareš (Ref. 47) presents tabulated values of bending moments and deflections for a series of plates in the form of *isosceles triangles* and *symmetric trapezoids* for linearly varying lateral pressures and for values of Poisson's ratio of 0.0 and 0.16. Tabulated values are also given for *skew* plates with uniform lateral loading and concentrated lateral loads for the support condition where two opposite edges are simply supported and two edges are free; the value of Poisson's ratio used was zero. In addition to many cases also included in Table 26, Marguerre and Woernle (Ref. 50) give results for line loading and uniform loading on a narrow strip across a rectangular plate. They also discuss the case of a rectangular plate supported within the span by elastic cross beams. Morley (Ref. 51) discusses solutions of problems involving *parallelogram*, or *skew*, plates and box structures. A few graphs and tables of results are given.

For plates with boundary shapes or restraints not discussed in the literature, we can only approximate an answer or resort to a direct numerical solution of the problem at hand. All numerical methods are approximate but can be carried to any degree of accuracy desired at the expense of time and computer costs. There are many numerical techniques used to solve plate problems, and the choice of a method for a given problem can be difficult. Leissa et al. (Ref. 56) have done a very complete and competent job of comparing and rating 9 approximate numerical methods on the basis of 11 different criteria. Szilard (Ref. 84) discusses both classical and numerical methods and tabulates many solutions.

Variable thickness. Petrina and Conway (Ref. 63) give numerical data for two sets of boundary conditions, three aspect ratios, and two nearly linear tapers in plate thickness. The loading was uniform and they found that the center deflection and center moment differed little from the same uniform-thickness case using the average thickness; the location and magnitude of maximum stress, however, did vary.

10.11 *Effect of large deflection; diaphragm stresses*

When the deflection becomes larger than about one-half the thickness, as may occur in thin plates, the middle surface becomes appreciably strained and the stress in it cannot be ignored. This stress, called *diaphragm* stress, or *direct* stress, enables the plate to carry part of the load as a diaphragm in direct tension. This tension may be balanced by radial tension at the edges if the edges are *held* or by circumferential compression if the edges are not horizontally restrained. In thin plates this circumferential compression may cause buckling.

When this condition of large deflection obtains, the plate is stiffer than indicated by the ordinary theory and the load-deflection and load-stress relations are nonlinear. Stresses for a given load are less and stresses for a given deflection are generally greater than the ordinary theory indicates.

Circular plates. Formulas for stress and deflection when middle surface stresses are taken into account are given below. These formulas should be used whenever the maximum deflection exceeds half the thickness if accurate results are desired. The table on page 407 gives the necessary constants for the several loadings and support conditions listed.

Let t = thickness of plate; a = outer radius of plate; q = unit load (pounds per square inch), assumed uniform over entire area; y = maximum deflection; σ_b = bending stress; σ_d = diaphragm stress; $\sigma = \sigma_b + \sigma_d$ = maximum stress due to flexure and diaphragm tension combined. Then the following formulas apply:

$$\frac{qa^4}{Et^4} = K_1\frac{y}{t} + K_2\left(\frac{y}{t}\right)^3 \qquad (1)$$

$$\frac{\sigma a^2}{Et^2} = K_3\frac{y}{t} + K_4\left(\frac{y}{t}\right)^2 \qquad (2)$$

First solve for y in Eq. 1 and then obtain the stresses from Eq. 2.

EXAMPLE

For the plate of Example 1 of Art. 10.2 (page 326) it is desired to determine the maximum deflection and maximum stress under a load of 10 lb/in^2.

Solution. If the linear theory held, the stresses and deflections would be directly proportional to the load, which would indicate a maximum stress of 9240(10)/3 = 30,800 lb/in^2 and a maximum deflection of 0.0883(10)/3 = 0.294 in. Since this deflection is much more than half the thickness, Eqs. 1 and 2 with the constants from case 1 in the table will be used to solve for the deflection and stress. From Eq. 1, we obtain

$$\frac{10(10^4)}{30(10^6)(0.2^4)} = \frac{1.016}{1-0.3}\frac{y}{t} + 0.376\left(\frac{y}{t}\right)^3$$

$$2.0833 = 1.4514\left(\frac{y}{t}\right) + 0.376\left(\frac{y}{t}\right)^3$$

Starting with a trial value for y somewhat less than 0.294 in, a solution is found when y = 0.219 in. From Eq. 2 the maximum stress is found to be 27,500 lb/in^2.

In addition to the limited data presented in the table, Mah (Ref. 71) has analyzed numerically and presented graphically data for solid circular plates under nine different combinations of loading and edge conditions and for annular plates under three such combinations.

Elliptical plates. Nash and Cooley (Ref. 72) present graphically the results of a uniform pressure on a clamped elliptical plate for $a/b = 2$. Their method of solution is presented in detail, and the numerical solution is compared with experimental results and with previous solutions they have referenced. Ng (Ref. 73) has tabulated the values of center deflection for clamped

Circular plates under uniform load producing large deflections

Case no., edge condition	Constants
1. Simply supported (neither fixed nor held)	$K_1 = \dfrac{1.016}{1-\nu}$ $\quad K_2 = 0.376$ $K_3 = \dfrac{1.238}{1-\nu}$ $\quad K_4 = 0.294$ (Ref. 5)
2. Fixed but not held (no edge tension)	$K_1 = \dfrac{5.33}{1-\nu^2}$ $\quad K_2 = 0.857$ (At center) $\quad K_3 = \dfrac{2}{1-\nu}$ $\quad K_4 = 0.50$ (At edge) $\quad K_3 = \dfrac{4}{1-\nu^2}$ $\quad K_4 = 0.0$ (Ref. 5)
3. Fixed and held	$K_1 = \dfrac{5.33}{1-\nu^2}$ $\quad K_2 = \dfrac{2.6}{1-\nu^2}$ (At center) $\quad K_3 = \dfrac{2}{1-\nu}$ $\quad K_4 = 0.976$ (At edge) $\quad K_3 = \dfrac{4}{1-\nu^2}$ $\quad K_4 = 0.476$ (Refs. 15 and 16)
4. Diaphragm without flexural stiffness, edge held	$K_1 = 0.0$ $\quad K_2 = 3.44$ (At center) $\quad K_3 = 0.0$ $\quad K_4 = 0.965$ (At edge) $\quad K_3 = 0.0$ $\quad K_4 = 0.748$ (At r from the center) $\quad y = \max y\left(1 - 0.9\dfrac{r^2}{a^2} - 0.1\dfrac{r^5}{a^5}\right)$ (Refs. 18 and 29)

elliptical plates on elastic foundations for ratios of a/b from 1 to 2 and for a wide range of foundation moduli. Large deflections are also graphed for two ratios a/b (1.5 and 2) for the same range of foundation moduli.

Rectangular plates. Analytical solutions for uniformly loaded rectangular plates with large deflections are given in Refs. 30 to 34, where the relations among load, deflection, and stress are expressed by numerical values of the dimensionless coefficients y/t, qb^4/Et^4, and $\sigma b^2/Et^2$. The values of these coefficients given in the table on page 408 are taken from these references and are for $\nu = 0.316$. In this table, a, b, q, E, y, and t have the same meaning as in Table 26; σ_d is the diaphragm stress, and σ is the total stress found by adding the diaphragm stress and the bending stress.

In Ref. 35 experimentally determined deflections are given and compared

Rectangular plates under uniform load producing large deflection

a/b	Edges and point of max σ	Coef.	qb^4/Et^4										
			0	12.5	25	50	75	100	125	150	175	200	250
1	Held, not fixed	y/t	0	0.430	0.650	0.930	1.13	1.26	1.37	1.47	1.56	1.63	1.77
	At center of plate	$\sigma_d b^2/Et^2$	0	0.70	1.60	3.00	4.00	5.00	6.10	7.00	7.95	8.60	10.20
		$\sigma b^2/Et^2$	0	3.80	5.80	8.70	10.90	12.80	14.30	15.60	17.00	18.20	20.50
1	Held and riveted	y/t	0	0.406	0.600	0.840	1.00	1.13	1.23	1.31	1.40	1.46	1.58
	At center of plate	$\sigma_d b^2/Et^2$	0	0.609	1.380	2.68	3.80	4.78	5.75	6.54	7.55	8.10	9.53
		$\sigma b^2/Et^2$	0	3.19	5.18	7.77	9.72	11.34	12.80	14.10	15.40	16.40	18.40
1	Held and fixed	y/t	0	0.165	0.25	0.59	0.80	0.95	1.08	1.19	1.28	1.38	1.54
	At center of long edges	$\sigma_d b^2/Et^2$	0	0.070	0.22	0.75	1.35	2.00	2.70	3.30	4.00	4.60	5.90
		$\sigma b^2/Et^2$	0	3.80	6.90	14.70	21.0	26.50	31.50	36.20	40.70	45.00	53.50
	At center of plate	$\sigma_d b^2/Et^2$	0	0.075	0.30	0.95	1.65	2.40	3.10	3.80	4.50	5.20	6.50
		$\sigma b^2/Et^2$	0	1.80	3.50	6.60	9.20	11.60	13.0	14.50	15.80	17.10	19.40
1.5	Held, not fixed	y/t	0	0.625	0.879	1.18	1.37	1.53	1.68	1.77	1.88	1.96	2.12
	At center of plate	$\sigma_d b^2/Et^2$	0	1.06	2.11	3.78	5.18	6.41	7.65	8.60	9.55	10.60	12.30
		$\sigma b^2/Et^2$	0	4.48	6.81	9.92	12.25	14.22	16.0	17.50	18.90	20.30	22.80
2 to ∞	Held, not fixed	y/t	0	0.696	0.946	1.24	1.44	1.60	1.72	1.84	1.94	2.03	2.20
	At center of plate	$\sigma_d b^2/Et^2$	0	1.29	2.40	4.15	5.61	6.91	8.10	9.21	10.10	10.90	13.20
		$\sigma b^2/Et^2$	0	4.87	7.16	10.30	12.60	14.60	16.40	18.00	19.40	20.90	23.60
1.5 to ∞	Held and fixed	y/t	0	0.28	0.51	0.825	1.07	1.24	1.40	1.50	1.63	1.72	1.86
	At center of long edges	$\sigma_d b^2/Et^2$	0	0.20	0.66	1.90	3.20	4.35	5.40	6.50	7.50	8.50	10.30
		$\sigma b^2/Et^2$	0	5.75	11.12	20.30	27.8	35.0	41.0	47.0	52.50	57.60	67.00

with those predicted by theory. In Ref. 74 a numerical solution for uniformly loaded rectangular plates with simply supported edges is discussed, and the results for a square plate are compared with previous approximate solutions. Graphs are presented to show how stresses and deflections vary across a square plate.

Parallelogram plates. Kennedy and Ng (Ref. 53) present several graphs of large elastic deflections and the accompanying stresses for uniformly loaded skew plates with clamped edges. Several aspect ratios and skew angles are represented.

10.12 *Plastic analysis of plates*

The onset of yielding in plates may occur before the development of appreciable diaphragm stress if the plate is relatively thick. For thinner plates, the nonlinear increase in stiffness due to diaphragm stresses is counteracted by the decrease in stiffness which occurs when the material starts to yield (Ref. 80). Save and Massonnet (Ref. 81) discuss the effect of the several yield criteria on the response of circular and rectangular plates under various loadings and give an extensive list of references. They also compare the results of theory with referenced experiments which have been performed. *Orthotropy* in plates can be caused by cold-forming the material or by the positioning of stiffeners. The effect of this orthotropic behavior on the yielding of circular plates is discussed by Save and Massonnet (Ref. 81) as well as by Markowitz and Hu (Ref. 82).

Crose and Ang (Ref. 83) describe an iterative solution scheme which first solves the elastic case and then increments the loading upward to allow a slow expansion of the yielded volume after it forms. The results of a test on a clamped plate are compared favorably with a theoretical solution.

10.13 *Ultimate strength*

Plates of brittle material fracture when the actual maximum tensile stress reaches the ultimate tensile strength of the material. A flat-plate modulus of rupture, analogous to the modulus of rupture of a beam, may be determined by calculating the (fictitious) maximum stress corresponding to the breaking load, using for this purpose the appropriate formula for elastic stress. This flat-plate modulus of rupture is usually greater than the modulus of rupture determined by testing a beam of rectangular section.

Plates of ductile material fail by excessive plastic deflection, as do beams of similar material. For a number of cases the load required to produce collapse has been determined analytically, and the results for some of the simple loadings are summarized as follows:

Circular plate; uniform load, edges simply supported:

$$W_u = \sigma_y(\tfrac{3}{2}\pi t^2) \quad \text{(Ref. 43)}$$

Circular plate; uniform load, fixed edges:

$$W_u = \sigma_y(2.814\pi t^2) \quad \text{(Ref. 43)}$$

(For collapse loads on partially loaded orthotropic annular plates see Refs. 81 and 82.)

Rectangular plate, length a, width b; uniform load, edges supported:

$$W_u = \beta\sigma_y t^2$$

where β depends on the ratio of b to a and has the following values (Ref. 44):

$b/a = 1$	0.9	0.8	0.7	0.6	0.5	0.4	0.3	0.2
$\beta = 5.48$	5.50	5.58	5.64	5.89	6.15	6.70	7.68	9.69

Plate of any shape and size, any type of edge support, concentrated load at any point:

$$W_u = \sigma_y(\tfrac{1}{2}\pi t^2) \quad \text{(Ref. 45)}$$

In each of the above cases W_u denotes the total load required to collapse the plate (pounds), t the thickness of the plate (inches), and σ_y the yield point of the material (pounds per square inch). Accurate prediction of W_u is hardly to be expected; the theoretical error in some of the formulas may range up to 30 percent, and few experimental data seem to be available.

REFERENCES

1. Roark, R. J.: Stresses Produced in a Circular Plate by Eccentric Loading and by a Transverse Couple, *Univ. Wis. Eng. Exp. Sta., Bull.* 74, 1932. The deflection formulas are due to Föppl. See Die Biegung einer kreisförmigen Platte, *Sitzungsber. math.-phys. Kl. K. B. Akad. Wiss. Münch.,* p. 155, 1912.
2. Michell, J. H.: The Flexure of Circular Plates, *Proc. Math. Soc. Lond.,* p. 223, 1901.
3. Morley, A.: "Strength of Materials," Longmans, Green & Co., Ltd., 1919.
4. Timoshenko, S.: Über die Biegung der allseitig unterstützten rechteckigen Platte unter Wirkung einer Einzellast, *Der Bauingenieur,* vol. 3, Jan. 31, 1922.
5. Prescott, J.: "Applied Elasticity," Longmans, Green & Co., Ltd., 1924.
6. Nadai, A.: Über die Spannungsverteilung in einer durch eine Einzelkraft belasteten rechteckigen Platte, *Der Bauingenieur,* vol. 2, Jan. 15, 1921.
7. Timoshenko, S., and J. M. Lessells: "Applied Elasticity," Westinghouse Technical Night School Press, 1925.
8. Wojtaszak, I. A.: Stress and Deflection of Rectangular Plates, *ASME Paper* A-71, *J. Appl. Mech.,* vol. 3, no. 2, 1936.
9. Westergaard, H. M., and A. Slater: Moments and Stresses in Slabs, *Proc. Am. Concr. Inst.,* vol. 17, 1921.
10. Wahl, A. M.: Strength of Semicircular Plates and Rings under Uniform External Pressure, *Trans. ASME,* vol. 54, no. 23, 1932.
11. Nadai, A.: Die Formänderungen und die Spannungen von durchlaufenden Platten, *Der Bauingenieur,* vol. 5, p. 102, 1924.
12. Nadai, A.: "Elastische Platten," J. Springer, Berlin, 1925.
13. Holl, D. L.: Analysis of Thin Rectangular Plates Supported on Opposite Edges, *Iowa Eng. Exp. Sta., Iowa State College, Bull.* 129, 1936.

14. Westergaard, H. M.: Stresses in Concrete Pavements Computed by Theoretical Analysis, *Public Roads,* U.S. Dept. of Agriculture, Bureau of Public Roads, vol. 7, no. 2, 1926.
15. Timoshenko, S.: "Vibration Problems in Engineering," p. 319, D. Van Nostrand Company, Inc., 1928.
16. Way, S.: Bending of Circular Plates with Large Deflection, *Trans. ASME,* vol. 56, no. 8, 1934 (see also discussion by E. O. Waters).
17. Sturm, R. G., and R. L. Moore: The Behavior of Rectangular Plates under Concentrated Load, *ASME Paper* A-75, *J. Appl. Mech.,* vol. 4, no. 2, 1937.
18. Hencky, H.: "Über den Spannungszustand in kreisrunder Platten mit verschwindender Biegungssteifigkeit," *Z. Math. Phys.,* vol. 63, p. 311, 1915.
19. Wahl, A. M.: Stresses and Deflections in Flat Circular Plates with Central Holes, *Trans. ASME Paper* APM-52-3, vol. 52(1), p. 29, 1930.
20. Flügge, W.: Kreisplatten mit linear veränderlichen Belastungen, *Bauingenieur,* vol. 10, no. 13, p. 221, 1929.
21. Timoshenko, S., and S. Woinowsky-Krieger: "Theory of Plates and Shells," 2d ed., McGraw-Hill Book Company, 1959.
22. Reissner, H.: Über die unsymmetrische Biegung dünner Kreisringplatte, *Ing.-Arch.,* vol. 1, p. 72, 1929.
23. Woinowsky-Krieger, S.: Berechnung der ringsum frei aufliegenden gleichseitigen Dreiecksplatte, *Ing.-Arch.,* vol. 4, p. 254, 1933.
24. Jensen, V. P.: Analysis of Skew Slabs, *Eng. Exp. Sta. Univ. Ill., Bull.* 332, 1941.
25. Evans, T. H.: Tables of Moments and Deflections for a Rectangular Plate Fixed at All Edges and Carrying a Uniformly Distributed Load, *ASME J. Appl. Mech.,* vol. 6, no. 1, March 1939.
26. Young, D.: Clamped Rectangular Plates with a Central Concentrated Load, *ASME Paper* A-114, *J. Appl. Mech.,* vol. 6, no. 3, 1939.
27. Almen, J. O., and A. Laszlo: The Uniform-section Disc Spring, *Trans. ASME,* vol. 58, p. 305, 1936.
28. Odley, E. G.: Deflections and Moments of a Rectangular Plate Clamped on all Edges and under Hydrostatic Pressure, *ASME J. Appl. Mech.,* vol. 14, no. 4, December 1947.
29. Stevens, H. H.: Behavior of Circular Membranes Stretched above the Elastic Limit by Air Pressure, *Exp. Stress Anal.,* vol. 2, no. 1, 1944.
30. Levy, S.: Bending of Rectangular Plates with Large Deflections, *Natl. Adv. Comm. Aeron., Tech. Note* 846, 1942.
31. Levy, S.: Square Plate with Clamped Edges under Normal Pressure Producing Large Deflections, *Natl. Adv. Comm. Aeron., Tech. Note* 847, 1942.
32. Levy, S., and S. Greenman: Bending with Large Deflection of a Clamped Rectangular Plate with Length-width Ratio of 1.5 under Normal Pressure, *Natl. Adv. Comm. Aeron., Tech. Note* 853, 1942.
33. Chi-Teh Wang: Nonlinear Large Deflection Boundary-value Problems of Rectangular Plates, *Natl. Adv. Comm. Aeron., Tech. Note* 1425, 1948.
34. Chi-Teh Wang: Bending of Rectangular Plates with Large Deflections, *Natl. Adv. Comm. Aeron., Tech. Note* 1462, 1948.
35. Ramberg, W., A. E. McPherson, and S. Levy: Normal Pressure Tests of Rectangular Plates, *Natl. Adv. Comm. Aeron., Rept.* 748, 1942.
36. Conway, H. D.: The Bending of Symmetrically Loaded Circular Plates of Variable Thickness, *ASME J. Appl. Mech.,* vol. 15 no. 1, March 1948.
37. Reissmann, Herbert: Bending of Clamped Wedge Plates, *ASME J. Appl. Mech.,* vol. 20, March 1953.
38. Bassali, W. A., and R. H. Dawoud: Bending of an Elastically Restrained Circular Plate under Normal Loading on a Sector, *ASME J. Appl. Mech.,* vol. 25, no. 1, March 1958.
39. Bassali, W. A., and M. Nassif: Stresses and Deflections in Circular Plate Loaded over a Segment, *ASME J. Appl. Mech.,* vol. 26, no. 1, March 1959.

40. Jurney, W. H.: Displacements and Stresses of a Laterally Loaded Semicircular Plate with Clamped Edges, *ASME J. Appl. Mech.,* vol. 26, no. 2, June 1959.
41. Conway, H. D.: Bending of Rectangular Plates Subjected to a Uniformly Distributed Lateral Load and to Tensile or Compressive Forces in the Plane of the Plate, *ASME J. Appl. Mech.,* vol. 16, no. 3, September 1949.
42. Morse, R. F., and H. D. Conway: The Rectangular Plate Subjected to Hydrostatic Tension and to Uniformly Distributed Lateral Load, *ASME J. Appl. Mech.,* vol. 18, no. 2, June 1951.
43. Hodge, P. G., Jr.: "Plastic Analysis of Structures," McGraw-Hill Book Company, 1959.
44. Shull, H. E., and L. W. Hu: Load-carrying Capacities of Simply Supported Rectangular Plates, *ASME J. Appl. Mech.,* vol. 30, no. 4, December 1963.
45. Zaid, M.: Carrying Capacity of Plates of Arbitrary Shape, *ASME J. Appl. Mech.,* vol. 25, no. 4, December 1958.
46. Márkus, G.: "Theorie und Berechnung rotationssymmetrischer Bauwerke," Werner-Verlag, 1967.
47. Bareś, R.: "Tables for the Analysis of Plates, Slabs and Diaphragms Based on the Elastic Theory," Bauverlag GmbH., 1969 (English transl. by Carel van Amerongen).
48. Heap, J.: Bending of Circular Plates Under a Variable Symmetrical Load, *Argonne Natl. Lab. Bull.* 6882, 1964.
49. Moody, W.: Moments and Reactions for Rectangular Plates, *Bur. Reclamation Eng. Monogr.* 27, 1960.
50. Marguerre, K., and H. Woernle: "Elastic Plates," Blaisdell Publishing Company, 1969.
51. Morley, L.: "Skew Plates and Structures," The Macmillan Company, 1963.
52. Hodge, P.: "Limit Analysis of Rotationally Symmetric Plates and Shells," Prentice-Hall, Inc., 1963.
53. Kennedy, J., and S. Ng: Linear and Nonlinear Analyses of Skewed Plates, *ASME J. Appl. Mech.,* vol. 34, no. 2, June 1967.
54. Monforton, G., and L. Schmit, Jr.: Finite Element Analysis of Skew Plates in Bending, *AIAA J.,* vol. 6, no. 6, June 1968.
55. Leissa, A., C. Lo, and F. Niedenfuhr: Uniformly Loaded Plates of Regular Polygonal Shape, *AIAA J.,* vol. 3, no. 3, March 1965.
56. Leissa, A., W. Clausen, L. Hulbert, and A. Hopper: A Comparison of Approximate Methods for the Solution of Plate Bending Problems, *AIAA J.,* vol. 7, no. 5, May 1969.
57. Stanek, F. J.: "Stress Analysis of Circular Plates and Cylindrical Shells," Dorrance & Co., Inc., 1970.
58. Griffel, W.: "Plate Formulas," Frederick Ungar Publishing Co., 1968.
59. Tuncel, Özcan: Circular Ring Plates Under Partial Arc Loading, *ASME J. Appl. Mech.,* vol. 31, no. 2, June 1964.
60. Lee, T. M.: Flexure of Circular Plate by Concentrated Force, Proc. *Am. Soc. Civil Eng., Eng. Mech. Div.,* vol. 94, no. 3, June 1968.
61. Essenburg, F., and S. T. Gulati: On the Contact of Two Axisymmetric Plates, *ASME J. Appl. Mech.,* vol. 33, no. 2, June 1966.
62. Slot, T., and W. J. O'Donnell: Effective Elastic Constants for Thick Perforated Plates with Square and Triangular Penetration Patterns, *ASME J. Eng. Ind.,* vol. 11, no. 4, November 1971.
63. Petrina, P., and H. D. Conway: Deflection and Moment Data for Rectangular Plates of Variable Thickness, *ASME J. Appl. Mech.,* vol. 39, no. 3, September 1972.
64. Conway, H. D.: Nonaxial Bending of Ring Plates of Varying Thickness, *ASME J. Appl. Mech.,* vol. 25, no. 3, September 1958.
65. Strock, R. R., and Yi-Yuan Yu: Bending of a Uniformly Loaded Circular Disk With a Wedge-Shaped Cross Section, *ASME J. Appl. Mech.,* vol. 30, no. 2, June 1963.
66. Conway, H. D.: The Ribbed Circular Plate, *ASME J. Appl. Mech.,* vol. 30, no. 3, September 1963.

67. Wempner, G. A.: The Conical Disc Spring, *Proc. 3rd Natl. Congr. Appl. Mech., ASME,* 1958.
68. Wempner, G. A.: Axisymmetric Deflections of Shallow Conical Shells, *Proc. Am. Soc. Civil Eng., Eng. Mech. Div.,* vol. 90, no. 2, April 1964.
69. Owens, J. H., and D. C. Chang: Belleville Springs Simplified, *Mach. Des.,* May 14, 1970.
70. Paul, B.: Collapse Loads of Rings and Flanges Under Uniform Twisting Moment and Radial Force, *ASME J. Appl. Mech.,* vol. 26, no. 2, June 1959.
71. Mah, G. B. J.: Axisymmetric Finite Deflection of Circular Plates, *Proc. Am. Soc. Civil Eng., Eng. Mech. Div.,* vol. 95, no. 5, October 1969.
72. Nash, W. A., and I. D. Cooley: Large Deflections of a Clamped Elliptical Plate Subjected to Uniform Pressure, *ASME J. Appl. Mech.,* vol. 26, no. 2, June 1959.
73. Ng, S. F.: Finite Deflection of Elliptical Plates on Elastic Foundations, *AIAA J.,* vol. 8, no. 7, July 1970.
74. Bauer, F., L. Bauer, W. Becker, and E. Reiss: Bending of Rectangular Plates With Finite Deflections, *ASME J. Appl. Mech.,* vol. 32, no. 4, December 1965.
75. Dundurs, J., and Tung-Ming Lee: Flexure by a Concentrated Force of the Infinite Plate on a Circular Support, *ASME J. Appl. Mech.,* vol. 30, no. 2, June 1963.
76. Amon, R., and O. E. Widera: Clamped Annular Plate under a Concentrated Force, *AIAA J.,* vol. 7, no. 1, January 1969.
77. Amon, R., O. E. Widera, and R. G. Ahrens: Problem of the Annular Plate, Simply Supported and Loaded with an Eccentric Concentrated Force, *AIAA J.,* vol. 8, no. 5, May 1970.
78. Widera, O. E., R. Amon, and P. L. Panicali: On the Problem of the Infinite Elastic Plate with a Circular Insert Reinforced by a Beam, *Nuclear Eng. Des.,* vol. 12, no. 3, June 1970.
79. Amon, R., O. E. Widera, and S. M. Angel: Green's Function of an Edge-Beam Reinforced Circular Plate, *Proc. CANCAM, Calgary,* May 1971.
80. Ohashi, Y., and S. Murakami: Large Deflection in Elastoplastic Bending of a Simply Supported Circular Plate Under a Uniform Load, *ASME J. Appl. Mech.,* vol. 33, no. 4, December 1966.
81. Save, M. A., and C. E. Massonnet: "Plastic Analysis and Design of Plates, Shells and Disks," North-Holland Publishing Company, 1972.
82. Markowitz, J., and L. W. Hu: Plastic Analysis of Orthotropic Circular Plates, *Proc. Am. Soc. Civil Eng., Eng. Mech. Div.,* vol. 90, no. 5, October 1964.
83. Crose, J. G., and A. H.-S. Ang: Nonlinear Analysis Method for Circular Plates, *Proc. Am. Soc. Civil Eng., Eng. Mech. Div.,* vol. 95, no. 4, August 1969.
84. Szilard, R.: "Theory and Analysis of Plates; Classical and Numerical Methods," Prentice-Hall, Inc., 1974.

ELEVEN

Columns and Other Compression Members

11.1 *Columns; common case*

The formulas and discussion of this article are based on the following assumptions: (1) The column is nominally straight and is subjected only to nominally concentric and axial end loads, and such crookedness and eccentricity as may occur are accidental and not greater than is consistent with standard methods of fabrication and ordinary conditions of service; (2) the column is homogeneous and of uniform cross section; (3) if the column is made up of several longitudinal elements, these elements are so connected as to act integrally; (4) there are no parts so thin as to fail by local buckling before the column as a whole has developed its full strength.

End conditions. The strength of a column is in part dependent on the *end conditions,* that is, the degree of end fixity or constraint. A column with ends that are supported and fixed, so that there can be neither lateral displacement nor change in slope at either end, is called *fixed-ended.* A column with ends that are supported against lateral displacement but not constrained against change in slope is called *round-ended.* A column with one end fixed and the other end neither laterally supported nor otherwise constrained is called *free-ended.* A column with both end surfaces that are flat and normal to the axis and that bear evenly against rigid loading surfaces is called *flat-ended.* A column with ends that bear against transverse pins is called *pin-ended.*

Truly fixed-ended and truly round-ended columns practically never occur in practice; the actual conditions are almost always intermediate. The greatest degree of fixity is found in columns with ends that are riveted or welded to relatively rigid parts that are also fixed. Theoretically a flat-ended column is equivalent to a fixed-ended column until the load reaches a certain critical value at which the column "kicks out" and bears only on one edge of each end surface instead of on the whole surface. Actually, flat-ended columns have a degree of end constraint considerably less than that required to produce fixity. The nearest approach to round-ended conditions is found in pin-ended columns subject to vibration or other imposed motion. The degree of end fixity may be expressed by the *coefficient of constraint* (explained following Eq. 1) or by the *free* or *effective* length, which is the length measured between points of counterflexure or the length of a round-ended column of equal strength.

Behavior. If sufficiently slender, a column will fail by elastic instability (see Chap. 14). In this case the maximum unit stress sustained is less than the proportional limit of the material; it depends on the modulus of elasticity, the slenderness ratio (ratio of the length of the column to the least radius of gyration of the section), and the end conditions and is independent of the strength of the material. Columns which fail in this way are called *long columns.*

Columns that are too short to fail by elastic instability are called *short columns;* such a column will fail when the maximum fiber stress due to direct compression and to the bending that results from accidental crookedness and eccentricity reaches a certain value. For structural steel this value is about equal to the tensile yield point; for light alloys it is about equal to the compressive yield strength; and for wood it lies between the flexural elastic limit and the modulus of rupture.

For a given material and given end conditions, there is a certain slenderness ratio which marks the dividing point between long and short columns; this is called the *critical slenderness ratio.*

Formulas for long columns. The unit stress at which a long column fails by elastic instability is given by the Euler formula

$$\frac{P}{A} = \frac{C\pi^2 E}{(L/r)^2} \tag{1}$$

where P = total load, A = area of section, E = modulus of elasticity, L/r = slenderness ratio, and C is the coefficient of constraint, which depends on end conditions. For round ends, $C = 1$; for fixed ends, $C = 4$; and for the end conditions that occur in practice, C can rarely be assumed greater than 2. It is generally not considered good practice to employ long columns in building and bridge construction, but they are used in aircraft; in Table 27 the Euler equations used in aeronautical design are given for a number of materials and end conditions. (Formulas for the loads producing elastic instability of uniform and tapered bars under a wide variety of conditions of loading and support are given in Table 34.)

TABLE 27 ***Standard column formulas***

Notation: Q = allowable load (pounds); P = ultimate load (pounds); A = section area of column (square inches); L = length of column (inches); r = least radius of gyration of column section (inches); σ_u = ultimate strength, and σ_y = yield point or yield strength of material (pounds per square inch); E = modulus of elasticity of material (pounds per square inch); m = factor of safety; and $(L/r)_{cr}$ = critical slenderness ratio

Material	Service	Type of section	End conditions assumed	Formulas for: Allowable unit load Q/A	Formulas for: Ultimate unit load P/A
Structural steel	Buildings	Structural shapes or fabricated	Ends riveted or welded, or columns continuous	$\frac{Q}{A} = \frac{1 - \frac{(L/r)^2}{2C_c^2}}{m}\sigma_y \quad \text{for } \frac{L}{r} < C_c$ $\frac{Q}{A} = \frac{149{,}000{,}000}{(L/r)^2} \quad \text{for } C_c < \frac{L}{r} < 200$ Here: $C_c = \sqrt{\frac{2\pi^2 E}{\sigma_y}}$ (Ref. 1) $m = \frac{5}{3} + \frac{3(L/r)}{8C_c} - \frac{(L/r)^3}{8C_c^3}$ For σ_y = 33,000 36,000 42,000 46,000 50,000 C_c = 131.7 126.1 116.7 111.6 107.0	$\frac{P}{A} = \frac{Q}{A} \times m$
Structural carbon steel $\sigma_y = 33{,}000$	Bridges	Structural shapes or fabricated	Ends riveted	$\frac{Q}{A} = \frac{\sigma_y/m}{1 + 0.25 \sec\left(\frac{0.75L}{2r}\sqrt{\frac{mQ}{EA}}\right)}$ (Ref. 4) Suggested $\sigma_y = 32{,}000$; suggested $m = 1.7$	$\frac{P}{A} = \frac{\sigma_y}{1 + 0.25 \sec\left(\frac{0.75L}{2r}\sqrt{\frac{P}{EA}}\right)}$ or (Ref. 4) $\frac{P}{A} = 25{,}600 - 0.425\left(\frac{L}{r}\right)^2 \text{ up to } \frac{L}{r} = 160$
				$\frac{Q}{A} = 15{,}000 - \frac{1}{4}\left(\frac{L}{r}\right)^2 \text{ up to } \frac{L}{r} = 140$ (Refs. 3, 4)	
				$\frac{Q}{A} = \frac{18{,}750}{1 + 0.25 \sec\left(\frac{0.75L}{2r}\sqrt{\frac{1.76Q}{EA}}\right)} \quad \text{for } \frac{L}{r} > 140$ (Ref. 3)	

			Ends pinned	$\frac{Q}{A} = 15{,}000 - \frac{1}{3}\left(\frac{L}{r}\right)^2$ up to $\frac{L}{r} = 140$ (Refs. 3 and 4) $\frac{Q}{A} = \frac{18{,}750}{1 + 0.25 \sec\left(\frac{0.875L}{2r}\sqrt{\frac{1.76Q}{EA}}\right)}$ for $\frac{L}{r} > 140$ (Ref. 3)	$\frac{P}{A} = \frac{\sigma_y}{1 + 0.25 \sec\left(\frac{0.85L}{2r}\sqrt{\frac{P}{EA}}\right)}$ or $\frac{P}{A} = 25{,}600 - 0.566\left(\frac{L}{r}\right)^2$ up to $\frac{L}{r} = 140$ (Ref. 4)
Structural silicon steel $\sigma_u = 80{,}000$ $\sigma_y = 45{,}000$	Bridges and general structural use	Structural shapes or fabricated	Ends riveted	$\frac{Q}{A} = 20{,}000 - 0.46\left(\frac{L}{r}\right)^2$ up to $\frac{L}{r} = 130$ (Ref. 3) $\frac{Q}{A} = \frac{25{,}000}{1 + 0.25 \sec\left(\frac{0.75L}{2r}\sqrt{\frac{1.8Q}{EA}}\right)}$ for $\frac{L}{r} > 130$	$\frac{P}{A} = 1.8\frac{Q}{A}$
				$\frac{Q}{A} = 27{,}000 - 80\frac{L}{r}$ max $\frac{Q}{A} = 23{,}000$ (Ref. 5)	
			Ends pinned	$\frac{Q}{A} = 20{,}000 - 0.61\left(\frac{L}{r}\right)^2$ up to $\frac{L}{r} = 130$ $\frac{Q}{A} = \frac{25{,}000}{1 + 0.25 \sec\left(\frac{0.875L}{2r}\sqrt{\frac{1.8Q}{EA}}\right)}$ for $\frac{L}{r} > 130$ (Ref. 3)	$\frac{P}{A} = 1.8\frac{Q}{A}$
Structural nickel steel $\sigma_y = 55{,}000$	Bridges and general structural use	Structural shapes or fabricated	Ends riveted	$\frac{Q}{A} = 24{,}000 - 0.66\left(\frac{L}{r}\right)^2$ up to $\frac{L}{r} = 120$ $\frac{Q}{A} = \frac{30{,}000}{1 + 0.25 \sec\left(\frac{0.75L}{2r}\sqrt{\frac{1.83Q}{EA}}\right)}$ for $\frac{L}{r} > 120$ (Ref. 3)	$\frac{P}{A} = 1.83\frac{Q}{A}$
				$\frac{Q}{A} = 24{,}000 - 98\frac{L}{r}$ (Ref. 6)	
			Ends pinned	$\frac{Q}{A} = 24{,}000 - 0.86\left(\frac{L}{r}\right)^2$ up to $\frac{L}{r} = 120$ $\frac{Q}{A} = \frac{30{,}000}{1 + 0.25 \sec\left(\frac{0.875L}{2r}\sqrt{\frac{1.83Q}{EA}}\right)}$ for $\frac{L}{r} > 120$ (Ref. 3)	$\frac{P}{A} = 1.83\frac{Q}{A}$

TABLE 27 *Standard column formulas* (*Cont.*)

Material	Service	Type of section	End conditions assumed	Formulas for					
				Allowable unit load Q/A			Ultimate unit load P/A		
				σ_y	Q/A	L/r			
High-strength steel	General	Structural shapes and fabricated	Pinned	33,000	$15{,}000 - 0.325\left(\frac{L}{r}\right)^2$	0–140	$\frac{P}{A} = 1.8\frac{Q}{A}$		
					$\frac{15{,}000}{0.5 + \frac{1}{15{,}860}\left(\frac{L}{r}\right)^2}$	140–200			
				45,000	$20{,}500 - 0.605\left(\frac{L}{r}\right)^2$	0–120			
					$\frac{20{,}500}{0.5 + \frac{1}{11{,}630}\left(\frac{L}{r}\right)^2}$	120–200			
				50,000	$22{,}500 - 0.738\left(\frac{L}{r}\right)^2$	0–110			
					$\frac{22{,}500}{0.5 + \frac{1}{10{,}460}\left(\frac{L}{r}\right)^2}$	110–200			
				55,000	$25{,}000 - 0.902\left(\frac{L}{r}\right)^2$	0–105			
					$\frac{25{,}000}{0.5 + \frac{1}{9510}\left(\frac{L}{r}\right)^2}$	105–200	(From Ref. 28. Copyright U.S. Steel Corp.)		
							σ_y	P/A	$\left(\frac{L'}{r}\right)_{cr}$
Low-carbon and low-alloy steel	Airplanes	Tubular, circular, or noncrippling shapes	Represented by C	$\frac{Q}{A} = \frac{P/A}{m}$ m is usually taken as 1.5 in aircraft design			36,000	$36{,}000 - 1.172\left(\frac{L'}{r}\right)^2$	122
							75,000	$79{,}500 - 51.9\left(\frac{L'}{r}\right)^{1.5}$	91
							103,000	$113{,}000 - 11.15\left(\frac{L'}{r}\right)^2$	73

					132,000; $145{,}000 - 18.36\left(\frac{L'}{r}\right)^2$; 63
					163,000; $179{,}000 - 27.95\left(\frac{L'}{r}\right)^2$; 56
					Here σ_y is tensile yield strength for 0.002 offset. Column yield strength is higher and is the first term in the formula. $L' = L/C$, where C is usually taken as 1.5 for riveted ends and as 2 for welded ends For $L'/r > (L'/r)_{cr}$, $P/A = 286{,}000{,}000/(L'/r)^2$ (Ref. 9)
Cast iron	Buildings	Hollow, round	Flat	$\frac{Q}{A} = 12{,}000 - 60\frac{L}{r}$ $\text{Max}\frac{Q}{A} = 10{,}000$; $\max\frac{L}{r} = 100$ (Ref. 3)	$\frac{P}{A} = 34{,}000 - 88\frac{L}{r}$
				$\frac{Q}{A} = 9000 - 40\frac{L}{r}$ $\text{Max}\frac{L}{r} = 70$ Min diameter = 6 in; min thickness = $\frac{1}{2}$ in (Ref. 7)	
Structural aluminum: 6061-T6, 6062-T6 $\sigma_y = 35{,}000$ $\sigma_u = 38{,}000$ $E = 10{,}000{,}000$	Building structures (nonwelded)	Structural shapes or fabricated	Partial constraint	$\frac{L}{r} < 10$ $\quad \frac{Q}{A} = 19{,}000$ $10 < \frac{L}{r} < 67$ $\quad \frac{Q}{A} = 20{,}400 - 135\frac{L}{r}$ $\frac{L}{r} > 67$ $\quad \frac{Q}{A} = \frac{51{,}000{,}000}{(L/r)^2}$ (Ref. 30, spec. A-7b, p. 42)	$\frac{P}{A} = 1.95\frac{Q}{A}$
	Bridge structures (nonwelded)	Structural shapes or fabricated	Partial constraint	$\frac{L}{r} < 9.2$ $\quad \frac{Q}{A} = 17{,}000$ $9.2 < \frac{L}{r} < 67$ $\quad \frac{Q}{A} = 18{,}100 - 120\frac{L}{r}$ $\frac{L}{r} > 67$ $\quad \frac{Q}{A} = \frac{45{,}000{,}000}{(L/r)^2}$ (Ref. 30, spec. A-7a, p. 6)	$\frac{P}{A} = 2.20\frac{Q}{A}$

TABLE 27 *Standard column formulas* (*Cont.*)

Material	Service	Type of section	End conditions assumed	Formulas for: Allowable unit load Q/A	Formulas for: Ultimate unit load P/A
Structural aluminum: 2014-T4 $F_{cy} = 35{,}000$; 2024-T3 $F_{cy} = 42{,}000$; 2024-T4 $F_{cy} = 40{,}000$; 6061-T6 $F_{cy} = 35{,}000$	Airplanes	Tubular, circular, or noncrippling shapes	Represented by $c = 1.5$ (rivited ends) to 2 (welded ends)	$\frac{Q}{A} = \frac{P/A}{m}$ m is usually taken as 1.5 in aircraft design	$\frac{P}{A} = F_{co}\left(1 - 0.385\frac{L'/r}{\pi\sqrt{E/F_{co}}}\right)$ up to $(L'/r)_{cr} = 1.732\pi\sqrt{E/F_{co}}$ $\frac{P}{A} = \frac{\pi^2 E}{(L'/r)^2}$ for $\frac{L'}{r} > \left(\frac{L'}{r}\right)_{cr}$ Here $L' = L/\sqrt{c}$ F_{co} = column yield stress $= F_{cy}\left(1 + \frac{F_{cy}}{200{,}000}\right)$ where F_{cy} = comp. yield strength at 0.2% offset (Ref. 9)
7075-T6 $F_{cy} = 66{,}000$	Airplanes	Tubular, circular, or noncrippling shapes	Represented by c (as above)		$\frac{P}{A} = F_{co}\left[1 - F_{co}\frac{(L'/r)^2}{4\pi^2 E}\right]$ up to $(L'/r)_{cr} = 1.414\pi\sqrt{E/F_{co}}$ $F_{co} = 1.075F_{cy}$ (Ref. 9)
Structural magnesium alloy (see Ref. 10)	General structural use	Structural shapes or fabricated	Ends riveted or with equivalent constraint	$\frac{Q}{A} = \frac{1}{2}\frac{P}{A}$ or less	$\frac{P}{A} = \frac{\sigma}{1 + f(kL/r)^2}$ not to exceed σ' Alloy / σ / f / σ': AMC 58S-T51: 160,900 / 0.00249 / 36,000 AMC 58S: 46,000 / 0.00072 / 22,000 AMC 57S: 34,300 / 0.00053 / 19,000 AMC 52S: 25,500 / 0.00040 / 16,000 AM 3S: 16,750 / 0.00026 / 11,000 Here $k = 0.5$
			Ends pinned	$\frac{Q}{A} = \frac{1}{2}\frac{P}{A}$ or less	Same as for riveted ends except $k = 1$
Wood (values are for timber of select structural grade)	General structural use under continuously dry conditions	Solid square or rectangular	Ends flat	$\frac{Q}{A} = \sigma\left[1 - \frac{1}{3}\left(\frac{L}{Kd}\right)^4\right]$ up to $\frac{L}{d} = K = 0.64\sqrt{\frac{E}{\sigma}}$ $\frac{Q}{A} = \frac{0.274E}{(L/d)^2}$ for $\frac{L}{d} > K$	$\frac{P}{A} = 4\frac{Q}{A}$

Hemlock $\sigma = 700$ $E = 1{,}100{,}000$ Longleaf yellow pine $\sigma = 1450$ $E = 1{,}600{,}000$ Southern cypress $\sigma = 1100$ $E = 1{,}200{,}000$ Douglas fir $\sigma = 1200$ $E = 1{,}600{,}000$				$\frac{Q}{A} = \sigma \quad \text{up to } \frac{L}{d} = 11$ Here σ = allowable compressive stress; d = least dimension of cross section $\frac{Q}{A} = 700\left[1 - 0.00000097\left(\frac{L}{d}\right)^4\right] \quad K = 24.2$ $\frac{Q}{A} = 1450\left[1 - 0.00000162\left(\frac{L}{d}\right)^4\right] \quad K = 21.23$ $\frac{Q}{A} = 1100\left[1 - 0.00000168\left(\frac{L}{d}\right)^4\right] \quad K = 21.10$ $\frac{Q}{A} = 1200\left[1 - 0.00000112\left(\frac{L}{d}\right)^4\right] \quad K = 23.35$ (Ref. 11)	
(*Note:* For additional data see Ref. 4 for Chap. 2)	General structural use under continuously dry conditions	Laminated with cover plates or boxed around solid core; square or rectangular	Ends flat	Ratio of Q/A to Q/A for solid column of same dimensions depends on L/d and is as follows: L/d = 6, 10, 14, 18, 22, 26 Ratio = 0.82, 0.77, 0.71, 0.65, 0.74, 0.82 (Ref. 11)	
		Solid, circular	Ends flat	Q/A is same as for square column of equal area. For tapered round column, d and A are taken as for a section a distance $\frac{1}{3}L$ from the smaller end. Q/A at small end must not exceed σ (Ref. 11)	
	Airplanes	Solid	Represented by c taken as 1 to 1.5	$\frac{Q}{A} = \frac{P/A}{m}$	$\frac{P}{A} = F_{cu}\left[1 - \frac{1}{3}\left(\frac{L'/r}{(L'/r)_{cr}}\right)^4\right]$ up to $\left(\frac{L'}{r}\right)_{cr}$ $\frac{P}{A} = \frac{10E_L}{(L'/r)^2} \quad \text{for } \left(\frac{L'}{r}\right) > \left(\frac{L'}{r}\right)_{cr}$ $\left(\frac{L'}{r}\right)_{cr} = \sqrt{\frac{15E_L}{F_{cu}}}$ Here: F_{cu} = ultimate compressive strength E_L = modulus of elasticity from static bending test $L' = \frac{L}{\sqrt{c}}$ (Ref. 25)

Formulas for short columns. It is not possible to calculate with accuracy the maximum stress produced in a short column by a nominally concentric load because of the large influence of the indeterminate crookedness and eccentricity. The maximum unit stress that a column will sustain, however, can be expressed by any of a number of formulas, each of which contains one or more terms that is empirically adjusted to secure conformity with test results. Of such formulas, those given below are the best known and provide the basis for most of the design formulas used in American practice. In these equations P denotes the load at failure, A the cross-sectional area, L the length, and r the least radius of gyration of the section; the meaning of other symbols used is explained in the discussion of each formula.

Secant formula.

$$\frac{P}{A} = \frac{\sigma}{1 + \frac{ec}{r^2} \sec\left(\frac{KL}{2r}\sqrt{\frac{P}{AE}}\right)} \tag{2}$$

This formula is adapted from the formula for stress due to eccentric loading (Art. 11.4, Eq. 27). Here σ denotes the maximum fiber stress at failure (usually taken as the yield point for steel and as the yield strength for light alloys); e denotes the *equivalent eccentricity* (that eccentricity which in a perfectly straight column would cause the same amount of bending as the actual eccentricity and crookedness); c denotes the distance from the central axis about which bending occurs to the extreme fiber on the concave or compression side of the bent column; and K is a numerical coefficient, dependent on end conditions, such that KL is the effective length of the column, or distance between points of inflection. The term ec/r^2 is called the *eccentric ratio,* and a value is assumed for this ratio which makes the formula agree with the results of tests on columns of the type under consideration. For example, tests on structural steel columns of conventional design indicate that the average value of the eccentric ratio is 0.25. In using the secant formula, P/A must be solved for by trial or by the use of prepared charts.

Parabolic formula.

$$\frac{P}{A} = \sigma - k\left(\frac{L}{r}\right)^2 \tag{3}$$

This is an empirical formula. The value of σ is the same as for the secant formula; the value of the coefficient k is adjusted so as to make the curve agree with the results of tests through the L/r range of most importance or make it tangent to the Euler curve for the material and end conditions in question.

Straight-line formula.

$$\frac{P}{A} = \sigma - k\left(\frac{L}{r}\right) \tag{4}$$

This also is an empirical formula. The value of σ is sometimes taken as the ultimate compressive strength of the material when tested in the form of a short prism, k then being taken so as to make the straight-line tangent to the corresponding Euler curve. But more frequently, both σ and k are empirically adjusted to make the straight line agree with the results of tests through the L/r range of most importance.

Rankine formula.

$$\frac{P}{A} = \frac{\sigma}{1 + \phi(L/r)^2} \tag{5}$$

This is a semirational formula. The value of σ is sometimes taken as the ultimate strength of the material and the value of ϕ as $\sigma/C\pi^2E$, thus making the formula agree with the results of tests on short prisms when L/r is very small and with Euler's equation when L/r is very large. More often σ and ϕ are adjusted empirically to make the equation agree with the results of tests through the L/r range of most importance.

By applying a proper safety factor, a formula for the *safe* unit load Q/A may be derived from any one of the preceding formulas for ultimate unit load. In Table 27 a number of the ultimate strength and working formulas most commonly used in American practice are given, together with specified restrictions as to the permissible slenderness ratio and values of the critical slenderness ratio $(L/r)'$ which marks the transition point between long and short columns. The use of the formulas is illustrated in Examples 1 and 2 below.

Calculation of stress. The best way to compute the probable value of the maximum fiber stress in a short column, caused by the imposition of a nominally concentric load that is less than the ultimate load, is to use the secant formula (Eq. 2) with an assumed value of e or ec/r^2. However, by transposing terms, any one of Eqs. 3 to 5 can be written so as to give the maximum stress σ in terms of the load P. Such procedure is logical only when σ is the fiber stress at failure and P the ultimate load; but if the maximum stress due to some load that is less than the ultimate load is thus computed, the result, although probably considerably in error, is almost sure to be greater than the true stress and hence the method errs on the side of safety. The stress in a transmission shaft due to a longitudinal compressive load is sometimes computed in this way, a straight-line formula being written in the form

$$\sigma = \frac{P/A}{1 - 0.0044\, L/r} \qquad \text{(Ref. 12)}$$

EXAMPLES

1. Figure 11.1 represents the cross section of a structural steel column composed of two 10-in, 35-lb channels placed 12 in back to back and latticed together. The length of the column is 349.3 in. It is required to determine the maximum load this column will carry and the safe

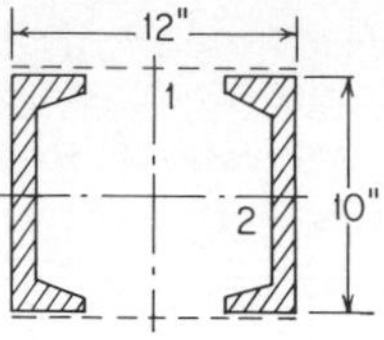

Fig. 11.1

load, assuming riveted ends and using appropriate formulas from Table 27. (This test of this column is reported in Ref. 4; it is there designated as no. 27, ser. *D*.)

Solution. The actual area of the section is 29.13 in^2; the least radius of gyration is that for axis 2 and is 3.34 in; and the slenderness ratio is $L/r = 349.3/3.34 = 104.5$. First the maximum unit load is found by the parabolic formula to be

$$\frac{P}{A} = 25{,}600 - 0.425(104.5)^2 = 25{,}600 - 4650 = 20{,}950 \text{ lb/in}^2$$

The maximum total load is therefore

$$P = 20.13(20{,}950) = 421{,}500 \text{ lb}$$

Solution will also be effected by the secant formula. This involves successive trials; and since the parabolic and secant formulas are known to be in close agreement, we begin with 21,000. We have

$$\frac{P}{A} = \frac{32{,}000}{1 + 0.25 \sec\left[\dfrac{(0.75)(349.3)}{(2)(3.34)}\sqrt{\dfrac{21{,}000}{30{,}000{,}000}}\right]} = 21{,}400 \text{ lb/in}^2$$

Evidently 21,000 is too small, and so we use slightly higher assumed values of P/A. After one or two trials, we find that $P/A = 21{,}300$ lb/in^2 or $P = 429{,}000$ lb.

The safe unit load is given by

$$\frac{Q}{A} = 15{,}000 - \tfrac{1}{4}(104.5)^2 = 12{,}270 \text{ lb/in}^2$$

Hence the safe load is

$$Q = 20.13(12{,}270) = 247{,}000 \text{ lb}$$

2. A piece of cypress piling, 18 ft long, is to be used as a flat-ended column. The section is circular, the diameter being 18 in at one end and 15 in at the other. It is required to find the safe load.

Solution. Assuming a uniform taper, the diameter at a distance from the smaller end equal to one-third the length is 16 in; the section area at that point is therefore 201 in^2, and in a square column of equal area $d = 14.2$ in. L/d is therefore $18(12)/14.2 = 15.2$, and since this is greater than 11 and less than K, the allowable unit load is calculated as follows:

$$\frac{Q}{A} = 1100[1 - 0.00000168(15.2)^4] = 1100(1 - 0.09) = 1000$$

The safe load is therefore $201(1000) = 201{,}000$ lb.

11.2 *Local buckling*

If a column is composed wholly or partially of thin material, local buckling may occur at a unit load less than that required to cause failure of the column as a whole. When such local buckling occurs at a unit stress less than the proportional limit, it represents elastic instability; the critical stress at which this occurs can be determined by mathematical analysis. Formulas for the critical stress at which bars and thin plates exhibit elastic instability, under various conditions of loading and support, are given in Tables 34 and 35 of Chap. 14. All such formulas are based upon assumptions as to homogeneity of material, regularity of form, and boundary conditions that are never realized in practice; the critical stress to be expected under any actual set of circumstances is nearly always less than that indicated by the corresponding theoretical formula and can be determined with certainty only by test. This is also true of the ultimate load that will be carried by such parts as buckle since elastic buckling is not necessarily attended by failure and thin flanges and webs, by virtue of the support afforded by attached parts, may carry a load considerably in excess of that at which buckling occurs (see Art. 11.6).

In the following paragraphs, the more important facts and relations that have been established concerning local buckling are stated, insofar as they apply to columns of more or less conventional design. In the formulas given, b represents the unsupported width of the part under consideration, t its thickness, σ_y the yield point or yield strength, and E and ν have their usual meanings.

Outstanding flanges. For a long flange having one edge fixed and the other edge free, the theoretical formula for buckling stress is

$$\sigma' = \frac{1.09E}{1 - \nu^2}\left(\frac{t}{b}\right)^2 \tag{6}$$

and for a flange having one edge simply supported and the other edge free, the corresponding formula is

$$\sigma' = \frac{0.416E}{1 - \nu^2}\left(\frac{t}{b}\right)^2 \tag{7}$$

(See Table 35.)

For the outstanding flange of a column, the edge condition is intermediate, the degree of constraint depending upon the torsional rigidity of the main member and on the way in which the flange is attached. The conclusions of the ASCE Column Research Committee (Ref. 4) on this point may be summed up as follows: For columns of structural steel having a proportional limit of 30,000 lb/in^2, an outstanding flange riveted between two angles, each having a thickness equal to that of the flange, will not fail by elastic buckling if b/t is less than 15, b being measured from the free edge of the flange to the first row of rivets; for wider flanges, the formula for buckling stress is

$$\sigma' = 0.4E\left(\frac{t}{b}\right)^2 \tag{8}$$

If the thickness of each supporting angle is twice that of the flange, elastic buckling will not occur if b/t is less than 20, b in this case being measured from the free edge of the flange to the toe of angle; for wider flanges, the formula for buckling stress is

$$\sigma' = 0.6E\left(\frac{t}{b}\right)^2 \tag{9}$$

The ultimate strength of an outstanding flange is practically equal to the area times the yield point up to a b/t ratio of 15; for wider flanges the ultimate load is not appreciably greater, and so there is no substantial gain in load-carrying capacity when the width of a flange is increased to more than 15 times the thickness. In Ref. 1 are given recommended limiting values of width/thickness ratios in terms of σ_y for webs, flanges, and other parts subject to buckling.

In the case of aluminum, the *allowable* unit stress on an outstanding flange may be found by the formulas

$$\text{(Allowable)} \quad \sigma = 15{,}000 - 123\,k\frac{b}{t} \qquad \text{when } k\frac{b}{t} < 81 \tag{10}$$

$$\text{(Allowable)} \quad \sigma = \frac{33{,}000{,}000}{\left(k\frac{b}{t}\right)^2} \qquad \text{when } k\frac{b}{t} > 81 \tag{11}$$

Here k is to be taken as 4 when the outstanding flange is one leg of an angle T or other section having relatively little torsional rigidity, and may be taken as 3 when the flange is part of or firmly attached to a heavy web or other part that offers relatively great edge constraint.

A formula (Ref. 13) for the ultimate strength of short compression members consisting of single angles, which takes into account both local and general buckling, is

$$\frac{P}{A} = \sigma \operatorname{Tanh}\left[K\left(\frac{t}{b}\right)^2\right] \tag{12}$$

where $K = 149.1 + 0.1(L/r - 47)^2$ and σ, which depends on L/r, has the following values:

$\frac{L}{r}$	0	20	40	60	80
σ	40,000	38,000	34,000	27,000	18,000

This formula is for an alloy (24ST) having a yield strength of 43,000 lb/in^2 and a modulus of elasticity of 10,500,000 lb/in^2, and is for round-ended columns ($c = 1$). A more general formula for thin sections other than angles is

$$\frac{P}{A} = \sigma \text{ Tanh } (Kt) \tag{12a}$$

Here $\sigma = \sigma_y(1 + B)/(1 + B + B^2)$, where $B = \sigma_y(L/r)^2/(c\pi^2 E)$, and $K = K_o(\sigma_y/\sigma)^{1/2}$, where K_o is a *shape factor,* the value of which is found from Eq. 12*a*, P/A being experimentally determined by testing columns of the section in question that have a slenderness ratio of about 20. For a closed box or "hat" section, $K_o = 15.6$; for a section with flat flanges with a width that is not more than 25 times the thickness, $K_o = 10.8$; for a section of oval form or having wholly or partially curved flanges, K_o ranges from 12 to 32 (Ref. 13). (An extensive discussion of design procedures and buckling formulas for aluminum columns and other structural elements is found in Ref. 8.)

For spruce and other wood of similar properties, Trayer and March (Chap. 14, Ref. 3) give as the formula for buckling stress

$$\sigma' = 0.07E\left(\frac{t}{b}\right)^2 \tag{13}$$

when the edge constraint is as great as normally can be expected in all-wood construction and

$$\sigma' = 0.044E\left(\frac{t}{b}\right)^2 \tag{14}$$

when conditions are such as to make the edge constraint negligible.

Thin webs. For a long thin web that is fixed along each edge, the theoretical formula for buckling stress is

$$\sigma' = \frac{5.73E}{1 - \nu^2}\left(\frac{t}{b}\right)^2 \tag{15}$$

and for a web that is simply supported along each edge, the corresponding formula is

$$\sigma' = \frac{3.29E}{1 - \nu^2}\left(\frac{t}{b}\right)^2 \tag{16}$$

(See Table 35.)

For structural steel columns, the conclusion of the ASCE Column Research Committee (Ref. 4) is that elastic buckling will not occur at b/t ratios less than 30. Tests made by the Bureau of Standards (Ref. 14) on steel members consisting of wide webs riveted between edge angles indicate that this conclusion is conservative and that b/t may be safely as great as 35 if b is taken as the width between rivet lines.

For aluminum columns, the same formulas for allowable stress on a thin web are suggested as are given previously for the outstanding flange (Eqs. 10 and 11) but with $k = 1.2$. (For discussion of the ultimate strength developed by a thin web, see Art. 11.6.)

Thin cylindrical tubes. For a thin cylindrical tube, the theoretical formula for the critical stress at which buckling occurs is

$$\sigma' = \frac{E}{\sqrt{3}\sqrt{1-\nu^2}}\frac{t}{R} \tag{17}$$

where R denotes the mean radius of the tube (see Table 35). Tests indicate that the critical stress actually developed is usually only 40 to 60 percent of this theoretical value. Lundquist (Chap. 14, Ref. 12) suggests the use of the formula

$$\sigma' = KE \tag{18}$$

where the nondimensional coefficient K, which depends on the ratio R/t and on the imperfections of the cylinder, is to be based on tests. From the data available, it would appear that even for very carefully prepared cylinders K should not be taken as more than 50 percent of the theoretical value (given by Eq. 17) when R/t is about 200 or more than 30 percent for values of R/t around 1000. Younger (Ref. 15) suggests as a conservative formula $\sigma' = 0.12Et/R$.

Wilson and Newmark (Ref. 16; Chap. 14, Ref. 13) on the basis of extensive tests, suggest as safe *design formulas* for steel columns

$$\text{(Allowable)} \qquad \sigma = 2{,}000{,}000\,\frac{t}{R}\left(\leq \frac{1}{2}\text{ proportional limit}\right) \tag{19}$$

for cold-drawn or machined tubes, and

$$\text{(Allowable)} \qquad \sigma = 1{,}600{,}000\,\frac{t}{R}\left(\leq \frac{1}{3}\text{ yield point}\right) \tag{20}$$

for fabricated columns without local indentations or lap girth seams. Wilson concludes that when $t/R \geq 0.015$, the full yield point of the steel will be developed before wrinkling occurs, assuming this yield point to be 33,000 lb/in^2 or less. These experiments and others on very thin tubes indicate that for a constant t/R ratio the wrinkling stress is less for a thin plate than for a thick plate owing to the fact that form irregularities occur on a relatively greater scale in the thinner material.

For the design of large fabricated tubular columns of structural steel, the Chicago Bridge and Iron Company has used a formula which takes into account both local and general buckling:

$$\text{(Allowable)} \qquad \frac{Q}{A} = XY \tag{21}$$

where $Y = 1$ $\qquad$ for $\dfrac{L}{R} \leq 60$

$$Y = \frac{21{,}600}{18{,}000 + (L/r)^2} \qquad \text{for } \frac{L}{r} > 60$$

$$X = 1{,}000{,}000 \frac{t}{R}\left[2 - \frac{2}{3}\left(100 \frac{t}{R}\right)\right] \qquad \text{for } \frac{t}{R} \leq 0.015$$

$$X = 15{,}000 \qquad \text{for } \frac{t}{R} \geq 0.015$$

Min $t = \frac{1}{4}$ in

This formula is based on tests made by Wilson and Newmark at University of Illinois (Ref. 16; Chap. 14, Ref. 13). Transverse bulkheads have no appreciable effect upon the critical stress at which buckling occurs unless they are very closely spaced; the same statement holds true of longitudinal stiffeners, but these may greatly increase the *ultimate strength* of a column (see Art. 11.6).

Attached plates. When the flanges or web of a column are formed by riveting a number of plates placed flat against one another, there is a possibility that the outer plate or plates will buckle between points of attachment if the unsupported length is too great compared with the thickness. If the full yield strength σ_y of an outer plate is to be developed, the ratio of unsupported length a to thickness t should not exceed the value indicated by the formula

$$\frac{a}{t} = 0.52\sqrt{\frac{E}{\sigma_y}} \tag{22}$$

(Ref. 17). Some specifications (Ref. 3) guard against the possibility of such buckling by limiting the maximum distance between rivets (in the direction of the stress) to 16 times the thickness of the thinnest outside plate and to 20 times the thickness of the thinnest inside plate; the ratio 16 is in agreement with Eq. 22.

Local buckling of latticed columns To guard against the possibility that the longitudinal elements of a latticed column will buckle individually between points of support, some specifications (Ref. 3) limit the slenderness ratio of such parts between points of attachment of lacing bars to 40 or to two-thirds the slenderness ratio of the column as a whole, whichever is less.

Lacing bars. In a column composed of channels or other structural shapes connected by lacing bars, the function of the lacing bars is to resist the transverse shear due to initial obliquity and that consequent upon such bending as may occur under load. The amount of this shear is conjectural since the obliquity is accidental and indeterminate.

Salmon (Ref. 17) shows that with the imperfections usually to be expected, the transverse shear will be at least 1 percent of the axial load. Moore and Talbot (Ref. 18) found that for certain experimental columns the shear amounted to from 1 to 3 percent of the axial load. Some specifications require that in buildings the lacing be designed to resist a shear equal to 2 percent of the axial load (Ref. 1), and that in bridges it be designed to resist a shear V given by

$$V = \frac{P}{100}\left[\frac{100}{(L/r) + 10} + \frac{L/r}{100}\right]$$

where P is the allowable axial load and r is the radius of gyration of the column section with respect to the central axis perpendicular to the plane of the lacing (Ref. 3).

The strength of individual lacing bars as columns has been investigated experimentally. For a bar of rectangular section with a single rivet at each end, the ultimate strength is given by

$$\frac{P}{A} = 25{,}000 - 50\,\frac{L}{r} \qquad \text{(Ref. 4)}$$

or

$$\frac{P}{A} = 21{,}400 - 45\,\frac{L}{r} \qquad \text{(Ref. 18)}$$

For bars of angle or channel section, these formulas are conservative. For flat bars used as double lacing, the crossed bars being riveted together, tests show that the effective L is about half the actual distance between end rivets. Some specifications (Refs. 1 and 3) require lacing bars of any section to be designed by the regular column formula, L being taken as the distance between end rivets for single lacing and as 70 percent of that distance for double lacing. There are additional limitations as to slope of lacing, minimum section, and method of riveting.

11.3 *Strength of latticed columns.*

Although it is customary to assume that a latticed column acts integrally and develops the full strength of the nominal section, tests show that when bending occurs in the plane of the lacing, the column is less stiff than would be the case if this assumption were valid. For a column so designed that buckling occurs in a plane normal to that of the lacing, this fact is unimportant; but in long open columns laced on all sides, such as are often used for derrick booms and other light construction, it may be necessary to take it into account.

For any assumed transverse loading, it is easy to calculate that part of the deflection of a latticed member which is due to strains in the lacing bars and thus to derive a value for what may be called the *reduced modulus of elasticity* KE. Such calculations agree reasonably well with the results of tests (see Ref. 4), but K, of course, varies with the nature of the assumed transverse loading

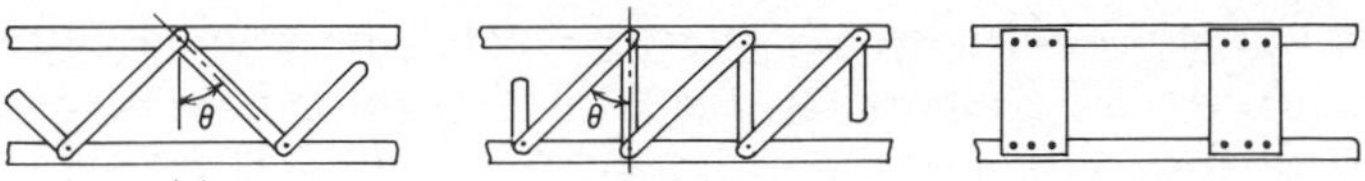

(a) (b) (c)

Fig. 11.2

or with the form of the assumed elastic curve, which amounts to the same thing. For uniformly distributed loading and end support, and for the type of lacing shown in Fig. 11.2*a*, K is given by

$$K = \frac{1}{1 + \dfrac{4.8I}{AL^2 \cos^2 \theta \sin \theta}} \tag{23}$$

where L = length of the column, I = moment of inertia of the column cross section about the principal axis which is normal to the plane of battens, and A = cross-sectional area of a single lacing bar. For double lacing, 2.4 should be used in place of 4.8. If KE is used in place of E, the effect of reduced stiffness on the strength of a long column will be approximately allowed for. The method is theoretically inexact mainly because the form of elastic curve assumed is not identical with that taken by the column, but the error due to this is small.

Timoshenko (Ref. 19) gives formulas based upon the assumption that the elastic curve of the column is a sinusoid, from which the following expressions for K may be derived: For the arrangement shown in Fig. 11.2*a*,

$$K = \frac{1}{1 + \dfrac{4.93I}{AL^2 \cos^2 \theta \sin \theta}} \tag{24}$$

For the arrangement shown in Fig. 11.2*b*,

$$K = \frac{1}{1 + \dfrac{4.93I}{A_1L^2 \cos^2 \theta \sin \theta} + \dfrac{4.93I}{A_2L^2 \tan \theta}} \tag{25}$$

where A_1 = cross-sectional area of each diagonal bar and A_2 = cross-sectional area of each transverse bar. For the channel and batten-plate arrangement shown in Fig. 11.2*c*,

$$K = \frac{1}{1 + \dfrac{\pi^2 I}{L^2}\left(\dfrac{ab}{12I_2} + \dfrac{a^2}{24I_1}\right)} \tag{26}$$

where a = center-to-center distance between battens; b = length of a batten between rivets; I_1 = moment of inertia of a single-channel section (or any similar section being used for each column leg) about its own centroidal axis normal to the plane of the battens, and I_2 = moment of inertia of a pair of batten plates (that is, $I_2 = 2tc^3/12$, where t is the batten-plate thickness and c is the batten-plate width measured parallel to the length of the column).

In all the preceding expressions for K, it is assumed that all parts have the same modulus of elasticity, and only the additional deflection due to longitudinal strain in the lacing bars and to secondary flexure of channels and batten plates is taken into account. For fairly long columns laced over practically the entire length, the values of K given by Eqs. 23 to 25 are probably sufficiently accurate. More elaborate formulas for shear deflection, in which direct shear stress in the channels, bending of the end portions of channels between stay plates, and rivet deformation, as well as longitudinal strains in lacing bars, are taken into account, are given in Ref. 4; these should be used when calculating the deflection of a short latticed column under direct transverse loading.

The use of K as a correction factor for obtaining a reduced value of E is convenient in designing long latticed columns; for short columns the correction is best made by replacing L in whatever column formula is selected by $\sqrt{(1/K)}L$.

11.4 *Eccentric loading; initial curvature*

When a round-ended column is loaded eccentrically with respect to one of the principal axes of the section (here called axis 1), the formula for the maximum stress produced is

$$\sigma = \frac{P}{A}\left\{1 + \frac{ec}{r^2}\sec\left[\frac{P}{4EA}\left(\frac{L}{r}\right)^2\right]^{1/2}\right\} \tag{27}$$

where e = eccentricity, c = distance from axis 1 to the extreme fiber on the side nearest the load, and r = radius of gyration of the section with respect to axis 1. (This equation may be derived from the formula for case 3e, Table 10 by putting $M_1 = Pe$.)

If a column with fixed ends is loaded eccentrically, as is assumed here, the effect of the eccentricity is merely to increase the constraining moments at the ends; the moment at midlength and the buckling load are not effected. If the ends are *partially* constrained, as by a frictional moment M, this constraint may be taken into account by considering the actual eccentricity e reduced to $e - M/P$. If a free-ended column is loaded eccentrically, as is assumed here, the formula for the maximum stress is

$$\sigma = \frac{P}{A}\left\{1 + \frac{ec}{r^2}\sec\left[\frac{P}{EA}\left(\frac{L}{r}\right)^2\right]^{1/2}\right\} \tag{28}$$

where the notation is the same as for Eq. 27.

When a round-ended column is loaded eccentrically with respect to *both* principal axes of the section (here called axes 1 and 2), the formula for the maximum stress is

$$\sigma = \frac{P}{A}\left\{1 + \frac{e_1c_1}{r_1^2}\sec\left[\frac{P}{4EA}\left(\frac{L}{r_1}\right)^2\right]^{1/2} + \frac{e_2c_2}{r_2^2}\sec\left[\frac{P}{4EA}\left(\frac{L}{r^2}\right)^2\right]^{1/2}\right\} \tag{29}$$

where the subscripts 1 and 2 have reference to axes 1 and 2 and the notation is otherwise the same as for Eq. 27. (The use of Eq. 27 is illustrated in the example below, which also shows the use of Eq. 23 in obtaining a reduced modulus of elasticity to use with a latticed column.)

If a round-ended column is initially curved in a plane that is perpendicular to principal axis 1 of the section, the formula for the maximum stress produced by concentric end loading is

$$\sigma = \frac{P}{A}\left(1 + \frac{dc}{r^2}\frac{8EA}{P(L/r)^2}\left\{\sec\left[\frac{P}{4EA}\left(\frac{L}{r}\right)^2\right]^{1/2} - 1\right\}\right) \tag{30}$$

where d = maximum initial deflection, c = distance from axis 1 to the extreme fiber on the concave side of the column, and r = radius of gyration of the section with respect to axis 1. If the column is initially curved in a plane that is not the plane of either of the principal axes 1 and 2 of the section, the formula for the maximum stress is

$$\sigma = \frac{P}{A}\left(1 + \frac{d_1c_1}{r_1^2}\frac{8EA}{P(L/r_1)^2}\left\{\sec\left[\frac{P}{4EA}\left(\frac{L}{r_1}\right)^2\right]^{1/2} - 1\right\}\right.$$
$$\left. + \frac{d_2c_2}{r_2^2}\frac{8EA}{P(L/r_2)^2}\left\{\sec\left[\frac{P}{4EA}\left(\frac{L}{r_2}\right)^2\right]^{1/2} - 1\right\}\right) \tag{31}$$

where d_1 = the component of the initial deflection perpendicular to the plane of axis 1, d_2 = the component of the initial deflection perpendicular to the plane of axis 2, and c_1, c_2, r_1, and r_2 each has reference to the axis indicated by the subscript.

Eccentrically loaded columns and columns with initial curvature can also be designed by the interaction formulas given in Art. 11.5.

EXAMPLE

The column described in Art. 11.1, Example 1, has single lacing, the bars being of rectangular section, $2\frac{1}{2}$ by $\frac{1}{4}$ in, and inclined at 45°. This column is loaded eccentrically, the load being applied on axis 2 but 2.40 in from axis 1. With respect to bending in the plane of the eccentricity, the column is round-ended. It is required to calculate the maximum fiber stress in the column when a load of 299,000 lb, or 14,850 lb/in^2, is thus applied.

Solution. For axis 1, r = 5.38 in, c = 6.03 in (measured), and e = 2.40 in. Since the bending due to the eccentricity is in the plane of the lacing, a reduced E is used. K is calculated by Eq. 23, where I = 583, $A = 2\frac{1}{2} \times \frac{1}{4} = 0.625$ in^2, L = 349.3 in, and θ = 45°. Therefore

$$K = \frac{1}{1 + \dfrac{(4.8)(583)}{(0.625)(349.3^2)(0.707^2)(0.707)}} = 0.94$$

and using the secant formula (Eq. 27), we have

$$\sigma = 14{,}850\left\{1 + \frac{(2.40)(6.03)}{5.38^2}\sec\left[\frac{14{,}850}{(4)(0.94)(30{,}000{,}000)}\left(\frac{349.3}{5.38}\right)^2\right]^{1/2}\right\}$$

$$= 25{,}300 \text{ lb/in}^2$$

(This column was actually tested under the loading just described, and the maximum stress (as determined by strain-gage measurements) was found to be 25, 250 lb/in^2. Such close agreement between measured and calculated stress must be regarded as fortuitous, however.)

11.5 *Columns under combined compression and bending*

A column bent by lateral forces or by couples presents essentially the same problem as a beam under axial compression, and the stresses produced can be found by the formulas of Table 10 provided the end conditions are determinable. Because these and other uncertainties generally preclude precise solution, it is common practice to rely upon some interaction formula, such as one of those given below. The column may be considered safe for the given loading when the relevant equations are satisfied.

The following notation is common to all the equations; other terms are defined as introduced:

F_a = allowable value of P/A for the member considered as a concentrically loaded column

F_b = allowable value of compressive fiber stress for the member considered as a beam under bending only

$f_a = P/A$ = average axial compressive stress due to the axial load P

f_b = computed maximum bending stress due to the transverse loads, applied couples, or a combination of these

L = unbraced length in plane of bending

L/r = slenderness ratio for buckling in that plane

For structural steel:

$$\frac{fa}{F_a} + \frac{C_m f_b}{(1 - f_a/F_e)\,F_b} \le 1 \qquad \text{when } \frac{f_a}{F_a} > 0.15$$

$$\frac{f_a}{F_a} + \frac{f_b}{F_b} \le 1 \qquad \text{when } \frac{f_a}{F_a} < 0.15$$

for sections between braced points, and

$$\frac{f_a}{0.6F_y} + \frac{f_b}{F_b} \le 1$$

for sections at braced points only. Here $F_e = 149{,}000{,}000/(L/r)^2$ and F_y = yield point of steel; $C_m = 0.85$ except that for restrained compression members in frames braced against joint translation and without transverse loading between joints; $C_m = 0.6 + 0.4(M_1/M_2)$, where M_1 is the smaller and M_2 the larger of the moments at the ends of the critical unbraced length of the member. M_1/M_2 is positive when the unbraced length is bent in single curvature and negative when it is bent in reverse curvature. For such members with transverse loading between joints, C_m may be determined by rational analysis, or the appropriate formula from Table 10 may be used. (Formulas are adapted from Ref. 1.)

For structural aluminum:

$$\frac{f_a}{F_a} + \frac{f_b}{F_b(1 - f_a/F_e)} \leq 1$$

Here $F_e = 51{,}000{,}000/(L/r)^2$ for building structures and $F_e = 45{,}000{,}000/(L/r)^2$ for bridge structures. (Formulas are taken from Ref. 30 with some changes of notation.)

For wood (solid rectangular):

when $L/d \leq \sqrt{0.3E/F_a}$

$$\frac{f_b}{F_b} + \frac{f_a}{F_a} \leq 1$$

when $L/d > \sqrt{0.3E/F_a}$:

1. Concentric end loads plus lateral loads,

$$\frac{f_b}{F_b - f_a} + \frac{f_a}{F_a} \leq 1$$

2. Eccentric end load,

$$\frac{1.25f_b}{F_b - f_a} + \frac{f_a}{F_a} \leq 1$$

3. Eccentric end load plus lateral loads,

$$\frac{f_{bl} + 1.25f_{be}}{F_b - f_a} + \frac{f_a}{F_a} \leq 1$$

Here d = dimension of the section in the plane of bending, f_{bl} = computed bending stress due to lateral loads, and f_{be} = computed bending stress due to the eccentric moment. (Formulas are taken from Ref. 26 with some changes of notation.)

11.6 *Thin plates with stiffeners*

Compression members and compression flanges of flexural members are sometimes made of a very thin sheet reinforced with attached stiffeners; this construction is especially common in airplanes, where both wings and fuselage are often of the "stressed-skin" type.

When a load is applied to such a combination, the portions of the plate not very close to the stiffeners buckle elastically at a very low unit stress, but those portions immediately adjacent to the stiffeners develop the same stress as do the latter, and portions a short distance from the stiffeners develop an intermediate stress. In calculating the part of any applied load that will be carried by the plate or in calculating the strength of the combination, it is convenient to make use of the concept of "effective," or "apparent," width, i.e., the width of that portion of the sheet which, if it developed the same stress as the stiffener, would carry the same load as is actually carried by the entire sheet.

For a flat, rectangular plate that is supported but not fixed along each of two opposite edges and subjected to a uniform shortening parallel to those edges, the theoretical expression (Ref. 20) for the effective width is

$$w = \frac{\pi t}{2\sqrt{3(1-\nu^2)}}\sqrt{\frac{E}{\sigma}} \tag{32}$$

where w = the effective width along each supported edge, t = the thickness of the plate, and σ = the unit stress at the supported edge. Since the maximum value of σ is σ_y (the yield point or yield strength), the maximum load that can be carried by the effective strip or by the whole plate (which amounts to the same thing) is

$$P = \frac{\pi t^2}{\sqrt{3(1-\nu^2)}}\sqrt{E\sigma_y} \tag{33}$$

This formula can be written

$$P = Ct^2\sqrt{E\sigma_y} \tag{34}$$

where C is an empirical constant to be determined experimentally for any given material and manner of support. Tests (Ref. 21) made on single plates of various metals, supported at the edges, gave values for C ranging from 1.18 to 1.67; its theoretical value from Eq. 33 (taking $\nu = 0.25$) is 1.87.

Sechler (Ref. 22) represents C as a function of $\lambda = (t/b)\sqrt{E/\sigma_y}$, where b is the panel width, and gives a curve showing experimentally determined values of C plotted against λ. The following table of corresponding values is taken from Sechler's corrected graph:

λ	0.02	0.05	0.1	0.15	0.2	0.3	0.4	0.5	0.6	0.8
C	2.0	1.76	1.62	1.50	1.40	1.28	1.24	1.20	1.15	1.10

The effective width at failure can be calculated by the relation

$$w = \frac{1}{2}Ct\sqrt{\frac{E}{\sigma_y}} = \frac{1}{2}Cb\lambda$$

In the case of a cylindrical panel loaded parallel to the axis, the effective width at failure can be taken as approximately equal to that for a flat sheet, but the increase in the buckling stress in the central portion of the panel due to curvature must be taken into account. Sechler shows that the contribution of this central portion to the strength of the panel may be allowed for by using for C in the formula $P = Ct^2 \sqrt{E\sigma_y}$, a value given by

$$C = C_f - 0.3C_f\lambda\eta + 0.3\eta$$

where $\lambda = (t/b)\sqrt{E/\sigma_y}$, $\eta = (b/r)\sqrt{E/\sigma_y}$, and C_f is the value of C for a flat sheet, as given by the above table.

The above formulas and experimental data refer to single sheets supported along each edge. In calculating the load carried by a flat sheet with longitudinal stiffeners at any given stiffener stress σ_s, the effective width corresponding to that stress is found by

$$w = b(0.25 + 0.91\lambda^2)$$

where $\lambda = (t/b)\sqrt{E/\sigma_s}$ and b = distance between the stiffeners (Ref. 23). The total load carried by n stiffeners and the supported plate is then

$$P = n(A_s + 2wt)\sigma_s$$

where A_s is the section area of one stiffener. When σ_s is the maximum unit load the stiffener can carry as a column, P becomes the ultimate load for the reinforced sheet.

In calculating the ultimate load on a curved sheet with stiffeners, the strength of each unit or panel may be found by adding to the buckling strength of the central portion of the panel the strength of a column made up of the stiffener and the effective width of the attached sheet, this effective width being found by

$$w = \tfrac{1}{2}C_f t\sqrt{\frac{E}{\sigma_c}},$$

where C_f is the flat-sheet coefficient corresponding to $\lambda = (t/b)\sqrt{E/\sigma_c}$ and σ_c is the unit load that the stiffener-and-sheet column will carry before failure, determined by an appropriate column formula. (For the type of thin section often used for stiffeners in airplane construction, σ_c may be found by Eq. 12 or 12*a*.) Since the unit load σ_c and the effective width w are interdependent (because of the effect of w on the column radius of gyration), it is necessary to assume a value of σ_c, calculate the corresponding w, and then ascertain if the value of σ_c is consistent (according to the column formula used) with this w. (This procedure may have to be repeated several times before agreement is reached.) Then, σ_c and w being known, the strength of the stiffener-and-sheet combination is calculated as

$$P = n\left[\sigma_c(A_s + 2wt) + (b - 2w)t\sigma'\right]$$

where n is the number of stiffeners, A_s is the section area of one stiffener, b is the distance between stiffeners (rivet line to rivet line) and σ' is the critical buckling stress for the central portion of the sheet, taken as $\sigma' = 0.3Et/r$ (r being the radius of curvature of the sheet).

Methods of calculating the strength of stiffened panels and thin columns subject to local and torsional buckling are being continually modified in the light of current study and experimentation. A more extensive discussion than is appropriate here can be found in books on airplane stress analysis, as well as in Refs. 8 and 9, and in the book "Specification for the Design of Cold-Formed Steel Structural Members" (in 5 parts including Commentary) published by the American Iron and Steel Institute.

11.7 *Short prisms under eccentric loading*

When a compressive or tensile load is applied eccentrically to a short prism (i.e., one so short that the effect of deflection is negligible), the resulting stresses are readily found by superposition. The eccentric load P is replaced by an equal axial load P' and by couples Pe_1 and Pe_2, where e_1 and e_2 denote the eccentricities of P with respect to the principal axes 1 and 2, respectively. The stress at any point, or the maximum stress, is then found by superposing the direct stress P'/A due to the axial load and the bending stresses due to the couples Pe_1 and Pe_2, these being found by the ordinary flexure formula (Art. 7.1).

If, however, the prism is composed of a material that can withstand compression only (masonry) or tension only (very thin shell), this method cannot be employed when the load acts *outside the kern* because the reversal of stress implied by the flexure formula cannot occur. By assuming a linear stress distribution and making use of the facts that the volume of the stress solid must equal the applied load P and that the center of gravity of the stress solid must lie on the line of action of P, formulas can be derived for the position of the neutral axis (line of zero stress) and for the maximum fiber stress in a prism of any given cross section. A number of such formulas are given in Table 28, together with the dimensions of the kern for each of the sections considered. For any section that is symmetrical about the axis of eccentricity, the maximum stress $K(P/A)$ is just twice the average stress P/A when the load is applied at the edge of the kern and increases as the eccentricity increases, becoming (theoretically) infinite when the load is applied at the extreme fiber. A prism made of material incapable of sustaining both tension and compression will fail completely when the resultant of the loads falls outside the boundary of any cross section, and will crack (under tension) or buckle (under compression) part way across any section through which the resultant of the loads passes at a point lying outside the kern.

For any section not shown in Table 28, a chart may be constructed showing the relation between e and x; this is done by assuming successive positions of the neutral axis (parallel to one principal axis) and solving for the corresponding eccentricity by the relation $b = I/M$, where b = distance from the neutral axis to the point of application of the load (assumed to be on the other principal axis), I = the moment of inertia, and M = the statical moment about the neutral axis of that part of the section carrying stress. The position of the neutral axis for any given eccentricity being known, the maximum stress can be found by the relation max $\sigma = Px/M$. These equations simply express the facts stated above—that the center of gravity of the stress solid lies on the line of action of P, and that the volume of the stress solid is equal to P. The procedure outlined is simple in principle but rather laborious when applied to any except the simpler type of section since both M and I may have to be determined by graphical integration.

The method of solution just outlined and all the formulas of Table 28 are based on the assumption that the load is applied on one of the principal axes of the section. If the load is applied outside the kern and on neither principal axis, solution is more difficult because neither the position nor the direction of the neutral axis corresponding to a given position of the load is known. The following graphical method, which involves successive trials, may be used for a section of any form.

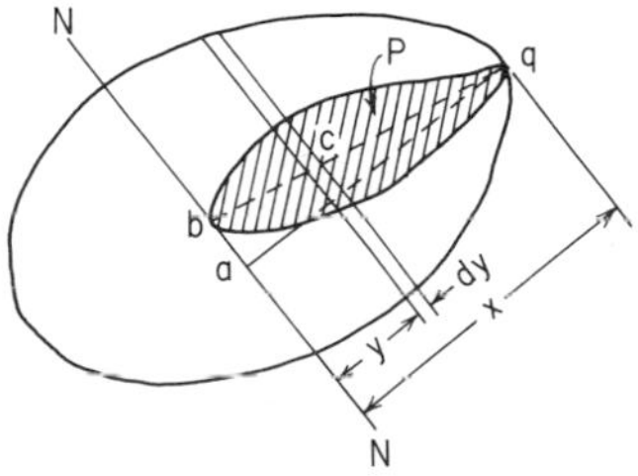

Fig. 11.3

Let a prism of any section (Fig. 11.3) be loaded at any point P. Guess the position of the neutral axis NN. Draw from NN to the most remote fiber q the perpendicular aq. That part of the section on the load side of NN is under compression, and the intensity of stress varies linearly from 0 at NN to σ at q. Divide the stressed area into narrow strips of uniform width dy running parallel to NN. The total stress on any strip acts at the center of that strip and is proportional to the area of the strip $w\,dy$ and to its distance y from NN. Draw the locus of the centers of the strips bcq and mark off a length of strip extending $\frac{1}{2}wy/x$ to each side of this locus. This portion of the strip, if it sustained a unit stress σ, would carry the same total load as does the whole strip when sustaining the actual unit stress $\sigma y/x$ and may be called the *effective portion* of the strip. The effective portions of all strips combine to form the *effective area*, shown as the shaded portion of Fig. 11.3. Now if the assumed position of NN is correct, the centroid of this effective

TABLE 28 ***Formulas for short prisms loaded eccentrically; stress reversal impossible***

NOTATION: m and n are dimensions of the kern which is shown shaded in each case; x is the distance from the most stressed fiber to the neutral axis; and A is the net area of the section. Formulas for x and for maximum stress assume the prism to be subjected to longitudinal load P acting outside the kern on one principal axis of the section and at a distance e from the other principal axis

Form of section, form of kern, and case no.	Formulas for m, n, x, and max σ	
1. Solid rectangular section 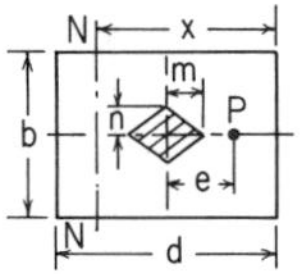	$m = \frac{1}{6}d \qquad n = \frac{1}{6}b$ $x = 3(\frac{1}{2}d - e)$ $\text{Max } \sigma = \frac{P}{A}\frac{4d}{3d - 6e}$	
2. Hollow rectangular section 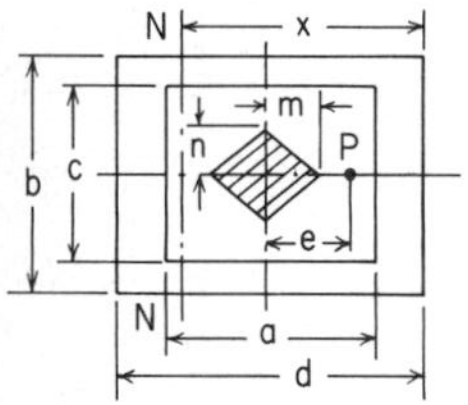	$m = \frac{1}{6}\frac{bd^3 - ca^3}{d(db - ac)} \qquad n = \frac{1}{6}\frac{db^3 - ac^3}{b(db - ac)}$	
	x satisfies $\frac{e}{d} = \frac{1}{2} - \frac{\frac{1}{6}bx^3 - \frac{1}{2}a^2c(\frac{1}{2}d - \frac{1}{6}a) - \frac{1}{2}acd(x - \frac{1}{2}a - \frac{1}{2}d)}{d[\frac{1}{2}bx^2 - ac(x - \frac{1}{2}d)]}$	if $x > \frac{1}{2}(a + d)$
	x satisfies $\frac{e}{d} = \frac{1}{2} - \frac{\frac{1}{6}bx^3 - \frac{1}{2}c(x - \frac{1}{2}d - \frac{1}{2}a)^2(\frac{1}{3}x + \frac{1}{6}d - \frac{1}{6}a)}{d[\frac{1}{2}bx^2 - \frac{1}{2}c(x - \frac{1}{2}d - \frac{1}{2}a)^2]}$	if $x < \frac{1}{2}(a + d)$
	$\text{Max } \sigma = \frac{P}{\frac{1}{2}bx - ac(x - \frac{1}{2}d)/x}$	if $x > \frac{1}{2}(a + d)$
	$\text{Max } \sigma = \frac{P}{\frac{1}{2}bx - (x - \frac{1}{2}d + \frac{1}{2}a)^2/2x}$	if $x < \frac{1}{2}(a + d)$
3. Thin-walled rectangular shell 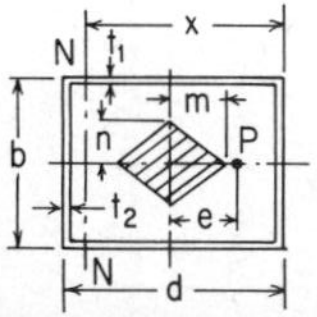	$m = \frac{1}{6}d\left(\frac{dt_1 + 3bt_2}{dt_1 + bt_2}\right) \qquad n = \frac{1}{6}b\left(\frac{bt_2 + 3dt_1}{dt_1 + bt_2}\right)$ $x = \frac{1}{2}\left(\frac{3}{2}d - 3e\right) + \sqrt{b\left(\frac{3}{2}d\frac{t_2}{t_1} - 3e\frac{t_2}{t_1}\right) + \frac{1}{4}\left(\frac{3}{2}d - 3e\right)^2}$ $\text{Max } \sigma = \frac{P}{xt_1 + bt_2}$	

4. Solid circular section

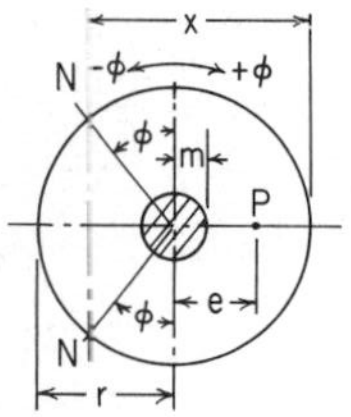

$m = \dfrac{r}{4}$

$x = r(1 - \sin\phi)$ where ϕ satisfies $\dfrac{e}{r} = \dfrac{\frac{1}{8}\pi - \frac{1}{4}\phi - \frac{5}{12}\sin\phi\cos\phi + \frac{1}{3}\sin^3\phi\cos\phi}{\cos\phi - \frac{1}{3}\cos^3\phi - \frac{1}{2}\pi\sin\phi + \phi\sin\phi}$

$\text{Max } \sigma = \dfrac{P}{A}\,\dfrac{\pi(1 - \sin\phi)}{\cos\phi - \frac{1}{3}\cos^3\phi - \frac{1}{2}\pi\sin\phi + \phi\sin\phi}$ or $\max \sigma = \dfrac{P}{A}K$ where K is given by following table:

$e/r =$	0.25	0.30	0.35	0.40	0.45	0.50	0.55	0.60	0.65	0.70	0.75	0.80	0.90
$K =$	2	2.20	2.43	2.70	3.10	3.55	4.20	4.92	5.90	7.20	9.20	13	80

5. Hollow circular section

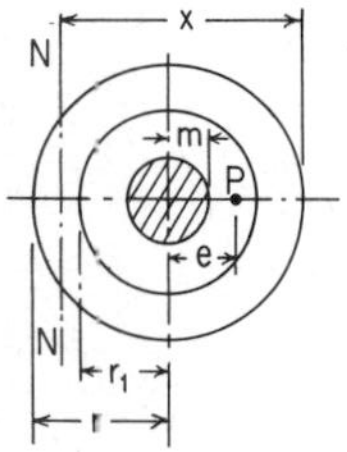

$m = \dfrac{\frac{1}{4}(r^2 + r_1^2)}{r}$

$\text{Max } \sigma = \dfrac{P}{A}K$ where K is given by following table:

r_1/r \ e/r	0.34	0.40	0.41	0.45	0.50	0.55	0.60	0.65	0.70	0.75	0.80	0.85	0.90	0.95
0.6	2.00	2.15		2.35	2.60	2.90	3.30	3.80	4.60	5.80	8.00			
0.8			2.00	2.10	2.25	2.40	2.64	2.90	3.35	4.00	5.00	8.00		
1.0					2.00	2.08	2.23	2.40	2.65	2.90	3.25	3.80	4.60	6.70

6. Thin-walled circular shell

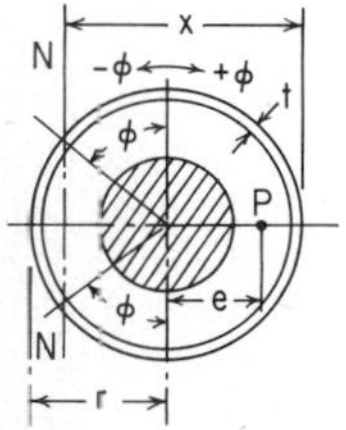

$m = \frac{1}{2}r$

$x = r(1 - \sin\phi)$ where ϕ satisfies $\dfrac{e}{r} = \dfrac{\frac{1}{2}\pi - \phi - \sin\phi\cos\phi}{2\cos\phi - \pi\sin\phi + 2\phi\sin\phi}$

$\text{Max } \sigma = \dfrac{P}{tr}\,\dfrac{1 - \sin\phi}{2\cos\phi - \pi\sin\phi + 2\phi\sin\phi}$ or $\max \sigma = \dfrac{P}{A}K$ where K is given by the above table (case 5) for $\dfrac{r_1}{r} = 1$

area will coincide with the point P and the maximum stress σ will then be equal to the load P divided by the effective area.

To ascertain whether or not the centroid of the effective area does coincide with P, trace its outline on stiff cardboard; then cut out the piece so outlined and balance it on a pin thrust through at P. If the piece balances in any position, P is, of course, the centroid. Obviously the chance of guessing the position of NN correctly at the first attempt is remote, and a number of trials are likely to be necessary. Each trial, however, enables the position of NN to be estimated more closely, and the method is less tedious than might be supposed.

For a solid rectangular section, Esling (Ref. 24) explains a special method of analysis and gives tabulated constants which greatly facilitate solution for this particular case. The coefficient K, by which the average stress P/A is multiplied to give the maximum stress σ, is given as a function of the eccentric ratios e_1/d and e_2/b, where the terms have the meaning shown by Fig. 11.4.

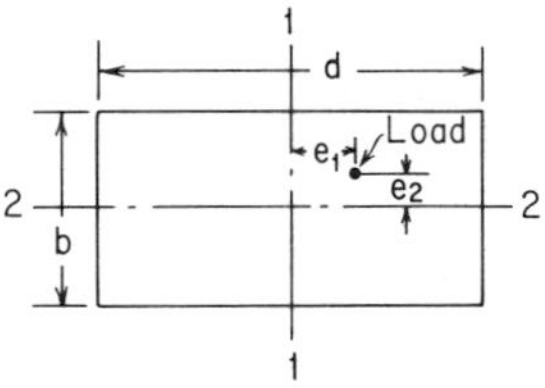

Fig. 11.4

The values of K, taken from Esling's paper, are as shown in the accompanying table.

e_2/b \ e_1/d	0	0.05	0.10	0.15	0.175	0.200	0.225	0.250	0.275	0.300	0.325	0.350	0.375	0.400
0	1.0	1.30	1.60	1.90	2.05	2.22	2.43	2.67	2.96	3.33	3.87	4.44	5.33	6.67
0.05	1.30	1.60	1.90	2.21	2.38	2.58	2.81	3.09	3.43	3.87	4.41	5.16	6.17	7.73
0.10	1.60	1.90	2.20	2.56	2.76	2.99	3.27	3.60	3.99	4.48	5.14	5.99	7.16	9.00
0.15	1.90	2.21	2.56	2.96	3.22	3.51	3.84	4.22	4.66	5.28	6.03	7.04	8.45	10.60
0.175	2.05	2.38	2.76	3.22	3.50	3.81	4.16	4.55	5.08	5.73	6.55	7.66	9.17	11.50
0.200	2.22	2.58	2.99	3.51	3.81	4.13	4.50	4.97	5.54	6.24	7.12	8.33	9.98	
0.225	2.43	2.81	3.27	3.84	4.16	4.50	4.93	5.48	6.05	6.83	7.82	9.13	10.90	
0.250	2.67	3.09	3.60	4.22	4.55	4.97	5.48	6.00	6.67	7.50	8.57	10.0	12.00	
0.275	2.96	3.43	3.99	4.66	5.08	5.54	6.05	6.67	7.41	8.37	9.55	11.1		
0.300	3.33	3.87	4.48	5.28	5.73	6.24	6.83	7.50	8.37	9.37	10.80			
0.325	3.87	4.41	5.14	6.03	6.55	7.12	7.82	8.57	9.55	10.80				
0.350	4.44	5.16	5.99	7.04	7.66	8.33	9.13	10.00	11.10					
0.375	5.33	6.17	7.16	8.45	9.17	9.98	10.90	12.00						
0.400	6.67	7.73	9.00	10.60	11.50									

By double linear interpolation, the value of K for any eccentricity within the limits of the table may readily be found.

EXAMPLE

A bridge pier of masonry, 80 ft high, is rectangular in section, measuring at the base 20 by 10 ft, the longer dimension being parallel to the track. This pier is subjected to a vertical load P (including its own weight) of 1500 tons, a horizontal braking load (parallel to the track) of 60 tons, and a horizontal wind load P_z (transverse to the track) of 50 tons. It is required to determine the maximum compressive stress at the base of the pier, first assuming that the masonry can sustain tension and, second, that it cannot.

Solution. For convenience in numerical work, the ton will be retained as the unit of force and the foot as the unit of distance.

(*a*) *Masonry can sustain tension:* Take $d = 20$ ft, and $b = 10$ ft, and take axes 1 and 2 as shown in Fig. 11.5. Then, with respect to axis 1, the bending moment $M_1 = 60 \times 80 = 4800$ ft-tons,

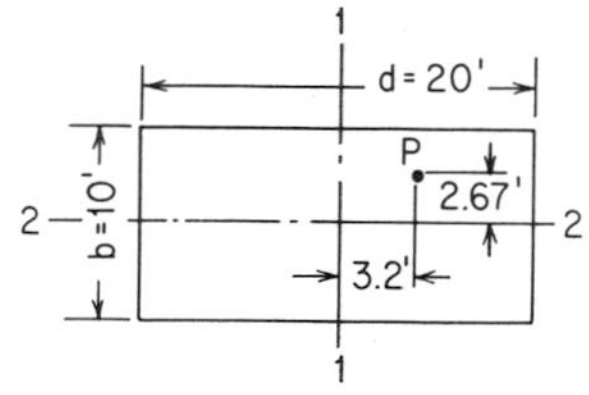

Fig. 11.5

and the section modulus $(I/c)_1 = \frac{1}{6}(10)(20^2) = 667$ ft^3. With respect to axis 2, the bending moment $M_2 = 50 \times 80 = 4000$ ft-tons, and the section modulus $(I/c)_2 = \frac{1}{6}(20)(10^2) = 333$ ft^3. The section area is $10 \times 20 = 200$ ft^2. The maximum stress obviously occurs at the corner where both bending moments cause compression and is

$$\sigma = \frac{1500}{200} + \frac{4800}{667} + \frac{4000}{333} = 7.5 + 7.2 + 12 = 26.7 \text{ tons/ft}^2$$

(*b*) *Masonry cannot sustain tension:* The resultant of the loads pierces the base section of the pier at point P, a distance $e_1 = 60 \times 80/1500 = 3.2$ ft from axis 1 and $e_2 = 50 \times 80/1500 = 2.67$ ft from axis 2. This resultant is resolved at point P into rectangular components, the only one of which causing compression is the vertical component, equal to 1500, with eccentricities e_1 and e_2. The eccentric ratios are $e_1/d = 0.16$ and $e_2/b = 0.267$. Referring to the tabulated values of K, linear interpolation between $e_1/d = 0.15$ and 0.175 at $e_2/b = 0.250$ gives $K = 4.35$. Similar interpolation at $e_2/b = 0.275$ gives $K = 4.83$. Linear interpolation between these values gives $K = 4.68$ as the true value at $e_2/b = 0.267$. The maximum stress is therefore

$$\sigma = KP/A = 4.68 \times \frac{1500}{200} = 35.1 \text{ tons/ft}^2$$

REFERENCES

1. Specification for the Design, Fabrication and Erection of Structural Steel for Buildings, American Institute of Steel Construction, 1961.
2. Recommended Building Code for Working Stresses in Building Materials, U.S. Bureau of Standards, 1926.
3. Specifications for Steel Railway Bridges, American Railway Association, 1950.
4. Final Report of the Special Committee on Steel Column Research, *Trans. Am. Soc. Civil Eng.*, vol. 98, p. 1376, 1933.
5. Specifications for George Washington Bridge, *Trans. Am. Soc. Civil Eng.*, vol. 97, p. 105, 1933.
6. Quebec Bridge Specifications, Board of Engineers.
7. New York City Building Code, 1937.
8. "Alcoa Structural Handbook," Aluminum Company of America, 1960.
9. "ANC Mil-Hdbk-5, Strength of Metal Aircraft Elements," Armed Forces Supply Support Center, March 1959.

10. "Designing with Magnesium," Aluminum Company of America, 1951.
11. "Wood Handbook," Forest Products Laboratory, U.S. Dept. of Agriculture, 1935.
12. Standard Formulas for Use in Determining the Size of Transmission Shafting, *ASME Code.*
13. Kilpatrick, S. A., and O. U. Schaefer: Stress Calculations for Thin Aluminum Alloy Sections, *Prod. Eng.*, February, March, April, May, 1936.
14. Johnston, R. S.: Compressive Strength of Column Web Plates and Wide Web Columns, *Tech. Paper Bur. Stand.* 327, 1926.
15. Younger, J. E.: "Structural Design of Metal Airplanes," McGraw-Hill Book Company, 1935.
16. Wilson, W. M.: Tests of Steel Columns, *Univ. Ill. Eng. Exp. Sta., Bull.* 292, 1937.
17. Salmon, E. H.: "Columns," Oxford Technical Publications, Henry Frowde and Hodder & Stoughton, 1921.
18. Talbot, A. N., and H. F. Moore: Tests of Built-up Steel and Wrought Iron Compression Pieces, *Trans. Am. Soc. Civil Eng.*, vol. 65, p. 202, 1909. (Also *Univ. Ill. Eng. Exp. Sta., Bull.* 44.)
19. Timoshenko, S.: "Strength of Materials," D. Van Nostrand Company, Inc., 1930.
20. von Kármán, Th., E. E. Sechler, and L. H. Donnell: The Strength of Thin Plates in Compression, *Trans. ASME*, vol. 54, no. 2, p. 53, 1932.
21. Schuman, L., and G. Black: Strength of Rectangular Flat Plates under Edge Compression, *Nat. Adv. Comm. Aeron., Rept.* 356, 1930.
22. Sechler, E. E.: A Preliminary Report on the Ultimate Compressive Strength of Curved Sheet Panels, *Guggenheim Aeron. Lab., Calif. Inst. Tech., Publ.* 36, 1937.
23. Sechler, E. E.: Stress Distribution in Stiffened Panels under Compression, *J. Aeron. Sci.*, vol. 4, no. 8, p. 320, 1937.
24. Esling, K. E.: A Problem Relating to Railway-bridge Piers of Masonry or Brickwork, *Proc. Inst. Civil Eng.*, vol. 165, p. 219, 1905–1906.
25. ANC 18 Bulletin, Design of Wood Aircraft Structures, Munitions Board Aircraft Committee, June 1951.
26. National Design Specification for Stress-grade Lumber and Its Fastenings, National Lumber Manufacturers Association, 1960.
27. "Standard Specifications for Highway Bridges," 7th ed., American Association of State Highway Officials, 1957.
28. Priest, H. Malcolm: "Design Manual for High Strength Steels," U.S. Steel Corporation, 1954.
29. National Building Code of Canada, 1953.
30. Suggested Specifications for Structures of Aluminum Alloys 6061-T6 and 6062-T6, *Proc. Am. Soc. Civil Eng.*, ST6, paper 3341, vol. 88, December, 1962.

TWELVE

Shells of Revolution; Pressure Vessels; Pipes

12.1 *Circumstances and general state of stress*

The discussion and formulas in this article apply to any vessel that is a figure of revolution. For convenience of reference, a line that represents the intersection of the wall and a plane containing the axis of the vessel is called a *meridian*, and a line representing the intersection of the wall and a plane normal to the axis of the vessel is called a *circumference*. Obviously the meridian through any point is perpendicular to the circumference through that point.

When a vessel of the kind under consideration is subjected to internal or external pressure, stresses are set up in the walls. The state of stress is triaxial; at any point there is a meridional stress σ_1 acting parallel to the meridian, a circumferential, or hoop, stress σ_2 acting parallel to the circumference, and a radial stress σ_3. In addition, there may be bending stresses and shear stresses. In consequence of these stresses, there will be meridional, circumferential, and radial strains and there may be a change in meridional slope. If there is symmetry of both the loading and the vessel about the axis, there will be no tendency for any circumference to depart from the circular form unless buckling occurs.

12.2 *Thin shells of revolution under distributed loadings producing membrane stresses only*

If the walls of the vessel are relatively thin (less than about one-tenth the radius) and have no abrupt changes in thickness, slope, or curvature and if the loading is uniformly distributed or smoothly varying and axisymmetric, the stresses σ_1 and σ_2 are practically uniform throughout the thickness of the wall. The stresses σ_1 and σ_2 are the only important ones present; the radial stress σ_3 and such bending stresses as occur are negligibly small. Table 29 gives formulas for the stresses and deformations under loadings such as those described previously for cylindrical, conical, spherical, and toroidal vessels as well as for general smooth figures of revolution as listed under case 4.

If two thin-walled shells are joined to produce a vessel, and if it is desired to have no bending stresses at the joint under uniformly distributed or smoothly varying loads, then it is necessary to choose shells for which the radial deformations and the rotations of the meridian are the same for each shell at the point of connection. For example, a *cylindrical* shell under uniform internal pressure will have a radial deformation of $qR^2(1 - \nu/2)/Et$ while a *hemispherical* head of equal thickness under the same pressure will have a radial deformation of $qR^2(1 - \nu)/2Et$; the meridian rotation ψ is zero in both cases. This mismatch in radial deformation will produce bending and shear stresses in the near vicinity of the joint. An examination of case 4a (Table 29) shows that if R_1 is infinite at $\theta = 90°$ for a smooth figure of revolution, the radial deformation and the rotation of the meridian will match those of the cylinder.

Flügge (Ref. 5) points out that the family of *cassinian* curves has the property just described. He also discusses in some detail the *ogival* shells, which have a constant radius of curvature R_1 for the meridian but for which R_2 is a variable. If R_2 is everywhere less than R_1, the ogival shell has a pointed top, as shown in Fig. 12.1*a*. If R_2 is infinite, as it is at point A in Fig. 12.1*b*,

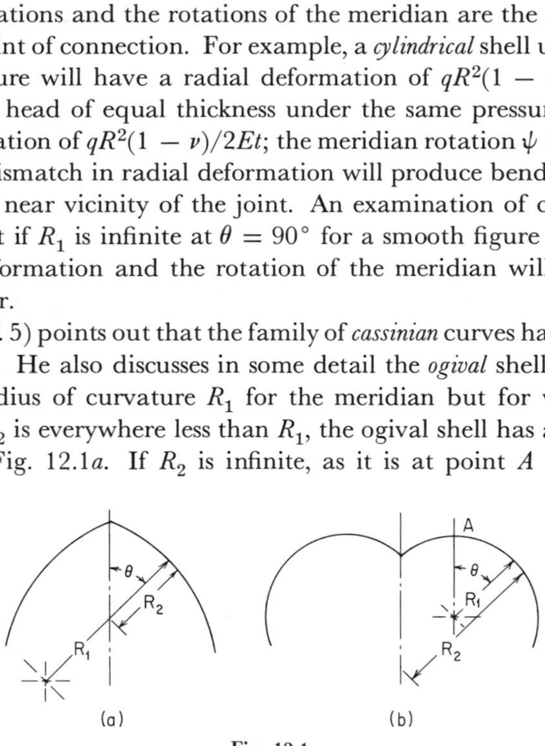

Fig. 12.1

the center of the shell must be supported to avoid large bending stresses although some bending stresses are still present in the vicinity of point A.

For very thin shells where bending stresses are negligible, a nonlinear membrane theory can provide more realistic values near the crown, point A. Rossettos and Sanders have carried out such a solution (Ref. 52). Chou and Johnson (Ref. 57) have examined large deflections of elastic toroidal membranes of a type used in some sensitive pressure measuring devices.

Baker, Kovalevsky, and Rish (Ref. 6) give formulas for toroidal segments, ogival shells, elliptical shells, and Cassini shells under various loadings; all these cases can be evaluated from case 4 of Table 29 once R_1 and R_2 are calculated. In addition to the axisymmetric shells considered in this chapter, Refs. 5, 6, 45, and 59 discuss in some detail the membrane stresses in nonaxisymmetric shells, such as barrel vaults, elliptic cylinders, and hyperbolic paraboloids.

EXAMPLES

1. A segment of a toroidal shell shown in Fig. 12.2 is to be used as a transition between a cylinder and a head closure in a thin-walled pressure vessel. In order to properly match

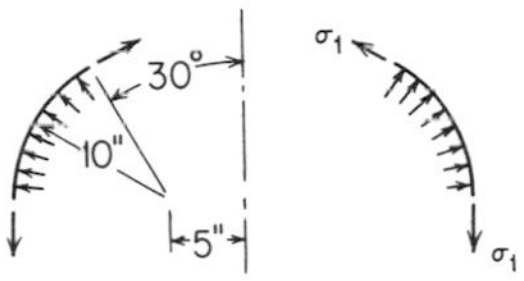

Fig. 12.2

the deformations, it is desired to know the change in radius and the rotation of the meridian at both ends of the toroidal segment under an internal pressure loading of 200 lb/in^2. Given: $E = 30(10^6)$ lb/in^2, $\nu = 0.3$, and the wall thickness $t = 0.1$ in.

Solution. Since this particular case is not included in Table 29, the general case 4a can be used. At the upper end $\theta = 30°$, $R_1 = 10$ in, and $R_2 = 10 + 5/\sin 30° = 20$ in; therefore,

$$\Delta R_{30°} = \frac{200(20^2)(0.5)}{2(30)(10^6)(0.1)}\left(2 - \frac{20}{10} - 0.3\right) = -0.002 \text{ in}$$

Since R_1 is a constant, $dR_1/d\theta = 0$ throughout the toroidal segment; therefore,

$$\psi_{30°} = \frac{200(20^2)}{2(30)(10^6)(0.1)(10)(0.577)}\left[3\frac{10}{20} - 5 + \frac{20}{10}(2 + 0)\right] = 0.00116 \text{ rad}$$

At the lower end, $\theta = 90°$, $R_1 = 10$ in, and $R_2 = 15$ in; therefore,

$$\Delta R_{90°} = \frac{200(15^2)(1)}{2(30)(10^6)(0.1)}\left(2 - \frac{15}{10} - 0.3\right) = 0.0015 \text{ in}$$

Since tan 90° = infinity and $dR_1/d\theta = 0$, $\psi_{90°} = 0$.

2. The truncated thin-walled cone shown in Fig. 12.3 is supported at its base by the membrane stress σ_1. The material in the cone weighs 0.10 lb/in^3, $t = 0.25$ in, $E = 10(10^6)$ lb/in^2, and Poisson's ratio is 0.3. Find the stress σ_1 at the base, the change in radius at the base, and the change in height of the cone if the cone is subjected to an acceleration parallel to its axis of 399g.

TABLE 29 *Formulas for membrane stresses and deformations in thin-walled pressure vessels*

NOTATION: P = axial load (pounds); p = unit load (pounds per linear inch); q and w = unit pressures (pounds per square inch); δ = density (pounds per cubic inch); σ_1 = meridional stress (pounds per square inch); σ_2 = circumferential, or hoop, stress (pounds per square inch); R_1 = radius of curvature of a meridian, a principal radius of curvature of the shell surface (inches); R_2 = length of the normal between the point on the shell and the axis of rotation, the second principal radius of curvature (inches); R = radius of curvature of a circumference (inches); ΔR = radial displacement of a circumference (inches); Δy = change in the height dimension y (inches); ψ = rotation of a meridian, positive when ΔR increases with y (radians); E = modulus of elasticity (pounds per square inch); and ν = Poisson's ratio. *Note:* y is considered positive upward where applicable.

Case no., form of vessel	Manner of loading	Formulas
1. Cylindrical 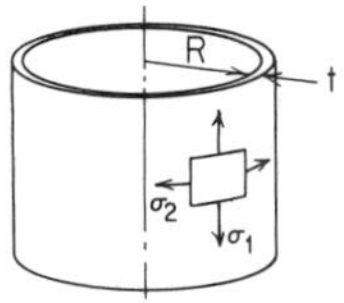 $\frac{R}{t} > 10$	1a. Uniform axial load, p lb/linear in 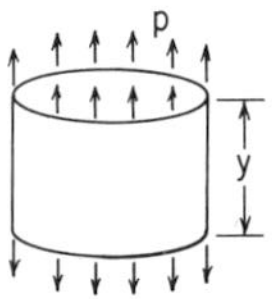	$\sigma_1 = \frac{p}{t}$ $\sigma_2 = 0$ $\Delta R = \frac{-p\nu R}{Et}$ $\Delta y = \frac{py}{Et}$ $\psi = 0$
	1b. Uniform radial pressure, q lb/in^2 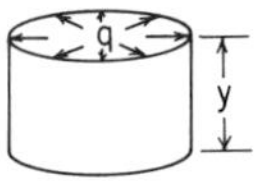	$\sigma_1 = 0$ $\sigma_2 = \frac{qR}{t}$ $\Delta R = \frac{qR^2}{Et}$ $\Delta y = \frac{-qR\nu y}{Et}$ $\psi = 0$
	1c. Uniform internal or external pressure, q lb/in^2 (ends capped)	At points away from the ends $\sigma_1 = \frac{qR}{2t}$ $\sigma_2 = \frac{qR}{t}$ $\Delta R = \frac{qR^2}{Et}\left(1 - \frac{\nu}{2}\right)$ $\Delta y = \frac{qRy}{Et}(0.5 - \nu)$ $\psi = 0$

	1d. Linearly varying radial pressure, q lb/in^2 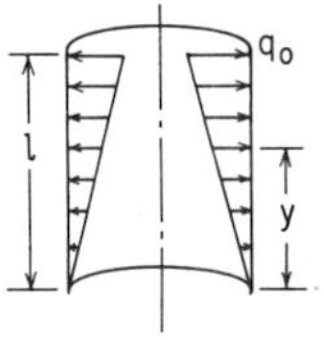	$q = \frac{q_o y}{l}$ (where y must be measured from a free end. If pressure starts away from the end, see Case 6 in Table 30) $\sigma_1 = 0$ $\sigma_2 = \frac{qR}{t} = \frac{q_o R y}{lt}$ $\Delta R = \frac{qR^2}{Et} = \frac{q_o R^2 y}{Etl}$ $\Delta y = \frac{-q_o R \nu y^2}{2Etl}$ $\psi = \frac{q_o R^2}{Etl}$
	1e. Own weight, δ lb/in^3; top edge support, bottom edge free	$\sigma_1 = \delta y$ $\sigma_2 = 0$ $\Delta R = \frac{-\delta \nu R y}{E}$ $\Delta y = \frac{\delta y^2}{2E}$ $\psi = \frac{-\delta \nu R}{E}$
2. Cone 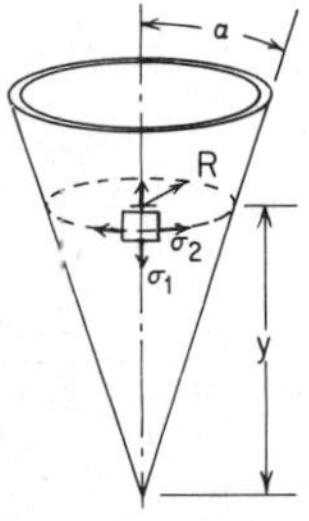 $\frac{R_2}{t} > 10$	2a. Uniform internal or external pressure, q lb/in^2; tangential edge support 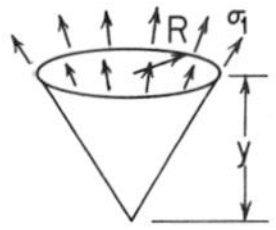	$\sigma_1 = \frac{qR}{2t \cos \alpha}$ $\sigma_2 = \frac{qR}{t \cos \alpha}$ $\Delta R = \frac{qR^2}{Et \cos \alpha}\left(1 - \frac{\nu}{2}\right)$ $\Delta y = \frac{qR^2}{4Et \sin \alpha}(1 - 2\nu - 3 \tan^2 \alpha)$ $\psi = \frac{3qR \tan \alpha}{2Et \cos \alpha}$

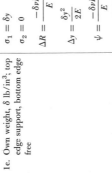

TABLE 29 *Formulas for membrane stresses and deformations in thin-walled pressure vessels* (*Cont.*)

Case no., form of vessel	Manner of loading	Formulas
	2b. Filled to depth d with liquid of density δ lb/in^3; tangential edge support 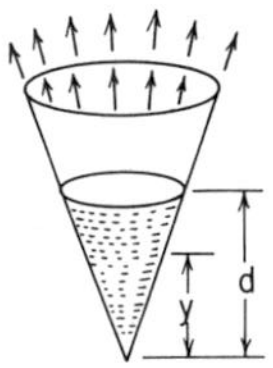	At any level y below the liquid surface, $y < d$ $\sigma_1 = \frac{\delta y \tan\alpha}{2t\cos\alpha}\left(d - \frac{2y}{3}\right)$ Max $\sigma_1 = \frac{3\delta d^2 \tan\alpha}{16t\cos\alpha}$ when $y = \frac{3d}{4}$ $\sigma_2 = \frac{y(d-y)\,\delta\tan\alpha}{t\cos\alpha}$ Max $\sigma_2 = \frac{\delta d^2\tan\alpha}{4t\cos\alpha}$ when $y = \frac{d}{2}$ $\Delta R = \frac{\delta y^2 \tan^2\alpha}{Et\cos\alpha}\left[d\left(1 - \frac{\nu}{2}\right) - y\left(1 - \frac{\nu}{3}\right)\right]$ $\Delta y = \frac{\delta y^2 \sin\alpha}{Et\cos^4\alpha}\left\{\frac{d}{4}(1-2\nu) - \frac{y}{9}(1-3\nu) - \sin^2\alpha\left[\frac{d}{2}(2-\nu) - \frac{y}{3}(3-\nu)\right]\right\}$ $\psi = \frac{\delta y \sin^2\alpha}{6Et\cos^3\alpha}(9d - 16y)$ At any level y above the liquid level $\sigma_1 = \frac{\delta d^3\tan\alpha}{6ty\cos\alpha}$ $\sigma_2 = 0$ $\Delta R = \frac{-\nu\,\delta d^3\tan^2\alpha}{6Et\cos\alpha}$ $\Delta y = \frac{\delta d^3\sin\alpha}{6Et\cos^4\alpha}\left[\frac{5}{6} - \nu(1 - \sin^2\alpha) + \ln\frac{y}{d}\right]$ $\psi = \frac{-\delta d^3\sin^2\alpha}{6Et\cos^3\alpha}\frac{1}{y}$
	2c. Own weight, δ lb/in^3; tangential top edge support	$\sigma_1 = \frac{\delta R}{2\sin\alpha\cos\alpha}$ $\sigma_2 = \delta R\tan\alpha$ $\Delta R = \frac{\delta R^2}{E\cos\alpha}\left(\sin\alpha - \frac{\nu}{2\sin\alpha}\right)$ $\Delta y = \frac{\delta R^2}{E\cos^2\alpha}\left(\frac{1}{4\sin^2\alpha} - \sin^2\alpha\right)$ $\psi = \frac{2\delta R}{E\cos^2\alpha}\left[\sin^2\alpha\left(1 + \frac{\nu}{2}\right) - \frac{1}{4}(1 + 2\nu)\right]$

	2d. Tangential loading only; resultant load = P 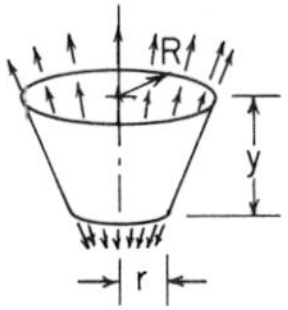	$\sigma_1 = \dfrac{P}{2\pi Rt \cos\alpha}$ $\sigma_2 = 0$ $\Delta R = \dfrac{-\nu P}{2\pi Et \cos\alpha}$ $\Delta y = \dfrac{P}{2\pi Et \sin\alpha \cos^2\alpha} \ln\dfrac{R}{r}$ $\psi = \dfrac{-P\sin\alpha}{2\pi ERt\cos^2\alpha}$	r must be finite to avoid infinite stress and $r/t > 10$ in order to be considered thin-walled
	2e. Uniform loading, w lb/in^2, on the horizontal projected area; tangential top edge support 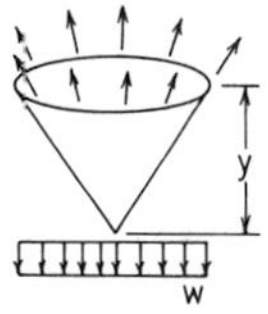	$\sigma_1 = \dfrac{wR}{2t\cos\alpha}$ $\sigma_2 = \dfrac{wR\sin^2\alpha}{t\cos\alpha}$ $\Delta R = \dfrac{wR^2}{Et\cos\alpha}\left(\sin^2\alpha - \dfrac{\nu}{2}\right)$ $\Delta y = \dfrac{wR^2}{2Et\cos^2\alpha}\left[\dfrac{1}{2\sin\alpha} + \nu(1 - \sin\alpha) - 2\sin^2\alpha\right]$ $\psi = \dfrac{wR\sin\alpha}{2Et\cos^2\alpha}(4\sin^2\alpha - 1 - 2\nu\cos^2\alpha)$	
3. Spherical 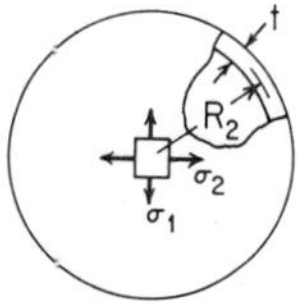 $\dfrac{R_2}{t} > 10$	3a. Uniform internal or external pressure, q lb/in^2; tangential edge support 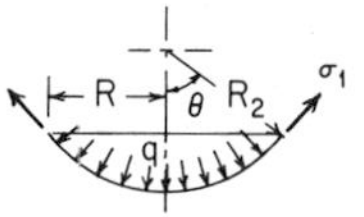	$\sigma_1 = \sigma_2 = \dfrac{qR_2}{2t}$ $\Delta R = \dfrac{qR_2^2(1-\nu)\sin\theta}{2Et}$ $\Delta R_2 = \dfrac{qR_2^2(1-\nu)}{2Et}$ $\Delta y = \dfrac{qR_2^2(1-\nu)(1-\cos\theta)}{2Et}$ $\psi = 0$	

TABLE 29 *Formulas for membrane stresses and deformations in thin-walled pressure vessels* (*Cont.*)

Case no., form of vessel	Manner of loading	Formulas
	3b. Filled to depth d with liquid of density δ lb/in^3; tangential edge support 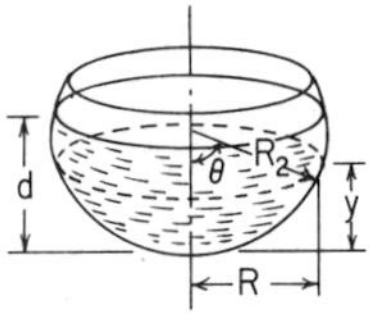 	At any level y below the liquid surface, $y < d$ $\sigma_1 = \frac{R_2^2}{6t}\left(3\frac{d}{R_2} - 1 + \frac{2\cos^2\theta}{1+\cos\theta}\right)$ $\sigma_2 = \frac{R_2^2}{6t}\left[3\frac{d}{R_2} - 5 + \frac{(3+2\cos\theta)2\cos\theta}{1+\cos\theta}\right]$ $\Delta R = \frac{R_2^3\sin\theta}{6Et}\left[3(1-\nu)\frac{d}{R_2} - 5 + \nu + 2\cos\theta\,\frac{3+(2-\nu)\cos\theta}{1+\cos\theta}\right]$ $\Delta y = \frac{\delta R_2^3}{6Et}\left\{3(1-\nu)\left[\frac{d}{R_2}(1-\cos\theta)+\cos\theta\right] - (2-\nu)\cos^2\theta + (1+\nu)\left(1 + 2\ln\frac{2}{1+\cos\theta}\right)\right\}$ $\psi = \frac{-\delta R_2^2}{Et}\sin\theta$ At any level y above the liquid level use case 3d with the load equal to the entire weight of the liquid
	3c. Own weight, δ lb/in^3; tangential top edge support 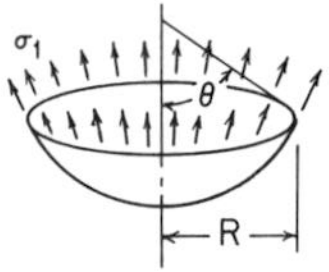 	$\sigma_1 = \frac{\delta R_2}{1+\cos\theta}$ $\quad\sigma_2 = -\delta R_2\left(\frac{1}{1+\cos\theta} - \cos\theta\right)$ Max tensile $\sigma_2 = \frac{\delta R_2}{2}$ at $\theta = 0$ $\sigma_2 = 0$ at $\theta = 51.83°$ $\Delta R = \frac{-\delta R_2^2\sin\theta}{E}\left(\frac{1+\nu}{1+\cos\theta} - \cos\theta\right)$ $\Delta y = \frac{\delta R_2^2}{E}\left[\sin^2\theta + (1+\nu)\ln\frac{2}{1+\cos\theta}\right]$ $\psi = \frac{-\delta R_2}{E}(2+\nu)\sin\theta$
	3d. Tangential loading only; resultant load $= P$ 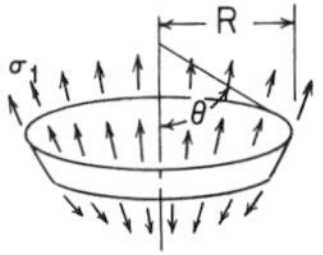 	$\sigma_1 = \frac{P}{2\pi R_2 t\sin^2\theta}$ $\quad\sigma_2 = -\sigma_1$ $\Delta R = \frac{-P(1+\nu)}{2\pi Et\sin\theta}$ $\Delta y = \frac{P(1+\nu)}{2\pi Et}\left[\ln\left(\tan\frac{\theta}{2}\right) - \ln\left(\tan\frac{\theta_o}{2}\right)\right]$ (*Note:* θ_o is the angle to the lower edge and cannot go to zero without local bending occurring in the shell) $\psi = 0$

	3e. Uniform loading, w lb/in^2, on the horizontal projected area; tangential top edge support 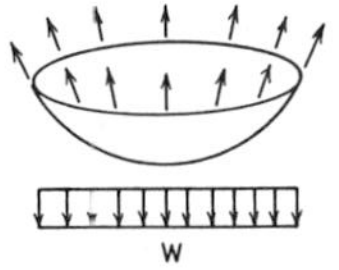 	For $\theta \leq 90°$ $\sigma_1 = \dfrac{wR_2}{2t}$ $\sigma_2 = \dfrac{wR_2}{2t} \cos 2\theta$ $\Delta R = \dfrac{wR_2^2 \sin \theta}{2Et} (\cos 2\theta - \nu)$ $\Delta y = \dfrac{wR_2^2}{2Et} [2(1 - \cos^3 \theta) + (1 + \nu)(1 - \cos \theta)]$ $\psi = \dfrac{-wR_2}{Et} (3 + \nu) \sin \theta \cos \theta$
4. Any smooth figure of revolution if R_2 is less than infinity 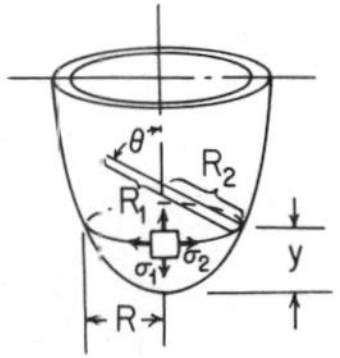 $\dfrac{R_2}{t} > 10$	4a. Uniform internal or external pressure, q lb/in^2; tangential edge support	$\sigma_1 = \dfrac{qR_2}{2t}$ $\sigma_2 = \dfrac{qR_2}{2t}\left(2 - \dfrac{R_2}{R_1}\right)$ $\Delta R = \dfrac{qR_2^2 \sin \theta}{2Et}\left(2 - \dfrac{R_2}{R_1} - \nu\right)$ $\psi = \dfrac{qR_2^2}{2EtR_1 \tan \theta}\left[3\dfrac{R_1}{R_2} - 5 + \dfrac{R_2}{R_1}\left(2 + \dfrac{dR_1}{d\theta} \tan \theta\right)\right]$
	4b. Filled to depth d with liquid of density δ lb/in^3; tangential edge support W = weight of liquid contained to a depth y	At any level below the liquid surface, $y < d$, $\sigma_1 = \dfrac{W}{2\pi R_2 t \sin^2 \theta} + \dfrac{\delta R_2(d - y)}{2t}$ $\sigma_2 = \dfrac{-W}{2\pi R_1 t \sin^2 \theta} + \dfrac{\delta R_2(d - y)}{2t}\left(2 - \dfrac{R_2}{R_1}\right)$ $\Delta R = \dfrac{R_2 \sin \theta}{E}(\sigma_2 - \nu\sigma_1)$ At any level y above the liquid level use case $4d$ with the load equal to the entire weight of the liquid

TABLE 29 *Formulas for membrane stresses and deformations in thin-walled pressure vessels* (Cont.)

Case no., form of vessel	Manner of loading	Formulas
	4c. Own weight, δ lb/in^3; tangential top edge support. W = weight of vessel below the level y	$\sigma_1 = \dfrac{W}{2\pi R_2 t \sin^2\theta}$ $\sigma_2 = \dfrac{W}{2\pi R_1 t \sin^2\theta} + \delta R_2 \cos\theta$ $\Delta R = \dfrac{R_2 \sin\theta}{E}(\sigma_2 - \nu\sigma_1)$
	4d. Tangential loading only, resultant load = P	$\sigma_1 = \dfrac{P}{2\pi R_2 t \sin^2\theta}$ $\sigma_2 = \dfrac{-P}{2\pi R_1 t \sin^2\theta}$ $\Delta R = \dfrac{-P}{2\pi Et \sin\theta}\left(\dfrac{R_2}{R_1} + \nu\right)$ $\psi = \dfrac{-P}{2\pi EtR_1 \sin^2\theta}\left[\dfrac{1}{\tan\theta}\left(1 + \dfrac{R_1}{R_2} - 2\dfrac{R_2}{R_1}\right) - \dfrac{R_2}{R_1}\dfrac{dR_1}{d\theta}\right]$
	4e. Uniform loading, w lb/in^2, on the horizontal projected area; tangential top edge support	For $\theta \le 90°$ $\sigma_1 = \dfrac{wR_2}{2t}$ $\sigma_2 = \dfrac{wR_2}{2t}\left(2\cos^2\theta - \dfrac{R_2}{R_1}\right)$ $\Delta R = \dfrac{wR_2^2 \sin\theta}{2Et}\left(2\cos^2\theta - \dfrac{R_2}{R_1} - \nu\right)$ $\psi = \dfrac{w}{2EtR_1 \tan\theta}\left[R_1R_2(4\cos^2\theta - 1 - 2\nu\sin^2\theta) - R_2^2(7 - 2\cos\theta) + \dfrac{R_2^3}{R_1}\left(2 + \dfrac{\tan\theta}{R_1}\dfrac{dR_1}{d\theta}\right)\right]$
5. Toroidal shell 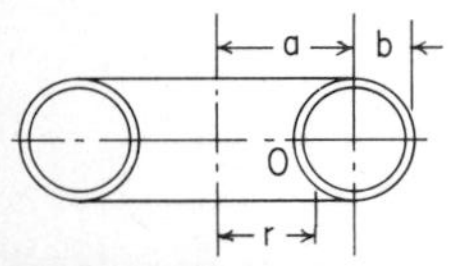	5a. Uniform internal or external pressure, q lb/in^2	$\sigma_1 = \dfrac{qb}{2t}\dfrac{r + a}{r}$ $\text{Max}\ \sigma_1 = \dfrac{qb}{2t}\dfrac{2a - b}{a - b}$ at point O $\sigma_2 = \dfrac{qb}{2t}$ (throughout) $\Delta r = \dfrac{qb}{2Et}[r - \nu(r + a)]$

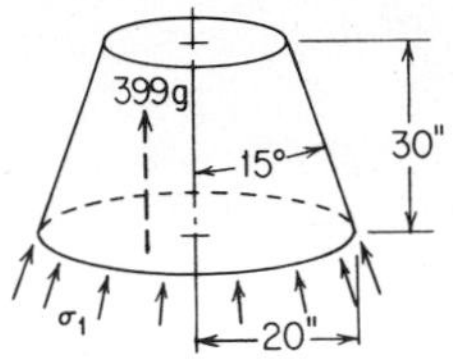

Fig. 12.3

Solution. Since the formulas for a cone loaded by its own weight are given only for a complete cone, superposition will have to be used. From Table 29, cases 2c and 2d will be applicable. First take a complete cone loaded by its own weight with its density multiplied by 400 to account for the acceleration. Since the vertex is up instead of down, a negative value can be used for δ. $R = 20$ in, $\delta = -40.0$, and $\alpha = 15°$; therefore,

$$\sigma_1 = \frac{-40(20)}{2 \sin 15° \cos 15°} = -1600 \text{ lb/in}^2$$

$$\Delta R = \frac{-40(20^2)}{10(10^6) \cos 15°}\left(\sin 15° - \frac{0.3}{2 \sin 15°}\right) = 0.000531 \text{ in}$$

$$\Delta y = \frac{-40(20^2)}{10(10^6) \cos^2 15°}\left(\frac{1}{4 \sin^2 15°} - \sin^2 15°\right) = -0.00628 \text{ in}$$

Next we find the radius of the top as 11.96 in and calculate the change in length and effective weight of the portion of the complete cone to be removed. $R = 11.96$ in, $\delta = -40.0$, and $\alpha = 15°$; therefore,

$$\Delta y = -0.00628\left(\frac{11.96}{20}\right)^2 = -0.00225 \text{ in}$$

The volume of the removed cone is

$$\frac{11.96}{\sin 15°}\left[\frac{11.96(2\pi)}{2}\right](0.25) = 434 \text{ in}^3$$

and the effective weight of the removed cone is 434(0.1)(400) = 17,360 lb.

Removing the load of 17,360 lb can be accounted for by using case 2d, where $P = 17{,}360$, $R = 20$ in, $r = 11.96$ in, $h = 30$ in, and $\alpha = 15°$:

$$\sigma_1 = \frac{17{,}360}{2\pi(20)(0.25) \cos 15°} = 572 \text{ lb/in}^2$$

$$\Delta R = \frac{-0.3(17{,}360)}{2\pi(10)(10^6)(0.25) \cos 15°} = -0.000343 \text{ in}$$

$$\Delta h = \frac{17{,}360 \ln(20/11.96)}{2\pi(10)(10^6) \sin 15° \cos^2 15°} = 0.000588 \text{ in}$$

Therefore, for the truncated cone,

$$\sigma_1 = -1600 + 572 = -1028 \text{ lb/in}^2$$
$$\Delta R = 0.000531 - 0.000343 = 0.000188 \text{ in}$$
$$\Delta h = -0.00628 + 0.00225 + 0.000588 = -0.00344 \text{ in}$$

12.3 *Thin shells of revolution under concentrated or discontinuous loadings producing bending and membrane stresses*

Table 30 gives formulas for forces, moments, and displacements for several *axisymmetric* loadings on both *long* and *short* thin-walled *cylindrical* shells having free ends. These expressions are based on differential equations similar in form to those used to develop the formulas for beams on elastic foundations in Chap. 7. To avoid excessive redundancy in the presentation, only the free-end cases are given in this chapter, but all of the loadings and boundary conditions listed in Tables 7 and 8 as well as the tabulated data in Tables 5 and 6 are directly applicable to cylindrical shells by substituting the shell parameters λ and D for the beam parameters β and EI, respectively. (This will be demonstrated in the examples which follow.) Since many loadings on cylindrical shells occur at the ends, note carefully on page 129 the modified numerators to be used in the equations in Table 7 for the condition when $a = 0$. A special application of this would be the situation where one end of a cylindrical shell is forced to increase a known amount in radius while maintaining zero slope at that same end. This reduces to an application of an externally created concentrated lateral displacement Δ_0 at $a = 0$ (Table 7, case 6) with the left end fixed. (See Example 4.)

Pao (Ref. 60) has tabulated influence coefficients for short cylindrical shells under edge loads with wall thicknesses varying according to $t = Cx^n$ for values of $n = \frac{1}{4}(\frac{1}{4})2$ and for values of t_1/t_2 of 2, 3, and 4. Various degrees of taper are considered by representing data for $k = 0.2(0.2)1.0$ where $k^4 = 3(1 - \nu^2)x_1^4/R^2t_1^2$. Stanek (Ref. 49) has tabulated similar coefficients for constant-thickness cylindrical shells.

A word of caution is in order at this point. The original differential equations used to develop the formulas presented in Table 30 were based on the assumption that radial deformations were small. If the magnitude of the radial deflection approaches the wall thickness, the accuracy of the equations declines. In addition, if axial loads are involved on a relatively short shell, the moments of these axial loads might have an appreciable effect if large deflections are encountered. The effects of these moments are not included in the expressions given.

EXAMPLES

1. A steel tube with a 4.180-in outside diameter and a 0.05-in wall thickness is free at both ends and is 6 in long. At a distance of 2 in from the left end a steel ring with a circular cross section is shrunk onto the outside of the tube such as to compress the tube radially inward a distance of 0.001 in. The maximum tensile stress in the tube is desired. Given: $E = 30(10^6)$ lb/in^2 and $\nu = 0.30$.

Solution. We calculate the following constants:

$$R = 2.090 - 0.025 = 2.065$$

$$\lambda = \left[\frac{3(1 - 0.3^2)}{2.065^2(0.05^2)}\right]^{\frac{1}{4}} = 4.00$$

$$D = \frac{30(10^6)(0.05^3)}{12(1 - 0.3^2)} = 344$$

Since $6/\lambda = 6/4.0 = 1.5$ in and the closest end of the tube is 2 in from the load, this can be considered a very long tube. From Table 30, case 15 indicates that both the maximum deflection and the maximum moment are under the load, so that

$$-0.001 = \frac{-p}{8(344)(4.00^3)} \qquad \text{or} \qquad p = 176 \text{ lb/in}$$

$$\text{Max } M = \frac{176}{4(4)} = 11.0 \text{ in-lb/in}$$

At the cross section under the load and on the inside surface, the following stresses are present:

$$\sigma_1 = 0$$

$$\sigma_1' = \frac{6M}{t^2} = \frac{6(11.0)}{0.05^2} = 26{,}400 \text{ lb/in}^2$$

$$\sigma_2 = \frac{yE}{R} + \nu\sigma_1 = \frac{-0.001(30)(10^6)}{2.065} = -14{,}500 \text{ lb/in}^2$$

$$\sigma_2' = 0.30(26{,}400) = 7920 \text{ lb/in}^2$$

The principal stresses on the inside surface are 26,400 and -6580 lb/in^2

2. Given the same tube and loading as in Example 1, except the tube is only 1.2 in long and the ring is shrunk in place 0.4 in from the left end, the maximum tensile stress is desired.

Solution. Since both ends are closer than $6/\lambda = 1.5$ in from the load, the free ends influence the behavior of the tube under the load. From Table 30, case 2 applies in this example, and since the deflection under the load is the given value from which to work, we must evaluate y at $x = a = 0.4$ in. Note that $\lambda l = 4.0(1.2) = 4.8$, $\lambda x = \lambda a = 4.0(0.4) = 1.6$, and $\lambda(l - a) = 4.0(1.2 - 0.4) = 3.2$. Also,

$$y = y_A F_1 + \frac{\psi_A}{2\lambda} + LT_y$$

$$y_A = \frac{-p}{2D\lambda^3} \frac{C_3 C_{a2} - C_4 C_{a1}}{C_{11}}$$

$$\psi_A = \frac{p}{2D\lambda^2} \frac{C_3 C_{a2} - C_4 C_{a1}}{C_{11}}$$

where $C_3 = -60.51809$ (from Table 5, under F_3 for $\beta x = 4.8$)
$C_{a2} = -12.94222$ (from Table 5, under F_2 for $\beta x = 3.2$)
$C_4 = -65.84195$ $\quad C_{a1} = -12.26569$ $\quad C_2 = -55.21063$ $\quad C_{11} = 3689.703$

Also $F_1(\text{at } x = a) = -0.07526$ and $F_2(\text{at } x = a) = 2.50700$; therefore,

$$y_A = \frac{-p}{2(344)(4.0^3)} \frac{-60.52(-12.94) - (-65.84)(-12.27)}{3689.7} = 0.154(10^{-6})p$$

$$\psi_A = \frac{p}{2(344)(4.0^2)} \frac{-55.21(-12.94) - 2(-60.52)(-12.27)}{3689.7} = -19.0(10^{-6})p$$

TABLE 30 *Shear, moment, slope, and deflection formulas for long and short thin-walled cylindrical shells under axisymmetric loading*

NOTATION: V_o, H, and p = unit loads (pounds per linear inch); q = unit pressure (pounds per square inch); M_o = unit applied couple (inch-pounds per linear inch); all loads are positive as shown. At a distance x (inches) from the left end, the following quantities are defined: V = meridional radial shear (pounds per inch), positive when acting outward on the right hand portion; M = meridional bending moment (inch-pounds per inch), positive when compressive on the outside; ψ = meridional slope (radians), positive when the deflection increases with x; y = radial deflection (inches), positive outward. σ_1 and σ_2 = meridional and circumferential membrane stresses (pounds per square inch), positive when tensile; σ_1' and σ_2' = meridional and circumferential bending stresses (pounds per square inch), positive when tensile on the outside; τ = meridional radial shear stress (pounds per square inch). E = modulus of elasticity (pounds per square inch); ν = Poisson's ratio; R = mean radius (inches); t = wall thickness (inches).

The following constants and functions are hereby defined in order to permit condensing the tabulated formulas which follow:

$$\lambda = \left[\frac{3(1-\nu^2)}{R^2t^2}\right]^{\frac{1}{4}} \qquad D = \frac{Et^3}{12(1-\nu^2)}$$

(*Note:* See page 94 for a definition of $\langle x-a\rangle^n$; also all hyperbolic and trigonometric functions of the argument $\langle x-a\rangle$ are also defined as zero if $x < a$)

$F_1 = \text{Cosh}\,\lambda x \cos\lambda x$
$F_2 = \text{Cosh}\,\lambda x \sin\lambda x + \text{Sinh}\,\lambda x \cos\lambda x$
$F_3 = \text{Sinh}\,\lambda x \sin\lambda x$
$F_4 = \text{Cosh}\,\lambda x \sin\lambda x - \text{Sinh}\,\lambda x \cos\lambda x$

$C_1 = \text{Cosh}\,\lambda l \cos\lambda l$
$C_2 = \text{Cosh}\,\lambda l \sin\lambda l + \text{Sinh}\,\lambda l \cos\lambda l$
$C_3 = \text{Sinh}\,\lambda l \sin\lambda l$
$C_4 = \text{Cosh}\,\lambda l \sin\lambda l - \text{Sinh}\,\lambda l \cos\lambda l$

$C_{11} = \text{Sinh}^2\,\lambda l - \sin^2\lambda l$
$C_{12} = \text{Cosh}\,\lambda l\,\text{Sinh}\,\lambda l + \cos\lambda l \sin\lambda l$
$C_{13} = \text{Cosh}\,\lambda l\,\text{Sinh}\,\lambda l - \cos\lambda l \sin\lambda l$
$C_{14} = \text{Sinh}^2\,\lambda l + \sin^2\lambda l$

$F_{a1} = \langle x-a\rangle^0\,\text{Cosh}\,\lambda\langle x-a\rangle \cos\lambda\langle x-a\rangle$
$F_{a2} = \text{Cosh}\,\lambda\langle x-a\rangle \sin\lambda\langle x-a\rangle + \text{Sinh}\,\lambda\langle x-a\rangle \cos\lambda\langle x-a\rangle$
$F_{a3} = \text{Sinh}\,\lambda\langle x-a\rangle \sin\lambda\langle x-a\rangle$
$F_{a4} = \text{Cosh}\,\lambda\langle x-a\rangle \sin\lambda\langle x-a\rangle - \text{Sinh}\,\lambda\langle x-a\rangle \cos\lambda\langle x-a\rangle$
$F_{a5} = \langle x-a\rangle^0 - F_{a1}$
$F_{a6} = 2\lambda(x-a)\langle x-a\rangle^0 - F_{a2}$

$C_{a1} = \text{Cosh}\,\lambda(l-a)\cos\lambda(l-a)$
$C_{a2} = \text{Cosh}\,\lambda(l-a)\sin\lambda(l-a) + \text{Sinh}\,\lambda(l-a)\cos\lambda(l-a)$
$C_{a3} = \text{Sinh}\,\lambda(l-a)\sin\lambda(l-a)$
$C_{a4} = \text{Cosh}\,\lambda(l-a)\sin\lambda(l-a) - \text{Sinh}\,\lambda(l-a)\cos\lambda(l-a)$
$C_{a5} = 1 - C_{a1}$
$C_{a6} = 2\lambda(l-a) - C_{a2}$

$A_1 = \frac{1}{2}e^{-\lambda a}\cos\lambda a$
$A_2 = \frac{1}{2}e^{-\lambda a}(\sin\lambda a - \cos\lambda a)$
$A_3 = -\frac{1}{2}e^{-\lambda a}\sin\lambda a$
$A_4 = \frac{1}{2}e^{-\lambda a}(\sin\lambda a + \cos\lambda a)$

$B_1 = \frac{1}{2}e^{-\lambda b}\cos\lambda b$
$B_2 = \frac{1}{2}e^{-\lambda b}(\sin\lambda b - \cos\lambda b)$
$B_3 = -\frac{1}{2}e^{-\lambda b}\sin\lambda b$
$B_4 = \frac{1}{2}e^{-\lambda b}(\sin\lambda b + \cos\lambda b)$

Numerical values of F_1, F_2, F_3, and F_4 for λx ranging from 0 to 6 are tabulated in Table 5; numerical values of C_{11}, C_{12}, C_{13}, and C_{14} are tabulated in Table 6

Short Shells with free ends

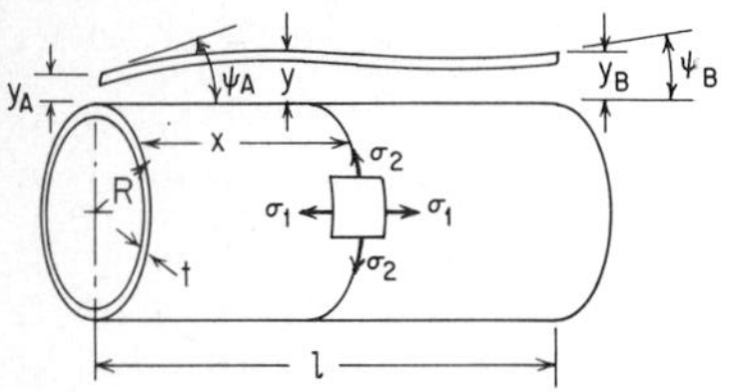

$R/t > 10$

Meridional radial shear $= V = -y_A 2D\lambda^3 F_2 - \psi_A 2D\lambda^2 F_3 + LT_V$

Meridional bending moment $= M = -y_A 2D\lambda^2 F_3 - \psi_A D\lambda F_4 + LT_M$ (*Note:* The load terms LT_V, LT_M, etc., are given for each of the following cases)

Meridional slope $= \psi = \psi_A F_1 - y_A \lambda F_4 + LT_\psi$

Radial deflection $= y = y_A F_1 + \dfrac{\psi_A}{2\lambda} F_2 + LT_y$

Circumferential membrane stress $= \sigma_2 = \dfrac{yE}{R} + \nu\sigma_1$

Meridional bending stress $= \sigma_1' = \dfrac{-6M}{t^2}$

Circumferential bending stress $= \sigma_2' = \nu\sigma_1'$

Meridional radial shear stress $= \tau = \dfrac{V}{t}$ (average value)

Loading and case no.	End deformations	Load terms or load and deformation equations	Selected values
1. Radial end load, V_o lb/in 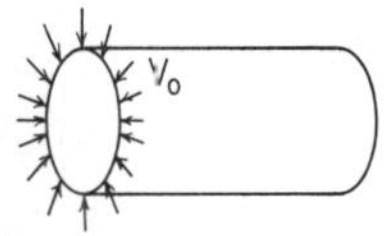 If $\lambda l > 6$, see case 8	$\psi_A = \dfrac{V_o}{2D\lambda^2}\dfrac{C_{14}}{C_{11}}$ $y_A = \dfrac{-V_o}{2D\lambda^3}\dfrac{C_{13}}{C_{11}}$ $\psi_B = \dfrac{V_o}{2D\lambda^2}\dfrac{2C_3}{C_{11}}$ $y_B = \dfrac{V_o}{2D\lambda^3}\dfrac{C_4}{C_{11}}$	$LT_V = -V_o F_1$ $LT_M = \dfrac{-V_o}{2\lambda} F_2$ $LT_\psi = \dfrac{-V_o}{2D\lambda^2} F_3$ $LT_y = \dfrac{-V_o}{4D\lambda^3} F_4$	$\sigma_1 = 0$ Max $\sigma_2 = \dfrac{y_A E}{R}$ Max $\psi = \psi_A$ Max $y = y_A$
2. Intermediate radial load, p lb/in 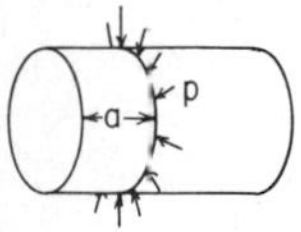 If $\lambda l > 6$, consider case 9	$\psi_A = \dfrac{p}{2D\lambda^2}\dfrac{C_2 C_{a2} - 2C_3 C_{a1}}{C_{11}}$ $y_A = \dfrac{-p}{2D\lambda^3}\dfrac{C_3 C_{a2} - C_4 C_{a1}}{C_{11}}$ $\psi_B = \psi_A C_1 - y_A \lambda C_4 - \dfrac{p}{2D\lambda^2} C_{a3}$ $y_B = y_A C_1 + \dfrac{\psi_A C_2}{2\lambda} - \dfrac{p}{4D\lambda^3} C_{a4}$	$LT_V = -p F_{a1}$ $LT_M = \dfrac{-p}{2\lambda} F_{a2}$ $LT_\psi = \dfrac{-p}{2D\lambda^2} F_{a3}$ $LT_y = \dfrac{-p}{4D\lambda^3} F_{a4}$	$\sigma_1 = 0$

TABLE 30 ***Shear, moment, slope, and deflection formulas for long and short thin-walled cylindrical shells under axisymmetric loading*** *(Cont.)*

Loading and case no.	End deformations	Load terms or load and deformation equations	Selected values
3. End moment, M_o in-lb/in 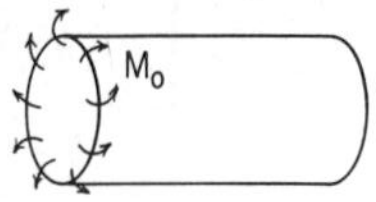 If $\lambda l > 6$, see case 10	$\psi_A = \frac{-M_o}{D\lambda}\frac{C_{12}}{C_{11}}$ $y_A = \frac{M_o}{2D\lambda^2}\frac{C_{14}}{C_{11}}$ $\psi_B = \frac{-M_o}{D\lambda}\frac{C_2}{C_{11}}$ $y_B = \frac{-M_o}{D\lambda^2}\frac{C_3}{C_{11}}$	$LT_V = -M_o\lambda F_4$ $LT_M = M_o F_1$ $LT_\psi = \frac{M_o}{2D\lambda}F_2$ $LT_y = \frac{M_o}{2D\lambda^2}F_3$	$\sigma_1 = 0$ Max $\sigma_2 = \frac{y_A E}{R}$ Max $M = M_o$ (at $x = 0$) Max $\psi = \psi_A$ Max $y = y_A$
4. Intermediate applied moment, M_o in-lb/in 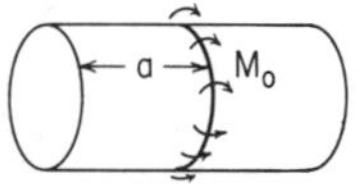 If $\lambda l > 6$, consider case 11	$\psi_A = \frac{-M_o}{D\lambda}\frac{C_2C_{a1} + C_3C_{a4}}{C_{11}}$ $y_A = \frac{M_o}{2D\lambda^2}\frac{2C_3C_{a1} + C_4C_{a4}}{C_{11}}$ $\psi_B = \psi_A C_1 - y_A\lambda C_4 + \frac{M_o}{2D\lambda}C_{a2}$ $y_B = y_A C_1 + \frac{\psi_A}{2\lambda}C_2 + \frac{M_o}{2D\lambda^2}C_{a3}$	$LT_V = -M_o\lambda F_{a4}$ $LT_M = M_o F_{a1}$ $LT_\psi = \frac{M_o}{2D\lambda}F_{a2}$ $LT_y = \frac{M_o}{2D\lambda^2}F_{a3}$	
5. Uniform radial pressure from a to l 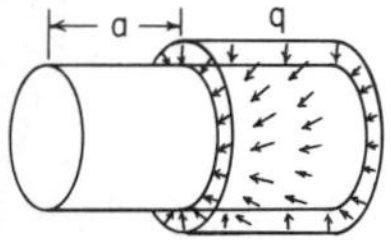 If $\lambda l > 6$, consider case 12	$\psi_A = \frac{q}{2D\lambda^3}\frac{C_2C_{a3} - C_3C_{a2}}{C_{11}}$ $y_A = \frac{-q}{4D\lambda^4}\frac{2C_3C_{a3} - C_4C_{a2}}{C_{11}}$ $\psi_B = \psi_A C_1 - y_A\lambda C_4 - \frac{q}{4D\lambda^3}C_{a4}$ $y_B = y_A C_1 + \frac{\psi_A}{2\lambda}C_2 - \frac{q}{4D\lambda^4}C_{a5}$	$LT_V = \frac{-q}{2\lambda}F_{a2}$ $LT_M = \frac{-q}{2\lambda^2}F_{a3}$ $LT_\psi = \frac{-q}{4D\lambda^3}F_{a4}$ $LT_y = \frac{-q}{4D\lambda^4}F_{a5}$	
6. Uniformly increasing pressure from a to l 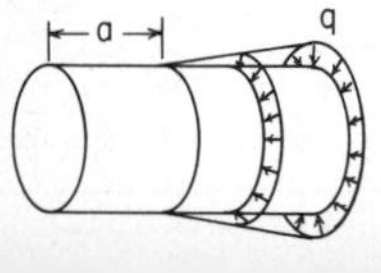	$\psi_A = \frac{-q}{4D\lambda^4(l-a)}\frac{2C_3C_{a3} - C_2C_{a4}}{C_{11}}$ $y_A = \frac{-q}{4D\lambda^5(l-a)}\frac{C_3C_{a4} - C_4C_{a3}}{C_{11}}$ $\psi_B = \psi_A C_1 - y_A\lambda C_4 - \frac{qC_{a5}}{4D\lambda^4(l-a)}$ [illegible]	$LT_V = \frac{-q}{2\lambda^2(l-a)}F_{a3}$ $LT_M = \frac{-q}{4\lambda^3(l-a)}F_{a4}$ $LT_\psi = \frac{-q}{4D\lambda^4(l-a)}F_{a5}$ [illegible]	$\sigma_1 = 0$ Max $\sigma_2 = \frac{y_B E}{R}$ Max $y = y_B$ Max $\psi = \psi_B$

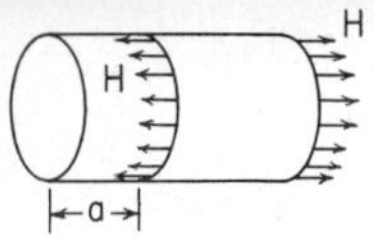	$y_A = \frac{-\nu HR}{Et}\frac{2C_3C_{a3} - C_3C_{a2}}{C_{11}}$ $\psi_B = \psi_A C_1 - y_A\lambda C_4 - \frac{\nu HR\lambda}{Et}C_{a4}$ $y_B = y_A C_1 + \frac{\psi_A C_2}{2\lambda} - \frac{\nu HR}{Et}C_{a5}$	$LT_M = \frac{-\nu H}{2\lambda^2 R}F_{a3}$ $LT_\psi = \frac{-\nu HR\lambda}{Et}F_{a4}$ $LT_y = \frac{-\nu HR}{Et}F_{a5}$	

Long shells with the left end free (right end more than $6/\lambda$ units of length from the closest load)

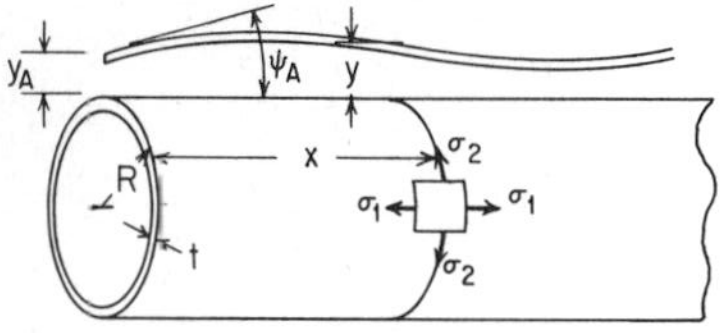

$R/t > 10$

Loading and case no.

Meridional radial shear $= V = -y_A 2D\lambda^3 F_2 - \psi_A 2D\lambda^2 F_3 + LT_V$

Meridional bending moment $= M = -y_A 2D\lambda^2 F_3 - \psi_A D\lambda F_4 + LT_M$ (*Note:* The load terms LT_V, LT_M, etc., are given where needed in the following cases)

Meridional slope $= \psi = \psi_A F_1 - y_A \lambda F_4 + LT_\psi$

Radial deflection $= y = y_A F_1 + \frac{\psi_A}{2\lambda}F_2 + LT_y$

Circumferential membrane stress $= \sigma_2 = \frac{yE}{R} + \nu\sigma_1$

Meridional bending stress $= \sigma_1' = \frac{-6M}{t^2}$

Circumferential bending stress $= \sigma_2' = \nu\sigma_1'$

Meridional radial shear stress $= \tau = \frac{V}{t}$ (average value)

Loading and case no.	End deformations	Load terms or load and deformation equations	Selected values
8. Radial end load, V_o lb/in 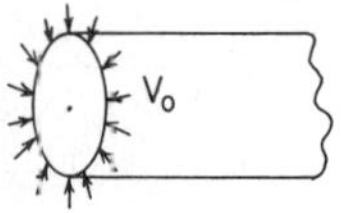	$\psi_A = \frac{V_o}{2D\lambda^2}$ $y_A = \frac{-V_o}{2D\lambda^3}$	$V = -V_o e^{-\lambda x}(\cos\lambda x - \sin\lambda x)$ $M = \frac{-V_o}{\lambda}e^{-\lambda x}\sin\lambda x$ $\psi = \frac{V_o}{2D\lambda^2}e^{-\lambda x}(\cos\lambda x + \sin\lambda x)$ $y = \frac{-V_o}{2D\lambda^3}e^{-\lambda x}\cos\lambda x$	Max $V = -V_o$ at $x = 0$ Max $M = -0.3225\frac{V_o}{\lambda}$ at $x = \frac{\pi}{4\lambda}$ Max $\psi = \psi_A$ Max $y = y_A$ $\sigma_1 = 0$ Max $\sigma_2 = \frac{-2V_o\lambda R}{t}$ at $x = 0$
9. Intermediate radial load. p lb/in 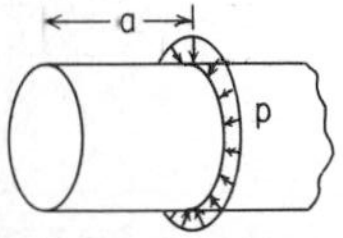 If $\lambda a > 3$, consider case 15	$\psi_A = \frac{-p}{D\lambda^2}A_2$ $y_A = \frac{-p}{D\lambda^3}A_1$	$LT_V = -pF_{a1}$ $LT_M = \frac{-p}{2\lambda}F_{a2}$ $LT_\psi = \frac{-p}{2D\lambda^2}F_{a3}$ $LT_y = \frac{-p}{4D\lambda^3}F_{a4}$	$\sigma_1 = 0$

TABLE 30 ***Shear, moment, slope, and deflection formulas for long and short thin-walled cylindrical shells under axisymmetric loading*** ***(Cont.)***

Loading and case no.	End deformations	Load terms or load and deformation equations	Selected values
10. End moment, M_o in-lb/in 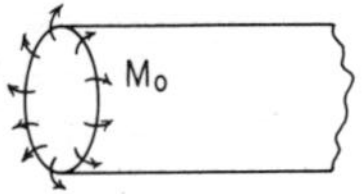	$\psi_A = \frac{-M_o}{D\lambda}$ $y_A = \frac{M_o}{2D\lambda^2}$	$V = -M_o\lambda e^{-\lambda x} \sin \lambda x$ $M = M_o e^{-\lambda x} (\cos \lambda x + \sin \lambda x)$ $\psi = \frac{-M_o}{D\lambda} e^{-\lambda x} \cos \lambda x$ $y = \frac{-M_o}{2D\lambda^2} e^{-\lambda x} (\sin \lambda x - \cos \lambda x)$	Max $V = -0.3225 M_o\lambda$ at $x = \frac{\pi}{4\lambda}$ Max $M = M_o$ at $x = 0$ Max $\psi = \psi_A$ Max $y = y_A$ $\sigma_1 = 0$ Max $\sigma_2 = \frac{2M_o\lambda^2 R}{t}$ at $x = 0$
11. Intermediate applied moment, M_a in-lb/in 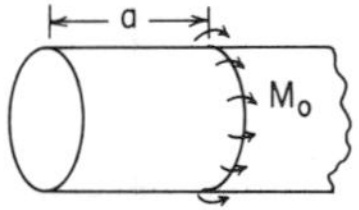 If $\lambda a > 3$, consider case 16	$\psi_A = \frac{-2M_o}{D\lambda} A_1$ $y_A = \frac{M_o}{D\lambda^2} A_4$	$LT_V = -M_o\lambda F_{a4}$ $LT_M = M_o F_{a1}$ $LT_\psi = \frac{M_o}{2D\lambda} F_{a2}$ $LT_y = \frac{M_o}{2D\lambda^2} F_{a3}$	$\sigma_1 = 0$
12. Uniform radial pressure from a to b 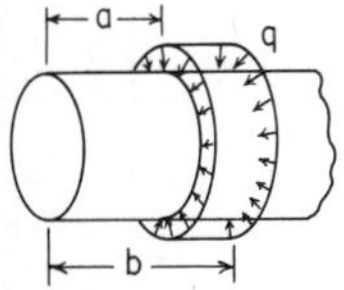	$\psi_A = \frac{-q}{D\lambda^3}(B_3 - A_3)$ $y_A = \frac{-q}{2D\lambda^4}(B_2 - A_2)$	$LT_V = \frac{-q}{2\lambda}(F_{a2} - F_{b2})$ $LT_M = \frac{-q}{2\lambda^2}(F_{a3} - F_{b3})$ $LT_\psi = \frac{-q}{4D\lambda^3}(F_{a4} - F_{b4})$ $LT_y = \frac{-q}{4D\lambda^4}(F_{a5} - F_{b5})$	$\sigma_1 = 0$
13. Uniformly increasing pressure from a to b 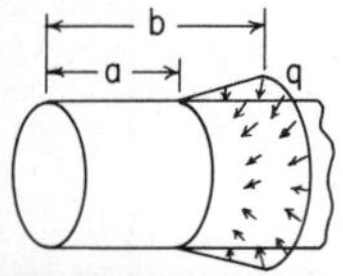	$\psi_A = \frac{q}{D}\left[\frac{B_4 - A_4}{2\lambda^4(b-a)} - \frac{B_3}{\lambda^3}\right]$ $y_A = \frac{q}{2D}\left[\frac{B_3 - A_3}{\lambda^5(b-a)} - \frac{B_2}{\lambda^4}\right]$	$LT_V = \frac{-q}{2}\left[\frac{F_{a3} - F_{b3}}{\lambda^2(b-a)} - \frac{F_{b2}}{\lambda}\right]$ $LT_M = \frac{-q}{2}\left[\frac{F_{a4} - F_{b4}}{2\lambda^3(b-a)} - \frac{F_{b3}}{\lambda^2}\right]$ $LT_\psi = \frac{-q}{4D}\left[\frac{F_{a5} - F_{b5}}{\lambda^4(b-a)} - \frac{F_{b4}}{\lambda^3}\right]$ $LT_y = \frac{-q}{4D}\left[\frac{F_{a6} - F_{b6}}{2\lambda^5(b-a)} + \frac{F_{b1}}{\lambda^4}\right]$	$\sigma_1 = 0$

14. Axial load along the portion from a to b 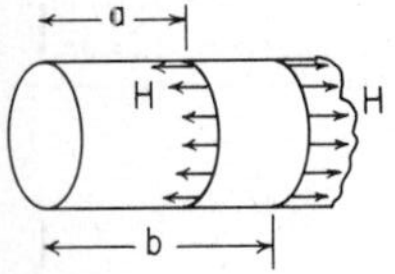	$\psi_A = \frac{-\nu H}{RD\lambda^3}(B_3 - A_3)$ $y_A = \frac{-\nu H}{2RD\lambda^4}(B_2 - A_2)$	$LT_V = \frac{-\nu H}{2R\lambda}(F_{a2} - F_{b2})$ $LT_M = \frac{-\nu H}{2R\lambda^2}(F_{a3} - F_{b3})$ $LT_\psi = \frac{-\nu H}{4RD\lambda^3}(F_{a4} - F_{b4})$ $LT_y = \frac{-\nu H}{4RD\lambda^4}(F_{a5} - F_{b5})$	$\sigma_1 = \frac{H}{t}\langle x - a\rangle^0 - \frac{H}{t}\langle x - b\rangle^0$

Very long shells (both ends more than $6/\lambda$ units of length from the nearest loading)

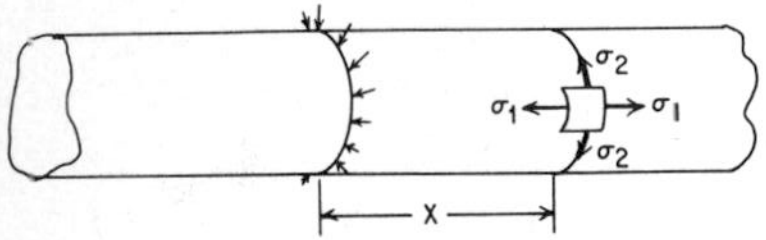

Circumferential membrane stress $= \sigma_2 = \frac{yE}{R} + \nu\sigma_1$

Meridional bending stress $= \sigma_1' = \frac{-6M}{t^2}$

Circumferential bending stress $= \sigma_2' = \nu\sigma_1'$

Meridional radial shear stress $= \tau = \frac{V}{t}$

Loading and case no.	Load and deformation equations	Selected values
15. Concentrated radial load, p (lb/linear in of circumference) 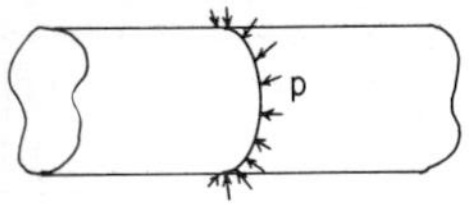	$V = \frac{-p}{2}e^{-\lambda x}\cos\lambda x$ $M = \frac{p}{4\lambda}e^{-\lambda x}(\cos\lambda x - \sin\lambda x)$ $\psi = \frac{p}{8D\lambda^2}e^{-\lambda x}\sin\lambda x$ $y = \frac{-p}{8D\lambda^3}e^{-\lambda x}(\cos\lambda x + \sin\lambda x)$	Max $V = \frac{-p}{2}$ at $x = 0$ $\quad \sigma_1 = 0$ Max $M = \frac{p}{4\lambda}$ at $x = 0$ Max $\psi = 0.0403\frac{p}{D\lambda^2}$ at $x = \frac{\pi}{4\lambda}$ Max $y = \frac{-p}{8D\lambda^3}$ at $x = 0$

TABLE 30 *Shear, moment, slope, and deflection formulas for long and short thin-walled cylindrical shells under axisymmetric loading* (*Cont.*)

Loading and case no.	Load terms or load and deformation equations	Selected values
16. Applied moment 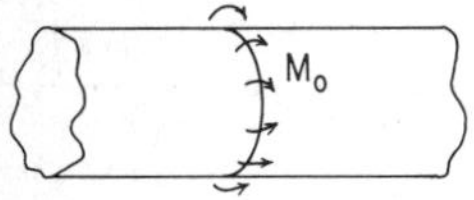	$V = \frac{-M_o\lambda}{2} e^{-\lambda x} (\cos \lambda x + \sin \lambda x)$ $M = \frac{M_o}{2} e^{-\lambda x} \cos \lambda x$ $\psi = \frac{-M_o}{4D\lambda} e^{-\lambda x} (\cos \lambda x - \sin \lambda x)$ $y = \frac{-M_o}{8D\lambda^2} e^{-\lambda x} \sin \lambda x$	$\text{Max } V = \frac{-M_o\lambda}{2}$ at $x = 0$ $\sigma_1 = 0$ $\text{Max } M = \frac{M_o}{2}$ at $x = 0$ $\text{Max } \psi = \frac{-M_o}{4D\lambda}$ at $x = 0$ $\text{Max } y = -0.0403 \frac{M_o}{D\lambda^2}$ at $x = \frac{\pi}{4\lambda}$
17. Uniform pressure over a band of width $2a$ 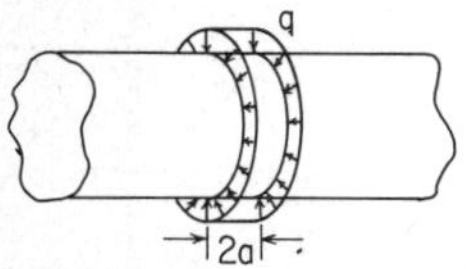	Superimpose cases 10 and 12 to make ψ_A (at $x = 0$) $= 0$	$\text{Max } M = \frac{q}{2\lambda^2} e^{-\lambda a} \sin \lambda a$ at $x = 0$ $\text{Max } y = \frac{-q}{4D\lambda^4}(1 - e^{-\lambda a} \cos \lambda a)$ at $x = 0$ $\sigma_1 = 0$

and $LT_y = 0$ since x is not greater than a. Substituting into the expression for y at $x = a$ gives

$$-0.001 = 0.154(10^{-6})p(-0.07526) - \frac{19.0(10^{-6})p(2.507)}{2(4.0)} = -5.96(10^{-6})p$$

or $p = 168$ lb/in, $y_A = 0.0000259$ in, and $\psi_A = -0.00319$ rad.

Although the position of the maximum moment depends upon the position of the load, the maximum moment in this case would be expected to be under the load since the load is some distance from the free end:

$$M = -y_A 2D\lambda^2 F_3 - \psi_A D\lambda F_4 + LT_M$$

and at $x = a$, $F_3 = 2.37456$, $F_4 = 2.64573$, and $LT_M = 0$ since x is not greater than a. Therefore,

$$\begin{aligned}\text{Max } M &= -(0.0000259)(2)(344)(4.0^2)(2.375) - (-0.00319)(344)(4.0)(2.646)\\ &= 10.92 \text{ in-lb/in}\end{aligned}$$

At the cross section under the load and on the inside surface the following stresses are present:

$$\sigma_1 = 0 \qquad \sigma_1' = \frac{6(10.92)}{0.05^2} = 26{,}200 \text{ lb/in}^2 \qquad \sigma_2 = \frac{-0.001(30)(10^6)}{2.065} = -14{,}500 \text{ lb/in}^2 \qquad \sigma_2' = 0.30(26{,}200) = 7{,}860 \text{ lb/in}^2$$

The small change in the maximum stress produced in this shorter tube points out how localized the effect of a load on a shell can be. Had the radial load been the same, however, instead of the radial deflection, a greater difference might have been noted and the stress σ_2 would have increased in magnitude instead of decreasing.

3. A cylindrical aluminum shell is 10 in long and 15 in in diameter and must be designed to carry an internal pressure of 300 lb/in² without exceeding a maximum tensile stress of 12,000 lb/in². The ends are capped with massive flanges, which are sufficiently clamped to the shell to effectively resist any radial or rotational deformation at the ends. Given: $E = 10(10^6)$ lb/in² and $\nu = 0.3$.

First solution. Case 1c from Table 29 and cases 1 and 3 or cases 8 and 10 from Table 30 can be superimposed to find the radial end load and the end moment which will make the slopes and deflections at both ends zero. Figure 12.4 shows the loadings applied to the shell.

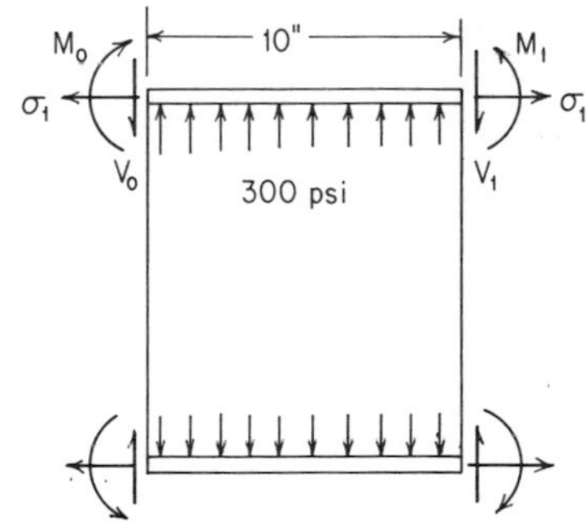

Fig. 12.4

First we evaluate the necessary constants:

$$R = 7.5 \text{ in} \qquad l = 10 \text{ in} \qquad D = \frac{10(10^6)t^3}{12(1 - 0.3^2)} = 915{,}800t^3$$

$$\lambda = \left[\frac{3(1 - 0.3^2)}{7.5^2 t^2}\right]^{\frac{1}{4}} = \frac{0.4694}{t^{0.5}} \qquad \lambda l = \frac{4.694}{t^{0.5}}$$

Since the thickness is unknown at this step in the calculation, we can only estimate whether the shell must be considered long or short, i.e., whether the loads at one end will have any influence on the deformations at the other. To make an estimate of this effect we can calculate the wall thickness necessary for just the internal pressure. From case 1c of Table 29, the value of the hoop stress $\sigma_2 = qR/t$ can be equated to 12,000 lb/in^2 and the expression solved for the thickness:

$$t = \frac{300(7.5)}{12{,}000} = 0.1875 \text{ in}$$

Using this value for t gives $\lambda l = 10.84$, which would be a very long shell.

For a trial solution the assumption will be made that the radial load and bending moment at the right end do not influence the deformations at the left end. Owing to the rigidity of the end caps, the radial deformation and the angular rotation of the left end will be set equal to zero. From Table 29, case 1c,

$$\sigma_1 = \frac{qR}{2t} = \frac{300(7.5)}{2t} = \frac{1125}{t} \qquad \sigma_2 = \frac{qR}{t} = \frac{2250}{t} \qquad \psi = 0$$

$$\Delta R = \frac{qR^2}{Et}\left(1 - \frac{\nu}{2}\right) = \frac{300(7.5^2)}{10(10^6)t}\left(1 - \frac{0.3}{2}\right) = \frac{0.001434}{t}$$

From Table 30, case 8,

$$y_A = \frac{-V_o}{2D\lambda^3} = \frac{-V_o}{2(915{,}800t^3)(0.4694/t^{\frac{1}{2}})^3} = \frac{-5.279(10^{-6})V_o}{t^{\frac{3}{2}}}$$

$$\psi_A = \frac{V_o}{2D\lambda^2} = \frac{2.478(10^{-6})V_o}{t^{\frac{5}{2}}}$$

From Table 30, case 10,

$$y_A = \frac{M_o}{2D\lambda^2} = \frac{2.478(10^{-6})M_o}{t^2}$$

$$\psi_A = \frac{-M_o}{D\lambda} = \frac{-2.326(10^{-6})M_o}{t^{\frac{5}{2}}}$$

Summing the radial deformations to zero gives

$$\frac{0.001434}{t} - \frac{5.279(10^{-6})V_o}{t^{\frac{3}{2}}} + \frac{2.478(10^{-6})M_o}{t^2} = 0$$

Similarly, summing the end rotations to zero gives

$$\frac{2.478(10^{-6})V_o}{t^2} - \frac{2.326(10^{-6})M_o}{t^{\frac{5}{2}}} = 0$$

Solving these two equations gives

$$V_o = 543t^{\frac{1}{2}} \qquad \text{and} \qquad M_o = 579t$$

A careful examination of the problem reveals that the maximum bending stress will occur at the end, and so the following stresses must be combined: From Table 29, case 1c,

$$\sigma_1 = \frac{1125}{t} \qquad \sigma_2 = \frac{2250}{t}$$

From Table 30, case 8,

$$\sigma_1 = 0 \qquad \sigma_1' = 0 \qquad \sigma_2' = 0$$

$$\sigma_2 = \frac{-2V_o\lambda R}{t} = \frac{-2(543t^{\frac{1}{2}})(0.4694/t^{\frac{1}{2}})(7.5)}{t} = \frac{-3826}{t}$$

From Table 30, case 10,

$$\sigma_1 = 0$$

$$\sigma_2 = \frac{2M_o\lambda^2 R}{t} = \frac{2(579t)(0.4694/t^{\frac{1}{2}})^2(7.5)}{t} = \frac{1913}{t}$$

and on the inside surface

$$\sigma_1' = \frac{6M_o}{t^2} = \frac{3473}{t}$$

$$\sigma_2' = \nu\sigma_1' = \frac{1042}{t}$$

Therefore, at the end of the cylinder the maximum longitudinal tensile stress is $1125/t + 3473/t = 4598/t$; similarly the maximum circumferential tensile stress is $2250/t - 3826/t + 1913/t + 1042/t = 1379/t$.

Since the allowable tensile stress was 12,000 lb/in^2, we can evaluate $4598/t = 12{,}000$ to obtain $t = 0.383$ in. This allows λl to be calculated as 7.59, which verifies the assumption that the shell can be considered a long shell for this loading and support.

Second solution. This loaded shell represents a case where both ends are fixed and a uniform radial pressure is applied over the entire length. Since the shell is considered long, we can find the expressions for R_A and M_A in Table 8, case 2, under the condition of the left end fixed and where the distance $a = 0$ and b can be considered infinite:

$$R_A = \frac{-2w}{\beta}(B_1 - A_1) \qquad \text{and} \qquad M_A = \frac{w}{\beta^2}(B_4 - A_4)$$

If $-V_o$ is substituted for R_A, M_o for M_A, λ for β, and D for EI, the solution should apply to the problem at hand. Care must be exercised when substituting for the distributed load w. A purely radial pressure would produce a radial deformation $\Delta R = qR^2/Et$, while the effect of the axial pressure on the ends reduces this to $\Delta R = qR^2(1 - \nu/2)/Et$. Therefore, for w we must substitute $-300(1 - \nu/2) = -255$ lb/in^2. Also note that for $a = 0$, $A_1 = A_4 = 0.5$, and for $b = \infty$, $B_1 = B_4 = 0$. Therefore,

$$V_o = \frac{-2(255)}{\lambda}(0 - 0.5) = \frac{255}{\lambda} = 543t^{\frac{1}{2}}$$

$$M_o = \frac{-255}{\lambda^2}(0 - 0.5) = \frac{127.5}{\lambda^2} = 579t$$

which verifies the results of the first solution.

If we examine case 2 of Table 7 under the condition of both ends fixed, we find the expression

$$M_o = M_A = \frac{w}{2\lambda^2}\,\frac{2C_3C_5 - C_4^2}{C_{11}}$$

Substituting for the several constants and reducing the expression to a simple form, we obtain

$$M_o = \frac{-w}{2\lambda^2}\,\frac{\text{Sinh}\,\lambda l - \sin\lambda l}{\text{Sinh}\,\lambda l + \sin\lambda l}$$

The hyperbolic sine of 7.59 is 989, and so for all practical purposes

$$M_o = \frac{-w}{2\lambda^2} = 579t$$

which, of course, is the justification for the formulas in Table 8.

4. A 2-in length of steel tube described in Example 1 is heated, and rigid plugs are inserted $\frac{1}{2}$ in into each end. The rigid plugs have a diameter equal to the inside diameter of the tube plus 0.004 in at room temperature. Find the longitudinal and circumferential stresses at the outside of the tube adjacent to the end of the plug and the diameter at midlength after the tube is shrunk onto the plugs.

Solution. The most straightforward solution would consist of assuming that the portion of the tube outside the rigid plug is, in effect, displaced radially a distance of 0.002 in and owing to symmetry the midlength has zero slope. A steel cylindrical shell $\frac{1}{2}$ in in length, fixed on the left end with a radial displacement of 0.002 in at $a = 0$ and with the right end guided, i.e., slope equal to zero, is the case to be solved.

From Example 1, $R = 2.065$ in, $\lambda = 4.00$, and $D = 344$; $\lambda l = 4.0(0.5) = 2.0$. From Table 7, case 6, for the left end fixed and the right end guided, we find the following expressions when $a = 0$:

$$R_A = \Delta_o 2EI\beta^3 \frac{C_4^2 + C_2^2}{C_{12}} \qquad \text{and} \qquad M_A = \Delta_o 2EI\beta^2 \frac{C_1C_4 - C_3C_2}{C_{12}}$$

Replace EI with D and β with λ; $\Delta_o = 0.002$, and from page 129 note that $C_4^2 + C_2^2 = 2C_{14}$ and $C_1C_4 - C_3C_2 = -C_{13}$. From Table 5, for $\lambda l = 2.00$, we find that $2C_{14} = 27.962$, $-C_{13} = -14.023$, and $C_{12} = 13.267$. Therefore

$$R_A = 0.002(2)(344)(4.0^3)\frac{27.962}{13.267} = 185.6 \text{ lb/in}$$

$$M_A = 0.002(2)(344)(4.0^2)\frac{-14.023}{13.267} = -23.27 \text{ in-lb/in}$$

To find the deflection at the midlength of the shell, which is the right end of the half-shell being used here, we solve for y at $x = 0.5$ in and $\lambda x = 4.0(0.5) = 2.0$. Note that $y_A = 0$ because the deflection of 0.002 in was forced into the shell just beyond the end in the solution being considered here. Therefore

$$y = \frac{M_A}{2D\lambda^2}F_3 + \frac{R_A}{2D\lambda^3}F_4 + \Delta_o F_{a1}$$

where from Table 5 at $\lambda x = 2.0$

$$F_3 = 3.298 \quad F_4 = 4.930 \qquad F_{a1} = F_1 = -1.566 \quad \text{since } a = 0$$

$$y_{x=0.5} = \frac{-23.27}{2(344)(4.0^2)}3.298 + \frac{185.6}{4(344)(4.0^3)}4.93 + 0.002(-1.566) = 0.00029 \text{ in}$$

For a partial check on the solution we can calculate the slope at midlength. From Table 7, case 6,

$$\theta = \frac{M_A}{2EI\beta}F_2 + \frac{R_A}{2EI\beta^2}F_3 - \Delta_o \beta F_{a4}$$

where $F_2 = 1.912$ and $F_{a4} = F_4$ since $a = 0$. Therefore,

$$\theta = \frac{-23.27}{2(344)(4.0)}1.912 + \frac{185.6}{2(344)(4.0^2)}3.298 - 0.002(4.0)(4.930) = 0.00000$$

Now from Table 30,

$$\sigma_1' = \frac{-6M}{t^2} = \frac{-6(-23.27)}{0.05^2} = 55{,}850 \text{ lb/in}^2$$

Since $\sigma_1 = 0$,

$$\sigma_2 = \frac{0.002(30)(10^6)}{2.065} = 29{,}060 \text{ lb/in}^2$$

$$\sigma_2' = 0.3(55{,}800) = 16{,}750 \text{ lb/in}^2$$

On the outside surface at the cross section adjacent to the plug the longitudinal stress is 55,850 lb/in^2 and the circumferential stress is 29,060 + 16,750 = 45,810 lb/in^2. Since a rigid plug is only hypothetical, the actual stresses present would be smaller when a solid but elastic plug is used. External clamping around the shell over the plugs would also be necessary to fulfill the assumed fixed-end condition. The stresses calculated are, therefore, maximum possible values and would be conservative.

Spherical shells. The format used to present the formulas for the finite-length cylindrical shells could be adapted for finite portions of open spherical and conical shells with both edge loads and loads applied within the shells if we were to accept the approximate solutions based on equivalent cylinders. Baker, Kovalevsky, and Rish (Ref. 6) present formulas based on this approximation for open spherical and conical shells under edge loads and edge displacements. For partial spherical shells under axisymmetric loading, Hetényi (Ref. 14) discusses the errors introduced by this approximation and compares it with a better approximate solution derived therein. Table 31, case 1, gives formulas based on Hetényi's work, and although it is estimated that the calculational effort is twice that of the simpler approximation, the errors in maximum stresses are decreased substantially, especially when the opening angle ϕ is much different from 90°.

Stresses and deformations due to edge loads decrease exponentially away from the loaded edges of axisymmetric shells, and consequently boundary conditions or other restraints are not important if they are far enough from the loaded edge. For example, the exponential term decreases to approximately 1 percent when the product of the spherical shell parameter β (see Table 31, case 1) and the angle ω (in radians) is greater than 4.5; similarly it reduces to approximately 5 percent at $\beta\omega = 3$. This means that a spherical shell with a radius/thickness ratio of 50, for which $\beta \approx 9$, can have an opening angle ϕ as small as $\frac{1}{3}$ rad, or 19°, and still be solved with formulas for case 1 with very little error. Figure 12.5 shows three shells, for which R/t is approximately 50, which would respond similarly to the edge loads M_o and Q_o. In fact, the conical portion of the shell in Fig. 12.5*c* could be extended much closer than 19° to the loaded edge since the conical portion near the junction of the cone and sphere would respond in a similar way to the sphere. (Hetényi discusses this in Ref. 14.)

Similar bounds on *nonspherical* but axisymmetric shells can be approximated by using closely matching equivalent spherical shells (Ref. 6). (We should

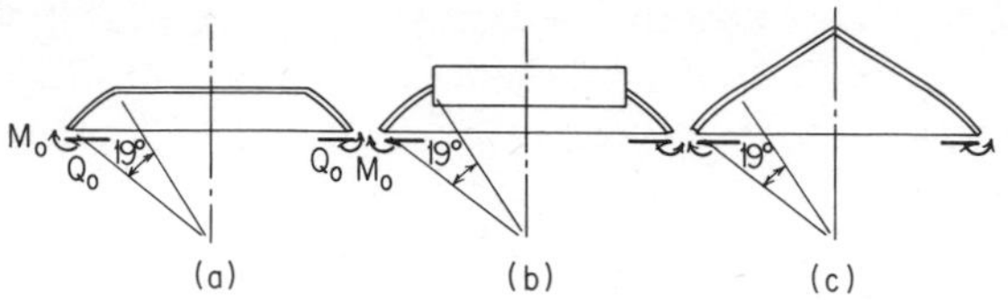

Fig. 12.5

note that the angle ϕ in Table 31, case 1, is not limited to a maximum of 90°, as will be illustrated in the examples at the end of this article.)

For shallow spherical shells where ϕ is small, Gerdeen and Niedenfuhr (Ref. 46) have developed influence coefficients for uniform pressure and for edge loads and moments. Shells with central cutouts are also included as are loads and moments on the edge of the cutouts. Many graphs as well as tabulated data are presented, which permits the solution of a wide variety of problems by superposition.

Conical shells. Exact solutions to the differential equations for both long and short thin-walled truncated conical shells are described in Refs. 30, 31, 64, and 65. Verifications of these expressions by tests are described in Ref. 32, and applications to reinforced cones are described in Ref. 33. In Table 31, cases 4 and 5 for long cones, where the loads at one end do not influence the displacements at the other, are based on the solution described by Taylor (Ref. 65) in which the Kelvin functions and their derivatives are replaced by asymptotic formulas involving negative powers of the conical shell parameter

$$k = 2\left[\frac{12(1-\nu^2)x^2}{t^2\tan^2\alpha}\right]^{\frac{1}{4}}$$

These asymptotic formulas will give three-place accuracy for the Kelvin functions for all values of $k > 5$. To appreciate this fully, one must understand that a truncated thin-walled cone with an R/t ratio of 10 at the small end, a semiapex angle of 80°, and a Poisson's ratio of 0.3 will have a value of $k = 4.86$. For problems where k is much larger than 5, fewer terms can be used in the series, but a few trial calculations will soon indicate the number of terms it is necessary to carry. If only displacements and stresses at the loaded edge are needed, the simpler forms of the expressions can be used. (See the example at the end of Art. 12.4.)

Baltrukonis (Ref. 64) obtains approximations for the influence coefficients which give the edge displacements for *short* truncated conical shells under axisymmetric edge loads and moments; this is done by using one-term asymptotic expressions for the Kelvin functions. Applying the multiterm asymptotic expressions suggested by Taylor to a short truncated conical shell leads to formulas that are too complicated to present in a reasonable form. Instead, in Table 31, case 6 tabulates numerical coefficients based upon this more accurate formulation but evaluated by a computer for the case where Poisson's ratio is 0.3. Because of limited space, only five values of k and four values of the length parameter $\mu_D = (k_A - k_B)/\sqrt{2}$ are presented. If μ_D is greater than 4, the end loads do not interact appreciably and the formulas from cases 4 and 5 may be used.

Tsui (Ref. 58) derives expressions for deformations of conical shells for which the thickness tapers linearly with distance along the meridian; influence coefficients are tabulated for a limited range of shell parameters. Blythe and Kyser (Ref. 50) give formulas for thin-walled conical shells loaded in torsion.

Toroidal shells. Simple closed-form solutions for toroidal shells are generally valid for a rather limited range of parameters, so that usually it is necessary to resort to numerical solutions. Osipova and Tumarkin (Ref. 18) present extensive tables of functions for the asymptotic method of solution of the differential equations for toroidal shells; this reference also contains an extensive bibliography of work on toroidal shells. Tsui and Massard (Ref. 43) tabulate the results of numerical solutions in the form of influence coefficients and influence functions for internal pressure and edge loadings on finite portions of segments of toroidal shells. Segments having *positive* and *negative* gaussian curvatures are considered; when both positive and negative curvatures are present in the same shell, the solutions can be obtained by matching slopes and deflections at the junction. References 29, 51, and 61 describe similar solutions.

Jordon (Ref. 53) works with the shell-equilibrium equations of a *deformed* shell to examine the effect of pressure on the stiffness of an axisymmetrically loaded toroidal shell.

Kraus (Ref. 44), in addition to an excellent presentation of the theory of thin elastic shells, devotes one chapter to numerical analysis under static loadings and another to numerical analysis under dynamic loadings. Comparisons are made among results obtained by finite-element methods, finite-difference methods, and analytic solutions. Numerical techniques, element sizes, and techniques of shell subdivision are discussed in detail. It would be impossible to list here all the references describing the finite-element computer programs available for solving shell problems, but Perrone (Ref. 62) has presented an excellent summary and Bushnell (Ref. 63) describes work on shells in great detail.

EXAMPLES

1. Two partial spheres of aluminum are to be welded together as shown in Fig. 12.6 to form a pressure vessel to withstand an internal pressure of 200 lb/in^2. The mean radius of each sphere is 2 ft, and the wall thickness is 0.5 in. Calculate the stresses at the seam. Given: $E = 10(10^6)$ lb/in^2 and $\nu = 0.33$.

Solution. The edge loading will be considered in three parts, as shown in Fig. 12.6*b*. The tangential edge force T will be applied to balance the internal pressure and, together with the pressure, will cause only membrane stresses and the accompanying change in circumferential radius ΔR; this loading will produce no rotation of the meridian. Owing to the symmetry of

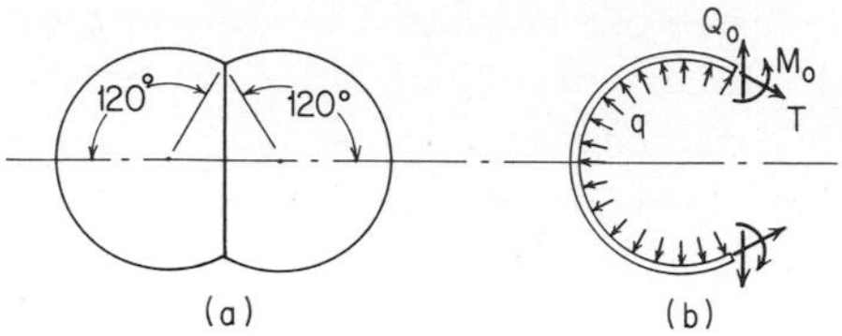

Fig. 12.6

the two shells, there is no resultant radial load on the edge, and so Q_o is added to eliminate that component of T. M_o is needed to ensure no edge rotation.

First apply the formulas from Table 29, case 3a:

$$\sigma_1 = \sigma_2 = \frac{qR_2}{2t} = \frac{200(24)}{2(0.5)} = 4800 \text{ lb/in}^2$$

$$\Delta R = \frac{qR_2^2(1-\nu)\sin\theta}{2Et} = \frac{200(24^2)(1-0.33)(\sin 120°)}{2(10)(10^6)(0.5)} = 0.00668 \text{ in}$$

$$T = \sigma_1 t = 4800(0.5) = 2400 \text{ lb/in}$$

$$\psi = 0$$

Next apply case 1a from Table 31:

$$Q_o = T\sin 30° = 2400(0.5) = 1200 \text{ lb/in}$$

$$\phi = 120°$$

$$\beta = \left[3(1-\nu^2)\left(\frac{R_2}{t}\right)^2\right]^{\frac{1}{4}} = \left[3(1-0.33^2)\left(\frac{24}{0.5}\right)^2\right]^{\frac{1}{4}} = 8.859$$

At the edge where $\omega = 0$,

$$K_1 = 1 - \frac{1-2\nu}{2\beta}\cot\phi = 1 - \frac{1-2(0.33)}{2(8.859)}\cot 120° = 1.011$$

$$K_2 = 1 - \frac{1+2\nu}{2\beta}\cot\phi = 1.054$$

$$\Delta R = \frac{Q_o R_2 \beta \sin^2\phi}{EtK_1}(1+K_1K_2)$$

$$= \frac{1200(24)(8.859)(\sin^2 120°)}{10(10^6)(0.5)(1.011)}[1+1.011(1.054)] = 0.0782 \text{ in}$$

$$\psi = \frac{Q_o 2\beta^2 \sin\phi}{EtK_1} = \frac{1200(2)(8.859^2)(\sin 120°)}{10(10^6)(0.5)(1.011)} = 0.0323 \text{ rad}$$

$$\sigma_1 = \frac{Q_o \cos\phi}{t} = \frac{1200\ (\cos 120°)}{0.5} = -1200 \text{ lb/in}^2$$

$$\sigma_1' = 0$$

$$\sigma_2 = \frac{Q_o\beta\sin\phi}{2t}\left(\frac{2}{K_1} + K_1 + K_2\right)$$

$$= \frac{1200(8.859)(\sin 120°)}{2(0.5)}\left(\frac{2}{1.011} + 1.011 + 1.054\right) = 37{,}200 \text{ lb/in}^2$$

$$\sigma_2' = \frac{-Q_o\beta^2\cos\phi}{K_1R_2} = \frac{-1200(8.859^2)(\cos 120°)}{1.011(24)} = 1{,}940 \text{ lb/in}^2$$

Now apply case 1b from Table 31:

$$\Delta R = \frac{M_o 2\beta^2 \sin\phi}{EtK_1} = 0.00002689M_o$$

$$\psi = \frac{M_o 4\beta^3}{EtR_2K_1} = \frac{M_o 4(8.859)^3}{10(10^6)(0.5)(24)(1.011)} = 0.00002292M_o$$

Since the combined edge rotation ψ must be zero,

$$0 = 0 + 0.0323 + 0.00002292M_o$$

or $M_o = -1{,}409$ in-lb/in

and $\Delta R = 0.00668 + 0.0782 + 0.00002689(-1409) = 0.04699$ in

$$\sigma_1 = 0$$

$$\sigma_1' = \frac{-6(-1409)}{0.05^2} = 33{,}800 \text{ lb/in}^2$$

$$\sigma_2 = \frac{M_o 2\beta^2}{R_2K_1t} = \frac{-1409(2)(8.859^2)}{24(1.011)(0.5)} = -18{,}200 \text{ lb/in}^2$$

$$M_2 = \frac{M_o}{2\beta K_1}[(1+\nu^2)(K_1+K_2) - 2K_2]$$

$$= \frac{-1409}{2(8.859)(1.011)}[(1+0.33^2)(1.011+1.054) - 2(1.054)] = -14.3 \text{ in-lb/in}$$

$$\sigma_2' = \frac{-6(-14.3)}{0.5^2} = 340 \text{ lb/in}^2$$

The superimposed stresses at the joint are, therefore,

$\sigma_1 = 4800 - 1200 + 0 = 3600$ lb/in² $\qquad \sigma_1' = 0 + 0 + 33{,}800 = 33{,}800$ lb/in²

$\sigma_2 = 4800 + 37{,}200 - 18{,}200 = 23{,}800$ lb/in² $\qquad \sigma_2' = 0 + 1940 + 340 = 2280$ lb/in²

The maximum stress is a tensile meridional stress of 37,400 lb/in² on the outside surface at the joint. A further consideration would be given to any stress concentrations due to the shape of the weld cross section.

2. To reduce the high stresses in Example 1, it is proposed to add to the joint a reinforcing ring of aluminum having a cross-sectional area A. Calculate the optimum area to use.

Solution. If the ring could be designed to expand in circumference by the same amount that the sphere does under membrane loading only, then all bending stresses could be eliminated. Therefore, let a ring be loaded radially with a load of $2Q_o$ and have the radius increase by 0.00668 in. Since $\Delta R/R = 2Q_oR/AE$, then

$$A = \frac{2Q_oR^2}{E\,\Delta R} = \frac{2(1200)(24^2)(\sin^2 60°)}{10(10^6)(0.00668)} = 15.5 \text{ in}^2$$

With this large an area required, the simple expression just given for $\Delta R/R$ based on a thin ring is not adequate; furthermore, there is not enough room to place such a ring external to the shell. An internal reinforcement seems more reasonable. If a 6-in-diameter hole is required for passage of the fluid, the internal reinforcing disk can have an outer radius of 20.78 in, an inner radius of 3 in, and a thickness t_1 to be determined. The loading on the disk is shown in Fig. 12.7. The change in the outer radius is desired.

From Table 32, case 1a, the effect of the 200 lb/in² internal pressure can be evaluated:

$$\Delta a = \frac{q}{E}\left(\frac{2ab^2}{a^2 - b^2}\right) = \frac{200}{10(10^6)} \frac{2(20.78)(3^2)}{20.78^2 - 3^2} = 0.0000177 \text{ in}$$

TABLE 31 ***Formulas for bending and membrane stresses and deformations in thin-walled pressure vessels***

NOTATION: Q_o and p = unit loads (pounds per linear inch); q = unit pressure (pounds per square inch); M_o = unit applied couple (inch-pounds per linear inch); ψ_o = applied edge rotation (radians); Δ_o = applied edge displacement (inches); all loads are positive as shown. V = meridional transverse shear (pounds per inch), positive as shown; M_1 and M_2 = meridional and circumferential bending moments, respectively (inch-pounds per inch), positive when compressive on the outside; ψ = change in meridional slope (radians), positive when the change is in the same direction as a positive V_1; ΔR = change in circumferential radius (inches), positive outward; σ_1 and σ_2 = meridional and circumferential membrane stresses (pounds per square inch), positive when tensile; σ_1' and σ_2' = meridional and circumferential bending stresses (pounds per square inch), positive when tensile on the outside. E = modulus of elasticity; ν = Poisson's ratio; and $D = Et^3/12(1-\nu^2)$

1. Partial spherical shells

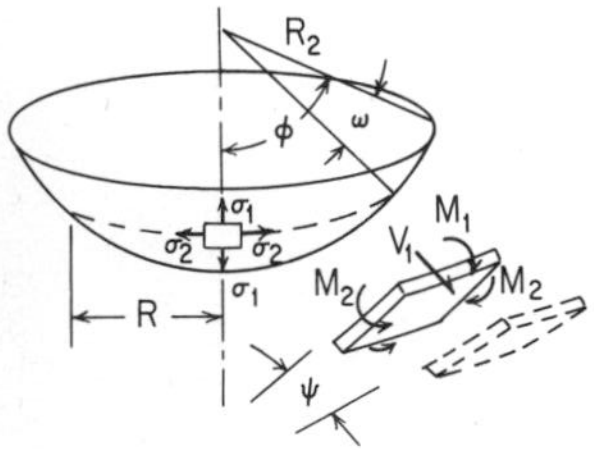

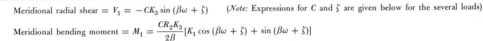

$$\beta = \left[3(1-\nu^2)\left(\frac{R_2}{t}\right)^2\right]^{\frac{1}{4}}$$

$$K_1 = 1 - \frac{1-2\nu}{2\beta}\cot(\phi-\omega)$$

$$K_2 = 1 - \frac{1+2\nu}{2\beta}\cot(\phi-\omega)$$

$$K_3 = \frac{e^{-\beta\omega}}{\sqrt{\sin(\phi-\omega)}}$$

$$\frac{R_2}{t} > 10$$

Meridional radial shear $= V_1 = -CK_3 \sin(\beta\omega + \zeta)$ (*Note:* Expressions for C and ζ are given below for the several loads)

Meridional bending moment $= M_1 = \dfrac{CR_2K_3}{2\beta}[K_1 \cos(\beta\omega+\zeta) + \sin(\beta\omega+\zeta)]$

Circumferential bending moment $= M_2 = \dfrac{CR_2K_3}{2\beta}\left\{\nu \sin(\beta\omega+\zeta) + \left[2 - \left(1-\dfrac{\nu}{2}\right)(K_1+K_2)\right]\cos(\beta\omega+\zeta)\right\}$

Change in meridional slope $= \psi = \dfrac{C2\beta^2K_3}{Et}\cos(\beta\omega+\zeta)$

Change in radius of circumference $= \Delta R = \dfrac{CR_2\beta K_3 \sin(\phi-\omega)}{Et}[\cos(\beta\omega+\zeta) - K_2\sin(\beta\omega+\zeta)]$

Meridional membrane stress $= \sigma_1 = \dfrac{-CK_3\cot(\phi-\omega)}{t}\sin(\beta\omega+\zeta)$

Meridional bending stress $= \sigma_1' = \dfrac{-6M_1}{t^2}$

Circumferential membrane stress $= \sigma_2 = \dfrac{C\beta K_3}{2t}[2\cos(\beta\omega+\zeta) - (K_1+K_2)\sin(\beta\omega+\zeta)]$

Circumferential bending stress $= \sigma_2' = \dfrac{-6M_2}{t^2}$

Meridional radial shear stress $= \tau = \dfrac{V_1}{t}$ (average value)

[*Note:* For reasonable accuracy $3/\beta < \phi < (\pi - 3/\beta)$ (see discussion)]

1a. Uniform radial force, Q_o lb/linear in at the edge

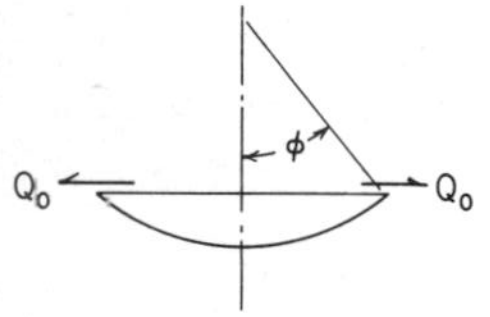

$$C = \frac{Q_o (\sin\phi)^{3/2}\sqrt{1+K_1^2}}{K_1} \quad \text{and} \quad \zeta = \tan^{-1}(-K_1) \quad \text{where } -\frac{\pi}{2} < \zeta < \frac{\pi}{2}$$

Max value of M_1 occurs at $\omega = \pi/4\beta$

At the edge, $\omega = 0$,

$$V_1 = Q_o \sin\phi \qquad M_1 = 0 \qquad \sigma_1' = 0 \qquad \sigma_1 = \frac{Q_o \cos\phi}{t}$$

$$\sigma_2 = \frac{Q_o \beta \sin\phi}{2t}\left(\frac{2}{K_1} + K_1 + K_2\right) = \max \sigma_2$$

$$M_2 = \frac{Q_o t^2 \beta^2 \cos\phi}{6K_1R_2} \qquad \sigma_2' = \frac{-Q_o \beta^2 \cos\phi}{K_1R_2}$$

$$\psi = \frac{Q_o 2\beta^2 \sin\phi}{EtK_1} \qquad \Delta R = \frac{Q_o R_2 \beta \sin^2\phi}{EtK_1}(1 + K_1K_2)$$

(Refs. 14 and 42)

1b. Uniform edge moment, M_o in-lb/linear in

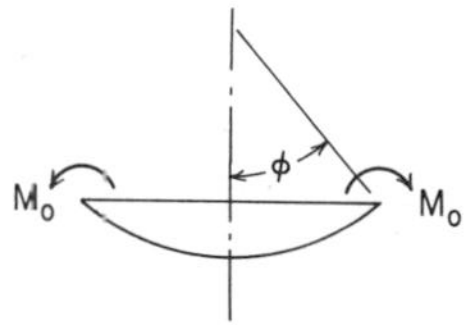

$$C = \frac{M_o 2\beta \sin\phi}{R_2K_1} \qquad \zeta = 0$$

At the edge, $\omega = 0$,

$$V_1 = 0 \qquad \sigma_1 = 0 \qquad M_1 = M_o \qquad \sigma_1' = \frac{-6M_o}{t^2}$$

$$\sigma_2 = \frac{M_o 2\beta^2}{R_2K_1t} \qquad M_2 = \frac{M_o}{2\beta K_1}[(1+\nu^2)(K_1+K_2) - 2K_2] \qquad \sigma_2' = \frac{-6M_2}{t^2}$$

$$\psi = \frac{M_o 4\beta^3}{EtR_2K_1} \qquad \Delta R = \frac{M_o 2\beta^2 \sin\phi}{EtK_1}$$

(Refs. 14 and 42)

1c. Radial displacement, Δ_o in; no edge rotation

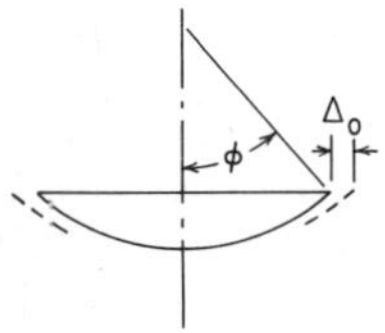

$$C = \frac{-\Delta_o Et}{R_2\beta K_2\sqrt{\sin\phi}} \qquad \zeta = \frac{\pi}{2} = 90°$$

At the edge, $\omega = 0$,

$$V_1 = \frac{\Delta_o Et}{R_2\beta K_2 \sin\phi} \qquad \sigma_1 = \frac{\Delta_o E \cos\phi}{R_2\beta K_2 \sin^2\phi}$$

$$\text{Resultant radial edge force} = \frac{\Delta_o Et}{R_2\beta K_2 \sin^2\phi}$$

$$M_1 = \frac{-\Delta_o Et}{2\beta^2K_2 \sin\phi} \qquad \sigma_1' = \frac{3\Delta_o E}{t\beta^2K_2 \sin\phi}$$

$$\sigma_2 = \frac{\Delta_o E(K_1+K_2)}{2R_2K_2 \sin\phi} \qquad M_2 = \frac{-\Delta_o Et\nu}{2\beta^2K_2 \sin\phi} \qquad \sigma_2' = \frac{3\Delta_o E\nu}{t\beta^2K_2 \sin\phi}$$

$$\psi = 0 \qquad \Delta R = \Delta_o$$

TABLE 31 *Formulas for bending and membrane stresses and deformations in thin-walled pressure vessels* (*Cont.*)

Case no., loading | **Formulas**

1d. Edge rotation, ψ_o rad; no edge displacement

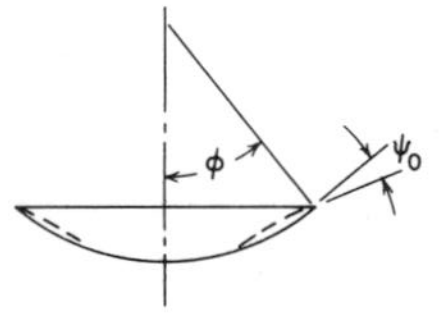

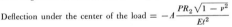

$$C = \frac{\psi_o Et\sqrt{(1+K_2^2)}\sin\phi}{2\beta^2 K_2} \qquad \zeta = \tan^{-1}\frac{1}{K_2} \qquad \text{where } 0 < \zeta < \pi$$

At the edge, $\omega = 0$,

$$V_1 = \frac{-\psi_o Et}{2\beta^2 K_2} \qquad \sigma_1 = \frac{-\psi_o E\cot\phi}{2\beta^2 K_2}$$

$$\text{Resultant radial edge force} = \frac{-\psi_o Et}{2\beta^2 K_2 \sin\phi}$$

$$M_1 = \frac{\psi_o EtR_2}{4\beta^3}\left(K_1 + \frac{1}{K_2}\right) \qquad \sigma_1' = \frac{-6M_1}{t^2}$$

$$\sigma_2 = \frac{-\psi_o E}{4\beta K_2}(K_1 - K_2) \qquad M_2 = \frac{\psi_o EtR_2}{8\nu\beta^3}\left[(1+\nu^2)(K_1+K_2) - 2K_2 + \frac{2\nu^2}{K_2}\right] \qquad \sigma_2' = \frac{-6M_2}{t^2}$$

$$\psi = \psi_o \qquad \Delta R = 0$$

2. Partial spherical shell, load P concentrated on small circular area of radius r_o at pole; any edge support

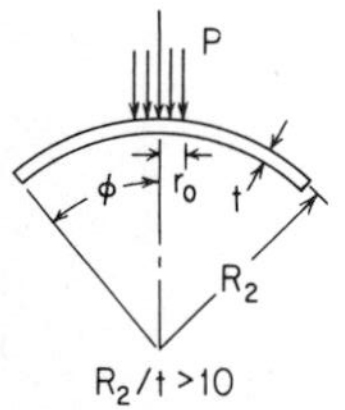

Note: $r_o' = \sqrt{1.6r_o^2 + t^2} - 0.675t$ if $r_o' < 0.5t$; $r_o' = r_o$ if $r_o > 0.5t$ (see Art. 10.1).

For $\phi > \sin^{-1}(1.65\sqrt{t/R_2})$

Note: The deflection for this case is measured locally relative to the undeformed shell. It does not include any deformations due to the edge supports or membrane stresses remote from the loading. The formulas for deflection and stress are applicable also to off-axis loads if no edge is closer than $\phi = \sin^{-1}(1.65\sqrt{t/R_2})$. If an edge were as close as half this angle, the results would be modified very little.

$$\text{Deflection under the center of the load} = -A\frac{PR_2\sqrt{1-\nu^2}}{Et^2}$$

$$\text{Max membrane stress under the center of the load } \sigma_1 = \sigma_2 = -B\frac{P\sqrt{1-\nu^2}}{t^2}$$

$$\text{Max bending stress under the center of the load } \sigma_1' = \sigma_2' = -C\frac{P(1+\nu)}{t^2}$$

Here A, B, and C are numerical coefficients that depend upon

$\mu = r_o'\left[\frac{12(1-\nu^2)}{R_2^2 t^2}\right]^{\frac{1}{4}}$ and have the values tabulated below

μ		0.1	0.2	0.4	0.6	0.8	1.0	1.2	1.4
A	0.433	0.431	0.425	0.408	0.386	0.362	0.337	0.311	0.286
B	0.217	0.215	0.212	0.204	0.193	0.181	0.168	0.155	0.143
C	∞	1.394	1.064	0.739	0.554	0.429	0.377	0.266	0.211

(Ref. 15)

3. Shallow spherical shell, point load P at the pole

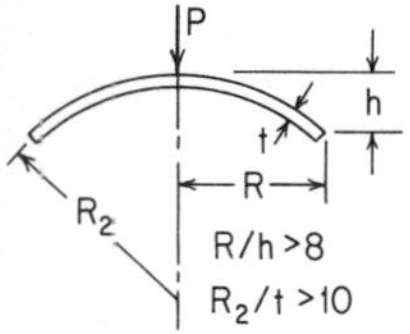

$h < \frac{R}{8}$

$t < \frac{R_2}{10}$

3a. Edge vertically supported and guided

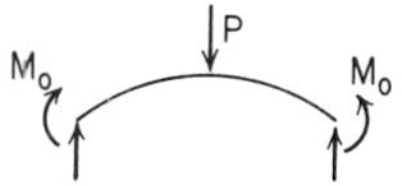

Max deflection $y = -A_1 \frac{PR^2}{16\pi D}$

Edge moment $M_o = -B_1 \frac{P}{4\pi}$

3b. Edge fixed and held

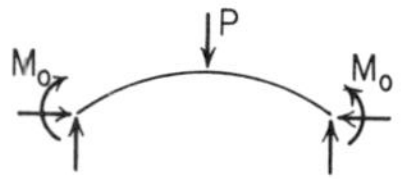

Max deflection $y = -A_2 \frac{PR^2}{16\pi D}$

Edge moment $M_o = -B_2 \frac{P}{4\pi}$

Here A and B are numerical coefficients that depend upon $\alpha = 2\left[\frac{3(1-\nu^2)h^2}{t^2}\right]^{\frac{1}{4}}$ and have the values tabulated below

α	0	1	2	3	4	5	6	7	8	9	10
A_1	1.000	0.996	0.935	0.754	0.406	0.321	0.210	0.148	0.111	0.085	0.069
B_1	1.000	0.995	0.932	0.746	0.498	0.324	0.234	0.192	0.168	0.153	0.140
A_2	1.000	0.985	0.817	0.515	0.320	0.220	0.161	0.122	0.095	0.075	0.061
B_2	1.000	0.975	0.690	0.191	−0.080	−0.140	−0.117	−0.080	−0.059	−0.034	−0.026

(Ref. 15)

4. Long conical shells with edge loads at the large end. Expressions are accurate if $R/t > 10$ and $k > 5$ everywhere in the region from the loaded end to the position where $\mu = 4$

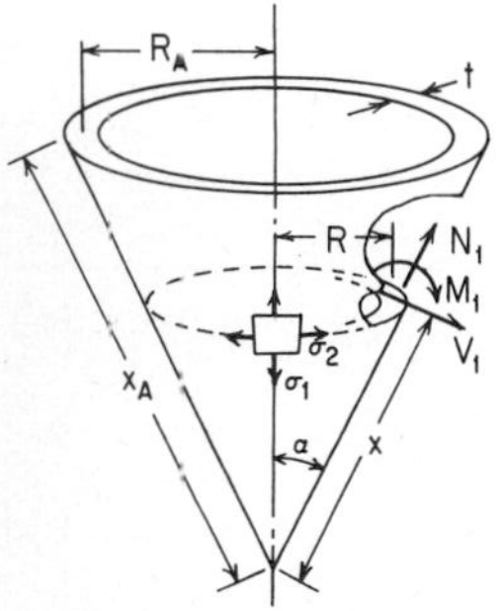

$$k = 2\left[\frac{12(1-\nu^2)x^2}{t^2 \tan^2 \alpha}\right]^{\frac{1}{4}}$$

$$\mu = \frac{k_A - k}{\sqrt{2}}$$

$$\beta = [12(1-\nu^2)]^{\frac{1}{2}}$$

$$l = 1 - \frac{1.326}{k} - \frac{0.218}{k^3} - \frac{0.317}{k^4}$$

$$m = \frac{1.326}{k} - \frac{0.820}{k^2} - \frac{0.218}{k^3}$$

$$s = 1 - \frac{1.679}{k} + \frac{1.233}{k^3} + \frac{0.759}{k^4}$$

$$f = \frac{1.679}{k} - \frac{3.633}{k^2} + \frac{1.233}{k^3}$$

$$C_1 = l_A(s_A - f_A) + m_A(s_A + f_A) + \frac{2\sqrt{2}\nu}{k_A}(l_A^2 + m_A^2)$$

(*Note:* The subscript A denotes that the quantity subscripted is evaluated at $x = x_A$)

$$\sigma_1 = \frac{N_1}{t} \qquad \sigma_2 = \frac{N_2}{t} \qquad \sigma_1' = \frac{-6M_1}{t^2} \qquad \sigma_2' = \frac{-6M_2}{t^2}$$

$$V_1 = \frac{N_1}{\tan \alpha}$$

$$F_1 = ml_A - lm_A$$
$$F_2 = ll_A + mm_A$$
$$F_3 = fs_A - sf_A$$
$$F_4 = ss_A + ff_A$$
$$F_5 = l(s_A - f_A) + m(s_A + f_A)$$
$$F_6 = l(s_A + f_A) - m(s_A - f_A)$$
$$F_7 = s(l_A - m_A) + f(l_A + m_A)$$
$$F_8 = s(l_A + m_A) - f(l_A - m_A)$$

$$F_9 = F_5 + \frac{2\sqrt{2}\nu}{k_A}F_2$$

$$F_{10} = F_6 - \frac{2\sqrt{2}\nu}{k_A}F_1$$

$$F_{11} = F_4 + \frac{\sqrt{2}\nu}{k_A}F_8$$

$$F_{12} = F_3 + \frac{\sqrt{2}\nu}{k_A}F_7$$

(Ref. 65)

TABLE 31 ***Formulas for bending and membrane stresses and deformations in thin-walled pressure vessels*** **(Cont.)**

For use at the loaded end where $x = x_A$,

$$F_{1A} = F_{3A} = 0 \qquad F_{2A} = 1 - \frac{2.652}{k_A} + \frac{3.516}{k_A^2} - \frac{2.610}{k_A^3} + \frac{0.038}{k_A^4}$$

$$F_{4A} = 1 - \frac{3.359}{k_A} + \frac{5.641}{k_A^2} - \frac{9.737}{k_A^3} + \frac{14.716}{k_A^4} \qquad F_{8A} = F_{5A} = 1 - \frac{3.359}{k_A} + \frac{7.266}{k_A^2} - \frac{10.068}{k_A^3} + \frac{5.787}{k_A^4}$$

$$F_{10A} = F_{7A} = F_{6A} = 1 - \frac{2.652}{k_A} + \frac{1.641}{k_A^2} - \frac{0.290}{k_A^3} - \frac{2.211}{k_A^4} \qquad F_{9A} = C_1 = F_{5A} + \frac{2\sqrt{2\nu}}{k_A}F_{2A}$$

Case no., loading	Formulas
4a. Uniform radial force, Q_A lb/linear in 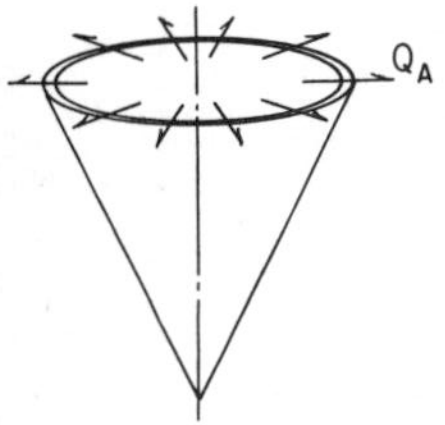	$N_1 = Q_A \sin\alpha \left(\frac{k_A}{k}\right)^{\frac{5}{2}} \frac{e^{-\mu}}{C_1}(F_9 \cos\mu - F_{10} \sin\mu)$ $N_2 = Q_A \sin\alpha \left(\frac{k_A}{k}\right)^{\frac{3}{2}} \frac{k_A}{\sqrt{2}} \frac{e^{-\mu}}{C_1}(F_{11} \cos\mu + F_{12} \sin\mu)$ $M_1 = Q_A \sin\alpha \left(\frac{k_A}{k}\right)^{\frac{3}{2}} \frac{k_A t}{\sqrt{2}\beta} \frac{e^{-\mu}}{C_1}\left[-\left(F_{12} - \frac{\sqrt{2\nu}}{k}F_{10}\right)\cos\mu + \left(F_{11} + \frac{\sqrt{2\nu}}{k}F_9\right)\sin\mu\right]$ $M_2 = Q_A \nu \sin\alpha \left(\frac{k_A}{k}\right)^{\frac{3}{2}} \frac{k_A t}{\sqrt{2}\beta} \frac{e^{-\mu}}{C_1}\left[-\left(F_{12} - \frac{\sqrt{2}}{\nu k}F_{10}\right)\cos\mu + \left(F_{11} + \frac{\sqrt{2}}{\nu k}F_9\right)\sin\mu\right]$ $\Delta R = \frac{Q_A x}{Et}\sin^2\alpha \left(\frac{k_A}{k}\right)^{\frac{3}{2}} \frac{k_A}{\sqrt{2}} \frac{e^{-\mu}}{C_1}\left[\left(F_{11} - \frac{\sqrt{2\nu}}{k}F_9\right)\cos\mu + \left(F_{12} + \frac{\sqrt{2\nu}}{k}F_{10}\right)\sin\mu\right]$ $\psi = \frac{Q_A x_A \beta}{Et^2}\sin\alpha \left(\frac{k_A}{k}\right)^{\frac{1}{2}} \frac{e^{-\mu}}{C_1}(F_{10}\cos\mu + F_9 \sin\mu)$ At the loaded end where $x = x_A$ $N_{1A} = Q_A \sin\alpha \qquad N_{2A} = Q_A \sin\alpha \frac{k_A}{\sqrt{2}C_1}\left(F_{4A} + \frac{\sqrt{2\nu}}{k_A}F_{8A}\right)$ $M_{1A} = 0 \qquad M_{2A} = Q_A \sin\alpha(1 - \nu^2)\frac{t}{\beta C_1}F_{10A}$ $\Delta R_A = \frac{Q_A x_A \sin^2\alpha}{Et}\frac{k_A}{\sqrt{2}C_1}\left(F_{4A} - \frac{4\nu^2}{k_A^2}F_{2A}\right) \qquad \psi_A = \frac{Q_A x_A \beta \sin\alpha}{Et^2 C_1}F_{10A}$

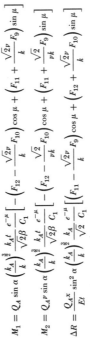

4b. Uniform edge moment, M_A in-lb/linear in

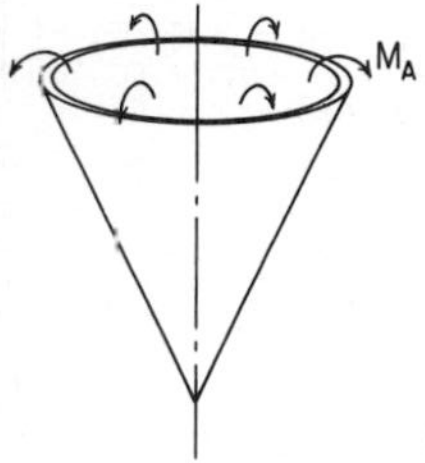

$$N_1 = M_A \left(\frac{k_A}{k}\right)^{\frac{5}{2}} \frac{2\sqrt{2}\beta}{tk_A} \frac{e^{-\mu}}{C_1}(F_1 \cos\mu - F_2 \sin\mu)$$

$$N_2 = M_A \left(\frac{k_A}{k}\right)^{\frac{3}{2}} \frac{\beta}{t} \frac{e^{-\mu}}{C_1}(F_7 \cos\mu - F_8 \sin\mu)$$

$$M_1 = M_A \left(\frac{k_A}{k}\right)^{\frac{3}{2}} \frac{e^{-\mu}}{C_1}\left[\left(F_8 + \frac{2\sqrt{2}\nu}{k}F_2\right)\cos\mu + \left(F_7 + \frac{2\sqrt{2}\nu}{k}F_1\right)\sin\mu\right]$$

$$M_2 = M_A \left(\frac{k_A}{k}\right)^{\frac{3}{2}} \nu \frac{e^{-\mu}}{C_1}\left[\left(F_8 + \frac{2\sqrt{2}}{\nu k}F_2\right)\cos\mu + \left(F_7 + \frac{2\sqrt{2}}{\nu k}F_1\right)\sin\mu\right]$$

$$\Delta R = M_A \left(\frac{k_A}{k}\right)^{\frac{3}{2}} \frac{\beta x \sin\alpha}{Et^2} \frac{e^{-\mu}}{C_1}\left[\left(F_7 - \frac{2\sqrt{2}\nu}{k}F_1\right)\cos\mu - \left(F_8 - \frac{2\sqrt{2}\nu}{k}F_2\right)\sin\mu\right]$$

$$\psi = M_A \left(\frac{k_A}{k}\right)^{\frac{1}{2}} \frac{2\sqrt{2}\beta^2 x_A}{Et^3 k_A} \frac{e^{-\mu}}{C_1}(F_2 \cos\mu + F_1 \sin\mu)$$

At the loaded end where $x = x_A$,

$$N_{1A} = 0 \qquad N_{2A} = M_A \frac{\beta}{tC_1} F_{2A}$$

$$M_{1A} = M_A \qquad M_{2A} = M_A \left[\nu + \frac{2\sqrt{2}(1 - \nu^2)}{k_A C_1} F_{2A}\right]$$

$$\Delta R_A = M_A \frac{\beta x_A \sin\alpha}{Et^2 C_1} F_{7A} \qquad \psi_A = M_A \frac{2\sqrt{2}\beta^2 x_A}{Et^3 k_A C_1} F_{2A}$$

TABLE 31 ***Formulas for bending and membrane stresses and deformations in thin-walled pressure vessels (Cont.)***

5. Long conical shells with edge loads at the small end. Expressions are accurate if $R/t > 10$ and $k > 5$ at the loaded end

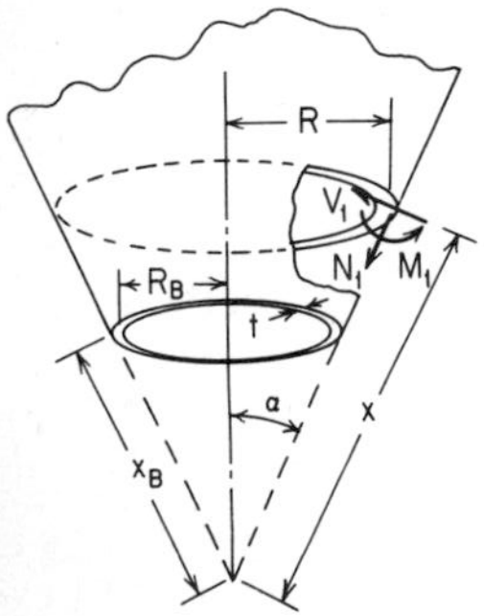

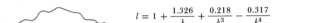

$$k = 2\left[\frac{12(1-\nu^2)x^2}{t^2\tan^2\alpha}\right]^{\frac{1}{4}}$$

$$\eta = \frac{k - k_B}{\sqrt{2}}$$

$$\beta = [12(1-\nu^2)]^{\frac{1}{2}}$$

$$l = 1 + \frac{1.326}{k} + \frac{0.218}{k^3} - \frac{0.317}{k^4}$$

$$m = -\frac{1.326}{k} - \frac{0.820}{k^2} + \frac{0.218}{k^3}$$

$$s = 1 + \frac{1.679}{k} - \frac{1.233}{k^3} + \frac{0.759}{k^4}$$

$$f = -\frac{1.679}{k} - \frac{3.633}{k^2} - \frac{1.233}{k^3}$$

$$D_1 = l_B(s_B - f_B) + m_B(s_B + f_B) - \frac{2\sqrt{2}\nu}{k_B}(l_B^2 + m_B^2)$$

(*Note:* The subscript B denotes that the quantity subscripted is evaluated at $x = x_B$)

$$\sigma_1 = \frac{N_1}{t} \qquad \sigma_2 = \frac{N_2}{t} \qquad \sigma_1' = \frac{-6M_1}{t^2} \qquad \sigma_2' = \frac{-6M_2}{t_2}$$

$$V_1 = \frac{N_1}{\tan\alpha}$$

$$G_1 = ml_B - lm_B$$
$$G_2 = ll_B + mm_B$$
$$G_3 = fs_B - sf_B$$
$$G_4 = ss_B + ff_B$$
$$G_5 = l(s_B - f_B) + m(s_B + f_B)$$
$$G_6 = l(s_B + f_B) - m(s_B - f_B)$$
$$G_7 = s(l_B - m_B) + f(l_B + m_B)$$
$$G_8 = s(l_B + m_B) - f(l_B - m_B)$$

$$G_9 = G_5 - \frac{2\sqrt{2}\nu}{k_B}G_2$$

$$G_{10} = G_6 + \frac{2\sqrt{2}\nu}{k_B}G_1$$

$$G_{11} = G_4 - \frac{\sqrt{2}\nu}{k_B}G_8$$

$$G_{12} = G_3 - \frac{\sqrt{2}\nu}{k_B}G_7$$

(Ref. 65)

For use at the loaded end where $x = x_B$,

$$G_{1B} = G_{3B} = 0$$

$$G_{4B} = 1 + \frac{3.359}{k_B} + \frac{5.641}{k_B^2} + \frac{9.737}{k_B^3} + \frac{14.716}{k_B^4}$$

$$G_{10B} = G_{7B} = G_{6B} = 1 + \frac{2.652}{k_B} + \frac{1.641}{k_B^2} + \frac{0.290}{k_B^3} - \frac{2.211}{k_B^4}$$

$$G_{2B} = 1 + \frac{2.652}{k_B} + \frac{3.516}{k_B^2} + \frac{2.610}{k_B^3} + \frac{0.038}{k_B^4}$$

$$G_{8B} = G_{5B} = 1 + \frac{3.359}{k_B} + \frac{7.266}{k_B^2} + \frac{10.068}{k_B^3} + \frac{5.787}{k_B^4}$$

$$G_{9B} = D_1 = G_{5B} - \frac{2\sqrt{2}\nu}{k_B}G_{2B}$$

Case no., loading	Formulas
5a. Uniform radial force, Q_B lb/linear in 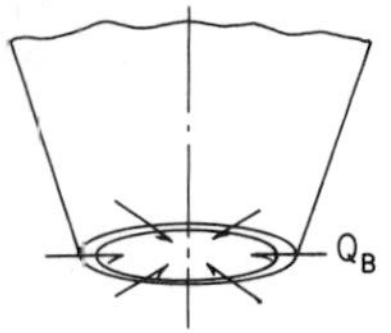	$N_1 = Q_B \sin\alpha \left(\frac{k_B}{k}\right)^{\frac{5}{2}} \frac{e^{-\eta}}{D_1}(G_9 \cos\eta - G_{10}\sin\eta)$ $N_2 = -Q_B \sin\alpha \left(\frac{k_B}{k}\right)^{\frac{3}{2}} \frac{k_B}{\sqrt{2}}\frac{e^{-\eta}}{D_1}(G_{11}\cos\eta + G_{12}\sin\eta)$ $M_1 = -Q_B \sin\alpha \left(\frac{k_B}{k}\right)^{\frac{3}{2}} \frac{k_B t}{\sqrt{2}\beta}\frac{e^{-\eta}}{D_1}\left[-\left(G_{12} + \frac{\sqrt{2}\nu}{k}G_{10}\right)\cos\eta + \left(G_{11} - \frac{\sqrt{2}\nu}{k}G_9\right)\sin\eta\right]$ $M_2 = -Q_B\nu \sin\alpha \left(\frac{k_B}{k}\right)^{\frac{3}{2}} \frac{k_B t}{\sqrt{2}\beta}\frac{e^{-\eta}}{D_1}\left[-\left(G_{12} + \frac{\sqrt{2}}{\nu k}G_{10}\right)\cos\eta + \left(G_{11} - \frac{\sqrt{2}}{\nu k}G_9\right)\sin\eta\right]$ $\Delta R = \frac{-Q_B x}{Et}\sin^2\alpha \left(\frac{k_B}{k}\right)^{\frac{3}{2}} \frac{k_B}{\sqrt{2}}\frac{e^{-\eta}}{D_1}\left[\left(G_{11} + \frac{\sqrt{2}\nu}{k}G_9\right)\cos\eta + \left(G_{12} - \frac{\sqrt{2}\nu}{k}G_{10}\right)\sin\eta\right]$ $\psi = \frac{Q_B x_B \beta}{Et^2}\sin\alpha \left(\frac{k_B}{k}\right)^{\frac{1}{2}} \frac{e^{-\eta}}{D_1}(G_{10}\cos\eta + G_9\sin\eta)$ At the loaded end where $x = x_B$, $N_{1B} = Q_B \sin\alpha \qquad N_{2B} = -Q_B\sin\alpha \frac{k_B}{\sqrt{2}D_1}\left(G_{4B} - \frac{\sqrt{2}\nu}{k_B}G_{8B}\right)$ $M_{1B} = 0 \qquad M_{2B} = Q_B \sin\alpha(1-\nu^2)\frac{t}{\beta D_1}G_{10B}$ $\Delta R_B = \frac{-Q_B x_B \sin^2\alpha}{Et}\frac{k_B}{\sqrt{2}D_1}\left(G_{4B} - \frac{4\nu^2}{k_B^2}G_{2B}\right) \qquad \psi_B = \frac{Q_B x_B \beta \sin\alpha}{Et^2 D_1}G_{10B}$

TABLE 31 ***Formulas for bending and membrane stresses and deformations in thin-walled pressure vessels (Cont.)***

Case no., loading	Formulas

5b. Uniform edge moment, M_B in-lb/linear inch.

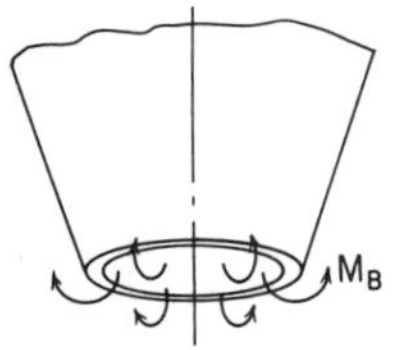

$$N_1 = -M_B\left(\frac{k_B}{k}\right)^{\frac{5}{2}} \frac{2\sqrt{2}\beta}{tk_B} \frac{e^{-\eta}}{D_1}(G_1 \cos\eta - G_2 \sin\eta)$$

$$N_2 = M_B\left(\frac{k_B}{k}\right)^{\frac{3}{2}} \frac{\beta}{t} \frac{e^{-\eta}}{D_1}(G_7 \cos\eta - G_8 \sin\eta)$$

$$M_1 = M_B\left(\frac{k_B}{k}\right)^{\frac{3}{2}} \frac{e^{-\eta}}{D_1}\left[\left(G_8 - \frac{2\sqrt{2}\nu}{k}G_2\right)\cos\eta + \left(G_7 - \frac{2\sqrt{2}\nu}{k}G_1\right)\sin\eta\right]$$

$$M_2 = M_B\nu\left(\frac{k_B}{k}\right)^{\frac{3}{2}} \frac{e^{-\eta}}{D_1}\left[\left(G_8 - \frac{2\sqrt{2}}{\nu k}G_2\right)\cos\eta + \left(G_7 - \frac{2\sqrt{2}}{\nu k}G_1\right)\sin\eta\right]$$

$$\Delta R = M_B\left(\frac{k_B}{k}\right)^{\frac{3}{2}} \frac{\beta x \sin\alpha}{Et^2} \frac{e^{-\eta}}{D_1}\left[\left(G_7 + \frac{2\sqrt{2}\nu}{k}G_1\right)\cos\eta - \left(G_8 + \frac{2\sqrt{2}\nu}{k}G_2\right)\sin\eta\right]$$

$$\psi = -M_B\left(\frac{k_B}{k}\right)^{\frac{1}{2}} \frac{2\sqrt{2}\beta^2 x_B}{Et^3 k_B} \frac{e^{-\eta}}{D_1}(G_2 \cos\eta + G_1 \sin\eta)$$

At the loaded end where $x = x_B$,

$$N_{1B} = 0 \qquad N_{2B} = M_B \frac{\beta}{tD_1} G_{7B}$$

$$M_{1B} = M_B \qquad M_{2B} = M_B\left[\nu - \frac{2\sqrt{2}(1-\nu^2)}{k_B D_1} G_{2B}\right]$$

$$\Delta R_B = M_B \frac{\beta x_B \sin\alpha}{Et^2 D_1} G_{7B} \qquad \psi_B = -M_B \frac{2\sqrt{2}\beta^2 x_B}{Et^3 k_B D_1} G_{2B}$$

6. Short conical shells. Expressions are accurate if $R/t > 10$ and $k_B > 5$

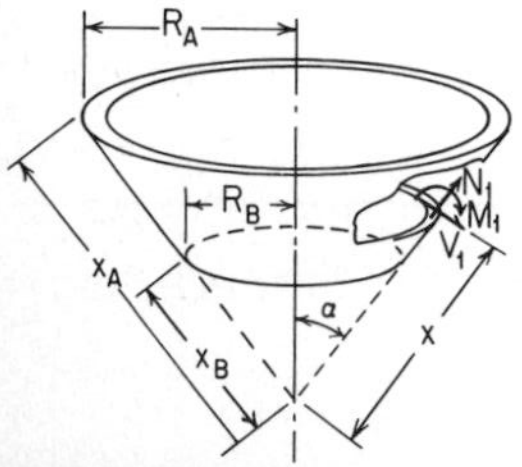

$$k = 2\left[\frac{12(1-\nu^2)x^2}{t^2 \tan^2\alpha}\right]^{\frac{1}{4}} \qquad \sigma_1 = \frac{N_1}{t} \qquad \sigma_2 = \frac{N_2}{t}$$

$$\mu_D = \frac{k_A - k_B}{\sqrt{2}} \qquad \sigma_1' = \frac{-6M_1}{t^2} \qquad \sigma_2' = \frac{-6M_2}{t^2}$$

$$\beta = [12(1-\nu^2)]^{\frac{1}{2}} \qquad V_1 = \frac{N_1}{\tan\alpha}$$

6a. Uniform radial force, Q_A lb/linear in at the large end

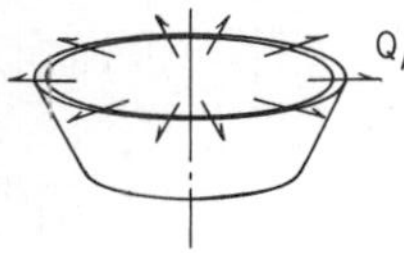

$$N_1 = Q_A \sin\alpha\, K_{N1} \qquad N_2 = Q_A \sin\alpha \frac{k_A}{\sqrt{2}} K_{N2}$$

$$M_1 = Q_A \sin\alpha \frac{k_A t}{\sqrt{2}\beta} K_{M1} \qquad M_2 = Q_A \nu \sin\alpha \frac{k_A t}{\sqrt{2}\beta} K_{M2}$$

$$\Delta R = \frac{Q_A x \sin^2\alpha}{Et} \frac{k_A}{\sqrt{2}} K_{\Delta R} \qquad \psi = \frac{Q_A x_A \beta \sin\alpha}{Et^2} K_\psi$$

For $R_B/t > 10$ and $k_B > 5$ and for $\nu = 0.3$, the following tables give the values of K at several locations along the shell $[\Omega = (x_A - x)/(x_A - x_B)]$

k_A	μ_D	0.4					0.8				
	Ω	0.000	0.250	0.500	0.750	1.000	0.000	0.250	0.500	0.750	1.000
10	K_{N1}	1.000	0.548	0.216	0.025	0.000	1.000	0.392	−0.029	−0.192	0.000
	K_{N2}	2.748	2.056	1.319	0.533	−0.307	1.952	1.371	0.720	−0.016	−0.862
	K_{M1}	0.000	0.047	0.054	0.034	0.000	0.000	0.106	0.114	0.060	0.000
	K_{M2}	3.279	3.445	3.576	3.690	3.801	1.468	1.672	1.780	1.848	1.941
	$K_{\Delta R}$	2.706	2.033	1.310	0.532	−0.307	1.909	1.354	0.721	−0.008	−0.862
	K_ψ	7.644	7.703	7.759	7.819	7.887	3.421	3.454	3.470	3.501	3.559

k_A	μ_D	1.6					3.2				
	Ω	0.000	0.250	0.500	0.750	1.000	0.000	0.250	0.500	0.750	1.000
10	K_{N1}	1.000	0.315	−0.185	−0.377	0.000	1.000	0.156	−0.301	−0.361	0.000
	K_{N2}	1.197	0.815	0.394	−0.082	−0.696	0.940	0.466	0.088	−0.097	−0.182
	K_{M1}	0.000	0.220	0.235	0.117	0.000	0.000	0.347	0.318	0.131	0.000
	K_{M2}	0.555	0.813	0.832	0.746	0.743	0.407	0.729	0.538	0.191	0.010
	$K_{\Delta R}$	1.155	0.802	0.402	−0.066	−0.696	0.898	0.459	0.101	−0.082	−0.182
	K_ψ	1.294	1.243	1.112	1.025	1.037	0.948	0.734	0.333	0.066	0.007

k_A	μ_D	0.4					0.8				
	Ω	0.000	0.250	0.500	0.750	1.000	0.000	0.250	0.500	0.750	1.000
20	K_{N1}	1.000	0.346	−0.053	−0.175	0.000	1.000	0.255	−0.193	−0.296	0.000
	K_{N2}	4.007	2.674	1.298	−0.123	−1.591	2.334	1.528	0.677	−0.224	−1.186
	K_{M1}	0.000	0.052	0.052	0.025	0.000	0.000	0.111	0.106	0.045	0.000
	K_{M2}	2.974	3.079	3.133	3.163	3.198	0.933	1.072	1.093	1.064	1.059
	$K_{\Delta R}$	3.985	2.666	1.299	−0.119	−1.591	2.313	1.523	0.681	−0.218	−1.186
	K_ψ	13.866	13.918	13.967	14.020	14.079	4.348	4.359	4.351	4.360	4.393

TABLE 31 *Formulas for bending and membrane stresses and deformations in thin-walled pressure vessels* (*Cont.*)

k_A	μ_D	1.6					3.2				
	Ω	0.000	0.250	0.500	0.750	1.000	0.000	0.250	0.500	0.750	1.000
20	K_{N1}	1.000	0.222	−0.247	−0.352	0.000	1.000	0.049	−0.274	−0.215	0.000
	K_{N2}	1.283	0.808	0.334	−0.139	−0.655	0.977	0.390	0.027	−0.094	−0.112
	K_{M1}	0.000	0.222	0.211	0.087	0.000	0.000	0.342	0.253	0.079	0.000
	K_{M2}	0.299	0.519	0.489	0.359	0.289	0.211	0.507	0.322	0.089	−0.002
	$K_{\Delta R}$	1.262	0.803	0.340	−0.132	−0.655	0.956	0.389	0.033	−0.089	−0.112
	K_{ψ}	1.392	1.313	1.160	1.065	1.061	0.983	0.692	0.259	0.032	−0.007

k_A	μ_D	0.4					0.8				
	Ω	0.000	0.250	0.500	0.750	1.000	0.000	0.250	0.500	0.750	1.000
40	K_{N1}	1.000	0.238	−0.192	−0.275	0.000	1.000	0.209	−0.238	−0.317	0.000
	K_{N2}	4.692	2.998	1.277	−0.471	−2.248	2.459	1.565	0.649	−0.289	−1.255
	K_{M1}	0.000	0.055	0.051	0.021	0.000	0.000	0.112	0.103	0.040	0.000
	K_{M2}	1.851	1.922	1.934	1.921	1.918	0.498	0.617	0.611	0.555	0.524
	$K_{\Delta R}$	4.682	2.995	1.279	−0.468	−2.248	2.448	1.563	0.652	−0.285	−1.255
	K_{ψ}	17.257	17.286	17.312	17.342	17.378	4.645	4.637	4.612	4.603	4.617

k_A	μ_D	1.6					3.2				
	Ω	0.000	0.250	0.500	0.750	1.000	0.000	0.250	0.500	0.750	1.000
40	K_{N1}	1.000	0.189	−0.253	−0.324	0.000	1.000	0.012	−0.245	−0.165	0.000
	K_{N2}	1.309	0.792	0.304	−0.153	−0.616	0.991	0.352	0.007	−0.087	−0.092
	K_{M1}	0.000	0.221	0.200	0.077	0.000	0.000	0.333	0.223	0.063	0.000
	K_{M2}	0.152	0.367	0.331	0.201	0.126	0.107	0.408	0.250	0.066	−0.001
	$K_{\Delta R}$	1.299	0.790	0.307	−0.150	−0.616	0.981	0.352	0.010	−0.085	−0.092
	K_{ψ}	1.419	1.324	1.160	1.061	1.048	0.994	0.665	0.229	0.023	−0.009

k_A	μ_D	0.4					0.8				
	Ω	0.000	0.250	0.500	0.750	1.000	0.000	0.250	0.500	0.750	1.000
80	K_{N1}	1.000	0.202	−0.235	−0.305	0.000	1.000	0.194	−0.248	−0.317	0.000
	K_{N2}	4.917	3.097	1.264	−0.584	−2.446	2.494	1.568	0.635	−0.306	−1.259
	K_{M1}	0.000	0.056	0.050	0.019	0.000	0.000	0.112	0.101	0.038	0.000
	K_{M2}	0.985	1.045	1.043	1.016	1.002	0.253	0.367	0.355	0.293	0.257
	$K_{\Delta R}$	4.912	3.096	1.265	−0.582	−2.446	2.489	1.567	0.636	−0.304	−1.259
	K_{ψ}	18.363	18.376	18.384	18.397	18.415	4.725	4.707	4.672	4.654	4.658

k_A	μ_D	1.6					3.2				
	Ω	0.000	0.250	0.500	0.750	1.000	0.000	0.250	0.500	0.750	1.000
80	K_{N1}	1.000	0.176	−0.252	−0.308	0.000	1.000	−0.002	−0.230	−0.145	0.000
	K_{N2}	1.318	0.781	0.290	−0.157	−0.593	0.997	0.334	0.000	−0.083	−0.084
	K_{M1}	0.000	0.220	0.194	0.073	0.000	0.000	0.327	0.208	0.056	0.000
	K_{M2}	0.076	0.292	0.258	0.132	0.059	0.053	0.363	0.221	0.057	−0.001
	$K_{\Delta R}$	1.313	0.780	0.291	−0.155	−0.593	0.992	0.334	0.001	−0.082	−0.084
	K_ψ	1.426	1.324	1.155	1.053	1.037	0.998	0.651	0.216	0.020	−0.009

k_A	μ_D	0.4					0.8				
	Ω	0.000	0.250	0.500	0.750	1.000	0.000	0.250	0.500	0.750	1.000
160	K_{N1}	1.000	0.192	−0.247	−0.312	0.000	1.000	0.189	−0.250	−0.315	0.000
	K_{N2}	4.978	3.121	1.257	−0.614	−2.492	2.505	1.566	0.627	−0.311	−1.253
	K_{M1}	0.000	0.056	0.050	0.019	0.000	0.000	0.112	0.100	0.038	0.000
	K_{M2}	0.500	0.557	0.552	0.522	0.504	0.127	0.240	0.227	0.164	0.127
	$K_{\Delta R}$	4.976	3.120	1.257	−0.613	−2.492	2.502	1.565	0.628	−0.310	−1.253
	K_ψ	18.662	18.665	18.663	18.666	18.675	4.746	4.723	4.683	4.660	4.659

k_A	μ_D	1.6					3.2				
	Ω	0.000	0.250	0.500	0.750	1.000	0.000	0.250	0.500	0.750	1.000
160	K_{N1}	1.000	0.171	−0.250	−0.300	0.000	1.000	−0.008	−0.223	−0.136	0.000
	K_{N2}	1.322	0.775	0.283	−0.158	−0.582	1.000	0.325	−0.003	−0.081	−0.080
	K_{M1}	0.000	0.219	0.191	0.071	0.000	0.000	0.324	0.201	0.053	0.000
	K_{M2}	0.038	0.255	0.223	0.100	0.028	0.027	0.341	0.207	0.054	0.000
	$K_{\Delta R}$	1.319	0.775	0.284	−0.157	−0.582	0.998	0.325	−0.003	−0.081	−0.080
	K_ψ	1.429	1.323	1.151	1.048	1.030	0.999	0.644	0.210	0.018	−0.010

6b. Uniform edge moment, M_A in-lb/linear in at the large end

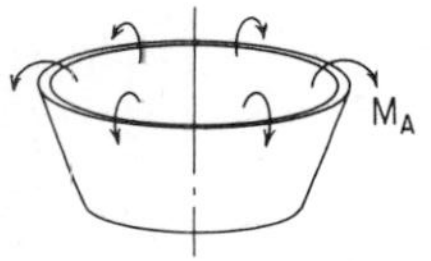

$$N_1 = M_A \frac{2\sqrt{2}\beta}{tk_A} K_{N1} \qquad N_2 = M_A \frac{\beta}{t} K_{N2}$$

$$M_1 = M_A K_{M1} \qquad M_2 = M_A \nu K_{M2}$$

$$\Delta R = M_A \frac{\beta x \sin\alpha}{Et^2} K_{\Delta R} \qquad \psi = M_A \frac{2\sqrt{2}\beta^2 x_A}{Et^3 k_A} K_\psi$$

For $R_B/t > 10$ and $k_B > 5$ and for $\nu = 0.3$, the following tables give the values of K at several locations along the shell $[\Omega = (x_A - x)/(x_A - x_B)]$

TABLE 31 *Formulas for bending and membrane stresses and deformations in thin-walled pressure vessels* (*Cont.*)

k_A	μ_D	0.4					0.8				
	Ω	0.000	0.250	0.500	0.750	1.000	0.000	0.250	0.500	0.750	1.000
10	K_{N1}	0.000	−0.584	−0.819	−0.647	0.000	0.000	−0.528	−0.772	−0.641	0.000
	K_{N2}	7.644	4.020	0.170	−3.934	−8.329	3.421	1.852	0.138	−1.780	−3.983
	K_{M1}	1.000	0.816	0.538	0.238	0.000	1.000	0.865	0.568	0.226	0.000
	K_{M2}	17.471	17.815	18.123	18.475	18.961	4.850	4.836	4.718	4.644	4.792
	$K_{\Delta R}$	7.644	4.069	0.240	−3.879	−8.329	3.421	1.896	0.203	−1.726	−3.983
	K_{ψ}	19.197	19.269	19.369	19.503	19.671	4.488	4.382	4.320	4.325	4.393

k_A	μ_D	1.6					3.2				
	Ω	0.000	0.250	0.500	0.750	1.000	0.000	0.250	0.500	0.750	1.000
10	K_{N1}	0.000	−0.370	−0.547	−0.476	0.000	0.000	−0.375	−0.389	−0.193	0.000
	K_{N2}	1.294	0.602	−0.004	−0.594	−1.323	0.948	0.127	−0.235	−0.240	−0.018
	K_{M1}	1.000	0.899	0.596	0.225	0.000	1.000	0.816	0.395	0.071	0.000
	K_{M2}	2.052	1.772	1.319	0.904	0.784	1.833	1.258	0.479	−0.058	−0.262
	$K_{\Delta R}$	1.294	0.633	0.043	−0.553	−1.323	0.948	0.158	−0.202	−0.224	−0.018
	K_{ψ}	1.226	0.915	0.674	0.553	0.547	0.971	0.425	0.064	−0.071	−0.091

k_A	μ_D	0.4					0.8				
	Ω	0.000	0.250	0.500	0.750	1.000	0.000	0.250	0.500	0.750	1.000
20	K_{N1}	0.000	−1.050	−1.435	−1.104	0.000	0.000	−0.660	−0.919	−0.722	0.000
	K_{N2}	13.866	7.110	0.149	−7.036	−14.461	4.348	2.246	0.072	−2.210	−4.638
	K_{M1}	1.000	0.832	0.517	0.191	0.000	1.000	0.858	0.530	0.181	0.000
	K_{M2}	15.905	15.971	15.909	15.860	15.967	3.421	3.286	2.988	2.703	2.617
	$K_{\Delta\psi}$	13.866	7.155	0.210	−6.989	−14.461	4.348	2.274	0.111	−2.179	−4.638
	K_{ψ}	34.745	34.799	34.881	34.998	35.145	5.643	5.504	5.415	5.395	5.431

k_A	μ_D	1.6					3.2				
	Ω	0.000	0.250	0.500	0.750	1.000	0.000	0.250	0.500	0.750	1.000
20	K_{N1}	0.000	−0.389	−0.519	−0.401	0.000	0.000	−0.357	−0.279	−0.091	0.000
	K_{N2}	1.392	0.587	−0.055	−0.611	−1.187	0.983	0.048	−0.240	−0.177	0.010
	K_{M1}	1.000	0.865	0.529	0.175	0.000	1.000	0.730	0.287	0.040	0.000
	K_{M2}	1.552	1.292	0.861	0.466	0.305	1.425	0.906	0.297	−0.011	−0.067
	$K_{\Delta\psi}$	1.392	0.604	−0.033	−0.594	−1.187	0.983	0.063	−0.228	−0.173	0.010
	K_{ψ}	1.286	0.943	0.690	0.571	0.559	0.992	0.369	0.019	−0.083	−0.094

k_A	μ_D	0.4					0.8				
	Ω	0.000	0.250	0.500	0.750	1.000	0.000	0.250	0.500	0.750	1.000
40	K_{N1}	0.000	−1.300	−1.755	−1.333	0.000	0.000	−0.698	−0.948	−0.726	0.000
	K_{N2}	17.257	8.734	0.088	−8.690	−17.610	4.645	2.339	0.024	−2.327	−4.743
	K_{M1}	1.000	0.842	0.508	0.168	0.000	1.000	0.352	0.514	0.165	0.000
	K_{M2}	10.272	10.179	9.917	9.658	9.578	2.290	2.125	1.780	1.441	1.298
	$K_{\Delta R}$	17.257	8.761	0.126	−8.661	−17.610	4.645	2.354	0.044	−2.311	−4.743
	K_{ψ}	43.230	43.228	43.252	43.312	43.399	6.015	5.352	5.741	5.700	5.714

k_A	μ_D	1.6					3.2				
	Ω	0.000	0.250	0.500	0.750	1.000	0.000	0.250	0.500	0.750	1.000
40	K_{N1}	0.000	−0.388	−0.494	−0.362	0.000	0.000	−0.341	−0.233	−0.065	0.000
	K_{N2}	1.419	0.565	−0.079	−0.603	−1.107	0.994	0.014	−0.231	−0.150	0.011
	K_{M1}	1.000	0.845	0.498	0.156	0.000	1.000	0.684	0.243	0.030	0.000
	K_{M2}	1.280	1.053	0.654	0.289	0.133	1.214	0.762	0.244	0.009	−0.025
	$K_{\Delta R}$	1.419	0.573	−0.069	−0.596	−1.107	0.994	0.021	−0.226	−0.148	0.011
	K_{ψ}	1.303	0.944	0.687	0.569	0.552	0.999	0.342	0.005	−0.084	−0.091

k_A	μ_D	0.4					0.8				
	Ω	0.000	0.250	0.500	0.750	1.000	0.000	0.250	0.500	0.750	1.000
80	K_{N1}	0.000	−1.380	−1.851	−1.397	0.000	0.000	−0.707	−0.948	−0.716	0.000
	K_{N2}	18.363	9.232	0.043	−9.211	−18.537	4.725	2.348	−0.001	−2.348	−4.720
	K_{M1}	1.000	0.845	0.504	0.160	0.000	1.000	0.848	0.506	0.160	0.000
	K_{M2}	5.934	5.791	5.465	5.141	5.004	1.656	1.489	1.138	0.791	0.636
	$K_{\Delta R}$	18.363	9.247	0.063	−9.196	−18.537	4.725	2.355	0.009	−2.341	−4.720
	K_{ψ}	46.004	45.959	45.939	45.956	46.000	6.116	5.941	5.817	5.765	5.767

k_A	μ_D	1.6					3.2				
	Ω	0.000	0.250	0.500	0.750	1.000	0.000	0.250	0.500	0.750	1.000
80	K_{N1}	0.000	−0.386	−0.479	−0.342	0.000	0.000	−0.331	−0.212	−0.055	0.000
	K_{N2}	1.426	0.551	−0.091	−0.596	−1.065	0.998	−0.001	−0.224	−0.138	0.010
	K_{M1}	1.000	0.834	0.483	0.148	0.000	1.000	0.660	0.224	0.027	0.000
	K_{M2}	1.140	0.937	0.558	0.211	0.062	1.107	0.697	0.223	0.017	−0.011
	$K_{\Delta R}$	1.426	0.555	−0.085	−0.592	−1.065	0.998	0.002	−0.222	−0.137	0.010
	K_{ψ}	1.309	0.942	0.683	0.565	0.547	1.002	0.329	−0.001	−0.083	−0.089

TABLE 31 *Formulas for bending and membrane stresses and deformations in thin-walled pressure vessels* (Cont.)

k_A	μ_D	0.4					0.8				
	Ω	0.000	0.250	0.500	0.750	1.000	0.000	0.250	0.500	0.750	1.000
160	K_{N1}	0.000	−1.401	−1.873	−1.409	0.000	0.000	−0.708	−0.943	−0.709	0.000
	K_{N2}	18.662	9.350	0.018	−9.342	−18.736	4.746	2.342	−0.014	−2.349	−4.690
	K_{M1}	1.000	0.845	0.502	0.158	0.000	1.000	0.845	0.502	0.157	0.000
	K_{M2}	3.507	3.352	3.012	2.672	2.519	1.329	1.166	0.817	0.470	0.314
	$K_{\Delta R}$	18.662	9.358	0.028	−9.334	−18.736	4.746	2.346	−0.009	−2.346	−4.690
	K_ψ	46.754	46.685	46.641	46.634	46.653	6.144	5.962	5.832	5.774	5.769

k_A	μ_D	1.6					3.2				
	Ω	0.000	0.250	0.500	0.750	1.000	0.000	0.250	0.500	0.750	1.000
160	K_{N1}	0.000	−0.385	−0.471	−0.332	0.000	0.000	−0.325	−0.202	−0.051	0.000
	K_{N2}	1.429	0.543	−0.096	−0.592	−1.044	0.999	−0.008	−0.221	−0.133	0.010
	K_{M1}	1.000	0.829	0.475	0.144	0.000	1.000	0.649	0.214	0.025	0.000
	K_{M2}	1.070	0.879	0.512	0.175	0.030	1.054	0.666	0.214	0.020	−0.005
	$K_{\Delta R}$	1.429	0.545	−0.093	−0.590	−1.044	0.999	−0.007	−0.220	−0.132	0.010
	K_ψ	1.311	0.940	0.680	0.562	0.543	1.003	0.323	−0.004	−0.083	−0.088

6c. Uniform radial force, Q_B lb/linear in at the small end

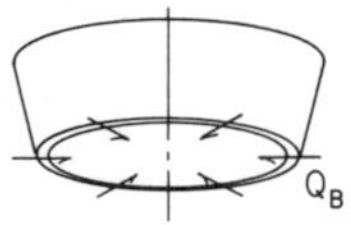

$$N_1 = Q_B \sin\alpha K_{N1} \qquad N_2 = -Q_B \sin\alpha \frac{k_B}{\sqrt{2}} K_{N2}$$

$$M_1 = -Q_B \sin\alpha \frac{k_B t}{\sqrt{2}\beta} K_{M1} \qquad M_2 = -Q_B \nu \sin\alpha \frac{k_B t}{\sqrt{2}\beta} K_{M2}$$

$$\Delta R = \frac{-Q_B x \sin^2\alpha}{Et} \frac{k_B}{\sqrt{2}} K_{\Delta R} \qquad \psi = \frac{Q_B x_B \beta \sin\alpha}{Et^2} K_\psi$$

For $R_B/t > 10$ and $k_B > 5$ and for $\nu = 0.3$, the following tables give the values of K at several locations along the shell $[\Omega = (x_A - x)/(x_A - x_B)]$

k_B	μ_D	0.4					0.8				
	Ω	0.000	0.250	0.500	0.750	1.000	0.000	0.250	0.500	0.750	1.000
5	K_{N1}	0.000	0.119	0.311	0.595	1.000	0.000	−0.013	0.090	0.384	1.000
	K_{N2}	0.520	0.817	1.151	1.534	1.974	−0.187	0.114	0.488	0.971	1.614
	K_{M1}	0.000	0.026	0.041	0.036	0.000	0.000	0.035	0.071	0.078	0.000
	K_{M2}	−2.063	−2.175	−2.320	−2.507	−2.746	−1.106	−1.209	−1.354	−1.585	−1.960
	$K_{\Delta R}$	0.520	0.827	1.178	1.584	2.059	−0.187	0.112	0.495	1.003	1.699
	K_ψ	2.404	2.442	2.487	2.535	2.583	1.289	1.328	1.383	1.451	1.519

k_B	μ_D	1.6					3.2				
	Ω	0.000	0.250	0.500	0.750	1.000	0.000	0.250	0.500	0.750	1.000
5	K_{N1}	0.000	−0.081	−0.079	0.142	1.000	0.000	−0.023	−0.058	−0.036	1.000
	K_{N2}	−0.220	−0.075	0.142	0.516	1.236	−0.028	−0.031	−0.018	0.128	1.037
	K_{M1}	0.000	0.036	0.107	0.157	0.000	0.000	0.013	0.067	0.178	0.000
	K_{M2}	−0.267	−0.291	−0.337	−0.509	−1.015	−0.002	0.008	0.029	−0.031	−0.775
	$K_{\Delta R}$	−0.220	−0.081	0.136	0.528	1.320	−0.028	−0.033	−0.023	0.125	1.122
	K_ψ	0.311	0.331	0.381	0.470	0.560	0.002	0.005	0.028	0.111	0.249

k_B	μ_D	0.4					0.8				
	Ω	0.000	0.250	0.500	0.750	1.000	0.000	0.250	0.500	0.750	1.000
15	K_{N1}	0.000	−0.098	0.021	0.379	1.000	0.000	−0.208	−0.152	0.228	1.000
	K_{N2}	−1.055	0.045	1.193	2.392	3.645	−0.906	−0.207	0.553	1.387	2.307
	K_{M1}	0.000	0.024	0.048	0.048	0.000	0.000	0.036	0.092	0.105	0.000
	K_{M2}	−3.078	−3.128	−3.182	−3.263	−3.396	−0.968	−0.977	−0.977	−1.028	−1.199
	$K_{\Delta R}$	−1.055	0.042	1.193	2.402	3.673	−0.906	−0.213	0.549	1.394	2.335
	K_ψ	10.762	10.822	10.889	10.960	11.028	3.383	3.423	3.484	3.560	3.625

k_B	μ_D	1.6					3.2				
	Ω	0.000	0.250	0.500	0.750	1.000	0.000	0.250	0.500	0.750	1.000
15	K_{N1}	0.000	−0.204	−0.198	0.136	1.000	0.000	−0.067	−0.142	−0.046	1.000
	K_{N2}	−0.437	−0.145	0.213	0.688	1.324	−0.051	−0.059	−0.022	0.232	1.024
	K_{M1}	0.000	0.055	0.160	0.205	0.000	0.000	0.030	0.135	0.276	0.000
	K_{M2}	−0.198	−0.164	−0.101	−0.127	−0.400	0.001	0.028	0.103	0.141	−0.284
	$K_{\Delta R}$	−0.437	−0.150	0.207	0.691	1.352	−0.051	−0.060	−0.026	0.231	1.052
	K_ψ	0.691	0.717	0.802	0.947	1.057	−0.003	0.007	0.091	0.328	0.586

k_B	μ_D	0.4					0.8				
	Ω	0.000	0.250	0.500	0.750	1.000	0.000	0.250	0.500	0.750	1.000
35	K_{N1}	0.000	−0.249	−0.174	0.237	1.000	0.000	−0.278	−0.225	0.189	1.000
	K_{N2}	−2.078	−0.445	1.219	2.914	4.643	−1.138	−0.292	0.588	1.509	2.473
	K_{M1}	0.000	0.020	0.049	0.054	0.000	0.000	0.036	0.096	0.110	0.000
	K_{M2}	−1.987	−1.988	−1.981	−1.999	−2.075	−0.512	−0.487	−0.440	−0.443	−0.569
	$K_{\Delta R}$	−2.078	−0.448	1.217	2.917	4.655	−1.138	−0.295	0.586	1.511	2.485
	K_ψ	16.215	16.255	16.301	16.351	16.397	4.174	4.199	4.244	4.307	4.355

TABLE 31 *Formulas for bending and membrane stresses and deformations in thin-walled pressure vessels* (Cont.)

k_B	μ_D	1.6					3.2				
	Ω	0.000	0.250	0.500	0.750	1.000	0.000	0.250	0.500	0.750	1.000
35	K_{N1}	0.000	−0.253	−0.231	0.148	1.000	0.000	−0.096	−0.182	−0.033	1.000
	K_{N2}	−0.513	−0.157	0.247	0.737	1.331	−0.064	−0.070	−0.016	0.277	1.014
	K_{M1}	0.000	0.062	0.176	0.214	0.000	0.000	0.040	0.167	0.303	0.000
	K_{M2}	−0.106	−0.050	0.048	0.059	−0.175	0.001	0.039	0.147	0.234	−0.123
	$K_{\Delta R}$	−0.513	−0.160	0.244	0.738	1.343	−0.064	−0.071	−0.018	0.277	1.026
	K_ψ	0.866	0.890	0.986	1.148	1.259	−0.007	0.011	0.141	0.472	0.785

k_B	μ_D	0.4					0.8				
	Ω	0.000	0.250	0.500	0.750	1.000	0.000	0.250	0.500	0.750	1.000
75	K_{N1}	0.000	−0.295	−0.231	0.198	1.000	0.000	−0.299	−0.243	0.184	1.000
	K_{N2}	−2.384	−0.583	1.234	3.068	4.919	−1.204	−0.309	0.605	1.542	2.504
	K_{M1}	0.000	0.019	0.050	0.056	0.000	0.000	0.037	0.098	0.112	0.000
	K_{M2}	−1.027	−1.013	−0.988	−0.987	−1.049	−0.254	−0.221	−0.163	−0.155	−0.271
	$K_{\Delta R}$	−2.384	−0.584	1.233	3.069	4.924	−1.204	−0.311	0.604	1.543	2.510
	K_ψ	17.954	17.975	18.002	18.034	18.061	4.448	4.463	4.500	4.554	4.592

k_B	μ_D	1.6					3.2				
	Ω	0.000	0.250	0.500	0.750	1.000	0.000	0.250	0.500	0.750	1.000
75	K_{N1}	0.000	−0.274	−0.241	0.156	1.000	0.000	−0.111	−0.199	−0.024	1.000
	K_{N2}	−0.543	−0.159	0.262	0.754	1.329	−0.070	−0.075	−0.011	0.297	1.008
	K_{M1}	0.000	0.066	0.183	0.216	0.000	0.000	0.045	0.181	0.313	0.000
	K_{M2}	−0.054	0.010	0.120	0.142	−0.082	0.000	0.044	0.171	0.278	−0.057
	$K_{\Delta R}$	−0.543	−0.160	0.261	0.755	1.335	−0.070	−0.075	−0.013	0.297	1.014
	K_ψ	0.947	0.969	1.069	1.238	1.348	−0.008	0.014	0.171	0.552	0.891

k_B	μ_D	0.4					0.8				
	Ω	0.000	0.250	0.500	0.750	1.000	0.000	0.250	0.500	0.750	1.000
155	K_{N1}	0.000	−0.307	−0.245	0.189	1.000	0.000	−0.306	−0.248	0.184	1.000
	K_{N2}	−2.464	−0.615	1.242	3.108	4.983	−1.227	−0.312	0.613	1.553	2.510
	K_{M1}	0.000	0.019	0.050	0.056	0.000	0.000	0.037	0.099	0.112	0.000
	K_{M2}	−0.511	−0.493	−0.464	−0.459	−0.517	−0.126	−0.090	−0.029	−0.018	−0.131
	$K_{\Delta R}$	−2.464	−0.616	1.241	3.108	4.985	−1.227	−0.313	0.612	1.554	2.513
	K_ψ	18.462	18.474	18.491	18.512	18.529	4.557	4.567	4.599	4.649	4.681

k_B	μ_D	1.6					3.2				
	Ω	0.000	0.250	0.500	0.750	1.000	0.000	0.250	0.500	0.750	1.000
155	K_{N1}	0.000	−0.283	−0.245	0.161	1.000	0.000	−0.119	−0.207	−0.019	1.000
	K_{N2}	−0.557	−0.159	0.269	0.762	1.327	−0.073	−0.077	−0.009	0.307	1.006
	K_{M1}	0.000	0.068	0.186	0.217	0.000	0.000	0.048	0.188	0.317	0.000
	K_{M2}	−0.027	0.040	0.155	0.181	−0.040	0.000	0.047	0.183	0.300	−0.028
	$K_{\Delta R}$	−0.557	−0.160	0.268	0.762	1.330	−0.073	−0.077	−0.009	0.307	1.008
	K_ψ	0.985	1.006	1.109	1.280	1.390	−0.009	0.015	0.187	0.594	0.945

6d. Uniform edge moment, M_B in-lb/linear in at the small end

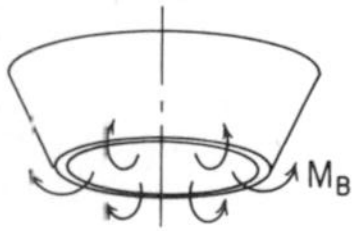

$$N_1 = -M_B \frac{2\sqrt{2}\beta}{tk_B} K_{N1} \qquad N_2 = M_B \frac{\beta}{t} K_{N2}$$

$$M_1 = M_B K_{M1} \qquad M_2 = M_B \nu K_{M2}$$

$$\Delta R = M_B \frac{\beta x \sin\alpha}{Et^2} K_{\Delta R} \qquad \psi = -M_B \frac{2\sqrt{2}\beta^2 x_B}{Et^3 k_B} K_\psi$$

For $R_B/t > 10$ and $k_B > 5$ and for $\nu = 0.3$, the following tables give the values of K at several locations along the shell $[\Omega = (x_A - x)/(x_A - x_B)]$

k_B	μ_D	0.4					0.8				
	Ω	0.000	0.250	0.500	0.750	1.000	0.000	0.250	0.500	0.750	1.000
5	K_{N1}	0.000	−0.188	−0.276	−0.228	0.000	0.000	−0.208	−0.332	−0.304	0.000
	K_{N2}	−2.694	−1.485	−0.125	1.420	3.200	−1.599	−0.958	−0.164	0.870	2.284
	K_{M1}	0.000	0.183	0.423	0.704	1.000	0.000	0.129	0.365	0.678	1.000
	K_{M2}	−10.343	−10.859	−11.439	−12.132	−12.997	−2.757	−2.976	−3.225	−3.620	−4.342
	$K_{\Delta R}$	−2.694	−1.517	−0.172	1.382	3.200	−1.599	−0.994	−0.220	0.818	2.284
	K_ψ	6.028	6.125	6.246	6.398	6.583	1.607	1.658	1.742	1.876	2.070

k_B	μ_D	1.6					3.2				
	Ω	0.000	0.250	0.500	0.750	1.000	0.000	0.250	0.500	0.750	1.000
5	K_{N1}	0.000	−0.107	−0.206	−0.244	0.000	0.000	−0.008	−0.045	−0.142	0.000
	K_{N2}	−0.458	−0.337	−0.149	0.236	1.183	−0.003	−0.041	−0.099	−0.094	0.904
	K_{M1}	0.000	0.069	0.263	0.594	1.000	0.000	0.005	0.060	0.308	1.000
	K_{M2}	−0.287	−0.287	−0.262	−0.367	−1.017	0.022	0.033	0.078	0.138	−0.645
	$K_{\Delta R}$	−0.458	−0.355	−0.184	0.195	1.183	−0.003	−0.043	−0.106	−0.118	0.904
	K_ψ	0.167	0.180	0.226	0.339	0.557	−0.013	−0.014	−0.007	0.045	0.264

TABLE 31 *Formulas for bending and membrane stresses and deformations in thin-walled pressure vessels* (*Cont.*)

k_B	μ_D	0.4					0.8				
	Ω	0.000	0.250	0.500	0.750	1.000	0.000	0.250	0.500	0.750	1.000
15	K_{N1}	0.000	−0.820	−1.130	−0.877	0.000	0.000	−0.525	−0.749	−0.604	0.000
	K_{N2}	−11.195	−5.799	−0.170	5.714	11.875	−3.656	−1.970	−0.133	1.903	4.193
	K_{M1}	0.000	0.177	0.477	0.791	1.000	0.000	0.145	0.456	0.798	1.000
	K_{M2}	−15.399	−15.591	−15.689	−15.809	−16.069	−2.403	−2.373	−2.214	−2.076	−2.131
	$K_{\Delta R}$	−11.195	−5.845	−0.234	5.664	11.875	−3.656	−2.000	−0.175	1.869	4.193
	K_ψ	26.923	27.076	27.255	27.469	27.713	4.201	4.254	4.352	4.515	4.733

k_B	μ_D	1.6					3.2				
	Ω	0.000	0.250	0.500	0.750	1.000	0.000	0.250	0.500	0.750	1.000
15	K_{N1}	0.000	−0.229	−0.370	−0.349	0.000	0.000	−0.023	−0.117	−0.257	0.000
	K_{N2}	−0.802	−0.514	−0.140	0.435	1.400	0.004	−0.084	−0.174	−0.068	0.994
	K_{M1}	0.000	0.109	0.392	0.754	1.000	0.000	0.012	0.133	0.512	1.000
	K_{M2}	−0.210	−0.126	0.084	0.285	0.256	0.024	0.040	0.142	0.388	0.425
	$K_{\Delta R}$	−0.802	−0.527	−0.161	0.415	1.400	0.004	−0.086	−0.180	−0.083	0.994
	K_ψ	0.367	0.386	0.472	0.671	0.982	−0.043	−0.043	−0.013	0.151	0.593

k_B	μ_D	0.4					0.8				
	Ω	0.000	0.250	0.500	0.750	1.000	0.000	0.250	0.500	0.750	1.000
35	K_{N1}	0.000	−1.225	−1.657	−1.261	0.000	0.000	−0.637	−0.877	−0.681	0.000
	K_{N2}	−16.494	−8.378	−0.109	8.324	16.931	−4.318	−2.244	−0.080	2.203	4.641
	K_{M1}	0.000	0.161	0.491	0.828	1.000	0.000	0.149	0.480	0.826	1.000
	K_{M2}	−9.936	−9.880	−9.662	−9.449	−9.408	−1.269	−1.148	−0.859	−0.573	−0.476
	$K_{\Delta R}$	−16.494	−8.408	−0.149	8.293	16.931	−4.318	−2.259	−0.101	2.187	4.641
	K_ψ	40.535	40.637	40.766	40.929	41.121	5.176	5.210	5.293	5.445	5.650

k_B	μ_D	1.6					3.2				
	Ω	0.000	0.250	0.500	0.750	1.000	0.000	0.250	0.500	0.750	1.000
35	K_{N1}	0.000	−0.280	−0.424	−0.371	0.000	0.000	−0.034	−0.155	−0.294	0.000
	K_{N2}	−0.926	−0.559	−0.121	0.495	1.427	0.007	−0.106	−0.199	−0.042	1.001
	K_{M1}	0.000	0.126	0.434	0.795	1.000	0.000	0.018	0.170	0.582	1.000
	K_{M2}	−0.112	0.005	0.283	0.577	0.678	0.015	0.033	0.173	0.516	0.753
	$K_{\Delta R}$	−0.926	−0.566	−0.132	0.486	1.427	0.007	−0.107	−0.203	−0.049	1.001
	K_ψ	0.458	0.478	0.581	0.813	1.158	−0.063	−0.062	−0.012	0.227	0.789

k_B	μ_D	0.4					0.8				
	Ω	0.000	0.250	0.500	0.750	1.000	0.000	0.250	0.500	0.750	1.000
75	K_{N1}	0.000	−1.351	−1.814	−1.370	0.000	0.000	−0.674	−0.914	−0.700	0.000
	K_{N2}	−18.098	−9.119	−0.059	9.090	18.334	−4.520	−2.312	−0.053	2.285	4.731
	K_{M1}	0.000	0.156	0.496	0.833	1.000	0.000	0.152	0.490	0.836	1.000
	K_{M2}	−5.133	−5.002	−4.692	−4.382	−4.257	−0.631	−0.486	−0.162	0.163	0.299
	$K_{\Delta R}$	−18.098	−9.134	−0.079	9.074	18.334	−4.520	−2.319	−0.063	2.277	4.731
	K_{ψ}	44.869	44.925	45.007	45.125	45.269	5.512	5.535	5.608	5.752	5.949

k_B	μ_D	1.6					3.2				
	Ω	0.000	0.250	0.500	0.750	1.000	0.000	0.250	0.500	0.750	1.000
75	K_{N1}	0.000	−0.302	−0.445	−0.378	0.000	0.000	−0.041	−0.174	−0.308	0.000
	K_{N2}	−0.977	−0.575	−0.111	0.517	1.431	0.008	−0.117	−0.209	−0.029	1.001
	K_{M1}	0.000	0.133	0.452	0.810	1.000	0.000	0.020	0.188	0.611	1.000
	K_{M2}	−0.057	0.073	0.378	0.705	0.850	0.009	0.029	0.189	0.578	0.885
	$K_{\Delta R}$	−0.977	−0.578	−0.116	0.513	1.431	0.008	−0.117	−0.211	−0.032	1.001
	K_{ψ}	0.500	0.520	0.631	0.877	1.238	−0.074	−0.072	−0.009	0.270	0.894

k_B	μ_D	0.4					0.8				
	Ω	0.000	0.250	0.500	0.750	1.000	0.000	0.250	0.500	0.750	1.000
155	K_{N1}	0.000	−1.387	−1.856	−1.397	0.000	0.000	−0.688	−0.927	−0.705	0.000
	K_{N2}	−18.534	−9.304	−0.032	9.288	18.665	−4.592	−2.332	−0.039	2.312	4.750
	K_{M1}	0.000	0.156	0.498	0.842	1.000	0.000	0.153	0.494	0.839	1.000
	K_{M2}	−2.554	−2.404	−2.070	−1.736	−1.589	−0.312	−0.161	0.174	0.511	0.659
	$K_{\Delta R}$	−18.534	−9.312	−0.042	9.281	18.665	−4.592	−2.336	−0.044	2.308	4.750
	K_{ψ}	46.135	46.165	46.222	46.315	46.434	5.645	5.662	5.730	5.870	6.063

k_B	μ_D	1.6					3.2				
	Ω	0.000	0.250	0.500	0.750	1.000	0.000	0.250	0.500	0.750	1.000
155	K_{N1}	0.000	−0.313	−0.454	−0.381	0.000	0.000	−0.044	−0.184	−0.314	0.000
	K_{N2}	−1.001	−0.581	−0.106	0.527	1.431	0.009	−0.122	−0.213	−0.022	1.001
	K_{M1}	0.000	0.137	0.460	0.817	1.000	0.000	0.022	0.197	0.624	1.000
	K_{M2}	−0.029	0.107	0.423	0.765	0.927	0.004	0.026	0.198	0.607	0.944
	$K_{\Delta R}$	−1.001	−0.583	−0.108	0.525	1.431	0.009	−0.122	−0.214	−0.024	1.001
	K_{ψ}	0.520	0.540	0.654	0.903	1.276	−0.080	−0.077	−0.008	0.293	0.948

TABLE 31 ***Formulas for bending and membrane stresses and deformations in thin-walled pressure vessels (Cont.)***

Case no., loading		Formulas

7. Toroidal shells

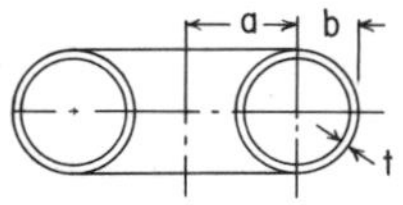

$\mu = \frac{b^2}{at}\sqrt{12(1-\nu^2)}$

$\frac{t}{b} < \frac{1}{10}$

7a. Split toroidal shell under axial load P (omega joint)

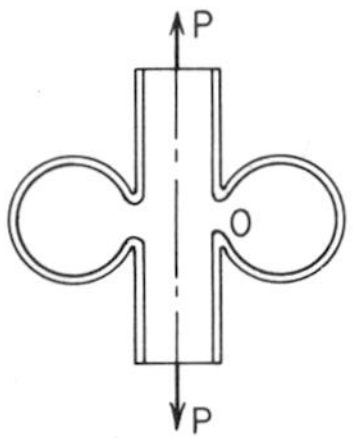

(Refs. 16 and 40)

For $4 < \mu < 40$,

$$\text{Stretch} = \frac{3.46Pb\sqrt{1-\nu^2}}{Et^2}$$

$$\text{Max } \sigma_2 = \frac{2.15P}{2\pi at}\left[\frac{ab(1-\nu^2)}{t^2}\right]^{\frac{1}{3}} \quad \text{at point 0}$$

$$\text{Max } \sigma_1' = \frac{2.99P}{2\pi at}\left[\frac{ab}{t^2\sqrt{1-\nu^2}}\right]^{\frac{1}{3}} \quad \text{near point 0}$$

If $\mu < 4$, the following values for stretch should be used, where $\Delta = \frac{Pb^3}{2aD}$:

μ	<1	1	2	3
Stretch	1.00Δ	0.95Δ	0.80Δ	0.70Δ

7b. Corrugated tube under axial load P

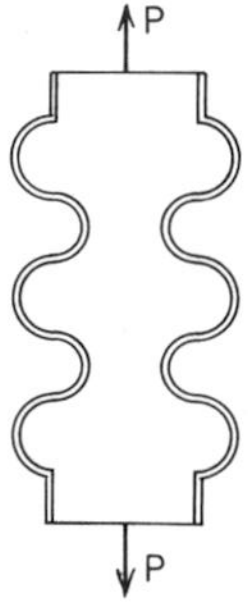

For $4 < \mu < 40$,

$$\text{Stretch} = \frac{0.577Pbn\sqrt{1-\nu^2}}{Et^2}$$ where n is the number of semicircular corrugations (five shown in figure)

$$\text{Max } \sigma_2 = \frac{0.925P}{2\pi at}\left[\frac{ab(1-\nu^2)}{t^2}\right]^{\frac{1}{3}}$$

$$\text{Max } \sigma_1' = \frac{1.63P}{2\pi at}\left[\frac{ab}{t^2\sqrt{1-\nu^2}}\right]^{\frac{1}{3}}$$ (Ref. 16)

If $\mu < 4$, let $\Delta = \frac{Pb^3n}{4aD}$ and use the tabulated values from case 7a

For U-shaped corrugations where a flat annular plate separates the inner and outer semicircles, see Ref. 41

	7c. Corrugated tube under internal pressure, q lb/in^2. If internal pressure on the ends must be carried by the walls, calculate the end load and use case 7b in addition (See Ref. 47 and Art. 12-5 for a discussion of a possible instability due to internal pressure in a long bellows)	For $4 < \mu < 40$, Stretch per semicircular corrugation $= \pm 2.45(1 - \nu^2)^{\frac{1}{3}} \left(\frac{a}{t}\right)^{\frac{4}{3}} \left(\frac{b}{t}\right)^{\frac{1}{3}} \frac{bq}{E}$ Total stretch = 0 if there are an equal number of inner and outer corrugations Max $\sigma_2 = 0.955q(1 - \nu^2)^{\frac{1}{6}} \left(\frac{ab}{t^2}\right)^{\frac{2}{3}}$ Max $\sigma_1' = 0.955q(1 - \nu^2)^{-\frac{1}{3}} \left(\frac{ab}{t^2}\right)^{\frac{2}{3}}$ (Ref. 16) If $\mu < 1$, the stretch per semicircular corrugation $= \pm 3.28(1 - \nu^2)\frac{b^4 q}{Et^3}$ For U-shaped corrugations, see Ref. 41
8. Cylindrical shells with open ends 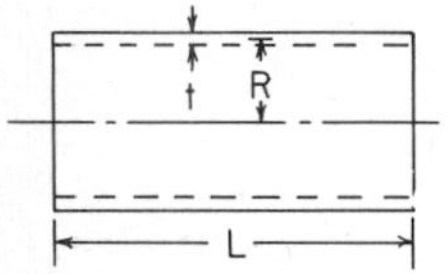	8a. Diametrically opposite and equal concentrated loads, P at midlength 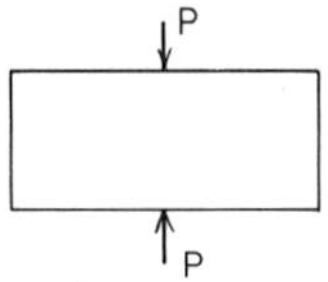	For $1 < L/R < 18$ and $R/t > 10$, Deflection under the load $= 6.5\frac{P}{Et}\left(\frac{R}{t}\right)^{\frac{3}{2}}\left(\frac{L}{R}\right)^{-\frac{3}{4}}$ For $L/R > 18$, the maximum stresses and deflections are approximately the same as for case 9a For loads at the extreme ends, the maximum stresses are approximately four times as great as for loading at midlength

TABLE 31 *Formulas for bending and membrane stresses and deformations in thin-walled pressure vessels* (*Cont.*)

Case no., loading		Formulas
9. Cylindrical shells with closed ends and end support 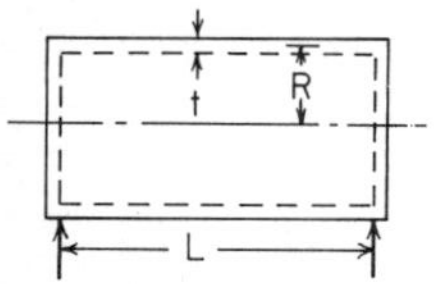	9a. Radial load P uniformly distributed over small area A, approximately square or round, located near midspan	Maximum stresses are circumferential stresses at center of loaded area and can be found from following table. Values given are for $L/R = 8$ but may be used for L/R ratios between 3 and 40. [Coefficients adapted from Bjilaard (Refs. 22, 23, 28)] (see table below)

R/t \ A/R^2	0.0004	0.0016	0.0036	0.0064	0.010	0.0144	0.0196	0.0256	0.0324	0.040	0.0576	0.090	0.160	0.25
						Values of $\sigma_2'(t^2/P)$								
300	1.475	1.11	0.906	0.780	0.678	0.600	0.522	0.450	0.390	0.348	0.264	0.186	0.120	0.078
100		1.44	1.20	1.044	0.918	0.840	0.750	0.666	0.600	0.540	0.444	0.342	0.240	0.180
50			1.44	1.254	1.11	1.005	0.900	0.840	0.756	0.720	0.600	0.480	0.360	0.264
15										0.990	0.888	0.780	0.600	0.468
						Values of $\sigma_2(Rt/P)$								
300	58	53.5	49	44.5	40	35.5	32	28	24	21	16	11	6	4
100		33.5	30.5	27.6	25	22.5	20	17.5	15	13	10	7	4.2	3.6
50					9.6	9	8.5	8.0	7.7	7.5	6.5	5.6	4.1	3.1
15										3.25	3.0	2.4	2.0	1.56

For A very small (nominal point loading) at point of load

$$\sigma_2 = \frac{0.4P}{t^2} \qquad \sigma_2' = \frac{2.4P}{t^2} \qquad y = \frac{P}{Et}\left[0.48\left(\frac{L}{R}\right)^{\frac{1}{2}}\left(\frac{R}{t}\right)^{1.22}\right]$$

(Approximate empirical formulas which are based on tests of Refs. 2 and 19)

For a more extensive presentation of Bjilaard's work in graphic form over a somewhat extended range of parameters, see Ref. 27

9b. Center load, P lb, concentrated on a very short length $2b$

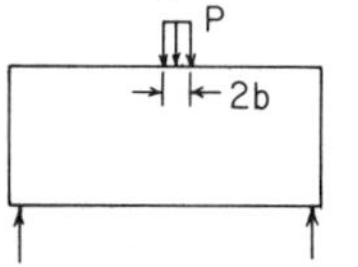

At the top center,

Max $\sigma_2 = -0.130BPR^{\frac{3}{4}}b^{-\frac{3}{2}}t^{-\frac{5}{4}}$

Max $\sigma_2' = -B^{-1}PR^{\frac{1}{4}}b^{-\frac{1}{2}}t^{-\frac{7}{4}}$

Max $\sigma_1 = -0.150B^3PR^{\frac{1}{4}}b^{-\frac{1}{2}}t^{-\frac{7}{4}}$

Deflection $= 0.0820B^5PR^{\frac{3}{4}}L^{\frac{1}{2}}t^{-\frac{9}{4}}E^{-1}$

where $B = [12(1-\nu^2)]^{\frac{1}{8}}$ (Ref. 13)

9c. Uniform load, p lb/in, over entire length of top element

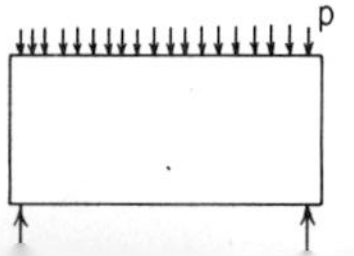

At the top center,

Max $\sigma_2 = -0.492BpR^{\frac{3}{4}}L^{-\frac{1}{2}}t^{-\frac{5}{4}}$

Max $\sigma_2' = -1.217B^{-1}pR^{\frac{1}{4}}L^{\frac{1}{2}}t^{-\frac{7}{4}}$

Max $\sigma_1 = -0.1188B^3pR^{\frac{1}{4}}L^{\frac{1}{2}}t^{-\frac{7}{4}}$

Deflection $= 0.0305B^5pR^{\frac{3}{4}}L^{\frac{3}{2}}t^{-\frac{9}{4}}E^{-1}$

where B is given in case 9b (Ref. 13)

Quarter span deflection $= 0.774$ midspan deflection

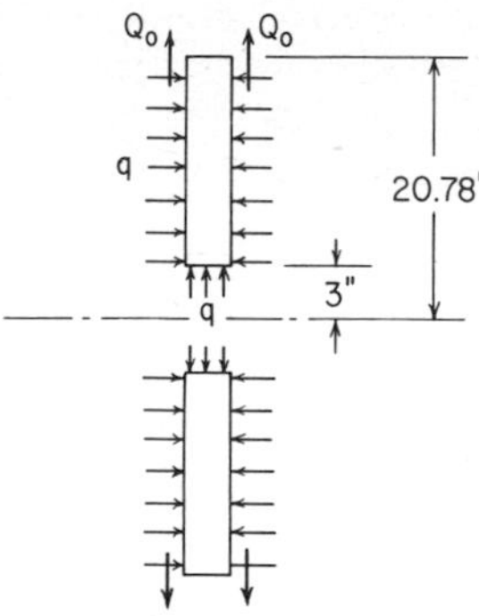

Fig. 12.7

From Table 32, case 1c, the effect of the loads Q_o can be determined if the loading is modeled as an outward pressure of $-2Q_o/t_1$. Therefore,

$$\Delta a = \frac{-qa}{E}\left(\frac{a^2+b^2}{a^2-b^2} - \nu\right) = \frac{2(1200)(20.78)}{t_1 10(10^6)}\left(\frac{20.78^2+3^2}{20.78^2-3^2} - 0.33\right) = \frac{0.00355}{t_1}$$

The longitudinal pressure of 200 lb/in^2 will cause a small lateral expansion in the outer radius of

$$\Delta a = \frac{200(0.33)(20.78)}{10(10^6)} = 0.000137 \text{ in}$$

Summing the changes in the outer radius to the desired value gives

$$0.00668 = 0.0000177 + 0.000137 + \frac{0.00355}{t_1} \qquad \text{or} \qquad t_1 = 0.545 \text{ in}$$

(Undoubtedly further optimization could be carried out on the volume of material required and the ease of welding the joint by varying the thickness of the disk and the size of the internal hole.)

3. A truncated cone of aluminum with a uniform wall thickness of 0.050 in and a semiapex angle of 55° has a radius of 2 in at the small end and 2.5 in at the large end. It is desired to know the radial loading at the small end which will increase the radius by half the wall thickness. Given: $E = 10(10^6)$ lb/in^2 and $\nu = 0.33$.

Solution. Evaluate the distances from the apex along a meridian to the two ends of the shell and then obtain the shell parameters:

$$x_A = \frac{2.5}{\sin 55^\circ} = 3.052 \text{ in}$$

$$x_B = \frac{2.0}{\sin 55^\circ} = 2.442 \text{ in}$$

$$k_A = 2\left[\frac{12(1-0.33^2)(3.052^2)}{0.050^2(\tan^2 55^\circ)}\right]^{\frac{1}{4}} = 23.64$$

$$k_B = 21.15$$

$$\mu_D = \frac{23.64 - 21.15}{2} = 1.76$$

$$\beta = [12(1-0.33^2)]^{\frac{1}{2}} = 3.27$$

From Table 31, case 6c, tabulated constants for shell forces, moments, and deformations can be found when a radial load is applied to the small end. For the present problem the value of $K_{\Delta R}$ at the small end ($\Omega = 1.0$) is needed when $\mu_D = 1.76$ and $k_B = 21.15$. Interpolation from the following data gives $K_{\Delta R} = 1.27$:

k_B	5.0				15.0				35.0			
μ_D	0.4	0.8	1.6	3.2	0.4	0.8	1.6	3.2	0.4	0.8	1.6	3.2
$K_{\Delta R}$	2.059	1.699	1.320	1.122	3.673	2.335	1.352	1.052	4.655	2.485	1.343	1.026

At $\Omega = 1.0$

Therefore,

$$\Delta R_B = \frac{-Q_B(2.442)(\sin^2 55°)}{10(10^6)(0.050)} \frac{21.15}{2}(1.27) = -0.00006225Q_B$$

Since $\Delta R_B = 0.050/2$ (half the thickness), $Q_B = -402$ lb/in (outward).

12.4 *Thin multielement shells of revolution*

The discontinuity stresses at the junctions of shells or shell elements due to changes in thickness or shape are not serious under static loading of ductile materials; however, they are serious under conditions of cyclic or fatigue loading. In Ref. 9, discontinuity stresses are discussed with a numerical example; also, allowable levels of the membrane stresses due to internal pressure are established, as well as allowable levels of membrane and bending stresses due to discontinuities under both static and cyclic loadings.

Langer (Ref. 10) discusses four modes of failure of a pressure vessel —bursting due to general yielding, ductile tearing at a discontinuity, brittle fracture, and creep rupture—and the way in which these modes are affected by the choice of material and wall thickness; he also compares pressure-vessel codes of several countries. Zaremba (Ref. 47) and Johns and Orange (Ref. 48) describe in detail the techniques for accurate deformation matching at the intersections of axisymmetric shells.

The following example illustrates the use of the formulas in Tables 29 to 31 to determine discontinuity stresses.

EXAMPLE

The vessel shown in quarter longitudinal section in Fig. 12.8*a* consists of a cylindrical shell ($R = 24$ in and $t = 0.633$ in) with conical ends ($\alpha = 45°$ and $t = 0.755$ in). The parts are welded together, and the material is steel, for which $E = 30(10^6)$ lb/in^2 and $\nu = 0.25$. It is required to determine the maximum stresses at the junction of the cylinder and cone due to an internal pressure of 300 lb/in^2. (This vessel corresponds to one for which the results of a supposedly precise analysis and experimentally determined stress values are available. See Ref. 17.)

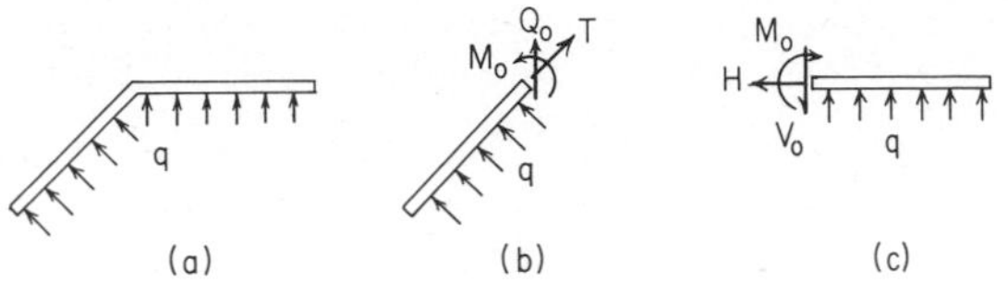

Fig. 12.8

Solution. For the cone, case 2a in Table 29 and cases 4a and 4b in Table 31 can be used: $R = 24$ in, $\alpha = 45°$, and $t = 0.755$ in. The following conditions exist at the end of the cone: From Table 29, case 2a, for the load T and pressure q,

$$\sigma_1 = \frac{300(24)}{2(0.755)(\cos 45°)} = 6740 \text{ lb/in}^2 \qquad T = 6740(0.755) = 5091 \text{ lb/in}$$

$$\sigma_2 = 13{,}480 \text{ lb/in}^2 \qquad \sigma_1' = 0 \qquad \sigma_2' = 0$$

$$\Delta R = \frac{300(24^2)}{30(10^6)(0.755)(\cos 45°)}\left(1 - \frac{0.25}{2}\right) = 0.00944 \text{ in}$$

$$\psi = \frac{3(300)(24)(1)}{2(30)(10^6)(0.755)(\cos 45°)} = 0.000674 \text{ rad}$$

From Table 31, case 4a, for the radial edge load Q_o,

$$x_A = \frac{24}{\sin 45°} = 33.94 \text{ in}$$

$$k_A = 2\left[\frac{12(1 - 0.25^2)(33.94^2)}{0.755^2(\tan^2 45°)}\right]^{\frac{1}{4}} = 24.56$$

$$\beta = [12(1 - 0.25^2)]^{\frac{1}{2}} = 3.354$$

Only values at $x = x_A$ are needed for this solution. Therefore, the series solutions for the constants can be used to give

$$F_{9A} = C_1 = 0.9005 \qquad F_{1A} = 0 \qquad F_{3A} = 0 \qquad F_{2A} = 0.8977$$

$$F_{4A} = 0.8720 \qquad F_{5A} = F_{8A} = 0.8746 \qquad F_{10A} = F_{7A} = F_{6A} = 0.8947$$

$$\Delta R_A = \frac{Q_o 33.94(0.5)(24.56)}{30(10^6)(0.755)(\sqrt{2})(0.9005)}\left[0.8720 - \frac{4(0.25^2)}{24.56^2}0.8977\right] = 12.59(10^{-6})Q_o$$

$$\psi_A = \frac{Q_o 33.94(3.354)(0.7071)}{30(10^6)(0.755^2)(0.9005)}(0.8947) = 4.677(10^{-6})Q_o$$

$$N_{1A} = 0.7071 Q_o \qquad M_{1A} = 0$$

$$N_{2A} = \frac{Q_o(0.7071)(24.56)}{2(0.9005)}\left[0.8720 + \frac{2(0.25)}{24.56}(0.8746)\right] = 12.063 Q_o$$

$$M_{2A} = \frac{Q_o(0.7071)(1 - 0.25^2)(0.755)}{3.354(0.9005)}(0.8947) = 0.1483 Q_o$$

From Table 31, case 4b, for the edge moment M_A,

$\Delta R_A = 4.677(10^{-6})M_o$ (same coefficient shown for ψ_A for the loading Q_o as would be expected from Castigliano's theorem)

$$\psi_A = \frac{M_o 2(2)(3.354^2)(33.94)}{30(10^6)(0.755^3)(24.56)}\left(\frac{0.8977}{0.9005}\right) = 3.395(10^{-6})M_o$$

$$N_{1A} = 0 \qquad N_{2A} = M_o\frac{3.354(0.8947)}{0.755(0.9005)} = 4.402 M_o$$

$$M_{1A} = M_o \qquad M_{2A} = M_o\left[0.25 + \frac{2(2)(1-0.25^2)(0.8977)}{24.56(0.9005)}\right] = 0.3576M_o$$

For the cylinder, case 1c in Table 29 and cases 8 and 10 in Table 30 can be used (it is assumed that the other end of the cylinder is far enough away so as to not affect the deformations and stresses at the cone-cylinder junction): $R = 24$ in; $t = 0.633$ in; $\lambda = [3(1-0.25^2)/24^2/0.633^2]^{\frac{1}{4}} = 0.3323$; and $D = 30(10^6)(0.633^3)/12(1-0.25^2) = 6.76(10^5)$. The following conditions exist at the end of the cylinder: From Table 29, case 1c, for the axial load H and the pressure q,

$$\sigma_1 = \frac{300(24)}{2(0.633)} = 5690 \text{ lb/in}^2 \qquad H = 5690(0.633) = 3600 \text{ lb/in}$$

$$\sigma_2 = 11{,}380 \text{ lb/in}^2 \qquad \sigma_1' = 0 \qquad \sigma_2' = 0$$

$$\Delta R = \frac{300(24^2)}{30(10^6)(0.633)}\left(1 - \frac{0.25}{2}\right) = 0.00796 \text{ in}$$

$$\psi = 0$$

From Table 30, case 8, for the radial end load V_o,

$$\psi_A = \frac{V_o}{2(6.76)(10^5)(0.3323^2)} = 6.698(10^{-6})V_o$$

$$\Delta R_A = y_A = \frac{-V_o}{2(6.76)(10^5)(0.3323^3)} = -20.16(10^{-6})V_o$$

$$\sigma_1 = 0 \qquad \sigma_2 = \frac{yE}{R} = \frac{-20.16(10^{-6})V_o(30)(10^6)}{24} = -25.20V_o$$

$$\sigma_1' = 0 \qquad \sigma_2' = 0$$

From Table 30, case 10, for the end moment M_o,

$$\psi_A = \frac{-M_o}{6.76(10^5)(0.3323)} = -4.452(10^{-6})M_o$$

$$\Delta R_A = y_A = \frac{M_o}{2(6.76)(10^5)(0.3323^2)} = 6.698(10^{-6})M_o$$

$$\sigma_1 = 0 \qquad \sigma_2 = \frac{2M_o\lambda^2 R}{t} = \frac{2M_o(0.3323^2)(24)}{0.633} = 8.373M_o$$

$$\sigma_1' = \frac{-6M_o}{t^2} = \frac{-6M_o}{0.633^2} = -14.97M_o \qquad \sigma_2' = \nu\sigma_1' = -3.74M_o$$

Summing the radial deflections for the end of the cone and equating to the sum for the cylinder gives

$$0.00944 + 12.59(10^{-6})Q_o + 4.677(10^{-6})M_o = 0.00796 - 20.16(10^{-6})V_o + 6.698(10^{-6})M_o$$

Doing the same with the meridian rotations gives

$$0.000674 + 4.677(10^{-6})Q_o + 3.395(10^{-6})M_o = 0 + 6.698(10^{-6})V_o - 4.452(10^{-6})M_o$$

Finally, equating the radial forces gives

$$Q_o + 5091 \cos 45° = V_o$$

Therefore,

$$Q_o = -2110 \text{ lb/in,}$$
$$V_o = 1490 \text{ lb/in,}$$
$$M_o = 2443 \text{ in-lb/in}$$

In the cylinder,

$$\sigma_1 = 5690 + 0 + 0 = 5690 \text{ lb/in}^2$$
$$\sigma_2 = 11{,}380 - 25.20(1490) + 8.373(2443) = -5712 \text{ lb/in}^2$$
$$\sigma_1' = 0 + 0 - 14.97(2443) = -36{,}570 \text{ lb/in}^2$$
$$\sigma_2' = 0 + 0 - 3.74(2443) = -9140 \text{ lb/in}^2$$

Combined hoop stress on the outside $= -5712 - 9140 = -14{,}852$ lb/in^2
Combined hoop stress on the inside $= -5712 + 9140 = 3428$ lb/in^2
Combined meridional stress on the outside $= 5690 - 36{,}570 = -30{,}880$ lb/in^2
Combined meridional stress on the inside $= 5690 + 36{,}570 = 42{,}260$ lb/in^2

Similarly, in the cone,

$$\sigma_1 = 6740 + \frac{0.7071(-2110)}{0.755} + 0 = 4764 \text{ lb/in}^2$$

$$\sigma_2 = 13{,}480 + \frac{12.063(-2110)}{0.755} + \frac{4.402(2443)}{0.755} = -5989 \text{ lb/in}^2$$

$$\sigma_1' = 0 + 0 - \frac{2443(6)}{0.755^2} = -25{,}715 \text{ lb/in}^2$$

$$\sigma_2' = 0 - \frac{0.1483(-2110)(6)}{0.755^2} - \frac{0.3576(2443)(6)}{0.755^2} = -5902 \text{ lb/in}^2$$

Combined hoop stress on the outside $= -5989 - 5902 = -11{,}891$ lb/in^2
Combined hoop stress on the inside $= -5989 + 5902 = -87$ lb/in^2
Combined meridional stress on the outside $= 4764 - 25{,}715 = -20{,}951$ lb/in^2
Combined meridional stress on the inside $= 4764 + 25{,}715 = 30{,}480$ lb/in^2

These stress values are in substantial agreement with the computed and experimental values cited in Refs. 17 and 26. Note that the radial deflections are much less than the wall thicknesses. See the last paragraph on page 456.

12.5 *Thin shells of revolution under external pressure*

All formulas given in Tables 29 and 31 for thin vessels under distributed pressure are for internal pressure, but they will apply equally to cases of external pressure if q is given a negative sign. The formulas in Table 30 for distributed pressure are for external pressure in order to correspond to similar loadings for beams on elastic foundations in Chap. 7. It should be noted with care that the application of external pressure may cause an instability failure due to stresses lower than the elastic limit, and in such a case the formulas in this chapter do not apply. This condition is discussed in Chap. 14, and formulas for the critical pressures or stresses producing instability are given in Table 35.

A vessel of moderate thickness may collapse under external pressure at stresses just below the yield point, its behavior being comparable to that of a short column. The problem of ascertaining the pressure that produces failure of this kind is of special interest in connection with cylindrical vessels and pipe. For external loading such as that in Table 29, case 1c, the external collapsing pressure can be given by

$$q' = \frac{t}{R} \frac{\sigma_y}{1 + (4\sigma_y/E)(R/t)^2} \qquad \text{(see Refs. 1, 7, and 8)}$$

In Refs. 8 and 9, charts are given for designing vessels under external pressure.

A special instability problem should be considered when designing long cylindrical vessels or even *relatively short corrugated tubes* under internal pressure. Haringx (Refs. 54 and 55) and Flügge (Ref. 5) have shown that vessels of this type will buckle laterally if the ends are restrained against longitudinal displacement and if the product of the internal pressure and the cross-sectional area reaches the Euler load for the column as a whole. For cylindrical shells this is seldom a critical factor, but for corrugated tubes or bellows this is recognized as a so-called *squirming instability*. To determine the Euler load for a bellows, an equivalent thin-walled circular cross section can be established which will have a radius equal to the mean radius of the bellows and a product Et, for which the equivalent cylinder will have the same axial deflection under end load as would the bellows. The overall bending moment of inertia I of the very thin equivalent cylinder can then be used in the expression $P_u = K\pi^2 EI/l^2$ for the Euler load. In a similar way Seide (Ref. 56) discusses the effect of pressure on the lateral bending of a bellows.

EXAMPLE

A corrugated-steel tube has a mean radius of 5 in, a wall thickness of 0.015 in, and 60 semicircular corrugations along its 40-in length. The ends are rigidly fixed, and the internal pressure necessary to produce a squirming instability is to be calculated. Given: $E = 30(10^6)$ lb/in^2 and $\nu = 0.3$.

Solution. Refer to Table 30, case 5b: $a = 5$ in, length = 40 in, $b = \frac{40}{120} = 0.333$ in, and $t = 0.015$ in

$$\mu = \frac{b^2}{at}\sqrt{12(1-\nu^2)} = \frac{0.333^2}{5(0.015)}\sqrt{12(1-0.3^2)} = 4.90$$

$$\text{Axial stretch} = \frac{-0.577Pbn\sqrt{1-\nu^2}}{Et^2} = \frac{0.577P(0.333)(60)(\sqrt{0.91})}{30(10^6)(0.015^2)} = -0.00163P$$

If a cylinder with a radius of 5 in and product E_1t_1 were loaded in compression with a load P, the stretch would be

$$\text{Stretch} = \frac{-Pl}{A_1E_1} = \frac{-P(40)}{2\pi 5 t_1 E_1} = -0.00163P$$

or

$$t_1E_1 = \frac{40}{2(5)(0.00163)} = 780.7 \text{ lb/in}$$

The bending moment of inertia of such a cylinder is $I_1 = \pi R^3 t_1$ (see Table 1, case 12). The Euler load for fixed ends is

$$P_{cr} = \frac{4\pi^2 E_1 I_1}{l^2}$$

or

$$P_{cr} = \frac{4\pi^2 E_1 \pi R^3 t_1}{l^2} = \frac{4\pi^3 5^3(780.7)}{40^2} = 7565 \text{ lb}$$

The internal pressure is therefore

$$q' = \frac{P_{cr}}{\pi R^2} = \frac{7565}{\pi 5^2} = 96.3 \text{ lb/in}^2$$

From Table 30, case 5c, the maximum stresses caused by this pressure are

$$\text{Max } \sigma_2 = 0.955(96.3)(0.91)^{\frac{1}{6}}\left[\frac{5(0.333)}{0.015^2}\right]^{\frac{2}{3}} = 34{,}400 \text{ lb/in}^2$$

$$\text{Max } \sigma_1' = 0.955(96.3)(0.91)^{-\frac{1}{3}}\left[\frac{5(0.333)}{0.015^2}\right]^{\frac{2}{3}} = 36{,}060 \text{ lb/in}^2$$

If the yield strength is greater than 36,000 lb/in^2, the corrugated tube should buckle laterally, that is, squirm, at an internal pressure of 96.3 lb/in^2.

12.6 *Thick shells of revolution*

If the wall thickness of a vessel is more than about one-tenth the radius, the meridional and hoop stresses cannot be considered uniform throughout the thickness of the wall and the radial stress cannot be considered negligible. These stresses in thick vessels, called *wall stresses,* must be found by formulas that are quite different from those used in finding membrane stresses in thin vessels.

It can be seen from the formulas for cases 1a and 1b of Table 32 that the stress σ_2 at the inner surface of a thick cylinder approaches q as the ratio of outer to inner radius approaches infinity. It is apparent, therefore, that if the stress is to be limited to some specified value σ, the pressure must never exceed $q = \sigma$, no matter how thick the wall is made. To overcome this limitation, the material at and near the inner surface must be put into a state of initial compression. This can be done by shrinking on one or more jackets (as explained in Art. 2.12 and in the example which follows) or by subjecting the vessel to a high internal pressure that stresses the inner part into the plastic range and, when removed, leaves residual compression there and residual tension in the outer part. This procedure is called *autofrettage,* or *self-hooping.* If many successive jackets are superimposed on the original tube by shrinking or wrapping, the resulting structure is called a *multilayer vessel.* Such a construction has certain advantages, but it should be noted that the formulas for hoop stresses are based on the assumption that an isotropic material is used. In a multilayered vessel the effective radial modulus of elasticity is less than the tangential modulus, and in consequence the hoop stress at and near the outer wall is less than the formula would indicate; therefore, the outer layers of material contribute less to the strength of the vessel than might be supposed.

Cases 1e and 1f in Table 32 represent radial body-force loading, which can be superimposed to give results for centrifugal loading, etc. (see Art. 15.2). Case 1f is directly applicable to thick-walled disks with embedded electrical conductors used to generate magnetic fields. In many such cases

TABLE 32 ***Formulas for thick-walled vessels under internal and external loading***

NOTATION: q = unit pressure (pounds per square inch); δ and δ_b = radial body forces (pounds per cubic inch); a = outer radius; b = inner radius; σ_1, σ_2, and σ_3 are normal stresses in the longitudinal, circumferential, and radial directions, respectively (positive when tensile); E = modulus of elasticity; ν = Poisson's ratio. Δa, Δb, and Δl are the changes in the radii a and b and in the length l, respectively. ε_1 = unit normal strain in the longitudinal direction

Case no., form of vessel	Case no., manner of loading	Formulas
1. Cylindrical disk or shell 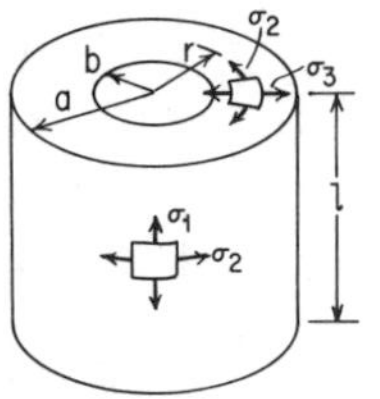	1a. Uniform internal radial pressure, q lb/in^2; longitudinal pressure zero or externally balanced; for a disk or a shell	$\sigma_1 = 0$ $\sigma_2 = \dfrac{qb^2(a^2 + r^2)}{r^2(a^2 - b^2)}$ max $\sigma_2 = q\dfrac{a^2 + b^2}{a^2 - b^2}$ at $r = b$ $\sigma_3 = \dfrac{-qb^2(a^2 - r^2)}{r^2(a^2 - b^2)}$ max $\sigma_3 = -q$ at $r = b$ Max shear stress $= \dfrac{\sigma_2 - \sigma_3}{2} = q\dfrac{a^2}{a^2 - b^2}$ at $r = b$ $\Delta a = \dfrac{q}{E}\dfrac{2ab^2}{a^2 - b^2}$ $\Delta b = \dfrac{qb}{E}\left(\dfrac{a^2 + b^2}{a^2 - b^2} + \nu\right)$ $\Delta l = \dfrac{-q\nu l}{E}\dfrac{2b^2}{a^2 - b^2}$
	1b. Uniform internal pressure, q lb/in^2, in all directions; ends capped; for a disk or a shell	$\sigma_1 = \dfrac{qb^2}{a^2 - b^2}$ (σ_2, σ_3, and the max shear stress are the same as for case 1a) $\Delta a = \dfrac{qa}{E}\dfrac{b^2(2 - \nu)}{a^2 - b^2}$ $\Delta b = \dfrac{qb}{E}\dfrac{a^2(1 + \nu) + b^2(1 - 2\nu)}{a^2 - b^2}$ $\Delta l = \dfrac{ql}{E}\dfrac{b^2(1 - 2\nu)}{a^2 - b^2}$
	1c. Uniform external radial pressure, q lb/in^2; longitudinal pressure zero or externally balanced; for a disk or a shell	$\sigma_1 = 0$ $\sigma_2 = \dfrac{-qa^2(b^2 + r^2)}{r^2(a^2 - b^2)}$ max $\sigma_2 = \dfrac{-q2a^2}{a^2 - b^2}$ at $r = b$ $\sigma_3 = \dfrac{-qa^2(r^2 - b^2)}{r^2(a^2 - b^2)}$ max $\sigma_3 = -q$ at $r = a$ Max shear stress $= \dfrac{\max \sigma_2}{2} = \dfrac{qa^2}{a^2 - b^2}$ at $r = b$ $\Delta a = \dfrac{-qa}{E}\left(\dfrac{a^2 + b^2}{a^2 - b^2} - \nu\right)$ $\Delta b = \dfrac{-q}{E}\dfrac{2a^2b}{a^2 - b^2}$ $\Delta l = \dfrac{q\nu l}{E}\dfrac{2a^2}{a^2 - b^2}$

1d. Uniform external pressure, q lb/in^2, in all directions; ends capped; for a disk or a shell	$\sigma_1 = \dfrac{-qa^2}{a^2 - b^2}$ (σ_2, σ_3, and the max shear stress are the same as for case 1c) $\Delta a = \dfrac{-qa}{E} \dfrac{a^2(1 - 2\nu) + b^2(1 + \nu)}{a^2 - b^2}$ $\Delta b = \dfrac{-qb}{E} \dfrac{a^2(2 - \nu)}{a^2 - b^2}$ $\Delta l = \dfrac{-ql}{E} \dfrac{a^2(1 - 2\nu)}{a^2 - b^2}$
1e. Uniformly distributed radial body force, δ lb/in^3, acting outward throughout the wall; for a disk only	$\sigma_1 = 0$ $\sigma_2 = \dfrac{\delta(2 + \nu)}{3(a + b)} \left[a^2 + ab + b^2 - (a + b)\left(\dfrac{1 + 2\nu}{2 + \nu}\right) r + \dfrac{a^2b^2}{r^2}\right]$ $\text{Max } \sigma_2 = \dfrac{\delta a^2}{3} \left[\dfrac{2(2 + \nu)}{a + b} + \dfrac{b}{a^2}(1 - \nu)\right]$ at $r = b$ $\sigma_3 = \dfrac{\delta(2 + \nu)}{3(a + b)} \left[a^2 + ab + b^2 - (a + b)r - \dfrac{a^2b^2}{r^2}\right]$ (*Note:* $\sigma_3 = 0$ at both $r = b$ and $r = a$.) $\text{Max shear stress} = \dfrac{\max \sigma_2}{2}$ at $r = b$ $\Delta a = \dfrac{\delta a^2}{3E} \left[1 - \nu + \dfrac{2(2 + \nu)b^2}{a(a + b)}\right]$ $\Delta b = \dfrac{\delta ab}{3E} \left[\dfrac{b}{a}(1 - \nu) + \dfrac{2a(2 + \nu)}{a + b}\right]$ $\varepsilon_1 = \dfrac{-\delta a\nu}{E} \left[\dfrac{2(a^2 + ab + b^2)}{3a(a + b)}(2 + \nu) - \dfrac{r}{a}(1 + \nu)\right]$
1f. Linearly varying radial body force from δ_b lb/in^3 outward at $r = b$ to zero at $r = a$; for a disk only	$\sigma_1 = 0$ $\sigma_2 = \delta_b \left[\dfrac{(7 + 5\nu)a^4 - 8(2 + \nu)ab^3 + 3(3 + \nu)b^4}{24(a - b)(a^2 - b^2)} - \dfrac{(1 + 2\nu)a}{3(a - b)} r + \dfrac{1 + 3\nu}{8(a - b)} r^2 + \dfrac{b^2a^2}{24r^2} \dfrac{(7 + 5\nu)a^2 - 8(2 + \nu)ab + 3(3 + \nu)b^2}{(a - b)(a^2 - b^2)}\right]$ $\sigma_3 = \delta_b \left[\dfrac{(7 + 5\nu)a^4 - 8(2 + \nu)ab^3 + 3(3 + \nu)b^4}{24(a - b)(a^2 - b^2)} - \dfrac{(2 + \nu)a}{3(a - b)} r + \dfrac{(3 + \nu)}{8(a - b)} r^2 + \dfrac{b^2a^2}{24r^2} \dfrac{(7 + 5\nu)a - 3(3 + \nu)b}{a^2 - b^2}\right]$ (*Note:* $\sigma_3 = 0$ at both $r = b$ and $r = a$) $\text{Max } \sigma_2 = \dfrac{b}{12} \dfrac{2a^4 + (1 + \nu)a^2(5a^2 - 12ab + 6b^2) - (1 - \nu)b^3(4a - 3b)}{(a - b)(a^2 - b^2)}$ at $r = b$

TABLE 32 *Formulas for thick-walled vessels under internal and external loading* (Cont.)

Case no., form of vessel	Case no., manner of loading	Formulas
		Max shear stress $= \dfrac{\max \sigma_2}{2}$ at $r = b$ $\Delta a = \dfrac{\delta_b a}{12E} \dfrac{(1-\nu)a^4 - 8(2+\nu)ab^3 + 3(3+\nu)b^4 + 6(1+\nu)a^2b^2}{(a-b)(a^2-b^2)}$ $\Delta b = \max \sigma_2 \dfrac{b}{E}$ $\epsilon_1 = \dfrac{-\delta_b \nu}{E}\left[\dfrac{(7+5\nu)a^4 - 8(2+\nu)ab^3 + 3(3+\nu)b^4}{12(a-b)(a^2-b^2)} - \dfrac{1+\nu}{a-b}\left(a - \dfrac{r}{2}\right)r\right]$
2. Spherical 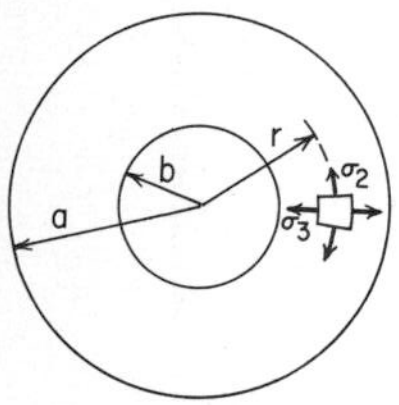 	2a. Uniform internal pressure, q lb/in^2	$\sigma_1 = \sigma_2 = \dfrac{qb^3}{2r^3}\dfrac{a^3+2r^3}{a^3-b^3}$ max $\sigma_1 = \max \sigma_2 = \dfrac{q}{2}\dfrac{a^3+2b^3}{a^3-b^3}$ at $r = b$ $\sigma_3 = \dfrac{-qb^3}{r^3}\dfrac{a^3-r^3}{a^3-b^3}$ max $\sigma_3 = -q$ at $r = b$ Max shear stress $= \dfrac{q3a^3}{4(a^3-b^3)}$ at $r = b$ The inner surface yields at $q = \dfrac{2\sigma_y}{3}\left(1 - \dfrac{b^3}{a^3}\right)$ (Ref. 20) $\Delta a = \dfrac{qa}{E}\dfrac{3(1-\nu)b^3}{2(a^3-b^3)}$ $\Delta b = \dfrac{qb}{E}\left[\dfrac{(1-\nu)(a^3+2b^3)}{2(a^3-b^3)} + \nu\right]$ (Ref. 3)
	2b. Uniform external pressure, q lb/in^2	$\sigma_1 = \sigma_2 = \dfrac{-qa^3}{2r^3}\dfrac{b^3+2r^3}{a^3-b^3}$ max $\sigma_1 = \max \sigma_2 = \dfrac{-q3a^3}{2(a^3-b^3)}$ at $r = b$ $\sigma_3 = \dfrac{-qa^3}{r^3}\dfrac{r^3-b^3}{a^3-b^3}$ max $\sigma_3 = -q$ at $r = a$ $\Delta a = \dfrac{-qa}{E}\left[\dfrac{(1-\nu)(b^3+2a^3)}{2(a^3-b^3)} - \nu\right]$ $\Delta b = \dfrac{-qb}{E}\dfrac{3(1-\nu)a^3}{2(a^3-b^3)}$ (Ref. 3)

the magnetic field varies linearly through the wall to zero at the outside. If there is a field at the outer turn, cases 1e and 1f can be superimposed in the necessary proportions.

The tabulated formulas for elastic wall stresses are accurate for both thin and thick vessels, but formulas for predicted yield pressures do not always agree closely with experimental results (Refs. 21, 34, 35, 37, and 39). The expression for q_y given in Table 32 is based on the minimum strain-energy theory of elastic failure. The expression for bursting pressure

$$q_u = 2\sigma_u \frac{a - b}{a + b}$$

commonly known as the *mean diameter formula*, is essentially empirical but agrees reasonably well with experiment for both thin and thick cylindrical vessels and is convenient to use. For very thick vessels the formula

$$q_u = \sigma_u \ln \frac{a}{b}$$

is preferable. Greater accuracy can be obtained by using with this formula a multiplying factor that takes into account the strain-hardening properties of the material (Refs. 10, 20, and 37). With the same objective, Faupel (Ref. 39) proposes (with different notation) the formula

$$q_u = \frac{2\sigma_y}{3} \ln \left[\frac{a}{b} \left(2 - \frac{\sigma_y}{\sigma_u} \right) \right]$$

A rather extensive discussion of bursting pressure is given in Ref. 38, which presents a tabulated comparison between bursting pressures as calculated by a number of different formulas and as determined by actual experiment.

EXAMPLE

At the powder chamber, the inner radius of a 3-in gun tube is 1.605 in and the outer radius is 2.425 in. It is desired to shrink a jacket on this tube so as to produce a radial pressure between the tube and jacket of 7600 lb/in^2. The outer radius of this jacket is 3.850 in. It is required to determine the difference between the inner radius of the jacket and the outer radius of the tube in order to produce the desired pressure, calculate the stresses in each part when assembled, and calculate the stresses in each part when the gun is fired, generating a powder pressure of 32,000 lb/in^2.

Solution. Using the formulas for Table 32, case 1c, it is found that for an external pressure of 7600, the stress σ_2 at the outer surface of the tube is $-19{,}430$, the stress σ_2 at the inner surface is $-27{,}050$, and the change in outer radius $\Delta a = -0.001385$; for an internal pressure of 7600, the stress σ_2 at the inner surface of the jacket is $+17{,}630$, the stress σ_2 at the outer surface is $+10{,}050$, and the change in inner radius $\Delta b = +0.001615$. (In making these calculations the inner radius of the jacket is assumed to be 2.425 in.) The initial difference between the inner radius of the jacket and the outer radius of the tube must be equal to the sum of the radial deformations they suffer, or $0.001385 + 0.001615 = 0.0030$; therefore the initial radius of the jacket should be $2.425 - 0.0030 = 2.422$ in.

The stresses produced by the powder pressure are calculated at the inner surface of the tube, at the common surface of tube and jacket ($r = 2.425$), and at the outer surface of the jacket.

These stresses are then superimposed on those found previously. The calculations are as follows: For the tube at the inner surface,

$$\sigma_2 = +32{,}000\left(\frac{3.85^2 + 1.605^2}{3.82^2 - 1.605^2}\right) = +45{,}450$$

$$\sigma_3 = -32{,}000$$

For tube and jacket at the interface,

$$\sigma_2 = +32{,}000\left(\frac{1.605^2}{2.425^2}\right)\left(\frac{3.85^2 + 2.425^2}{3.85^2 - 1.605^2}\right) = +23{,}500$$

$$\sigma_3 = -32{,}000\left(\frac{1.605^2}{2.425^2}\right)\left(\frac{3.85^2 - 2.425^2}{3.85^2 - 1.605^2}\right) = -10{,}200$$

For the jacket at the outer surface,

$$\sigma_2 = +32{,}000\left(\frac{1.605^2}{3.85^2}\right)\left(\frac{3.85^2 + 3.85^2}{3.85^2 - 1.605^2}\right) = +13{,}500$$

These are the stresses due to the powder pressure. Superimposing the stresses due to the shrinkage, we have as the resultant stresses:

At inner surface of tube,

$$\sigma_2 = -27{,}050 + 45{,}450 = +18{,}400 \text{ lb/in}^2$$
$$\sigma_3 = 0 - 32{,}000 = -32{,}000 \text{ lb/in}^2$$

At outer surface of tube,

$$\sigma_2 = -19{,}430 + 23{,}500 = +4070 \text{ lb/in}^2$$
$$\sigma_3 = -7600 - 10{,}200 = -17{,}800 \text{ lb/in}^2$$

At inner surface of jacket,

$$\sigma_2 = +17{,}630 + 23{,}500 = +41{,}130 \text{ lb/in}^2$$
$$\sigma_3 = -7600 - 10{,}200 = -17{,}800 \text{ lb/in}^2$$

At outer surface of jacket,

$$\sigma_2 = +10{,}050 + 13{,}500 = +23{,}550 \text{ lb/in}^2$$

12.7 *Pipe on supports at intervals*

For a pipe or cylindrical tank supported at intervals on saddles or pedestals and filled or partly filled with liquid, the stress analysis is difficult and the results are rendered uncertain by doubtful boundary conditions. Certain conclusions arrived at from a study of tests (Refs. 11 and 12) may be helpful in guiding design:

1. For a circular pipe or tank supported at intervals and held circular at the supports by rings or bulkheads, the ordinary theory of flexure is applicable if the pipe is completely filled.

2. If the pipe is only partially filled, the cross section at points between supports becomes out of round and the distribution of longitudinal fiber stress is neither linear nor symmetrical across the section. The highest stresses occur for the half-full condition; then the maximum longitudinal compressive stress

and the maximum circumferential bending stresses occur at the ends of the horizontal diameter, the maximum longitudinal tensile stress occurs at the bottom, and the longitudinal stress at the top is practically zero. According to theory (Ref. 4), the greatest of these stresses is the longitudinal compression, which is equal to the maximum longitudinal stress for the full condition divided by

$$K = \left(\frac{L}{R}\sqrt{\frac{t}{R}}\right)^{\frac{1}{2}}$$

where R = pipe radius, t = thickness, and L = span. The maximum circumferential stress is about one-third of this. Tests (Ref. 11) on a pipe having $K = 1.36$ showed a longitudinal stress that is somewhat less and a circumferential stress that is considerably greater than indicated by this theory.

3. For an unstiffened pipe resting in saddle supports, there are high local stresses, both longitudinal and circumferential, adjacent to the tips of the saddles. These stresses are less for a large saddle angle β (total angle subtended by arc of contact between pipe and saddle) than for a small angle, and for the ordinary range of dimensions they are practically independent of the thickness of the saddle, i.e., its dimension parallel to the pipe axis. For a pipe that fits the saddle well, the maximum value of these localized stresses will probably not exceed that indicated by the formula

$$\sigma_{\max} = k\frac{P}{t^2}\ln\frac{R}{t}$$

where P = total saddle reaction, R = pipe radius, t = pipe thickness, and k = coefficient given by

$$k = 0.02 - 0.00012(\beta - 90)$$

where β is in degrees. This stress is almost wholly due to circumferential bending and occurs at points about 15° above the saddle tips.

4. The maximum value of P the pipe can sustain is about 2.25 times the value that will produce a maximum stress equal to the yield point of the pipe material, according to the formula given above.

5. For a pipe supported in flexible slings instead of on rigid saddles, the maximum local stresses occur at the points of tangency of sling and pipe section; in general, they are less than the corresponding stresses in the saddle-supported pipe but are of the same order of magnitude.

REFERENCES

1. Southwell, R. V.: On the Collapse of Tubes by External Pressure, *Philos. Mag.*, vol. 29, p. 67, 1915.
2. Roark, R. J.: The Strength and Stiffness of Cylindrical Shells under Concentrated Loading, *ASME J. Appl. Mech.*, vol. 2, no. 4, p. A-147, 1935.
3. Timoshenko, S.: "Theory of Plates and Shells," Engineering Societies Monograph, McGraw-Hill Book Company, 1940.

4. Schorer, H.: Design of Large Pipe Lines, *Trans. Am. Soc. Civil Eng.*, vol. 98, p. 101, 1933.
5. Flügge, W.: "Stresses in Shells," Springer-Verlag, 1960.
6. Baker, E. H., L. Kovalevsky, and F. L. Rish: "Structural Analysis of Shells," McGraw Hill Book Company, 1972.
7. Saunders, H. E., and D. F. Windenburg: Strength of Thin Cylindrical Shells Under External Pressure, *Trans. ASME*, vol. 53, p. 207, 1931.
8. Jasper, T. M., and J. W. W. Sullivan: The Collapsing Strength of Steel Tubes, *Trans. ASME*, vol. 53, p. 219, 1931.
9. American Society of Mechanical Engineers: Rules for Construction of Nuclear Power Plant Components, Sec. III; Rules for Construction of Pressure Vessels, Division 1, and Division 2, Sec. VIII; ASME Boiler and Pressure Vessel Code, 1971.
10. Langer, B. F.: Design-stress Basis for Pressure Vessels, *Exp. Mech., J. Soc. Exp. Stress Anal.*, vol. 11, no. 1, January 1971.
11. Hartenberg, R. S.: The Strength and Stiffness of Thin Cylindrical Shells on Saddle Supports, doctoral dissertation, University of Wisconsin, 1941.
12. Wilson, W. M., and E. D. Olson: Tests on Cylindrical Shells, *Eng. Exp. Sta., Univ. Ill. Bull.* 331, 1941.
13. Odqvist, F. K. G.: Om Barverkan Vid Tunna Cylindriska Skal Ock Karlvaggar, *Proc. Roy. Swed. Inst. for Eng. Res.*, No. 164, 1942.
14. Hetényi, M.: Spherical Shells Subjected to Axial Symmetrical Bending, vol. 5 of the "Publications," International Association for Bridge and Structural Engineers, 1938.
15. Reissner, E.: Stresses and Small Displacements of Shallow Spherical Shells, II, *J. Math. Phys.*, vol. 25, No. 4, 1947.
16. Clark, R. A.: On the Theory of Thin Elastic Toroidal Shells, *J. Math. Phys.*, vol. 29, no. 3, 1950.
17. O'Brien, G. J., E. Wetterstrom, M. G. Dykhuizen, and R. G. Sturm: Design Correlations for Cylindrical Pressure Vessels with Conical or Toriconical Heads, *Weld. Res. Suppl.*, vol. 15, no. 7, p. 336, 1950.
18. Osipova, L. N., and S. A. Tumarkin: "Tables for the Computation of Toroidal Shells," P. Noordhoff Ltd., 1965 (English transl. by M. D. Friedman).
19. Roark, R. J.: Stresses and Deflections in Thin Shells and Curved Plates due to Concentrated and Variously Distributed Loading, *Natl. Adv. Comm. Aeron., Tech. Note* 806, 1941.
20. Svensson, N. L.: The Bursting Pressure of Cylindrical and Spherical Vessels, *ASME J. Appl. Mech.*, vol. 25, no. 1, March, 1958.
21. Durelli, A. J., J. W. Dally, and S. Morse: Experimental Study of Thin-wall Pressure Vessels, *Proc. Soc. Exp. Stress Anal.*, vol. 18, no. 1, 1961.
22. Bjilaard, P. P.: Stresses from Local Loadings in Cylindrical Pressure Vessels, *Trans. ASME*, vol. 77, no. 6, August 1955 (also in Ref. 28).
23. Bjilaard, P. P.: Stresses from Radial Loads in Cylindrical Pressure Vessels, *Weld. J.*, vol. 33, December 1954 (also in Ref. 28).
24. Yuan, S. W., and L. Ting: On Radial Deflections of a Cylinder Subjected to Equal and Opposite Concentrated Radial Loads, *ASME J. Appl. Mech.*, vol. 24, no. 6, June 1957.
25. Ting, L., and S. W. Yuan: On Radial Deflection of a Cylinder of Finite Length with Various End Conditions, *J. Aeron. Sci.*, vol. 25, 1958.
26. Final Report, Purdue University Project, Design Division, Pressure Vessel Research Committee, Welding Research Council, 1952.
27. Wichman, K. R., A. G. Hopper, and J. L. Mershon: "Local Stresses in Spherical and Cylindrical Shells Due to External Loadings," *Weld. Res. Counc. Bull. No. 107*, August 1965.
28. von Kármán, Th., and Hsue-shen Tsien: Pressure Vessel and Piping Design, *ASME Collected Papers* 1927–1959.
29. Galletly, G. D.: Edge Influence Coefficients for Toroidal Shells of Positive; Also Negative Gaussian Curvature, *ASME J. Eng. Ind.*, vol. 82, February 1960.

30. Wenk, Edward, Jr., and C. E. Taylor: Analysis of Stresses at the Reinforced Intersection of Conical and Cylindrical Shells, *U.S. Dept. of the Navy, David W. Taylor Model Basin, Rep.* 826, March 1953.
31. Taylor, C. E., and E. Wenk, Jr.: Analysis of Stresses in the Conical Elements of Shell Structures, *Proc. 2d U.S. Natl. Congr. Appl. Mech.,* 1954.
32. Borg, M. F.: Observations of Stresses and Strains Near Intersections of Conical and Cylindrical Shells, *U.S. Dept. of the Navy, David W. Taylor Model Basin, Rept.* 911, March 1956.
33. Raetz, R. V., and J. G. Pulos: A Procedure for Computing Stresses in a Conical Shell Near Ring Stiffeners or Reinforced Intersections, *U.S. Dept of the Navy, David W. Taylor Model Basin, Rept.* 1015, April 1958.
34. Narduzzi, E. D., and Georges Welter: High-pressure Vessels Subjected to Static and Dynamic Loads, *Weld. J. Res. Suppl.,* 1954.
35. Dubuc, J., and Georges Welter: Investigation of Static and Fatigue Resistance of Model Pressure Vessels, *Weld. J. Res. Suppl.,* July 1956.
36. Kooistra, L. F., and M. M. Lemcoe: Low Cycle Fatigue Research on Full-size Pressure Vessels, *Weld. J.,* July, 1962.
37. Weil, N. A.: Bursting Pressure and Safety Factors for Thin-walled Vessels, *J. Franklin Inst.,* February 1958.
38. Brownell, L. E., and E. H. Young: "Process Equipment Design: Vessel Design," John Wiley & Sons, Inc., 1959.
39. Faupel, J. H.: Yield and Bursting Characteristics of Heavy-wall Cylinders, *Trans. ASME,* vol. 78, no. 5, 1956.
40. Dahl, N. C.: Toroidal-shell Expansion Joints, *ASME J. Appl. Mech.,* vol. 20, 1953.
41. Laupa, A., and N. A. Weil: Analysis of U-shaped Expansion Joints, *ASME J. Appl. Mech.,* vol. 29, no. 1, March 1962.
42. Baker, B. R., and G. B. Cline, Jr.: Influence Coefficients for Thin Smooth Shells of Revolution Subjected to Symmetric Loads, *ASME J. Appl. Mech.,* vol. 29, no. 2, June 1962.
43. Tsui, E. Y. W., and J. M. Massard: Bending Behavior of Toroidal Shells, *Proc. Am. Soc. Civil Eng., J. Eng. Mech. Div.,* vol. 94, no. 2, April 1968.
44. Kraus, H.: "Thin Elastic Shells," John Wiley & Sons, Inc., 1967.
45. Pflüger, A.: "Elementary Statics of Shells," 2d ed., McGraw-Hill Information Systems Company, 1961 (English transl. by E. Galantay).
46. Gerdeen J. C., and F. W. Niedenfuhr: Influence Numbers for Shallow Spherical Shells of Circular Ring Planform, *Proc. 8th Midwestern Mech. Conf., Development in Mechanics,* vol. 2, part 2, Pergamon Press, 1963.
47. Zaremba, W. A.: Elastic interactions at the Junction of an Assembly of Axi-symmetric Shells, *J. Mech. Eng. Sci.,* vol. 1, no. 3, 1959.
48. Johns, R. H., and T. W. Orange: Theoretical Elastic Stress Distributions Arising from Discontinuities and Edge Loads in Several Shell-type Structures, *NASA Tech. Rept.* R-103, 1961.
49. Stanek, F. J.: "Stress Analysis of Circular Plates and Cylindrical Shells," Dorrance & Co., Inc., 1970.
50. Blythe, W., and E. L. Kyser: A Flügge-Vlasov Theory of Torsion for Thin Conical Shells, *ASME J. Appl. Mech.,* vol. 31, no. 3, September 1964.
51. Payne, D. J.: Numerical Analysis of the Axi-symmetric Bending of a Toroidal Shell, *J. Mech. Eng. Sci.,* vol. 4, no. 4, 1962.
52. Rossettos, J. N., and J. L. Sanders Jr.: Toroidal Shells Under Internal Pressure in the Transition Range, *AIAA J.,* vol. 3, no. 10, October 1965.
53. Jordan, P. F.: Stiffness of Thin Pressurized Shells of Revolution, *AIAA J.,* vol. 3, no. 5, May 1965.

54. Haringx, J. A.: Instability of Thin-walled Cylinders Subjected to Internal Pressure, *Philips Res. Rept.* 7, 1952.
55. Haringx, J. A.: Instability of Bellows Subjected to Internal Pressure, *Philips Res. Rept.* 7, 1952.
56. Seide, Paul: The Effect of Pressure on the Bending Characteristics of an Actuator System, *ASME J. Appl. Mech.,* vol. 27, no. 3, September 1960.
57. Chou, Seh-Ieh, and M. W. Johnson, Jr.: On the Finite Deformation of an Elastic Toroidal Membrane, *Proc. 10th Midwestern Mech. Conf.,* 1967.
58. Tsui, E. Y. W.: Analysis of Tapered Conical Shells, *Proc. 4th U.S. Natl. Congr. Appl. Mech.,* 1962.
59. Fischer, L.: "Theory and Practice of Shell Structures," Wilhelm Ernst & Sohn KG, 1968.
60. Pao, Yen-Ching: Influence Coefficients of Short Circular Cylindrical Shells with Varying Wall Thickness, *AIAA J.,* vol. 6, no. 8, August 1968.
61. Turner, C. E.: Study of the Symmetrical Elastic Loading of Some Shells of Revolution, with Special Reference to Toroidal Elements, *J. Mech. Eng. Sci.,* vol. 1, no. 2, 1959.
62. Perrone, N.: Compendium of Structural Mechanics Computer Programs, *Computers and Structures,* vol. 2, no. 3, April 1972. (Also available as *NTIS paper* N71-32026, April 1971.)
63. Bushnell, D.: Stress, Stability, and Vibration of Complex, Branched Shells of Revolution, *AIAA/ASME/SAE 14th Structures, Struct. Dynam. & Mater. Conf.,* Williamsburg, Va., March, 1973.
64. Baltrukonis, J. H.: Influence Coefficients for Edge-Loaded Short, Thin, Conical Frustums, *ASME J. Appl. Mech.,* vol. 26, no. 2, June 1959.
65. Taylor, C. E.: Simplification of the Analysis of Stress in Conical Shells, *Univ. Ill., TAM Rept.* 385, April 1974.

THIRTEEN

Bodies under Direct Bearing and Shear Stress

13.1 *Stress due to pressure between elastic bodies*

The stresses caused by the pressure between elastic bodies are of importance in connection with the design or investigation of ball and roller bearings, trunnions, expansion rollers, track stresses, etc. Hertz (Ref. 1) developed the mathematical theory for the surface stresses and deformations produced by pressure between curved bodies, and the results of his analysis are supported by experiment. Formulas based on this theory give the maximum compressive stresses, which occur at the center of the surfaces of contact, but not the maximum shear stresses, which occur in the interiors of the compressed parts, nor the maximum tensile stress, which occurs at the boundary of the contact area and is normal thereto.

Both surface and subsurface stresses were studied by Belajef (Refs. 28 and 29), and some of his results are cited in Ref. 6. A tabulated summary of surface and subsurface stresses, greatly facilitating calculation, is given in Ref. 33. For a cylinder on a plane and for crossed cylinders Thomas and Hoersch (Ref. 2) investigated mathematically surface compression and internal shear and checked the calculated value of the latter experimentally. The stresses due to the pressure of a sphere on a plate (Ref. 3) and of a cylinder on a plate (Ref. 4) have also been investigated by photoelasticity. The deformation and contact area for a ball in a race were measured by Whittemore and Petrenko (Ref. 8) and compared with the theoretical values. Some recent investigations have considered the influence of tangential combined with normal loading (Refs. 35 and 47 to 49).

In Table 33 are given formulas for the elastic stress and deformation produced by pressure between bodies of various forms, and for the dimensions of the circular, elliptical, or rectangular area of contact formed by the compressed surfaces. Except where otherwise indicated, these equations are based on Hertz's theory, which assumes the length of the cylinder and dimensions of the plate to be infinite. For a very short cylinder and for a plate having a width less than five or six times that of the contact area or a thickness less than five or six times the depth to the point of maximum shear stress, the actual stresses may vary considerably from the values indicated by the theory (see Refs. 4, 45, and 50). Tu (Ref. 50) discusses the stresses and deformations for a plate pressed between two identical elastic spheres with no friction; graphs are also presented. Pu and Hussain (Ref. 51) consider the unbonded contact between a flat plate and an elastic half-space when a normal load is applied to the plate. Graphs of the contact radii are presented for a concentrated load and two distributed loadings on circular areas.

Hertz (Ref. 1) based his work on the assumption that the contact area was small compared with the radius of the ball or cylinder; Goodman and Keer (Ref. 52) compare the work of Hertz with a solution which permits the contact area to be larger, such as the case when the negative radius of one surface is only very slightly larger (1.01 to 1) than the positive radius of the other. Cooper (Ref. 53) presents some reformulated hertzian coefficients in a more easily interpolated form and also points out some numerical errors in the coefficients originally published by Hertz. Dundurs and Stippes (Ref. 54) discuss the effect of Poisson's ratio on contact stress problems.

The use of the formulas of Table 33 is illustrated in the example at the end of this article. The general formula for case 4 can be used, as in the example, for any contact-stress problems involving any geometrically regular bodies except parallel cylinders, but for bearing calculations use should be made of charts such as those given in Refs. 33 and 34, which not only greatly facilitate calculations but provide for influences not taken into account in the formulas.

Because of the very small area involved in what initially approximates a point or line contact, contact stresses for even light loads are very high; but as the formulas show, the stresses do not increase in proportion to the loading. Furthermore, because of the facts that the stress is highly localized and triaxial, the actual stress intensity can be very high without producing apparent damage. In order to make use of the Hertz formulas for purposes of design or safe-load determination, it is necessary to know the relationship between theoretical stresses and likelihood of failure, whether from excessive deformation or fracture. In discussing this relationship, it is convenient to refer to the computed stress as the Hertz stress, whether the elastic range has been exceeded or not. Some of the available information showing the Hertz stress corresponding to loadings found to be safe and to loadings that produced excessive deformations or fracture may be summarized as follows.

Static or near static conditions.

Cylinder. The American Railway Engineering Association gives as the allowable loading for a steel cylinder on a flat steel plate the formulas

$$p = \frac{\sigma_y - 13{,}000}{20{,}000} 600d \qquad \text{for } d < 25 \text{ in}$$

and

$$p = \frac{\sigma_y - 13{,}000}{20{,}000} 3000\sqrt{d} \qquad \text{for } 25 < d < 125 \text{ in}$$

Here (and in subsequent equations) p is the load per linear inch in pounds, d is the diameter of the cylinder in inches, and σ_y is the tensile yield point of the steel in the roller or plate, whichever is lower. If σ_y is taken as 32,000 lb/in^2, the Hertz stress corresponding to this loading is constant at 76,200 lb/in^2 for any diameter up to 25 in and decreases as $d^{-\frac{1}{4}}$ to 50,900 at $d = 125$ in.

Wilson (Refs. 7, 11, and 32) carried out several series of static and slow-rolling tests on large rollers. From static tests on rollers of medium-grade cast steel having diameters of 120 to 720 in, he concluded that the load per linear inch required to produce appreciable permanent set could be represented by the empirical formula $p = 500 + 110d$, provided the bearing plates were 3 in thick or more. He found that p increased with the axial length of the roller up to a length of 6 in, after which it remained practically constant (Ref. 32). Slow-rolling tests (Ref. 11) undertaken to determine the load required to produce a permanent elongation or spread of 0.001 in/in in the bearing plate led to the empirical formula

$$p = (18{,}000 + 120d)\frac{\sigma_y - 13{,}000}{23{,}000}$$

for rollers with $d > 120$ in. Wilson's tests indicated that the average pressure on the area of contact required to produce set was greater for small rollers than for large rollers, and that there was little difference in bearing capacity under static and slow-rolling conditions, though the latter showed more tendency to produce surface deterioration.

Jensen (Ref. 4), making use of Wilson's test results and taking into account the three-dimensional aspect of the problem, proposed for the load-producing set the formula

$$p = 1 + \frac{1.78}{1 + d^2/800L^2} \frac{\sigma_y^2 d}{E}$$

where L is the length of the cylinder in inches and E is the modulus of elasticity in pounds per square inch. For values of the ratio d/L from 0.1 to 10, the corresponding Hertz stress ranges from $1.66\sigma_y$ to $1.72\sigma_y$.

Whittemore (Ref. 8) found that the elastic limit load for a flexible roller

TABLE 33 ***Formulas for stress and strain due to pressure on or between elastic bodies***

NOTATION: P = total load (pounds); p = load per unit length (pounds per linear inch); a = radius of circular contact area for case 1; b = width of rectangular contact area for case 2; c = major semiaxis and d = minor semiaxis of elliptical contact area for cases 3 and 4; y = relative motion of approach along the axis of loading of two points, one in each of the two contacting bodies, remote from the contact zone; ν = Poisson's ratio; E = modulus of elasticity. All dimensions are in inches, and all forces are in pounds. Subscripts 1 and 2 refer to bodies 1 and 2, respectively. To simplify expressions let

$$C_E = \frac{1-\nu_1^2}{E_1} + \frac{1-\nu_2^2}{E_2}$$

Conditions and case no.	Formulas	
1. Sphere 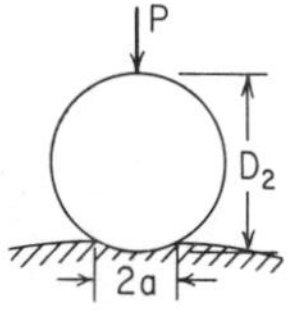	$a = 0.721\sqrt[3]{PK_DC_E}$ $\text{Max } \sigma_c = 1.5\dfrac{P}{\pi a^2}$ $= 0.918\sqrt[3]{\dfrac{P}{K_D^2C_E^2}}$ $y = 1.040\sqrt[3]{\dfrac{P^2C_E^2}{K_D}}$ If $E_1 = E_2 = E$ and $\nu_1 = \nu_2 = 0.3$, then $a = 0.881\sqrt[3]{\dfrac{PK_D}{E}}$ $\text{Max } \sigma_c = 0.616\sqrt[3]{\dfrac{PE^2}{K_D^2}}$ $y = 1.55\sqrt[3]{\dfrac{P^2}{E^2K_D}}$ Max $\sigma_t \approx 0.133$ (max σ_c) radially at the edge of contact area Max $\tau \approx \frac{1}{3}$ (max σ_c) at a point on the load line a distance $a/2$ below the contact surface (Approximate stresses from Refs. 3 and 6)	1a. Sphere on a flat plate  $K_D = D_2$ 1b. Sphere on a sphere 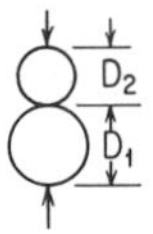 $K_D = \dfrac{D_1D_2}{D_1 + D_2}$ 1c. Sphere in a spherical socket 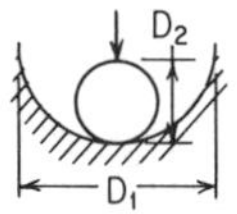 $K_D = \dfrac{D_1D_2}{D_1 - D_2}$

2. Cylinder of length L large as compared with D; p = load per linear in $= P/L$

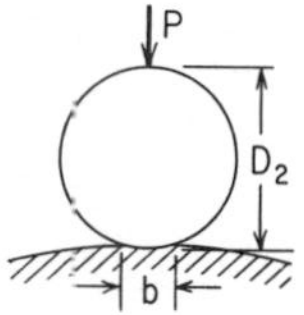

$$b = 1.60\sqrt{pK_DC_E}$$

$$\text{Max } \sigma_c = 0.798\sqrt{\frac{p}{K_DC_E}}$$

If $E_1 = E_2 = E$ and $\nu_1 = \nu_2 = 0.3$, then

$$b = 2.15\sqrt{\frac{pK_D}{E}}$$

$$\text{Max } \sigma_c = 0.591\sqrt{\frac{pE}{K_D}}$$

For a cylinder between two flat plates

$$\Delta D_2 = \frac{4p(1-\nu^2)}{\pi E}\left(\frac{1}{3} + \ln\frac{2D}{b}\right) \qquad \text{Ref. 5}$$

For two cylinders the distance between centers is reduced by

$$\frac{2p(1-\nu^2)}{\pi E}\left(\frac{2}{3} + \ln\frac{2D_1}{b} + \ln\frac{2D_2}{b}\right) \qquad \text{Ref. 31}$$

2a. Cylinder on a flat plate

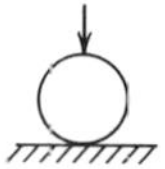

$K_D = D_2$

Max $\tau \approx \frac{1}{3}(\text{max } \sigma_c)$ at a depth of $0.4b$ below the surface of the plane

2b. Cylinder on a cylinder

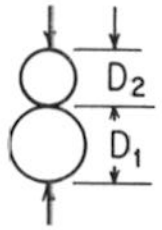

$$K_D = \frac{D_1D_2}{D_1 + D_2}$$

2c. Cylinder in a cylindrical socket

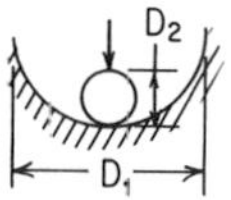

$$K_D = \frac{D_1D_2}{D_1 - D_2}$$

3. Cylinder on a cylinder; axes at right angles

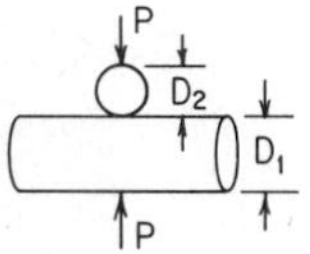

$$c = \alpha\sqrt[3]{PK_DC_E}$$

$$d = \beta\sqrt[3]{PK_DC_E}$$

$$\text{Max } \sigma_c = \frac{1.5P}{\pi cd}$$

$$y = \lambda\sqrt[3]{\frac{P^2C_E^2}{K_D}}$$

Max $\tau \approx \frac{1}{3}(\text{max } \sigma_c)$

$K_D = \dfrac{D_1D_2}{D_1 + D_2}$ and α, β, and λ depend upon $\dfrac{D_1}{D_2}$ as shown

D_1/D_2	1	1.5	2	3	4	6	10
α	0.908	1.045	1.158	1.350	1.505	1.767	2.175
β	0.908	0.799	0.732	0.651	0.602	0.544	0.481
λ	0.825	0.818	0.804	0.774	0.747	0.702	0.641

TABLE 33 ***Formulas for stress and strain due to pressure on or between elastic bodies (Cont.)***

Conditions and case no.	Formulas

4. General case of two bodies in contact; P = total load

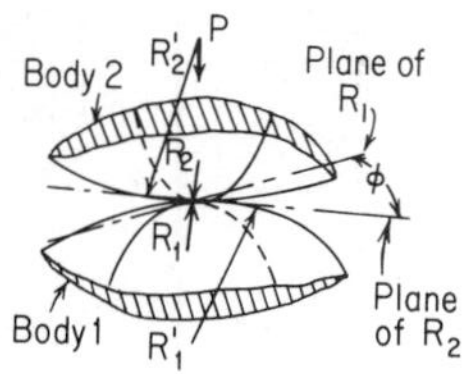

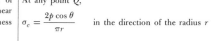

At point of contact minimum and maximum radii of curvature are R_1 and R'_1 for body 1, and R_2 and R'_2 for body 2. Then $1/R_1$ and $1/R'_1$ are principal curvatures of body 1, and $1/R_2$ and $1/R'_2$ of body 2; and in each body the principal curvatures are mutually perpendicular. The radii are positive if the center of curvature lies within the given body, i.e., the surface is convex, and negative otherwise. The plane containing curvature $1/R_1$ in body 1 makes with the plane containing curvature $1/R_2$ in body 2 the angle ϕ. Then:

$$c = \alpha \sqrt[3]{PK_DC_E} \qquad d = \beta \sqrt[3]{PK_DC_E} \qquad \max \sigma_c = \frac{1.5P}{\pi cd} \qquad \text{and} \qquad y = \lambda \sqrt[3]{\frac{P^2C_E^2}{K_D}} \quad \text{where } K_D = \frac{1.5}{1/R_1 + 1/R_2 + 1/R'_1 + 1/R'_2}$$

and α, β, and λ are given by the following table in which

$$\cos\theta = \frac{K_D}{1.5}\sqrt{\left(\frac{1}{R_1} - \frac{1}{R'_1}\right)^2 + \left(\frac{1}{R_2} - \frac{1}{R'_2}\right)^2 + 2\left(\frac{1}{R_1} - \frac{1}{R'_1}\right)\left(\frac{1}{R_2} - \frac{1}{R'_2}\right)\cos 2\phi}$$

$\cos\theta$	0.00	0.10	0.20	0.30	0.40	0.50	0.60	0.70	0.75	0.80	0.85	0.90	0.92	0.94	0.96	0.98	0.99
α	1.000	1.070	1.150	1.242	1.351	1.486	1.661	1.905	2.072	2.292	2.600	3.093	3.396	3.824	4.508	5.937	7.774
β	1.000	0.936	0.878	0.822	0.769	0.717	0.664	0.608	0.578	0.544	0.507	0.461	0.438	0.412	0.378	0.328	0.287
λ	0.750	0.748	0.743	0.734	0.721	0.703	0.678	0.644	0.622	0.594	0.559	0.510	0.484	0.452	0.410	0.345	0.288

(Ref. 8)

5. Rigid knife edge across edge of semi-infinite plate; load p lb/linear in $= P/t$ where t is plate thickness

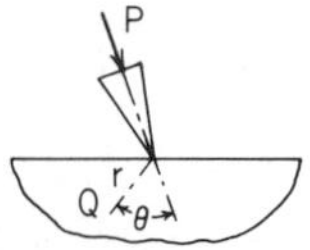

At any point Q,

$$\sigma_c = \frac{2p\cos\theta}{\pi r} \qquad \text{in the direction of the radius } r$$

(Ref. 6)

6. Rigid block of width $2b$ across edge of semi-infinite plate; load p lb/linear in $= P/t$ where t is plate thickness

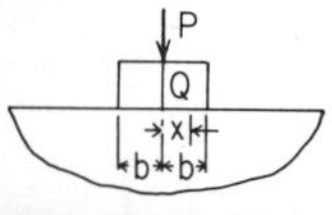

At any point Q on surface of contact,

$$\sigma_c = \frac{p}{\pi\sqrt{b^2 - x^2}}$$

(For loading on block of finite width and influence of distance of load from corner see Ref. 45)

(Ref. 6)

7. Uniform pressure q lb/in^2 over length L across edge of semi-infinite plate 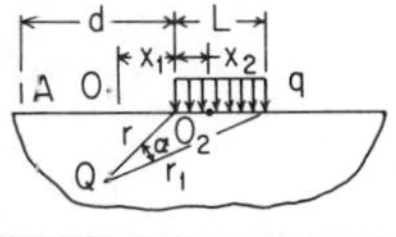	At any point O_1 outside loaded area, $y = \frac{2q}{\pi E}\left[(L + x_1)\ln\frac{d}{L + x_1} - x_1 \ln\frac{d}{x_1}\right] + qL\frac{1-\nu}{\pi E}$ At any point O_2 inside loaded area, $y = \frac{2q}{\pi E}\left[(L - x_2)\ln\frac{d}{L - x_2} + x_2 \ln\frac{d}{x_2}\right] + qL\frac{1-\nu}{\pi E}$ Where y = deflection relative to a remote point A a distance d from edge of loaded area At any point Q, $\sigma_c = 0.318q\ (\alpha + \sin\alpha)$ $\tau = 0.318q \sin\alpha$ (Ref. 6)
8. Rigid cylindrical die of radius R on surface of semi-infinite body; total load P lb 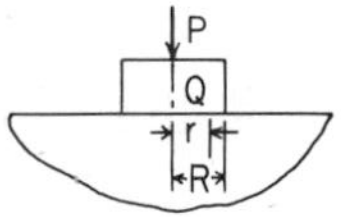	$y = \frac{P(1-\nu^2)}{2RE}$ At any point Q on surface of contact $\sigma_c = \frac{P}{2\pi R\sqrt{R^2 - r^2}}$ Max $\sigma_c = \infty$ at edge (theoretically) Min $\sigma_c = \frac{P}{2\pi R^2}$ at center (Ref. 6)
9. Uniform pressure q lb/in^2 over circular area of radius R on surface of semi-infinite body 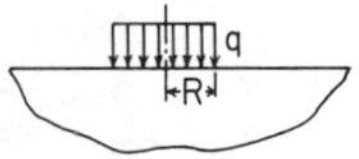	Max $y = \frac{2qR(1-\nu^2)}{E}$ at center y at edge $= \frac{4qR(1-\nu^2)}{\pi E}$ Max $\tau = 0.33q$ at point $0.638R$ below center of loaded area
10. Uniform pressure q lb/in^2 over square area of sides $2b$ on the surface of semi-infinite body 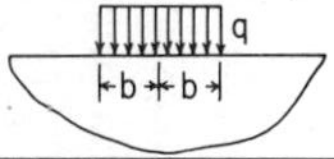	Max $y = \frac{2.24qb(1-\nu^2)}{E}$ at center $y = \frac{1.12qb(1-\nu^2)}{E}$ at corners Average $y = \frac{1.90qb(1-\nu^2)}{E}$ (Ref. 6)

of hardened steel (tensile strength about 265,000 lb/in^2) tested between slightly hardened races corresponded to a Hertz stress of about 436,000 lb/in^2. The roller failed before the races.

Sphere. Tests reported in Whittemore and Petrenko (Ref. 8) gave, for balls 1, $1\frac{1}{4}$, and $1\frac{1}{2}$ in in diameter, tested between straight races, Hertz stresses of 239,000, 232,000, and 212,000 lb/in^2, respectively, at loads producing a permanent strain of 0.0001. The balls were of steel having sclerescope hardness of 60 to 68, and the races were of approximately the same hardness. The critical strain usually occurred first in the races.

From the results of crushing tests of a sphere between two similar spheres, SKF derived the empirical formula $P = 1960(8d)^{1.75}$, where P is the crushing load in pounds and d is the diameter of the sphere in inches. The test spheres were made of steel believed to be of hardness 64 to 66 Rockwell C, and the formula corresponds to a Hertz stress of about $4{,}000{,}000 \times d^{-\frac{1}{12}}$.

Knife edge. Knife-edge pivots are widely used in scales and balances, and if accuracy is to be maintained, the bearing loads must not cause excessive deformation. It is impossible for a truly sharp edge to bear against a flat plane without suffering plastic deformation, and so pivots are not designed on the supposition that the contact stresses will be elastic; instead, the maximum load per inch consistent with the requisite degree of accuracy in weighing is determined by experience or by testing. In Wilson et al. (Ref. 9), the National Bureau of Standards is quoted as recommending that for heavy service the load per linear inch should not exceed 5000 lb/in for high-carbon steels or 6000 for special alloy steels; for light service the values can be increased to 6000 and 7000, respectively. In the tests described in Ref. 9, the maximum load that could be sustained without damage—the so-called *critical load*—was defined as the load per linear inch that produced an increase in the edge width of 0.0005 in or a sudden increase in the load rate of vertical deformation. The two methods gave about the same results when the bearing was harder than the pivot, as it should be for good operation. The conclusions drawn from the reported tests may be summarized as follows.

The bearing value of a knife-edge or pivot varies approximately with the wedge angle for angles of 30 to 120°, the bearing value of a flat pivot varies approximately with the width of the edge for widths of 0.004 to 0.04 in, and the bearing value of pivots increases with the hardness for variations in hardness of 45 to 60 on the Rockwell C scale. Successive applications of a load less than the critical load will cause no plastic flow; the edge of a pivot originally sharp will increase in width with the load, but no further plastic deformation is produced by successive applications of the same or smaller loads. The application of a load greater than the critical load will widen the edge at the first application, but additional applications of the same load will not cause additional flow; the average unit pressure on 90° pivots having a hardness represented by Rockwell C numbers of 50 to 60 is about 400,000 to 500,000 lb/in^2 at the critical load. This critical unit pressure appears to be independent of the width of the edge but increases with the pivot angle and the hardness of the material (Ref. 9).

These tests and the quoted recommendations relate to applications involving heavy loads (thousands of pounds) and reasonable accuracy. For light loads and extreme accuracy, as in analytical balances, the pressures are limited to much smaller values. Thus, in Ref. 39, on the assumptions that an originally sharp edge indents the bearing and that the common surface becomes cylindrical, it is stated that the radius of the loaded edge must not exceed 0.25 μ (approximately 0.00001 in) if satisfactory accuracy is to be attained, and that the corresponding loading would be about 35,000 lb/in^2 of contact area.

Dynamic Conditions. If the motion involved is a true rolling motion without any slip, then under conditions of slow motion (expansion rollers, bascules, etc.) the stress conditions are comparable with those produced by static loading. This is indicated by a comparison of the conclusions reached in Ref. 7, where the conditions are truly static, with those reached in Ref. 11, where there is a slow-rolling action. If there is even a slight amount of slip, however, the conditions are very much more severe and failure is likely to occur through mechanical wear. The only guide to proper design against wear is real or simulated service testing (Refs. 24 and 41).

When the motion involved is at high speed and produces cyclic loading, as in ball and roller bearings, fatigue is an important consideration. A great many tests have been made to determine the fatigue properties of bearings, especially ball bearings, and such tests have been carried out to as many as 1 billion cycles and with Hertz stresses up to 750,000 lb/in^2 (Ref. 37). The number of cycles to damage (either spalling or excessive deformation) has been found to be inversely proportional to the cube of the load for point contact (balls) and to the fourth power for line contact; this would be inversely proportional to the ninth and eighth powers, respectively, of the Hertz stress. Styri (Ref. 40) found the cycles to failure to vary as the ninth power of the Hertz stress and was unable to establish a true endurance limit. Some of these tests show that ball bearings can run for a great number of cycles at very high stresses; for example, $\frac{1}{2}$-in balls of SAE 52,100 steel (RC 63 to 64) withstood 17,500,000 cycles at a stress of 174,000 lb/in^2 before 10 percent failures occurred, and withstood 700,000,000 cycles at that stress before 90 percent failures occurred.

One difficulty in correlating different tests on bearings is the difference in criteria for judging damage; some experimenters have defined failure as a certain permanent deformation, others as visible surface damage through spalling. Palmgren (Ref. 36) states that a permanent deformation at any one contact point of rolling element and bearing ring combined equal to 0.0001 times the diameter of the rolling element has no significant influence on the functioning of the bearing. In the tests of Ref. 37, spalling of the surface was taken as the sign of failure; this spalling generally originated on planes of maximum shear stress below the surface.

It is apparent from the foregoing discussion that the practical design of parts that sustain direct bearing must be based largely on experience since this alone affords a guide as to whether, at any given load and number of stress cycles, there is enough deformation or surface damage to interfere with proper functioning. The rated capacities of bearings and gears are furnished by the manufacturers, with proper allowance indicated for the conditions of service and recommendations as to proper lubrication. Valid and helpful conclusions, however, can often be drawn from a comparison of service records with calculated stresses.

EXAMPLE

A ball 1.50 in in diameter, in a race which has a diameter of 10 in and a groove radius of 0.80 in, is subjected to a load of 2000 lb. It is required to find the dimensions of the contact area, the combined deformation of ball and race at the contact, and the maximum compressive stress.

Solution. The formulas and table of case 4 (Table 33) are used. The race is taken as body 1 and the ball as body 2; hence $R_1 = -0.80$, $R_1' = -5$, and $R_2 = R_2' = 0.75$ in. Taking $E_1 = E_2 = 30{,}000{,}000$ and $\nu_1 = \nu_2 = 0.3$, we have

$$C_E = \frac{2(1 - 0.09)}{30(10^6)} = 6.067(10^{-8})$$

$$K_D = \frac{1.5}{-1.25 + 1.33 - 0.20 + 1.33} = 1.233$$

$$\cos\theta = \frac{1.233}{1.5}\sqrt{(-1.25 + 0.20)^2 + 0 + 0} = 0.863$$

From the table, by interpolation,

$$\alpha = 2.634 \qquad \beta = 0.495 \qquad \lambda = 0.546$$

Then

$$c = 2.634\sqrt[3]{2000(1.233)(6.067)(10^{-8})} = 0.140 \text{ in}$$

$$d = 0.495\sqrt[3]{2000(1.233)(6.067)(10^{-8})} = 0.0263 \text{ in}$$

$$\text{Max } \sigma_c = \frac{1.5(2000)}{0.140(0.0263)} = 259{,}000 \text{ lb/in}^2$$

$$y = 0.546\sqrt[3]{\frac{2000^2(6.067^2)(10^{-16})}{1.233}} = 0.00125 \text{ in}$$

Therefore the contact area is an ellipse with a major axis of 0.280 in and a minor axis of 0.0526 in.

13.2 *Rivets and riveted joints*

Although the actual state of stress in a riveted joint is complex, it is customary—and experience shows it is permissible—to ignore such considerations as stress concentration at the edges of rivet holes, unequal division of load among rivets, and nonuniform distribution of shear stress across the section of the rivet and of the bearing stress between rivet and plate. Simplifying assumptions are made, which may be summarized as follows: (1) The applied load is assumed to be transmitted entirely by the rivets, friction between the connected plates being ignored; (2) when the center of gravity of the rivets is on the line of action of the load, all the rivets of the joint are assumed to carry equal parts of the load if they are of the same size, or to be loaded proportionally to their respective section areas if they are of different sizes; (3) the shear stress is assumed to be uniformly distributed across the rivet section; (4) the bearing stress between plate and rivet is assumed to be uniformly distributed over an area equal to the rivet diameter times the plate thickness; (5) the stress in a tension member is assumed to be uniformly distributed over the net area; and (6) the stress in a compression member is assumed to be uniformly distributed over the gross area.

The design of riveted joints on the basis of these assumptions is the accepted practice, although none of them is strictly correct and methods of stress calculation that are supposedly more accurate have been proposed (Ref. 12).

Details of design and limitations. The possibility of secondary failure duc to secondary causes, such as the shearing or tearing out of a plate between the rivet and the edge of a plate or between adjacent rivets, the bending or insufficient upsetting of long rivets, or tensile failure along a zigzag line when rivets are staggered, is guarded against in standard specifications (Ref. 13) by detailed rulcs for edge clearance, maximum grip of rivets, maximum pitch, and computing the net width of riveted parts. Provision is made for the use of high-strength bolts in place of rivets under certain circumstances (Ref. 42). Joints may be made by welding instead of riveting, but the use of welding in conjunction with riveting is not approved on new work; the division of the load as between the welds and the rivets would be indeterminate.

Tests on riveted joints. In general, tests on riveted joints show that although under working loads the stress conditions may be considerably at variance with the usual assumptions, the ultimate strength may be closely predicted by calculations based thereon. Some of the other conclusions drawn from such tests may be summarized as follows.

In either lap or double-strap butt joints in very wide plates, the unit tensile strength developed by the net section is greater than that developed by the plate itself when tested in full width and is practically equal to that developed by narrow tension specimens cut from the plate. The rivets in lap joints are as strong relative to undriven rivets tested in shear as are the rivets in butt joints. Lap joints bend sufficiently at stresses below the usual design stresses to cause opening of caulked joints (Ref. 14).

Although it is frequently specified that rivets shall not be used in tension, tests show that hot-driven buttonhead rivets develop a strength in direct tension greater than the strength of the rod from which they are made, and that they may be relied upon to develop this strength in every instance. Although the initial tension in such rivets due to cooling usually amounts to 70 percent or more of the yield strength, this initial tension does not reduce the ability of the rivets to resist an applied tensile load (see also Art. 2.12). Unless a joint is subjected to reversals of primary load, the use of rivets in tension appears to be justified; but when the primary load producing shear in the rivets is reversed, the reduction in friction due to simultaneous rivet tension may permit slip to occur, with possible deleterious effects (Ref. 15).

With respect to the form of the rivet head, the rounded or button-head type is standard; but countersunk rivets are often used, and tests show that these develop the same ultimate strength, although they permit much more slip and deformation at working loads than do the buttonhead rivets (Ref. 16).

In designing riveted joints in very thin metals, especially the light alloys, it may be necessary to take into account factors that are not usually considered in ordinary structural-steel work, such as the radial stresses caused at the hole edges by closing pressure and the buckling of the plates under rivet pressure (Ref. 17).

Eccentric loading. When the rivets of a joint are so arranged that the center of gravity G of the group lies, not on the line of action of the load, but at a distance e therefrom, the load P can be replaced by an equal and parallel load P' acting through G and a couple Pe. The load on any one of the n rivets is then found by *vectorially* adding the load P/n due to P' and the load Q due to the couple Pe. This load Q acts normal to the line from G to the rivet and is given by the equation $Q = PeA_1r_1/J$, where A_1 is the area of the rivet in question, r_1 is its distance from G, and $J = \Sigma Ar^2$ for all the rivets of the group. When all rivets are of the same size, as is usually the case, the formula becomes $Q = Per_1/\Sigma r^2$. Charts and tables are available which greatly facilitate the labor of the calculation involved, and which make possible direct design of the joint without recourse to trial and error (Ref. 18). (The direct procedure, as outlined previously, is illustrated in the following example.)

The stiffness or resistance to angular displacement of a riveted joint determines the degree of fixity that should be assumed in the analysis of beams with riveted ends or of rectangular frames. Tests (Ref. 19) have shown that although joints made with wide gusset plates are practically rigid, joints made by simply riveting through clip angles are not even approximately so. A method of calculating the elastic constraint afforded by riveted joints of different types, based on an extensive series of tests, has been proposed by Rathbun (Ref. 20). Brombolich (Ref. 55) describes the use of a finite-element-analysis procedure to determine the effect of yielding, interference fits, and load sequencing on the stresses near fastener holes.

EXAMPLE

Figure 13.1 represents a lap joint in which three 1-in rivets are used to connect a 15-in channel to a plate. The channel is loaded eccentrically as shown. It is required to determine the maximum shear stress in the rivets. (This is not, of course, a properly designed joint intended to develop the full strength of the channel. It represents a simple arrangement of rivets assumed for the purpose of illustrating the calculation of rivet stress due to a moment.)

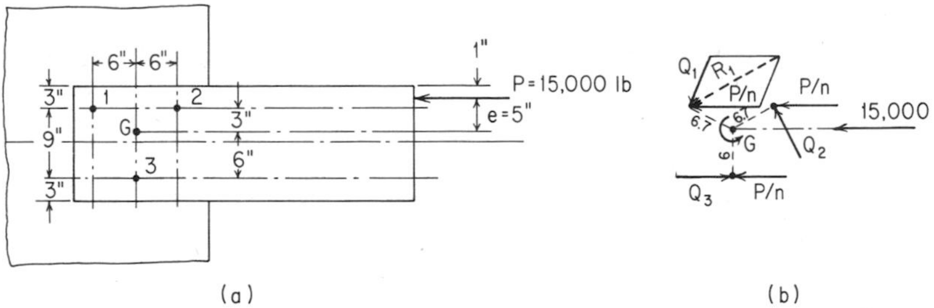

Fig. 13.1

Solution. The center of gravity of the rivet group is found to be at G. The applied load is replaced by an equal load through G and a couple equal to

$$15{,}000 \times 5 = 75{,}000 \text{ in-lb}$$

as shown in Fig. 13.1*b*. The distances r_1, r_2, and r_3 of rivets 1, 2, and 3, respectively, from G are as shown; the value of Σr^2 is 126. The loads on the rivets due to the couple of 75,000 lb are therefore

$$Q_1 = Q_2 = \frac{(75{,}000)(6.7)}{126} = 3990 \text{ lb}$$

$$Q_3 = \frac{(75{,}000)(6)}{126} = 3570 \text{ lb}$$

These loads act on the rivets in the directions shown. In addition, each rivet is subjected to a load in the direction of P' of $P/n = 5000$ lb. The resultant load on each rivet is then found by graphically (or algebraically) solving for the resultant of Q and P/n as shown. These resultant loads are $R_1 = R_2 = 8750$ lb; $R_3 = 1430$ lb. The maximum shear stress occurs in rivets 1 and 2 and is $\tau = 8750/0.785 = 11{,}150$ lb/in^2.

13.3 *Miscellaneous cases*

In most instances, the stress in bodies subjected to direct shear or pressure is calculated on the basis of simplifying assumptions such as are made in analyzing a riveted joint. Design is based on rules justified by experience rather than exact theory, and a full discussion does not properly come within the scope of this book. However, a brief consideration of a number of cases is given here; a more complete treatment of these cases may be found in books on machine and structural design and in the references cited.

Pins and bolts. These are designed on the basis of shear and bearing stress calculated in the same way as for rivets. In the case of pins bearing on wood, the allowable bearing stress must be reduced to provide for nonuniformity of pressure when the length of bolt is more than five or six times its diameter. When the pressure is inclined to the grain, the safe load is found by the formula

$$N = \frac{PQ}{P\sin^2\theta + Q\cos^2\theta}$$

where N is the safe load for the case in question, P is the safe load applied parallel to the grain, Q is the safe load applied transverse to the grain, and θ is the angle N makes with the direction of the grain (Ref. 21).

Hollow pins and rollers are usually much too thick-walled to be analyzed as circular rings by the formulas of Table 17. The loading is essentially as

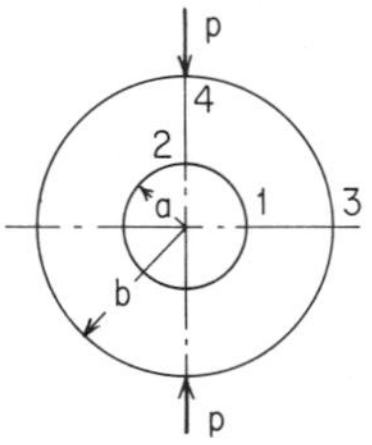

Fig. 13.2

shown in Figure 13.2, and the greatest circumferential stresses, which occur at points 1 to 4, may be found by the formula

$$\sigma = K\frac{2p}{\pi b}$$

where the numerical coefficient K depends on the ratio a/b and has the following values [a plus sign for K indicates tensile stress and a minus sign compressive stress (Ref. 30)]:

Point	a/b							
	0	0.1	0.2	0.3	0.4	0.5	0.6	0.7
1	−5.0	−5.05	−5.30	−5.80	−7.00	−9.00	−12.9	−21.4
2	+3.0	+3.30	+3.80	+4.90	+7.00	+10.1	+16.0	+31.0
3	0	+0.06	+0.20	+1.0	+1.60	+3.0	+5.8	+13.1
4	+0.5	+0.40	0	−0.50	−1.60	−3.8	−8.4	−19.0

Gear teeth. Gear teeth may be investigated by considering the tooth as a cantilever beam, the critical stress being the tensile bending stress at the base. This stress can be calculated by the modified Heywood formula for a very short cantilever beam (Art. 7.10) or by a combination of the modified Lewis formula and stress concentration factor given for case 21 in Table 37 (see also Refs. 22 to 24). The allowable stress is reduced according to speed of operation by one of several empirical formulas (Ref. 24). Under certain conditions, the bearing stress between teeth may become important (especially as this stress affects wear), and this stress may be calculated by the formula for case 2b, Table 33. The total deformation of the tooth, the result of direct compression at the point of contact and of beam deflection and shear, may be calculated by the formula of case 2b and the methods of Art. 7.1 (Ref. 23).

Keys. Keys are designed for a total shearing force $F = T/r$ (Fig. 13.3), where T represents the torque transmitted. The shear stress is assumed to be uniformly distributed over the horizontal section AB, and the bearing stress is assumed to be uniformly distributed over half the face. These assumptions lead to the following formulas: $\tau = F/Lb$; $\sigma_b = 2F/tL$ on the sides; and $\sigma_b = 2Ft/b^2L$ on top and bottom. Here L is the length of the key; in conventional design $4b < L < 16b$. As usually made, $b \geq t$; hence the bearing stress on the sides is greater than that on the top and bottom.

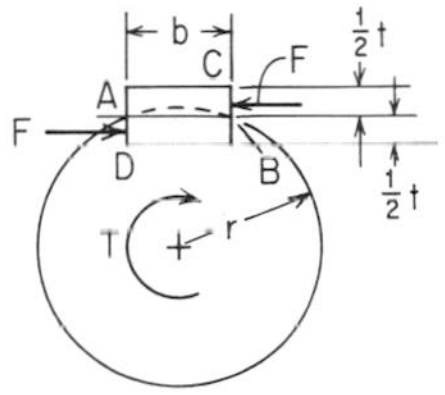

Fig. 13.3

Photoelastic analysis of the stresses in square keys shows that the shear stress is not uniform across the breadth b but is greatest at A and B, where it may reach a value of from two to four times the average value (Ref. 25). Undoubtedly the shear stress also varies in intensity along the length of the key. The bearing stresses on the surfaces of the key are also nonuniform, that on the sides being greatest near the common surface of shaft and hub, and that on the top and bottom being greatest near the corners C and D. When conservative working stresses are used, however, and the proportions of the key are such as have been found satisfactory in practice, the approximate methods of stress calculation that have been indicated result in satisfactory design.

Fillet welds. These are successfully designed on the basis of uniform distribution of shear stress on the longitudinal section of least area, although analysis and tests show that there is considerable variation in the intensity of shear stress along the length of the fillet (Refs. 26 and 27). (Detailed recommendations for the design of welded structural joints are given in Ref. 13.)

Screwthreads. The strength of screwthreads is of great importance in the design of joints, where the load is transferred to a bolt or stud by a nut. A major consideration is the load distribution. The load is not transferred uniformly along the engaged thread length; both mathematical analysis and tests show that the maximum load per linear inch of thread, which occurs near the loaded face of the nut, is several times as great as the average over the engaged length. This ratio, called the *thread-load concentration factor* and denoted by H, is often 2, 3, or even 4 (Ref. 43). The maximum load per linear inch on a screwthread is therefore the total load divided by the helical length of the engaged screwthread times H. The maximum stress due to this loading can be computed by the Heywood-Kelley-Pedersen formula for a short cantilever, as given in Art. 7.10. It is important to note that in some cases the values of k_f given in the literature are for loading through a nut, and so include H, while in other cases (as in rotating-beam tests) this influence is absent. Because of the combined effects of reduced area, nonuniform load distribution, and stress concentration, the efficiency of a bolted joint under reversed repeated loading is likely to be quite small. In Ref. 49 of Chap. 2, values from 18 percent (for a 60,000-lb/in^2 steel with rolled threads) to 6.6 percent (for a 200,000-lb/in^2 steel with machine-cut threads) are cited.

The design of bolted connections has received much study, and an extensive discussion and bibliography are given by Heywood (Chap. 2, Ref. 49) and in some of the papers of Ref. 54 of Chap. 2.

REFERENCES

1. Hertz, H.: "Gesammelte Werke," vol. I, Leipzig, 1895.
2. Thomas, H. R., and V. A. Hoersch: Stresses Due to the Pressure of One Elastic Solid Upon Another, *Eng. Exp. Sta. Univ. Ill., Bull.* 212, 1930.
3. Oppel, G.: The Photoelastic Investigation of Three-Dimensional Stress and Strain Conditions, *Natl. Adv. Comm. Aeron., Tech. Memo.* 824, 1937.
4. Jensen, V. P.: Stress Analysis by Means of Polarized Light with Special Reference to Bridge Rollers, *Bull. Assoc. State Eng. Soc.,* October 1936.
5. Föppl, A.: "Technische Mechanik," 4th ed., vol. 5, p. 350.
6. Timoshenko, S., and J. N. Goodier: "Theory of Elasticity," 2d ed., Engineering Societies Monograph, McGraw-Hill Book Company, 1951.
7. Wilson, W. M.: The Bearing Value of Rollers, *Eng. Exp. Sta. Univ. Ill., Bull.* 263, 1934.
8. Whittemore, H. L., and S. N. Petrenko: Friction and Carrying Capacity of Ball and Roller Bearings, *Tech. Paper Bur. Stand.,* No. 201, 1921.
9. Wilson, W. M., R. L. Moore, and F. P. Thomas: Bearing Value of Pivots for Scales, *Eng. Exp. Sta. Univ. Ill., Bull.* 242, 1932.
10. Manual of the American Railway Engineering Association, 1936.
11. Wilson, W. M.: Rolling Tests of Plates, *Eng. Exp. Sta. Univ. Ill., Bull.* 191, 1929.
12. Hrenikoff, A.: Work of Rivets in Riveted Joints, *Trans. Am. Soc. Civil Eng.,* vol. 99, p. 437, 1934.
13. Specifications for the Design, Fabrication and Erection of Structural Steel for Buildings, and Commentary, American Institute of Steel Construction, 1969.
14. Wilson, W. M., J. Mather, and C. O. Harris: Tests of Joints in Wide Plates, *Eng. Exp. Sta. Univ. Ill., Bull.* 239, 1931.
15. Wilson, W. M., and W. A. Oliver: Tension Tests of Rivets, *Eng. Exp. Sta. Univ. Ill., Bull.* 210, 1930.
16. Kommers, J. B.: Comparative Tests of Button Head and Countersunk Riveted Joints, *Bull. Eng. Exp. Sta. Univ. Wis.,* vol. 9, no. 5, 1925.

17. Hilbes, W.: Riveted Joints in Thin Plates, *Natl. Adv. Comm. Aeron., Tech. Memo.* 590.
18. Dubin, E. A.: Eccentric Riveted Connections, *Trans. Am. Soc. Civil Eng.*, vol. 100, p. 1086, 1935.
19. Wilson, W. M., and H. F. Moore: Tests to Determine the Rigidity of Riveted Joints of Steel Structures, *Eng. Exp. Sta. Univ. Ill., Bull.* 104, 1917.
20. Rathbun, J. C.: Elastic Properties of Riveted Connections, *Trans. Am. Soc. Civil Eng.*, vol. 101, p. 524, 1936.
21. "Wood Handbook," Forest Products Laboratory, U.S. Dept. of Agriculture, 1974.
22. Baud, R. V., and R. E. Peterson: Loads and Stress Cycles in Gear Teeth, *Mech. Eng.*, vol. 51, p. 653, 1929.
23. Timoshenko, S., and R. V. Baud: The Strength of Gear Teeth, *Mech. Eng.*, vol. 48, p. 1108, 1926.
24. Dynamic Loads on Gear Teeth, *ASME Res. Pub.*, 1931.
25. Solakian, A. G., and G. B. Karelitz: Photoelastic Study of Shearing Stress in Keys and Keyways, *ASME J. Appl. Mech.*, vol. 54, no. 11, p. 97, 1932.
26. Troelsch, H. W.: Distributions of Shear in Welded Connections, *Trans. Am. Soc. Civil Eng.*, vol. 99, p. 409, 1934.
27. Report of Structural Steel Welding Committee of the American Bureau of Welding, 1931.
28. Belajef, N. M.: On the Problem of Contact Stresses, *Bull. Eng. Ways Commun.*, St. Petersburg, 1917.
29. Belajef, N. M.: Computation of Maximal Stresses Obtained from Formulas for Pressure in Bodies in Contact, *Bull. Eng. Ways Commun.*, Leningrad, 1929.
30. Horger, O. J.: Fatigue Tests of Some Manufactured Parts, *Proc. Soc. Exp. Stress Anal.*, vol. 3, no. 2, p. 135, 1946.
31. Radzimovsky, E. I.: Stress Distribution and Strength Conditions of Two Rolling Cylinders Pressed Together, *Eng. Exp. Sta. Univ. Ill., Bull.* 408, 1953.
32. Wilson, W. M.: Tests on the Bearing Value of Large Rollers, *Univ. Ill., Eng. Exp. Sta., Bull.* 162, 1927.
33. New Departure, Division of General Motors Corp.: "Analysis of Stresses and Deflections," Bristol, Conn., 1946.
34. Lundberg, G., and H. Sjovall: Stress and Deformation in Elastic Contacts, Institution of Theory of Elasticity and Strength of Materials, Chalmers University of Technology, Gothenburg, 1958.
35. Smith, J. O., and C. K. Lin: Stresses Due to Tangential and Normal Loads on an Elastic Solid with Application to Some Contact Stress Problems, *ASME J. Appl. Mech.*, vol. 20, no. 2, June 1953.
36. Palmgren, Arvid: "Ball and Roller Bearing Engineering," 3d ed., SKF Industries Inc., 1959.
37. Butler, R. H., H. R. Bear, and T. L. Carter: Effect of Fiber Orientation on Ball Failure, *Natl. Adv. Comm. Aeron., Tech. Note* 3933 (also *Tech. Note* 3930).
38. Selection of Bearings, Timken Roller Bearing Co.
39. Corwin, A. H.: "Techniques of Organic Chemistry," 3d ed., vol 1, part 1, Interscience Publishers, Inc., 1959.
40. Styri, Haakon: Fatigue Strength of Ball Bearing Races and Heat Treated Steel Specimens, *Proc. ASTM*, vol. 51, 1951.
41. Burwell, J. T., Jr. (ed.): "Mechanical Wear," American Society for Metals, 1950.
42. Specifications for Assembly of Structural Joints Using High Strength Steel Bolts, distributed by American Institute of Steel Construction; approved by Research Council on Riveted and Bolted Structural Joints of the Engineering Foundation; endorsed by American Institute of Steel Construction and Industrial Fasteners Institute.
43. Sopwith, D. G.: The Distribution of Load in Screw Threads, *Proc. Inst. Mech. Eng.*, vol. 159, 1948.
44. Lundberg, Gustaf: Cylinder Compressed between Two Plane Bodies, reprint courtesy of SKF Industries Inc., 1949.

45. Hiltscher, R., and G. Florin: Spalt- und Abreisszugspannungen in rechteckingen Scheiben, die durch eine Last in verschiedenem Abstand von einer Scheibenecke belastet sind, *Die Bautech.,* vol. 12, 1963.
46. MacGregor, C. W. (ed.): "Handbook of Analytical Design for Wear," IBM Corp., Plenum Press, 1964.
47. Hamilton, G. M., and L. E. Goodman: The Stress Field Created by a Circular Sliding Contact, *ASME J. Appl. Mech.,* vol. 33, no. 2, June 1966.
48. Goodman, L. E.: Contact Stress Analysis of Normally Loaded Rough Spheres, *ASME J. Appl. Mech.,* vol. 29, no. 3, September 1962.
49. O'Connor, J. J.: Compliance Under a Small Torsional Couple of an Elastic Plate Pressed Between Two Identical Elastic Spheres, *ASME J. Appl. Mech.,* vol. 33, no. 2, June 1966.
50. Tu, Yih-O: A Numerical Solution for an Axially Symmetric Contact Problem, *ASME J. Appl. Mech.,* vol. 34, no. 2, June 1967.
51. Pu, S. L., and M. A. Hussain: Note on the Unbonded Contact Between Plates and an Elastic Half Space, *ASME J. Appl. Mech.,* vol. 37, no. 3, September 1970.
52. Goodman, L. E., and L. M. Keer: The Contact Stress Problem for an Elastic Sphere Indenting an Elastic Cavity, *Int. J. Solids Struct.,* vol. 1, 1965.
53. Cooper, D. H.: Hertzian Contact-Stress Deformation Coefficients, *ASME J. Appl. Mech.,* vol. 36, no. 2, June 1969.
54. Dundurs, J., and M. Stippes: Role of Elastic Constants in Certain Contact Problems, *ASME J. Appl. Mech.,* vol. 37, no. 4, December 1970.
55. Brombolich, L. J.: Elastic-Plastic Analysis of Stresses Near Fastener Holes, *McDonnell Aircraft Co., MCAIR* 73-002, January 1973.

FOURTEEN

Elastic Stability

14.1 *General considerations*

Failure through elastic instability has been discussed briefly in Art. 2.13, where it was pointed out that it may occur when the bending or twisting effect of an applied load is proportional to the deformation it produces. In this chapter, formulas for the critical load or critical unit stress at which such failure occurs are given for a wide variety of members and conditions of loading.

Such formulas can be derived mathematically by integrating the differential equation of the elastic curve or by equating the strain energy of bending to the work done by the applied load in the corresponding displacement of its point of application, the form of the elastic curve being assumed when unknown. Of all possible forms of the curve, that which makes the critical load a minimum is the correct one; but almost any reasonable assumption (consistent with the boundary conditions) can be made without gross error resulting, and for this reason the strain-energy method is especially adapted to the approximate solution of difficult cases. A very thorough discussion of the general problem, with detailed solutions of many specified cases, is given in Timoshenko (Ref. 1), from which many of the formulas in this chapter are taken. Formulas for many cases are also given in Refs. 35 and 36; in addition Ref. 35 contains many graphs of numerically evaluated coefficients.

At one time, most of the problems involving elastic stability were of academic interest only since engineers were reluctant to use compression members so slender as to fail by buckling at elastic stresses and danger of corrosion interdicted the use of very thin material in exposed structures. The requirements for minimum-weight construction in the fields of aerospace and transportation, however, have given great impetus to the theoretical and experimental investigation of elastic stability and to the use of parts for which it is a governing design consideration.

There are certain definite advantages in lightweight construction, in which stability determines strength. One is that since elastic buckling may occur without damage, part of a structure—such as the skin of an airplane wing or web of a deep beam—may be used safely at loads that cause local buckling, and under these circumstances the resistance afforded by the buckled part is definitely known. Furthermore, members such as Euler columns may be loaded experimentally to their maximum capacity without damage or permanent deformation and subsequently incorporated in a structure.

14.2 *Buckling of bars*

In Table 34 formulas are given for the critical loads on columns, beams, and shafts. In general, the theoretical values are in good agreement with test results as long as the assumed conditions are reasonably well satisfied. It is to be noted that even slight changes in the amount of end constraint have a marked effect on the critical loads, and therefore it is important that such constraint be closely estimated. Slight irregularities in form and small accidental eccentricities are less likely to be important in the case of columns than in the case of thin plates. For latticed columns or columns with tie plates, a reduced value of E may be used, calculated as shown in Art. 11.3. Formulas for the elastic buckling of bars may be applied to conditions under which the proportional limit is exceeded if a reduced value of E corresponding to the actual stress is used (Ref. 1), but the procedure requires a stress-strain diagram for the material and, in general, is not practical.

In Table 34, cases 1 to 3, the tabulated buckling coefficients are worked out for various combinations of concentrated and distributed axial loads. Tensile end loads are included so that the effect of axial end restraint under axial loading within the column length can be considered (see the example at the end of this article). Carter and Gere (Ref. 46) present graphs of buckling coefficients for columns with single tapers for various end conditions, cross sections, and degrees of taper. Culver and Preg (Ref. 47) investigate and tabulate buckling coefficients for singly tapered beam-columns in which the effect of torsion, including warping restraint, is considered for the case where the loading is by end moments in the stiffer principal plane.

Kitipornchai and Trahair describe (Ref. 55) the lateral stability of singly tapered cantilever and doubly tapered simple I-beams, including the effect of warping restraint; experimental results are favorably compared with numerical solutions. Morrison (Ref. 57) considers the effect of lateral restraint of the tensile flange of a beam under lateral buckling; example calculations are presented. Massey and McGuire (Ref. 54) present graphs of buckling coefficients for both stepped and tapered cantilever beams; good agreement with experiments is reported. Tables of lateral stability constants for laminated timber beams are presented in Fowler (Ref. 53) along with two design examples.

Clark and Hill (Ref. 52) derive a general expression for the lateral stability of unsymmetrical I-beams with boundary conditions based on both bending and warping supports; tables of coefficients as well as nomographs are presented. Anderson and Trahair (Ref. 56) present tabulated lateral buckling coefficients for uniformly loaded and end-loaded cantilevers and center- and uniformly loaded simply supported beams having unsymmetric I-beam cross sections; favorable comparisons are made with extensive tests on cantilever beams.

The Southwell plot is a graph in which the lateral deflection of a column or any other linearly elastic member undergoing a manner of loading which will produce buckling is plotted versus the lateral deflection divided by the load; the slope of this line gives the critical load. For columns and some frameworks, significant deflections do occur within the range where small-deflection theory is applicable. If the initial imperfections are such that experimental readings of lateral deflection must be taken beyond the small-deflection region, then the Southwell procedure is not adequate. Roorda (Ref. 93) discusses the extension of this procedure into the nonlinear range.

Bimetallic beams. Burgreen and Manitt (Ref. 48) and Burgreen and Regal (Ref. 49) discuss the analysis of bimetallic beams and point out some of the difficulties in predicting the *snap-through instability* of these beams under changes in temperature. The thermal expansion of the support structure is an important design factor.

Rings and arches. Austin (Ref. 50) tabulates in-plane buckling coefficients for circular, parabolic, and catenary arches for pinned and fixed ends as well as for the three-hinged case; he considers cases where the cross section varies with the position in the span as well as the usual case of a uniform cross section. Uniform loads, unsymmetric distributed loads, and concentrated center loads are considered, and the stiffening effect of tying the arch to the girder with columns is also evaluated. (The discussion referenced with the paper gives an extensive bibliography of work on arch stability.)

A thin ring shrunk by cooling and inserted into a circular cavity usually will yield before buckling unless the radius/thickness ratio is very large and the elastic-limit stress is high. Chicurel (Ref. 51) derives approximate solutions to this problem when the effect of friction is considered. He suggests a conservative expression for the *no-friction* condition: $P_o/AE = 2.67(k/r)^{1.2}$, where P_o is the prebuckling hoop compressive force, A is the hoop cross-sectional area, E is the modulus of elasticity, k is the radius of gyration of the cross section, and r is the radius of the ring.

TABLE 34 ***Formulas for elastic stability of bars, rings, and beams***

NOTATION: P' = critical load (pounds); p' = critical unit load (pounds per linear inch); T' = critical torque (inch-pounds); M' = critical bending moment (inch-pounds); E = modulus of elasticity (pounds per square inch); and I = moment of inertia of cross section about central axis perpendicular to plane of buckling

Reference number, form of bar, and manner of loading and support

1a. Stepped straight bar under end load P_1 and intermediate load P_2; upper end free, lower end fixed $P_1' = K_1 \frac{\pi^2 E_1 I_1}{l^2}$ where K_1 is tabulated below

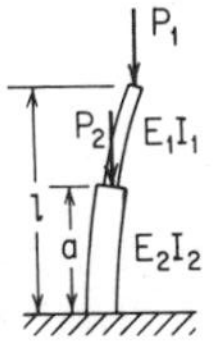

E_2I_2/E_1I_1	1.000					1.500					2.000				
a/l → / P_2/P_1 ↓	$\frac{1}{6}$	$\frac{1}{3}$	$\frac{1}{2}$	$\frac{2}{3}$	$\frac{5}{6}$	$\frac{1}{6}$	$\frac{1}{3}$	$\frac{1}{2}$	$\frac{2}{3}$	$\frac{5}{6}$	$\frac{1}{6}$	$\frac{1}{3}$	$\frac{1}{2}$	$\frac{2}{3}$	$\frac{5}{6}$
0.0	0.250	0.250	0.250	0.250	0.250	0.279	0.312	0.342	0.364	0.373	0.296	0.354	0.419	0.471	0.496
0.5	0.249	0.243	0.228	0.208	0.187	0.279	0.306	0.317	0.306	0.279	0.296	0.350	0.393	0.399	0.372
1.0	0.248	0.237	0.210	0.177	0.148	0.278	0.299	0.295	0.261	0.223	0.296	0.345	0.370	0.345	0.296
2.0	0.246	0.222	0.178	0.136	0.105	0.277	0.286	0.256	0.203	0.158	0.295	0.335	0.326	0.267	0.210
4.0	0.242	0.195	0.134	0.092	0.066	0.274	0.261	0.197	0.138	0.099	0.294	0.314	0.257	0.184	0.132
8.0	0.234	0.153	0.088	0.056	0.038	0.269	0.216	0.132	0.084	0.057	0.290	0.266	0.174	0.112	0.076

1b. Stepped straight bar under end load P_1 and intermediate load P_2; both ends pinned $P_1' = K_1 \frac{\pi^2 E_1 I_1}{l^2}$ where K_1 is tabulated below

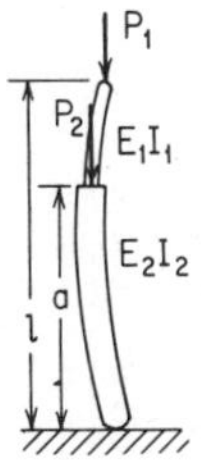

E_2I_2/E_1I_1	1.00					1.50					2.00				
a/l → / P_2/P_1 ↓	$\frac{1}{6}$	$\frac{1}{3}$	$\frac{1}{2}$	$\frac{2}{3}$	$\frac{5}{6}$	$\frac{1}{6}$	$\frac{1}{3}$	$\frac{1}{2}$	$\frac{2}{3}$	$\frac{5}{6}$	$\frac{1}{6}$	$\frac{1}{3}$	$\frac{1}{2}$	$\frac{2}{3}$	$\frac{5}{6}$
0.0	1.000	1.000	1.000	1.000	1.000	1.010	1.065	1.180	1.357	1.479	1.014	1.098	1.297	1.633	1.940
0.5	0.863	0.806	0.797	0.789	0.740	0.876	0.872	0.967	1.091	1.098	0.884	0.908	1.069	1.339	1.452
1.0	0.753	0.672	0.663	0.646	0.584	0.769	0.736	0.814	0.908	0.870	0.776	0.769	0.908	1.126	1.153
2.0	0.594	0.501	0.493	0.473	0.410	0.612	0.557	0.615	0.676	0.613	0.621	0.587	0.694	0.850	0.814
4.0	0.412	0.331	0.325	0.307	0.256	0.429	0.373	0.412	0.442	0.383	0.438	0.397	0.470	0.566	0.511
8.0	0.254	0.197	0.193	0.180	0.147	0.267	0.225	0.248	0.261	0.220	0.272	0.240	0.284	0.336	0.292

1c. Stepped straight bar under end load P_1 and intermediate load P_2; upper end guided, lower end fixed $P_1' = K_1 \frac{\pi^2 E_1 I_1}{l^2}$ where K_1 is tabulated below

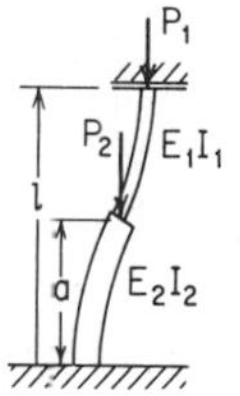

E_2I_2/E_1I_1	1.00					1.50					2.00				
a/l \ P_2/P_1	$\frac{1}{6}$	$\frac{1}{3}$	$\frac{1}{2}$	$\frac{2}{3}$	$\frac{5}{6}$	$\frac{1}{6}$	$\frac{1}{3}$	$\frac{1}{2}$	$\frac{2}{3}$	$\frac{5}{6}$	$\frac{1}{6}$	$\frac{1}{3}$	$\frac{1}{2}$	$\frac{2}{3}$	$\frac{5}{6}$
0.0	1.000	1.000	1.000	1.000	1.000	1.113	1.208	1.237	1.241	1.309	1.184	1.367	1.452	1.461	1.565
0.5	0.986	0.904	0.792	0.711	0.672	1.105	1.117	1.000	0.897	0.885	1.177	1.288	1.192	1.063	1.063
1.0	0.972	0.817	0.650	0.549	0.507	1.094	1.026	0.830	0.697	0.669	1.171	1.206	1.000	0.832	0.805
2.0	0.937	0.671	0.472	0.377	0.339	1.073	0.872	0.612	0.482	0.449	1.156	1.047	0.745	0.578	0.542
4.0	0.865	0.480	0.304	0.231	0.204	1.024	0.642	0.397	0.297	0.270	1.126	0.794	0.486	0.358	0.327
8.0	0.714	0.299	0.176	0.130	0.114	0.910	0.406	0.232	0.169	0.151	1.042	0.511	0.284	0.203	0.182

1d. Stepped straight bar under end load P_1 and intermediate load P_2; upper end pinned, lower end fixed $P_1' = K_1 \frac{\pi^2 E_1 I_1}{l^2}$ where K_1 is tabulated below

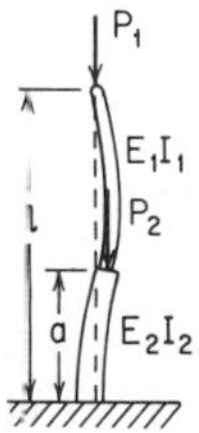

E_2I_2/E_1I_1	1.000					1.500					2.000				
a/l \ P_2/P_1	$\frac{1}{6}$	$\frac{1}{3}$	$\frac{1}{2}$	$\frac{2}{3}$	$\frac{5}{6}$	$\frac{1}{6}$	$\frac{1}{3}$	$\frac{1}{2}$	$\frac{2}{3}$	$\frac{5}{6}$	$\frac{1}{6}$	$\frac{1}{3}$	$\frac{1}{2}$	$\frac{2}{3}$	$\frac{5}{6}$
0.0	2.048	2.048	2.048	2.048	2.048	2.241	2.289	2.338	2.602	2.976	2.369	2.503	2.550	2.983	3.838
0.5	1.994	1.814	1.711	1.700	1.590	2.208	2.071	1.991	2.217	2.344	2.344	2.286	2.196	2.570	3.066
1.0	1.938	1.613	1.464	1.450	1.290	2.167	1.869	1.727	1.915	1.918	2.313	2.088	1.915	2.250	2.525
2.0	1.820	1.300	1.130	1.111	0.933	2.076	1.535	1.355	1.506	1.390	2.250	1.742	1.518	1.796	1.844
4.0	1.570	0.918	0.773	0.753	0.594	1.874	1.107	0.941	1.042	0.891	2.097	1.277	1.065	1.270	1.184
8.0	1.147	0.569	0.469	0.454	0.343	1.459	0.697	0.582	0.643	0.514	1.727	0.812	0.664	0.796	0.686

TABLE 34 *Formulas for elastic stability of bars, rings, and beams* (*Cont.*)

Reference number, form of bar, and manner of loading and support

1e. Stepped straight bar under end load P_1 and intermediate load P_2; both ends fixed $P_1' = K_1 \frac{\pi^2 E_1 I_1}{l^2}$ where K_1 is tabulated below

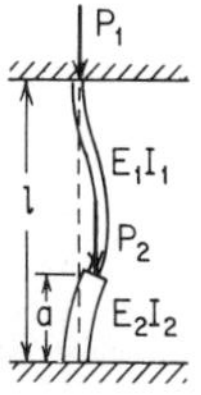

E_2I_2/E_1I_1	1.000					1.500					2.000				
P_2/P_1 \ a/l	$\frac{1}{6}$	$\frac{1}{3}$	$\frac{1}{2}$	$\frac{2}{3}$	$\frac{5}{6}$	$\frac{1}{6}$	$\frac{1}{3}$	$\frac{1}{2}$	$\frac{2}{3}$	$\frac{5}{6}$	$\frac{1}{6}$	$\frac{1}{3}$	$\frac{1}{2}$	$\frac{2}{3}$	$\frac{5}{6}$
0.0	4.000	4.000	4.000	4.000	4.000	4.389	4.456	4.757	5.359	5.462	4.657	4.836	5.230	6.477	6.838
0.5	3.795	3.298	3.193	3.052	2.749	4.235	3.756	3.873	4.194	3.795	4.545	4.133	4.301	5.208	4.787
1.0	3.572	2.779	2.647	2.443	2.094	4.065	3.211	3.254	3.411	2.900	4.418	3.568	3.648	4.297	3.671
2.0	3.119	2.091	1.971	1.734	1.414	3.679	2.459	2.459	2.452	1.968	4.109	2.766	2.782	3.136	2.496
4.0	2.365	1.388	1.297	1.088	0.857	2.921	1.659	1.649	1.555	1.195	3.411	1.882	1.885	2.008	1.523
8.0	1.528	0.826	0.769	0.623	0.479	1.943	1.000	0.992	0.893	0.671	2.334	1.138	1.141	1.158	0.854

2a. Stepped straight bar under tensile end load P_1 and intermediate load P_2; upper end free, lower end fixed $P_2' = K_2 \frac{\pi^2 E_1 I_1}{l^2}$ where K_2 is tabulated below

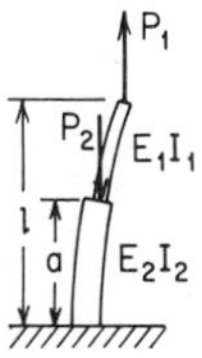

E_2I_2/E_1I_1	1.000					1.500					2.000				
P_1/P_2 \ a/l	$\frac{1}{6}$	$\frac{1}{3}$	$\frac{1}{2}$	$\frac{2}{3}$	$\frac{5}{6}$	$\frac{1}{6}$	$\frac{1}{3}$	$\frac{1}{2}$	$\frac{2}{3}$	$\frac{5}{6}$	$\frac{1}{6}$	$\frac{1}{3}$	$\frac{1}{2}$	$\frac{2}{3}$	$\frac{5}{6}$
0	9.00	2.25	1.00	0.56	0.36	13.50	3.38	1.50	0.84	0.54	18.00	4.50	2.00	1.25	0.72
0.125	15.55	3.75	1.48	0.74	0.44	21.87	5.36	2.19	1.11	0.65	27.98	6.92	2.89	1.48	0.87
0.250	21.33	5.30	2.19	1.03	0.55	29.51	7.36	3.13	1.53	0.82	37.30	9.31	4.02	2.02	1.10
0.375	29.02	7.25	3.13	1.52	0.74	39.89	9.97	4.37	2.21	1.10	50.10	12.52	5.52	2.86	1.46
0.500	40.50	10.12	4.46	2.31	1.08	55.66	13.92	6.16	3.28	1.60	69.73	17.43	7.73	4.18	2.12

2b. Stepped straight bar under tensile end load P_1 and intermediate load P_2; both ends pinned $P_2' = K_2 \dfrac{\pi^2 E_1 I_1}{l^2}$ where K_2 is tabulated below

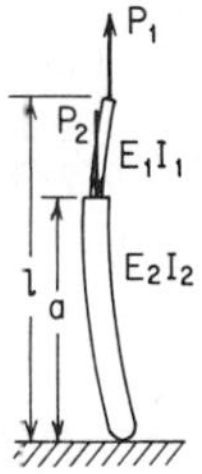

E_2I_2/E_1I_1	1.000					1.500					2.000				
a/l \ P_1/P_2	$\frac{1}{6}$	$\frac{1}{3}$	$\frac{1}{2}$	$\frac{2}{3}$	$\frac{5}{6}$	$\frac{1}{6}$	$\frac{1}{3}$	$\frac{1}{2}$	$\frac{2}{3}$	$\frac{5}{6}$	$\frac{1}{6}$	$\frac{1}{3}$	$\frac{1}{2}$	$\frac{2}{3}$	$\frac{5}{6}$
0	2.60	1.94	1.89	1.73	1.36	2.77	2.24	2.47	2.54	2.04	2.86	2.41	2.89	3.30	2.72
0.125	3.51	2.49	2.43	2.14	1.62	3.81	2.93	3.26	3.18	2.43	3.98	3.21	3.89	4.19	3.24
0.250	5.03	3.41	3.32	2.77	1.99	5.63	4.15	4.64	4.15	2.99	5.99	4.65	5.75	5.52	3.98
0.375	7.71	5.16	4.96	3.76	2.55	8.98	6.61	7.26	5.63	3.82	9.80	7.67	9.45	7.50	5.09
0.500	12.87	9.13	8.00	5.36	3.48	15.72	12.55	12.00	7.96	5.18	17.71	15.45	16.00	10.54	6.87

2c. Stepped straight bar under tensile end load P_1 and intermediate load P_2; upper end guided, lower end fixed $P_2' = K_2 \dfrac{\pi^2 E_1 I_1}{l^2}$ where K_2 is tabulated below

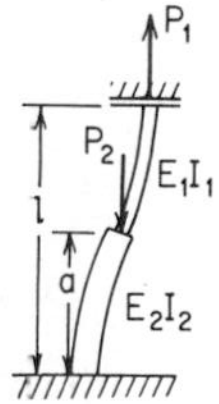

E_2I_2/E_1I_1	1.000					1.500					2.000				
a/l \ P_1/P_2	$\frac{1}{6}$	$\frac{1}{3}$	$\frac{1}{2}$	$\frac{2}{3}$	$\frac{5}{6}$	$\frac{1}{6}$	$\frac{1}{3}$	$\frac{1}{2}$	$\frac{2}{3}$	$\frac{5}{6}$	$\frac{1}{6}$	$\frac{1}{3}$	$\frac{1}{2}$	$\frac{2}{3}$	$\frac{5}{6}$
0	10.40	3.08	1.67	1.19	1.03	14.92	4.23	2.21	1.55	1.37	19.43	5.37	2.73	1.88	1.65
0.125	15.57	4.03	2.03	1.40	1.18	21.87	5.57	2.71	1.82	1.57	27.98	7.07	3.36	2.21	1.90
0.250	21.33	5.37	2.54	1.67	1.38	29.52	7.40	3.42	2.20	1.84	37.32	9.34	4.26	2.68	2.24
0.375	29.02	7.26	3.31	2.08	1.67	39.90	9.97	4.50	2.76	2.24	50.13	12.53	5.61	3.39	2.73
0.500	40.51	10.12	4.53	2.72	2.10	55.69	13.91	6.21	3.66	2.84	69.76	17.43	7.76	4.52	3.47

TABLE 34 *Formulas for elastic stability of bars, rings, and beams* (*Cont.*)

Reference number, form of bar, and manner of loading and support

2d. Stepped straight bar under tensile end load P_1 and intermediate load P_2; upper end pinned, lower end fixed $P_2' = K_2 \frac{\pi^2 E_1 I_1}{l^2}$ where K_2 is tabulated below

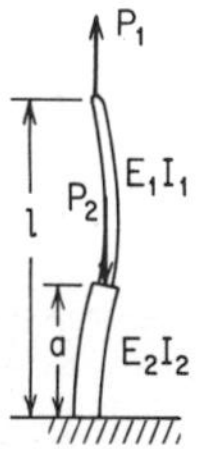

E_2I_2/E_1I_1	1.000					1.500					2.000				
P_1/P_2 \ a/l	$\frac{1}{6}$	$\frac{1}{3}$	$\frac{1}{2}$	$\frac{2}{3}$	$\frac{5}{6}$	$\frac{1}{6}$	$\frac{1}{3}$	$\frac{1}{2}$	$\frac{2}{3}$	$\frac{5}{6}$	$\frac{1}{6}$	$\frac{1}{3}$	$\frac{1}{2}$	$\frac{2}{3}$	$\frac{5}{6}$
0	13.96	5.87	4.80	4.53	3.24	18.66	7.33	6.04	6.58	4.86	23.26	8.64	6.98	8.40	6.48
0.125	20.21	7.93	6.50	5.84	3.91	27.12	10.12	8.43	8.71	5.86	33.71	12.06	9.92	11.51	7.81
0.250	28.58	11.35	9.64	7.68	4.87	38.32	14.82	13.13	11.50	7.27	47.36	17.85	15.96	15.30	9.65
0.375	41.15	17.64	15.82	10.15	6.26	55.43	23.67	23.44	14.96	9.30	68.40	28.88	30.65	19.66	12.28
0.500	62.90	31.73	23.78	13.58	8.42	85.64	44.28	34.97	19.80	12.40	106.27	55.33	45.81	25.83	16.27

2e. Stepped straight bar under tensile end load P_1 and intermediate load P_2; both ends fixed $P_2' = K_2 \frac{\pi^2 E_1 I_1}{l^2}$ where K_2 is tabulated below

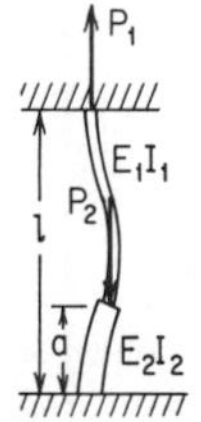

E_2I_2/E_1I_1	1.000					1.500					2.000				
P_1/P_2 \ a/l	$\frac{1}{6}$	$\frac{1}{3}$	$\frac{1}{2}$	$\frac{2}{3}$	$\frac{5}{6}$	$\frac{1}{6}$	$\frac{1}{3}$	$\frac{1}{2}$	$\frac{2}{3}$	$\frac{5}{6}$	$\frac{1}{6}$	$\frac{1}{3}$	$\frac{1}{2}$	$\frac{2}{3}$	$\frac{5}{6}$
0	16.19	8.11	7.54	5.79	4.34	21.06	9.93	9.89	8.34	6.09	25.75	11.44	11.55	10.87	7.78
0.125	21.83	10.37	9.62	6.86	5.00	28.74	12.93	13.03	9.92	7.05	35.28	15.06	15.55	12.96	9.01
0.250	30.02	14.09	12.86	8.34	5.91	39.81	17.99	18.36	12.09	8.35	48.88	21.25	22.98	15.79	10.69
0.375	42.72	20.99	17.62	10.47	7.19	57.14	27.66	26.02	15.17	10.20	70.23	33.29	34.36	19.79	13.11
0.500	64.94	36.57	24.02	13.70	9.16	86.23	50.39	35.09	19.86	13.07	102.53	61.71	45.87	25.86	16.85

3a. Uniform straight bar under end load P and a uniformly distributed load p over a lower portion of the length; several end conditions. $(pa)' = K\frac{\pi^2 EI}{l^2}$ where K is tabulated below (a negative value for P/pa means the end load is tensile)

End conditions	Upper end free, lower end fixed 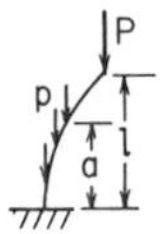				Both ends pinned 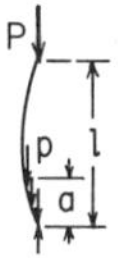				Upper end pinned, lower end fixed 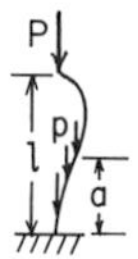				Both ends fixed 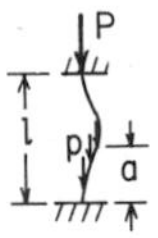			
P/pa \ a/l	$\frac{1}{4}$	$\frac{1}{2}$	$\frac{3}{4}$	1	$\frac{1}{4}$	$\frac{1}{2}$	$\frac{3}{4}$	1	$\frac{1}{4}$	$\frac{1}{2}$	$\frac{3}{4}$	1	$\frac{1}{4}$	$\frac{1}{2}$	$\frac{3}{4}$	1
−0.25		11.31	5.18	2.38	9.03	5.32	4.25	3.30		27.9	17.4	11.3		31.3	19.4	13.4
0.00	12.74	3.185	1.413	0.795	3.52	2.53	2.22	1.88	22.2	9.46	7.13	5.32	25.3	13.0	9.78	7.56
0.25	0.974	0.825	0.614	0.449	1.97	1.59	1.46	1.30	6.83	4.70	3.98	3.30	11.2	7.50	6.25	5.20
0.50	0.494	0.454	0.383	0.311	1.34	1.15	1.08	0.98	3.76	3.03	2.71	2.37	6.75	5.18	4.54	3.94
1.00	0.249	0.238	0.218	0.192	0.81	0.73	0.70	0.66	1.97	1.75	1.64	1.51	3.69	3.17	2.91	2.65

3b. Uniform straight bar under end load P and a uniformly distributed load p over an upper portion of the length; several end conditions. $(pa)' = K\frac{\pi^2 EI}{l^2}$ where K is tabulated below (a negative value for P/pa means the end load is tensile)

End conditions	Upper end free, lower end fixed 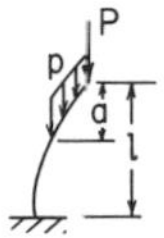			Both ends pinned 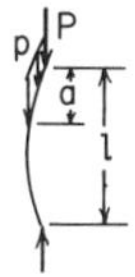			Upper end pinned, lower end fixed 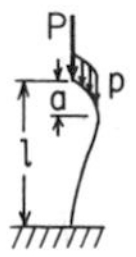			Both ends fixed 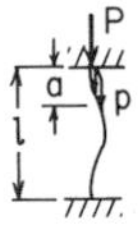		
P/pa \ a/l	$\frac{1}{4}$	$\frac{1}{2}$	$\frac{3}{4}$	$\frac{1}{4}$	$\frac{1}{2}$	$\frac{3}{4}$	$\frac{1}{4}$	$\frac{1}{2}$	$\frac{3}{4}$	$\frac{1}{4}$	$\frac{1}{2}$	$\frac{3}{4}$
−0.25	0.481	0.745	1.282	1.808	2.272	2.581	4.338	5.937	7.385	5.829	7.502	9.213
0.00	0.327	0.440	0.600	1.261	1.479	1.611	2.904	3.586	4.160	4.284	5.174	5.970
0.25	0.247	0.308	0.380	0.963	1.088	1.159	2.164	2.529	2.815	3.384	3.931	4.383
0.50	0.198	0.236	0.276	0.778	0.859	0.903	1.720	1.943	2.111	2.796	3.164	3.453
1.00	0.142	0.161	0.179	0.561	0.603	0.624	1.215	1.323	1.400	2.073	2.273	2.419

TABLE 34 *Formulas for elastic stability of bars, rings, and beams* (*Cont.*)

Reference number, form of bar, and manner of loading and support

3c. Uniform straight bar under end load P and a distributed load of maximum value p at the bottom linearly decreasing to zero at a distance a from the bottom. $(pa)' = K \frac{\pi^2 EI}{l^2}$ where K is tabulated below (a negative value for P/pa means the end load is tensile)

End conditions	Upper end free, lower end fixed 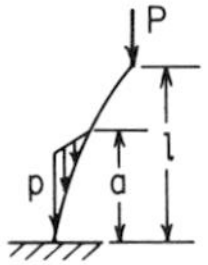				Both ends pinned 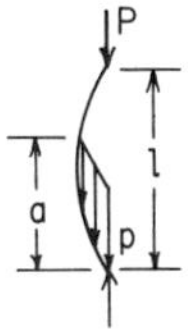				Upper end pinned, lower end fixed 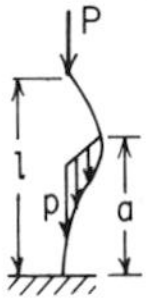				Both ends fixed 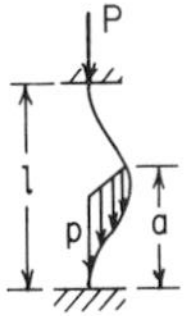			
a/l → / P/pa ↓	$\frac{1}{4}$	$\frac{1}{2}$	$\frac{3}{4}$	1	$\frac{1}{4}$	$\frac{1}{2}$	$\frac{3}{4}$	1	$\frac{1}{4}$	$\frac{1}{2}$	$\frac{3}{4}$	1	$\frac{1}{4}$	$\frac{1}{2}$	$\frac{3}{4}$	1
−0.250						58.9	41.1	30.4								
−0.125			26.7	15.5	31.9	15.7	12.0	9.41			62.1	43.8		113.0	70.2	48.7
0.000	52.4	13.1	5.80	3.26	9.66	6.31	5.32	4.72		30.3	20.6	16.1		38.9	27.8	21.9
0.125	1.98	1.85	1.58	1.29	4.65	3.66	3.29	3.03	15.2	11.7	9.73	8.50	27.3	18.9	15.6	13.4
0.250	0.995	0.961	0.887	0.787	2.98	2.54	2.35	2.22	7.90	6.92	6.18	5.66	14.9	12.1	10.6	9.53
0.500	0.499	0.490	0.471	0.441	1.72	1.56	1.49	1.43	4.02	3.77	3.54	3.36	7.73	6.95	6.43	6.00
1.000	0.250	0.248	0.243	0.235	0.93	0.88	0.86	0.84	2.03	1.96	1.90	1.85	3.93	3.73	3.57	3.44

4. Uniform straight bar under end load P; both ends hinged and bar elastically supported by lateral pressure p proportional to deflection ($p = ky$, where k = lateral force in pounds per inch per inch of deflection)

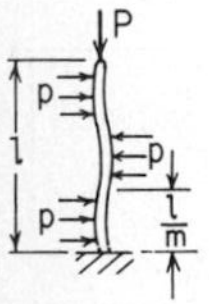

$$P' = \frac{\pi^2 EI}{l^2}\left(m^2 + \frac{kl^4}{m^2\pi^2 EI}\right)$$

where m represents the number of half-waves in which the bar buckles and is equal to the lowest integer greater than

$$\frac{1}{2}\left(\sqrt{1 + \frac{4l^2}{\pi^2}\sqrt{\frac{k}{EI}}} - 1\right)$$

(Ref. 1)

deflection toward the harder foundation); these are also called unattached foundations

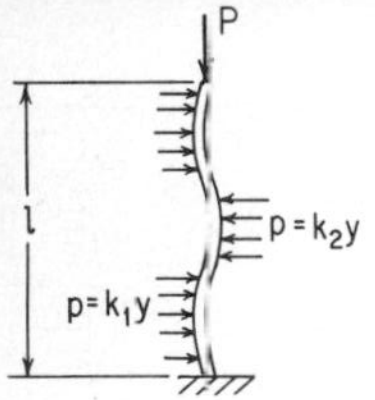

$$P' = \frac{\pi^2 EI}{l^2}\left(m^2 + \frac{k_2 l^4}{m^2 \pi^4 EI}\phi^\alpha\right) \quad \text{where } \phi = \frac{k_1}{k_2} \text{ and } \alpha \text{ depends upon } m \text{ as given below}$$

m	α
1	1
2	$1 + \phi\left(0.23 - 0.017 l^2 \sqrt{k_2/EI}\right)$
3	$0.75 - 0.56\phi$

This is an emperical expression which closely fits numerical solutions found in Ref. 45 and is valid only over the range $0 \leq l^2\sqrt{k_2/EI} \leq 120$. Solutions for P' are carried out for values of $m = 1, 2$, and 3, and the lowest one governs

6. Straight bar, middle portion uniform, end portions tapered and alike; end load; I = moment of inertia of cross section of middle portion; I_0 = moment of inertia of end cross sections; I_x = moment of inertia of section x

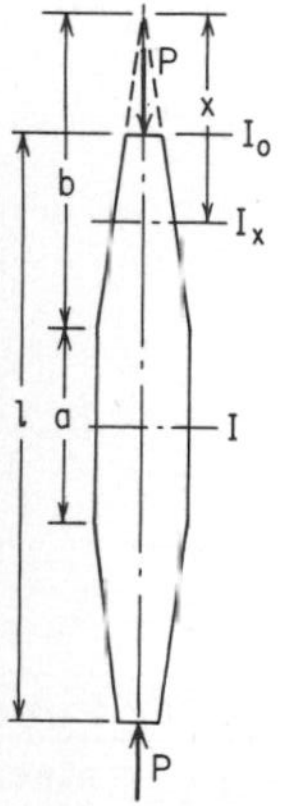

(For singly tapered columns see Ref. 46.)

6a. $I_x = I\frac{x}{b}$

for example, rectangular section tapering uniformly in width

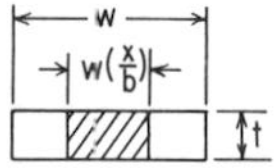

$P' = \frac{KEI}{l^2}$ where K depends on $\frac{I_0}{I}$ and $\frac{a}{l}$ and may be found from the following table:

I_0/I → / a/l ↓	K for ends hinged 0	0.01	0.10	0.2	0.4	0.6	0.8	K for ends fixed 0.2	0.4	0.6	0.8
0	5.78	5.87	6.48	7.01	7.86	8.61	9.27	20.36	26.16	31.04	35.40
0.2	7.04	7.11	7.58	7.99	8.59	9.12	9.53	22.36	27.80	32.20	36.00
0.4	8.35	8.40	8.63	8.90	9.19	9.55	9.68	23.42	28.96	32.92	36.36
0.6	9.36	9.40	9.46	9.73	9.70	9.76	9.82	25.44	30.20	33.80	36.84
0.8	9.80	9.80	9.82	9.82	9.83	9.85	9.86	29.00	33.08	35.80	37.84

(Ref. 5)

6b. $I_x = I\left(\frac{x}{b}\right)^2$

for example, section of four slender members latticed together

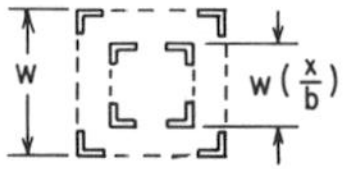

$P' = \frac{KEI}{l^2}$ where K may be found from the following table:

I_0/I → / a/l ↓	K for ends hinged 0	0.01	0.10	0.2	0.4	0.6	0.8	K for ends fixed 0.2	0.4	0.6	0.8
0	1.00	3.45	5.40	6.37	7.61	8.51	9.24	18.94	25.54	30.79	35.35
0.2	1.56	4.73	6.67	7.49	8.42	9.04	9.50	21.25	27.35	32.02	35.97
0.4	2.78	6.58	8.08	8.61	9.15	9.48	9.70	22.91	28.52	32.77	36.34
0.6	6.25	8.62	9.25	9.44	9.63	9.74	9.82	24.29	29.69	33.63	36.80
0.8	9.57	9.71	9.79	9.81	9.84	9.85	9.86	27.67	32.59	35.64	37.81

(Ref. 5)

TABLE 34 *Formulas for elastic stability of bars, rings, and beams* (*Cont.*)

Reference number, form of bar, and manner of loading and support

6c. $I_x = I\left(\frac{x}{b}\right)^3$

for example, rectangular section tapering uniformly in thickness

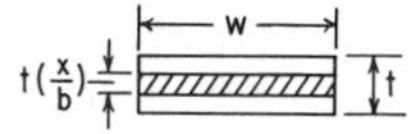

$P' = \frac{KEI}{l^2}$ where K may be found from the following table:

a/l \ I_0/I	K for ends hinged 0.01	0.10	0.2	0.4	0.6	0.8	K for ends fixed 0.2	0.4	0.6	0.8
0	2.55	5.01	6.14	7.52	8.50	9.23	18.48	25.32	30.72	35.32
0.2	3.65	6.32	7.31	8.38	9.02	9.50	20.88	27.20	31.96	35.96
0.4	5.42	7.84	8.49	9.10	9.46	9.69	22.64	28.40	32.72	36.32
0.6	7.99	9.14	9.39	9.62	9.74	9.81	23.96	29.52	33.56	36.80
0.8	9.63	9.77	9.81	9.84	9.85	9.86	27.24	32.44	35.60	37.80

(Ref. 5)

6d. $I_x = I\left(\frac{x}{b}\right)^4$

for example, end portions pyramidal or conical

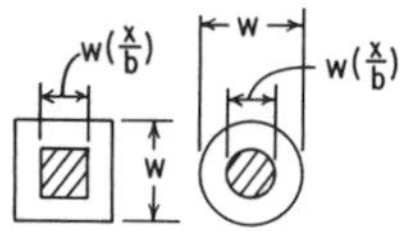

$P' = \frac{KEI}{l^2}$ where K may be found from the following table:

a/l \ I_0/I	K for ends hinged 0.01	0.10	0.2	0.4	0.6	0.8	K for ends fixed 0.2	0.4	0.6	0.8
0	2.15	4.81	6.02	7.48	8.47	9.23	18.23	25.23	30.68	35.33
0.2	3.13	6.11	7.20	8.33	9.01	9.49	20.71	27.13	31.94	35.96
0.4	4.84	7.68	8.42	9.10	9.45	9.69	22.49	28.33	32.69	36.32
0.6	7.53	9.08	9.38	9.62	9.74	9.81	23.80	29.46	33.54	36.78
0.8	9.56	9.77	9.80	9.84	9.85	9.86	27.03	32.35	35.56	37.80

(Ref. 5)

7. Uniform straight bar under end loads P and end twisting couples T; cross section of bar has same I for all central axes; both ends hinged

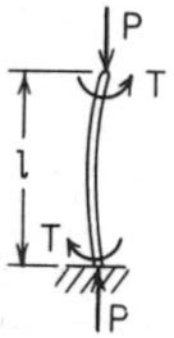

Critical combination of P and T is given by

$$\frac{T^2}{4(EI)^2} + \frac{P}{EI} = \frac{\pi^2}{l^2}$$ (Ref. 1)

If $P = 0$, the formula gives critical twisting moment T' which, acting alone, would cause buckling
If for a given value of T the formula gives a negative value for P, $T > T'$ and P represents tensile load required to prevent buckling
For thin circular tube of diameter D and thickness t under torsion only, critical shear stress

$$\tau = \frac{\pi ED}{l(1 - \nu)}\left(1 - \frac{t}{D} + \frac{1}{3}\frac{t^2}{D^2}\right)$$ for helical buckling (Ref. 2)

---

8. Uniform circular ring under uniform radial pressure p lb/in; mean radius of ring r

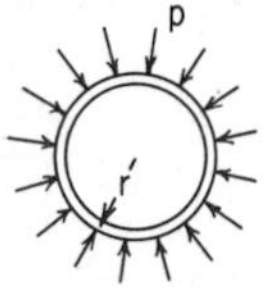

$$p' = \frac{3EI}{r^3}$$ (Ref. 1)

---

9. Uniform circular arch under uniform radial pressure p lb/in; mean radius r; ends hinged

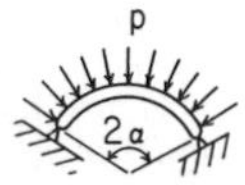

$$p' = \frac{EI}{r^3}\left(\frac{\pi^2}{\alpha^2} - 1\right)$$ (Ref. 1)

(For symmetrical arch of any form under central concentrated loading, see Ref. 40; for parabolic and catenary arches, see Ref. 50)

---

10. Uniform circular arch under uniform radial pressure p lb/in; mean radius r; ends fixed

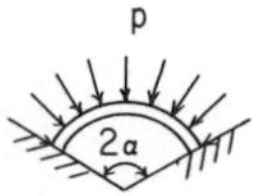

$$p' = \frac{EI}{r^3}(k^2 - 1)$$ (Ref. 1)

where k depends on α and is found by trial from the equation: $k \tan \alpha \cot k\alpha = 1$ or from the following table:

$\alpha =$	15°	30°	45°	60°	75°	90°	120°	180°
$k =$	17.2	8.62	5.80	4.37	3.50	3.00	2.36	2.00

(For parabolic and catenary arches, see Ref. 50)

TABLE 34 *Formulas for elastic stability of bars, rings, and beams* (*Cont.*)

Reference number, form of bar, and manner of loading and support	
11. Straight uniform beam of narrow rectangular section under pure bending 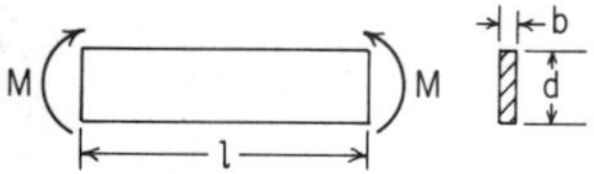	For ends held vertical but not fixed in horizontal plane: $M' = \dfrac{\pi b^3 d \sqrt{EG\left(1 - 0.63\dfrac{b}{d}\right)}}{6l}$ For ends held vertical and fixed in horizontal plane: $M' = \dfrac{2\pi b^3 d \sqrt{EG\left(1 - 0.63\dfrac{b}{d}\right)}}{6l}$ (Refs. 1, 3, 4)
12. Straight uniform cantilever beam of narrow rectangular section under end load applied at a point a distance a above (a positive) or below (a negative) centroid of section 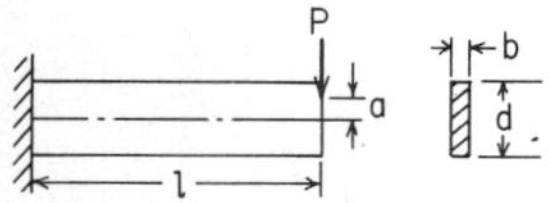	$P' = \dfrac{0.669 b^3 d \sqrt{\left(1 - 0.63\dfrac{b}{d}\right)EG}}{l^2}\left[1 - \dfrac{a}{2l}\sqrt{\dfrac{E}{G\left(1 - 0.63\dfrac{b}{d}\right)}}\right]$ For a load W uniformly distributed along the beam the critical load, $W' = 3P'$ (approximately) (For tapered and stepped beams, see Ref. 54) (Refs. 1, 3, 4)
13. Straight uniform beam of narrow rectangular section under center load applied at a point a distance a above (a positive) or below (a negative) centroid of section: ends of beam simply supported and constrained against twisting 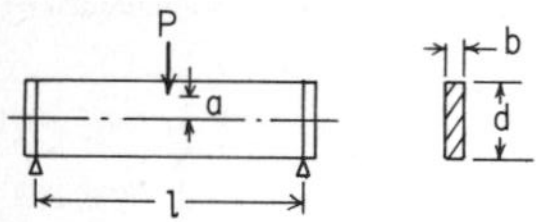	$P' = \dfrac{2.82 b^3 d \sqrt{\left(1 - 0.63\dfrac{b}{d}\right)EG}}{l^2}\left[1 - \dfrac{1.74a}{l}\sqrt{\dfrac{E}{G\left(1 - 0.63\dfrac{b}{d}\right)}}\right]$ For a uniformly distributed load, the critical load $W' = 1.67P'$ (approximately) If P is applied at an intermediate point, a distance C from one end, its critical value is practically the same as for central loading if $0.4l < C < 0.5l$; if $C < 0.4l$, the critical load is given approximately by multiplying the P' for central loading by $0.36 + 0.28\dfrac{l}{C}$ If the ends of the beam are fixed and the load P is applied at the centroid of the middle cross section, $P' = \dfrac{4.43 b^3 d}{l^2}\sqrt{\left(1 - 0.63\dfrac{b}{d}\right)EG}$ (Refs. 1, 3, 4)

14. Straight uniform I-beam under pure bending; d = depth center to center of flange; ends constrained against twisting 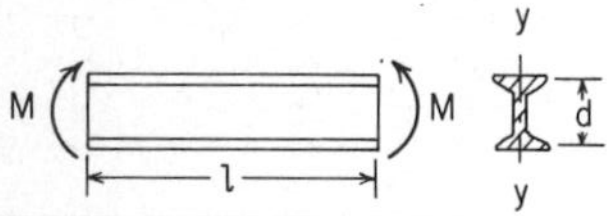 	$M' = \frac{\pi \sqrt{EI_y KG}}{l} \sqrt{1 + \pi^2 \frac{I_f E d^2}{2KGl^2}}$ where I_y is the moment of inertia of the cross section about its vertical axis of symmetry, I_f is the moment of inertia of one flange about this axis, and KG is the torsional rigidity of the section (see Table 20, case 26) (For tapered I-beams, see Ref. 47.) (Refs. 1, 3)
15. Straight uniform cantilever beam of I section under end load applied at centroid of cross section; d = depth center to center of flanges 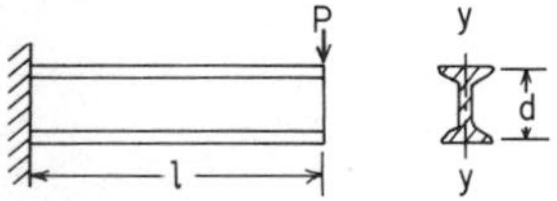 	$P' = \frac{m \sqrt{EI_y KG}}{l^2}$ where m is approximately equal to $4.01 + 11.7 \sqrt{\frac{I_f E d^2}{2KGl^2}}$ and I_y, I_f, and KG have the same significance as in case 14 (For unsymmetric I-beams, see Refs. 52 and 56; for tapered I-beams, see Ref. 55.) (Refs. 1, 3)
16. Straight uniform I-beam loaded at centroid of middle section; ends simply supported and constrained against twisting 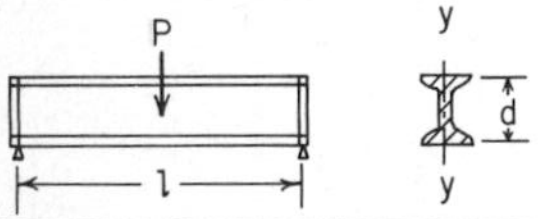 	$P' = \frac{m \sqrt{EI_y KG}}{l^2}$ where m is approximately equal to $16.93 + 45 \left(\frac{I_f E d^2}{2KGl^2}\right)^{0.8}$ and I_y, I_f, and KG have same significance as in case 14. (For unsymmetric I-beams, see Refs. 52 and 56; for tapered I-beams, see Ref. 55.) (Refs. 1, 3)

EXAMPLE

A 4-in steel pipe is to be used as a column to carry 8000 lb of transformers centered axially on a platform 20 ft above the foundation. The factor of safety *FS* is to be determined for the following conditions, based on elastic buckling of the column.

(*a*) The platform is supported only by the pipe fixed at the foundation.

(*b*) A $3\frac{1}{2}$-in steel pipe is to be slipped into the top of the 4-in pipe a distance of 4 in, welded in place, and extended 10 ft to the ceiling above, where it will extend through a close-fitting hole in a steel plate.

(*c*) This condition is the same as in (*b*) except that the $3\frac{1}{2}$-in pipe will be welded solidly into a heavy steel girder passing 10 ft above the platform.

Solution. A 4-in steel pipe has a cross-sectional area of 3.174 in^2 and a bending moment of inertia of 7.233 in^4. For a $3\frac{1}{2}$-in pipe these are 2.68 in^2 and 4.788 in^4, respectively.

(*a*) This case is a column fixed at the bottom and free at the top with an end load only. In Table 34, case 1a, for $I_2/I_1 = 1.00$ and $P_2/P_1 = 0$, K_1 is given as 0.25. Therefore,

$$P_1' = 0.25\frac{\pi^2 30(10^6)(7.233)}{240^2} = 9295 \text{ lb}$$

$$FS = \frac{9295}{8000} = 1.162$$

(*b*) This case is a column fixed at the bottom and pinned at the top with a load at a distance of two-thirds the 30-ft length from the bottom: $I_1 = 4.788$ in^4, $I_2 = 7.233$ in^4, and $I_2/I_1 = 1.511$. In Table 34, case 2d, for $E_2I_2/E_1I_1 = 1.5$, $P_1/P_2 = 0$, and $a/l = \frac{2}{3}$, K_2 is given as 6.58. Therefore,

$$P_2' = 6.58\frac{\pi^2 30(10^6)(4.788)}{360^2} = 72{,}000 \text{ lb}$$

$$FS = \frac{72{,}000}{8000} = 9$$

(*c*) This case is a column fixed at both ends and subjected to an upward load on top and a downward load at the platform. The upward load depends to some extent on the stiffness of the girder to which the top is welded, and so we can only bracket the actual critical load. If we assume the girder is infinitely rigid and permits no vertical deflection of the top, the elongation of the upper 10 ft would equal the reduction in length of the lower 20 ft. Equating these deformations gives

$$\frac{P_1(10)(12)}{2.68(30)(10^6)} = \frac{(P_1 - P_2)(20)(12)}{3.174(30)(10^6)} \qquad \text{or} \qquad P_1 = 0.628P_2$$

From Table 34, case 2e, for $E_2I_2/E_1I_1 = 1.5$ and $a/l = \frac{2}{3}$, we find the following values of K_2 for the several values of P_1/P_2:

$\frac{P_1}{P_2}$	0	0.125	0.250	0.375	0.500
K_2	8.34	9.92	12.09	15.17	19.86

By extrapolation, for $P_1/P_2 = 0.628$, $K_2 = 26.5$.

If we assume the girder provides no vertical load but does prevent rotation of the top, then $K_2 = 8.34$. Therefore, the value of P_2 ranges from 91,200 to 289,900 lb, and the factor of safety lies between 11.4 and 36.2. A reasonable estimate of the rotational and vertical stiffness of the girder will allow a good estimate to be made of the actual factor of safety from the values calculated.

14.3 *Buckling of flat and curved plates*

In Table 35 formulas are given for the critical loads and critical stresses on plates and thin-walled members. Because of the greater likelihood of serious geometrical irregularities and their greater relative effect, the critical stresses actually developed by such members usually fall short of the theoretical values by a wider margin than in the case of bars. The discrepancy is generally greater for pure compression (thin tubes under longitudinal compression or external pressure) than for tension and compression combined (thin tubes under torsion or flat plates under edge shear), and increases with the thinness of the material. The critical stress or load indicated by any one of the theoretical formulas should therefore be regarded as an upper limit, approached more or less closely according to the closeness with which the actual shape of the member approximates the geometrical form assumed. In Table 35 the approximate discrepancy to be expected between theory and experiment is indicated wherever the data available have made this possible.

The majority of the theoretical analyses of the stability of plates and shells require a numerical evaluation of the resulting equations. Considering the variety of shapes and combinations of shapes as well as the multiplicity of boundary conditions and loading combinations, it is not possible in the limited space available to present anything like a comprehensive coverage of plate and shell buckling. As an alternative, Table 35 contains many of the simpler loadings and shapes. The following paragraphs and the References contain some, but by no means all, of the more easily acquired sources giving results in tabular or graphic form that can be applied directly to specific problems.

Rectangular plates. Stability coefficients for *orthotropic* rectangular plates with several combinations of boundary conditions and several ratios of the bending stiffnesses parallel to the sides of the plate are tabulated in Shuleshko (Ref. 60); these solutions were obtained by reducing the problem of plate buckling to that of an isotropic bar that is in a state of vibration and under tension. Srinivas and Rao (Ref. 63) evaluate the effect of shear deformation on the stability of simply supported rectangular plates under edge loads parallel to one side; the effect becomes noticeable for $h/b > 0.05$ and is greatest when the loading is parallel to the short side.

Skew plates. Ashton (Ref. 61) and Durvasula (Ref. 64) consider the buckling of skew (parallelogram) plates under combinations of edge compression, edge tension, and edge shear. Since the loadings evaluated are generally parallel to orthogonal axes and not to both sets of the plate edges, we would not expect to find the particular case desired represented in the tables of coefficients; the general trend of results is informative.

Circular plates. Vijayakumar and Joga Rao (Ref. 58) describe a technique for solving for the radial buckling loads on a *polar orthotropic annular plate.* They give graphs of stability coefficients for a wide range of rigidity ratios and for the several combinations of free, simply supported, and fixed inner and outer edges for the radius ratio (outer to inner) 2:1. Two loadings are presented: outer edge only under uniform compression, and inner and outer edges under equal uniform compression.

Amon and Widera (Ref. 59) present graphs showing the effect of an edge beam on the stability of a circular plate of uniform thickness.

Sandwich plates. There is a great amount of literature on the subject of sandwich construction. References 38 and 100 and the publications listed in Ref. 39 provide initial sources of information.

14.4 *Buckling of shells*

Baker, Kovalevsky, and Rish (Ref. 97) discuss the stability of unstiffened, orthotropic composite, stiffened, and sandwich shells. They present data based on theory and experiment which permits the designer to choose a loading or pressure with a 90 percent probability of no stability failure; the work is extensively referenced.

Stein (Ref. 95) discusses some recent comparisons of theory with experimentation in shell buckling. Rabinovich (Ref. 96) describes in some detail the work in structural mechanics, including shell stability, in the U.S.S.R. from 1917 to 1957.

In recent years, there have been increasing development and application of the finite-element method for the numerical solution of shell problems. Navaratna, Pian, and Witmer (Ref. 94) describe a finite-element method of solving axisymmetric shell problems where the element considered is either a conical frustum or a frustum with a curved meridian; examples are presented of cylinders with uniform or tapered walls under axial load, a truncated hemisphere under axial tension, and a conical shell under torsion. Bushnell (Ref. 99) presents a very general finite-element program for shell analysis, and Perrone (Ref. 98) gives a compendium of such programs.

Cylindrical and conical shells. In general, experiments to determine the axial loads required to buckle cylindrical shells yield results that are between one-half and three-fourths of the classical buckling loads predicted by theory. The primary causes of these discrepancies are the deviations from a true cylindrical form in most manufactured vessels and the inability to accurately define the boundary conditions. Hoff (Refs. 67 and 68) shows that removing the in-plane shear stress at the boundary of a simply supported cylindrical shell under axial compression can reduce the theoretical buckling load by a factor of 2 from that predicted by the more usual boundary conditions associated with a simply supported edge. Baruch, Harari, and Singer (Ref. 84) find similar low-buckling loads for simply supported conical shells under axial load but for a different modification of the boundary support. Tani and Yamaki (Ref. 83) carry out further work on this problem, including the effect of clamped edges.

The random nature of manufacturing deviations leads to the use of the statistical approach, as mentioned previously (Ref. 97) and as Hausrath and Dittoe have done for conical shells (Ref. 77). Weingarten, Morgan, and Seide (Ref. 80) have developed empirical expressions for lower bounds of stability coefficients for cylindrical and conical shells under axial compression with references for the many data they present.

McComb, Zender, and Mikulas (Ref. 44) discuss the effects of internal pressure on the bending stability of very thin-walled cylindrical shells. Internal pressure has a stabilizing effect on axially and/or torsionally loaded cylindrical and conical shells. This subject is discussed in several references: Seide (Ref. 75), Weingarten (Ref. 76), and Weingarten, Morgan, and Seide (Ref. 82) for conical and cylindrical shells; Ref. 97 contains much information on this subject as well.

Axisymmetric snap-buckling of conical shells is discussed by Newman and Reiss (Ref. 73), which leads to the concept of the Belleville spring for the case of shallow shells. (See also Art. 10.8.)

External pressure as a cause of buckling is examined by Singer (Ref. 72) for cones, Newman and Reiss (Ref. 73), and Yao and Jenkins (Ref. 69) for elliptic cylinders. External pressure caused by pretensioned filament winding on cylinders is analyzed by Mikulas and Stein (Ref. 66); they point out that material compressibility in the thickness direction is important in this problem.

The combination of external pressure and axial loads on cylindrical and conical shells is very thoroughly examined and referenced by Radkowski (Ref. 79) and Weingarten and Seide (Ref. 81). The combined loading on orthotropic and stiffened conical shells is discussed by Singer (Ref. 74).

Attempts to manufacture nearly perfect shells in order to test the theoretical results have led to the construction of thin-walled shells by electroforming; Sendelbeck and Singer (Ref. 85) and Arbocz and Babcock (Ref. 91) describe the results of such tests.

A very thorough survey of buckling theory and experimentation for conical shells of constant thickness is presented by Seide (Ref. 78).

Spherical shells. Experimental work is described by Loo and Evan-Iwanowski on the effect of a concentrated load at the apex of a spherical cap (Ref. 90) and the effect of multiple concentrated loads (Ref. 89). Carlson, Sendelbeck, and Hoff (Ref. 70) report on the experimental study of buckling of electroformed complete spherical shells; they report experimental critical pressures of up to 86 percent of those predicted by theory and the correlation of flaws with lower test pressures.

Burns (Ref. 92) describes tests of static and dynamic buckling of thin spherical caps due to external pressure; both elastic and plastic buckling are considered and evaluated in these tests. Wu and Cheng (Ref. 71) discuss

TABLE 35 *Formulas for elastic stability of plates and shells*

NOTATION: E = modulus of elasticity; ν = Poisson's ratio; and t = thickness for all plates and shells. All dimensions are in inches, all forces are in pounds, and all angles are in radians. Compression is positive; tension is negative

Form of plate or shell and manner of loading	Manner of support	Formulas for critical unit compressive stress σ', unit shear stress τ', load P', bending moment M', or unit external pressure q' at which elastic buckling occurs
1. Rectangular plate under equal uniform compression on two opposite edges b 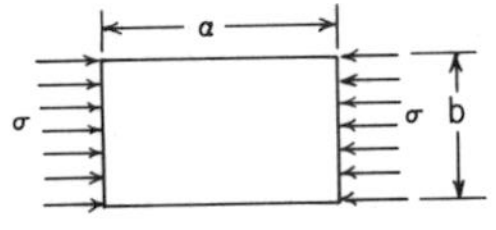 $\frac{b}{t} > 10$	1a. All edges simply supported	$\sigma' = K\frac{E}{1-\nu^2}\left(\frac{t}{b}\right)^2$ Here K depends on ratio $\frac{a}{b}$ and may be found from the following table: $\frac{a}{b}$ = 0.2, 0.3, 0.4, 0.6, 0.8, 1.0, 1.2, 1.4, 1.6, 1.8, 2.0, 2.2, 2.4, 2.7, 3, ∞ K = 22.2, 10.9, 6.92, 4.23, 3.45, 3.29, 3.40, 3.68, 3.45, 3.32, 3.29, 3.32, 3.40, 3.32, 3.29, 3.29 (For unequal end compressions, see Ref. 33) (Refs. 1, 6)
	1b. All edges clamped	$\sigma' = K\frac{E}{1-\nu^2}\left(\frac{t}{b}\right)^2$ $\frac{a}{b}$ = 1, 2, 3, ∞ K = 7.7, 6.7, 6.4, 5.73 (Refs. 1, 6, 7)
	1c. Edges b simply supported, edges a clamped	$\sigma' = K\frac{E}{1-\nu^2}\left(\frac{t}{b}\right)^2$ $\frac{a}{b}$ = 0.4, 0.5, 0.6, 0.7, 0.8, 1.0, 1.2, 1.4, 1.6, 1.8, 2.1, ∞ K = 7.76, 6.32, 5.80, 5.76, 6.00, 6.32, 5.80, 5.76, 6.00, 5.80, 5.76, 5.73 (Refs. 1, 6)
	1d. Edges b simply supported, one edge a simply supported, other edge a free	$\sigma' = K\frac{E}{1-\nu^2}\left(\frac{t}{b}\right)^2$ $\frac{a}{b}$ = 0.5, 1.0, 1.2, 1.4, 1.6, 1.8, 2.0, 2.5, 3.0, 4.0, 5.0 K = 3.62, 1.18, 0.934, 0.784, 0.687, 0.622, 0.574, 0.502, 0.464, 0.425, 0.416 (Ref. 1)
	1e. Edges b simply supported, one edge a clamped, other edge a free	$\sigma' = K\frac{E}{1-\nu^2}\left(\frac{t}{b}\right)^2$ $\frac{a}{b}$ = 1, 1.1, 1.2, 1.3, 1.4, 1.5, 1.6, 1.7, 1.8, 1.9, 2.0, 2.2, 2.4 K = 1.40, 1.28, 1.21, 1.16, 1.12, 1.10, 1.09, 1.09, 1.10, 1.12, 1.14, 1.19, 1.21 (Ref. 1)

1f. Edges b clamped, edges a simply supported

$$\sigma' = K\frac{E}{1-\nu^2}\left(\frac{t}{b}\right)^2$$

$\frac{a}{b} =$	0.6	0.8	1.0	1.2	1.4	1.6	1.7	1.8	2.0	2.5	3.0
$K =$	11.0	7.18	5.54	4.80	4.48	4.39	4.39	4.26	3.99	3.72	3.63

(Ref. 1)

2. Rectangular plate under uniform compression (or tension) σ_x on edges b and uniform compression (or tension) σ_y on edges a

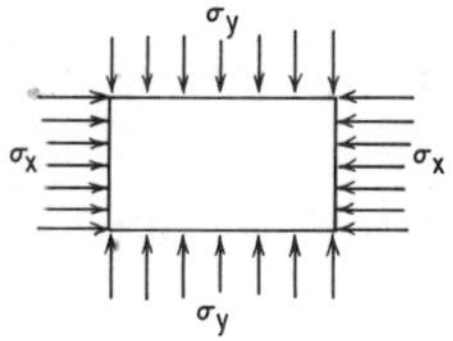

2a. All edges simply supported

$$\sigma'_x\frac{m^2}{a^2} + \sigma'_y\frac{n^2}{b^2} = 0.823\frac{E}{1-\nu^2}t^2\left(\frac{m^2}{a^2}+\frac{n^2}{b^2}\right)^2$$

Here m and n signify the number of half-waves in the buckled plate in the x and y directions, respectively. To find σ'_y for a given σ_x, take $m = 1$, $n = 1$ if $C\left(1 - 4\frac{a^4}{b^4}\right) < \sigma_x < C\left(5 + 2\frac{a^2}{b^2}\right)$, where $C = \frac{0.823Et^2}{(1-\nu^2)a^2}$.

If σ_x is too large to satisfy this inequality, take $n = 1$ and m to satisfy: $C\left(2m^2 - 2m + 1 + 2\frac{a^2}{b^2}\right) < \sigma_x < C\left(2m^2 + 2m + 1 + 2\frac{a^2}{b^2}\right)$. If σ_x is too small to satisfy the first inequality, take $m = 1$ and n to satisfy: $C\left[1 - n^2(n-1)^2\frac{a^4}{b^4}\right] > \sigma_x > C\left[1 - n^2(n+1)^2\frac{a^4}{b^4}\right]$

(Refs. 1, 6)

2b. All edges clamped

$$\sigma_x + \frac{a^2}{b^2}\sigma_y = 1.1\frac{Et^2a^2}{1-\nu^2}\left(\frac{3}{a^4}+\frac{3}{b^4}+\frac{2}{a^2b^2}\right)$$

(This equation is approximate and is most accurate when the plate is nearly square and σ_x and σ_y nearly equal

(Ref. 1)

3. Rectangular plate under linearly varying stress on edges b (bending or bending combined with tension or compression)

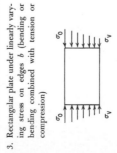

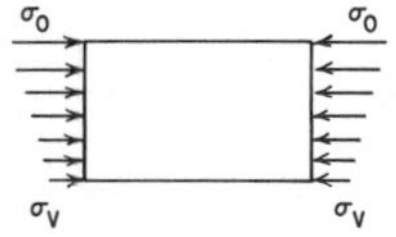

3a. All edges simply supported

$$\sigma'_o = K\frac{E}{1-\nu^2}\left(\frac{t}{b}\right)^2$$

Here K depends on $\frac{a}{b}$ and on $\alpha = \frac{\sigma_v}{\sigma_o - \sigma_v}$ and may be found from the following table:

	$\frac{a}{b} =$ 0.4	0.5	0.6	0.667	0.75	0.8	0.9	1.0	1.5
$\alpha = 0.5$	$K =$ 23.9	21.1	19.8	19.7	19.8	20.1	21.1	21.1	19.8
0.75	15.4		10.6		9.5	9.2		9.1	9.5
1.00	12.4		8.0		6.9	6.7		6.4	6.9
1.25	10.95		6.8		5.8	5.7		5.4	5.8
1.50	8.9		5.3		5.0	4.9		4.8	5.0
∞ (pure compression)	6.92		4.25			3.45		3.29	3.57

(Refs. 1, 6)

TABLE 35 *Formulas for elastic stability of plates and shells* (*Cont.*)

Form of plate or shell and manner of loading	Manner of support	Formulas for critical unit compressive stress σ', unit shear stress τ', load P', bending moment M', or unit external pressure q' at which elastic buckling occurs
4. Rectangular plate under uniform shear on all edges 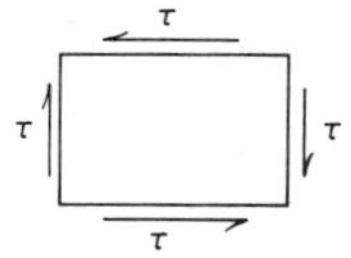	4a. All edges simply supported	$\tau' = K\dfrac{E}{1-\nu^2}\left(\dfrac{t}{b}\right)^2$ $\frac{a}{b} =$ 1.0, 1.2, 1.4, 1.5, 1.6, 1.8, 2.0, 2.5, 3.0, ∞ $K =$ 7.75, 6.58, 6.00, 5.84, 5.76, 5.59, 5.43, 5.18, 5.02, 4.40 (Refs. 1, 6, 8, 22)
	4b. All edges clamped	$\tau' = K\dfrac{E}{1-\nu^2}\left(\dfrac{t}{b}\right)^2$ $\frac{a}{b} =$ 1, 2, ∞ $K =$ 12.7, 9.5, 7.38 Test results indicate a value for K of about 4.1 for very large values of $\frac{a}{b}$ (Ref. 9) (For continuous panels, see Ref. 30)
5. Rectangular plate under uniform shear on all edges; compression (or tension) σ_x on edges b; compression (or tension) σ_y on edges a; a/b very large 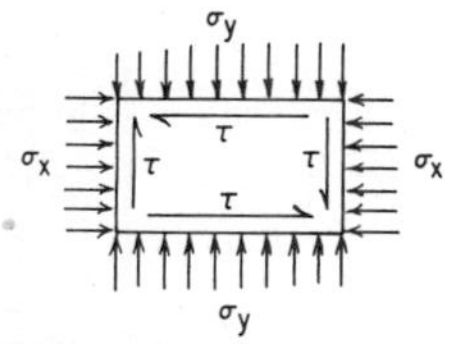	5a. All edges simply supported	$\tau' = \sqrt{C^2\left(2\sqrt{1-\dfrac{\sigma_y}{C}}+2-\dfrac{\sigma_x}{C}\right)\left(2\sqrt{1-\dfrac{\sigma_y}{C}}+6-\dfrac{\sigma_x}{C}\right)}$ where $C = \dfrac{0.823}{1-\nu^2}\left(\dfrac{t}{b}\right)^2 E$ (Refs. 1, 6, 23, and 31)
	5b. All edges clamped	$\tau' = \sqrt{C^2\left(2.31\sqrt{4-\dfrac{\sigma_y}{C}}+\dfrac{4}{3}-\dfrac{\sigma_x}{C}\right)\left(2.31\sqrt{4-\dfrac{\sigma_y}{C}}+8-\dfrac{\sigma_x}{C}\right)}$ where $C = \dfrac{0.823}{1-\nu^2}\left(\dfrac{t}{b}\right)^2 E$ (σ_x and σ_y are negative when tensile) (Ref. 6)
6. Rectangular plate under uniform shear on all edges and bending stresses on edges b 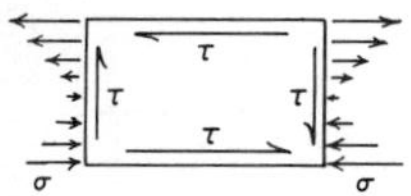	6a. All edges simply supported	$\sigma' = K\dfrac{E}{1-\nu^2}\left(\dfrac{t}{b}\right)^2$ Here K depends on $\frac{\tau}{\tau'}$ (ratio of actual shear stress to shear stress that, acting alone, would be critical) and on $\frac{a}{b}$. K varies less than 10 percent for values of $\frac{a}{b}$ from 0.5 to 1, and for $\frac{a}{b} = 1$ is approximately as follows: $\frac{\tau}{\tau'} =$ 0, 0.2, 0.3, 0.4, 0.5, 0.6, 0.7, 0.8, 0.9, 1.0 $K =$ 21.1, 20.4, 19.6, 18.5, 17.7, 16.0, 14.0, 11.9, 8.20, 0 (Refs. 1, 10)

7. Rectangular plate under concentrated center loads on two opposite edges	7a. All edges simply supported	$P = \dfrac{\tau}{3}\dfrac{Et^3}{(1-\nu^2)b}$ $\left(\text{for } \dfrac{a}{b} > 2\right)$		(Ref. 1)
	7b. Edges b simply supported, edges a clamped	$P = \dfrac{2\pi}{3}\dfrac{Et^3}{(1-\nu^2)b}$ $\left(\text{for } \dfrac{a}{b} > 2\right)$		(Ref. 1)
8. Rhombic plate under uniform compression on all edges	8a. All edges simply supported	$\sigma' = K\dfrac{Et^2}{a^2(1-\nu^2)}$ $\begin{array}{c\|cccccc} \alpha & 0° & 9° & 18° & 27° & 36° & 45° \\ \hline K & 1.645 & 1.678 & 1.783 & 1.983 & 2.338 & 2.898 \end{array}$		(Ref. 65)
9. Polygon plate under uniform compression on all edges N = number of sides	9a. All edges simply supported	$\sigma' = K\dfrac{Et^2}{a^2(1-\nu^2)}$ $\begin{array}{c\|cccccc} N & 3 & 4 & 5 & 6 & 7 & 8 \\ \hline K & 4.393 & 1.645 & 0.916 & 0.597 & 0.422 & 0.312 \end{array}$		(Ref. 65)
10. Parabolic and semielliptic plates under uniform compression on all edges	10a. All edges simply supported	$\sigma' = K\dfrac{Et^2}{a^2(1-\nu^2)}$		
	10b. All edges fixed	where K is tabulated below for the several shapes and boundary conditions for $\nu = \frac{1}{3}$:		(Ref. 62)

	Square	Semiellipse	Parabola	Triangle
Simply supported	1.65	1.86	2.50	3.82
Fixed	4.36	5.57	7.22	10.60

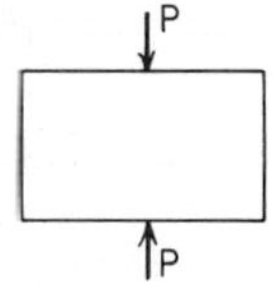

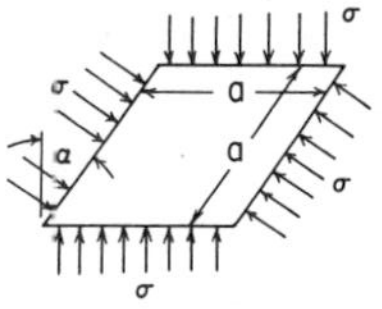

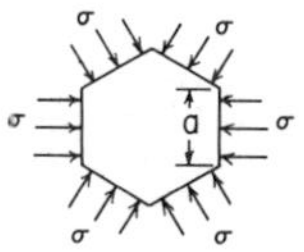

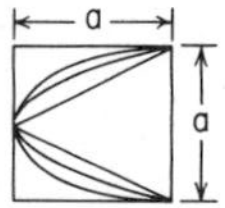

TABLE 35 *Formulas for elastic stability of plates and shells* (*Cont.*)

Form of plate or shell and manner of loading	Manner of support	Formulas for critical unit compressive stress σ', unit shear stress τ', load P', bending moment M', or unit external pressure q' at which elastic buckling occurs
11. Isotropic circular plate under uniform radial edge compression 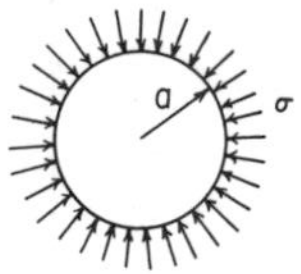	11a. Edges simply supported	$\sigma' = 0.35\frac{E}{1-\nu^2}\left(\frac{t}{a}\right)^2$ (Ref. 1)
	11b. Edges clamped	$\sigma' = 1.22\frac{E}{1-\nu^2}\left(\frac{t}{a}\right)^2$ (Ref. 1) For elliptical plate with major semiaxis a, minor semiaxis b, $\sigma' = K\frac{E}{1-\nu^2}\left(\frac{t}{b}\right)^2$, where K has values as follows: $\frac{a}{b} = 1.0$ 1.1 1.2 1.3 2.0 5.0 $K = 1.22$ 1.13 1.06 1.01 0.92 0.94 (Ref. 21)
12. Circular plate with concentric hole under uniform radial compression on outer edge 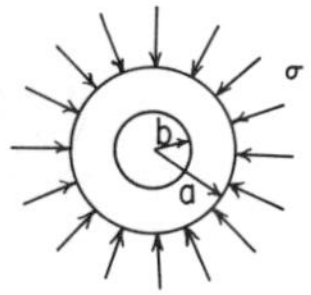 $\frac{a}{t} > 10$	12a. Outer edge simply supported, inner edge free	$\sigma' = K\frac{E}{1-\nu^2}\left(\frac{t}{a}\right)^2$ Here K depends on $\frac{b}{a}$ and is given approximately by following table: $\frac{b}{a} = 0$ 0.1 0.2 0.3 0.4 0.5 0.6 0.7 0.8 0.9 $K = 0.35$ 0.33 0.30 0.27 0.23 0.21 0.19 0.18 0.17 0.16 (Ref. 1)
	12b. Outer edge clamped, inner edge free	$\sigma' = K\frac{E}{1-\nu^2}\left(\frac{t}{a}\right)^2$ Here K depends on $\frac{b}{a}$ and is given approximately by following table: $\frac{b}{a} = 0$ 0.1 0.2 0.3 0.4 0.5 $K = 1.22$ 1.17 1.11 1.21 1.48 2.07 (Ref. 1)
13. Curved panel under uniform compression on curved edges b (b = width of panel measured on arc; r = radius of curvature) 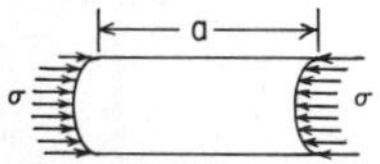 $\frac{b}{t} > 10$	13a. All edges simply supported	$\sigma' = \frac{1}{6}\frac{E}{1-\nu^2}\left[\sqrt{12(1-\nu^2)\left(\frac{t}{r}\right)^2 + \left(\frac{\pi t}{b}\right)^4} + \left(\frac{\pi t}{b}\right)^2\right]$ (*Note:* With $a > b$, the solution does not depend upon a.) or $\sigma' = 0.6E\frac{t}{r}$ if $\frac{b}{r}$ (central angle of curve) is less than $\frac{1}{2}$ and b and a are nearly equal (Refs. 1 and 6) (For compression combined with shear, see Refs. 28 and 34.)

14. Curved panel under uniform shear on all edges 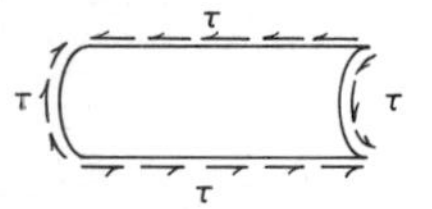	14a. All edges simply supported	$\tau' = 0.1E\frac{t}{r} + 5E\left(\frac{t}{b}\right)^2$ (Refs. 6, 27, 29)
	14b. All edges clamped	$\tau' = 0.1E\frac{t}{r} + 7.5E\left(\frac{t}{b}\right)^2$ (Ref. 6) Tests show $\tau' = 0.075E\frac{t}{r}$ for panels curved to form quadrant of a circle (Ref. 11) (See also Refs. 27, 29)
15. Thin-walled circular tube under uniform longitudinal compression (radius of tube $= r$) 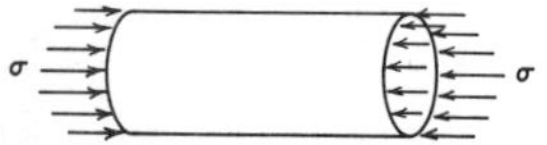 $\frac{r}{t} > 10$	15a. Ends not constrained	$\sigma' = \frac{1}{\sqrt{3}}\frac{E}{\sqrt{1-\nu^2}}\frac{t}{r}$ (Refs. 6, 12, 13, 24) Most accurate for very long tubes, but applicable if length is several times as great as $1.72\sqrt{rt}$, which is the length of a half-wave of buckling. Tests indicate an actual buckling strength of from 40 to 60 percent of this theoretical value, or $\sigma' = 0.3Et/r$ approximately
16. Thin-walled circular tube under a transverse bending moment M (radius of tube $= r$) 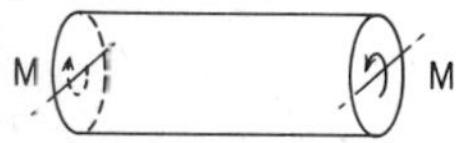 $\frac{r}{t} > 10$	16a. No constraint	$M' = K\frac{E}{1-\nu^2}rt^2$ Here the theoretical value of K for pure bending and long tubes is 0.99. The average value of K determined by tests is 1.14, and the minimum value is 0.72. Except for very short tubes, length effect is negligible and a small transverse shear produces no appreciable reduction in M'. A very short cylinder under transverse (beam) shear may fail by buckling at neutral axis when shear stress there reaches a value of about $1.25\tau'$ for case 17a (Refs. 6, 14, 15)
17. Thin-walled circular tube under a twisting moment T that produces a uniform circumferential shear stress: $\tau = \frac{T}{2\pi r^2 t}$ (length of tube $= l$; radius of tube $= r$) 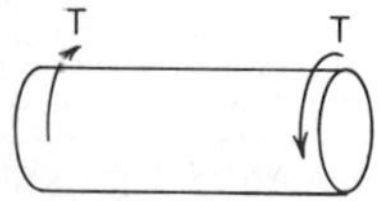 $\frac{r}{t} > 10$	17a. Ends hinged, i.e., wall free to change angle with cross section, but circular section maintained	$\tau' = \frac{E}{1-\nu^2}\left(\frac{t}{l}\right)^2(1.27 + \sqrt{9.64 + 0.466H^{1.5}})$ where $H = \sqrt{1-\nu^2}\frac{l^2}{tr}$ Tests indicate that the actual buckling stress is from 60 to 75 percent of this theoretical value, with the majority of the data points nearer 75 percent. (Refs. 6, 16, 18, 25)
	17b. Ends clamped, i.e., wall held perpendicular to cross section and circular section maintained	$\tau' = \frac{E}{1-\nu^2}\left(\frac{t}{l}\right)^2(-2.39 + \sqrt{96.9 + 0.605H^{1.5}})$ where H is given in part 17a. The statement in part a regarding actual buckling stress applies here as well. (Refs. 6, 16, 18, 25)

TABLE 35 *Formulas for elastic stability of plates and shells* (*Cont.*)

Form of plate or shell and manner of loading	Manner of support	Formulas for critical unit compressive stress σ', unit shear stress τ', load P', bending moment M', or unit external pressure q' at which elastic buckling occurs
18. Thin-walled circular tube under uniform longitudinal compression σ and uniform circumferential shear τ due to torsion (case 15 combined with case 17) 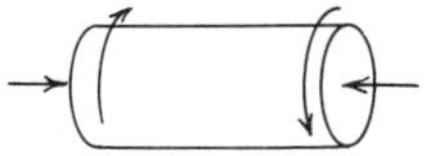 $\frac{r}{t} > 10$	18a. Edges hinged as in case 17a. 18b. Edges clamped as in case 17b.	The equation $1 - \frac{\sigma'}{\sigma'_o} = \left(\frac{\tau'}{\tau'_o}\right)^n$ holds, where σ' and τ' are the critical compressive and shear stresses for the combined loading, σ'_o is the critical compressive stress for the cylinder under compression alone (case 15), and τ'_o is the critical shear stress for the cylinder under torsion alone (case 17a or 17b according to end conditions). Tests indicate that n is approximately 3. If σ is tensile, then σ' should be considered negative. (Ref. 6) (See also Ref. 26. For square tube, see Ref. 32)
19. Thin tube under uniform lateral external pressure (radius of tube $= r$) 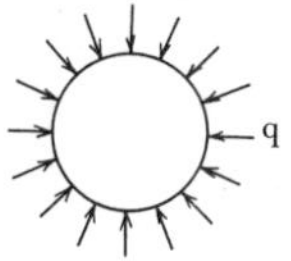 $\frac{r}{t} > 10$	19a. Very long tube with free ends; length l	$q' = \frac{1}{4}\frac{E}{1-\nu^2}\frac{t^3}{r^3}$ Applicable when $l > 4.90r\sqrt{\frac{r}{t}}$ (Ref. 19)
	19b. Short tube, of length l, ends held circular, but not otherwise constrained, or long tube held circular at intervals l	$q' = 0.807\frac{Et^2}{lr}\sqrt[4]{\left(\frac{1}{1-\nu^2}\right)^3\frac{t^2}{r^2}}$ approximate formula (Ref. 19)
20. Thin tube with closed ends under uniform external pressure, lateral and longitudinal (length of tube $= l$; radius of tube $= r$) 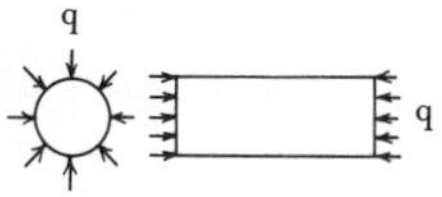 $\frac{r}{t} > 10$	20a. Ends held circular	$q' = \frac{E\frac{t}{r}}{1 + \frac{1}{2}\left(\frac{\pi r}{nl}\right)^2}\left\{\frac{1}{n^2\left[1 + \left(\frac{nl}{\pi r}\right)^2\right]^2} + \frac{n^2t^2}{12r^2(1-\nu^2)}\left[1 + \left(\frac{\pi r}{nl}\right)^2\right]^2\right\}$ (Refs. 19, 20) where n = number of lobes formed by the tube in buckling. To determine q' for tubes of a given t/r, plot a group of curves, one curve for each integral value of n of 2 or more, with l/r as ordinates and q' as abscissa; that curve of the group which gives the least value of q' is then used to find the q' corresponding to a given l/r. If $\left(\frac{l}{r}\right)^2\left(\frac{r}{t}\right) > 300$, the critical pressure can be approximated by $q' = \frac{0.92E}{\left(\frac{l}{r}\right)\left(\frac{r}{t}\right)^{2.5}}$. (Ref. 81) Values of experimentally determined critical pressures range 20% above and below the theoretical values given by the expressions above. A recommended probable minimum critical pressure is $0.80q'$.

angle 2α, when 2α = arc AB/r) 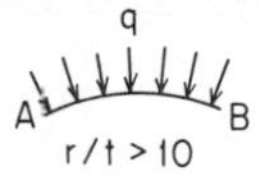	supported (i.e., hinged)	$q = \frac{\ }{12r^3(1-\nu^2)}$ (Ref. 1)
	21b. Curved edges free, straight edges at A and B clamped	Here k is found from the equation $k \tan \alpha \cot k\alpha = 1$ and has the following values: $q' = \frac{Et^3(k^2-1)}{12r^3(1-\nu^2)}$ $\alpha =$ 15° 30° 60° 90° 120° 150° 180° (Ref. 1) $k =$ 17.2 8.62 4.37 3.0 2.36 2.07 2.0
22. Thin sphere under uniform external pressure (radius of sphere $= r$) 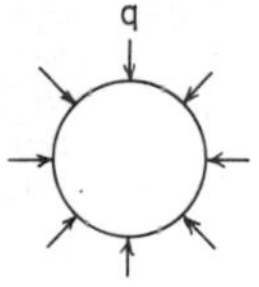 r/t >10	22a. No constraint	$q' = \frac{2Et^2}{r^2\sqrt{3(1-\nu^2)}}$ (for ideal case) (Refs. 1, 37) $q' = \frac{0.365Et^2}{r^2}$ (probable actual minimum q') For spherical cap, half-central angle ϕ between 20 and 60°, R/t between 400 and 2000, $q' = [1 - 0.00875(\phi° - 20°)]\left(1 - 0.000175\frac{R}{t}\right)(0.3E)\left(\frac{t}{R}\right)^2$ (Empirical formula, Ref. 43)
23. Thin truncated conical shell with closed ends under external pressure (both lateral and longitudinal pressure) 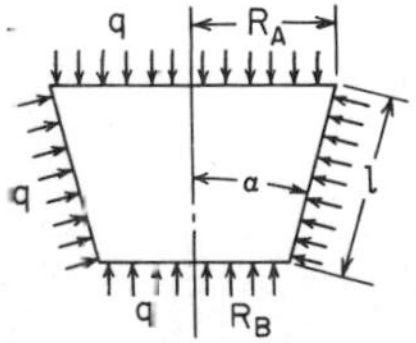 $R_B/t > 10$	23a. Ends held circular	q' can be found from the formulas of case 20a if the slant length of the cone is substituted for the length of the cylinder and if the average radius of curvature of the wall of the cone normal to the meridian $(R_A + R_B)/2 \cos \alpha$ is substituted for the radius of the cylinder. The same recommendation of a probable minimum critical pressure of $0.8q'$ is made from the examination of experimental data for cones. (Refs. 78, 81)
24. Thin truncated conical shell under axial load 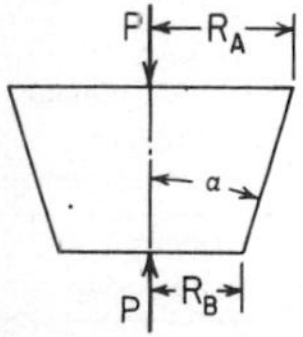 $R_B/t > 10$	24a. Ends held circular	$P' = \frac{2\pi Et^2 \cos^2 \alpha}{\sqrt{3(1-\nu^2)}}$ (theoretical) Tests indicate an actual buckling strength of from 40 to 60% of the above theoretical value, or $P' = 0.3(2\pi Et^2 \cos^2 \alpha)$ approximately. (Ref. 78) In Ref. 77 it is stated that $P' = 0.277(2\pi Et^2 \cos^2 \alpha)$ will give 95% confidence in at least 90% of the cones carrying more than this critical load. This is based on 170 tests.

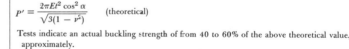

TABLE 35 *Formulas for elastic stability of plates and shells* (*Cont.*)

<table>
<tr><th>Form of plate or shell and manner of loading</th><th>Manner of support</th><th>Formulas for critical unit compressive stress σ', unit shear stress τ', load P', bending moment M', or unit external pressure q' at which elastic buckling occurs</th></tr>
<tr><td>25. Thin truncated conical shell under combined axial load and internal pressure
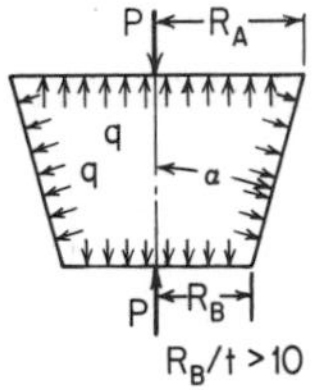
</td><td>25a. Ends held circular</td><td>$P' - q\pi R_B^2 = K_A 2\pi E t^2 \cos^2 \alpha$
The probable minimum values of K_A are tabulated for several values of $K_p = \frac{q}{E}\left(\frac{R_B}{t \cos \alpha}\right)^2$.
$k_B = 2\left[\frac{12(1 - \nu^2)R_B^2}{t^2 \tan^2 \alpha \sin^2 \alpha}\right]^{1/4}$
<table><tr><th>K_p</th><th>0.00</th><th>0.25</th><th>0.50</th><th>1.00</th><th>1.50</th><th>2.00</th><th>3.00</th></tr><tr><td>for $k_B < 150$</td><td>0.30</td><td>0.52</td><td>0.60</td><td>0.68</td><td>0.73</td><td>0.76</td><td>0.80</td></tr><tr><td>for $k_B > 150$</td><td>0.20</td><td>0.36</td><td>0.48</td><td>0.60</td><td>0.64</td><td>0.66</td><td>0.69</td></tr></table>(Ref. 78)</td></tr>
<tr><td>26. Thin truncated conical shell under combined axial load and external pressure
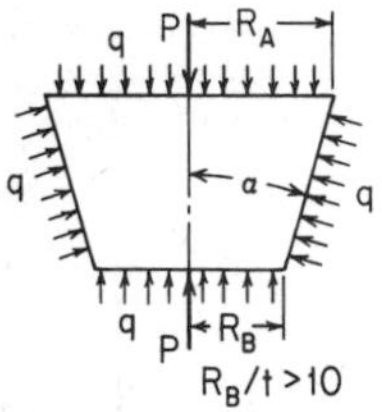
</td><td>26a. Ends held circular</td><td>The following conservative interaction formula may be used for design. It is applicable equally to theoretical values or to minimum probable values of critical load and pressure.
$$\frac{P'}{P'_{\text{case 24}}} + \frac{q'}{q'_{\text{case 23}}} = 1$$
This expression can be used for cylinders if the angle α is set equal to zero or use is made of cases 15 and 20. For small values of $P'/P'_{\text{case 24}}$ the external pressure required to collapse the shell is greater than that required to initiate buckling. See Ref. 78.</td></tr>
<tr><td>27. Thin truncated conical shell under torsion
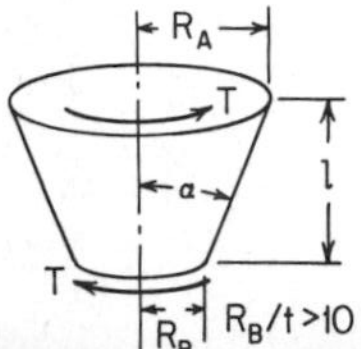
</td><td>27a. Ends held circular</td><td>Let $T = \tau' 2\pi r_e^2 t$ and for τ' use the formulas for thin-walled circular tubes, case 17, substituting for the radius r of the tube the equivalent radius r_e, where $r_e = R_B \cos \alpha \left\{1 + \left[\frac{1}{2}\left(1 + \frac{R_A}{R_B}\right)\right]^{1/2} - \left[\frac{1}{2}\left(1 + \frac{R_A}{R_B}\right)\right]^{-1/2}\right\}$. l and t remain the axial length and wall thickness respectively.
(Ref. 17)</td></tr>
</table>

in detail the buckling due to circumferential hoop compression which is developed when a truncated spherical shell is subjected to an axisymmetric tensile load.

Toroidal shells. Stein and McElman (Ref. 86) derive nonlinear equations of equilibrium and buckling equations for segments of toroidal shells; segments that are symmetric with the equator are considered for both inner and outer diameters, as well as segments centered at the crown. Sobel and Flügge (Ref. 87) tabulate and graph the minimum buckling external pressures on full toroidal shells. Almroth, Sobel, and Hunter (Ref. 88) compare favorably the theory in Ref. 87 with experiments they performed.

Corrugated tubes or bellows. An instability can develop when a corrugated tube or bellows is subjected to an internal pressure with the ends partially or totally restrained against axial displacement. (This instability can also occur in very long cylindrical vessels under similar restraints.) For a discussion and an example of this effect, see Art. 12.5.

REFERENCES

1. Timoshenko, S.: "Theory of Elastic Stability," Engineering Societies Monograph, McGraw-Hill Book Company, 1936.
2. Schwerin, E.: Die Torsionstabilität des dünnwandigen Rohres, *Z. angew. Math. Mech.*, vol. 5, no. 3, p. 235, 1925.
3. Trayer, G. W., and H. W. March: Elastic Instability of Members Having Sections Common in Aircraft Construction, *Natl. Adv. Comm. Aeron., Rept.* 382, 1931.
4. Dumont, C., and H. N. Hill: The Lateral Instability of Deep Rectangular Beams, *Natl. Adv. Comm. Aeron., Tech. Note* 601, 1937.
5. Dinnik, A.: Design of Columns of Varying Cross-section, *Trans. ASME,* vol. 54, no. 18, p. 165, 1932.
6. Heck, O. S., and H. Ebner: Methods and Formulas for Calculating the Strength of Plate and Shell Construction as Used in Airplane Design, *Natl. Adv. Comm. Aeron., Tech. Memo.* 785, 1936.
7. Maulbetsch, J. L.: Buckling of Compressed Rectangular Plates with Built-in Edges, *ASME J. Appl. Mech.,* vol. 4, no. 2, June 1937.
8. Southwell, R. V., and S. W. Skan: On the Stability under Shearing Forces of a Flat Elastic Strip, *Proc. R. Soc. Lond., Ser. A,* vol. 105, p. 582, 1924.
9. Bollenrath, F.: Wrinkling Phenomena of Thin Flat Plates Subjected to Shear Stresses, *Natl. Adv. Comm. Aeron., Tech. Memo.* 601, 1931.
10. Way, S.: Stability of Rectangular Plates under Shear and Bending Forces, *ASME J. Appl. Mech.,* vol. 3, no. 4, December 1936.
11. Smith, G. M.: Strength in Shear of Thin Curved Sheets of Alclad, *Natl. Adv. Comm. Aeron., Tech. Note* 343, 1930.
12. Lundquist, E. E.: Strength Tests of Thin-walled Duralumin Cylinders in Compression, *Natl. Adv. Comm. Aeron., Rept.* 473, 1933.
13. Wilson, W. M., and N. M. Newmark: The Strength of Thin Cylindrical Shells as Columns, *Eng. Exp. Sta. Univ. Ill., Bull.* 255, 1933.
14. Lundquist, E. E.: Strength Tests of Thin-walled Duralumin Cylinders in Pure Bending, *Natl. Adv. Comm. Aeron., Tech. Note* 479, 1933.
15. Lundquist, E. E.: Strength Tests of Thin-walled Duralumin Cylinders in Combined Transverse Shear and Bending, *Natl. Adv. Comm. Aeron., Tech. Note* 523, 1935.

16. Donnell, L. H.: Stability of Thin-walled Tubes under Torsion, *Natl. Adv. Comm. Aeron., Tech. Rept.* 479, 1933.
17. Seide, P.: On the Buckling of Truncated Conical Shells in Torsion, *ASME, J. Appl. Mech.*, vol. 29, no. 2, June 1962.
18. Ebner, H.: Strength of Shell Bodies—Theory and Practice, *Natl. Adv. Comm. Aeron., Tech. Memo.* 838, 1937.
19. Saunders, H. E., and D. F. Windenberg: Strength of Thin Cylindrical Shells under External Pressure, *Trans. ASME,* vol. 53, no. 15, p. 207, 1931.
20. von Mises, R.: Der kritische Aussendruck zylindrischer Rohre, *Z. Ver. Dtsch. Ing.*, vol. 58, p. 750, 1914.
21. Woinowsky-Krieger, S.: The Stability of a Clamped Elliptic Plate under Uniform Compression, *ASME J. Appl. Mech.*, vol. 4, no. 4, December 1937.
22. Stein, M., and J. Neff: Buckling Stresses in Simply Supported Rectangular Flat Plates in Shear, *Natl. Adv. Comm. Aeron., Tech. Note* 1222, 1947.
23. Batdorf, S. B., and M. Stein: Critical Combinations of Shear and Direct Stress for Simply Supported Rectangular Flat Plates, *Natl. Adv. Comm. Aeron., Tech. Note* 1223, 1947.
24. Batdorf, S. B., M. Schildcrout, and M. Stein: Critical Stress of Thin-walled Cylinders in Axial Compression, *Natl. Adv. Comm. Aeron., Tech. Note* 1343, 1947.
25. Batdorf, S. B., M. Stein, and M. Schildcrout: Critical Stress of Thin-walled Cylinders in Torsion, *Natl. Adv. Comm. Aeron., Tech. Note* 1344, 1947.
26. Batdorf, S. B., M. Stein, and M. Schildcrout: Critical Combinations of Torsion and Direct Axial Stress for Thin-walled Cylinders, *Natl. Adv. Comm. Aeron., Tech. Note* 1345, 1947.
27. Batdorf, S. B., M. Schildcrout, and M. Stein: Critical Shear Stress of Long Plates with Transverse Curvature, *Natl. Adv. Comm. Aeron., Tech. Note* 1346, 1947.
28. Batdorf, S. B., M. Schildcrout, and M. Stein, Critical Combinations of Shear and Longitudinal Direct Stress for Long Plates with Transverse Curvature, *Natl. Adv. Comm. Aeron., Tech. Note* 1347, 1947.
29. Batdorf, S. B., M. Stein, and M. Schildcrout: Critical Shear Stress of Curved Rectangular Panels, *Natl. Adv. Comm. Aeron., Tech. Note* 1348, 1947.
30. Budiansky, B., R. W. Connor, and M. Stein: Buckling in Shear of Continuous Flat Plates, *Natl. Adv. Comm. Aeron., Tech. Note* 1565, 1948.
31. Peters, R. W.: Buckling Tests of Flat Rectangular Plates under Combined Shear and Longitudinal Compression, *Natl. Adv. Comm. Aeron., Tech. Note* 1750, 1948.
32. Budiansky, B., M. Stein, and A. C. Gilbert: Buckling of a Long Square Tube in Torsion and Compression, *Natl. Adv. Comm. Aeron., Tech. Note* 1751, 1948.
33. Libove, C., S. Ferdman, and J. G. Reusch: Elastic Buckling of a Simply Supported Plate under a Compressive Stress that Varies Linearly in the Direction of Loading, *Natl. Adv. Comm. Aeron., Tech. Note* 1891, 1949.
34. Schildcrout, M., and M. Stein: Critical Combinations of Shear and Direct Axial Stress for Curved Rectangular Panels, *Natl. Adv. Comm. Aeron., Tech. Note* 1928, 1949.
35. Pflüger, A.: "Stabilitätsprobleme der Elastostatik," Springer-Verlag, 1964.
36. Gerard, G., and Herbert Becker: Handbook of Structural Stability, *Natl. Adv. Comm. Aeron., Tech. Notes* 3781–3786 inclusive, and D163, 1957–1959.
37. von Kármán, Th., and Hsue-shen Tsien: The Buckling of Spherical Shells by External Pressure, Pressure Vessel and Piping Design, *ASME Collected Papers* 1927–1959.
38. Cheng, Shun: On the Theory of Bending of Sandwich Plates, *Proc. 4th U.S. Natl. Congr. Appl. Mech.*, 1962.
39. U.S. Forest Products Laboratory: List of Publications on Structural Sandwich, Plastic Laminates, and Wood-base Aircraft Components, 1962.
40. Lind, N. C.: Elastic Buckling of Symmetrical Arches, *Univ. Ill., Eng. Exp. Sta. Tech. Rept.* 3, 1962.
41. Goodier, J. N., and N. J. Hoff (eds.): "Structural Mechanics," Proc. 1st Symp. Nav. Struct. Mech., Pergamon Press, 1960.

42. Collected Papers on Instability of Shell Structures, *Natl. Aeron. Space Admin., Tech. Note D*-1510, 1962.
43. Kloppel, K., and O. Jungbluth: Beitrag zum Durchschlagproblem dünnwandiger Kugelschalen, *Der Stahlbau,* 1953.
44. McComb, H. G. Jr., G. W. Zender, and M. M. Mikulas, Jr.: The Membrane Approach to Bending Instability of Pressurized Cylindrical Shells (in Ref. 42), p. 229.
45. Burkhard, A., and W. Young: Buckling of a Simply-Supported Beam between Two Unattached Elastic Foundations, *AIAA J.,* vol. 11, no. 3, March 1973.
46. Gere, J. M., and W. O. Carter: Critical Buckling Loads for Tapered Columns, *Trans. Am. Soc. Civil Eng.,* vol. 128, pt. 2, 1963.
47. Culver, C. G., and S. M. Preg, Jr.: Elastic Stability of Tapered Beam-Columns, *Proc. Am. Soc. Civil Eng.,* vol. 94, no. ST2, February 1968.
48. Burgreen, D., and P. J. Manitt: Thermal Buckling of a Bimetallic Beam, *Proc. Am. Civil Eng.,* vol. 95, no. EM2, April 1969.
49. Burgreen, D., and D. Regal: Higher Mode Buckling of Bimetallic Beam, *Proc. Am. Soc. Civil Eng.,* vol. 97, no. EM4, August 1971.
50. Austin, W. J.: In-Plane Bending and Buckling of Arches, *Proc. Am. Soc. Civil Eng.,* vol. 97, no. ST5, May 1971. Discussion by R. Schmidt, D. A. DaDeppo, and K. Forrester: *ibid.,* vol. 98, no. ST1, January 1972.
51. Chicurel, R.: Shrink Buckling of Thin Circular Rings, *ASME J. Appl. Mech.,* vol. 35, no. 3, September 1968.
52. Clark, J. W., and H. N. Hill: Lateral Buckling of Beams, *Proc. Am. Soc. Civil Eng.,* vol. 86, no. ST7, July 1960.
53. Fowler, D. W.: Design of Laterally Unsupported Timber Beams, *Proc. Am. Soc. Civil Eng.,* vol. 97, no. ST3, March 1971.
54. Massey, C., and P. J. McGuire: Lateral Stability of Nonuniform Cantilevers, *Proc. Am. Soc. Civil Eng.,* vol. 97, no. EM3, June 1971.
55. Kitipornchai, S., and N. S. Trahair: Elastic Stability of Tapered I-Beams, *Proc. Am. Soc. Civil Eng.,* vol. 98, no. ST3, March 1972.
56. Anderson, J. M., and N. S. Trahair, Stability of Monosymmetric Beams and Cantilevers, *Proc. Am. Soc. Civil Eng.,* vol. 98, no. ST1, January 1972.
57. Morrison, T. G.: Lateral Buckling of Constrained Beams, *Proc. Am. Soc. Civil Eng.,* vol. 98, no. ST3, March 1972.
58. Vijayakumar, K., and C. V. Joga Rao: Buckling of Polar Orthotropic Annular Plates, *Proc. Am. Soc. Civil Eng.,* vol. 97, no. EM3, June 1971.
59. Amon, R., and O. E. Widera: Stability of Edge-Reinforced Circular Plate, *Proc. Am. Soc. Civil Eng.,* vol. 97, no. EM5, October 1971.
60. Shuleshko, P.: Solution of Buckling Problems by Reduction Method, *Proc. Am. Soc. Civil Eng.,* vol. 90, no. Em3, June 1964.
61. Ashton, J. E.: Stability of Clamped Skew Plated Under Combined Loads, *ASME J. Appl. Mech.,* vol. 36, no. 1, March 1969.
62. Robinson, N. I.: Buckling of Parabolic and Semi-Elliptic Plates, *AIAA J.,* vol. 7, no. 6, June 1969.
63. Srinivas, S., and A. K. Rao: Buckling of Thick Rectangular Plates, *AIAA J.,* vol. 7, no. 8, August 1969.
64. Durvasula, S.: Buckling of Clamped Skew Plates, *AIAA J.,* vol. 8, no. 1, January 1970.
65. Roberts, S. B.: Buckling and Vibrations of Polygonal and Rhombic Plates, *Proc. Am. Soc. Civil Eng.,* vol. 97, no. EM2, April 1971.
66. Mikulas, M. M., Jr., and M. Stein, Buckling of a Cylindrical Shell Loaded by a Pre-Tensioned Filament Winding, *AIAA J.,* vol. 3, no. 3, March 1965.
67. Hoff, N. J.: Low Buckling Stresses of Axially Compressed Circular Cylindrical Shells of Finite Length, *ASME J. Appl. Mech.,* vol. 32, no. 3, September 1965.

68. Hoff, N. J., and L. W. Rehfield: Buckling of Axially Compressed Circular Cylindrical Shells at Stresses Smaller Than the Classical Critical Value, *ASME J. Appl. Mech.*, vol. 32, no. 3, September 1965.
69. Yao, J. C., and W. C. Jenkins: Buckling of Elliptic Cylinders under Normal Pressure, *AIAA J.*, vol. 8, no. 1, January 1970.
70. Carlson, R. L., R. L. Sendelbeck, and N. J. Hoff: Experimental Studies of the Buckling of Complete Spherical Shells, Experimental Mechanics, *J. Soc. Exp. Stress Anal.*, vol. 7, no. 7, July 1967.
71. Wu, M. T., and Shun Cheng: Nonlinear Asymmetric Buckling of Truncated Spherical Shells, *ASME J. Appl. Mech.*, vol. 37, no. 3, September 1970.
72. Singer, J.: Buckling of Circular Conical Shells under Axisymmetrical External Pressure, *J. Mech. Eng. Sci.*, vol. 3, no. 4, 1961.
73. Newman, M., and E. L. Reiss: Axisymmetric Snap Buckling of Conical Shells (in Ref. 42), p. 45.
74. Singer, J.: Buckling of Orthotropic and Stiffened Conical Shells (in Ref. 42), p. 463.
75. Seide, P.: On the Stability of Internally Pressurized Conical Shells under Axial Compression, *Proc. 4th. U.S. Natl. Cong. Appl. Mech.*, June 1962.
76. Weingarten, V. I.: Stability of Internally Pressurized Conical Shells under Torsion, *AIAA J.*, vol. 2, no. 10, October 1964.
77. Hausrath, A. H., and F. A. Dittoe; Development of Design Strength Levels for the Elastic Stability of Monocoque Cones under Axial Compression (in Ref. 42), p. 45.
78. Seide, P.: A Survey of Buckling Theory and Experiment for Circular Conical Shells of Constant Thickness (in Ref. 42), p. 401.
79. Radkowski, P. P.: Elastic Instability of Conical Shells under Combined Loading (in Ref. 42), p. 427.
80. Weingarten, V. I., E. J. Morgan, and P. Seide: Elastic Stability of Thin-Walled Cylindrical and Conical Shells under Axial Compression, *AIAA J.*, vol. 3, no. 3, March 1965.
81. Weingarten, V. I., and P. Seide: Elastic Stability of Thin-Walled Cylindrical and Conical Shells under Combined External Pressure and Axial Compression, *AIAA J.*, vol. 3, no. 5, May 1965.
82. Weingarten, V. I., E. J. Morgan, and P. Seide: Elastic Stability of Thin-Walled Cylindrical and Conical Shells under Combined Internal Pressure and Axial Compression, *AIAA J.*, vol. 3, no. 6, June 1965.
83. Tani, J., and N. Yamaki: Buckling of Truncated Conical Shells under Axial Compression, *AIAA J.*, vol. 8, no. 3, March 1970.
84. Baruch, M., O. Harari, and J. Singer: Low Buckling Loads of Axially Compressed Conical Shells, *ASME J. Appl. Mech.*, vol. 37, no. 2, June 1970.
85. Sendelbeck, R. L., and J. Singer: Further Experimental Studies of Buckling of Electroformed Conical Shells, *AIAA J.*, vol. 8, no. 8, August 1970.
86. Stein, M., and J. A. McElman: Buckling of Segments of Toroidal Shells, *AIAA J.*, vol. 3, no. 9, September 1965.
87. Sobel, L. H., and W. Flügge: Stability of Toroidal Shells under Uniform External Pressure, *AIAA J.*, vol. 5, no. 3, March 1967.
88. Almroth, B. O., L. H. Sobel, and A. R. Hunter: An Experimental Investigation of the Buckling of Toroidal Shells, *AIAA J.*, vol. 7, no. 11, November 1969.
89. Loo, Ta-Cheng, and R. M. Evan-Iwanowski: Interaction of Critical Pressures and Critical Concentrated Loads Acting on Shallow Spherical Shells, *ASME J. Appl. Mech.*, vol. 33, no. 3, September 1966.
90. Loo, Ta-Cheng, and R. M. Evan-Iwanowski: Experiments on Stability on Spherical Caps, *Proc. Am. Soc. Civil Eng.*, vol. 90, no. EM3, June 1964.
91. Arbocz, J., and C. D. Babcock, Jr.: The Effect of General Imperfections on the Buckling of Cylindrical Shells, *ASME J. Appl. Mech.*, vol. 36, no. 1, March 1969.
92. Burns, J. J. Jr.: Experimental Buckling of Thin Shells of Revolution, *Proc. Am. Soc. Civil Eng.*, vol. 90, no. EM3, June 1964.

93. Roorda, J.: Some Thoughts on the Southwell Plot, *Proc. Am. Soc. Civil Eng.*, vol. 93, no. EM6, December 1967.
94. Navaratna, D. R., T. H. H. Pian, and E. A. Witmer: Stability Analysis of Shells of Revolution by the Finite-Element Method, *AIAA J.*, vol. 6, no. 2, February 1968.
95. Stein, M.: Some Recent Advances in the Investigation of Shell Buckling, *AIAA J.*, vol. 6, no. 12, December 1968.
96. Rabinovich, I. M. (ed.): "Structural Mechanics in the U.S.S.R. 1917–1957," Pergamon Press, 1960 (English transl. edited by G. Herrmann).
97. Baker, E. H., L. Kovalevsky, and F. L. Rish: "Structural Analysis of Shells," McGraw-Hill Book Company, 1972.
98. Perrone, N.: Compendium of Structural Mechanics Computer Programs, *Comput. & Struct.*, vol. 2, no. 3, April 1972. (Available from NTIS as N71-32026, April 1971.)
99. Bushnell, D.: Stress, Stability, and Vibration of Complex, Branched Shells of Revolution, *AIAA/ASME/SAE 14th Struct., Struct. Dynam. & Mater. Conf.*, March, 1973.
100. "Structural Sandwich Composites," MIL-HDBK-23, U.S. Dept. of Defense, 1968.

FIFTEEN

Dynamic and Temperature Stresses

15.1 *Dynamic loading; general conditions*

Dynamic loading was defined in Chap. 1 as any loading during which the parts of the body cannot be considered to be in static equilibrium. It was further pointed out that two kinds of dynamic loading can be distinguished: (1) that in which the body has imposed upon it a particular kind of motion involving known accelerations, and (2) impact, of which sudden loading may be considered a special case. In the following articles, specific cases of each kind of dynamic loading will be considered.

15.2 *Body in a known state of motion*

The acceleration a of each particle of mass dm being known, the effective force on each particle is $dm \times a$, directed like a. If to each particle a force equal and opposite to the effective force were applied, equilibrium would result. If then such reversed effective forces are assumed to be applied to all the constituent particles of the body, the body may be regarded as being in equilibrium under these forces and the actual forces (loads and reactions) that act upon it, and the resulting stresses can be found exactly as for a body at rest. The reversed effective forces are *imaginary* forces exerted *on* the particles but are equal to and directed like the actual reactions the particles exert on whatever gives them their acceleration, i.e., in general, on the rest of the body. Since these reactions are due to the inertia of the particles, they are called *inertia forces,* and the body may be thought of as loaded by these inertia forces. Similarly, any attached mass will exert on a body inertia forces equal and opposite to the forces which the body has to exert on the attached mass to accelerate it.

The results of applying this method of analysis to a number of more or less typical problems are given below. In all cases, in finding the accelerations of the particles, it has been assumed that the effect of deformation could be ignored; i.e., the acceleration of each particle has been found as though the body were rigid. For convenience, stresses, bending moments, and shears due to inertia forces only are called *inertia* stresses, moments, and shears; they are calculated as though the body were outside the field of gravitation. Stresses, moments, and shears due to balanced forces (including gravity) may be superimposed thereon. The gravitational acceleration constant g is given as 32.2 ft/s/s, or 386.4 in/s/s, depending upon the units used for the imposed acceleration; for other systems of units the appropriate values of g should be used.

1. A slender uniform rod of weight W lb, length L in, section area A in^2, and modulus of elasticity E lb/in^2 is given a motion of translation with an acceleration of a ft/s/s parallel to its axis by a pull (push) applied at one end. The maximum tensile (compressive) stress occurs at the loaded end and is $\sigma = Wa/32.2A$ lb/in^2. The elongation (shortening) due to the inertia stresses is

$$e = \frac{1}{2}\frac{W}{32.2}\frac{aL}{AE} \text{ in}$$

2. The rod described in problem 1 is given a motion of translation with an acceleration of a ft/s/s normal to its axis by forces applied at each end. The maximum inertia bending moment occurs at the middle of the bar and is $M = \frac{1}{8}WaL/32.2$ in-lb. The maximum inertia vertical (transverse) shear occurs at the ends and is $V = \frac{1}{2}Wa/32.2$ lb.

3. The rod described in problem 1 is made to rotate about an axis through one end normal to its length at a uniform angular velocity of ω rad/s. The maximum tensile inertia stress occurs at the pinned end and is

$$\sigma = \frac{1}{2}\frac{W}{386.4}\frac{L\omega^2}{A} \text{ lb/in}^2$$

The elongation due to inertia stresses is

$$e = \frac{1}{3}\frac{W}{386.4}\frac{L^2\omega^2}{AE} \text{ in}$$

4. The rod described in problem 1 is pinned at the lower end and allowed to swing down under the action of gravity from an initially vertical position.

When the rod reaches a position where it makes with the vertical the angle θ, it is subjected to a positive bending moment (owing to its weight and the inertia forces) which has its maximum value at a section a distance $\frac{1}{3}L$ from the pinned end. This maximum value is $M = \frac{1}{27}WL \sin\theta$ in-lb. The maximum positive inertia shear occurs at the pinned end and is $V = \frac{1}{4}W \sin\theta$ lb. The maximum negative inertia shear occurs at a section a distance $\frac{2}{3}L$ from the pinned end and is $V = -\frac{1}{12}W \sin\theta$ lb. The axial force at any section x in from the pinned end is given by

$$H = \frac{3W}{2}\left(1 - \frac{x^2}{L^2}\right) - \frac{W\cos\theta}{2}\left(5 - 2\frac{x}{L} - 3\frac{x^2}{L^2}\right)$$

and becomes tensile near the free end when θ exceeds 41.4°. (This case represents approximately the conditions existing when a chimney or other slender structure topples over, and the bending moment M explains the tendency of such a structure to break near the one-third point while falling.)

5. The rod described in problem 1 is pinned at the lower end and, while in the vertical position, has imposed upon its lower end a horizontal acceleration of a ft/s/s. The maximum inertia bending moment occurs at a section a distance $\frac{1}{3}L$ from the lower end and is $M = \frac{1}{27}WLa/32.2$ in-lb. The maximum inertia shear is in the direction of the acceleration, is at the lower end, and is $V = \frac{1}{4}Wa/32.2$ lb. The maximum inertia shear in the opposite direction occurs at a section a distance $\frac{2}{3}L$ from the lower end and is $V = \frac{1}{12}Wa/32.2$ lb. (This case represents approximately the conditions existing when a chimney or other slender structure without anchorage is subjected to an earthquake shock.)

6. A uniform circular ring of mean radius R in and weight δ lb/in³, having a thickness in the plane of curvature that is very small compared with R, rotates about its own axis with a uniform angular velocity of ω rad/s. The ring is subjected to a uniform tangential tensile inertia stress

$$\sigma = \frac{\delta R^2\omega^2}{386.4}\,\text{lb/in}^2$$

7. A solid homogeneous circular disk of uniform thickness (or a solid cylinder) of radius R in, Poisson's ratio ν, and density δ lb/in³ rotates about its own axis with a uniform angular velocity of ω rad/s. At any point a distance r in from the center there is a radial tensile inertia stress

$$\sigma_r = \frac{1}{8}\frac{\delta\omega^2}{386.4}[(3+\nu)(R^2 - r^2)]\ \text{lb/in}^2 \qquad (1)$$

and a tangential tensile inertia stress

$$\sigma_t = \frac{1}{8}\frac{\delta\omega^2}{386.4}[(3+\nu)R^2 - (1+3\nu)r^2]\ \text{lb/in}^2 \qquad (2)$$

The maximum radial stress and maximum tangential stress are equal, occur at the center, and are

$$\text{Max } \sigma_r = \text{max } \sigma_t = \frac{1}{8}\frac{\delta\omega^2}{386.4}(3 + \nu)R^2 \text{ lb/in}^2 \tag{3}$$

8. A homogeneous annular disk of uniform thickness outer radius R in, and density δ lb/in^3, with a central hole of radius R_0 in, rotates about its own axis with a uniform angular velocity of ω rad/s. At any point a distance r in from the center there is a radial tensile inertia stress

$$\sigma_r = \frac{3 + \nu}{8}\frac{\delta\omega^2}{386.4}\left(R^2 + R_0^2 - \frac{R^2R_0^2}{r^2} - r^2\right) \text{ lb/in}^2 \tag{4}$$

and a tangential tensile inertia stress

$$\sigma_t = \frac{1}{8}\frac{\delta\omega^2}{386.4}\left[(3 + \nu)\left(R^2 + R_0^2 + \frac{R^2R_0^2}{r^2}\right) - (1 + 3\nu)r^2\right] \text{ lb/in}^2 \tag{5}$$

The maximum radial stress occurs at $r = \sqrt{RR_0}$ and is

$$\text{Max } \sigma_r = \frac{3 + \nu}{8}\frac{\delta\omega^2}{386.4}(R - R_0)^2 \text{ lb/in}^2 \tag{6}$$

and the maximum tangential stress occurs at the perimeter of the hole and is

$$\text{Max } \sigma_t = \frac{1}{4}\frac{\delta\omega^2}{386.4}[(3 + \nu)R^2 + (1 - \nu)R_0^2] \text{ lb/in}^2 \tag{7}$$

The change in the outer radius is

$$\Delta R = \frac{1}{4}\frac{\delta\omega^2}{386.4}\frac{R}{E}[(1 - \nu)R^2 + (3 + \nu)R_0^2] \text{ in} \tag{8}$$

and the change in the inner radius is

$$\Delta R_0 = \frac{1}{4}\frac{\delta\omega^2}{386.4}\frac{R_0}{E}[(3 + \nu)R^2 + (1 - \nu)R_0^2] \text{ in} \tag{9}$$

If there are radial pressures or pulls distributed uniformly along either the inner or outer perimeter of the disk, such as a radial pressure from the shaft or a centrifugal pull from parts attached to the rim, the stresses due thereto can be found by the formula for thick cylinders (Table 32) and superimposed upon the inertia stresses given by the preceding formulas.

9. A homogeneous circular disk of conical section (Fig. 15.1) of density δ lb/in^3 rotates about its own axis with a uniform angular velocity of N rpm. At any point a distance r in from the center, the tensile inertia stresses σ_r and σ_t are given by

$$\sigma_r = TK_r + Ap_1 + Bp_2 \text{ lb/in}^2 \tag{10}$$
$$\sigma_t = TK_t + Aq_1 + Bq_2 \text{ lb/in}^2 \tag{11}$$

where $T = 0.0000282R^2N^2\delta$ (or for steel, $T = 0.000008R^2N^2$); K_r, K_t, p_1, p_2, q_1, and q_2 are given by the following table; and A and B are constants which

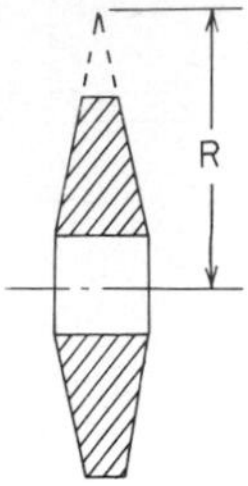

Fig. 15.1

may be found by setting σ_r equal to its known or assumed values at the inner and outer perimeters and solving the resulting equations simultaneously for A and B, as in the example on page 571. [See papers by Hodkinson and Rushing (Refs. 1 and 2) from which Eqs. 8 and 9 and the tabulated coefficients are taken.]

Tabulated values of coefficients

r/R	K_r	K_t	p_1	q_1	p_2	q_2
0.00	0.1655	0.1655	1.435	1.435	∞	∞
0.05	0.1709	0.1695	1.475	1.497	−273.400	288.600
0.10	0.1753	0.1725	1.559	1.518	−66.620	77.280
0.15	0.1782	0.1749	1.627	1.565	−28.680	36.550
0.20	0.1794	0.1763	1.707	1.617	−15.540	21.910
0.25	0.1784	0.1773	1.796	1.674	−9.553	14.880
0.30	0.1761	0.1767	1.898	1.738	−6.371	10.890
0.35	0.1734	0.1757	2.015	1.809	−4.387	8.531
0.40	0.1694	0.1739	2.151	1.890	−3.158	6.915
0.45	0.1635	0.1712	2.311	1.983	−2.328	5.788
0.50	0.1560	0.1675	2.501	2.090	−1.743	4.944
0.55	0.1465	0.1633	2.733	2.217	−1.309	4.301
0.60	0.1355	0.1579	3.021	2.369	−0.9988	3.816
0.65	0.1229	0.1525	3.390	2.556	−0.7523	3.419
0.70	0.1094	0.1445	3.860	2.794	−0.5670	3.102
0.75	0.0956	0.1370	4.559	3.111	−0.4161	2.835
0.80	0.0805	0.1286	5.563	3.557	−0.2971	2.614
0.85	0.0634	0.1193	7.263	4.276	−0.1995	2.421
0.90	0.0442	0.1100	10.620	5.554	−0.1203	2.263
0.95	0.0231	0.0976	20.645	8.890	−0.0555	2.140
1.00	0.0000	0.0840	∞	∞	−0.0000	2.051

10. A homogeneous circular disk of hyperbolic section (Fig. 15.2) of density δ lb/in^3 rotates about its own axis with uniform angular velocity ω rad/s. The equation $t = cr^a$ defines the section, where if t_1 = thickness at radius r_1 and t_2 at radius r_2,

$$a = \frac{\log_e (t_1/t_2)}{\log_e (r_1/r_2)}$$

and $$\log_e c = \log_e t_1 - a \log_e r_1 = \log_e t_2 - a \log_e r_2$$

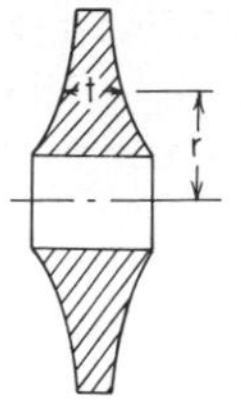

Fig. 15.2

(For taper toward the rim, a is negative; and for uniform t, $a = 0$.) At any point a distance r in from the center the tensile inertia stresses σ_r and σ_t are

$$\sigma_r = \frac{E}{1-\nu^2}[(3+\nu)Fr^2 + (m_1+\nu)Ar^{m_1-1} + (m_2+\nu)Br^{m_2-1}] \text{ lb/in}^2 \qquad (12)$$

$$\sigma_t = \frac{E}{1-\nu^2}[(1+3\nu)Fr^2 + (1+m_1\nu)Ar^{m_1-1} + (1+m_2\nu)Br^{m_2-1}] \text{ lb/in}^2 \qquad (13)$$

where $F - \dfrac{-(1-\nu^2)\delta\omega^2/386.4}{E[8+(3+\nu)a]}$

$$m_1 = -\frac{a}{2} - \sqrt{\frac{a^2}{4} - a\nu + 1}$$

$$m_2 = -\frac{a}{2} + \sqrt{\frac{a^2}{4} - a\nu + 1}$$

A and B are constants, found by setting σ_r equal to its known or assumed values at the inner and outer perimeters and solving the two resulting equations simultaneously for A and B. [Equations 12 and 13 are taken from Stodola (Ref. 3) with some changes in notation.]

11. A homogeneous circular disk with section bounded by curves and straight lines (Fig. 15.3) rotates about its own axis with a uniform angular velocity N rpm. The disk is imagined divided into annular rings of such width that each ring can be regarded as having a section with hyperbolic outline, as in problem 10. For each ring, a is calculated by the formulas of problem 10, using the inner and outer radii and the corresponding thick-

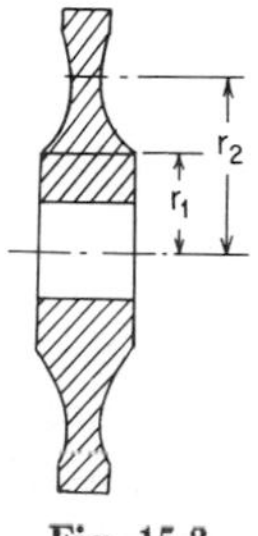

Fig. 15.3

nesses. Then, if r_1 and r_2 represent, respectively, the inner and outer radii of any ring, the tangential stresses σ_{t_1} and σ_{t_2} at the inner and outer boundaries of the ring are related to the corresponding radial stresses σ_{r_1} and σ_{r_2} as follows:

$$\sigma_{t_1} = Ar_2^2 - B\sigma_{r_1} + C\sigma_{r_2} \text{ lb/in}^2 \tag{14}$$

$$\sigma_{t_2} = Dr_2^2 - E\sigma_{r_1} + F\sigma_{r_2} \text{ lb/in}^2 \tag{15}$$

where $B = -\dfrac{m_2 K^{m_1-1} - m_1 K^{m_2-1}}{K^{m_2-1} - K^{m_1-1}}$

$$K = \frac{r_1}{r_2}$$

$$E = -\frac{m_2 - m_1}{K^{m_2-1} - K^{m_1-1}}$$

$$C = \frac{E}{K^{a+2}}$$

$$F = B + a$$

$$A = -\frac{7.956(N/1000)^2}{8 + 3.3a}[1.9K^2 + 3.3(K^2B - C)]$$

$$D = -\frac{7.956(N/1000)^2}{8 + 3.3a}[1.9 + 3.3(K^2E - F)]$$

$$m_1 = -\frac{a}{2} - \sqrt{\frac{a^2}{4} - 0.3a + 1}$$

$$m_2 = -\frac{a}{2} + \sqrt{\frac{a^2}{4} - 0.3a + 1}$$

The preceding formulas, which are given by Loewenstein (Ref. 4) are directly applicable to steel, for which the values $\nu = 0.3$ and $\delta = 0.28$ lb/in^3 have been assumed.

Two values of σ_r are known or can be assumed, viz., the values at the inner and outer perimeters of the disk. Then, by setting the tangential stress at the outer boundary of each ring equal to the tangential stress at the inner boundary of the adjacent larger ring, one equation in σ_r will be obtained for each common ring boundary. In this case the modulus of elasticity is the same for adjacent rings and the radial stress σ_r at the boundary is common to both rings, and so the tangential stresses can be equated instead of the tangential strains (Eq. 15 for the smaller ring equals Eq. 14 for the larger ring). Therefore there are as many equations as there are unknown boundary radial stresses, and hence the radial stress at each boundary can be found. The tangential stresses can then be found by Eqs. 14 and 15, and then the stresses at any point in a ring can be found by using, in Eq. 14, the known values of σ_{t_1} and σ_{r_1} and substituting for σ_{r_2} the unknown radial stress σ_r, and for r_2 the corresponding radius r.

A fact of importance with reference to turbine disks or other rotating bodies is that geometrically similar disks of different sizes will be equally stressed at corresponding points when running at the same *peripheral* velocity. Furthermore, for any given peripheral velocity, the axial and radial dimensions of a rotating body may be changed independently of each other and in any ratio without affecting the stresses at similarly situated points.

EXAMPLE

The conical steel disk shown in section in Fig. 15.4 rotates at 2500 rpm. To its rim it has attached buckets, the aggregate mass of which amounts to $w = 0.75$ lb/linear in of rim; this mass may be considered to be centered 30 in from the axis. It is desired to determine the stresses at a point 7 in from the axis.

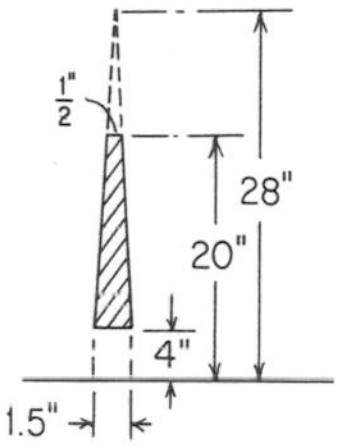

Fig. 15.4

Solution. From the dimensions of the section, R is found to be 28 in. The values of r/R for the inner and outer perimeters and for the circumference $r = 7$ are calculated, and the corresponding coefficients K_r, K_t, etc., are determined from the table by graphic interpolation. The results are tabulated here for convenience:

	r/R	K_r	K_t	p_1	q_1	p_2	q_2
Inner rim	0.143	0.1780	0.1747	1.616	1.558	−32.5	40.5
Outer rim	0.714	0.1055	0.1425	4.056	2.883	−0.534	3.027
$r = 7$ in	0.25	0.1784	0.1773	1.796	1.674	−9.553	14.88

The attached mass exerts on the rim outward inertia forces which will be assumed to be uniformly distributed; the amount of force per linear inch is

$$p = \frac{\omega}{g}\omega^2 r = \frac{0.75}{386.4}(261.5^2)(30) = 3980 \text{ lb/linear in}$$

Therefore at the outer rim $\sigma_r = 7960$.

It is usual to design the shrink fit so that in operation the hub pressure is a few hundred pounds; it will be assumed that the radial stress at the inner rim $\sigma_r = -700$. The value of $T = 0.000008(28^2)(2500^2) = 39{,}200$. Having two values of σ_r, Eq. 10 can now be written

$$-700 = (39{,}200)(0.1780) + A(1.616) + B(-32.5) \qquad \text{(inner rim)}$$

$$7960 = (39{,}200)(0.1055) + A(4.056) + B(-0.534) \qquad \text{(outer rim)}$$

The solution gives

$$A = 973 \qquad B = 285$$

The stresses at $r = 7$ are now found by Eqs. 10 and 11 to be

$$\sigma_r = (39{,}200)(0.1784) + (973)(1.796) + (285)(-9.553) = 6020 \text{ lb/in}^2$$
$$\sigma_t = (39{,}200)(0.1773) + (973)(1.674) + (285)(14.88) = 12{,}825 \text{ lb/in}^2$$

Bursting speed. The formulas given above for stresses in rotating disks presuppose *elastic* conditions; when the elastic limit is exceeded, plastic yielding tends to equalize the stress intensity along a diametral plane. Because of this, the average stress σ_a on such a plane is perhaps a better criterion of margin of safety against bursting than is the maximum stress computed for elastic conditions. For a solid disk of uniform thickness (case 7),

$$\sigma_a = 0.000863\delta\omega^2R^2$$

For a pierced disk (case 8),

$$\sigma_a = \frac{0.000863\delta\omega^2(R^3 - R_0^3)}{R - R_0}$$

Tests (Refs. 12 and 13) have shown that for some materials, rupture occurs in both solid and pierced disks when σ_a, computed for the original dimensions, becomes equal to the ultimate tensile strength of the material as determined by a conventional test. On the other hand, some materials fail at values of σ_a as low as 61.5 percent of the ultimate strength, and the lowest values have been observed in tests of solid disks. The ratio of σ_a at failure to the ultimate strength does not appear to be related in any consistent way to the ductility of the material; it seems probable that it depends on the form of the stress-strain diagram. In none of the tests reported did the weakening effect of a central hole prove to be nearly as great as the formulas for elastic stress would seem to indicate.

15.3 *Impact and sudden loading*

When a force is suddenly applied to an elastic body (as by a blow), a wave of stress is propagated, which travels through the body with a velocity

$$V = \sqrt{\frac{386.4E}{\delta}} \text{ in/s} \tag{16}$$

where E is the modulus of elasticity of the material in pounds per square inch and δ is the density of the material in pounds per cubic inch.

Bar with free ends. When one end of an unsupported uniform elastic bar is subjected to longitudinal impact from a rigid body moving with velocity v in/s, a wave of compressive stress of intensity

$$\sigma = \frac{v}{V}E = v\sqrt{\frac{\delta E}{386.4}} \text{ lb/in}^2 \tag{17}$$

is propagated. The intensity of stress is seen to be independent of the mass of the moving body, but the length of the stressed zone, or volume of material simultaneously subjected to this stress, does depend on the mass of the moving body. If this mass is infinite (or very large compared with that of the bar), the wave of compression is reflected back from the free end of the bar as a wave of tension and returns to the struck end after a period $t_1 = 2L/V$ s, where L is the length of the bar in inches and the period t_1 is the duration of contact between bar and body.

If the impinging body is very large compared with the bar (so that its mass may be considered infinite), the bar, after breaking contact, moves with a velocity $2v$ in the direction of the impact and is free of stress. If the mass of the impinging body is μ times the mass of the bar, the average velocity of the bar after contact is broken is

$$\mu v\,(1 - e^{-2/\mu})$$

and it is left vibrating with a stress of intensity

$$\sigma = \frac{v}{V} E\, e^{-\beta t_1}\ \text{lb/in}^2$$

where $\beta = 19.67A\sqrt{\delta E/M}$, A being the section area of the bar in square inches and M the mass of the moving body in pounds.

Bar with one end fixed. If one end of a bar is fixed, the wave of compressive stress resulting from impact on the free end is reflected back unchanged from the fixed end and combines with advancing waves to produce a maximum stress very nearly equal to

$$\text{Max}\ \sigma = \frac{v}{V} E \left(1 + \sqrt{\mu + \frac{2}{3}}\right) \tag{18}$$

where, as before, μ denotes the ratio of the mass of the moving body to the mass of the bar. The total time of contact is approximately

$$t_1 = \frac{L}{V}\left[\pi\sqrt{\mu + \frac{1}{2}} - \frac{1}{2}\right] \text{s}$$

[The above formulas are taken from the paper by Donnell (Ref. 5); see also Ref. 17.]

Sudden loading. If a dead load is suddenly transferred to the free end of a bar, the other end being fixed, the resulting state of stress is characterized by waves, as in the case of impact. The space-average value of the pull exerted by the bar on the load is not half the maximum tension, as is usually assumed, but is somewhat greater than that, and therefore the maximum stress that results from sudden loading is somewhat less than twice that which results from static loading. Love (Ref. 6) shows that if μ (the ratio of the mass of the load to that of the bar) is 1, sudden loading causes 1.63 times as much stress as static loading; for $\mu = 2$, the ratio is 1.68; for $\mu = 4$, it is 1.84; and it approaches 2 as a limit as μ increases. It can be seen that the ordinary assumption that sudden loading causes twice as much stress and deflection as static loading is always a safe one to make.

Moving load on beam. If a constant *force* moves at uniform speed across a beam with simply supported ends, the maximum deflection produced exceeds the static deflection that the same force would produce. If v represents the velocity of the force, l the span, and ω the lowest natural vibration frequency of the (unloaded) beam, then theoretically the maximum value of the ratio of dynamic to static deflection is 1.74; it occurs for $v = \omega l/1.64\pi$ and at the instant when the force has progressed a distance $0.757l$ along the span (Refs. 15 and 16).

If a constant *mass* W moves across a simple beam of relatively negligible mass, then the maximum ratio of dynamic to static deflection is $[1 + (v^2/g)(Wl/3EI)]$.* (Note that consistant units must be used in the preceding equations.)

Vibration. A very important type of dynamic loading occurs when an elastic body vibrates under the influence of a periodic impulse. This occurs whenever a rotating or reciprocating mass is unbalanced and also under certain conditions of fluid flow. The most serious situation arises when the impulse synchronizes (or nearly synchronizes) with the natural period of vibration, and it is of the utmost importance to guard against this condition of resonance (or near resonance). There is always some resistance to vibration, whether natural or introduced; this is called *damping* and tends to prevent vibrations of excessive amplitude. In the absence of effective damping, the amplitude y for near resonance vibration will much exceed the deflection y_s that would be produced by the same force under static conditions. The ratio y/y_s, called the *relative amplification factor,* in the absence of damping, is $1/[1 - (f/f_n)^2]$, where f is the frequency of the forcing impulse and f_n is the natural frequency of the elastic system. Obviously, it is necessary to know at least approximately the natural period of vibration of a member in order to guard against resonance.

Thomson (Ref. 19) describes in detail analytical and numerical techniques for determining resonant frequencies for systems with single and multiple degrees of freedom; he also describes methods and gives numerous examples for torsional and lateral vibrations of rods and beams. Huang (Ref. 22) has tabulated the first five resonant frequencies as well as deflections, slopes, bending moments, and shearing forces for each frequency at intervals of $0.02l$ for uniform beams; these are available for six combinations of boundary conditions. He has also included the first five resonant frequencies for all combinations of 7 different amounts of correction for rotary inertia and 10 different amounts of correction for lateral shear deflection; many mode shapes for these corrections are also included. In Table 36 the resonant frequencies and nodal locations are listed for the several boundary conditions with no corrections for rotary inertia or shear deflection. (Corrections for rotary

*From Timoshenko's "Vibration Problems in Engineering," copyright 1955, D. Van Nostrand Company, Inc., Princeton, N.J.

inertia and shear deflection have a relatively small effect on the fundamental frequency but a proportionally greater effect on the higher modes.)

Leissa (Ref. 20) has compiled, compared, and in some cases extended most of the known work on the vibration of plates; where possible, mode shapes are given in addition to the many resonant frequencies. Table 36 lists only a very few simple cases. Similarly, Leissa (Ref. 21) has done an excellent job of reporting the known work on the vibration of shells. Since, in general, this work must involve three additional variables—the thickness/radius ratio, length/radius ratio, and Poisson's ratio—no results are included here.

A simple but close approximation for the fundamental frequency of a uniform thin plate of arbitrary shape having any combination of fixed, partially fixed, or simply supported boundaries is given by Jones in Ref. 23. The equation

$$f = \frac{1.2769}{2\pi}\sqrt{\frac{g}{\delta_{max}}}$$

is based on his work where δ_{max} is the maximum static deflection produced by the weight of the plate and any uniformly distributed mass attached to the plate and vibrating with it. It is based on the expression for the fundamental frequency of a clamped elliptical plate but, as Jones points out with several examples of triangular, rectangular, and circular plates having various combinations of boundary conditions, it should hold equally well for all uniform plates having no free boundaries. In the 16 examples he presents, the maximum error in frequency is about 3 percent.

15.4 *Impact and sudden loading; approximate formulas*

If it is assumed that the stresses due to impact are distributed throughout any elastic body exactly as in the case of static loading, then it can be shown that the vertical deformation d_i and stress σ_i produced in any such body (bar, beam, truss, etc.) by the vertical impact of a body falling from a height of h in are greater than the deformation d and stress σ produced by the weight of the same body applied as a static load in the ratio

$$\frac{d_i}{d} = \frac{\sigma_i}{\sigma} = 1 + \sqrt{1 + 2\frac{h}{d}} \qquad (19)$$

If $h = 0$, we have the case of sudden loading, and $d_i/d = \sigma_i/\sigma = 2$, as usually assumed.

If the impact is horizontal instead of vertical, the impact deformation and stress are given by

$$\frac{d_i}{d} = \frac{\sigma_i}{\sigma} = \sqrt{\frac{v^2}{386.4d}} \qquad (20)$$

TABLE 36 *Natural frequencies of vibration for continuous members*

NOTATION: f = natural frequency (cycles per second); K_n = constant where n refers to the mode of vibration; g = gravitational acceleration (units consistent with length dimensions); E = modulus of elasticity; I = area moment of inertia; $D = Et^3/12(1 - v^2)$

Case no. and description	Natural frequencies
1. Uniform beam; both ends simply supported	
1a. Center load W, beam weight negligible	$f_1 = \dfrac{6.93}{2\pi}\sqrt{\dfrac{EIg}{Wl^3}}$ Ref. 22
1b. Uniform load w per unit length including beam weight	$f_n = \dfrac{K_n}{2\pi}\sqrt{\dfrac{EIg}{wl^4}}$

Mode	K_n	Nodal positions/l					
1	9.87	0.0	1.00				
2	39.5	0.0	0.50	1.00			
3	88.8	0.0	0.33	0.67	1.00		
4	158	0.0	0.25	0.50	0.75	1.00	
5	247	0.0	0.20	0.40	0.60	0.80	1.00

Case no. and description	Natural frequencies
1c. Uniform load w per unit length plus a center load W	$f_1 = \dfrac{6.93}{2\pi}\sqrt{\dfrac{EIg}{Wl^3 + 0.486\,wl^4}}$ approximately
2. Uniform beam; both ends fixed	
2a. Center load W, beam weight negligible	$f_1 = \dfrac{13.86}{2\pi}\sqrt{\dfrac{EIg}{Wl^3}}$
2b. Uniform load w per unit length including beam weight	$f_n = \dfrac{K_n}{2\pi}\sqrt{\dfrac{EIg}{wl^4}}$

Mode	K_n	Nodal position/l					
1	22.4	0.0	1.00				
2	61.7	0.0	0.50	1.00			
3	121	0.0	0.36	0.64	1.00		
4	200	0.0	0.28	0.50	0.72	1.00	
5	299	0.0	0.23	0.41	0.59	0.77	1.00

(Ref. 22)

Case no. and description	Natural frequencies
2c. Uniform load w per unit length plus a center load W	$f_1 = \dfrac{13.86}{2\pi}\sqrt{\dfrac{EIg}{Wl^3 + 0.383\,wl^4}}$ approximately
3. Uniform beam; left end fixed, right end free (cantilever)	
3a. Right end load W, beam weight negligible	$f_1 = \dfrac{1.732}{2\pi}\sqrt{\dfrac{EIg}{Wl^3}}$
3b. Uniform load w per unit length including beam weight	$f_n = \dfrac{K_n}{2\pi}\sqrt{\dfrac{EIg}{wl^4}}$

Mode	K_n	Nodal position/l				
1	3.52	0.0				
2	22.0	0.0	0.783			
3	61.7	0.0	0.504	0.868		
4	121	0.0	0.358	0.644	0.905	
5	200	0.0	0.279	0.500	0.723	0.926

(Ref. 22)

<table>
<tr><td></td><td>3c. Uniform load w per unit length plus an end load W</td><td>$f_1 = \frac{1.732}{2\pi}\sqrt{\frac{EIg}{Wl^3 + 0.236wl^4}}$ approximately</td></tr>
<tr><td>4. Uniform beam; both ends free</td><td>4a. Uniform load w per unit length including beam weight</td><td>$f_n = \frac{K_n}{2\pi}\sqrt{\frac{EIg}{wl^4}}$
<table>
<tr><th>Mode</th><th>K_n</th><th colspan="6">Nodal position/l</th></tr>
<tr><td>1</td><td>22.4</td><td>0.224</td><td>0.776</td><td></td><td></td><td></td><td></td></tr>
<tr><td>2</td><td>61.7</td><td>0.132</td><td>0.500</td><td>0.868</td><td></td><td></td><td></td></tr>
<tr><td>3</td><td>121</td><td>0.095</td><td>0.356</td><td>0.644</td><td>0.905</td><td></td><td></td></tr>
<tr><td>4</td><td>200</td><td>0.074</td><td>0.277</td><td>0.500</td><td>0.723</td><td>0.926</td><td></td></tr>
<tr><td>5</td><td>299</td><td>0.060</td><td>0.226</td><td>0.409</td><td>0.591</td><td>0.774</td><td>0.940</td></tr>
</table>
(Ref. 22)</td></tr>
<tr><td>5. Uniform beam; left end fixed, right end hinged</td><td>5a. Uniform load w per unit length including beam weight</td><td>$f_n = \frac{K_n}{2\pi}\sqrt{\frac{EIg}{wl^4}}$
<table>
<tr><th>Mode</th><th>K_n</th><th colspan="6">Nodal position/l</th></tr>
<tr><td>1</td><td>15.4</td><td>0.0</td><td>1.000</td><td></td><td></td><td></td><td></td></tr>
<tr><td>2</td><td>50.0</td><td>0.0</td><td>0.557</td><td>1.000</td><td></td><td></td><td></td></tr>
<tr><td>3</td><td>104</td><td>0.0</td><td>0.386</td><td>0.692</td><td>1.000</td><td></td><td></td></tr>
<tr><td>4</td><td>178</td><td>0.0</td><td>0.295</td><td>0.529</td><td>0.765</td><td>1.000</td><td></td></tr>
<tr><td>5</td><td>272</td><td>0.0</td><td>0.239</td><td>0.428</td><td>0.619</td><td>0.810</td><td>1.000</td></tr>
</table>
(Ref. 22)</td></tr>
<tr><td>6. Uniform beam; left end hinged, right end free</td><td>6a. Uniform load w per unit length including beam weight</td><td>$f_n = \frac{K_n}{2\pi}\sqrt{\frac{EIg}{wl^4}}$
<table>
<tr><th>Mode</th><th>K_n</th><th colspan="6">Nodal position/l</th></tr>
<tr><td>1</td><td>15.4</td><td>0.0</td><td>0.736</td><td></td><td></td><td></td><td></td></tr>
<tr><td>2</td><td>50.0</td><td>0.0</td><td>0.446</td><td>0.853</td><td></td><td></td><td></td></tr>
<tr><td>3</td><td>104</td><td>0.0</td><td>0.308</td><td>0.617</td><td>0.898</td><td></td><td></td></tr>
<tr><td>4</td><td>178</td><td>0.0</td><td>0.235</td><td>0.471</td><td>0.707</td><td>0.922</td><td></td></tr>
<tr><td>5</td><td>272</td><td>0.0</td><td>0.190</td><td>0.381</td><td>0.571</td><td>0.763</td><td>0.937</td></tr>
</table>
(Ref. 22)</td></tr>
<tr><td rowspan="3">7. Uniform bar or spring vibrating along its longitudinal axis; upper end fixed, lower end free</td><td>7a. Weight W at lower end, bar weight negligible</td><td>$f_1 = \frac{1}{2\pi}\sqrt{\frac{kg}{W}}$ for a spring where k is the spring constant
$f_1 = \frac{1}{2\pi}\sqrt{\frac{AEg}{Wl}}$ for a bar where A is the area</td></tr>
<tr><td>7b. Uniform load w per unit length including bar weight.</td><td>$f_n = \frac{K_n}{2\pi}\sqrt{\frac{AEg}{Wl^2}}$ where $K_1 = 1.57$ $K_2 = 4.71$ $K_3 = 7.85$</td></tr>
<tr><td>7c. Uniform load w per unit length plus a load W at the lower end</td><td>$f_1 = \frac{1}{2\pi}\sqrt{\frac{kg}{W + wl/3}}$ approximately for a spring where k is the spring constant
$f_1 = \frac{1}{2\pi}\sqrt{\frac{AEg}{Wl + wl^2/3}}$ approximately for a bar where A is the area</td></tr>
</table>

TABLE 36 ***Natural frequencies of vibration for continuous members (Cont.)***

Case no. and description		Natural frequencies
8. Uniform shaft or bar in torsional vibration; one end fixed, the other end free	8a. Concentrated end mass of J mass moment of inertia, shaft weight negligible	$f_1 = \frac{1}{2\pi}\sqrt{\frac{GK}{Jl}}$ where G is the shear modulus of elasticity and K is the torsional stiffness constant (see Chapter 9)
	8b. Uniform distribution of mass moment of inertia along shaft; J_s = total distributed mass moment of inertia	$f_n = \frac{K_n}{2\pi}\sqrt{\frac{GK}{J_s l}}$ where $K_1 = 1.57$ $K_2 = 4.71$ $K_3 = 7.85$
	8c. Uniformly distributed inertia plus a concentrated end mass	$f_1 = \frac{1}{2\pi}\sqrt{\frac{GK}{(J + J_s/3)l}}$ approximately
9. String vibrating laterally under a tension T with both ends fixed	9a. Uniform load w per unit length including own weight	$f = \frac{K_n}{2\pi}\sqrt{\frac{Tg}{wl^2}}$ where $K_1 = \pi$ $K_2 = 2\pi$ $K_3 = 3\pi$
10. Circular flat plate of uniform thickness t and radius r; edge fixed	10a. Uniform load w per unit area including own weight	$f = \frac{K_n}{2\pi}\sqrt{\frac{Dg}{wr^4}}$ where $K_1 = 10.2$ fundamental $K_2 = 21.3$ one nodal diameter $K_3 = 34.9$ two nodal diameters $K_4 = 39.8$ one nodal circle (Ref. 20)
11. Circular flat plate of uniform thickness t and radius r; edge simply supported	11a. Uniform load w per unit area including own weight; $\nu = 0.3$	$f = \frac{K_n}{2\pi}\sqrt{\frac{Dg}{wr^4}}$ where $K_1 = 4.99$ fundamental $K_2 = 13.9$ one nodal diameter $K_3 = 25.7$ two nodal diameters $K_4 = 29.8$ one nodal circle (Ref. 20)
12. Circular flat plate of uniform thickness t and radius r; edge free	12a. Uniform load w per unit area including own weight; $\nu = 0.33$	$f = \frac{K_n}{2\pi}\sqrt{\frac{Dg}{wr^4}}$ where $K_1 = 5.25$ two nodal diameters $K_2 = 9.08$ one nodal circle $K_3 = 12.2$ three nodal diameters $K_4 = 20.5$ one nodal diameter and one nodal circle (Ref. 20)

<table>
<tr><td>13. Circular flat plate of uniform thickness t and radius r; edge simply supported with an additional edge constraining moment $M = \beta\psi$ per unit circumference where ψ is the edge rotation</td><td>13a. Uniform load w per unit area including own weight; $\nu = 0.3$</td><td>$f = \frac{K_n}{2\pi}\sqrt{\frac{Dg}{wr^4}}$ where K_n is tabulated for various degrees of edge stiffness in the form of $\beta a/D$.
<table>
<tr><th rowspan="2">$\beta a/D$</th><th colspan="4">K_n</th></tr>
<tr><th>Fundamental</th><th>1 nodal diameter</th><th>2 nodal diameters</th><th>1 nodal circle</th></tr>
<tr><td>∞</td><td>10.2</td><td>21.2</td><td>34.8</td><td>39.7</td></tr>
<tr><td>1</td><td>10.2</td><td>21.2</td><td>34.8</td><td>39.7</td></tr>
<tr><td>0.1</td><td>10.0</td><td>20.9</td><td>34.2</td><td>39.1</td></tr>
<tr><td>0.01</td><td>8.76</td><td>18.6</td><td>30.8</td><td>35.2</td></tr>
<tr><td>0.001</td><td>6.05</td><td>15.0</td><td>26.7</td><td>30.8</td></tr>
<tr><td>0</td><td>4.93</td><td>13.9</td><td>25.6</td><td>29.7</td></tr>
</table>
(Ref. 20)</td></tr>
<tr><td>14 Elliptical flat plate of major radius a, minor radius b, and thickness t; edge fixed</td><td>14a. Uniform load w per unit area including own weight</td><td>$f = \frac{K_1}{2\pi}\sqrt{\frac{Dg}{wa^4}}$ where K_1 is tabulated for various ratios of $\frac{a}{b}$
$a/b =$ 1.0 1.1 1.2 1.5 2.0 3.0
$K_1 =$ 10.2 11.3 12.6 17.0 27.8 57.0
(Ref. 20)</td></tr>
<tr><td>15 Rectangular flat plate with short edge a, long edge b, and thickness t; all edges fixed</td><td>15a. Uniform load w per unit area including own weight</td><td>$f = \frac{K_1}{2\pi}\sqrt{\frac{Dg}{wa^4}}$ where K_1 is tabulated for various ratios of a/b
$a/b =$ 1 0.9 0.8 0.6 0.4 0.2 0
$K_1 =$ 36.0 32.7 29.9 25.9 23.6 22.6 22.4
(Ref. 20)</td></tr>
<tr><td>16. Rectangular flat plate with short edge a, long edge b, and thickness t; all edges simply supported</td><td>16a. Uniform load w per unit area including own weight</td><td>$f = \frac{K_n}{2\pi}\sqrt{\frac{Dg}{wa^4}}$ where $K_n = \pi^2\left[m_a^2 + \left(\frac{a}{b}\right)^2 m_b^2\right]$
<table>
<tr><td></td><td>$a/b =$</td><td>1.0</td><td>0.8</td><td>0.6</td><td>0.4</td><td>0.2</td><td>0.0</td></tr>
<tr><td>$(m_a = 1, m_b = 1)$</td><td>$K_1 =$</td><td>19.7</td><td>16.2</td><td>13.4</td><td>11.5</td><td>10.3</td><td>9.87</td></tr>
<tr><td>$(m_a = 1, m_b = 2)$</td><td>$K_2 =$</td><td>49.3</td><td>35.1</td><td>24.1</td><td>16.2</td><td>11.5</td><td></td></tr>
<tr><td>$(m_a = 2, m_b = 1)$</td><td>$K_3 =$</td><td>49.3</td><td>45.8</td><td></td><td></td><td></td><td></td></tr>
<tr><td>$(m_a = 1, m_b = 3)$</td><td>$K_3 =$</td><td></td><td></td><td>41.9</td><td>24.1</td><td>13.4</td><td></td></tr>
</table>
(Ref. 20)</td></tr>
<tr><td>17. Rectangular flat plate with two edges a fixed, one edge b fixed, and one edge b simply supported</td><td>17a. Uniform load w per unit area including own weight</td><td>$f = \frac{K_1}{2\pi}\sqrt{\frac{Dg}{wa^4}}$ where K_1 is tabulated for various ratios of $\frac{a}{b}$
$a/b =$ 3.0 2.0 1.6 1.2 1.0 0.8 0.6 0.4 0.2 0
$K_1 =$ 213 99 67 42.4 33.1 25.9 20.8 17.8 16.2 15.8
(Ref. 20)</td></tr>
</table>

where, as before, d is the deformation the weight of the moving body would produce if applied as a static load in the direction of the velocity and v is the velocity of impact in inches per second.

Energy losses. The above approximate formulas are derived on the assumptions that impact strains the elastic body in the same way (though not in the same degree) as static loading and that all the kinetic energy of the moving body is expended in producing this strain. Actually, on impact, some kinetic energy is dissipated; and this loss, which can be found by equating the momentum of the entire system before and after impact, is most conveniently taken into account by multiplying the available energy (measured by h or by v^2) by a factor K, the value of which is as follows for a number of simple cases involving members of uniform section:

1. A moving body of mass M strikes axially one end of a bar of mass M_1, the other end of which is fixed. Then

$$K = \frac{1 + \frac{1}{3}\frac{M_1}{M}}{\left(1 + \frac{1}{2}\frac{M_1}{M}\right)^2}$$

If there is a body of mass M_2 attached to the struck end of the bar, then

$$K = \frac{1 + \frac{1}{3}\frac{M_1}{M} + \frac{M_2}{M}}{\left(1 + \frac{1}{2}\frac{M_1}{M} + \frac{M_2}{M}\right)^2}$$

2. A moving body of mass M strikes transversely the center of a simple beam of mass M_1. Then

$$K = \frac{1 + \frac{17}{35}\frac{M_1}{M}}{\left(1 + \frac{5}{8}\frac{M_1}{M}\right)^2}$$

If there is a body of mass M_2 attached to the beam at its center, then

$$K = \frac{1 + \frac{17}{35}\frac{M_1}{M} + \frac{M_2}{M}}{\left(1 + \frac{5}{8}\frac{M_1}{M} + \frac{M_2}{M}\right)^2}$$

3. A moving body of mass M strikes transversely the end of a cantilever beam of mass M_1. Then

$$K = \frac{1 + \frac{33}{140}\frac{M_1}{M}}{\left(1 + \frac{3}{8}\frac{M_1}{M}\right)^2}$$

If there is a body of mass M_2 attached to the beam at the struck end, then

$$K = \frac{1 + \frac{33}{140}\frac{M_1}{M} + \frac{M_2}{M}}{\left(1 + \frac{3}{8}\frac{M_1}{M} + \frac{M_2}{M}\right)^2}$$

4. A moving body of mass M strikes transversely the center of a beam with fixed ends and of mass M_1. Then

$$K = \frac{1 + \frac{13}{35}\frac{M_1}{M}}{\left(1 + \frac{1}{2}\frac{M_1}{M}\right)^2}$$

If there is a body of mass M_2 attached to the beam at the center, then

$$K = \frac{1 + \frac{13}{35}\frac{M_1}{M} + \frac{M_2}{M}}{\left(1 + \frac{1}{2}\frac{M_1}{M} + \frac{M_2}{M}\right)^2}$$

15.5 *Remarks on stress due to impact*

It is improbable that in any actual case of impact the stresses can be calculated accurately by any of the methods or formulas given above. Equation 18, for instance, is supposedly very nearly precise if the conditions assumed are realized, but those conditions—perfect elasticity of the bar, rigidity of the moving body, and simultaneous contact of the moving body with all points on the end of the rod—are obviously unattainable. On the one hand, the damping of the initial stress wave by elastic hysteresis in the bar and the diminution of the intensity of that stress wave by the cushioning effect of the actually nonrigid moving body would serve to make the actual maximum stress less than the theoretical value; on the other hand, uneven contact between the moving body and the bar would tend to make the stress conditions nonuniform across the section and would probably increase the maximum stress.

The formulas given in Art. 15.4 are based upon an admittedly false assumption, viz., that the distribution of stress and strain under impact loading is the same as under static loading. It is known, for instance, that the elastic curve of a beam under impact is different from that under static loading. Such a difference exists in any case, but it is less marked for low than for high velocities of impact, and Eqs. 19 and 20 probably give reasonably accurate values for the deformation and stress (especially the deformation) resulting from the impact of a relatively heavy body moving at low velocity. The lenitive effect of the inertia of the body struck and of attached bodies, as expressed by K, is greatest when the masses of these parts are large compared with that of the moving body. When this is the case, impact can be serious only if the velocity is relatively high, and under such circumstances the formulas probably give only a rough indication of the actual stresses and deformations to be expected. (See Ref. 18.)

15.6 *Temperature stresses*

Whenever the expansion or contraction that would normally result from the heating or cooling of a body is prevented, stresses are developed that are called *thermal,* or *temperature, stresses.* It is convenient to distinguish two different sets of circumstances under which thermal stresses occur: (1) The form of the body and the temperature conditions are such that there would be no stresses except for the *constraint of external forces;* in any such case, the stresses may be found by determining the shape and dimensions the body would assume if unconstrained and then calculating the stresses produced by forcing it back to its original shape and dimensions (see Art. 6.2, Example 2). (2) The form of the body and the temperature conditions are such that stresses are produced in the *absence of external constraint* solely because of the incompatibility of the natural expansions or contractions of the different parts of the body.

A number of representative examples of each type of thermal stress will now be considered.[1] In all instances the modulus of elasticity E and the coefficient of thermal expansion α are assumed to be constant for the temperature range involved and the increment or difference in temperature ΔT is assumed to be positive; when ΔT is negative, the stress produced is of the opposite kind. Also, it is assumed that the compressive stresses produced do not produce buckling and that yielding does not occur; if either buckling or yielding is indicated by the stress levels found, then the solution must be modified by appropriate methods discussed in previous chapters.

Stresses due to external constraint.

1. A uniform straight bar is subjected to a temperature change ΔT throughout while held at the ends; the resulting unit stress is $\Delta T \alpha E$ (compression). (For other conditions of end restraint see Chap. 7, Table 4, cases 1q to 12q.)

[1] Most of the formulas here given are taken from the papers by Goodier (Refs. 7 and 14), Maulbetsch (Ref. 8), and Kent (Ref. 9).

2. A uniform flat plate is subjected to a temperature change ΔT throughout while held at the edges; the resulting unit stress is $\Delta T \alpha E/(1 - \nu)$ (compression).

3. A solid body of any form is subjected to a temperature change ΔT throughout while held to the same form and volume; the resulting stress is $\Delta T \alpha E/(1 - 2\nu)$, (compression).

4. A uniform bar of rectangular section has one face at a uniform temperature T and the opposite face at a uniform temperature $T + \Delta T$, the temperature gradient between these faces being linear. The bar would normally curve in the arc of the circle of radius $d/\Delta T \alpha$, where d is the distance between the hot and cold faces. If the ends are fixed, the bar will be held straight by end couples $EI\,\Delta T \alpha/d$, and the maximum resulting bending stress will be $\frac{1}{2}\Delta T \alpha E$ (compression on the hot face; tension on the cold face). [For many other conditions of end restraint and partial heating, see Table 3, cases 6a to 6f; Table 4, cases 1r to 12r; Table 7, case 7; Table 8, case 7; Table 10, cases 6a to 6f; and Table 11, cases 6a to 6f (all in Chap. 7).]

5. A flat plate of uniform thickness t and of any shape has one face at a uniform temperature T and the other face at a uniform temperature $T + \Delta T$, the temperature gradient between the faces being linear. The plate would normally assume a spherical curvature with radius $t/\Delta T\alpha$. If the edges are fixed, the plate will be held flat by uniform edge moments and the maximum resulting bending stress will be $\frac{1}{2}\Delta T \alpha E/(1 - \nu)$ (compression on the hot face; tension on the cold face). [For many other conditions of edge restraint and axisymmetric partial heating, see Table 24, cases 8a to 8h (Chap. 10); a more general treatment of the solid circular plate is given in Table 24, case 15 (Chap. 10).]

6. If the plate described in 5 is circular, no stress is produced by supporting the edges in a direction normal to the plane of the plate.

7. If the plate described in 5 has the shape of an equilateral triangle of altitude a (sides $2a/\sqrt{3}$) and the edges are rigidly supported so as to be held in a plane, the supporting reactions will consist of a uniform load $\frac{1}{8}\Delta T \alpha E t^2/a$ lb/linear in along each edge against the hot face and a concentrated load $\sqrt{3}\,\Delta T \alpha E t^2/12$ lb at each corner against the cold face. The maximum resulting bending stress is $\frac{3}{4}\Delta T \alpha E$ at the corners (compression on the hot face; tension on the cold face). There are also high shear stresses near the corners (Ref. 8).

8. If the plate described in 5 is square, no simple formula is available for the reactions necessary to hold the edges in their original plane. The maximum bending stress occurs near the edges, and its value approaches $\frac{1}{2}\Delta T \alpha E$. There are also high shear stresses near the corners (Ref. 8).

Stresses due to internal constraint.

9. Part or all of the surface of a solid body is suddenly subjected to a temperature change ΔT; a compressive stress $\Delta T \alpha E/(1 - \nu)$ is developed in the surface layer of the heated part (Ref. 7).

10. A thin circular disk at uniform temperature has the temperature changed ΔT throughout a comparatively small central circular portion of radius a. Within the heated part there are radial and tangential compressive stresses $\sigma_r = \sigma_t = \frac{1}{2}\Delta T\,\alpha E$. At points outside the heated part a distance r from the center of the disk but still close to the central portion, the stresses are $\sigma_r = \frac{1}{2}\Delta T\,\alpha E a^2/r^2$ (compression) and $\sigma_t = \frac{1}{2}\Delta T\,\alpha E a^2/r^2$ (tension); at the edge of the heated portion, there is a maximum shear stress $\frac{1}{2}\Delta T\,\alpha E$ (Ref. 7).

11. If the disk of 10 is heated uniformly throughout a small central portion of elliptical instead of circular outline, the maximum stress is the tangential stress at the ends of the ellipse and is $\sigma_t = \Delta T\,\alpha E/[1 + (b/a)]$, where a is the major and b the minor semiaxis of the ellipse (Ref. 7).

12. If the disk of 10 is heated symmetrically about its center and uniformly throughout its thickness so that the temperature is a function of the distance r from the center only, the radial and tangential stresses at any point a distance r_1 from the center are

$$\sigma_{r_1} = \alpha E\left(\frac{1}{R^2}\int_0^R Tr\,dr - \frac{1}{r_1^2}\int_0^{r_1} Tr\,dr\right)$$

$$\sigma_{t_1} = \alpha E\left(-T + \frac{1}{R^2}\int_0^R Tr\,dr + \frac{1}{r_1^2}\int_0^{r_1} Tr\,dr\right)$$

where R is the radius of the disk and T is the temperature at any point a distance r from the center minus the temperature of the coldest part of the disk. [In the preceding expressions, the negative sign denotes compressive stress (Ref. 7).]

13. A rectangular plate or strip $ABCD$ (Fig. 15.5) is heated along a transverse line FG uniformly throughout the thickness and across the width so

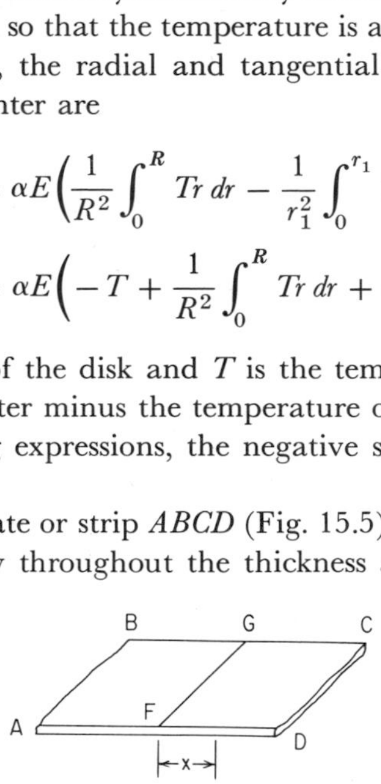

Fig. 15.5

that the temperature varies only along the length with x. At FG the temperature is T_1; the minimum temperature in the plate is T_0. At any point along the edges of the strip where the temperature is T, a tensile stress $\sigma_x = E\alpha(T - T_0)$ is developed; this stress has its maximum value at F and G, where it becomes $E\alpha(T_1 - T_0)$. Halfway between F and G, a compressive stress σ_y of equal intensity is developed (Ref. 7).

14. The plate of 13 is heated as described except that the lower face of the plate is cooler than the upper face, the maximum temperature there being T_2 and the temperature gradient through the thickness being linear. The maximum tensile stress at F and G is

$$\sigma_x = \frac{1}{2} E\alpha \left[T_1 + T_2 - 2T_0 + \frac{1-\nu}{3+\nu}(T_1 - T_2) \right] \quad \text{(Ref. 7)}$$

15. A long hollow cylinder with thin walls has the outer surface at the uniform temperature T and the inner surface at the uniform temperature $T + \Delta T$. The temperature gradient through the thickness is linear. At points remote from the ends, the maximum circumferential stress is $\frac{1}{2}\Delta T \alpha E/(1 - \nu)$ (compression at the inner surface; tension at the outer surface) and the longitudinal stress is $\frac{1}{2}\Delta T \alpha E/(1 - \nu)$ (compression at the inside; tension at the outside). (These formulas apply to a thin tube of any cross section.) At the ends, if these are free, the maximum tensile stress in a tube of circular section is about 25 percent greater than the value given by the formula (Ref. 7).

16. A hollow cylinder with thick walls of inner radius b and outer radius c has the outer surface at the uniform temperature T and the inner surface at the uniform temperature $T + \Delta T$; the temperature gradient is not linear. The maximum stresses, which are circumferential and which occur at the inner and outer surfaces, are

(Outer surface)

$$\sigma_t = \frac{\Delta T \alpha E}{2(1-\nu)\log_e(c/b)} \left(1 - \frac{2b^2}{c^2 - b^2} \log_e \frac{c}{b}\right) \qquad \text{tension}$$

(Inner surface)

$$\sigma_t = \frac{\Delta T \alpha E}{2(1-\nu)\log_e(c/b)} \left(1 - \frac{2c^2}{c^2 - b^2} \log_e \frac{c}{b}\right) \qquad \text{compression}$$

At the inner and outer surfaces, the longitudinal stresses are equal to the tangential stresses (Ref. 7).

17. If the thick tube of 16 has the temperature of the outer surface raised at the uniform rate of $m°$/s, then after a steady state of heat flow has been reached the maximum tangential stresses are

(Outer surface)

$$\sigma_t = \frac{E\alpha m}{8A(1-\nu)} \left(3b^2 - c^2 - \frac{4b^4}{c^2 - b^2} \log_e \frac{c}{b}\right) \qquad \text{compression}$$

(Inner surface)

$$\sigma_t = \frac{E\alpha m}{8A(1-\nu)} \left(b^2 + c^2 - \frac{4b^2c^2}{c^2 - b^2} \log_e \frac{c}{b}\right) \qquad \text{tension}$$

where A is the coefficient of thermal diffusivity equal to the coefficient of thermal conductivity divided by the product of the density of the material and its specific heat. (For steel, A may be taken as 0.027 in^2/s at moderate temperatures.) [At the inner and outer surfaces, the longitudinal stresses are equal to the tangential stresses (Ref. 9).]

18. A solid rod of circular section is heated or cooled symmetrically with

respect to its axis, the condition being uniform along the length, so that the temperature is a function of r (the distance from the axis) only. The stresses are equal to those given by the formulas for 12 divided by $(1 - \nu)$ (Ref. 7).

19. If the solid rod of 18 has the temperature of its convex surface raised at the uniform rate of $m°/\text{s}$, then after a steady state of heat flow has been reached the radial, tangential, and longitudinal stresses at any point a distance r from the center are

$$\sigma_r = \frac{E\alpha m}{1-\nu}\,\frac{c^2 - r^2}{16A}$$

$$\sigma_t = \frac{E\alpha m}{1-\nu}\,\frac{c^2 - 3r^2}{16A}$$

$$\sigma_x = \frac{E\alpha m}{1-\nu}\,\frac{c^2 - 2r^2}{8A}$$

Here A has the same meaning as in 17 and c is the radius of the shaft. [A negative result indicates compression, a positive result tension (Ref. 9).]

20. A solid sphere of radius c has the surface temperature increased at the uniform rate of $m°/\text{s}$. The radial and tangential stresses produced at any point a distance r from the center are

$$\sigma_r = \frac{E\alpha m}{15A(1-\nu)}(c^2 - r^2)$$

$$\sigma_t = \frac{E\alpha m}{15A(1-\nu)}(c^2 - 2r^2)$$

[A negative result indicates compression, a positive result tension (Ref. 9).]

21. If the sphere is hollow, with outer radius c and inner radius b, the stresses at any point are

$$\sigma_r = \frac{E\alpha m}{15A(1-\nu)}\left(-r^2 - \frac{5b^3}{r} + \phi - \psi\right)$$

$$\sigma_t = \frac{E\alpha m}{15A(1-\nu)}\left(-2r^2 - \frac{5b^3}{2r} + \phi + \frac{\psi}{2}\right)$$

where

$$\phi = \frac{c^5 + 5c^2b^3 - 6b^5}{c^3 - b^3}$$

$$\psi = \frac{c^5b^3 - 6c^3b^5 + 5c^2b^6}{r^3(c^3 - b^3)}$$

[A negative result indicates compression, a positive result tension (Ref. 9).]

Other problems involving thermal stress, the solutions of which cannot be expressed by simple formulas, are considered in the references cited above and in Refs. 3 and 10; charts for the solution of thermal stresses in tubes are given in Ref. 11.

REFERENCES

1. Hodkinson, B.: Rotating Discs of Conical Profile, *Engineering,* vol. 115, p. 1, 1923.
2. Rushing, F. C.: Determination of Stresses in Rotating Disks of Conical Profile, *Trans. ASME,* vol. 53, p. 91, 1931.
3. Stodola, A.: "Steam and Gas Turbines," 6th ed., McGraw-Hill Book Company, 1927 (transl. by L. C. Loewenstein).
4. Loewenstein, L. C.: "Marks' Mechanical Engineers' Handbook," McGraw-Hill Book Company, 1930.
5. Donnell, L. H.: Longitudinal Wave Transmission and Impact, *Trans. ASME,* vol. 52, no. 1, p. 153, 1930.
6. Love, A. E. H.: "Mathematical Theory of Elasticity," 2d ed., Cambridge University Press, 1906.
7. Goodier, J. N.: Thermal Stress, *ASME J. Appl. Mech.,* vol. 4, no. 1, March 1937.
8. Maulbetsch, J. L.: Thermal Stresses in Plates, *ASME J. Appl. Mech.,* vol. 2, no. 4, December 1935.
9. Kent, C. H.: Thermal Stresses in Spheres and Cylinders Produced by Temperatures Varying with Time, *Trans. ASME,* vol. 54, no. 18, p. 185, 1932.
10. Timoshenko, S.: "Theory of Elasticity," Engineering Societies Monograph, McGraw-Hill Book Company, 1934.
11. Barker, L. H.: The Calculation of Temperature Stresses in Tubes, *Engineering,* vol. 124, p. 443, 1927.
12. Robinson, E. L.: Bursting Tests of Steam-turbine Disk Wheels, *Trans. ASME,* vol. 66, no. 5, p. 373, 1944.
13. Holms, A. G., and J. E. Jenkins: Effect of Strength and Ductility on Burst Characteristics of Rotating Disks, *Natl. Adv. Comm. Aeron., Tech. Note* 1667, 1948.
14. Goodier, J. N.: Thermal Stress and Deformation, *ASME J. Appl. Mech.,* vol. 24, no. 3, September, 1957.
15. Eichmann, E. S.: Note on the Maximum Effect of a Moving Force on a Simple Beam, *ASME J. Appl. Mech.,* vol. 20, no. 4, December, 1953.
16. Ayre, R. S., L. S. Jacobsen, and C. S. Hsu: Transverse Vibration of 1 and 2-span Beams under Moving Mass-Load, *Proc. 1st U.S. Natl. Congr. Appl. Mech.,* 1952.
17. Burr, Arthur H.: Longitudinal and Torsional Impact in a Uniform Bar with a Rigid Body at One End, *ASME J. Appl. Mech.,* vol. 17, no. 2, June 1950.
18. Schwieger, Horst: A Simple Calculation of the Transverse Impact on Beams and Its Experimental Verification, *J. Soc. Exp. Mech.,* vol. 5, no. 11, November 1965.
19. Thomson, W. T.: "Vibration Theory and Applications," Prentice-Hall, Inc., 1965.
20. Leissa, A. W.: Vibration of Plates, NASA SP-160, National Aeronautics and Space Administration, 1969.
21. Leissa, A. W.: Vibration of Shells, NASA SP-288, National Aeronautics and Space Administration, 1973.
22. Huang, T. C.: Eigenvalues and Modifying Quotients of Vibration of Beams, and Eigenfunctions of Vibration of Beams, *Univ. Wis. Eng. Exp. Sta. Repts. Nos.* 25 *and* 26, 1964.
23. Jones, R.: An Approximate Expression for the Fundamental Frequency of Vibration of Elastic Plates, *J. Sound Vib.,* vol. 38, no. 4, February 1975.

Miscellaneous Tables

Factors of Stress Concentration

Properties of Materials

TABLE 37 ***Factors of stress concentration for elastic stress*** **(k)**

The elastic stress concentration factor k is the ratio of the maximum stress in the stress raiser to the nominal stress computed by the ordinary strength-of-materials formulas, using the dimensions of the net cross section unless defined otherwise in specific cases.

For those data presented in the form of equations, the equations have been developed to fit as closely as possible the many data points given in the literature referenced in each case. Over the majority of the ranges specified for the variables, the curves fit the data points with much less than a 5 percent error.

It is not possible to tabulate all the available values of stress concentration factors found in the literature, but the following list of topics and sources of data will be helpful. All the following references have extensive bibliographies.

Fatigue stress concentration factors (k_f); see the 4th edition of this book, Ref. 23.

Stress concentration factors for rupture (k_r); see the 4th edition of this book, Ref. 23.

Stress concentration factors pertaining to odd-shaped holes in plates, multiple holes arranged in various patterns, and reinforced holes under multiple loads; see Refs. 1 and 24.

Stress concentrations around holes in pressure vessels; see Ref. 1.

For a discussion of the effect of stress concentration on the response of machine elements and structures, see Ref. 25.

Type of form irregularity or stress raiser	Stress condition and manner of loading	Factor of stress concentration k for various dimensions
1. Two U notches in a member of rectangular section 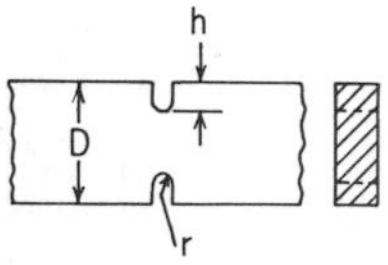	1a. Elastic stress, axial tension	$k = K_1 + K_2\left(\frac{2h}{D}\right) + K_3\left(\frac{2h}{D}\right)^2 + K_4\left(\frac{2h}{D}\right)^3$ where (see sub-table below) (Refs. 1 to 10) For the semicircular notch ($h/r = 1$) $k = 3.065 - 3.370\left(\frac{2h}{D}\right) + 0.647\left(\frac{2h}{D}\right)^2 + 0.658\left(\frac{2h}{D}\right)^3$

	$0.1 \le h/r \le 2.0$	$2.0 \le h/r \le 50.0$
K_1	$0.850 + 2.628\sqrt{h/r} - 0.413h/r$	$0.833 + 2.069\sqrt{h/r} - 0.009h/r$
K_2	$-1.119 - 4.826\sqrt{h/r} + 2.575h/r$	$2.732 - 4.157\sqrt{h/r} + 0.176h/r$
K_3	$3.563 - 0.514\sqrt{h/r} - 2.402h/r$	$-8.859 + 5.327\sqrt{h/r} - 0.320h/r$
K_4	$-2.294 + 2.713\sqrt{h/r} + 0.240h/r$	$6.294 - 3.239\sqrt{h/r} + 0.154h/r$

1b. Elastic stress, in-plane bending

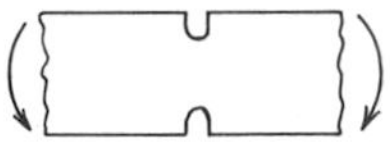

$$k = K_1 + K_2\left(\frac{2h}{D}\right) + K_3\left(\frac{2h}{D}\right)^2 + K_4\left(\frac{2h}{D}\right)^3$$

where

	$0.25 \le h/r \le 2.0$	$2.0 \le h/r \le 50.0$
K_1	$0.723 + 2.845\sqrt{h/r} - 0.504h/r$	$0.833 + 2.069\sqrt{h/r} - 0.009h/r$
K_2	$-1.836 - 5.746\sqrt{h/r} + 1.314h/r$	$0.024 - 5.383\sqrt{h/r} + 0.126h/r$
K_3	$7.254 - 1.885\sqrt{h/r} + 1.646h/r$	$-0.856 + 6.460\sqrt{h/r} - 0.199h/r$
K_4	$-5.140 + 4.785\sqrt{h/r} - 2.456h/r$	$0.999 - 3.146\sqrt{h/r} + 0.082h/r$

(Refs. 1, 3, 8, 11, and 12)

For the semicircular notch ($h/r = 1$)

$$k = 3.065 - 6.269\left(\frac{2h}{D}\right) + 7.015\left(\frac{2h}{D}\right)^2 - 2.812\left(\frac{2h}{D}\right)^3$$

1c. Elastic stress, out-of-plane bending

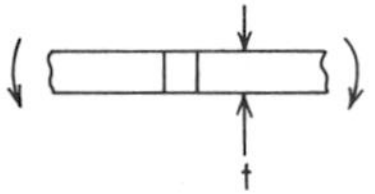

$$k = K_1 + K_2\left(\frac{2h}{D}\right) + K_3\left(\frac{2h}{D}\right)^2 + K_4\left(\frac{2h}{D}\right)^3$$

where for $0.25 \le h/r \le 4.0$ and h/t is large

$K_1 = 1.031 + 0.831\sqrt{h/r} + 0.014h/r$
$K_2 = -1.227 - 1.646\sqrt{h/r} + 0.117h/r$
$K_3 = 3.337 - 0.750\sqrt{h/r} + 0.469h/r$
$K_4 = -2.141 + 1.566\sqrt{h/r} - 0.600h/r$

(Refs. 1, 9, 13, and 14)

For the semicircular notch ($h/r = 1$)

$$k = 1.876 - 2.756\left(\frac{2h}{D}\right) + 3.056\left(\frac{2h}{D}\right)^2 - 1.175\left(\frac{2h}{D}\right)^3$$

2. Two V notches in a member of rectangular section

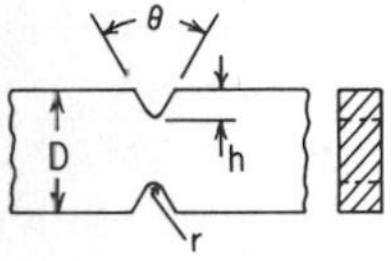

2a. Elastic stress, axial tension

The stress concentration factor for the V notch, k_θ, is the smaller of the values

$$k_\theta = k_U$$

or

$$k_\theta = 1.11k_U - \left[0.0275 + 0.000145\theta + 0.0164\left(\frac{\theta}{120}\right)^8\right]k_U^2 \qquad \text{for } \frac{2h}{D} = 0.40 \text{ and } \theta \le 120°$$

or

$$k_\theta = 1.11k_U - \left[0.0275 + 0.00042\theta + 0.0075\left(\frac{\theta}{120}\right)^8\right]k_U^2 \qquad \text{for } \frac{2h}{D} = 0.667 \text{ and } \theta \le 120°$$

where k_U is the stress concentration factor for a U notch, case 1a, when the dimensions h, r, and D are the same as for the V notch and θ is the notch angle in degrees.

(Refs. 1 and 15)

TABLE 37 ***Factors of stress concentration for elastic stress (k) (Cont.)***

Type of form irregularity or stress raiser	Stress condition and manner of loading	Factor of stress concentration k for various dimensions
3. One U notch in a member of rectangular section 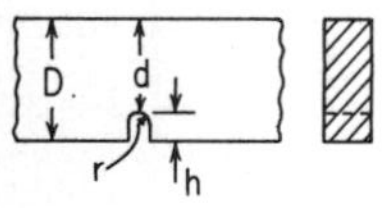	3a. Elastic stress, axial tension 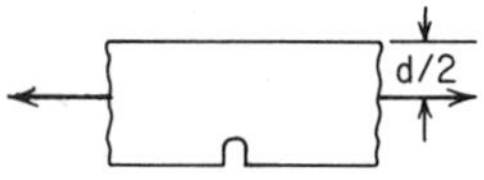	$k = K_1 + K_2\left(\frac{h}{D}\right) + K_3\left(\frac{h}{D}\right)^2 + K_4\left(\frac{h}{D}\right)^3$ where for $0.5 \le h/r \le 4.0$ $K_1 = 0.721 + 2.394\sqrt{h/r} - 0.127h/r$ $K_2 = 1.978 - 11.489\sqrt{h/r} + 2.211h/r$ $K_3 = -4.413 + 18.751\sqrt{h/r} - 4.596h/r$ $K_4 = 2.714 - 9.655\sqrt{h/r} + 2.512h/r$ (Refs. 16 and 17) For the semicircular notch ($h/r = 1$) $k = 2.988 - 7.300\left(\frac{h}{D}\right) + 9.742\left(\frac{h}{D}\right)^2 - 4.429\left(\frac{h}{D}\right)^3$
	3b. Elastic stress, in-plane bending 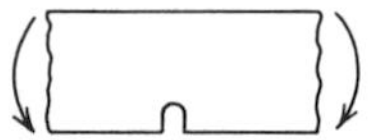	$k = K_1 + K_2\left(\frac{h}{D}\right) + K_3\left(\frac{h}{D}\right)^2 + K_4\left(\frac{h}{D}\right)^3$ where for $0.5 \le h/r \le 4.0$ $K_1 = 0.721 + 2.394\sqrt{h/r} - 0.127h/r$ $K_2 = -0.426 - 8.827\sqrt{h/r} + 1.518h/r$ $K_3 = 2.161 + 10.968\sqrt{h/r} - 2.455h/r$ $K_4 = -1.456 - 4.535\sqrt{h/r} + 1.064h/r$ (Refs. 17 and 18) For the semicircular notch ($h/r = 1$) $k = 2.988 - 7.735\left(\frac{h}{D}\right) + 10.674\left(\frac{h}{D}\right)^2 - 4.927\left(\frac{h}{D}\right)^3$
4. One V notch in a member of rectangular section 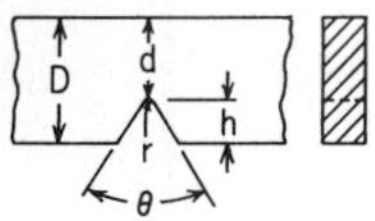	4b. Elastic stress, in-plane bending 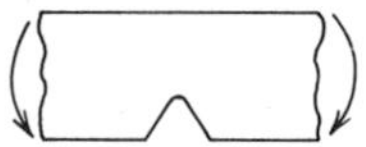	The stress concentration factor for the V notch, k_θ, is the smaller of the values $k_\theta = k_U$ or $k_\theta = 1.11k_U - \left[0.0275 + 0.1125\left(\frac{\theta}{150}\right)^4\right]k_U^2$ for $\theta \le 150°$ where k_U is the stress concentration factor for a U notch, case 3b, when the dimensions h, r, and D are the same as for the V notch and θ is the notch angle in degrees. (Ref. 18)

5. Square shoulder with fillet in a member of rectangular section

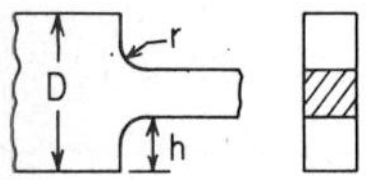

5a. Elastic stress, axial tension

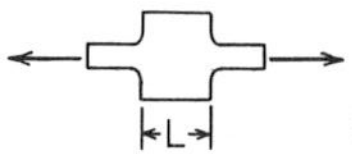

$$k = K_1 + K_2\left(\frac{2h}{D}\right) + K_3\left(\frac{2h}{D}\right)^2 + K_4\left(\frac{2h}{D}\right)^3$$

where $\frac{L}{D} > \frac{3}{[r/(D - 2h)]^{\frac{1}{4}}}$ and where

	$0.1 \leq h/r \leq 2.0$	$2.0 \leq h/r \leq 20.0$
K_1	$1.007 + 1.000\sqrt{h/r} - 0.031h/r$	$1.042 + 0.982\sqrt{h/r} - 0.036h/r$
K_2	$-0.114 - 0.585\sqrt{h/r} + 0.314h/r$	$-0.074 - 0.156\sqrt{h/r} - 0.010h/r$
K_3	$0.241 - 0.992\sqrt{h/r} - 0.271h/r$	$-3.418 + 1.220\sqrt{h/r} - 0.005h/r$
K_4	$-0.134 + 0.577\sqrt{h/r} - 0.012h/r$	$3.450 - 2.046\sqrt{h/r} + 0.051h/r$

(Refs. 1, 8, 11, and 19)

For cases where $\frac{L}{D} < \frac{3}{[r/(D - 2h)]^{\frac{1}{4}}}$ see Refs. 1, 21, and 22.

5b. Elastic stress, in-plane bending

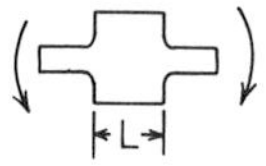

$$k = K_1 + K_2\left(\frac{2h}{D}\right) + K_3\left(\frac{2h}{D}\right)^2 + K_4\left(\frac{2h}{D}\right)^3$$

where $\frac{L}{D} > \frac{0.8}{[r/(D - 2h)]^{\frac{1}{4}}}$ and where

	$0.1 \leq h/r \leq 2.0$	$2.0 \leq h/r \leq 20.0$
K_1	$1.007 + 1.000\sqrt{h/r} - 0.031h/r$	$1.042 + 0.982\sqrt{h/r} - 0.036h/r$
K_2	$-0.270 - 2.404\sqrt{h/r} + 0.749h/r$	$-3.599 + 1.619\sqrt{h/r} - 0.431h/r$
K_3	$0.677 + 1.133\sqrt{h/r} - 0.904h/r$	$6.084 - 5.607\sqrt{h/r} + 1.158h/r$
K_4	$-0.414 + 0.271\sqrt{h/r} + 0.186h/r$	$-2.527 + 3.006\sqrt{h/r} - 0.691h/r$

(Refs. 1, 11, and 20)

For cases where $\frac{L}{D} < \frac{0.8}{[r/(D - 2h)]^{\frac{1}{4}}}$ see Refs. 1 and 20.

TABLE 37 ***Factors of stress concentration for elastic stress*** (k) ***(Cont.)***

Type of form irregularity or stress raiser	Stress condition and manner of loading	Factor of stress concentration k for various dimensions
6. Circular hole in an infinite plate 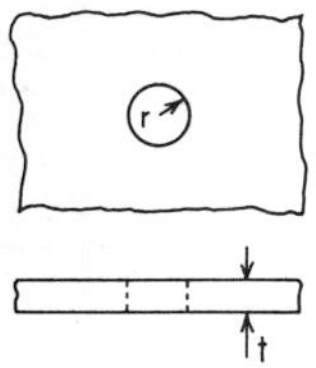	6a. Elastic stress, in-plane normal stress 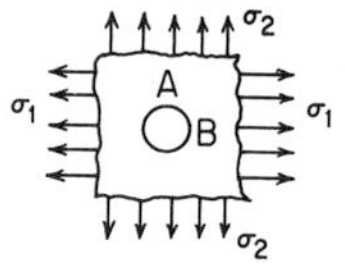	(a1) Uniaxial stress, $\sigma_2 = 0$ $\sigma_A = 3\sigma_1 \qquad \sigma_B = -\sigma_1$ (a2) Biaxial stress, $\sigma_2 = \sigma_1$ $\sigma_A = \sigma_B = 2\sigma_1$ (a3) Biaxial stress, $\sigma_2 = -\sigma_1$ (pure shear) $\sigma_A = -\sigma_B = 4\sigma_1$
	6b. Elastic stress, out-of-plane bending 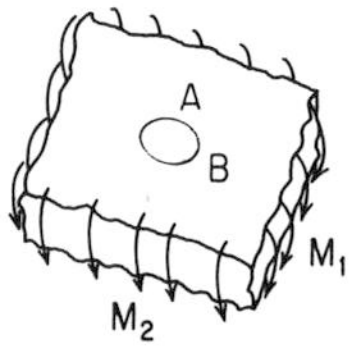	(b1) Simple bending, $M_2 = 0$ $\sigma_A = k\dfrac{6M_1}{t^2}$ where $k = 1.79 + \dfrac{0.25}{0.39 + (2r/t)} + \dfrac{0.81}{1 + (2r/t)^2} - \dfrac{0.26}{1 + (2r/t)^3}$ (b2) Cylindrical bending, $M_2 = \nu M_1$ $\sigma_A = k\dfrac{6M_1}{t^2}$ where $k = 1.85 + \dfrac{0.509}{0.70 + (2r/t)} - \dfrac{0.214}{1 + (2r/t)^2} + \dfrac{0.335}{1 + (2r/t)^3}$ for $\nu = 0.3$ (b3) Isotropic bending, $M_2 = M_1$ $\sigma_A = k\dfrac{6M_1}{t^2}$ where $k = 2$ (independent of r/t) (Refs. 1 and 26 to 30)
7. Central circular hole in a member of rectangular cross section 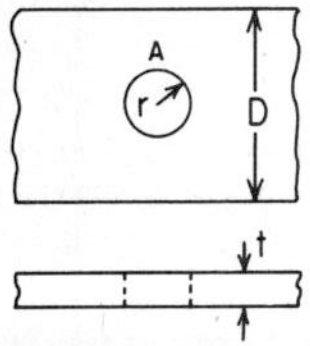	7a. Elastic stress, axial tension 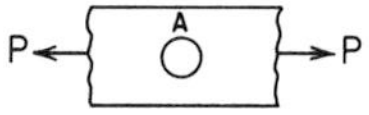	$\sigma_{max} = \sigma_A = k\sigma_{nom}$ where $\sigma_{nom} = \dfrac{P}{t(D - 2r)}$ $k = 3.00 - 3.13\left(\dfrac{2r}{D}\right) + 3.66\left(\dfrac{2r}{D}\right)^2 - 1.53\left(\dfrac{2r}{D}\right)^3$ (Refs. 5 and 25)

7b. Elastic stress, in-plane bending 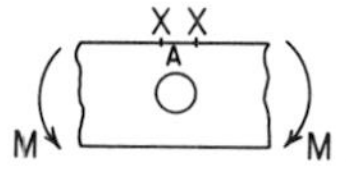 	The maximum stress at the edge of the hole is $$\sigma_A = k\sigma_{nom}$$ where $\sigma_{nom} = \dfrac{12Mr}{t[D^3 - (2r)^3]}$ (at the edge of the hole) $k = 2$ (independent of r/D) The maximum stress at the edge of the plate is not directly above the hole but is found a short distance away on either side, points X. $$\sigma_X = \sigma_{nom}$$ where $\sigma_{nom} = \dfrac{6MD}{t[D^3 - (2r)^3]}$ (at the edge of the plate) (Refs. 1, 25, and 31)
7c. Elastic stress, out-of-plane bending 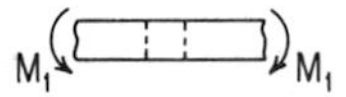 	(c1) Simple bending, $M_2 = 0$ $\sigma_{max} = \sigma_A = k\dfrac{6M_1}{t^2(D - 2r)}$ where $k = \left[1.79 + \dfrac{0.25}{0.39 + (2r/t)} + \dfrac{0.81}{1 + (2r/t)^2} - \dfrac{0.26}{1 + (2r/t)^3}\right]\left[1 - 1.04\left(\dfrac{2r}{D}\right) + 1.22\left(\dfrac{2r}{D}\right)^2\right]$ for $\dfrac{2r}{D} < 0.3$ (c2) Cylindrical bending (plate action), $M_2 = \nu M_1$ $\sigma_{max} = \sigma_A = k\dfrac{6M_1}{t^2(D - 2r)}$ where $k = \left[1.85 + \dfrac{0.509}{0.70 + (2r/t)} - \dfrac{0.214}{1 + (2r/t)^2} - \dfrac{0.335}{1 + (2r/t)^3}\right]\left[1 - 1.04\left(\dfrac{2r}{D}\right) + 1.22\left(\dfrac{2r}{D}\right)^2\right]$ for $\dfrac{2r}{D} < 0.3$ and $\nu = 0.3$ (Refs. 1 and 27 to 29)

TABLE 37 *Factors of stress concentration for elastic stress (k)* (*Cont.*)

Type of form irregularity or stress raiser	Stress condition and manner of loading	Factor of stress concentration k for various dimensions
8. Off-center circular hole in a member of rectangular cross section 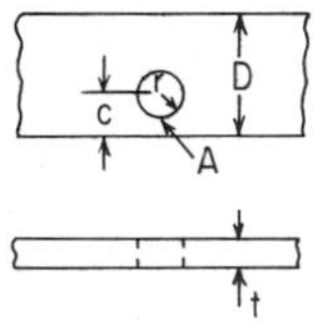 	8a. Elastic stress, axial tension 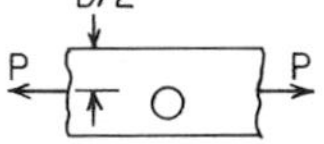 	$\sigma_{max} = \sigma_A = k\sigma_{nom}$ where $\sigma_{nom} = \dfrac{P}{Dt}\dfrac{\sqrt{1-(r/c)^2}}{1-(r/c)}\dfrac{1-(c/D)}{1-(c/D)[2-\sqrt{1-(r/c)^2}]}$ $k = 3.00 - 3.13\left(\dfrac{r}{c}\right) + 3.66\left(\dfrac{r}{c}\right)^2 - 1.53\left(\dfrac{r}{c}\right)^3$ (Refs. 1 and 32)
	8b. Elastic stress, in-plane bending 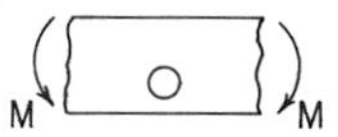 	$\sigma_{max} = \sigma_A = k\sigma_{nom}$ where $\sigma_{nom} = \dfrac{12M}{tD^3}\left(\dfrac{D}{2} - c + r\right)$ and $k = 3.0$ if $r/c < 0.05$ or $k = K_1 + K_2\left(\dfrac{2c}{D}\right) + K_3\left(\dfrac{2c}{D}\right)^2 + K_4\left(\dfrac{2c}{D}\right)^3$ where for $0.05 \leq r/c \leq 0.5$ $K_1 = 3.022 - 0.422r/c + 3.556(r/c)^2$ $K_2 = -0.569 + 2.664r/c - 4.397(r/c)^2$ $K_3 = 3.138 - 18.367r/c + 28.093(r/c)^2$ $K_4 = -3.591 + 16.125r/c - 27.252(r/c)^2$ (Ref. 31)
9. Elliptical hole in an infinite plate 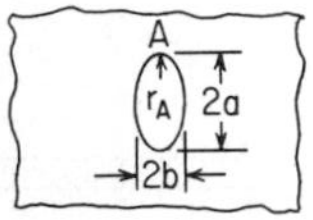 $r_A = \dfrac{b^2}{a}$	9a. Elastic stress, in-plane normal stress 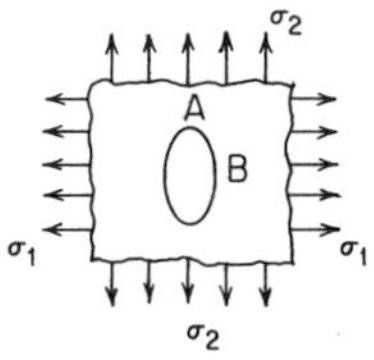 	(a1) Uniaxial stress, $\sigma_2 = 0$ $\sigma_A = \left(1 + \dfrac{2a}{b}\right)\sigma_1$ or $\sigma_A = \left(1 + 2\sqrt{\dfrac{a}{r_A}}\right)\sigma_1$ $\sigma_B = -\sigma_1$ (a2) Biaxial stress, $\sigma_2 = \sigma_1$ $\sigma_A = 2\dfrac{a}{b}\sigma_1$ $\sigma_B = 2\dfrac{b}{a}\sigma_1$ (a3) Biaxial stress, $\sigma_2 = -\sigma_1$ $\sigma_A = 2\left(1 + \dfrac{a}{b}\right)\sigma_1$ $\sigma_B = -2\left(1 + \dfrac{b}{a}\right)\sigma_1$

10. Central elliptical hole in a member of rectangular cross section 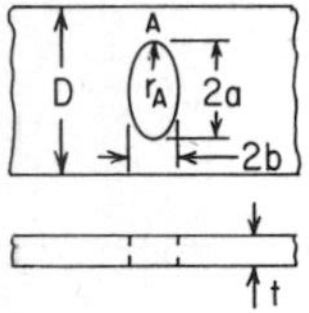 $r_A = \frac{b^2}{a}$	10a. Elastic stress, axial tension 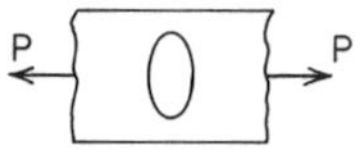	$\sigma_{max} = \sigma_A = k\sigma_{nom}$ where $\sigma_{nom} = \frac{P}{t(D - 2a)}$ $k = K_1 + K_2\left(\frac{2a}{D}\right) + K_3\left(\frac{2a}{D}\right)^2 + K_4\left(\frac{2a}{D}\right)^3$ where for $0.5 \le a/b \le 10.0$ $K_1 = 1.000 + 0.000\sqrt{a/b} + 2.000a/b$ $K_2 = -0.351 - 0.021\sqrt{a/b} - 2.483a/b$ $K_3 = 3.621 - 5.183\sqrt{a/b} + 4.494a/b$ $K_4 = -2.270 + 5.204\sqrt{a/b} - 4.011a/b$ (Refs. 33 to 37)
	10b. Elastic stress, in-plane bending 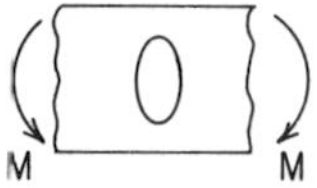	The maximum stress at the edge of the hole is $\sigma_A = k\sigma_{nom}$ where $\sigma_{nom} = \frac{12Ma}{t[D^3 - (2a)^3]}$ (at the edge of the hole) $k = K_1 + K_2\left(\frac{2a}{D}\right) + K_3\left(\frac{2a}{D}\right)^2$ where for $1.0 \le a/b \le 2.0$ and $0.4 \le 2a/D \le 1.0$ $K_1 = 3.465 - 3.739\sqrt{a/b} + 2.274a/b$ $K_2 = -3.841 + 5.582\sqrt{a/b} - 1.741a/b$ $K_3 = 2.376 - 1.843\sqrt{a/b} - 0.534a/b$ (Refs 1, 36, and 37)
11. Off-center elliptical hole in a member of rectangular cross section 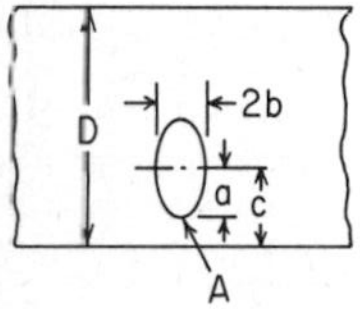	11a. Elastic stress, axial tension 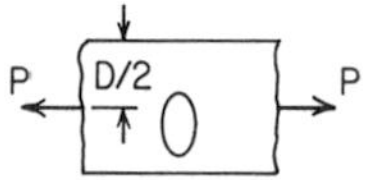	$\sigma_{max} = \sigma_A = k\sigma_{nom}$ The expression for σ_{nom} from case 8a can be used by substituting a/c for r/c. Use the expression for k from case 10a by substituting a/c for $2a/D$.

TABLE 37 *Factors of stress concentration for elastic stress* (k) *(Cont.)*

Type of form irregularity or stress raiser	Stress condition and manner of loading	Factor of stress concentration k for various dimensions
12. Rectangular hole with round corners in an infinite plate 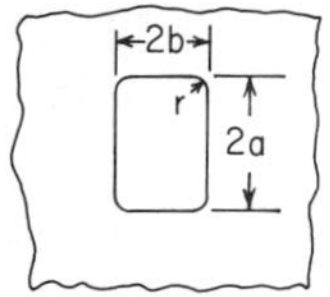	12a. Elastic stress, axial tension 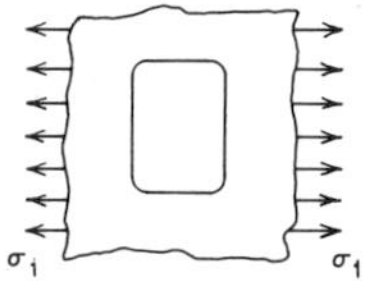	$\sigma_{max} = k\sigma_1$ and $k = K_1 + K_2\left(\frac{b}{a}\right) + K_3\left(\frac{b}{a}\right)^2 + K_4\left(\frac{b}{a}\right)^3$ where for $0.2 \le r/b \le 1.0$ and $0.3 \le b/a \le 1.0$ $K_1 = 14.815 - 15.774\sqrt{r/b} + 8.149r/b$ $K_2 = -11.201 - 9.750\sqrt{r/b} + 9.600r/b$ $K_3 = 0.202 + 38.622\sqrt{r/b} - 27.374r/b$ $K_4 = 3.232 - 23.002\sqrt{r/b} + 15.482r/b$ (Refs. 38 to 40)
13. Lateral slot with circular ends in a member of rectangular section 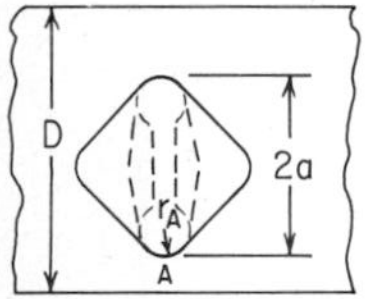 The equivalent ellipse has a width $2b_{eq}$ where $b_{eq} = \sqrt{r_A a}$	13a. Elastic stress, axial tension	A very close approximation to the maximum stress σ_A can be obtained by using the maximum stress for the given loading with the actual slot replaced by an ellipse having the same overall dimension normal to the loading direction, $2a$, and the same end radius r_A. See cases 9a, 10a, and 11a.
	13b. Elastic stress, in-plane bending	As above, but see cases 9b, 10b, and 11b.
14. Reinforced circular hole in a wide plate 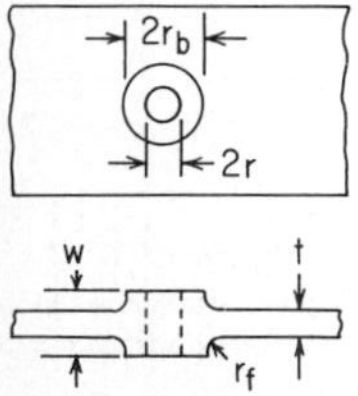	14a. Elastic stress, axial tension 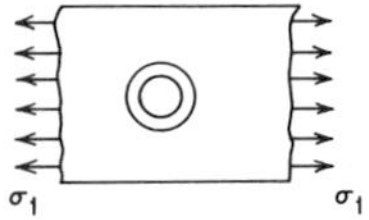	$\sigma_{max} = k\sigma_1$ where for $r_f \ge 0.6t$ and $w \ge 3t$ $k = 1.0 + \frac{1.66}{1 + A} - \frac{2.182}{(1 + A)^2} + \frac{2.521}{(1 + A)^3}$ A is the ratio of the transverse area of the added reinforcement to the transverse area of the hole: $A = \frac{(r_b - r)(w - t) + 0.429r_f^2}{rt}$ (Ref. 41)

15. U notch in a circular shaft

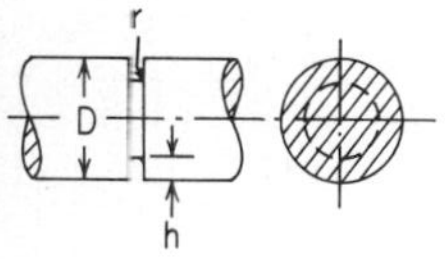

15a. Elastic stress, axial tension

$$k = K_1 + K_2\left(\frac{2h}{D}\right) + K_3\left(\frac{2h}{D}\right)^2 + K_4\left(\frac{2h}{D}\right)^3$$

where

	$0.25 \le h/r \le 2.0$	$2.0 \le h/r \le 50.0$
K_1	$0.455 + 3.354\sqrt{h/r} - 0.769h/r$	$0.935 + 1.922\sqrt{h/r} + 0.004h/r$
K_2	$3.129 - 15.955\sqrt{h/r} + 7.404h/r$	$0.537 - 3.708\sqrt{h/r} + 0.040h/r$
K_3	$-6.909 + 29.286\sqrt{h/r} - 16.104h/r$	$-2.538 + 3.438\sqrt{h/r} - 0.012h/r$
K_4	$4.325 - 16.685\sqrt{h/r} + 9.469h/r$	$2.066 - 1.652\sqrt{h/r} - 0.031h/r$

(Refs. 1, 9, and 42)

For the semicircular notch ($h/r = 1$)

$$k = 3.04 - 5.42\left(\frac{2h}{D}\right) + 6.27\left(\frac{2h}{D}\right)^2 - 2.89\left(\frac{2h}{D}\right)^3$$

15b. Elastic stress, bending

$$k = K_1 + K_2\left(\frac{2h}{D}\right) + K_3\left(\frac{2h}{D}\right)^2 + K_4\left(\frac{2h}{D}\right)^3$$

where

	$0.25 \le h/r \le 2.0$	$2.0 \le h/r \le 50.0$
K_1	$0.455 + 3.354\sqrt{h/r} - 0.769h/r$	$0.935 + 1.922\sqrt{h/r} + 0.004h/r$
K_2	$0.891 - 12.721\sqrt{h/r} + 4.593h/r$	$-0.552 - 5.327\sqrt{h/r} + 0.086h/r$
K_3	$0.286 + 15.481\sqrt{h/r} - 6.392h/r$	$0.754 + 6.281\sqrt{h/r} - 0.121h/r$
K_4	$-0.632 - 6.115\sqrt{h/r} + 2.568h/r$	$-0.138 - 2.876\sqrt{h/r} + 0.031h/r$

(Refs. 1 and 9)

For the semicircular notch ($h/r = 1$)

$$k = 3.04 - 7.236\left(\frac{2h}{D}\right) + 9.375\left(\frac{2h}{D}\right)^2 - 4.179\left(\frac{2h}{D}\right)^3$$

15c. Elastic stress, torsion

$$k = K_1 + K_2\left(\frac{2h}{D}\right) + K_3\left(\frac{2h}{D}\right)^2 + K_4\left(\frac{2h}{D}\right)^3$$

where

	$0.25 \le h/r \le 2.0$	$2.0 \le h/r \le 50.0$
K_1	$1.245 + 0.264\sqrt{h/r} + 0.491h/r$	$1.651 + 0.614\sqrt{h/r} + 0.040h/r$
K_2	$-3.030 + 3.269\sqrt{h/r} - 3.633h/r$	$-4.794 - 0.314\sqrt{h/r} - 0.217h/r$
K_3	$7.199 - 11.286\sqrt{h/r} + 8.318h/r$	$8.457 - 0.962\sqrt{h/r} + 0.389h/r$
K_4	$-4.414 + 7.753\sqrt{h/r} - 5.176h/r$	$-4.314 + 0.662\sqrt{h/r} - 0.212h/r$

(Refs. 1, 9, and 43 to 46)

For the semicircular notch ($h/r = 1$)

$$k = 2.000 - 3.394\left(\frac{2h}{D}\right) + 4.231\left(\frac{2h}{D}\right)^2 - 1.837\left(\frac{2h}{D}\right)^3$$

TABLE 37 *Factors of stress concentration for elastic stress* (k) *(Cont.)*

Type of form irregularity or stress raiser	Stress condition and manner of loading	Factor of stress concentration k for various dimensions
16. V notch in a circular shaft 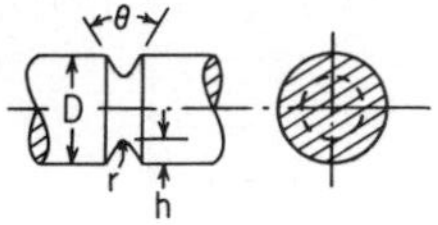	16c. Elastic stress, torsion	The stress concentration factor for the V notch, k_θ, is the smaller of the values $k_\theta = k_U$ or $k_\theta = 1.065k_U - \left[0.022 + 0.137\left(\frac{\theta}{135}\right)^2\right](k_U - 1)k_U$ for $\frac{r}{D - 2h} \le 0.01$ and $\theta \le 135°$ where k_U is the stress concentration factor for a U notch, case 15c, when the dimensions h, r, and D are the same as for the V notch and θ is the notch angle in degrees. (Refs. 1 and 44)
17. Square shoulder with fillet in circular shaft 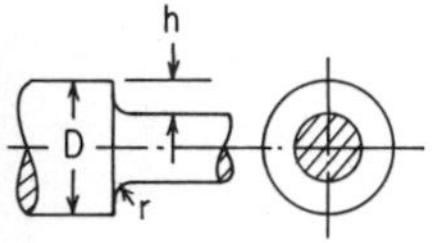	17a. Elastic stress, axial tension	$k = K_1 + K_2\left(\frac{2h}{D}\right) + K_3\left(\frac{2h}{D}\right)^2 + K_4\left(\frac{2h}{D}\right)^3$ where $0.25 \le h/r \le 2.0$ / $2.0 \le h/r \le 20.0$ K_1: $0.927 + 1.149\sqrt{h/r} - 0.086h/r$ / $1.225 + 0.831\sqrt{h/r} - 0.010h/r$ K_2: $0.011 - 3.029\sqrt{h/r} + 0.948h/r$ / $-1.831 - 0.318\sqrt{h/r} - 0.049h/r$ K_3: $-0.304 + 3.979\sqrt{h/r} - 1.737h/r$ / $2.236 - 0.522\sqrt{h/r} + 0.176h/r$ K_4: $0.366 - 2.098\sqrt{h/r} + 0.875h/r$ / $-0.630 + 0.009\sqrt{h/r} - 0.117h/r$ (Refs. 1, 19, and 47)
	17b. Elastic stress, bending	$k = K_1 + K_2\left(\frac{2h}{D}\right) + K_3\left(\frac{2h}{D}\right)^2 + K_4\left(\frac{2h}{D}\right)^3$ where $0.25 \le h/r \le 2.0$ / $2.0 \le h/r \le 20.0$ K_1: $0.927 + 1.149\sqrt{h/r} - 0.086h/r$ / $1.225 + 0.831\sqrt{h/r} - 0.010h/r$ K_2: $0.015 - 3.281\sqrt{h/r} + 0.837h/r$ / $-3.790 + 0.958\sqrt{h/r} - 0.257h/r$ K_3: $0.847 + 1.716\sqrt{h/r} - 0.506h/r$ / $7.374 - 4.834\sqrt{h/r} + 0.862h/r$ K_4: $-0.790 + 0.417\sqrt{h/r} - 0.246h/r$ / $-3.809 + 3.046\sqrt{h/r} - 0.595h/r$ (Refs. 1, 20, and 48)
	17c. Elastic stress, torsion	$k = K_1 + K_2\left(\frac{2h}{D}\right) + K_3\left(\frac{2h}{D}\right)^2 + K_4\left(\frac{2h}{D}\right)^3$ where for $0.25 \le h/r \le 4.0$ $K_1 = 0.953 + 0.680\sqrt{h/r} - 0.053h/r$ $K_2 = -0.493 - 1.820\sqrt{h/r} + 0.517h/r$ $K_3 = 1.621 + 0.908\sqrt{h/r} - 0.529h/r$ $K_4 = -1.081 + 0.232\sqrt{h/r} + 0.065h/r$ (Refs. 1, 19, 46, 49, and 50)

18. Radial hole in a hollow or solid circular shaft 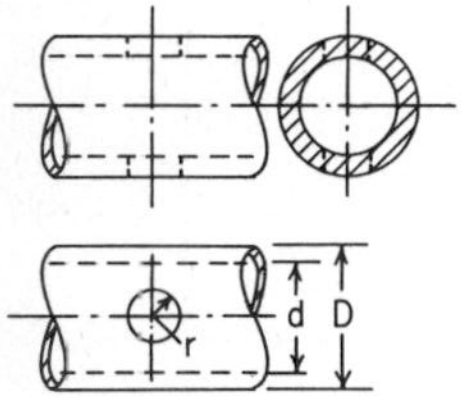 For a solid shaft $d = 0$	18a. Elastic stress, axial tension	$\sigma_{max} = k\dfrac{4P}{\pi(D^2 - d^2)}$ where $k = K_1 + K_2\left(\dfrac{2r}{D}\right) + K_3\left(\dfrac{2r}{D}\right)^2 + K_4\left(\dfrac{2r}{D}\right)^3$ and where for $d/D \leq 0.9$ and $2r/D \leq 0.45$ $K_1 = 3.000$ $K_2 = 2.773 + 1.529d/D - 4.379(d/D)^2$ $K_3 = -0.421 - 12.782d/D + 22.781(d/D)^2$ $K_4 = 16.841 + 16.678d/D - 40.007(d/D)^2$ (Refs. 1, 43, 51, and 52)
	18b. Elastic stress, bending when hole is farthest from bending axis	$\sigma_{max} = k\dfrac{32MD}{\pi(D^4 - d^4)}$ where $k = K_1 + K_2\left(\dfrac{2r}{D}\right) + K_3\left(\dfrac{2r}{D}\right)^2 + K_4\left(\dfrac{2r}{D}\right)^3$ and where for $d/D \leq 0.9$ and $2r/D \leq 0.3$ $K_1 = 3.000$ $K_2 = -6.690 - 1.620d/D + 4.432(d/D)^2$ $K_3 = 44.739 + 10.724d/D - 19.927(d/D)^2$ $K_4 = -53.307 - 25.998d/D + 43.258(d/D)^2$ (Refs. 1 and 51 to 54)
	18c. Elastic stress, torsion	$\tau_{max} = k\dfrac{16TD}{\pi(D^4 - d^4)}$ where $k = K_1 + K_2\left(\dfrac{2r}{D}\right) + K_3\left(\dfrac{2r}{D}\right)^2 + K_4\left(\dfrac{2r}{D}\right)^3$ and where for $d/D \leq 0.9$ and $2r/D \leq 0.4$ $K_1 = 4.000$ $K_2 = -6.793 + 1.133d/D - 0.126(d/D)^2$ $K_3 = 38.382 - 7.242d/D + 6.495(d/D)^2$ $K_4 = -44.576 - 7.428d/D + 58.656(d/D)^2$ (Refs. 1 and 51 to 53)
19. Multiple U notches in a member of rectangular section 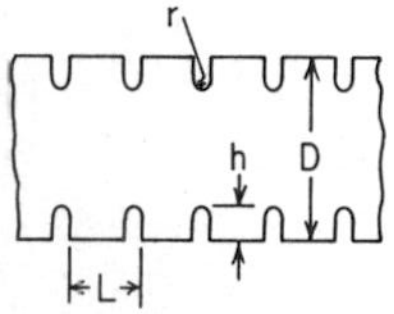	19a. Elastic stress, axial tension, semicircular notches only, i.e., $h = r$	The stress concentration factor for the multiple semicircular U notches, k_M, is the *smaller* of the values $k_M = k_U$ or $k_M = \left\{1.1 - \left[0.88 - 1.68\left(\dfrac{2r}{D}\right)\right]\dfrac{2r}{L} + \left[1.3\left(0.5 - \dfrac{2r}{D}\right)^2\right]\left(\dfrac{2r}{L}\right)^3\right\}k_U$ for $\dfrac{2r}{L} < 1$ where k_U is the stress concentration factor for a single pair of semicircular U notches, case 1a. (Refs. 1, 55, and 56)

TABLE 37 *Factors of stress concentration for elastic stress (k)* (*Cont.*)

Type of form irregularity or stress raiser	Stress condition and manner of loading	Factor of stress concentration k for various dimensions
20. Infinite row of circular holes in an infinite plate 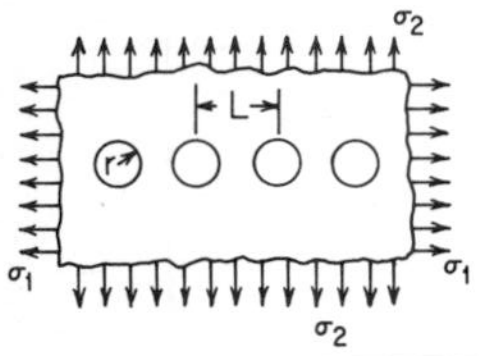	20a. Elastic stress, axial tension parallel to the row of holes	$\sigma_{max} = k\sigma_1$, $\sigma_2 = 0$ where $k = 3.0 - 1.061\left(\frac{2r}{L}\right) - 2.136\left(\frac{2r}{L}\right)^2 + 1.877\left(\frac{2r}{L}\right)^3$ (Refs. 1 and 57)
	20b. Elastic stress, axial tension normal to the row of holes	$\sigma_{max} = k\sigma_2$, $\sigma_1 = 0$ where $k = 3.0 - 3.057\left(\frac{2r}{L}\right) + 0.214\left(\frac{2r}{L}\right)^2 + 0.843\left(\frac{2r}{L}\right)^3$ (Refs. 1 and 57)
21. Gear tooth 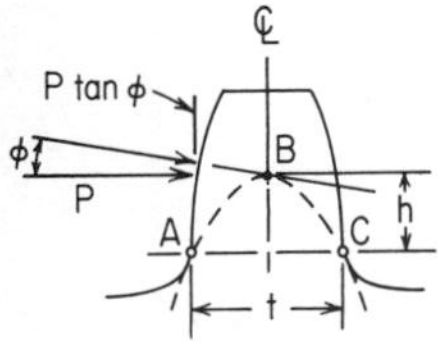 *A* and *C* are points of tangency of the inscribed parabola *ABC* with tooth profile, b = tooth width normal to plane of figure, r = minimum radius of tooth fillet.	Elastic stress, bending plus some compression	For 14.5° pressure angle: $k = 0.22 + \left(\frac{t}{r}\right)^{0.2}\left(\frac{t}{h}\right)^{0.4}$ For 20° pressure angle: $k = 0.18 + \left(\frac{t}{r}\right)^{0.15}\left(\frac{t}{h}\right)^{0.45}$ $k = (\text{max } \sigma_t \text{ by photoelastic analysis}) \div \left(\text{calculated max } \sigma_t = \frac{6Ph}{bt^2} - \frac{P \tan \phi}{bt}\right)$ (Ref. 58) (Alternatively, the maximum stress can be found by the formula for a short cantilever, p. 187.)
22. Square or filleted corner in tension 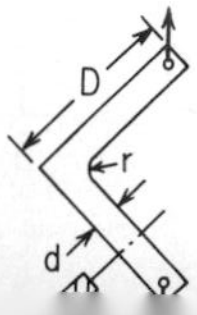	Elastic stress	$\frac{D}{d} = 5.5$ $\frac{r}{d} = 0.125$ 0.15 0.20 0.25 0.30 0.40 0.50 0.70 1.00 $k = 2.50$ 2.30 2.03 1.88 1.70 1.53 1.40 1.26 1.20 (Ref. 17)

23. U-shaped member

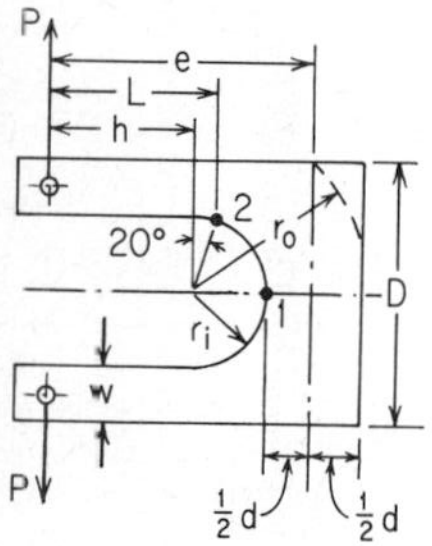

Elastic stress, as shown

k_1 is ratio of actual to nominal bending stress at point 1, and k_2 is this ratio at point 2. Nominal bending stress $= Pey/I$ at point 1 and Ply/I at point 2, where I/y = section modulus at the section in question

Outer corners	Dimension ratios and values of k				
	$\frac{e}{r_i} = \frac{e}{w} = \frac{e}{d}$	k_1		k_2	
Square	4.5	1.24		1.24	
	3.5	1.20		1.24	
	2.5	1.30		1.20	
	1.5	1.24		1.61	
	$\frac{e}{2r_i} = \frac{e}{2w} = \frac{e}{d}$				
Square	2.5	1.50		1.29	
	2.0	1.52		1.33	
	1.5	1.53		1.22	
	1.0	1.46		1.75	
	$\frac{d}{r_i} = \frac{d}{w}$	$h = \frac{3}{4}D$		$h = \frac{1}{4}D$	
		k_1	k_2	k_1	k_2
Square	2.0	1.50	1.29	1.53	1.22
	1.5	1.34	1.10	1.37	1.40
	1.25	1.29	1.23	1.33	1.41
	1.0	1.24	1.24	1.30	1.20
	0.75	1.21	1.10	1.24	1.22
	$\frac{r_0}{r_i} = \frac{r_0}{d} = \frac{r_0}{w}$				
Rounded to radius r_0	2.75	1.24	1.24	1.30	1.20
	2.37	1.18	1.21	1.18	1.22
	2.12	1.16	1.22	1.21	1.31
	2.0	1.27	1.42	1.31	1.56
	$\frac{d}{r_i} = \frac{w}{r_i}$				
Square	7.0	2.29	1.93	2.38	2.38
	3.0	1.72	1.59	1.76	1.62
	1.67	1.49	1.37	1.41	1.52
	1.0	1.24	1.24	1.30	1.20
	$\frac{d}{r_i}$; $\frac{d}{w}$				
Square	5 ; 1.67	2.33	1.73	2.32	2.00
	3 ; 1.80	1.82	1.30	1.75	1.56
	2 ; 2.0	1.50	1.29	1.53	1.22

(Ref. 59)

References for Table 37

1. Peterson, R. E.: "Stress Concentration Factors," John Wiley & Sons, Inc., 1974.
2. Isida, M.: On the Tension of the Strip with Semicircular Notches, *Trans. Jap. Soc. Mech. Eng.,* vol. 19, no. 83, 1953.
3. Ling, C-B.: On Stress-Concentration Factor in a Notched Strip, *ASME J. Appl. Mech.,* vol. 35, no. 4, December 1968.
4. Flynn, P. D., and A. A. Roll: A Comparison of Stress-concentration Factors in Hyperbolic and U-shaped Grooves, *Exp. Mech., J. Soc. Exp. Stress Anal.,* vol. 7, no. 6, June 1967.
5. Flynn, P. D.: Photoelastic Comparison of Stress Concentrations Due to Semicircular Grooves and a Circular Hole in a Tension Bar, *ASME J. Appl. Mech.,* vol. 36, no. 4, December 1969.
6. Slot, T., and D. F. Mowbray: A Note on Stress-Concentration Factors for Symmetric U-shaped Notches in Tension Strips, *ASME J. Appl. Mech.,* vol. 36, no. 4, December 1969.
7. Kikukawa, M.: Factors of Stress Concentration for Notched Bars under Tension and Bending, *Proc., 10th Int. Congr. Appl. Mech.,* Stresa, Italy, 1960.
8. Frocht, M.: Factors of Stress Concentration Photoelastically Determined, *ASME J. Appl. Mech.,* vol. 2, no. 2, June 1935.
9. Neuber, H.: Notch Stress Theory, *Tech. Rep.* AFML-TR-65-225, July 1965.
10. Baratta, F. I.: Comparison of Various Formulae and Experimental Stress-Concentration Factors for Symmetrical U-notched Plates, *J. Strain Anal.,* vol. 7, no. 2, 1972.
11. Wilson, I. H., and D. J. White: Stress-Concentration Factors for Shoulder Fillets and Grooves in Plates, *J. Strain Anal.,* vol. 8, no. 1, 1973.
12. Ling, C-B.: On the Stresses in a Notched Strip, *ASME J. Appl. Mech.,* vol. 19, no. 2, June, 1952.
13. Lee, G. H.: The Influence of Hyperbolic Notches on the Transverse Flexure of Elastic Plates, *Trans. ASME,* vol. 62, 1940.
14. Shioya, S.: The Effect of Square and Triangular Notches with Fillets on the Transverse Flexure of Semi-Infinite Plates, *Z. angew. Math. Mech.,* vol. 39, 1959.
15. Appl, F. J., and D. R. Koerner: Stress Concentration Factors for U-shaped, Hyperbolic and Rounded V-shaped Notches, *ASME Pap.* 69-DE-2, Engineering Society Library, United Engineering Center, New York, 1969.
16. Cole, A. G., and A. F. C. Brown: Photoelastic Determination of Stress Concentration Factors Caused by a Single U-Notch on One Side of a Plate in Tension, *J. Roy. Aeronaut. Soc.,* vol. 62, 1958.
17. Roark, R. J., R. S. Hartenberg, and R. Z. Williams: Influence of Form and Scale on Strength, *Univ. Wis. Eng. Exp. Sta. Bull.,* 1938.
18. Leven, M. M., and M. M. Frocht: Stress Concentration Factors for a Single Notch in a Flat Bar in Pure and Central Bending, *Proc. Soc. Exp. Stress Anal.,* vol. 11, no. 2, 1953.
19. Fessler, H., C. C. Rogers, and P. Stanley: Shouldered Plates and Shafts in Tension and Torsion, *J. Strain Anal.,* vol. 4, no. 3, 1969.
20. Hartman, J. B., and M. M. Leven: Factors of Stress Concentration for the Bending Case of Fillets in Flat Bars and Shafts with Central Enlarged Section, *Proc. Soc. Exp. Stress Anal.,* vol. 19, no. 1, 1951.
21. Kumagai, K., and H. Shimada: The Stress Concentration Produced by a Projection under Tensile Load, *Bull. Jap. Soc. Mech. Eng.,* vol. 11, 1968.
22. Derecho, A. T., and W. H. Munse: Stress Concentration at External Notches in Members Subjected to Axial Loading, *Univ. Ill. Eng. Exp. Sta. Bull. 494,* 1968.
23. Roark, R. J.: "Formulas for Stress and Strain," 4th ed., McGraw-Hill Book Company, 1965.
24. Savin, G. N.: "Stress Concentration Around Holes," Pergamon Press, 1961.
25. Heywood, R. B.: "Designing by Photoelasticity," Chapman & Hall, Ltd., 1952.
26. Goodier, J. N.: Influence of Circular and Elliptical Holes on Transverse Flexure of Elastic Plates, *Phil. Mag.,* vol. 22, 1936.
27. Goodier, J. N., and G. H. Lee: An Extension of the Photoelastic Method of Stress Measurement to Plates in Transverse Bending, *Trans. ASME,* vol. 63, 1941.

28. Drucker, D. C.: The Photoelastic Analysis of Transverse Bending of Plates in the Standard Transmission Polariscope, *Trans. ASME,* vol. 64, 1942.
29. Dumont, C.: Stress Concentration Around an Open Circular Hole in a Plate Subjected to Bending Normal to the Plane of the Plate, *NACA Tech. Note* 740, 1939.
30. Reissner, E.: The Effect of Transverse Shear Deformation on the Bending of Elastic Plates, *Trans. ASME,* vol. 67, 1945.
31. Isida, M.: On the Bending of an Infinite Strip with an Eccentric Circular Hole, *Proc. 2d Jap. Congr. Appl. Mech.,* 1952.
32. Sjöström, S.: On the Stresses at the Edge of an Eccentrically Located Circular Hole on a Strip under Tension, *Aeronaut. Res. Inst. Rept.* 36, Sweden, 1950.
33. Isida, M.: On the Tension of a Strip with a Central Elliptic Hole, *Trans. Jap. Soc. Mech. Eng.,* vol. 21, 1955.
34. Durelli, A. J., V. J. Parks, and H. C. Feng: Stresses Around an Elliptical Hole in a Finite Plate Subjected to Axial Loading, *ASME J. Appl. Mech.,* vol. 33, no. 1, March 1966.
35. Jones, N., and D. Hozos: A Study of the Stresses Around Elliptical Holes in Flat Plates, *ASME J. Eng. Ind.,* vol. 93, no. 2, May 1971.
36. Isida, M.: Form Factors of a Strip with an Elliptic Hole in Tension and Bending, *Sci. Pap. Fac. Eng., Tokushima Univ.,* vol. 4, 1953.
37. Frocht, M. M. and M. M. Leven: Factors of Stress Concentration for Slotted Bars in Tension and Bending, *ASME J. Appl. Mech.,* vol. 18, no. 1, March 1951.
38. Brock, J. S.: The Stresses Around Square Holes with Rounded Corners, *J. Ship Res.,* October 1958.
39. Sobey, A. J.: Stress Concentration Factors for Rounded Rectangular Holes in Infinite Sheets, *Aeronaut. Res. Counc.* R&M 3407, Her Majesty's Stationery Office, 1963.
40. Heller, S. R., J. S. Brock, and R. Bart: The Stresses Around a Rectangular Opening with Rounded Corners in a Uniformly Loaded Plate, *Proc. 3d U.S. Nat. Congr. Appl. Mech.,* 1958.
41. Seika, M. and A. Amano: The Maximum Stress in a Wide Plate with a Reinforced Circular Hole under Uniaxial Tension—Effects of a Boss with Fillet, *ASME J. Appl. Mech.,* vol. 34, no. 1, March 1967.
42. Cheng, Y. F.: Stress at Notch Root of Shafts under Axially Symmetric Loading, *Exp. Mech., J. Soc. Exp. Stress Anal.,* vol. 10, no. 12, December 1970.
43. Leven, M. M.: Quantitative Three-Dimensional Photoelasticity, *Proc. Soc. Exp. Stress Anal.,* vol. 12, no. 2, 1955.
44. Rushton, K. R.: Stress Concentrations Arising in the Torsion of Grooved Shafts, *J. Mech. Sci.,* vol. 9, 1967.
45. Hamada, M., and H. Kitagawa: Elastic Torsion of Circumferentially Grooved Shafts, *Bull. Jap. Soc. Mech. Eng.,* vol. 11, 1968.
46. Matthews, G. J., and C. J. Hooke: Solution of Axisymmetric Torsion Problems by Point Matching, *J. Strain Anal.,* vol. 6, 1971.
47. Allison, I. M.: The Elastic Concentration Factors in Shouldered Shafts, Part III: Shafts Subjected to Axial Load, *Aeronaut. Q.,* vol. 13, 1962.
48. Allison, I. M.: The Elastic Concentration Factors in Shouldered Shafts, Part II: Shafts Subjected to Bending, *Aeronaut. Q.,* vol. 12, 1961.
49. Allison, I. M.: The Elastic Concentration Factors in Shouldered Shafts, *Aeronaut. Q.,* vol. 12, 1961.
50. Rushton, K. R.: Elastic Stress Concentrations for the Torsion of Hollow Shouldered Shafts Determined by an Electrical Analogue, *Aeronaut. Q.,* vol. 15, 1964.
51. Jessop, H. T., C. Snell, and I. M. Allison: The Stress Concentration Factors in Cylindrical Tubes with Transverse Circular Holes, *Aeronaut. Q.,* vol. 10, 1959.
52. British Engineering Science Data, 65004, Engineering Science Data Unit, 4 Hamilton Place, London W1, 1965.
53. Thum, A. and W. Kirmser: Uberlagerte Wechselbeanspruchungen, ihre Erzeugung und ihr Einfluss auf die Dauerbarkeit und Spannungsausbildung quergebohrten Wellen, *VDI-Forschungsh.* 419, vol. 14b, 1943.

54. Fessler, H. and E. A. Roberts: Bending Stresses in a Shaft with a Transverse Hole, "Selected Papers on Stress Analyses, Stress Analysis Conference, Delft, 1959," Reinhold Publishing Co., 1961.
55. Atsumi, A.: Stress Concentrations in a Strip under Tension and Containing an Infinite Row of Semicircular Notches, *Q. J. Mech. Appl. Math.,* vol. 11, pt. 4, 1958.
56. Durelli, A. J., R. L. Lake, and E. Phillips: Stress Concentrations Produced by Multiple Semi-circular Notches in Infinite Plates under Uniaxial State of Stress, *Proc. Soc. Exp. Stress Anal.,* vol. 10, no. 1, 1952.
57. Schulz, K. J.: On the State of Stress in Perforated Strips and Plates, *Proc. Neth. Roy. Acad. Sci.,* vols. 45 to 48, 1942 to 1945.
58. Dolan, T. J., and E. L. Broghamer: A Photo-elastic Study of Stresses in Gear Tooth Fillets, *Univ. Ill. Eng. Exp. Sta. Bull.* 335, 1942.
59. Mantle, J. B., and T. J. Dolan: A Photoelastic Study of Stresses in U-shaped Members, *Proc. Soc. Exp. Stress Anal.,* vol. 6, no. 1, 1948.

TABLE 38 *Representative properties of some important structural materials*

1. Metals

Material	δ, weight, lb/in³	γ, coef. of thermal exp., °F × 10⁵	E, million lb/in²	G, million lb/in²	Poisson's ratio	Strength properties—thousand lb. per sq. in.: Tensile properties: Ultimate strength (σ_u)	Tensile: Elastic limit	Tensile: Yield pt. or yield strength	Compressive properties: Ultimate strength	Compressive: Yield strength	Shear properties: Ultimate strength	Shear: Yield strength	Modulus of rupture in cross-bending	Endurance strength (Rot. beam, 10^7 cycles)
Aluminum, cast, pure	0.0976	1.30	9	3.7	0.36	11	5							
Aluminum, cast, 220-T4	0.093	1.36	9.5	3.55	0.33	42	...	22	...	23	30	...	...	11
Aluminum, wrought, 2014-T6	0.101	1.28	10.6	4	0.33	68	...	60	...	62	39	35	...	20
Aluminum, wrought, 6061-T6	0.098	1.30	10	3.75	0.3	38	...	35	...	35	24	20	...	17
Beryllium copper	0.297	0.93	19	7	...	100–200	110–150	140	...	...	100–130	70–100	...	40
Brass, naval	0.304	1.18	15	5.5	...	57–75	...	25–50	...	...	40–45	...	...	$0.35\sigma_u$
Bronze, phosphor, A.S.T.M. B159	0.320	0.99	15	6.5	...	100–150	60–110	...	70–110	50–85	70–110	50–85	...	$0.32\sigma_u$
Cast iron, gray, No. 20	0.251	0.60	14	...	0.25	20	...	...	90	...	32	...	46	10
Cast iron, gray, No. 30	0.260	0.60	15.2	...	0.25	30	...	...	115	...	44	...	57	14.5
Cast iron, gray, No. 40	0.260	0.60	18.3	...	0.25	40	...	...	130	...	51	...	66	19
Cast iron, gray, No. 60	0.270	0.60	19	...	0.25	60	...	...	180	...	72	...	100	24
Cast iron, malleable	0.266	0.75	26	8.8	0.25	50–65	...	32–45	200	...	49	...	62	32
Cast iron, nodular	0.257	0.66	23.5	...	0.25	60–100	...	45–65	200					
Magnesium, AZ80A-T5	0.065	1.60	6.5	2.4		55	...	38	...	17	24	...	...	16
Titanium, pure	0.163	0.53	15.5	5.8	0.34	65–80	...	55–70	...	...	...	...	...	$0.6\sigma_u$
Titanium, alloy, 5 Al, 2.5 Sn	0.161	0.57	17	6.2	0.33	115	...	110	...	110	100	...	...	$0.6\sigma_u$
Steel for bridges and buildings, A.S.T.M. A7-61T:														
All shapes	0.283	0.65	29	11.5	0.27	60–75	...	33	...	33	...	17	...	$0.5\sigma_u$
Plates $t < 1.5$	0.283	0.65	29	11.5	0.27	60–72	...	33	...	33	...	17	...	$0.5\sigma_u$
Plates $t > 1.5$	0.283	0.65	29	11.5	0.27	60–75	...	33	...	33	...	17	...	$0.5\sigma_u$
High-strength low-alloy structural steel, A.S.T.M. A242-63T														
Most shapes	0.283	0.65	29	11.5	0.27	70	...	50	...	50	...	25	...	$0.5\sigma_u$
Plates $t < 0.75$	0.283	0.65	29	11.5	0.27	70	...	50	...	50	...	25	...	$0.5\sigma_u$
Plates $0.75 < t < 1.5$	0.283	0.65	29	11.5	0.27	67	...	46	...	46	...	23	...	$0.5\sigma_u$
Plates $1.5 < t < 4$	0.283	0.65	29	11.5	0.27	63	...	42	...	42	...	21	...	$0.5\sigma_u$
High-strength steel castings for structural purposes, A.S.T.M. A148-60 (7 grades)	0.283	0.83	29	11.5	0.27	80–175	..	40–145	...	...	...	...	...	$0.4\sigma_u$
Steel, spring, carbon, S.A.E. 1095	0.28	...	30	...	...	170–220	125–170	...	...	...	...	...	...	$0.36\sigma_u$
Steel, spring, alloy, S.A.E. 4068	0.28	...	30	...	...	200–270	175–240							
Steel, ball bearings, S.A.E. 52100	0.28	...	30	...	...	326								
Steel, stainless (0.08–0.2 C, 17 Cr, 7 Ni) ¼ hard	0.28	0.96	28	12.5	...	125	..	78	...	67	...	...	...	$0.35\sigma_u$
Same, full hard	0.28	0.96	26.6	12.0	...	185	..	150	...	99	...	...	...	$0.35\sigma_u$

TABLE 38 *Representative properties of some important structural materials* (*Cont.*)

2. Timber

Species	Weight lb/ft^3	γ, coeff. thermal exp., °F.		E, lb/in^2	Compressive strength			Shear strength with grain	Bending strength	
		With grain	Across grain		With grain: Ultimate	With grain: Elastic limit	Across grain elastic limit		Modulus of rupture (rect. section)	Fiber stress at elastic limit
Ash (white)	41	0.0000053	. . .	1,680,000	7280	5580	1510	1920	14,600	8,900
Birch (sweet, yellow)	44	0.0000011	0.000016	2,070,000	8310	6200	1250	2020	16,700	10,100
Elm (American)	35	. . .	. . .	1,340,000	5520	4030	850	1510	11,800	7,600
Hickory (true)	51	. . .	. . .	2,180,000	8970		2310	2140	19,700	10,900
Maple (sugar)	44	0.0000012	. . .	1,830,000	7830	5390	1810	2430	15,800	9,500
Oak (red)	44	0.0000019	0.00002	1,810,000	6920	4610	1260	1830	14,400	8,400
Oak (white)	48	0.0000027	0.00003	1,620,000	7040	4350	1410	1890	13,900	7,900
Fir (Douglas)	36	. . .	. . .	1,920,000	7420	6450	910	1140	11,700	8,100
Hemlock (Eastern)	30	. . .	. . .	1,200,000	5410	4020	800	1060	8,900	6,100
Spruce (Sitka)	26	. . .	. . .	1,570,000	5610	4780	710	1150	10,200	6,700
Cypress (southern)	32	. . .	. . .	1,440,000	6360	4740	900	1000	10,600	7,200
Pine (southern long-leaf)	40	0.000003	0.000019	1,990,000	8440	6150	1190	1500	14,700	9,300

G is approximately one-sixteenth E. The endurance limit in reversed bending is approximately 28 percent of the modulus of rupture.

Above values are taken from "Wood Handbook" and are based on tests of small specimens of select, clear, seasoned wood at 12 percent moisture content.

TABLE 38 *Representative properties of some important structural materials* (*Cont.*)

3. Concrete and Masonry

Material	Weight, lb/ft³	γ, coeff. thermal exp., °F.	E, lb/in²	ν	Ultimate strength values: Compression	Tension	Shear	Modulus of rupture in bending
Concrete	150	0.0000060	...	...	σ_c	$\frac{1}{10}\sigma_c$	$\frac{1}{2}\sigma_c$	$200 + 0.09\,\sigma_c$
1:1½:3, $w/c = 6.5$	...	...	3,500,000	0.15	3,500			
1:2½:3½, $w/c = 7.5$	...	...	3,000,000	0.13	2,500			
1:3:5, $w/c = 9.0$	...	...	2,500,000	0.10	1,500			
(w/c = water/cement ratio, gal per sack).	...	...	(at 30 days)	...	(at 30 days)			
Brick	...	0.000003	...	...	σ_c			
Soft	120	...	1,500,000	...	2,000	...	...	400
Medium	...	...	...	...	3,500	...	...	600
Hard	144	...	3,500,000	...	5,000	...	...	900
Vitrified	...	...	...	...	10,000	...	...	1,800
Brick masonry	120							
1:3 Portland cement mortar	...	...	...	...	$0.30\sigma_c$			
1:3 lime mortar	...	...	...	...	$0.15\sigma_c$			
(wall or pier height = 15 × thickness)								
Granite	168	0.0000036	7,000,000	0.28	25,000	...	...	2,500
Limestone	166	0.0000028	6,000,000	0.21	8,000–16,000	...	...	700–1,500
Marble	175	0.0000038	8,000,000	0.26	12,000	...	...	1,200
Sandstone	156	0.0000052	2,500,000	0.28	6,000	...	...	600

References for Table 38

The data in Table 38 are merely representative of a few among the almost infinite number of structural materials, and cannot even suggest the great variation introduced in metals by such factors as heat treatment, cold working, and temperature, and in wood and concrete by moisture, density, duration of loading, defects, etc. Far more detailed information is to be found in the following references; in the publications of the American Society for Testing and Materials, the American Society for Metals, the Portland Cement Association, and the United States Forest Products Laboratory; and in various trade publications.

1. "Alcoa Structural Handbook," Aluminum Co. of America, 1960.
2. "Material Properties Handbook," Vol. 1, "Aluminum Alloys," NATO Advisory Group for Aeronautical Research and Development, 1958.
3. ANC Mil-Hdbk-5, Strength of Metal Aircraft Elements, Armed Force Supply Support Center, March, 1959.
4. Mechanical Properties of Metals and Alloys, *Nat. Bur. Standards Circ.* C447, 1943.
5. "Handbook of Metals," Vol. 1, "Properties and Selection of Metals," 8th ed. American Society for Metals, 1961.
6. Watter, Michael, and Rush A. Lincoln: "Strength of Stainless Steel Structural Members as Function of Design," Allegheny Ludlum Steel Corp., 1950.
7. American Society for Testing and Materials, Book of Standards.
8. Kinsey, H. V.: "The Mechanical and Engineering Properties of Commercially Available Titanium Alloys," NATO Advisory Group for Aeronautical Research and Development, 1957.
9. "Materials in Design Engineering" (Materials Selector Issues), Reinhold Publishing Corporation.
10. "Wood Handbook," U.S. Forest Products Laboratory, 1955.
11. ANC 18 Bulletin, Design of Wood Aircraft Structures, Munitions Board Aircraft Committee, June, 1951.
12. Withey, M. O., and George W. Washa: "Materials of Construction," John Wiley & Sons, Inc., 1954.
13. Magnesium Mill Products, Dow Metal Products Co., 1961.
14. "Metallic Materials and Elements for Flight Vehicular Structures," Government Publications 19T and 20T, Catalog Nos. D 7.6/2.5/3/ and D 7.6/2.5/3/ch.1.

Name Index

Subject Index